ADVANCED MAGNETOHYDRODYNAMICS
With Applications to Laboratory and Astrophysical Plasmas

Following on from the companion volume *Principles of Magnetohydrodynamics*, this textbook analyzes the applications of plasma physics to thermonuclear fusion and plasma astrophysics from the single viewpoint of MHD. This approach turns out to be ever more powerful when applied to streaming plasmas (the vast majority of visible matter in the Universe), toroidal plasmas (the most promising approach to fusion energy), and nonlinear dynamics (where it all comes together with modern computational techniques and extreme transonic and relativistic plasma flows).

The textbook interweaves theory and explicit calculations of waves and instabilities of streaming plasmas in complex magnetic geometries. It is ideally suited to advanced undergraduate and graduate courses in plasma physics and astrophysics.

J. P. (HANS) GOEDBLOED is an Advisor of the FOM-Institute for Plasma Physics "Rijnhuizen," and Professor Emeritus of theoretical plasma physics at Utrecht University. He has been a Visiting Scientist at laboratories in the Soviet Union, the United States, Brazil and Europe. He has taught at Campinas, Rio de Janeiro, São Paulo, MIT, K.U. Leuven and regularly at Amsterdam Free University and Utrecht University. For many years he coordinated a large-scale computational effort with the Dutch Science Organization on Fast Changes in Complex Flows involving scientists of different disciplines.

RONY KEPPENS is a Professor at the Centre for Plasma-Astrophysics, K.U. Leuven, affiliated with the FOM-Institute for Plasma Physics "Rijnhuizen," and a Professor at Utrecht University. He headed numerical plasma dynamics teams at Rijnhuizen and Leuven and frequently lectures on computational methods in astrophysics. His career started with research at the National Center for Atmospheric Research, Boulder, and the Kiepenheuer Institute for Solar Physics, Freiburg. His expertise ranges from solar physics to high energy astrophysics and includes parallel computing, grid-adaptivity and visualization of large-scale simulations.

STEFAAN POEDTS is full Professor in the department of mathematics at K.U. Leuven. He graduated in Leuven, was a postdoctoral researcher at the Max-Planck-Institut für Plasmaphysik, Garching, a senior researcher at the FOM-Institute for Plasma Physics "Rijnhuizen," and a research associate at the Centre for Plasma Astrophysics, K.U. Leuven. His research interests include solar astrophysics, space weather, thermonuclear fusion and MHD stability. He teaches basic math courses, advanced courses on plasma physics of the Sun and numerical simulation, and is currently president of the European Solar Physics Division of the EPS.

ADVANCED MAGNETOHYDRODYNAMICS

With Applications to Laboratory and Astrophysical Plasmas

J. P. (HANS) GOEDBLOED

*FOM-Institute for Plasma Physics "Rijnhuizen"
and Astronomical Institute, Utrecht University*

RONY KEPPENS

*Centre for Plasma-Astrophysics, Katholieke Universiteit Leuven,
FOM-Institute for Plasma Physics "Rijnhuizen"
and Astronomical Institute, Utrecht University*

STEFAAN POEDTS

Centre for Plasma-Astrophysics, Katholieke Universiteit Leuven

CAMBRIDGE UNIVERSITY PRESS

CAMBRIDGE UNIVERSITY PRESS
Cambridge, New York, Melbourne, Madrid, Cape Town, Singapore,
São Paulo, Delhi, Dubai, Tokyo

Cambridge University Press
The Edinburgh Building, Cambridge CB2 2RU, UK

Published in the United States of America by Cambridge University Press, New York

www.cambridge.org
Information on this title: www.cambridge.org/9780521705240

© J. P. Goedbloed, R. Keppens and S. Poedts 2010

This publication is in copyright. Subject to statutory exception
and to the provisions of relevant collective licensing agreements,
no reproduction of any part may take place without the written
permission of Cambridge University Press.

First published 2010

Printed in the United Kingdom at the University Press, Cambridge

A catalogue record for this publication is available from the British Library

Library of Congress Cataloging in Publication data
Goedbloed, J. P., 1940–
Advanced magnetohydrodynamics : with applications to laboratory and
astrophysical plasmas / J. P. (Hans) Goedbloed, Rony Keppens, Stefaan Poedts.
p.   cm.
ISBN 978-0-521-87957-6 (hardback)
1. Magnetohydrodynamics.   2. Plasma astrophysics.   I. Keppens, Rony.
II. Poedts, Stefaan, 1962–   III. Title.
QC718.5.M36G638   2010
538′.6–dc22
2010000317

ISBN 978-0-521-87957-6 Hardback
ISBN 978-0-521-70524-0 Paperback

Cambridge University Press has no responsibility for the persistence or
accuracy of URLs for external or third-party Internet websites referred to
in this publication, and does not guarantee that any content on such
websites is, or will remain, accurate or appropriate.

To Antonia, 陆蓉 (Rong Lu), and Micheline

Contents

Preface   *page* xiii

Part III   Flow and dissipation   1

12   Waves and instabilities of stationary plasmas   3
   12.1   Laboratory and astrophysical plasmas   3
      12.1.1   Grand vision: magnetized plasma on all scales   3
      12.1.2   Differences between laboratory and astrophysical plasmas   6
      12.1.3   Plasmas with background flow   12
   12.2   Spectral theory of stationary plasmas   13
      12.2.1   Basic equations   13
      12.2.2   Frieman–Rotenberg formulation   16
      12.2.3   Self-adjointness of the generalized force operator* 22
      12.2.4   Energy conservation and stability   27
   12.3   Solution paths in the complex ω plane   35
      12.3.1   Opening up the boundaries   35
      12.3.2   Approach to eigenvalues   40
   12.4   Literature and exercises   47

13   Shear flow and rotation   49
   13.1   Spectral theory of plane plasmas with shear flow   49
      13.1.1   Gravito-MHD wave equation for plane plasma flow   49
      13.1.2   Kelvin–Helmholtz instabilities in interface plasmas   55
      13.1.3   Continua and oscillation theorem $\mathcal{R}$ for real eigenvalues   59
      13.1.4   Complex eigenvalues and the alternator   65
   13.2   Case study: flow-driven instabilities in diffuse plasmas   71
      13.2.1   Rayleigh–Taylor instabilities of magnetized plasmas   73
      13.2.2   Kelvin–Helmholtz instabilities of ordinary fluids   76
      13.2.3   Gravito-MHD instabilities of stationary plasmas   85

   13.2.4 Oscillation theorem $\mathcal{C}$ for complex eigenvalues 91

  13.3 Spectral theory of rotating plasmas 93

   13.3.1 MHD wave equation for cylindrical flow 93

   13.3.2 Local stability* 98

   13.3.3 WKB approximation 102

  13.4 Rotational instabilities 104

   13.4.1 Rigid rotation of incompressible plasmas 104

   13.4.2 Magneto-rotational instability: local analysis 112

   13.4.3 Magneto-rotational instability: numerical solutions 118

  13.5 Literature and exercises 123

14 Resistive plasma dynamics 127

  14.1 Plasmas with dissipation 127

   14.1.1 Conservative versus dissipative dynamical systems 127

   14.1.2 Stability of force-free magnetic fields: a trap 128

  14.2 Resistive instabilities 135

   14.2.1 Basic equations 135

   14.2.2 Tearing modes 138

   14.2.3 Resistive interchange modes 149

  14.3 Resistive spectrum 150

   14.3.1 Resistive wall mode 150

   14.3.2 Spectrum of homogeneous plasma 155

   14.3.3 Spectrum of inhomogeneous plasma 158

  14.4 Reconnection 162

   14.4.1 Reconnection in 2D Harris sheet 162

   14.4.2 Petschek reconnection 168

   14.4.3 Kelvin–Helmholtz induced tearing instabilities 169

   14.4.4 Extended MHD and reconnection 171

  14.5 Literature and exercises 175

15 Computational linear MHD 177

  15.1 Spatial discretization techniques 178

   15.1.1 Basic concepts for discrete representations 180

   15.1.2 Finite difference methods 182

   15.1.3 Finite element method 186

   15.1.4 Spectral methods 196

   15.1.5 Mixed representations 201

  15.2 Linear MHD: boundary value problems 204

   15.2.1 Linearized MHD equations 204

   15.2.2 Steady solutions to linearly driven problems 206

   15.2.3 MHD eigenvalue problems 209

15.2.4 Extended MHD examples 211
15.3 Linear algebraic methods 217
    15.3.1 Direct and iterative linear system solvers 217
    15.3.2 Eigenvalue solvers: the QR algorithm 220
    15.3.3 Inverse iteration for eigenvalues and eigenvectors 221
    15.3.4 Jacobi–Davidson method 222
15.4 Linear MHD: initial value problems 225
    15.4.1 Temporal discretizations: explicit methods 225
    15.4.2 Disparateness of MHD time scales 233
    15.4.3 Temporal discretizations: implicit methods 234
    15.4.4 Applications: linear MHD evolutions 236
15.5 Concluding remarks 240
15.6 Literature and exercises 241

Part IV Toroidal plasmas 245

16 Static equilibrium of toroidal plasmas 247
16.1 Axi-symmetric equilibrium 247
    16.1.1 Equilibrium in tokamaks 247
    16.1.2 Magnetic field geometry 252
    16.1.3 Cylindrical limits 256
    16.1.4 Global confinement and parameters 260
16.2 Grad–Shafranov equation 269
    16.2.1 Derivation of the Grad–Shafranov equation 269
    16.2.2 Large aspect ratio expansion: internal solution 271
    16.2.3 Large aspect ratio expansion: external solution 277
16.3 Exact equilibrium solutions 284
    16.3.1 Poloidal flux scaling 284
    16.3.2 Soloviev equilibrium 289
    16.3.3 Numerical equilibria* 293
16.4 Extensions 299
    16.4.1 Toroidal rotation 299
    16.4.2 Gravitating plasma equilibria* 301
    16.4.3 Challenges 302
16.5 Literature and exercises 304

17 Linear dynamics of static toroidal plasmas 307
17.1 "Ad more geometrico" 307
    17.1.1 Alfvén wave dynamics in toroidal geometry 307
    17.1.2 Coordinates and mapping 308
    17.1.3 Geometrical–physical characteristics 309

17.2 Analysis of waves and instabilities in toroidal geometry   315
       17.2.1 Spectral wave equation   315
       17.2.2 Spectral variational principle   318
       17.2.3 Alfvén and slow continuum modes   319
       17.2.4 Poloidal mode coupling   322
       17.2.5 Alfvén and slow ballooning modes   326
17.3 Computation of waves and instabilities in tokamaks   334
       17.3.1 Ideal MHD versus resistive MHD in computations   334
       17.3.2 Edge localized modes   340
       17.3.3 Internal modes   344
       17.3.4 Toroidal Alfvén eigenmodes and MHD spectroscopy   347
17.4 Literature and exercises   352

18 Linear dynamics of stationary toroidal plasmas* 355
18.1 Transonic toroidal plasmas   355
18.2 Axi-symmetric equilibrium of transonic stationary states* 357
       18.2.1 General equations and toroidal rescalings* 357
       18.2.2 Elliptic and hyperbolic flow regimes* 365
       18.2.3 Expansion of the equilibrium in small toroidicity* 366
18.3 Equations for the continuous spectrum* 374
       18.3.1 Reduction for straight-field-line coordinates* 374
       18.3.2 Continua of poloidally and toroidally rotating plasmas* 378
       18.3.3 Analysis of trans-slow continua for small toroidicity* 385
18.4 Trans-slow continua in tokamaks and accretion disks* 392
       18.4.1 Tokamaks and magnetically dominated accretion disks* 393
       18.4.2 Gravity dominated accretion disks* 396
       18.4.3 A new class of transonic instabilities   397
18.5 Literature and exercises* 402

Part V   Nonlinear dynamics   405

19 Computational nonlinear MHD   407
19.1 General considerations for nonlinear conservation laws   408
       19.1.1 Conservative versus primitive variable formulations   408
       19.1.2 Scalar conservation law and the Riemann problem   415
       19.1.3 Numerical discretizations for a scalar conservation law 420
       19.1.4 Finite volume treatments   430
19.2 Upwind-like finite volume treatments for 1D MHD   433
       19.2.1 The Godunov method   434
       19.2.2 A robust shock-capturing method: TVDLF   440
       19.2.3 Approximate Riemann solver type schemes   446

19.2.4 Simulating 1D MHD Riemann problems 451
19.3 Multi-dimensional MHD computations 454
19.3.1 $\nabla \cdot \mathbf{B} = 0$ condition for shock-capturing schemes 455
19.3.2 Example nonlinear MHD scenarios 461
19.3.3 Alternative numerical methods 466
19.4 Implicit approaches for extended MHD simulations 473
19.4.1 Alternating direction implicit strategies 474
19.4.2 Semi-implicit methods 475
19.4.3 Simulating ideal and resistive instability developments 481
19.4.4 Global simulations for tokamak plasmas 482
19.5 Literature and exercises 484

20 Transonic MHD flows and shocks 487
20.1 Transonic MHD flows 487
20.1.1 Flow in laboratory and astrophysical plasmas 487
20.1.2 Characteristics in space and time 488
20.2 Shock conditions 490
20.2.1 Special case: gas dynamic shocks 492
20.2.2 MHD discontinuities without mass flow 498
20.2.3 MHD discontinuities with mass flow 500
20.2.4 Slow, intermediate and fast shocks 505
20.3 Classification of MHD shocks 507
20.3.1 Distilled shock conditions 507
20.3.2 Time reversal duality 513
20.3.3 Angular dependence of MHD shocks 520
20.3.4 Observational considerations of MHD shocks 527
20.4 Stationary transonic flows 529
20.4.1 Modeling the solar wind–magnetosphere boundary 530
20.4.2 Modeling the solar wind by itself 531
20.4.3 Example astrophysical transonic flows 534
20.5 Literature and exercises 540

21 Ideal MHD in special relativity 543
21.1 Four-dimensional space-time: special relativistic concepts 544
21.1.1 Space-time coordinates and Lorentz transformations 544
21.1.2 Four-vectors in flat space-time and invariants 547
21.1.3 Relativistic gas dynamics and stress–energy tensor 551
21.1.4 Sound waves and shock relations in relativistic gases 556
21.2 Electromagnetism and special relativistic MHD 564
21.2.1 Electromagnetic field tensor and Maxwell's equations 564
21.2.2 Stress–energy tensor for electromagnetic fields 569

		21.2.3 Ideal MHD in special relativity	570
		21.2.4 Wave dynamics in a homogeneous plasma	572
		21.2.5 Shock conditions in relativistic MHD	577
	21.3	Computing relativistic magnetized plasma dynamics	580
		21.3.1 Numerical challenges from relativistic MHD	583
		21.3.2 Example astrophysical applications	584
	21.4	Literature and exercises	588

Appendices			591
A	**Vectors and coordinates**		591
	A.1	Vector identities	591
	A.2	Vector expressions in orthogonal coordinates	592
	A.3	Vector expressions in non-orthogonal coordinates	600
References			604
Index			629

Preface

This book, together with the preceding *Principles of Magnetohydrodynamics* (to be referred to as Volume [1]), describes the two main applications of plasma physics, laboratory research on thermonuclear fusion energy and plasma-astrophysics of the solar system, stars, accretion disks, etc., from the single viewpoint of magnetohydrodynamics (MHD). This provides effective methods and insights for the interpretation of plasma phenomena on virtually all scales, ranging from the laboratory to the Universe. The key issue is understanding the complexities of plasma dynamics in extended magnetic structures. In Volume [1], the classical MHD model was developed in great detail without omitting steps in the derivations. This necessitated restriction to ideal dissipationless plasmas, in static equilibrium and with inhomogeneity in one direction. In the present volume on *Advanced Magnetohydrodynamics* [2], these restrictions are relaxed one by one: introducing stationary background flows, resistivity and reconnection, two-dimensional toroidal geometry, linear and nonlinear computational techniques and transonic flows and shocks. These topics transform the subject into a vital new area with many applications in laboratory, space and astrophysical plasmas.

The two volumes now consist of five parts:

   I *Plasma physics preliminaries* (Volume [1], Chapters 1–3),

  II *Basic magnetohydrodynamics* (Volume [1], Chapters 4–11),

 III *Flow and dissipation* (Volume [2], Chapters 12–15),

 IV *Toroidal plasmas* (Volume [2], Chapters 16–18),

  V *Nonlinear dynamics* (Volume [2], Chapters 19–21).

Inevitably, with the chosen distinction of topics for Volume [1] (mostly ideal linear phenomena described by self-adjoint linear operators) and topics for Volume [2] (mostly non-ideal, toroidal and nonlinear phenomena), the difference between "basic" and "advanced" levels of magnetohydrodynamics could not be strictly maintained. The logical order required inclusion of some advanced topics in Volume [1], whereas some topics that now appear in Volume [2] (like stationary flows and toroidal effects) really belong to the "principles" of MHD. Difficult parts or asides with tedious derivations, that may be skipped on first reading, are again indicated by a star (*) or put in small print in between triangles ($\triangleright \cdots \triangleleft$).

An overview of the subject matter of the different chapters of the two volumes may help the reader to find his way.

Contents of Volume [1]:

- Chapter 1 gives an introduction to laboratory fusion and astrophysical plasmas, and formulates provisional microscopic and macroscopic definitions of the plasma state.

- Chapter 2 discusses the three complementary points of view of single particle motion, kinetic theory, and fluid description. The corresponding theoretical models provide the opportunity to introduce some of the basic concepts of plasma physics.

- Chapter 3 gives the "derivation" of the macroscopic equations from the kinetic (Boltzmann) equation. Quotation marks because a fully satisfactory derivation can not be given at present in view of the largely unknown contribution of turbulent transport processes. The presentation given is meant to provide some idea on the limitations of the macroscopic view point.

- Chapter 4 defines the MHD model and introduces the concept of scale independence. The central importance of the conservation laws is discussed at length. Based on this, the similarities and differences of laboratory and astrophysical plasmas are articulated in terms of a number of generic boundary value problems.

- Chapter 5 derives the basic MHD waves and describes their properties, with an eye on their role in spectral analysis and computational MHD. The theory of characteristics is introduced as a way to describe the propagation of nonlinear disturbances.

- Chapter 6 treats the subject of waves and instabilities from the unifying point of view of spectral theory. The force operator formulation and the energy principle are extensively discussed. The analogy with quantum mechanics is pointed out and exploited. The difficult extension to interface systems is treated in detail.

- Chapter 7 applies the spectral analysis developed in Chapter 6 to inhomogeneous plasmas in a plane slab. The wave equation for gravito-MHD waves is derived and solved in various limits. Here, all the intricacies of the subject enter: continuous spectra, damping of Alfvén waves, local instabilities, etc. The analogy between helioseismology and MHD spectroscopy in tokamaks is shown to hold great promise for the investigation of plasma dynamics.

- Chapter 8 introduces the enormous variety of magnetic phenomena in astrophysics, in particular for the solar system (dynamo, solar wind, magnetospheres, etc.), and provides basic examples of plasma dynamics worked out in later chapters.

- Chapter 9 is the cylindrical counterpart of Chapter 7, with a wave equation describing the various waves and instabilities. It presents the stability analysis of diffuse cylindrical plasmas (classical pinches and present tokamaks) from the spectral perspective.

- Chapter 10 solves the initial value problem for one-dimensional inhomogeneous MHD and the associated damping due to the continuous spectrum.

- Chapter 11 discusses resonant absorption and phase mixing in the context of heating mechanisms of solar and stellar coronae. Sunspot seismology is introduced as another example of MHD spectroscopy.

Contents of Volume [2]:

- Chapter 12 initiates the most urgent extension of the theory presented in Volume [1]: waves and instabilities in plasmas with stationary background flows, a theme of common interest for laboratory fusion and astrophysical plasma research. The old problem of how to find the complex eigenvalues of stationary plasmas is solved by means of a new method of constructing solution paths in the complex plane.

- Chapter 13 applies the new theory of Chapter 12 to the two classical topics of shear flow in plane plasma slabs, including the Kelvin–Helmholtz instability, and to rotation in cylindrical plasmas, including the magneto-rotational instability.

- Chapter 14 treats the considerable modification of plasma dynamics when resistivity is introduced in the MHD description, both in the linear domain of spectral theory and in the nonlinear domain of reconnection.

- Chapter 15 introduces the basic techniques of computational MHD, the discretization techniques, the methods of time stepping, etc. It thus provides the modern techniques needed to solve for the dynamics of plasmas in complicated magnetic geometries.

- Chapter 16 presents the classical theory of static equilibrium of toroidal plasmas, a topic of central interest in fusion research of tokamaks. Both analytical theory and numerical solutions are presented.

- Chapter 17 concerns the spectral theory of waves and instabilities in toroidal equilibria, again a central topic in tokamak research. Because of this important application, this part of MHD spectral theory is the most developed one, also with respect to comparison with experimental data. This activity is rightly called *MHD spectroscopy*.

- Chapter 18 introduces the theory of transonic equilibria and spectral theory of those equilibria, a subject of huge interest, but still in its infancy.

- Chapter 19 presents the counterpart of Chapter 15 by introducing the numerical methods for nonlinear MHD, in particular for plasmas with large background flows, applied in the last two chapters.

- Chapter 20 discusses the MHD shock conditions from a new perspective, scale independence leading to time reversal duality, and it introduces some of the important areas of application of nonlinear MHD, viz. astrophysical winds and transonic flows.

- Chapter 21 introduces special relativistic MHD, in particular the linear waves and nonlinear shocks that occur at relativistic speeds. The books ends with applications to astrophysical phenomena, like relativistic jets, and thus completes the panorama of the tremendously exciting field of magnetohydrodynamics dominated by flows.

It is impossible to include all topics that actually belong to the field of advanced MHD. Fortunately, books or chapters of books exist on most of those topics, like *dynamos* (Moffatt [337], Ortolani & Schnack [357], Ferriz-Mas & Núñez [133], Rüdiger & Hollerbach [397]); *chaos* (White [483]); *stellarators* (Freidberg [141]); *spheromaks* (Bellan [31]); *anomalous transport* (Balescu [21], Yoshizawa, Itoh & Itoh [492]); *MHD turbulence* (Biskamp [47]).

We wish to acknowledge Guido Huysmans, Jelle Kaastra, Giovanni Lapenta, Sasha Lifschitz, Zakaria Meliani, Gábor Tóth, Ronald Van der Linden and Henk van der Vorst for constructive comments on selected chapters, Jan Willem Blokland for his input on the exercises of various chapters, and Bram Achterberg, Hubert Baty, Sander Beliën, Nicolas Bessolaz, Tom Bogdan, Fabien Casse, Paul Charbonneau, Peter Delmont, Dan D'Ippolito, Jeff Freidberg, Ricardo Galvão, Marcel Goossens, Giel Halberstadt, Tony Hearn, Bart van der Holst, Hanno Holties, Wolfgang Kerner, Rob Kleibergen, Max Kuperus, Keith MacGregor, Daniel Mueller, Valery Nakariakov, Ronald Nijboer, Eric Priest, Jan Rem, Ilia Roussev, Paulo Sakanaka and Karel Schrijver for fruitful collaborations and exchange of ideas. We also thank our copy-editor, Frances Nex, for very careful and efficient editing of our text.

The first author is particularly indebted to the management of the FOM-Institute for Plasma Physics "Rijnhuizen", Aart Kleyn, Niek Lopes Cardozo, Noud Oomens and Jan Kranenbarg, for having provided optimum conditions to work on this book. We also wish to thank Simon Capelin of Cambridge University Press for his support and patience over all those years of preparation of this second volume (a project agreed to be completed in less than two years after the first one), thus accepting the universal validity of the circle theorem.

Circle theorem: The actual time to complete a project is precisely π times the best estimate of the time that one foresees at the beginning of it.
Proof: Standing at the disk of the unknown, the best estimate is based on how long it takes to reach the other side, the actual time spent involves encircling it so as to really enclose it from all sides. That path is precisely π times longer; QED.

Finally, a frequently asked question is: "Will there be a third volume?" Yes, there will be, and you, the serious students of these two volumes who realized that these are just introductions to an enormous field of largely unexplored territory, are going to write it. Remember, with plasmas making up 90% of all (so far visible) matter of the Universe, and plasma physics under-represented in the physics curriculum of the universities, there is no doubt that there will be completely unexpected discoveries for you in store. Nature is on your side!

Part III

Flow and dissipation

12

Waves and instabilities of stationary plasmas

12.1 Laboratory and astrophysical plasmas

12.1.1 Grand vision: magnetized plasma on all scales

In Chapter 1 of the preceding Volume [1] we pointed out that, since more than 90% of visible matter in the Universe is plasma, the dynamics of plasmas and the associated magnetic fields are an important constituent of the description of nature. In Chapter 4[1], we then showed that *the equations of magnetohydrodynamics (MHD) are scale-independent*: the scales of length, density and magnetic field strength of a magnetically confined plasma may be divided out. This simple fact has the amazing consequence that the macroscopic dynamics of plasmas in both laboratory fusion devices (tokamaks, stellarators, etc.) and astrophysical objects (stellar coronae, accretion disks, spiral arms of galaxies, etc.) may be described by the same equations, viz. the equations of MHD. We encountered several examples of this before, in Volume [1]. In the present Volume [2], we will continue the investigation of this common field of research by means of the new "wide-angle MHD telescope".

Figure 12.1 shows two representative, but very different, examples from science and technology, viz. the design drawing of the international tokamak experimental reactor ITER, presently under construction, and an image made by the Hubble Space Telescope of the Pinwheel Galaxy M101. The consequence of scale-independence is that the most obvious difference of the two configurations, their length scale indicated next to the figure, is actually irrelevant for the description of macroscopic plasma dynamics!

▷ **Scale-dependent models** To avoid misunderstanding: small-scale kinetic or two-fluid effects like electron inertia [20], described by the scale-dependent model of Hall-MHD, can have macroscopic consequences like reconnection and waves (see Section 14.5), which may even be detectable by spacecrafts flying through the bow shock of the magnetosphere; see Stasiewicz [418]. Likewise, in the description of hot plasmas in thermonuclear confinement experiments, kinetic effects exhibit a bewildering range of dynamical phenomena on many spatial and temporal scales presenting a challenge to the computational modeling of these plasmas by different, scale-dependent, fluid closures; see Schnack *et al.* [403]. ◁

For our present purpose, the Hubble Space Telescope picture is somewhat misleading since it only shows the stars and dust. Roughly an equal amount of plasma should be present in the plasma component of galaxies (not counting the plasma interiors of the stars themselves), and much more mass should be present in the dark matter component. According to a recent review by Fukugita [150], for the Universe as a whole the balance is shifted significantly towards plasma: ten times more mass is present in plasmas than in stars (again, not counting the fact that stars themselves are mostly plasma). Since we have no clue about the physics of dark matter, it might be advisable to first investigate the plasma component with all techniques that are presently available. Recalling our critical discussion of the standard view of nature, which does not articulate the distinction between neutral gas and plasma, as schematically represented in Figure 1.8 [1], one would expect on the contrary that the abundance of plasma ($\equiv$ abundance of magnetic fields $\equiv$ global anisotropic dynamics) should play a much more prominent role in the description of the Universe than it has done up till now.

In fact, there are many signs that astrophysics is beginning to fill in this gap. For example, when Land and Magueijo [292] established that there is a small but statistically significant anisotropy, with a preferred axis, in the cosmic background radiation as observed with the WMAP satellite, the far-going implications for cosmology were immediately realized. A number of researchers, e.g. Hutsemékers [234] and Longo [316], started to speculate that, amongst other more exotic possibilities, a large-scale cosmic magnetic field might be involved.

As another example, Kaastra *et al.* [251], and several other researchers (see the review by Peterson and Fabian [368]), have recently pointed out that magnetic fields may play an important role in the dynamics of clusters of galaxies. From X-ray spectra obtained from the XMM-Newton satellite, they conclude that magnetized plasmas in huge magnetic loops, of similar spatial structure to those in stellar coronae, may be responsible for the temperature decrement observed for cooling plasma flows in those clusters.

Also, recently, filaments of warm hot intergalactic matter (WHIM) connecting clusters of galaxies have been detected unequivocally by Werner *et al.* [478] by means of X-ray images obtained from the same satellite. This discovery appears to agree with dark matter simulations that ascribe this "cosmic web" mainly to dark matter, but it would come as no surprise if the filamentary structure were associated with a magnetized plasma component as well. The bookkeeping of the gravitational effects ascribed to dark matter might well change in the direction of a larger contribution of plasma.

Whatever the final outcome of these debates will be, it is probable that plasma

(a)

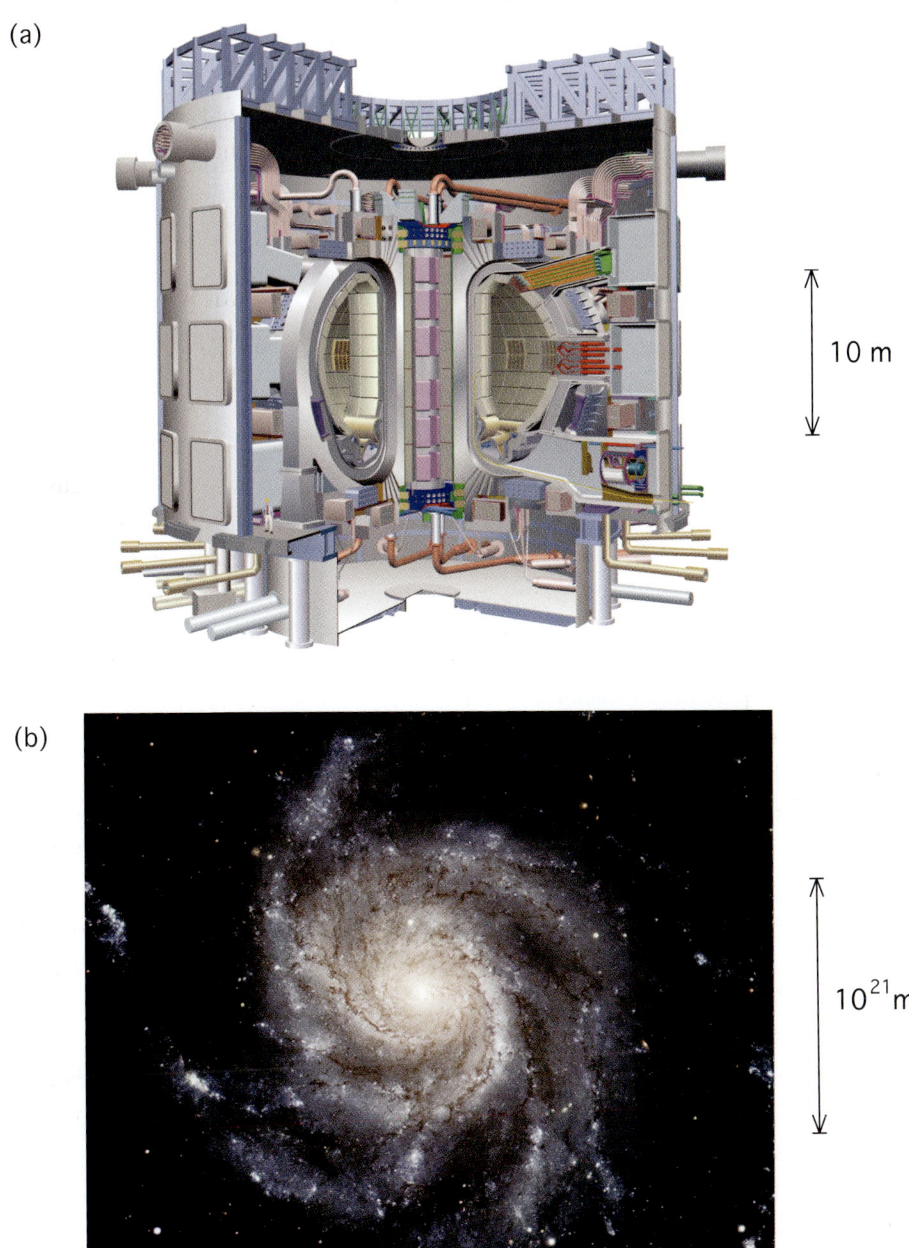

Fig. 12.1 Magnetized plasmas in the laboratory and in astrophysics: (a) the international tokamak experimental reactor ITER; (b) the Pinwheel Galaxy M101 (HST, NASA-ESA).

and, hence, magnetic fields will become much more central for our understanding of the dynamics of the Universe at large than presently accounted for.

12.1.2 Differences between laboratory and astrophysical plasmas

Although scale-independence of the MHD equations permits analysis of global plasma dynamics in laboratory and astrophysical plasmas by the same techniques, the important differences of the parameters that govern overall force balance should not be lost sight of. For example, the parameter $\beta \equiv 2\mu_0 p / B^2$ is small for tokamak plasmas and usually large for astrophysical plasmas, so that plasma dynamics in tokamaks is always dominated by magnetic fields whereas this may not be the case for astrophysical plasmas.

Roughly speaking, one could distinguish the two kinds of plasma configurations on the basis of the following global equilibrium characteristics.

(a) Tokamaks are *magneto*-hydrodynamic plasmas, with a magnetic field that is approximately a force-free field (FFF),

$$\mathbf{j} \times \mathbf{B} \approx 0 \qquad \textit{(FFF to leading order)} \qquad (12.1)$$

$$= \nabla p \ \sim \beta \ll 1 \qquad \textit{(important correction)}.$$

Consequently, the equilibrium is nearly exclusively determined by the magnetic field geometry, but the pressure corrections are essential since they determine the power output of a future fusion reactor.

(b) Most astrophysical objects are *hydro*-magnetic plasmas, with sizeable flows, and the gravitational acceleration is usually not negligible,

$$\rho \mathbf{v} \cdot \nabla \mathbf{v} + \nabla p + \rho \nabla \Phi \approx 0 \qquad \textit{(Keplerian flow to leading order)} \qquad (12.2)$$

$$= \mathbf{j} \times \mathbf{B} \ \sim \beta^{-1} \ll 1 \quad \textit{(important correction)}.$$

Consequently, gravity and rotation usually dominate over the magnetic terms, but the latter may be crucial for the growth or damping of instabilities (as for the Parker instability, discussed below, and the magneto-rotational instability which even operates when the magnetic field is infinitesimal, see Section 13.4.2).

It is well known that a force-free magnetic field cannot be extended indefinitely, as follows from the virial theorem (see Shafranov [409], p. 106). Eventually, the magnetic pressure has to be balanced by something. In tokamaks, equilibrium is due to balancing of the Lorentz forces on the plasma by mechanical forces on the induction coils, which have to be firmly fixed to the laboratory by "nuts and bolts". (Without those, the configuration would simply fly apart: a magnetic field of 5 T exerts a pressure of $B^2/(2\mu_0) \approx 10^7 \,\mathrm{N\,m^{-2}} \approx 100 \,\mathrm{atm}$.) The mechanical counterpart for accretion disks or galaxies is balancing of the centrifugal acceleration by the gravitational pull of the central objects, which may include a black hole. The implications of this difference for stability are much more wide-ranging than generally realized, as will be illustrated by contrasting "intuition" developed on tokamak stability to some major instabilities operating in astrophysical plasmas.

Interchanges in tokamaks and Parker instability in galaxies

To appreciate the issue, let us pronounce some general features of tokamak stability theory, based on the results from the quasi-cylindrical approximation presented in Section 9.4 [1] and anticipating the exact toroidal representation to be developed in Chapter 17. For the present purpose, the difference between the cylindrical approximation in terms of r, θ and z and the toroidal representation in terms of ψ (the poloidal magnetic flux, the "radial" coordinate), ϑ (the poloidal angle) and φ (the toroidal angle) may be ignored. Without exaggeration, it may then be said that the wide variety of MHD instabilities operating in tokamaks, represented by normal modes of the form

$$f(\psi, \vartheta, \varphi, t) = \sum_m \tilde{f}_m(\psi)\, e^{i(m\vartheta + n\varphi - \omega t)}, \tag{12.3}$$

is unstable only for (approximately) perpendicular wave vectors,

$$\mathbf{k}_0 \perp \mathbf{B} \quad \Rightarrow \quad -i\mathbf{B} \cdot \nabla \sim m + nq \approx 0. \tag{12.4}$$

The reason is the enormous field line bending energy of the Alfvén waves,

$$W_A \approx \tfrac{1}{2} \int \left[(\mathbf{k}_0 \cdot \mathbf{B})^2 \, |\mathbf{n} \cdot \boldsymbol{\xi}|^2 + \cdots \right] dV \gg 0, \tag{12.5}$$

so that field line localization ($k_\parallel \ll k_\perp$) is necessary to eliminate this term and to get instability from the different higher order terms due to, e.g. pressure gradients and currents. The Ansatz (12.4) is made in virtually all tokamak stability calculations, like in the derivation of the Mercier criterion [331] involving *interchanges* on rational magnetic surfaces, of *ballooning modes* [91] involving localization about rational magnetic field lines, of *internal kink modes*, of *neo-classical tearing modes*, of *external kink modes*, etc.; see Sections 17.2 and 17.3. All involve localization about rational magnetic surfaces, either inside the plasma or in an outer vacuum. Hence, it became a kind of "intuition" in tokamak physics to assume that this is a general truth about plasma instabilities.

In contrast, some major instabilities in astrophysical plasmas turn out to operate under precisely the opposite conditions:

$$\mathbf{k}_0 \parallel \mathbf{B} \quad \Rightarrow \quad -i\mathbf{B} \cdot \nabla \sim m + nq \approx 1. \tag{12.6}$$

These include the *Parker instability* operating in spiral arms of galaxies [365] and the *magneto-rotational instability* [467, 83, 18] which is held responsible for the turbulent dissipation in accretion disks about a compact object. Both have their largest growth rates when the wave vector $\mathbf{k}_0$ is about parallel to the magnetic field, and certainly not perpendicular! (It is most peculiar that this apparent contradiction with stability of laboratory plasmas went unnoticed so far.) How is the above argument about the dominance of field line bending energy circumvented for astrophysical plasmas?

In order to answer that question, let us compare how the two entirely different pairs of equilibrium conditions (12.1) and (12.2), and the associated pairs of instability conditions (12.4) and (12.6), appear in the analysis of the gravitational interchange [171]. This instability has played an important role in modeling both the stability of laboratory plasmas (where gravity is used as just a way to simulate magnetic field line curvature) and the Parker instability [365] which is concerned with instability due to genuine gravity in spiral arms of galaxies.

To that end, we recapitulate the major conclusions on the gravitational interchange from Sections 7.5.2 and 7.5.3 [1], Eqs. (7.199), (7.206) and (7.212). The stability criterion for gravitational interchanges of a plane plasma slab reads:

$$- \rho N_{\mathrm{B}}^2 \equiv \rho' g + \frac{\rho^2 g^2}{\gamma p} \leq \tfrac{1}{4} B^2 \varphi'^2 \,, \tag{12.7}$$

where N_{B} is the Brunt–Väisäläa frequency and φ' is the magnetic shear. Without magnetic shear, stability just appears to depend on the square of the Brunt–Väisäläa frequency: $N_{\mathrm{B}}^2 \geq 0$, which amounts to the Schwarzschild criterion for convective stability when expressed in terms of the equilibrium temperature gradient. This criterion is obtained from the marginal equation of motion ($\omega^2 = 0$) in the limit of small parallel wave number ($k_\parallel \to 0$). However, when these two limits are interchanged ($k_\parallel = 0$ and $\omega^2 \to 0$), an entirely different criterion is obtained:

$$- \rho N_{\mathrm{M}}^2 \equiv \rho' g + \frac{\rho^2 g^2}{\gamma p + B^2} \leq 0 \,, \tag{12.8}$$

where N_{M} is the magnetically modified Brunt–Väisäläa frequency. The apparent discrepancy between these stability criteria was resolved by Newcomb [348] who noted that there is a cross-over of two branches of the local dispersion equation with the solutions

$$\omega_1^2 = (k_0^2/k_{\mathrm{eff}}^2)\, N_{\mathrm{M}}^2 \qquad \qquad \textit{(pure interchanges)} \,, \tag{12.9}$$

$$\omega_2^2 = \frac{N_{\mathrm{B}}^2}{N_{\mathrm{M}}^2} \frac{\gamma p}{\gamma p + B^2} \frac{1}{\rho} (\mathbf{k}_0 \cdot \mathbf{B})^2 \quad \textit{(quasi-interchanges)} \,, \tag{12.10}$$

where the last mode is the first to become unstable when the density gradient is increased. The first expression holds for $k_\parallel = 0$, where the factor $k_{\mathrm{eff}}^2 \equiv k_0^2 + n^2 \pi^2 / a^2$ indicates clustering of the modes, $\omega_1^2 \to 0$, when the vertical mode number becomes large, $n \to \infty$ (this n should not be confused with the toroidal mode number n introduced above), and the second expression is only valid for $k_\parallel \ll k_\perp$. Hence, *field line bending is small in both cases.*

In cylindrical and toroidal plasmas, magnetic field line curvature is unavoidable and interchange instabilities then arise when the negative pressure gradient (associated with confinement) exceeds the shear of the magnetic field lines, analogous to the gravitational interchange criterion (12.7). This is expressed by the criteria of Suydam [427], Eq. (9.118)[1], and Mercier [331], Eq. (17.98). For cylindrical plasmas without magnetic shear, expressions were derived for the growth rates of interchanges and quasi-interchanges [474, 160, 177], analogous to Eqs. (12.9) and (12.10) with the following replacements:

$$N_{\mathrm{M}}^2 \to \frac{2B_\theta^2}{\rho r B^2}\left(p' + \frac{\gamma p}{\gamma p + B^2}\frac{2B_\theta^2}{r}\right), \qquad N_{\mathrm{B}}^2 \to \frac{2B_\theta^2}{\rho r B^2}p'. \tag{12.11}$$

As illustrated in Figs. 9.15 and 9.11 [1], when p' becomes negative (violation of the shearless limit of Suydam's criterion) first the quasi-interchanges become unstable and the pure interchanges become unstable when $p' \le -\gamma p\,(\gamma p + B^2)^{-1}\,(2B_\theta^2)/r$, in agreement with the expression for the z-pinch derived by Kadomtsev [252].

It would appear that the analogy between plasmas with curved magnetic fields and gravitational plasmas is perfect: instability only occurs at the interchange value $k_\parallel = 0$ or close to it. However (we now complete the analysis of the gravito-MHD waves started in Section 7.3.3[1]), the *Parker instability* operates under precisely opposite conditions ($k_\perp \approx 0$). For an exponential atmosphere, its growth rate is given by expanding the expression (7.112)[1]:

$$\omega^2 \approx \left(1 + \frac{\rho N_{\mathrm{B}}^{'2}}{k_{\mathrm{eff}}^2 B^2}\right)\frac{\gamma p}{\gamma p + B^2}\frac{1}{\rho}\,(k_0 B)^2. \tag{12.12}$$

This looks similar to the expression (12.10) for the quasi-interchanges, which gives the growth rate at $k_\parallel \approx 0$ for *localized modes* ($n \to \infty$), but it is actually completely different since the expression (12.12) for the Parker instability requires $k_\perp \approx 0$ and only yields instability for *global modes* ($n \approx 1$). This is so because the criterion for the Parker instability, $k_{\mathrm{eff}}^2 B^2 + \rho N_{\mathrm{B}}^2 < 0$, cannot be satisfied for $n \to \infty$, since $k_{\mathrm{eff}} \to \infty$ then. In other words, it is very well possible to have a global instability when the field line bending energy (12.5) is not small at all! This is also the case for the magneto-rotational instability (see Section 13.4.2).

Hence, MHD instabilities occur in astrophysical plasmas under conditions that do not allow instability in laboratory plasmas. The reason is the stabilizing "backbone" of a large toroidal magnetic field in the latter. Estimating orders of magnitudes for an equilibrium with inhomogeneity length scale L, the Parker instability requires

$$N_{\mathrm{B}}^2 \sim -\frac{B^2}{\rho L^2}, \quad \text{with } \beta \sim 1. \tag{12.13}$$

(Note that a sizeable magnetic field is required, actually violating the simplified order of magnitude estimate $\beta \gg 1$ of Eq. (12.2)(b).) In contrast, the corresponding term of Eq. (12.11)(b) for curvature–pressure gradient driven interchanges in quasi-cylindrical/toroidal equilibria requires

$$\frac{2B_\theta^2}{\rho r B^2}\, p' \sim -\epsilon^2 \beta \cdot \frac{B^2}{\rho L^2}\,, \quad \text{with } \epsilon \equiv a/R_0 \ll 1\,, \ \beta \sim \epsilon^2\,. \tag{12.14}$$

In the first case, the driving force of the instability can compete with the field line bending contributions (12.5). In the latter case, because of the small factor $\epsilon^2 \beta \sim \epsilon^4$, this is impossible so that pressure-driven interchanges in cylindrical and toroidal plasmas never occur for $\mathbf{k}_0 \parallel \mathbf{B}$. Consequently, tokamak "intuition" focusing on rational magnetic surfaces and field lines as exclusively determining stability may be misleading for astrophysical plasmas.

The two different view points can be reconciled as follows. Whereas the condition (12.4) for tokamak instability automatically leads to study of the degeneracy and couplings of the Alfvén and slow continua close to marginal stability ($\omega \approx 0$), an entirely different path to avoid the stabilizing contribution (12.5) of the Alfvén waves is exploited by the Parker instabilities. These are actually modified slow magneto-acoustic waves avoiding the coupling to the Alfvén waves by remaining orthogonal to them: the polarization (expressed by the eigenvector $\boldsymbol{\xi}$) of the Parker (slow) modes is parallel to $\mathbf{B}$ (flow along the magnetic field is essential), whereas the polarization of the Alfvén waves is mainly perpendicular to $\mathbf{B}$. This orthogonality is clearly exhibited by Fig. 12.2, which shows the complete low-frequency part of the spectrum of modes for a gravitating plasma slab with exponential dependence on height of the density, magnetic field and pressure, for different values of the angle ϑ between the horizontal wave vector $\mathbf{k}_0$ and the magnetic field.

The exponentially stratified equilibrium was analyzed in Section 7.3.2 [1], resulting in the dispersion equation (7.116) with solutions shown in Fig. 7.10 for fixed angle ϑ. These solutions are now shown in Fig. 12.2(a) for all directions of $\mathbf{k}_0$. At $\vartheta = 0$ ($\mathbf{k}_0 \parallel \mathbf{B}_0$), the Parker instability has its largest growth rate, whereas around $\vartheta = \frac{1}{2}\pi$ ($\mathbf{k}_0 \perp \mathbf{B}_0$), the interchanges and quasi-interchanges operate. These two ranges correspond to two different instability mechanisms: in the $k_\parallel \approx 0$ range, coupling of *local* (high n) slow and Alfvén modes causes interchange or quasi-interchange instability, whereas in the range $k_\perp \approx 0$, *global* (low n) instability of the slow magneto-sonic branch, viz. the Parker instability, occurs. In the intermediate range, there is a smooth transformation from the Parker instability to interchanges via modes that we have termed *quasi-Parker instabilities*.

The local interchange and quasi-interchange instabilities are modified substantially by the introduction of magnetic shear,

$$\mathbf{B} = B_0 \mathrm{e}^{-\frac{1}{2}\alpha x}\left[\sin(\lambda x)\mathbf{e}_y + \cos(\lambda x)\mathbf{e}_z\right]. \tag{12.15}$$

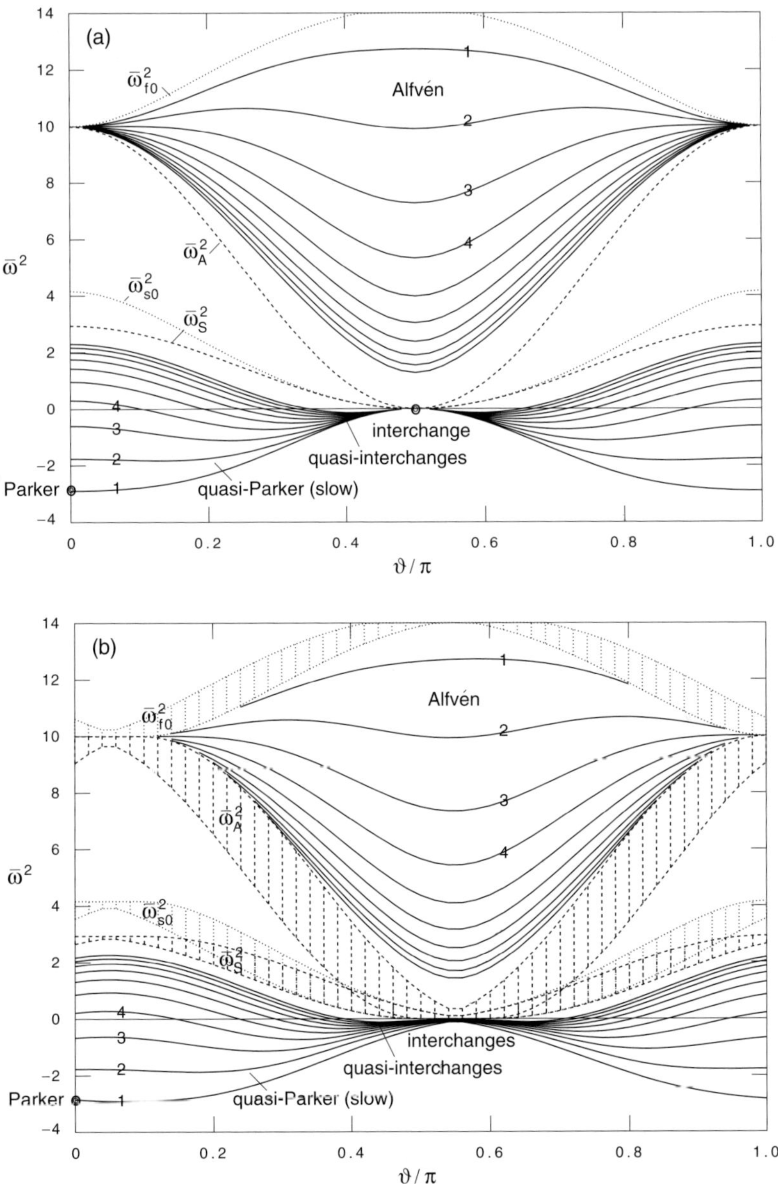

Fig. 12.2 Spectrum of slow (quasi-Parker) instabilities, connecting the Parker instability to the quasi-interchanges, and Alfvén waves for different angles between $\mathbf{k}_0$ and $\mathbf{B}_0$ for exponential atmosphere with (a) uni-directional field ($\bar{\lambda} = 0$), (b) magnetic shear ($\bar{\lambda} = 0.3$) and genuine continua $\bar{\omega}_A^2$ and $\bar{\omega}_S^2$; $\bar{\alpha} = 20$, $\beta = 0.5$, $\bar{k}_0^2 = 10$, $\bar{q} = n\pi$ ($n = 1, 2, \ldots, 10$).

Except for modifying the stability properties, it also leads to the bands of continuous spectra ω_A^2 and ω_S^2 separating the Alfvén waves from the gravitational instabilities, as shown in Fig. 12.2(b), obtained by solving the full implicit eigenvalue problem (7.91), (7.93)[1] with the "shooting" method. Clearly, the consideration of magnetic shear is essential for the analysis of local stability criteria, as usual in tokamak stability studies, but not so important for the Parker instability.

In conclusion, a full spectral analysis of gravitational instabilities exhibits the existence of a large class of instabilities, called *quasi-Parker instabilities*, that smoothly connect the Parker instability (operating at $\mathbf{k} \parallel \mathbf{B}$) of astrophysical plasmas to the quasi-interchanges and interchanges of laboratory plasmas (operating at $\mathbf{k} \perp \mathbf{B}$). Eventually (stretching the imagination now), such full spectral studies of the gravitational waves in galaxies could lead to *MHD spectroscopy of galactic plasmas*, i.e. determination of the internal characteristics of the galactic plasma by computing and observing the spatial distribution of the modes (since the frequencies themselves are unobservable on a human time scale).

12.1.3 Plasmas with background flow

We have come now to a point in our exposition where we have to face a basic omission of the theory expounded so far. In Volume [1] on *Principles of Magnetohydrodynamics*, we first presented "Plasma physics preliminaries" in Part I and then developed "Basic magnetohydrodynamics" on equilibrium, waves and instabilities in Part II mainly from the idealized picture of a plasma at rest in *static equilibrium*. This view on plasmas from laboratory fusion research has led to many fruitful insights, but it fails to account for the dynamics of the vast majority of astrophysical plasmas where this assumption is simply wrong. Also, in fusion research, the ultimate goal of an energy producing machine requires the presence of substantial plasma flows caused by the injection of neutral beams to heat the plasma fuel and by divertors to get rid of the exhaust and impurities. Hence, the study of the *influence of background flows on equilibrium, waves and instabilities* has become an important, and common, research theme for the study of both laboratory and astrophysical plasmas.

The subject of background plasma flow implies that the hydrodynamics component of MHD has to be taken more seriously. Hence, the remainder of this chapter, which is the first of Part III on "Flow and dissipation", is devoted to the urgent modifications of MHD spectral theory by the consequences of equilibrium plasma flows. This involves accounting for the effects of the Doppler shift on the spectra, eliminating misconceptions about supposed non-self-adjointness of the operators, construction of a method to compute eigenvalues in the complex plane, and application to plane shear flow and rotation.

In the later chapters of Part III, we will extend the exposition to include dissipation, in particular resistivity, which may be considered as the counterpart of viscosity in hydrodynamics (Chapter 14), and to discuss in detail the numerical solution techniques needed for spectral calculations of ideal and dissipative plasmas (Chapter 15). Only after these basic techniques have been elaborated will the other urgent extension of Volume [1] be presented, viz. spectral theory of toroidal plasmas like tokamaks, in Part IV on "Toroidal plasmas", since this demands moving from the solution of ordinary to the solution of partial differential equations. Chapters 16 and 17 are devoted to the description of toroidal equilibrium and stability for the standard picture of static plasmas, and Chapter 18 again to the substantial extension for stationary toroidal plasmas. Finally, in Part V on "Nonlinear dynamics", the analysis of moving plasmas will be extended to the nonlinear domain where separation of equilibrium and stability is no longer feasible. After presenting the numerical solution techniques needed to analyze the nonlinear dynamics (Chapter 19), the subject of moving plasmas is resumed with the analysis of transonic flow and shocks (Chapter 20), and finally to the extreme speeds of relativistic plasma flows encountered in the wide variety of explosions occurring in the final phases of stellar and galactic evolution (Chapter 21).

12.2 Spectral theory of stationary plasmas

12.2.1 Basic equations

To describe the dynamics of plasmas with stationary flow, we start from the set of nonlinear ideal MHD equations (see Volume [1], Chapter 4) for the density ρ, the velocity $\mathbf{v}$, the pressure p and the magnetic field $\mathbf{B}$:

$$\frac{\partial \rho}{\partial t} + \nabla \cdot (\rho \mathbf{v}) = 0 \,, \tag{12.16}$$

$$\rho \left(\frac{\partial \mathbf{v}}{\partial t} + \mathbf{v} \cdot \nabla \mathbf{v} \right) + \nabla p - \mathbf{j} \times \mathbf{B} - \rho \mathbf{g} = 0 \,, \qquad \mathbf{j} = \nabla \times \mathbf{B} \,, \tag{12.17}$$

$$\frac{\partial p}{\partial t} + \mathbf{v} \cdot \nabla p + \gamma p \nabla \cdot \mathbf{v} = 0 \,, \tag{12.18}$$

$$\frac{\partial \mathbf{B}}{\partial t} - \nabla \times (\mathbf{v} \times \mathbf{B}) = 0 \,, \qquad \nabla \cdot \mathbf{B} = 0 \,. \tag{12.19}$$

This set is to be complemented with appropriate initial and boundary conditions, e.g. the BCs for model I (plasma confined inside a rigid wall):

$$\mathbf{n} \cdot \mathbf{v} = 0 \,, \qquad \mathbf{n} \cdot \mathbf{B} = 0 \qquad \textit{(at the wall)} \,, \tag{12.20}$$

as discussed in Section 4.6.1 [1]. The gravitational acceleration $\mathbf{g} = -\nabla \Phi$ is considered to be caused by a fixed external gravitational field Φ. We will consistently exploit scale independence (see Section 4.1.2 [1]) and drop factors μ_0.

Most of Volume [1] was concerned with *static equilibria* ($\mathbf{v} = 0$),

$$\nabla p = \mathbf{j} \times \mathbf{B} + \rho \mathbf{g}, \qquad \mathbf{j} = \nabla \times \mathbf{B}, \qquad \nabla \cdot \mathbf{B} = 0, \tag{12.21}$$

and perturbations of these equilibria described by the equation of motion with the force operator $\mathbf{F}$ acting on the plasma displacement vector $\boldsymbol{\xi}$:

$$\mathbf{F}(\boldsymbol{\xi}) = \rho \frac{\partial^2 \boldsymbol{\xi}}{\partial t^2}. \tag{12.22}$$

The spectral problem for normal mode solutions $\hat{\boldsymbol{\xi}}(\mathbf{r}) \exp(-i\omega t)$,

$$\mathbf{F}(\hat{\boldsymbol{\xi}}) = -\rho \omega^2 \hat{\boldsymbol{\xi}}, \tag{12.23}$$

was particularly effective since the linear operator $\mathbf{F}$ is self-adjoint, so that the eigenvalues ω^2 are real.

All this changes when *stationary equilibria* ($\mathbf{v} \neq 0$) are considered. In the first place, none of the MHD equations (12.16)–(12.19) is trivially satisfied now, so that the description of stationary equilibria requires four differential equations:

$$\nabla \cdot (\rho \mathbf{v}) = 0, \tag{12.24}$$

$$\rho \mathbf{v} \cdot \nabla \mathbf{v} + \nabla p - \mathbf{j} \times \mathbf{B} - \rho \mathbf{g} = 0, \qquad \mathbf{j} = \nabla \times \mathbf{B}, \tag{12.25}$$

$$\mathbf{v} \cdot \nabla p + \gamma p \nabla \cdot \mathbf{v} = 0, \tag{12.26}$$

$$\nabla \times (\mathbf{v} \times \mathbf{B}) = 0, \qquad \nabla \cdot \mathbf{B} = 0. \tag{12.27}$$

Note that the presence of a background flow not only enlarges the set of equilibrium solutions with an additional function $\mathbf{v}(\mathbf{r})$, but it also enlarges the freedom of choice for the functions $\rho(\mathbf{r})$, $p(\mathbf{r})$ and $\mathbf{B}(\mathbf{r})$. Paradoxically, though more equations are to be satisfied, more solutions are permitted. To convince yourself of this basic fact, consider a variant of Fig. 6.1 [1] with two balls on opposite sides of the top of a hill: not an equilibrium. However, if the two balls are constrained by a wire connecting them, all of a sudden infinitely many equilibrium positions are obtained. In other words, constraints produce a more intricate energy landscape.

Before we turn to the implications for the waves and instabilities, let us consider the effects of flow on the two generic classes of equilibria, of plane plasma slabs and cylindrical flux tubes, that were studied in Chapters 7 and 9 of Volume [1]. For the stationary equilibrium of a plane slab with gravity, $\mathbf{g} = -g\mathbf{e}_x$, and the other inhomogeneities also in the vertical direction, one easily convinces oneself that the equilibrium equations (12.24), (12.26), (12.27) are trivially satisfied whereas the momentum equation (12.25) is unchanged with respect to the static case:

$$(p + \tfrac{1}{2}B^2)' = -\rho g \qquad \left(' \equiv \frac{d}{dx}\right). \tag{12.28}$$

Hence, the functions $v_y(x)$ and $v_z(x)$ are entirely free, whereas $\rho(x)$, $p(x)$, $B_y(x)$ and $B_z(x)$ are constrained by Eq. (12.28). This is nice since we can now study the effects of plasma flow on the *same* equilibria as studied for the static case.

For cylindrical equilibria, Eqs. (12.24), (12.26) and (12.27) are also trivially satisfied but the momentum equation has the important additional contribution $\rho\mathbf{v}\cdot\nabla\mathbf{v}$ compared to the static case. In the evaluation of this expression for cylindrical coordinates (using Appendix A), one should account for the non-vanishing derivatives of the unit vectors, $\partial\mathbf{e}_r/\partial\theta = \mathbf{e}_\theta$ and $\partial\mathbf{e}_\theta/\partial\theta = -\mathbf{e}_r$, resulting in the nasty vector expression (A.50). This yields the *centrifugal acceleration*

$$- \mathbf{v} \cdot \nabla\mathbf{v} = (v_\theta^2/r)\mathbf{e}_r \,, \tag{12.29}$$

so that the equilibrium equation becomes

$$(p + \tfrac{1}{2}B^2)' = (\rho v_\theta^2 - B_\theta^2)/r - \rho\Phi' \qquad \left(' \equiv \frac{d}{dr}\right). \tag{12.30}$$

For the present discussion, the gravitational potential is better ignored (cylindrical gravity fields are not very physical) but it is included here since we will need it later, in Section 13.4.2, when discussing the cylindrical limit of accretion disks. Hence, only the translational component $v_z(x)$ can be chosen arbitrarily, the remaining quantities $\rho(x)$, $p(x)$, $v_\theta(x)$, $B_\theta(x)$ and $B_z(x)$ are constrained by the equilibrium equation (12.30). The presence of the rotational component v_θ essentially changes the cylindrical equilibrium with respect to the static case. Moreover, the computation of waves and instabilities becomes much more complicated than for the plane slab: translation and rotation are physically quite different phenomena.

Finally, the Eulerian perturbations ρ_1, $\mathbf{v}_1$, p_1, $\mathbf{B}_1$ of stationary equilibria are described by the following linear differential equations:

$$\left(\frac{\partial}{\partial t} + \mathbf{v} \cdot \nabla\right)\rho_1 + \rho_1\nabla \cdot \mathbf{v} + \nabla \cdot (\rho\mathbf{v}_1) = 0\,, \tag{12.31}$$

$$\left(\frac{\partial}{\partial t} + \mathbf{v} \cdot \nabla\right)\mathbf{v}_1 + \rho\mathbf{v}_1 \cdot \nabla\mathbf{v} + \rho_1(\mathbf{v} \cdot \nabla\mathbf{v} + \nabla\Phi)$$
$$+ \nabla p_1 + \mathbf{B} \times (\nabla \times \mathbf{B}_1) - (\nabla \times \mathbf{B}) \times \mathbf{B}_1 = 0\,, \tag{12.32}$$

$$\left(\frac{\partial}{\partial t} + \mathbf{v} \cdot \nabla\right)p_1 + \gamma p_1\nabla \cdot \mathbf{v} + \mathbf{v}_1 \cdot \nabla p + \gamma p\nabla \cdot \mathbf{v}_1 = 0\,, \tag{12.33}$$

$$\left(\frac{\partial}{\partial t} + \mathbf{v} \cdot \nabla\right)\mathbf{B}_1 - \mathbf{B}_1 \cdot \nabla\mathbf{v} + \mathbf{B}_1\nabla \cdot \mathbf{v}$$
$$- \mathbf{B} \cdot \nabla\mathbf{v}_1 + \mathbf{B}\nabla \cdot \mathbf{v}_1 + \mathbf{v}_1 \cdot \nabla\mathbf{B} = 0\,, \qquad \nabla \cdot \mathbf{B}_1 = 0\,, \tag{12.34}$$

where the unsubscripted variables ρ, $\mathbf{v}$, p, $\mathbf{B}$ should satisfy the equilibrium conditions (12.24)–(12.27).

As in Volume [1], instead of the above seven primitive variables of the Eulerian perturbations, we will exploit a Lagrangian representation in terms of the three components of a displacement vector $\boldsymbol{\xi}$. This reduction will be elaborated in the following subsection. The Lagrangian representation implies a significant simplification for the analysis of ideal plasmas, but it has three implicit shortcomings:

(a) it cannot be generalized to dissipative plasmas;

(b) it involves second order derivatives instead of first order ones;

(c) it ignores the *entropy modes*.

Because of the first two items, we will exploit the Eulerian representation in Chapter 14 on resistive plasma dynamics and Chapter 15 on computational linear MHD. Concerning the third item, recall the discussion of the similar apparent loss of a degree of freedom for static plasmas in Section 5.1 [1]. This may be elucidated by the entropy evolution equation, obtained from Eqs. (12.16) and (12.18):

$$\frac{\mathrm{D}S}{\mathrm{D}t} = 0 \quad \Rightarrow \quad \left(\frac{\partial}{\partial t} + \mathbf{v} \cdot \nabla\right)S_1 + \mathbf{v}_1 \cdot \nabla S = 0. \tag{12.35}$$

where $\mathrm{D}/\mathrm{D}t$ denotes the Lagrangian time derivative. In the absence of a plasma displacement $\boldsymbol{\xi}$, i.e. a velocity perturbation, the Lagrangian representation will automatically deduce from the first expression that $S_{1\mathrm{L}} = 0$ and, hence, also $S_{1\mathrm{E}} = 0$ (since the two are connected by $\boldsymbol{\xi}$, see Eq. (12.54) below). On the other hand, from the second expression, the Eulerian representation will permit $S_{1\mathrm{E}} \neq 0$, even in the absence of a plasma displacement (when $\mathbf{v}_1 = 0$ as well). This corresponds to so-called non-holonomic initial data, i.e. the freedom to choose initial perturbations that do not arise from perturbations satisfying the ideal MHD constraints. In Section 6.1 [1], we have called this "cheating on the initial data" since it also applies to the other variables, like the magnetic field (see Fig. 6.5 [1]). In other words: although, admittedly, the Lagrangian representation misses out on this class of perturbations, one should consider this an advantage since those modes are actually inconsistent with the ideal MHD model. In particular, associated with the possible vanishing of the operator $\partial/\partial t + \mathbf{v} \cdot \nabla$ acting on S_1 in Eq. (12.35), a continuous spectrum of non-holonomic *Eulerian entropy continuum* modes is obtained, that is rightly ignored in the Lagrangian representation. It is important to distinguish these rather flimsy modes from the physically significant Lagrangian *flow continuum modes* of hydrodynamics (introduced by Case [77] in 1960) that are related to the vanishing of the same operator, but for variables that are expressible in terms of $\boldsymbol{\xi}$. We will come back to this in Section 13.1.3.

12.2.2 *Frieman–Rotenberg formulation*

Recall the discussion of MHD spectral theory in Volume [1], Chapter 6. That discussion was based on the reduction from a representation of the perturbations in terms of the primitive, Eulerian, variables ρ_1, $\mathbf{v}_1$, p_1 and $\mathbf{B}_1$ to a representation in terms of the plasma displacement vector $\boldsymbol{\xi}$ alone. This led to the powerful force operator formalism, with spectral representations in complete analogy with the mathematics of quantum mechanics and, hence, the possible transfer of methods and insights from that part of physics. Can this be generalized for plasmas with background flow?

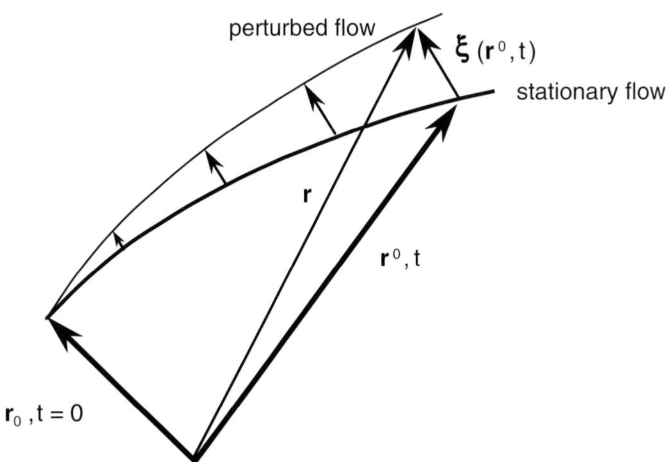

Fig. 12.3 Displacement vector field for plasma with stationary background flow. (Adapted from Frieman and Rotenberg [147].)

Clearly, the crucial part is to construct a displacement vector $\boldsymbol{\xi}$ that connects the perturbed flow with the unperturbed flow (Fig. 12.3) and to specify coordinates that exploit the fact that the stationary equilibrium is actually independent of time. This would not be the case if the perturbed flow were described, in a truly Lagrangian fashion, by the initial positions $\mathbf{r}_0$ of the fluid elements. Instead, a quasi-Lagrangian representation will be exploited, where the position vector $\mathbf{r}$ of a fluid element of the perturbed flow is connected to the position vector $\mathbf{r}^0$ of that same element on the unperturbed flow:

$$\mathbf{r} = \mathbf{r}(\mathbf{r}^0, t) = \mathbf{r}^0 + \boldsymbol{\xi}(\mathbf{r}^0, t)\,. \tag{12.36}$$

In the coordinates $(\mathbf{r}^0, t)$, the equilibrium variables are *time-independent*,

$$\rho^0 \equiv \rho^0(\mathbf{r}^0)\,, \quad \mathbf{v}^0 \equiv \mathbf{v}^0(\mathbf{r}^0)\,, \quad p^0 \equiv p^0(\mathbf{r}^0)\,, \quad \mathbf{B}^0 \equiv \mathbf{B}^0(\mathbf{r}^0)\,, \tag{12.37}$$

by definition satisfying the equilibrium equations (12.24)–(12.27), with a superscript ^{0} on all variables and $\nabla \rightarrow \nabla^0 \equiv \partial/\partial\mathbf{r}^0$. In the end, we will drop these superscripts again, but for now we keep them in order to distinguish perturbed and unperturbed quantities. It has to be demonstrated that the perturbations are expressible in terms of $\boldsymbol{\xi}(\mathbf{r}^0, t)$ alone. Needless to say, $\boldsymbol{\xi}$ is always considered to be small compared to the length scales of the pertinent magnetic geometry. A nonlinear generalization, without that restriction, was made by Newcomb [349, 350]. Here, we will exploit the linear counterparts of his expressions.

To systematically exploit the coordinates $(\mathbf{r}^0, t)$, we need to express the gradient operator at the perturbed position in terms of that at the unperturbed position:

$$\nabla \equiv \frac{\partial}{\partial \mathbf{r}} = \frac{\partial \mathbf{r}^0}{\partial \mathbf{r}} \cdot \frac{\partial}{\partial \mathbf{r}^0} \equiv (\nabla \mathbf{r}^0) \cdot \nabla^0 = (\nabla(\mathbf{r} - \boldsymbol{\xi})) \cdot \nabla^0 = (\mathbf{I} - \nabla \boldsymbol{\xi}) \cdot \nabla^0 .$$

Newcomb's exact expression for the gradient operator is obtained by iteration,

$$\nabla \boldsymbol{\xi} = \nabla^0 \boldsymbol{\xi} - (\nabla^0 \boldsymbol{\xi}) \cdot \nabla^0 \boldsymbol{\xi} + (\nabla^0 \boldsymbol{\xi}) \cdot (\nabla^0 \boldsymbol{\xi}) \cdot \nabla^0 \boldsymbol{\xi} - \cdots = (\mathbf{I} + \nabla^0 \boldsymbol{\xi})^{-1} \cdot \nabla^0 \boldsymbol{\xi}$$

$$\Rightarrow \quad \nabla = (\mathbf{I} + \nabla^0 \boldsymbol{\xi})^{-1} \cdot \nabla^0 \approx \nabla^0 - (\nabla^0 \boldsymbol{\xi}) \cdot \nabla^0 , \tag{12.38}$$

where the latter approximation holds for $\boldsymbol{\xi}$ small. The Lagrangian time derivative is expressed by

$$\frac{\mathrm{D}}{\mathrm{D}t} \equiv \frac{\partial}{\partial t}\bigg|_{\mathbf{r}^0} + \mathbf{v}^0 \cdot \nabla^0 , \tag{12.39}$$

so that the velocity at the perturbed position becomes

$$\mathbf{v}(\mathbf{r}^0 + \boldsymbol{\xi}, t) \equiv \frac{\mathrm{D}\mathbf{r}}{\mathrm{D}t} = \frac{\mathrm{D}\mathbf{r}^0}{\mathrm{D}t} + \frac{\mathrm{D}\boldsymbol{\xi}}{\mathrm{D}t} = \mathbf{v}^0 + \mathbf{v}^0 \cdot \nabla^0 \boldsymbol{\xi} + \frac{\partial \boldsymbol{\xi}}{\partial t} , \tag{12.40}$$

where $\mathbf{v}^0 = \mathbf{v}^0(\mathbf{r}^0)$ is the equilibrium velocity. As compared to the expression (6.18)[1] for static equilibria, the velocity perturbation now contains the important additional contribution $\mathbf{v}^0 \cdot \nabla^0 \boldsymbol{\xi}$ due to the *gradient parallel to the velocity field*. This gives rise to a Doppler shift and possibly, depending on the geometry, curvature contributions.

Similar to the procedure for static equilibria of Section 6.1 [1], we will integrate the equations (12.16), (12.18) and (12.19) for ρ, p and $\mathbf{B}$ to get the exact perturbed quantities, that we will expand to first order in $\boldsymbol{\xi}$, and then substitute the result in Eq. (12.17) for $\mathbf{v}$ to get the linearized equation of motion in terms of $\boldsymbol{\xi}$. This could be done by direct substitution of the expressions (12.38)–(12.40), but it is more instructive to exploit the local conservation laws derived in Section 4.3.3 [1], which were based on the kinematic expressions (4.88)–(4.90) for the time evolution of the line, surface and volume elements $d\mathbf{r}$, $d\boldsymbol{\sigma}$ and $d\tau$. Again, we could integrate those by substitution of the above expressions, but there is no need to do this since the coordinate transformation (12.36) already contains the answers.

▷ **Kinematic transformation** of the line, surface and volume elements $d\mathbf{r}$, $d\boldsymbol{\sigma}$ and $d\tau$ (at position $\mathbf{r}$) to $d\mathbf{r}^0$, $d\boldsymbol{\sigma}^0$ and $d\tau^0$ (at position $\mathbf{r}^0$) just involves the coordinate transformation (12.36), which gives rise to the transpose $\mathbf{R}^{\mathrm{T}}$ of the connection matrix $\mathbf{R}$ with elements

$$r_{ij} \equiv \frac{\partial r_i}{\partial r_j^0} = \delta_{ij} + \frac{\partial \xi_i}{\partial r_j^0} \quad\Rightarrow\quad \mathbf{R}^{\mathrm{T}} \equiv \nabla^0 \mathbf{r} = \mathbf{I} + \nabla^0 \boldsymbol{\xi} \,, \tag{12.41}$$

and the associated Jacobian with Levi-Civita symbols ϵ_{jmn} defined in Appendix A.3,

$$J \equiv \frac{\partial(r_1, r_2, r_3)}{\partial(r_1^0, r_2^0, r_3^0)} \equiv \det(r_{ij}) = \tfrac{1}{6}\,\epsilon_{ikl}\,\epsilon_{jmn} r_{ij} r_{km} r_{ln}$$

$$\equiv \det\left(\mathbf{I} + \nabla^0 \boldsymbol{\xi}\right) \approx 1 + \nabla^0 \cdot \boldsymbol{\xi} \,. \tag{12.42}$$

These expressions immediately provide the transformations of line and volume element, Eqs. (12.45) and (12.47) below. As always, transformation of the surface element $d\boldsymbol{\sigma}$ ($\equiv d\mathbf{r}_\alpha \times d\mathbf{r}_\beta$) is a bit more complicated since it involves the inverse $\mathbf{R}^{-1}$. This requires evaluation of the matrix $\mathbf{C}$ of cofactors c_{ij}, defined as the determinants of the minors of $\mathbf{R}$ (obtained by taking out the ith row and jth column) multiplied by $(-1)^{i+j}$, e.g.,

$$c_{23} = -\begin{vmatrix} r_{11} & r_{12} \\ r_{31} & r_{32} \end{vmatrix} = -r_{11}\,r_{32} + r_{12}\,r_{31} \,.$$

The inverse $\mathbf{R}^{-1}$ is related to the transpose $\mathbf{C}^{\mathrm{T}}$ through

$$\mathbf{R}^{-1} = (1/J)\,\mathbf{C}^{\mathrm{T}} \quad\Rightarrow\quad c_{ki} r_{kj} = J\delta_{ij} \,, \tag{12.43}$$

whereas the cofactors also have the following properties:

$$c_{ij} = \tfrac{1}{2}\epsilon_{ikl}\,\epsilon_{jmn} r_{km} r_{ln} = \frac{\partial J}{\partial r_{ij}} \,, \qquad \frac{\partial c_{ij}}{\partial r_j^0} = 0 \,. \tag{12.44}$$

This provides the *transformations of line, surface and volume element:*

$$d\mathbf{r} = \mathbf{R} \cdot d\mathbf{r}^0 = d\mathbf{r}^0 \cdot (\mathbf{I} + \nabla^0 \boldsymbol{\xi}) \,, \tag{12.45}$$

$$d\boldsymbol{\sigma} = \mathbf{C} \cdot d\boldsymbol{\sigma}^0 \,, \tag{12.46}$$

$$d\tau = J d\tau^0 = \det\left(\mathbf{I} + \nabla^0 \boldsymbol{\xi}\right) d\tau^0 \approx (1 + \nabla^0 \cdot \boldsymbol{\xi})\, d\tau^0 \,. \tag{12.47}$$

as one may check by substitution and use of the properties (12.44). $\triangleleft$

Integration of the local mass, entropy and magnetic flux conservation equations,

$$\frac{\mathrm{D}}{\mathrm{D}t}(dM) \equiv \frac{\mathrm{D}}{\mathrm{D}t}(\rho d\tau) = 0 \qquad (\textit{mass})\,, \tag{12.48}$$

$$\frac{\mathrm{D}}{\mathrm{D}t}(S) \equiv \frac{\mathrm{D}}{\mathrm{D}t}(p\rho^{-\gamma}) = 0 \qquad (\textit{entropy})\,, \tag{12.49}$$

$$\frac{\mathrm{D}}{\mathrm{D}t}(d\psi) \equiv \frac{\mathrm{D}}{\mathrm{D}t}(\mathbf{B} \cdot d\boldsymbol{\sigma}) = 0 \qquad (\textit{magnetic flux})\,, \tag{12.50}$$

is now straightforward:

$$\rho d\tau = \rho^0 d\tau^0 \quad\Rightarrow\quad \rho = \frac{\rho^0}{J} \approx \rho^0 - \rho^0 \nabla^0 \cdot \boldsymbol{\xi} \,, \tag{12.51}$$

$$p\rho^{-\gamma} = p^0(\rho^0)^{-\gamma} \quad \Rightarrow \quad p = \frac{p^0}{J^\gamma} \approx p^0 - \gamma p^0 \nabla^0 \cdot \boldsymbol{\xi} , \tag{12.52}$$

$$B_i d\sigma_i = B_i^0 d\sigma_i^0 \quad \Rightarrow \quad \mathbf{B} = \frac{1}{J} \mathbf{R} \cdot \mathbf{B}^0 \approx (1 - \nabla^0 \cdot \boldsymbol{\xi}) \, \mathbf{B}^0 \cdot (\mathbf{I} + \nabla^0 \boldsymbol{\xi})$$

$$\approx \mathbf{B}^0 + \mathbf{B}^0 \cdot \nabla^0 \boldsymbol{\xi} - \mathbf{B}^0 \nabla^0 \cdot \boldsymbol{\xi} . \tag{12.53}$$

This completes the first half of the program, viz. construction of the Lagrangian representation of density, pressure and magnetic field perturbations. Since

$$f_{1L} = f_{1E} + \boldsymbol{\xi} \cdot \nabla^0 f^0 , \tag{12.54}$$

the Lagrangian expressions (12.51)–(12.53) are fully compatible with the earlier obtained Eulerian counterparts (6.19)–(6.21)[1] for static equilibria,

$$\rho_{1E} = - \nabla^0 \cdot (\rho^0 \boldsymbol{\xi}) , \tag{12.55}$$

$$\pi \equiv p_{1E} = - \gamma p^0 \nabla^0 \cdot \boldsymbol{\xi} - \boldsymbol{\xi} \cdot \nabla^0 p^0 , \tag{12.56}$$

$$\mathbf{Q} \equiv \mathbf{B}_{1E} = \nabla^0 \times (\boldsymbol{\xi} \times \mathbf{B}^0) = \mathbf{B}^0 \cdot \nabla^0 \boldsymbol{\xi} - \mathbf{B}^0 \nabla^0 \cdot \boldsymbol{\xi} - \boldsymbol{\xi} \cdot \nabla^0 \mathbf{B}^0 . \tag{12.57}$$

The abbreviations π and $\mathbf{Q}$ will be used in the reductions below.

It remains to substitute the expressions obtained in the equation of motion,

$$\rho \frac{D\mathbf{v}}{Dt} + \nabla p - (\nabla \times \mathbf{B}) \times \mathbf{B} - \rho \mathbf{g} = 0 \qquad (momentum) . \tag{12.58}$$

This involves a number of steps that are put in small print below.

▷ **Reduction of the equation of motion** involves the following contributions, with the full perturbed expressions on the LHS, while in the first order expansions on the RHS the superscripts ^{0} on the unperturbed quantities are consistently dropped:

$$\rho \frac{D\mathbf{v}}{Dt} \approx (\rho - \rho \nabla \cdot \boldsymbol{\xi})\left(\frac{\partial}{\partial t} + \mathbf{v} \cdot \nabla\right)\left(\mathbf{v} + \mathbf{v} \cdot \nabla \boldsymbol{\xi} + \frac{\partial \boldsymbol{\xi}}{\partial t}\right)$$

$$\approx \rho \mathbf{v} \cdot \nabla \mathbf{v} - (\nabla \cdot \boldsymbol{\xi}) \rho \mathbf{v} \cdot \nabla \mathbf{v} + \rho (\mathbf{v} \cdot \nabla)^2 \boldsymbol{\xi} + 2\rho \mathbf{v} \cdot \nabla \frac{\partial \boldsymbol{\xi}}{\partial t} + \rho \frac{\partial^2 \boldsymbol{\xi}}{\partial t^2}$$

$$= \rho \mathbf{v} \cdot \nabla \mathbf{v} + \boldsymbol{\xi} \cdot \nabla (\rho \mathbf{v} \cdot \nabla \mathbf{v})$$

$$- \nabla \cdot (\boldsymbol{\xi} \rho \mathbf{v} \cdot \nabla \mathbf{v} - \rho \mathbf{v} \mathbf{v} \cdot \nabla \boldsymbol{\xi}) + 2\rho \mathbf{v} \cdot \nabla \frac{\partial \boldsymbol{\xi}}{\partial t} + \rho \frac{\partial^2 \boldsymbol{\xi}}{\partial t^2} , \tag{12.59}$$

$$\nabla p \approx [\nabla - (\nabla \boldsymbol{\xi}) \cdot \nabla] (p + \pi + \boldsymbol{\xi} \cdot \nabla p)$$

$$\approx \nabla p - (\nabla \boldsymbol{\xi}) \cdot \nabla p + \nabla(\boldsymbol{\xi} \cdot \nabla p) + \nabla \pi$$

$$= \nabla p + \boldsymbol{\xi} \cdot \nabla \nabla p + \nabla \pi , \tag{12.60}$$

$$-(\nabla \times \mathbf{B}) \times \mathbf{B} \approx - \left\{ [\nabla - (\nabla \boldsymbol{\xi}) \cdot \nabla] \times (\mathbf{B} + \mathbf{Q} + \boldsymbol{\xi} \cdot \nabla \mathbf{B}) \right\} \times (\mathbf{B} + \mathbf{Q} + \boldsymbol{\xi} \cdot \nabla \mathbf{B})$$

$$\approx -(\nabla \times \mathbf{B}) \times \mathbf{B} - (\nabla \times \mathbf{B}) \times (\boldsymbol{\xi} \cdot \nabla \mathbf{B}) - (\nabla \times (\boldsymbol{\xi} \cdot \nabla \mathbf{B})) \times \mathbf{B}$$

$$+ \left[((\nabla \boldsymbol{\xi}) \cdot \nabla) \times \mathbf{B} \right] \times \mathbf{B} + \mathbf{B} \times (\nabla \times \mathbf{Q}) - (\nabla \times \mathbf{B}) \times \mathbf{Q}$$

$$= -(\nabla \times \mathbf{B}) \times \mathbf{B} - \boldsymbol{\xi} \cdot \nabla\big[(\nabla \times \mathbf{B}) \times \mathbf{B}\big]$$

$$+ \mathbf{B} \times (\nabla \times \mathbf{Q}) - (\nabla \times \mathbf{B}) \times \mathbf{Q}, \tag{12.61}$$

$$\rho\nabla\Phi \approx (\rho - \rho\nabla \cdot \boldsymbol{\xi})[\nabla - (\nabla\boldsymbol{\xi}) \cdot \nabla](\Phi + \boldsymbol{\xi} \cdot \nabla\Phi)$$

$$\approx \rho\nabla\Phi - \rho(\nabla \cdot \boldsymbol{\xi})\nabla\Phi - \rho(\nabla\boldsymbol{\xi}) \cdot \nabla\Phi + \rho\nabla(\boldsymbol{\xi} \cdot \nabla\Phi)$$

$$= \rho\nabla\Phi + \boldsymbol{\xi} \cdot \nabla(\rho\nabla\Phi) - (\nabla\Phi)\nabla \cdot (\rho\boldsymbol{\xi}). \tag{12.62}$$

The derivation of the last line of Eq. (12.59) involves the equilibrium condition (12.24). Adding the expressions (12.59)–(12.62), the four first terms on the RHSs cancel because of the equilibrium condition (12.25) and the four second terms cancel as well because they represent the Lagrangian perturbation $\boldsymbol{\xi} \cdot \nabla$ of the same equilibrium condition. The remaining terms constitute the desired perturbation of the equation of motion. ◁

This finally yields the linearized equation of motion for perturbations of stationary equilibria that was first derived by Frieman and Rotenberg [147]:

$$\mathbf{G}(\boldsymbol{\xi}) - 2\rho\mathbf{v} \cdot \nabla\frac{\partial\boldsymbol{\xi}}{\partial t} - \rho\frac{\partial^2\boldsymbol{\xi}}{\partial t^2} = 0, \tag{12.63}$$

where $\mathbf{G}$ is the *generalized force operator,*

$$\mathbf{G}(\boldsymbol{\xi}) \equiv \mathbf{F}(\boldsymbol{\xi}) + \nabla \cdot (\boldsymbol{\xi}\rho\mathbf{v} \cdot \nabla\mathbf{v} - \rho\mathbf{v}\mathbf{v} \cdot \nabla\boldsymbol{\xi}), \tag{12.64}$$

involving the standard force operator expression (6.23) [1] of static equilibria,

$$\mathbf{F}(\boldsymbol{\xi}) \equiv \nabla(\gamma p\nabla \cdot \boldsymbol{\xi}) - \mathbf{B} \times (\nabla \times \mathbf{Q})$$

$$+ \nabla(\boldsymbol{\xi} \cdot \nabla p) + \mathbf{j} \times \mathbf{Q} + (\nabla\Phi)\nabla \cdot (\rho\boldsymbol{\xi}). \tag{12.65}$$

We here introduce the new notation $\mathbf{G}$ for the generalized force operator (Frieman and Rotenberg indicate that quantity also by $\mathbf{F}$), so that we can use the operator $\mathbf{F}$ as a convenient abbreviation for the RHS terms of Eq. (12.65).

Because of the equilibrium, the last term of $\mathbf{G}$ may be transformed to

$$- \nabla \cdot (\rho\mathbf{v}\mathbf{v} \cdot \nabla\boldsymbol{\xi}) \overset{(12.24)}{=} -\rho(\mathbf{v} \cdot \nabla)^2\boldsymbol{\xi}, \tag{12.66}$$

which yields an equivalent representation for the linearized equation of motion:

$$\mathbf{F}(\boldsymbol{\xi}) + \nabla \cdot (\boldsymbol{\xi}\rho\mathbf{v} \cdot \nabla\mathbf{v}) - \rho\Big(\frac{\partial}{\partial t} + \mathbf{v} \cdot \nabla\Big)^2\boldsymbol{\xi} = 0. \tag{12.67}$$

For plane shear flow, where $\mathbf{v} \cdot \nabla\mathbf{v} = 0$, the only change with respect to the static problem is the appearance of the *Doppler shift* operator $\mathbf{v} \cdot \nabla$. However, since this term varies from place to place, the relationship between the waves and instabilities of plane static and those of plane stationary plasmas is not as straightforward as it might appear (see Section 13.1). For more general flow fields, like rotations in a cylinder, the equation of motion becomes much more involved since $\mathbf{v} \cdot \nabla\mathbf{v}$ yields the centrifugal acceleration (12.29) and the operator $\mathbf{v} \cdot \nabla$ gives Coriolis contributions to the frequency in addition to the Doppler shift (see Section 13.4).

▷ **Alternative definition** Some authors [350, 199] exploit an alternative definition of the generalized force operator, extracting the term (12.66). This yields the following more compact form for the equation of motion:

$$\tilde{\mathbf{G}}(\xi) \equiv \mathbf{G}(\xi) + \rho(\mathbf{v} \cdot \nabla)^2 \xi \equiv \mathbf{F}(\xi) + \nabla \cdot (\xi \rho \mathbf{v} \cdot \nabla \mathbf{v}) = \rho \frac{\mathrm{D}^2 \xi}{\mathrm{D}t^2}. \qquad (12.68)$$

We will stick to the original definition of Frieman and Rotenberg [147], though, because it gives the more expedient expression of the spectral equation below.   ◁

For normal modes $\hat{\xi}(\mathbf{r}) \exp(-\mathrm{i}\omega t)$, the associated *spectral equation* reads:

$$\mathbf{G}(\hat{\xi}) + 2\mathrm{i}\rho\omega \mathbf{v} \cdot \nabla \hat{\xi} + \rho\omega^2 \hat{\xi} = 0, \qquad (12.69)$$

or, equivalently, from Eq. (12.67),

$$\mathbf{F}(\hat{\xi}) + \nabla \cdot (\hat{\xi} \rho \mathbf{v} \cdot \nabla \mathbf{v}) + \rho(\omega + \mathrm{i}\mathbf{v} \cdot \nabla)^2 \hat{\xi} = 0. \qquad (12.70)$$

This becomes an eigenvalue problem by supplementing appropriate BCs, e.g.

$$\mathbf{n} \cdot \hat{\xi} = 0 \qquad \textit{(at the wall)} \qquad (12.71)$$

for model I (plasma confined inside a rigid wall). This eigenvalue problem is now a *quadratic* one (involving both ω and ω^2), in contrast to the static spectral problem (12.23) which is linear in the eigenvalue ω^2. This implies that the eigenvalues ω are no longer restricted to the real and imaginary axes but may be genuinely complex, so that *overstable modes* occur. This represents a major complication in the theory of waves and instabilities of plasmas with background flow, as will be extensively illustrated in the following sections. From now on, we will drop the hat on $\hat{\xi}$ (we need it for a different purpose in the following section) leaving it understood that this time-independent part of ξ is meant if normal modes are being considered.

12.2.3 Self-adjointness of the generalized force operator*

In order to put the spectral theory of stationary plasmas on a firm mathematical basis, we need to study the adjointness properties of the basic spectral equation (12.69). As a first step, we will prove that the generalized force operator $\mathbf{G}$ itself is actually self-adjoint. The proof will be analogous that of self-adjointness of the operator $\mathbf{F}$ for static equilibria of Section 6.2.3 [1], but it cannot be copied blindly since the static equilibrium relations were used, so that we need to carefully retrace our steps and instead use the new stationary equilibrium relations (12.24)–(12.27).

▷ **Transformation #1 of the inhomogeneity terms** The four terms $\sim \nabla p$, $\mathbf{j}$, $\nabla\Phi$, $\mathbf{v} \cdot \nabla \mathbf{v}$ of the generalized force operator $\mathbf{G}$ may be transformed as follows:

$$\nabla(\xi \cdot \nabla p) = (\nabla p \times \nabla) \times \xi + (\nabla p)\nabla \cdot \xi + \xi \cdot \nabla\nabla p$$

$$\overset{(12.25)}{=} (\mathbf{B}\mathbf{j} \cdot \nabla - \mathbf{j}\mathbf{B} \cdot \nabla) \times \xi - \rho(\nabla\Phi \times \nabla) \times \xi$$

$$- \left[\rho(\mathbf{v} \cdot \nabla \mathbf{v}) \times \nabla\right] \times \boldsymbol{\xi} + (\nabla p) \nabla \cdot \boldsymbol{\xi} + \boldsymbol{\xi} \cdot \nabla \nabla p$$

$$= \ \mathbf{B} \times (\mathbf{j} \cdot \nabla \boldsymbol{\xi}) - \mathbf{j} \times (\mathbf{B} \cdot \nabla \boldsymbol{\xi}) - \rho(\nabla \boldsymbol{\xi}) \cdot \nabla \Phi + \rho(\nabla \Phi) \nabla \cdot \boldsymbol{\xi}$$

$$- \rho(\nabla \boldsymbol{\xi}) \cdot (\mathbf{v} \cdot \nabla \mathbf{v}) + \rho(\mathbf{v} \cdot \nabla \mathbf{v}) \nabla \cdot \boldsymbol{\xi} + (\nabla p) \nabla \cdot \boldsymbol{\xi} + \boldsymbol{\xi} \cdot \nabla \nabla p \, ,$$

$$\mathbf{j} \times \mathbf{Q} \ = \ \mathbf{j} \times (\mathbf{B} \cdot \nabla \boldsymbol{\xi}) - \mathbf{j} \times \mathbf{B} \nabla \cdot \boldsymbol{\xi} - \boldsymbol{\xi} \cdot \nabla (\mathbf{j} \times \mathbf{B}) - \mathbf{B} \times (\boldsymbol{\xi} \cdot \nabla \mathbf{j}) \, ,$$

$$(\nabla \Phi) \nabla \cdot (\rho \boldsymbol{\xi}) \ = \ \rho(\nabla \Phi) \nabla \cdot \boldsymbol{\xi} + (\nabla \Phi)(\nabla \rho) \cdot \boldsymbol{\xi} \, ,$$

$$\nabla \cdot (\rho \boldsymbol{\xi} \mathbf{v} \cdot \nabla \mathbf{v}) \ = \ \boldsymbol{\xi} \cdot \nabla (\rho \mathbf{v} \cdot \nabla \mathbf{v}) + \rho(\mathbf{v} \cdot \nabla \mathbf{v}) \nabla \cdot \boldsymbol{\xi} \, .$$

Hence, the sum of the four inhomogeneity terms becomes

$$\nabla(\boldsymbol{\xi} \cdot \nabla p) + \mathbf{j} \times \mathbf{Q} + (\nabla \Phi) \nabla \cdot (\rho \boldsymbol{\xi}) + \nabla \cdot (\rho \boldsymbol{\xi} \mathbf{v} \cdot \nabla \mathbf{v})$$

$$= \ - \mathbf{B} \times (\nabla \times (\mathbf{j} \times \boldsymbol{\xi})) + (\nabla p - 2\mathbf{j} \times \mathbf{B} + 2\rho \nabla \Phi + 2\rho \mathbf{v} \cdot \nabla \mathbf{v}) \nabla \cdot \boldsymbol{\xi}$$

$$+ \boldsymbol{\xi} \cdot \nabla (\nabla p - \mathbf{j} \times \mathbf{B} + \rho \mathbf{v} \cdot \nabla \mathbf{v}) - \rho(\nabla \boldsymbol{\xi}) \cdot \nabla \Phi + (\nabla \Phi)(\nabla \rho) \cdot \boldsymbol{\xi}$$

$$- \rho(\nabla \boldsymbol{\xi}) \cdot (\mathbf{v} \cdot \nabla \mathbf{v})$$

$$\overset{(12.25)}{=} - \mathbf{B} \times (\nabla \times (\mathbf{j} \times \boldsymbol{\xi})) - (\nabla p) \nabla \cdot \boldsymbol{\xi} - \rho \nabla(\boldsymbol{\xi} \cdot \nabla \Phi) - \rho(\nabla \boldsymbol{\xi}) \cdot (\mathbf{v} \cdot \nabla \mathbf{v}), \ (12.72)$$

which provides the revised expression of $\mathbf{G}$ exploited below. $\triangleleft$

This transformation yields an *equivalent form of the generalized force operator*:

$$\mathbf{G}(\boldsymbol{\xi}) \ = \ \nabla(\gamma p \nabla \cdot \boldsymbol{\xi}) - \mathbf{B} \times \left[\nabla \times \mathbf{Q} + \nabla \times (\mathbf{j} \times \boldsymbol{\xi}) \right] - (\nabla p) \nabla \cdot \boldsymbol{\xi}$$

$$- \rho \nabla(\boldsymbol{\xi} \cdot \nabla \Phi) - \rho(\nabla \boldsymbol{\xi}) \cdot (\mathbf{v} \cdot \nabla \mathbf{v}) - \nabla \cdot (\rho \mathbf{v} \mathbf{v} \cdot \nabla \boldsymbol{\xi}) , \quad (12.73)$$

where the first two terms correspond to the compressional and magnetic wave contributions of homogeneous plasmas (Alfvén and magneto-sonic waves), the next four terms correspond to the wide variety of current, pressure gradient, gravitation and velocity gradient driven instabilities of inhomogeneous plasmas (interchanges, kinks, gravitational and rotational instabilities, and all their combinations), whereas the last term gives the contribution (12.66) discussed above.

As in Volume [1], we exploit an inner product with real vectors $\boldsymbol{\xi}$ and $\boldsymbol{\eta}$, satisfying the BCs $\mathbf{n} \cdot \boldsymbol{\xi} = 0$ and $\mathbf{n} \cdot \boldsymbol{\eta} = 0$, and associated Eulerian magnetic perturbations $\mathbf{Q}$ and $\mathbf{R}$. This inner product is split into a simple part involving the homogeneity terms $\mathbf{G}_1$ and a complicated part with the inhomogeneity terms $\mathbf{G}_2$,

$$\boldsymbol{\eta} \cdot \mathbf{G} = \boldsymbol{\eta} \cdot \mathbf{G}_1 + \boldsymbol{\eta} \cdot \mathbf{G}_2 \, . \qquad (12.74)$$

The required transformation to a symmetric expression plus a divergence is straightforward for the first part:

$$\boldsymbol{\eta} \cdot \mathbf{G}_1 \equiv \boldsymbol{\eta} \cdot \left[\nabla(\gamma p \nabla \cdot \boldsymbol{\xi}) - \mathbf{B} \times (\nabla \times \mathbf{Q}) \right]$$

$$= - \gamma p (\nabla \cdot \boldsymbol{\xi})(\nabla \cdot \boldsymbol{\eta}) - \mathbf{Q} \cdot \mathbf{R} + \nabla \cdot \left[\boldsymbol{\eta} \gamma p \nabla \cdot \boldsymbol{\xi} + (\mathbf{B} \boldsymbol{\eta} - \boldsymbol{\eta} \mathbf{B}) \cdot \mathbf{Q} \right] .$$

$$(12.75)$$

The transformation of the second part starts from the original definition (12.64) of the operator $\mathbf{G}$ and then proceeds by inserting the transformed form (12.73) of $\mathbf{G}$:

$$\boldsymbol{\eta} \cdot \mathbf{G}_2 \equiv \boldsymbol{\eta} \cdot \left[\nabla(\boldsymbol{\xi} \cdot \nabla p) + \mathbf{j} \times \mathbf{Q} + (\nabla\Phi) \nabla \cdot (\rho\boldsymbol{\xi}) \right.$$
$$\left. + \nabla \cdot (\boldsymbol{\xi}\rho\mathbf{v} \cdot \nabla\mathbf{v} - \rho\mathbf{v}\mathbf{v} \cdot \nabla\boldsymbol{\xi}) \right] \tag{12.76}$$

$$= \boldsymbol{\eta} \cdot \left[-\mathbf{B} \times (\nabla \times (\mathbf{j} \times \boldsymbol{\xi})) - (\nabla p) \nabla \cdot \boldsymbol{\xi} - \rho\nabla(\boldsymbol{\xi} \cdot \nabla\Phi) \right.$$
$$\left. - \rho(\nabla\boldsymbol{\xi}) \cdot (\mathbf{v} \cdot \nabla\mathbf{v}) - \nabla \cdot (\rho\mathbf{v}\mathbf{v} \cdot \nabla\boldsymbol{\xi}) \right]. \tag{12.77}$$

The substantial algebra involved in reworking the expression (12.77) to a mirror image of the expression (12.76) plus a divergence is put in small print below.

▷ **Transformation #2 of the inhomogeneity terms** The first term of Eq. (12.77) is transformed by using the equality

$$-\boldsymbol{\eta} \cdot \mathbf{B} \times [\nabla \times (\mathbf{j} \times \boldsymbol{\xi})] = \boldsymbol{\xi} \cdot \mathbf{j} \times \mathbf{R} + \nabla \cdot [\mathbf{j}\,\mathbf{B} \cdot (\boldsymbol{\xi} \times \boldsymbol{\eta}) + \boldsymbol{\xi}\,\boldsymbol{\eta} \cdot (\mathbf{j} \times \mathbf{B})], \tag{12.78}$$

derived in the corresponding derivation of Volume [1]. Next, the following steps are taken:

$$\boldsymbol{\eta} \cdot \mathbf{G}_2 \equiv \boldsymbol{\xi} \cdot \mathbf{j} \times \mathbf{R} + \nabla \cdot [\mathbf{j}\,\mathbf{B} \cdot (\boldsymbol{\xi} \times \boldsymbol{\eta})] - \boldsymbol{\eta} \cdot (\nabla p - \mathbf{j} \times \mathbf{B})\nabla \cdot \boldsymbol{\xi} + \boldsymbol{\xi} \cdot \nabla(\boldsymbol{\eta} \cdot \mathbf{j} \times \mathbf{B})$$
$$- \rho\boldsymbol{\eta} \cdot \nabla(\boldsymbol{\xi} \cdot \nabla\Phi) - \rho(\boldsymbol{\eta} \cdot \nabla\boldsymbol{\xi}) \cdot (\mathbf{v} \cdot \nabla\mathbf{v}) - [\nabla \cdot (\rho\mathbf{v}\mathbf{v} \cdot \nabla\boldsymbol{\xi})] \cdot \boldsymbol{\eta}$$

$$= \boldsymbol{\xi} \cdot \left[\nabla(\boldsymbol{\eta} \cdot \nabla p) + \mathbf{j} \times \mathbf{R} + (\nabla\Phi) \nabla \cdot (\rho\boldsymbol{\eta}) \right]$$
$$- \rho(\boldsymbol{\eta} \cdot \nabla\boldsymbol{\xi}) \cdot (\mathbf{v} \cdot \nabla\mathbf{v}) - [\nabla \cdot (\rho\mathbf{v}\mathbf{v} \cdot \nabla\boldsymbol{\xi})] \cdot \boldsymbol{\eta}$$
$$+ \nabla \cdot [\mathbf{j}\,\mathbf{B} \cdot (\boldsymbol{\xi} \times \boldsymbol{\eta}) - (\nabla p - \mathbf{j} \times \mathbf{B}) \cdot \boldsymbol{\eta}\boldsymbol{\xi} - \rho\boldsymbol{\eta}\boldsymbol{\xi} \cdot \nabla\Phi]$$

$$\overset{(12.25)}{=} \boldsymbol{\xi} \cdot \left[\nabla(\boldsymbol{\eta} \cdot \nabla p) + \mathbf{j} \times \mathbf{R} + (\nabla\Phi) \nabla \cdot (\rho\boldsymbol{\eta}) \right]$$
$$- \rho(\boldsymbol{\eta} \cdot \nabla\boldsymbol{\xi}) \cdot (\mathbf{v} \cdot \nabla\mathbf{v}) - [\nabla \cdot (\rho\mathbf{v}\mathbf{v} \cdot \nabla\boldsymbol{\xi})] \cdot \boldsymbol{\eta}$$
$$+ \nabla \cdot [\mathbf{j}\,\mathbf{B} \cdot (\boldsymbol{\xi} \times \boldsymbol{\eta}) + (\nabla p - \mathbf{j} \times \mathbf{B}) \cdot (\boldsymbol{\xi}\boldsymbol{\eta} - \boldsymbol{\eta}\boldsymbol{\xi}) + \rho(\mathbf{v} \cdot \nabla\mathbf{v}) \cdot \boldsymbol{\xi}\boldsymbol{\eta}]$$

$$= \boldsymbol{\xi} \cdot \left[\nabla(\boldsymbol{\eta} \cdot \nabla p) + \mathbf{j} \times \mathbf{R} + (\nabla\Phi) \nabla \cdot (\rho\boldsymbol{\eta}) \right.$$
$$\left. + \nabla \cdot (\boldsymbol{\eta}\rho\mathbf{v} \cdot \nabla\mathbf{v} - \rho\mathbf{v}\mathbf{v} \cdot \nabla\boldsymbol{\eta}) \right]$$
$$+ \nabla \cdot [\mathbf{j}\,\mathbf{B} \cdot (\boldsymbol{\xi} \times \boldsymbol{\eta}) + (\nabla p - \mathbf{j} \times \mathbf{B}) \cdot (\boldsymbol{\xi}\boldsymbol{\eta} - \boldsymbol{\eta}\boldsymbol{\xi})$$
$$- \rho\mathbf{v}((\mathbf{v} \cdot \nabla\boldsymbol{\xi}) \cdot \boldsymbol{\eta} - (\mathbf{v} \cdot \nabla\boldsymbol{\eta}) \cdot \boldsymbol{\xi})]. \tag{12.79}$$

The latter expression has the required form of a mirror image of Eq. (12.76) plus a divergence. A symmetric expression is obtained by just averaging Eqs. (12.76) and (12.79):

$$\boldsymbol{\eta} \cdot \mathbf{G}_2 \equiv \tfrac{1}{2}\boldsymbol{\eta} \cdot \left[\nabla(\boldsymbol{\xi} \cdot \nabla p) + \mathbf{j} \times \mathbf{Q} + (\nabla\Phi) \nabla \cdot (\rho\boldsymbol{\xi}) \right.$$
$$\left. + \nabla \cdot (\boldsymbol{\xi}\rho\mathbf{v} \cdot \nabla\mathbf{v} - \rho\mathbf{v}\mathbf{v} \cdot \nabla\boldsymbol{\xi}) \right]$$
$$+ \tfrac{1}{2}\boldsymbol{\xi} \cdot \left[\nabla(\boldsymbol{\eta} \cdot \nabla p) + \mathbf{j} \times \mathbf{R} + (\nabla\Phi) \nabla \cdot (\rho\boldsymbol{\eta}) \right.$$
$$\left. + \nabla \cdot (\boldsymbol{\eta}\rho\mathbf{v} \cdot \nabla\mathbf{v} - \rho\mathbf{v}\mathbf{v} \cdot \nabla\boldsymbol{\eta}) \right]$$
$$+ \tfrac{1}{2}\nabla \cdot [\mathbf{j}\,\mathbf{B} \cdot (\boldsymbol{\xi} \times \boldsymbol{\eta}) + (\nabla p - \mathbf{j} \times \mathbf{B}) \cdot (\boldsymbol{\xi}\boldsymbol{\eta} - \boldsymbol{\eta}\boldsymbol{\xi})$$
$$- \rho\mathbf{v}((\mathbf{v} \cdot \nabla\boldsymbol{\xi}) \cdot \boldsymbol{\eta} - (\mathbf{v} \cdot \nabla\boldsymbol{\eta}) \cdot \boldsymbol{\xi})]$$

$$
\begin{aligned}
= \;& -(\tfrac{1}{2}\nabla p)\cdot(\boldsymbol{\xi}\nabla\cdot\boldsymbol{\eta}+\boldsymbol{\eta}\nabla\cdot\boldsymbol{\xi})-\tfrac{1}{2}\mathbf{j}\cdot(\boldsymbol{\eta}\times\mathbf{Q}+\boldsymbol{\xi}\times\mathbf{R})\\
&+\tfrac{1}{2}(\nabla\Phi)\cdot[\,\boldsymbol{\eta}\nabla\cdot(\rho\boldsymbol{\xi})+\boldsymbol{\xi}\nabla\cdot(\rho\boldsymbol{\eta})\,]\\
&-\tfrac{1}{2}\rho(\mathbf{v}\cdot\nabla\mathbf{v})\cdot(\boldsymbol{\xi}\cdot\nabla\boldsymbol{\eta}+\boldsymbol{\eta}\cdot\nabla\boldsymbol{\xi})+\rho(\mathbf{v}\cdot\nabla\boldsymbol{\xi})\cdot(\mathbf{v}\cdot\nabla\boldsymbol{\eta})\\
&+\nabla\cdot[\,\boldsymbol{\eta}\,\boldsymbol{\xi}\cdot\nabla p+\tfrac{1}{2}\mathbf{j}\,\mathbf{B}\cdot(\boldsymbol{\xi}\times\boldsymbol{\eta})-\tfrac{1}{2}(\mathbf{j}\times\mathbf{B})\cdot(\boldsymbol{\xi}\boldsymbol{\eta}-\boldsymbol{\eta}\boldsymbol{\xi})\\
&+\tfrac{1}{2}\rho(\mathbf{v}\cdot\nabla\mathbf{v})\cdot(\boldsymbol{\eta}\boldsymbol{\xi}+\boldsymbol{\xi}\boldsymbol{\eta})-\rho\mathbf{v}(\mathbf{v}\cdot\nabla\boldsymbol{\xi})\cdot\boldsymbol{\eta}\,]\,. \quad (12.80)
\end{aligned}
$$

This finally has the requisite form of a symmetric expression plus a divergence. $\lhd$

Adding Eqs. (12.75) and (12.80) yields the required form for the expression $\boldsymbol{\eta}\cdot\mathbf{G}$. By using Gauss' theorem, the integral of this expression may be transformed to the sum of a symmetric volume integral and a surface integral. The latter is simplified by exploiting the equality

$$
(\mathbf{j}\times\mathbf{B})\cdot(\boldsymbol{\xi}\eta_n-\boldsymbol{\eta}\xi_n)=j_n\,\mathbf{B}\cdot(\boldsymbol{\xi}\times\boldsymbol{\eta})\,,
$$

and introducing the Eulerian perturbation of the total pressure,

$$
\Pi\equiv(p+\tfrac{1}{2}B^2)_{1\mathrm{E}}\equiv\pi+\mathbf{B}\cdot\mathbf{Q}=-\gamma p\nabla\cdot\boldsymbol{\xi}-\boldsymbol{\xi}\cdot\nabla p+\mathbf{B}\cdot\mathbf{Q}\,. \quad (12.81)
$$

This gives the following general expression:

$$
\begin{aligned}
\int\boldsymbol{\eta}\cdot&\,\mathbf{G}(\boldsymbol{\xi})\,dV\\
=\;&-\int\Big[\gamma p\,(\nabla\cdot\boldsymbol{\xi})\,\nabla\cdot\boldsymbol{\eta}+\mathbf{Q}\cdot\mathbf{R}+\tfrac{1}{2}(\nabla p)\cdot(\boldsymbol{\xi}\,\nabla\cdot\boldsymbol{\eta}+\boldsymbol{\eta}\,\nabla\cdot\boldsymbol{\xi})\\
&\qquad+\tfrac{1}{2}\mathbf{j}\cdot(\boldsymbol{\eta}\times\mathbf{Q}+\boldsymbol{\xi}\times\mathbf{R})-\tfrac{1}{2}(\nabla\Phi)\cdot[\,\boldsymbol{\eta}\,\nabla\cdot(\rho\boldsymbol{\xi})+\boldsymbol{\xi}\,\nabla\cdot(\rho\boldsymbol{\eta})\,]\\
&\qquad+\tfrac{1}{2}\rho(\mathbf{v}\cdot\nabla\mathbf{v})\cdot(\boldsymbol{\xi}\cdot\nabla\boldsymbol{\eta}+\boldsymbol{\eta}\cdot\nabla\boldsymbol{\xi})-\rho(\mathbf{v}\cdot\nabla\boldsymbol{\xi})\cdot(\mathbf{v}\cdot\nabla\boldsymbol{\eta})\Big]\,dV\\
&-\int\Big[\eta_n\Pi(\boldsymbol{\xi})-\tfrac{1}{2}\rho(\mathbf{v}\cdot\nabla\mathbf{v})\cdot(\boldsymbol{\eta}\xi_n+\boldsymbol{\xi}\eta_n)\\
&\qquad+\rho v_n(\mathbf{v}\cdot\nabla\boldsymbol{\xi})\cdot\boldsymbol{\eta}-B_n\,\boldsymbol{\eta}\cdot\mathbf{Q}-j_n\,\mathbf{B}\cdot(\boldsymbol{\xi}\times\boldsymbol{\eta})\Big]\,dS\,, \quad (12.82)
\end{aligned}
$$

from which one may construct the quadratic forms for the different model problems of Section 4.6 [1]. For model I (plasma inside a rigid wall), as well as for models II and II* (configurations with a plasma–plasma or plasma–vacuum interface), the last three terms of the surface integral vanish because the equilibrium requires $v_n=0$, $B_n=0$, $j_n=0$ at a wall (model I), as well as at a stationary interface (models II and II*). The extra condition $v_n=0$ (compared to static plasmas) insures that equilibria have *stationary* interfaces. It remains to demonstrate symmetry of the remaining terms of the surface integral for the different models.

First, restricting the discussion to model I, due to the symmetry of the second volume integral of Eq. (12.82) only a surface integral survives in the pertinent difference expression, which gives the important *pre-self-adjointness relation*

$$\int \left[\boldsymbol{\eta} \cdot \mathbf{G}(\boldsymbol{\xi}) - \boldsymbol{\xi} \cdot \mathbf{G}(\boldsymbol{\eta}) \right] dV = -\int \left[\eta_n \Pi(\boldsymbol{\xi}) - \xi_n \Pi(\boldsymbol{\eta}) \right] dS \quad \textit{(no BCs)}, \quad (12.83)$$

that we will frequently exploit. In model I, the latter integral also vanishes because of the BCs on the normal components of $\boldsymbol{\xi}$ and $\boldsymbol{\eta}$ at the wall:

$$\int \left[\boldsymbol{\eta} \cdot \mathbf{G}(\boldsymbol{\xi}) - \boldsymbol{\xi} \cdot \mathbf{G}(\boldsymbol{\eta}) \right] dV = 0 \quad \textit{(BCs satisfied)}. \quad (12.84)$$

This proves that the generalized force operator $\mathbf{G}$ is indeed self-adjoint for model I perturbations, QED.

Next, we extend the discussion to model II*, where two plasmas are separated by an interface S with a tangential discontinuity. We distinguish the quantities of the two plasmas by putting a hat on one of them, indicate the discontinuities at the interface by the notation $[\![f]\!] \equiv \hat{f} - f$, and let the normal $\mathbf{n}$ point into the plasma with the hat. The equilibrium pressure balance BC then reads

$$[\![p + \tfrac{1}{2}B^2]\!] = 0 \quad \textit{(on S)}. \quad (12.85)$$

The terms of the surface integral of Eq. (12.82) now need to be converted into a symmetric expression by means of the BCs on the perturbations at the perturbed position of the interface. Those BCs are the same as given by Eqs. (6.144) and (6.147)[1] for static plasmas *(prove that!)*,

$$[\![\xi_n]\!] = 0 \quad \textit{(on S)}, \quad (12.86)$$

$$[\![\Pi + \xi_n \mathbf{n} \cdot \nabla(p + \tfrac{1}{2}B^2)]\!] = 0 \quad \textit{(on S)}. \quad (12.87)$$

Recall that the latter BC is obtained by evaluating pressure balance at $\mathbf{r}^0 + \xi_n \mathbf{n}$ (rather than at $\mathbf{r}^0 + \boldsymbol{\xi}$), since the tangential component of $\boldsymbol{\xi}$ is not continuous in general (see Fig. 6.18[1]). This gives the following generalization of Eq. (12.82) for the integral over the combined volume $V_{\text{all}} \equiv V + \hat{V}$ of the two plasmas:

$$\int \boldsymbol{\eta} \cdot \mathbf{G}(\boldsymbol{\xi}) \, dV_{\text{all}}$$

$$= -\int \left[\gamma p (\nabla \cdot \boldsymbol{\xi}) \nabla \cdot \boldsymbol{\eta} + \mathbf{Q} \cdot \mathbf{R} + \tfrac{1}{2}(\nabla p) \cdot (\boldsymbol{\xi} \nabla \cdot \boldsymbol{\eta} + \boldsymbol{\eta} \nabla \cdot \boldsymbol{\xi}) \right.$$

$$+ \tfrac{1}{2}\mathbf{j} \cdot (\boldsymbol{\eta} \times \mathbf{Q} + \boldsymbol{\xi} \times \mathbf{R}) - \tfrac{1}{2}(\nabla \Phi) \cdot [\boldsymbol{\eta} \nabla \cdot (\rho \boldsymbol{\xi}) + \boldsymbol{\xi} \nabla \cdot (\rho \boldsymbol{\eta})]$$

$$\left. + \tfrac{1}{2}\rho(\mathbf{v} \cdot \nabla \mathbf{v}) \cdot (\boldsymbol{\xi} \cdot \nabla \boldsymbol{\eta} + \boldsymbol{\eta} \cdot \nabla \boldsymbol{\xi}) - \rho(\mathbf{v} \cdot \nabla \boldsymbol{\xi}) \cdot (\mathbf{v} \cdot \nabla \boldsymbol{\eta}) \right] dV$$

$$- \int \left[\eta_n \xi_n \mathbf{n} \cdot [\![\nabla(p + \tfrac{1}{2}B^2)]\!] + [\![\tfrac{1}{2}\rho(\mathbf{v} \cdot \nabla \mathbf{v}) \cdot (\boldsymbol{\xi}\eta_n + \boldsymbol{\eta}\xi_n)]\!] \right] dS$$

$$- \int \left[\gamma \hat{p} (\nabla \cdot \hat{\boldsymbol{\xi}}) \nabla \cdot \hat{\boldsymbol{\eta}} + \hat{\mathbf{Q}} \cdot \hat{\mathbf{R}} + \tfrac{1}{2}(\nabla \hat{p}) \cdot (\hat{\boldsymbol{\xi}} \nabla \cdot \hat{\boldsymbol{\eta}} + \hat{\boldsymbol{\eta}} \nabla \cdot \hat{\boldsymbol{\xi}}) \right.$$

$$+ \tfrac{1}{2}\hat{\mathbf{j}} \cdot (\hat{\boldsymbol{\eta}} \times \hat{\mathbf{Q}} + \hat{\boldsymbol{\xi}} \times \hat{\mathbf{R}}) - \tfrac{1}{2}(\nabla\hat{\Phi}) \cdot [\hat{\boldsymbol{\eta}}\,\nabla \cdot (\hat{\rho}\hat{\boldsymbol{\xi}}) + \hat{\boldsymbol{\xi}}\,\nabla \cdot (\hat{\rho}\hat{\boldsymbol{\eta}})]$$

$$+ \tfrac{1}{2}\hat{\rho}(\hat{\mathbf{v}} \cdot \nabla\hat{\mathbf{v}}) \cdot (\hat{\boldsymbol{\xi}} \cdot \nabla\hat{\boldsymbol{\eta}} + \hat{\boldsymbol{\eta}} \cdot \nabla\hat{\boldsymbol{\xi}}) - \hat{\rho}(\hat{\mathbf{v}} \cdot \nabla\hat{\boldsymbol{\xi}}) \cdot (\hat{\mathbf{v}} \cdot \nabla\hat{\boldsymbol{\eta}})\Big]\, d\hat{V}\,.$$

$$(12.88)$$

Since the surface integral is now symmetric as well, this again leads to the equality (12.84), with V replaced by V_{all}. Hence, the generalized force operator $\mathbf{G}$ is self-adjoint for model II* equilibria, QED.

The proof of self-adjointness of the operator $\mathbf{G}$ for model II equilibria (plasma–vacuum configurations) is left as an exercise for the reader. The resulting quadratic forms are the same as for model II* in the limit $\hat{\rho} \to 0\,, \hat{p} \to 0\,, \hat{\mathbf{j}} \to 0$ and $\hat{\mathbf{v}} \to 0\,.$ This limit also applies to BC (12.87), but BC (12.86) needs to be replaced by

$$\mathbf{n} \cdot \nabla \times (\boldsymbol{\xi} \times \hat{\mathbf{B}}) = \hat{\mathbf{B}} \cdot \nabla\xi_n - \mathbf{n} \cdot (\nabla\hat{\mathbf{B}}) \cdot \mathbf{n}\,\xi_n = \hat{Q}_n \qquad (on\ S)\,, \qquad (12.89)$$

as shown in Eqs. (6.140) and (6.141)[1].

▷ **Exercise** Prove self-adjointness of the operator $\mathbf{G}$ for model II equilibria. To that end, describe the vacuum magnetic field perturbations with the vector potential $\hat{\mathbf{A}}$ and exploit the plasma–interface BCs of Section 6.6.1 [1]. If you open up the outer wall, you may also derive the pre-self-adjointness relation for model III perturbations.   ◁

12.2.4 Energy conservation and stability

Of course, the fact that the operator $\mathbf{G}$ is self-adjoint, or symmetric, does not imply that spectral theory of MHD waves and instabilities for plasmas with background flow is now at the same level as spectral theory for static plasmas. The important difference is the appearance of *the gradient operator parallel to the velocity*, $\rho\mathbf{v}\cdot\nabla$, in the second term of the equation of motion (12.63). Since

$$\nabla \cdot (\rho\mathbf{v}\boldsymbol{\xi} \cdot \boldsymbol{\eta}) \overset{(12.24)}{=} \rho\mathbf{v} \cdot \nabla(\boldsymbol{\xi} \cdot \boldsymbol{\eta}) = \rho(\mathbf{v} \cdot \nabla\boldsymbol{\xi}) \cdot \boldsymbol{\eta} + \rho(\mathbf{v} \cdot \nabla\boldsymbol{\eta}) \cdot \boldsymbol{\xi}\,,$$

that operator *is anti-symmetric*:

$$\int \Big[\boldsymbol{\eta} \cdot (\rho\mathbf{v} \cdot \nabla\boldsymbol{\xi}) + \boldsymbol{\xi} \cdot (\rho\mathbf{v} \cdot \nabla\boldsymbol{\eta})\Big]\, dV = \int \rho v_n\, \boldsymbol{\xi} \cdot \boldsymbol{\eta}\, dS = 0\,. \qquad (12.90)$$

Here the surface integral vanishes because $v_n = 0$, both at a wall (model I) and at stationary interfaces (models II and II*). The energy conservation equation is obtained from the equation of motion (12.63) by dotting it with $-\partial\boldsymbol{\xi}/\partial t$, and integrating over the plasma volume:

$$-\int \frac{\partial\boldsymbol{\xi}}{\partial t} \cdot \mathbf{G}(\boldsymbol{\xi})\, dV + 2\int \frac{\partial\boldsymbol{\xi}}{\partial t} \cdot \Big(\rho\mathbf{v} \cdot \nabla\frac{\partial\boldsymbol{\xi}}{\partial t}\Big)\, dV + \int \rho\, \frac{\partial\boldsymbol{\xi}}{\partial t} \cdot \frac{\partial^2\boldsymbol{\xi}}{\partial t^2}\, dV = 0\,,$$

where the second integral vanishes by virtue of the anti-symmetry (12.90). Also exploiting the self-adjointness of the operator $\mathbf{G}$ then gives

$$\frac{d}{dt}\left[-\tfrac{1}{2}\int \boldsymbol{\xi}\cdot\mathbf{G}(\boldsymbol{\xi})\,dV + \tfrac{1}{2}\int \rho\left|\frac{\partial\boldsymbol{\xi}}{\partial t}\right|^2 dV \right] = 0$$

$$\Rightarrow \quad H = \int \mathcal{H}\,dV = \text{const}\,, \quad \mathcal{H} \equiv \tfrac{1}{2}\rho\left|\frac{\partial\boldsymbol{\xi}}{\partial t}\right|^2 - \tfrac{1}{2}\boldsymbol{\xi}\cdot\mathbf{G}(\boldsymbol{\xi})\,, \quad (12.91)$$

so that the total energy, or Hamiltonian H, of the perturbations is conserved.

▷ **Hamiltonian formulations** Some of the most powerful and beautiful parts of physics are the Lagrangian and Hamiltonian formulations of classical mechanics. In particular, the formulation of a Lagrangian from which the equations of motion can be derived by means of Hamilton's principle is the most concise description of a dynamical system; see the chapter on Lagrangian and Hamiltonian formulations for continuous systems and fields by Goldstein [184]. One may consider a branch of physics to have become part of the classical curriculum if one succeeds in constructing the appropriate Lagrangian. For nonlinear ideal MHD, this was accomplished by Cotsaftis [95] and Newcomb [349] in 1962.

For linear ideal MHD, Frieman and Rotenberg [147] showed that the equation of motion (12.63) can be derived from Hamilton's principle (see Section 6.4.2 [1]), involving the volume integral of the Lagrangian density $\mathcal{L}$:

$$\delta\int_{t_1}^{t_2}\left[\int \mathcal{L}\,dV\right]dt = 0\,, \quad \mathcal{L} \equiv \tfrac{1}{2}\rho\left|\frac{\partial\boldsymbol{\xi}}{\partial t}\right|^2 - \rho\left(\mathbf{v}\cdot\nabla\frac{\partial\boldsymbol{\xi}}{\partial t}\right)\cdot\boldsymbol{\xi} + \tfrac{1}{2}\boldsymbol{\xi}\cdot\mathbf{G}(\boldsymbol{\xi})\,. \quad (12.92)$$

By defining the canonical momentum $\boldsymbol{\pi}$ (not to be confused with the Eulerian pressure perturbation), one obtains the corresponding Hamiltonian density,

$$\boldsymbol{\pi} \equiv \rho\left(\frac{\partial\boldsymbol{\xi}}{\partial t} + \mathbf{v}\cdot\nabla\boldsymbol{\xi}\right) \quad\Rightarrow\quad \mathcal{H} = \frac{1}{2\rho}\left(\boldsymbol{\pi} - \rho\mathbf{v}\cdot\nabla\boldsymbol{\xi}\right)^2 - \tfrac{1}{2}\boldsymbol{\xi}\cdot\mathbf{G}(\boldsymbol{\xi})\,, \quad (12.93)$$

which yields Hamilton's equations:

$$\frac{\partial\xi_i}{\partial t} = \frac{\partial\mathcal{H}}{\partial\pi_i} \qquad\qquad\Rightarrow\quad \frac{\partial\boldsymbol{\xi}}{\partial t} = \frac{1}{\rho}\left(\boldsymbol{\pi} - \rho\mathbf{v}\cdot\nabla\boldsymbol{\xi}\right)\,,$$

$$\frac{\partial\pi_i}{\partial t} = \frac{\partial}{\partial r_j}\left(\frac{\partial\mathcal{H}}{\partial(\partial\xi_i/\partial r_j)}\right) - \frac{\partial\mathcal{H}}{\partial\xi_i} \quad\Rightarrow\quad \frac{\partial\boldsymbol{\pi}}{\partial t} = \mathbf{G}(\boldsymbol{\xi}) - \rho\mathbf{v}\cdot\nabla\left(\pi/\rho - \mathbf{v}\cdot\nabla\boldsymbol{\xi}\right)\,.$$
$$(12.94)$$

The first equation just rephrases the definition of the canonical momentum, and the latter equation reproduces the equation of motion (12.63), QED.

Next, Newcomb [349] integrated the nonlinear MHD equations, and constructed the associated Lagrangian, by replacing the above quasi-Lagrangian description by a truly Lagrangian description in terms of the displacement vector field $\boldsymbol{\xi}$ with respect to the initial positions ($\mathbf{r}_0$ in Fig. 12.3). Later, this approach was put in the framework of modern Hamiltonian methods [14] and elaborated by a growing number of authors: Holm, Marsden, Ratiu and Weinstein [227], Morrison and Greene [340], Hameiri [211, 212], etc. ◁

So far, we have exploited real displacement vectors $\boldsymbol{\xi}$ and a real inner product. For normal modes $\boldsymbol{\xi}(\mathbf{r})\exp(-i\omega t)$, it is evident that we should revert to a complex inner product. Therefore, as for static plasmas (see Section 6.2.2[1]), we now assume a Hilbert space with an inner product and a finite norm defined by

$$\langle \boldsymbol{\xi}, \boldsymbol{\eta} \rangle \equiv \tfrac{1}{2} \int \rho \, \boldsymbol{\xi}^* \cdot \boldsymbol{\eta} \, dV , \qquad I[\boldsymbol{\xi}] \equiv \|\boldsymbol{\xi}\|^2 \equiv \langle \boldsymbol{\xi}, \boldsymbol{\xi} \rangle < \infty . \tag{12.95}$$

With this inner product, it is convenient to absorb a factor $-\mathrm{i}$ in a *revised definition of the gradient operator parallel to the velocity*,

$$U \equiv -\mathrm{i}\rho \mathbf{v} \cdot \nabla , \tag{12.96}$$

so that the equation of motion (12.63) and the spectral equation (12.69) become

$$\mathbf{G}(\boldsymbol{\xi}) - 2\mathrm{i}U \frac{\partial \boldsymbol{\xi}}{\partial t} - \rho \frac{\partial^2 \boldsymbol{\xi}}{\partial t^2} = 0 \quad \Rightarrow \quad \mathbf{G}(\boldsymbol{\xi}) - 2\omega \, U\boldsymbol{\xi} + \rho \omega^2 \boldsymbol{\xi} = 0 . \tag{12.97}$$

Both operators $\rho^{-1}\mathbf{G}$ and $\rho^{-1}U$ are then self-adjoint (or Hermitian) in the Hilbert space defined:

$$\langle \boldsymbol{\eta}, \rho^{-1}\mathbf{G}(\boldsymbol{\xi}) \rangle \equiv \tfrac{1}{2} \int \boldsymbol{\eta}^* \cdot \mathbf{G}(\boldsymbol{\xi}) \, dV = \tfrac{1}{2} \int \boldsymbol{\xi} \cdot \mathbf{G}(\boldsymbol{\eta}^*) \, dV \equiv \langle \rho^{-1}\mathbf{G}(\boldsymbol{\eta}), \boldsymbol{\xi} \rangle ,$$

$$\langle \boldsymbol{\eta}, \rho^{-1}U\boldsymbol{\xi} \rangle \equiv \tfrac{1}{2} \int \boldsymbol{\eta}^* \cdot U\boldsymbol{\xi} \, dV = \tfrac{1}{2} \int \boldsymbol{\xi} \cdot (U\boldsymbol{\eta})^* \, dV \equiv \langle \rho^{-1}U\boldsymbol{\eta}, \boldsymbol{\xi} \rangle ,$$
$$\tag{12.98}$$

where the latter equality follows from the anti-symmetry relation (12.90). We conclude that the complexity of the spectral problem in plasmas with stationary flow is not due to non-self-adjointness (as frequently stated in the literature), but to the fact that the eigenvalue problem is quadratic, with two operators, so that ω is complex and $\boldsymbol{\xi}$ has six (rather than three) real components.

▷ **Eigenvalue problem in the Hamiltonian formulation** As a bonus of the introduction of the canonical momentum, for normal modes Hamilton's equations (12.94) become

$$\begin{pmatrix} \rho^{-1}U & -\rho^{-1}\mathcal{G} + (\rho^{-1}U)^2 \\ 1 & \rho^{-1}U \end{pmatrix} \begin{pmatrix} \mathrm{i}\rho^{-1}\boldsymbol{\pi} \\ \boldsymbol{\xi} \end{pmatrix} = \omega \begin{pmatrix} \mathrm{i}\rho^{-1}\boldsymbol{\pi} \\ \boldsymbol{\xi} \end{pmatrix} , \tag{12.99}$$

where $\mathcal{G}\boldsymbol{\xi} \equiv \mathbf{G}(\boldsymbol{\xi})$. In contrast to the original spectral problem (12.97), this is now a *standard linear eigenvalue problem* in the six free components of $\mathrm{i}\rho^{-1}\boldsymbol{\pi}$ and $\boldsymbol{\xi}$. ◁

The Hamiltonian (12.91) is properly represented by $H \equiv K + W$. This suggests associating the following expression for the *kinetic energy* K with the dynamical variable $\boldsymbol{\xi}$:

$$K \equiv \tfrac{1}{2} \int \rho \frac{\partial \boldsymbol{\xi}^*}{\partial t} \cdot \frac{\partial \boldsymbol{\xi}}{\partial t} \, dV_{\text{all}} , \tag{12.100}$$

so that the boundedness of I turns out to be related to the physical condition of finite kinetic energy,

$$K[\boldsymbol{\xi}] \equiv \|\partial \boldsymbol{\xi}/\partial t\|^2 = |\omega|^2 I < \infty . \tag{12.101}$$

The corresponding expression for the *potential energy* W may be obtained from the manifestly symmetric quadratic form (12.88) by substituting $\boldsymbol{\eta} \equiv \boldsymbol{\xi}^*$. However, a more compact expression (without the need for complex conjugate doubling of some of the terms) is obtained from the sum of Eqs. (12.75) and (12.76) and integration by parts to eliminate the second derivatives of the equilibrium quantities. This gives:

$$ W \equiv -\tfrac{1}{2} \int \boldsymbol{\xi}^* \cdot \mathbf{G}(\boldsymbol{\xi})\, dV_{\mathrm{all}} = W^p[\boldsymbol{\xi}] + W^s[\xi_n] + W^{\hat{p}}[\hat{\boldsymbol{\xi}}] , \tag{12.102} $$

where

$$ \begin{aligned} W^p \equiv \tfrac{1}{2} \int \Big[& \gamma p |\nabla \cdot \boldsymbol{\xi}|^2 + |\mathbf{Q}|^2 \\ & + (\nabla p) \cdot \boldsymbol{\xi}\, \nabla \cdot \boldsymbol{\xi}^* + \mathbf{j} \cdot (\boldsymbol{\xi}^* \times \mathbf{Q}) - (\nabla \Phi) \cdot \boldsymbol{\xi}^*\, \nabla \cdot (\rho \boldsymbol{\xi}) \\ & + \rho(\mathbf{v} \cdot \nabla \mathbf{v}) \cdot (\boldsymbol{\xi} \cdot \nabla \boldsymbol{\xi}^*) - \rho |\mathbf{v} \cdot \nabla \boldsymbol{\xi}|^2 \Big]\, dV, \end{aligned} \tag{12.103} $$

$$ W^s \equiv \tfrac{1}{2} \int \Big[|\xi_n|^2\, \mathbf{n} \cdot \big[\![\nabla(p + \tfrac{1}{2}B^2)]\!\big] + \xi_n \big[\![\rho(\mathbf{v} \cdot \nabla \mathbf{v}) \cdot \boldsymbol{\xi}^*]\!\big] \Big]\, dS , \tag{12.104} $$

$$ W^{\hat{p}} \equiv \begin{cases} \tfrac{1}{2} \int |\hat{\mathbf{Q}}|^2\, d\hat{V} \quad \left(\equiv W^v \right) & (\textit{model II}) \\[2mm] \tfrac{1}{2} \int \Big[\gamma \hat{p} |\nabla \cdot \hat{\boldsymbol{\xi}}|^2 + \cdots (\text{as Eq. (12.103)}) \cdots \Big]\, d\hat{V} & (\textit{model II*}) \end{cases} \tag{12.105} $$

For model I (plasma–wall) equilibria, $W^s = 0$ because $\xi_n = 0$ at the wall, and the term $W^{\hat{p}}$ is evidently absent. The energy expressions for model II (plasma–vacuum) equilibria may be obtained from the ones for model II* (plasma–plasma) equilibria by taking the limit $\hat{\rho} \to 0$, $\hat{p} \to 0$, $\hat{\mathbf{j}} \to 0$ and $\hat{\mathbf{v}} \to 0$. As noticed at the end of Section 12.2.3, this also applies to the BC (12.87) on the perturbed pressure, but the BC (12.86) on $\hat{\xi}_n$ needs to be replaced by the condition (12.89) on $\hat{Q}_n$.

Compared to static equilibria, the background velocity $\mathbf{v}$ adds two terms to the volume integrals W^p and $W^{\hat{p}}$, one due to the centrifugal acceleration and another one due to the squared gradient operator parallel to $\mathbf{v}$ (last two terms of Eq. (12.103)). The surface integral W^s also has an extra centrifugal contribution (second term of Eq. (12.104)). For cylindrical plasmas in the absence of gravity, according to Eqs. (12.29) and (12.30), this term cancels with the rotational contribution so that only the magnetic curvature $-B_\theta^2/r$ remains. For toroidal equilibria, the second term of Eq. (12.104) contributes both a normal (perpendicular to the surface) and a geodesic (inside the surface) curvature of the flow lines (see Section 17.1.3 for the similar definitions for magnetic field lines).

The relationship between the quadratic forms and the spectral equation (12.97) becomes much clearer when the latter is subjected to a similar operation as above (in the derivation of Eq. (12.91)), dotting it with $\boldsymbol{\xi}^*$ and integrating over the volume.

This yields a quadratic equation for the eigenvalues:

$$I\omega^2 - 2V\omega - W = 0 , \tag{12.106}$$

where the three coefficients

$$I \equiv \tfrac{1}{2} \int \rho \boldsymbol{\xi}^* \cdot \boldsymbol{\xi}\, dV , \quad V \equiv \tfrac{1}{2} \int \boldsymbol{\xi}^* \cdot U \boldsymbol{\xi}\, dV , \quad W \equiv -\tfrac{1}{2} \int \boldsymbol{\xi}^* \cdot \mathbf{G}(\boldsymbol{\xi})\, dV \tag{12.107}$$

are *real* because of the proved self-adjointness of the operators $\rho^{-1}U$ and $\rho^{-1}\mathbf{G}$. (The meaning of the ambiguous symbol V, for the quadratic form as well as for the volume, should be clear from the context.) The solutions of the quadratic equation (12.106),

$$\omega = \frac{V \pm \sqrt{V^2 + IW}}{I} , \tag{12.108}$$

are physically significant relationships between the eigenvalues ω and the quadratic forms I, V and W, but they do not immediately provide a solution of the eigenvalue problem. Recall that, in contrast, for static plasmas (Section 6.4.3[1]), the relationship $\omega^2 = W/I$ could easily be transformed into the Rayleigh–Ritz variational principle $\delta \Lambda = 0$, where the eigenfunctions $\boldsymbol{\xi}$ yield the stationary values ω^2 of the Rayleigh quotient $\Lambda \equiv W[\boldsymbol{\xi}]/I[\boldsymbol{\xi}]$. This also provided the connection with the energy principle $W[\boldsymbol{\xi}] > 0$ for stability (Section 6.4.4[1]). We now need to account for the modifications of this spectral method by the presence of the operator U, and the associated quadratic form V, for stationary plasmas.

Splitting the eigenvalue parameter into real and imaginary parts, $\omega = \sigma + i\nu$, and also the eigenvalue equation (12.106),

$$(\sigma^2 - \nu^2)I - 2\sigma V - W = 0 , \qquad \nu(\sigma I - V) = 0 , \tag{12.109}$$

it is evident that the solutions divide into *stable waves* ($\nu = 0$) and *instabilities* ($\nu \neq 0$) with different energy content:

$$\begin{cases} K = \sigma^2 I & \Rightarrow \quad H = K + W = 2\sigma(\sigma I - V) \quad (\nu = 0) , \\[2mm] K = (\sigma^2 + \nu^2)I & \Rightarrow \quad H = K + W = 0 \qquad\qquad\quad (\nu \neq 0) . \end{cases} \tag{12.110}$$

Hence, for stable oscillations a finite amount of energy $K + W$ should be provided, but for exponentially growing instabilities the increase of kinetic energy that is accompanied by the decrease of potential energy is compatible with $K + W = 0$. As we will see, this does not imply that $W[\boldsymbol{\xi}] > 0$ is necessary for stability, nor that $W[\boldsymbol{\xi}] < 0$ is sufficient for instability.

Normalizing the quadratic forms V and W to correspond, in the terminology of quantum mechanics, to the expectation values (or averages) of the operators $\rho^{-1}U$

and $-\rho^{-1}\mathbf{G}$,

$$\overline{V} \equiv \frac{V}{I} \equiv \langle \rho^{-1}U \rangle, \qquad \overline{W} \equiv \frac{W}{I} \equiv \langle -\rho^{-1}\mathbf{G} \rangle, \qquad (12.111)$$

the expressions for the real and imaginary parts of ω following from Eqs. (12.109) become:

$$\begin{cases} \sigma = \overline{V} \pm \sqrt{\overline{V}^2 + \overline{W}}, & \nu = 0 & \textit{(stable waves)}, \\[2mm] \sigma = \overline{V}, & \nu = \pm\sqrt{-\overline{V}^2 - \overline{W}} \quad \Rightarrow \quad |\omega|^2 = -\overline{W} \;\textit{(instabilities)}. \end{cases}$$
$$(12.112)$$

This shows that the marginal state is no longer at the origin $\sigma = \nu = 0$ of the ω-plane (as for static plasmas), but shifted to the point $\sigma = \overline{V}$, $\nu = 0$, where ω becomes complex for instabilities, whereas stable waves divide into faster and slower ones there. This point is no longer at a fixed position in the ω-plane, but it depends on the solution $\boldsymbol{\xi}$ itself. Equation (12.112) implies that unstable modes are located at a distance $|\omega| \equiv \sqrt{\sigma^2 + \nu^2} = \sqrt{-\overline{W}}$ to the origin of the ω-plane and at a horizontal distance $\sigma = \overline{V}$ to the imaginary axis (Fig. 12.4). Hence, for every complex eigenvalue ω, a corresponding eigenvalue ω^* can be found. Moreover, since stationary equilibria are invariant under reflection of the background velocity, i.e. $\overline{V} \to -\overline{V}$, the pairs $-\omega$ and $-\omega^*$ will also be eigenvalues. (For separable systems, this involves consideration of wave numbers of opposite sign.)

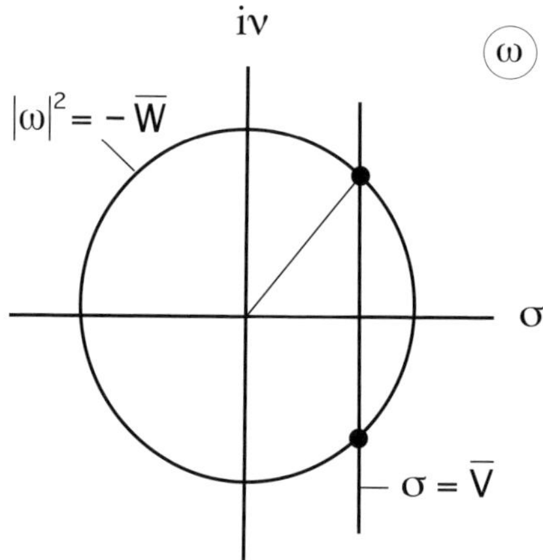

Fig. 12.4 Unstable modes ω and ω^* at the intersections of the circle $|\omega|^2 = -\overline{W}$ (with real radius for negative potential energy) and the vertical line $\sigma = \overline{V}$ (average Doppler–Coriolis shift).

The most important (but not the only) physical effect of the operator U behind these spectral properties is the *Doppler shift* of eigenfrequencies. For plane waves,

$\xi \sim \exp{(\mathbf{ik} \cdot \mathbf{r})}$, the relationship is immediate: $\overline{V} \equiv \langle -\mathrm{i}\mathbf{v} \cdot \nabla \rangle = \langle \mathbf{k} \cdot \mathbf{v} \rangle$. However, for inhomogeneous plasma configurations with curved flow fields (e.g. rotating cylindrical plasmas, see Sections 13.3 and 13.4, the differential operator U also acts on the unit vectors so that $\overline{V}$ will contain contributions from the Coriolis acceleration as well. Hence, we will call $\overline{V}$ the *average Doppler–Coriolis shift*.

With respect to stability, the standard criterion $\overline{W}[\xi] > 0$ is still sufficient for the stability of stationary plasmas, but the above expressions provide the following, much sharper, criteria involving the square of the average Doppler–Coriolis shift.

A *necessary and sufficient criterion for stability* of stationary plasmas is that

$$\overline{W}[\xi] > -\overline{V}^2[\xi] \tag{12.113}$$

for all ξ that are bound in norm and satisfy the appropriate boundary conditions. Vice versa, *sufficient for instability* is that

$$\overline{W}[\xi] < -\overline{V}^2[\xi] \tag{12.114}$$

for a particular choice of ξ. The proviso "bound in norm" is crucial here since it has been shown that stationary rotating plasmas may have unstable continuous spectra [222, 213, 351, 173]. Hence, these criteria refer to discrete modes only.

The stability criterion (12.113) should not be interpreted to imply that the effect of the average Doppler–Coriolis shift $\overline{V}$ is stabilizing. In fact, the opposite is true since the expression (12.103) for W^p contains the negative term $-\rho|\mathbf{v} \cdot \nabla\xi|^2$. Splitting this term off, we obtain the following alternative definition $\widetilde{W}$ for the potential energy (corresponding to the alternative definition (12.68) for $\widetilde{G}$), which is identical to that of static plasmas if the centrifugal term $\sim \rho\mathbf{v} \cdot \nabla\mathbf{v}$ is neglected (as appropriate for plane shear flows):

$$\widetilde{W} \equiv W + \Delta\,, \qquad \Delta \equiv \tfrac{1}{2}\int \rho|\mathbf{v} \cdot \nabla\xi|^2\, dV\,, \tag{12.115}$$

so that the stability criterion (12.113) becomes

$$\overline{W} + \overline{V}^2 \equiv \overline{\widetilde{W}} + \overline{V}^2 - \overline{\Delta} > 0\,. \tag{12.116}$$

Through Schwarz' inequality,

$$\left[\int \mathbf{f}^* \cdot \mathbf{g}\, dV\right]^2 \le \int |\mathbf{f}|^2\, dV \int |\mathbf{g}|^2\, dV\,, \quad \text{with } \mathbf{f} \equiv \sqrt{\rho}\,\xi\,, \ \mathbf{g} \equiv -\mathrm{i}\sqrt{\rho}\,\mathbf{v} \cdot \nabla\xi\,,$$

$$\Rightarrow \ \overline{V}^2 - \overline{\Delta} \equiv \langle -\mathrm{i}\mathbf{v} \cdot \nabla \rangle^2 - \langle (-\mathrm{i}\mathbf{v} \cdot \nabla)^2 \rangle \le 0\,, \tag{12.117}$$

the net result is zero or negative: in general, *plane shear flows are destabilizing*. This is the reason that such flows may exhibit the Kelvin–Helmholtz instability (Section 13.1.2).

▷ **Modified stability criterion** Similar to the σ-stability modification of the energy principle for static plasmas by Goedbloed and Sakanaka [183], discussed in Section 6.5.3 [1], the stability criterion (12.113) for stationary plasmas may also be relaxed as follows (see Fig. 12.5). A *necessary and sufficient criterion for ϵ-stability* of stationary plasmas is that

$$\overline{W}^{\epsilon}[\boldsymbol{\xi}] \equiv \overline{W}[\boldsymbol{\xi}] + \epsilon^2 > -\overline{V}^2[\boldsymbol{\xi}] \tag{12.118}$$

for all $\boldsymbol{\xi}$ that are bound in norm and satisfy the appropriate boundary conditions. (For obvious reasons, the terminology σ-stability of Volume [1] is changed to ϵ-stability here.) This stability criterion considers modes that exponentially grow as $\exp(\epsilon t)$ stable if ϵ does not surpass a certain pre-fixed value. The motivation could be (1) that the ideal MHD model is not appropriate anyway for very long time scales when dissipation becomes important, (2) that one is interested only in phenomena observed during a limited time interval of a

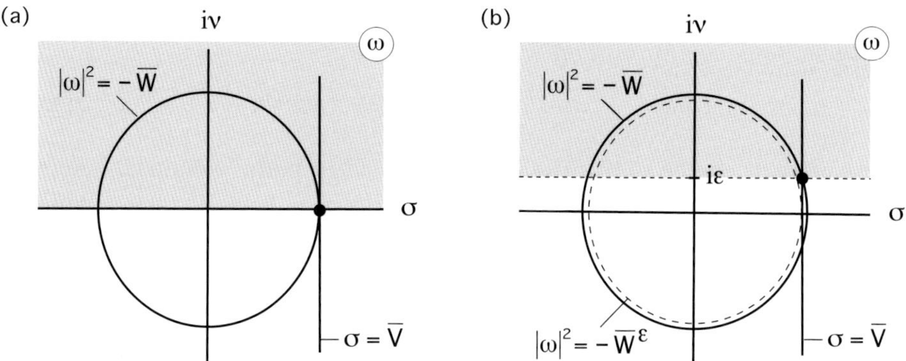

Fig. 12.5 Unstable part of the ω-plane (grey) and marginal discrete mode (dot) with respect to (a) ordinary stability, and (b) ϵ-stability.

certain experiment or astrophysical observation, (3) that the available numerical program is not accurate enough to calculate modes that grow too slowly. (Note that, if one demands an accuracy of $\epsilon = 10^{-6}$ in the growth rate, one should be able to compute the energy to an accuracy of $\epsilon^2 = 10^{-12}$!) In practice, all these limitations always occur.

Most importantly, the method of ϵ-stability avoids the neighborhood of the continua, which for static plasmas just complicate the analysis of the marginal point $\omega^2 = 0$, but now the analysis of the whole real ω-axis, already for plane shear flow (Section 13.1.3), not even considering the mentioned complex unstable continua for rotating plasmas. Finally, the proofs of sufficiency and necessity of the energy principle for static plasmas, due to Laval, Mercier and Pellat [298], can be adapted to stability of stationary plasmas by appropriate modification of the energy functional, and to ϵ-stability by replacing W by W^{ϵ}.   ◁

Having arrived at this point, it becomes clear that we should leave the esoteric realm of spectral theory and descend into the low lands of explicit calculations for specific equilibria in order to really get a grip on the waves and instabilities of stationary plasmas. This is what we will do in Chapter 13 for plane shear flows and for rotating cylindrical plasmas. The general solution method is introduced in the following section.

12.3 Solution paths in the complex ω plane

12.3.1 Opening up the boundaries

Recall the "simplicity" of spectral theory for static plasmas, which was based on the fact that the eigenvalue only appears as ω^2, so that ω itself is either real (for stable modes) or imaginary (for instabilities). With flow, because of the Doppler–Coriolis shift operator U, the eigenvalues are no longer confined to the real and imaginary axes so that the spectral problem becomes really complex. Do we now have to search randomly for eigenvalues in the complex plane? Of course not: we have the powerful expressions (12.112) at our disposal telling us that the eigenvalues are, again, to be found in particular locations, viz. on the real ω-axis (for stable waves) and on a path, or paths, where the real part of the average Doppler–Coriolis shifted frequency vanishes (for instabilities). Evidently, we have to exploit these expressions to determine those locations.

We start from the basic spectral equation (12.97), repeated here for convenience,

$$\mathbf{G}(\boldsymbol{\xi}) - 2\omega\, U\boldsymbol{\xi} + \rho\omega^2\boldsymbol{\xi} = 0 \,. \tag{12.119}$$

In order to solve this equation for a particular configuration, we have to be more explicit about the actual magnetic geometry of the confined plasma and the associated BCs. In fact, the formulation (12.71) of the BC on the normal component ξ_n of $\boldsymbol{\xi}$ at the wall does not show that, in general, such a "wall" may be multiply connected (like for the toroidal geometries considered in controlled fusion research) or, rather, consist of two walls (like the plane plasma slab considered in the following chapter). Moreover, for cylindrical and toroidal geometries, instead of an inner wall the magnetic axis will provide a limiting BC by demanding regularity of the solution there. The essential aspects of these conditions are well represented by considering the model I BCs for plasma confined between two "walls", one at $x = x_1$ (which may be the magnetic axis) and another one at $x = x_2$:

$$\left\{ \begin{array}{ll} \xi(x_1) = 0\,, \quad \text{or regular at the magnetic axis} & \textit{(left BC)} \\[2mm] \xi(x_2) = 0 & \textit{(right BC)} \end{array} \right. \tag{12.120}$$

Here and in the following, we drop the subscript on ξ_n and simply write $\xi \equiv \boldsymbol{\xi} \cdot \mathbf{n}$. The coordinate x may represent the vertical coordinate of a plane gravitating plasma slab (Section 13.1), or the radial coordinate r of a cylindrical plasma (Section 13.3), or the poloidal magnetic flux ψ of a toroidal plasma (Chapters 16–18). The associated differences in the metric coefficients, which are important for explicit calculations, are considered in detail at the places cited.

Next, notice that the spectral differential equation (12.119) can be solved, in principle, for *any* complex value $\omega = \sigma + \mathrm{i}\nu$ if one of the BCs (12.120) is ignored. Thus, the basic first step in many numerical computations of MHD spectra is to turn the eigenvalue problem (EVP) of Eqs. (12.119) and (12.120) into a one-sided boundary value problem (BVP) by integrating Eq. (12.119) starting from the left and dropping the right BC (12.120)(b), or starting from the right and dropping the left BC (12.120)(a). (This one-sided BVP is usually indicated as an initial value problem (IVP), but we will avoid that terminology to avoid confusion with the temporal IVP associated with the spectrum, as expanded in Chapter 10[1].) The corresponding solutions will be called $\boldsymbol{\xi}^\ell$ and $\boldsymbol{\xi}^r$. Taking the integration of the (partial) differential equation for granted (the necessary numerical techniques are detailed in Chapter 15), the problem of this section can be summarized as follows. *How do we construct a path, or paths, in the complex ω-plane from the solutions $\boldsymbol{\xi}^\ell$, or $\boldsymbol{\xi}^r$, such that the eigenvalues are encountered with* 100% *certainty?*

Let us reconsider the derivation of the quadratic equation (12.106), from which the central properties (12.112) for the eigenvalues were obtained. Again dotting Eq. (12.119) with $\boldsymbol{\xi}^*$ and integrating over volume, but this time not assuming that both BCs (12.120) are satisfied, instead of Eq. (12.106) we obtain

$$I\omega^2 - 2V\omega - W_1 - iW_2 = 0 , \tag{12.121}$$

where I and V are still real, but W has an imaginary contribution $W_2 \neq 0$. For complex ω, this quadratic gives solutions similar to the expressions (12.112)(b),

$$\sigma = \overline{V} + \frac{\overline{W}_2}{2\nu} , \quad \nu = \pm\left[-\overline{V}^2 - \overline{W}_1 + \left(\frac{\overline{W}_2}{2\nu}\right)^2\right]^{1/2} \quad\Rightarrow\quad |\omega|^2 = -\overline{W}_1 + \frac{\sigma}{\nu}\overline{W}_2 . \tag{12.122}$$

These equalities now hold for any solution of the left or right BVP (except at the continuous spectra). Clearly, W_2 may be considered as some kind of measure for the distance to eigenvalues. This contribution can be computed from the pre-self-adjointness relation (12.83), obtained in the proof of self-adjointness of $\mathbf{G}$:

$$W_2 \equiv -\tfrac{1}{2}i(W - W^*) \equiv \tfrac{1}{4}i \int \left[\boldsymbol{\xi}^* \cdot \mathbf{G}(\boldsymbol{\xi}) - \boldsymbol{\xi} \cdot \mathbf{G}(\boldsymbol{\xi}^*)\right] dV$$

$$= -\tfrac{1}{4}i \int \left[\boldsymbol{\xi}^*\Pi(\boldsymbol{\xi}) - \boldsymbol{\xi}\Pi(\boldsymbol{\xi}^*)\right] dS \overset{(1D)}{=} \tfrac{1}{2}\left(\xi_1\Pi_2 - \xi_2\Pi_1\right)\Big|_{x_1}^{x_2} . \tag{12.123}$$

For simplicity, the surface integral over the left and right boundaries is reduced in the last step for a one-dimensional (1D), separable, system (as analyzed in Chapter 13), where the complex variables are split into real and imaginary parts,

$$\xi = \xi_1 + i\xi_2 , \qquad \Pi = \Pi_1 + i\Pi_2 . \tag{12.124}$$

It is important to realize though that the boundary values appearing in W_2 are obtained in *any* numerical iteration procedure, either by solving a system of ordinary differential equation (ODEs) in the 1D case exploiting finite differences (see Section 15.1.2), or by solving a partial differential equation (PDE) in the 2D or 3D case with finite elements (see Section 15.1.3). For eigenvalues, these boundary contributions vanish so that W becomes real. How do we extend this desirable property from point eigenvalues to the paths in the complex ω-plane we are looking for?

Consider the following complex function, mapping the ω-plane onto itself:

$$Q^\ell(\omega) \equiv \omega - \overline{V}^\ell \equiv \omega - \langle \rho^{-1}U \rangle^\ell \equiv \omega - \frac{\int \boldsymbol{\xi}^{\ell*} \cdot U\boldsymbol{\xi}^\ell \, dV}{\int \rho |\boldsymbol{\xi}^\ell|^2 \, dV} . \tag{12.125}$$

It is defined everywhere in the complex ω-plane (except at the continua) for solutions $\boldsymbol{\xi}^\ell(\mathbf{r}; \sigma + i\nu)$ of the left BVP. A similar expression may be defined for the right solutions $\boldsymbol{\xi}^r(\mathbf{r}; \sigma + i\nu)$. For definiteness, we will restrict the discussion to the left solutions. Notice that, since $\overline{V}^\ell$ is real for any value of ω, two paths may be constructed from Q^ℓ, viz.

$$\begin{cases} \operatorname{Im} Q^{\mathrm{L}} = \nu & = 0 \quad \Rightarrow \quad \text{path } \mathcal{P}_{\mathrm{s}}^{\mathrm{L}} \text{ of stable solutions} \\[2mm] \operatorname{Re} Q^{\mathrm{L}} = \sigma - \overline{V}^{\mathrm{L}} = 0 \quad \Rightarrow \quad \text{path } \mathcal{P}_{\mathrm{u}}^{\mathrm{L}} \text{ of unstable solutions} \end{cases} , \tag{12.126}$$

where we now indicate the solutions on the paths by upper case superscripts. In fact, according to the conditions (12.112) for eigenvalues, *the straight line* $\operatorname{Im} Q^{\mathrm{L}} = 0$ *and the curve(s)* $\operatorname{Re} Q^{\mathrm{L}} = 0$ *provide the only possible locations where eigenvalues may be found:* with respect to stability, they play the same role as the real and imaginary axes in the static case. Hence, unstable eigenvalues are found on the path(s) $\mathcal{P}_{\mathrm{u}}^{\mathrm{L}}$ in the complex ω-plane that consists of the points $\sigma_{\mathrm{u}} + i\nu_{\mathrm{u}}$ determined by solving the following nonlinear equation for σ_{u} for every value of ν_{u}:

$$\sigma_{\mathrm{u}} = \overline{V}[\boldsymbol{\xi}^{\mathrm{L}}(\sigma_{\mathrm{u}} + i\nu_{\mathrm{u}})] \equiv \frac{\int \boldsymbol{\xi}^{\mathrm{L}*} \cdot U\boldsymbol{\xi}^{\mathrm{L}} \, dV}{\int \rho |\boldsymbol{\xi}^{\mathrm{L}}|^2 \, dV} \quad \Rightarrow \quad \mathcal{P}_{\mathrm{u}}^{\mathrm{L}} \equiv \{\sigma_{\mathrm{u}}^{\mathrm{L}}(\nu_{\mathrm{u}})\} . \tag{12.127}$$

This amounts to determining the zeros of the function $\sigma_{\mathrm{u}} - \overline{V}^{\mathrm{L}}(\sigma_{\mathrm{u}})$, which physically corresponds to *vanishing of the solution-averaged Doppler–Coriolis shifted real part of the frequency.* Since the velocities of the stationary flow are bounded, this equation always has at least one solution for every value of ν_{u}.

Returning now to the expression (12.123) for the imaginary contribution W_2 of the energy, notice that, for the above solutions $\sigma_{\mathrm{u}}(\nu_{\mathrm{u}})$ for the path $\mathcal{P}_{\mathrm{u}}^{\mathrm{L}}$, the first relation (12.122) yields $\sigma_{\mathrm{u}} - \overline{V} = \overline{W}_2/(2\nu_{\mathrm{u}}) = 0$. Hence, we obtain an alternative, even more powerful, expression for the determination of the path $\mathcal{P}_{\mathrm{u}}^{\mathrm{L}}$:

$$\begin{aligned} W_2^{\mathrm{L}} &= \tfrac{1}{2} \int \left(\xi_1^{\mathrm{L}} \Pi_2^{\mathrm{L}} - \xi_2^{\mathrm{L}} \Pi_1^{\mathrm{L}} \right) dS_2 \\[2mm] &\overset{(1\mathrm{D})}{=} \tfrac{1}{2} \left[\xi_1^{\mathrm{L}}(x_2) \Pi_2^{\mathrm{L}}(x_2) - \xi_2^{\mathrm{L}}(x_2) \Pi_1^{\mathrm{L}}(x_2) \right] = 0 \quad \Rightarrow \quad \mathcal{P}_{\mathrm{u}}^{\mathrm{L}} \equiv \{\sigma_{\mathrm{u}}^{\mathrm{L}}(\nu_{\mathrm{u}})\} . \end{aligned} \tag{12.128}$$

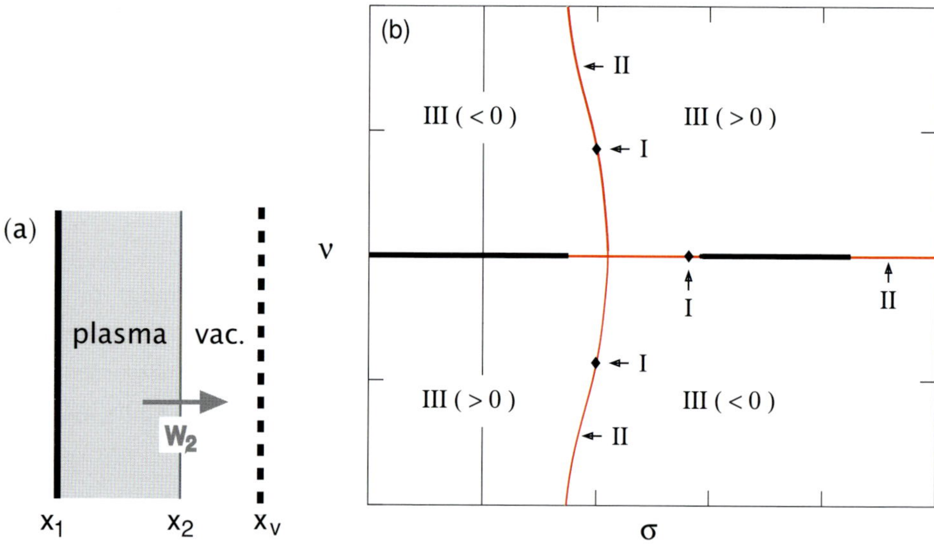

Fig. 12.6 Schematic presentation of the three BVPs for left solutions, $\xi(x_1) = 0$. (a) Model I: wall on plasma ($x_v = x_2$), model II: plasma–vacuum system ($x_2 < x_v < \infty$), model III: system coupled to external device (at $x = x_v$) injecting/extracting energy; (b) corresponding EVs (I) and solution paths (II, in red) for the closed system and external solution areas (III, with $W_2 > 0$ or $W_2 < 0$) in the complex ω-plane for the open system; thick black lines indicate continuous spectra.

Consequently, *the energy of the solutions of the left BVP is real,* not only for the eigenvalues, but *all along the path(s)* $\mathcal{P}_u^L$. This provides additional physical justification for the present method of exploiting the paths $\mathcal{P}_s^L$ and $\mathcal{P}_u^L$ for the iterative determination of the eigenvalues of stationary equilibria. For the numerical determination of the path(s), Eq. (12.128) is to be preferred over Eq. (12.127) since it does not require the additional operation of integration but just exploits the boundary values of ξ and Π. Those are directly available in whatever numerical scheme is exploited for the solution of the one-sided BVP for the spectral equation. Note though that Eq. (12.128), like Eq. (12.127), is a highly nonlinear equation for the determination of the path $\mathcal{P}_u^L$.

It will be noticed that the construction of solution paths by means of solutions of the one-sided BVP closely corresponds to the three confined plasma models I, II and III introduced in Section 4.6 [1]. This is illustrated in Fig. 12.6 for left solutions. Integrating the spectral equation, starting from the left, the right BC will only be satisfied for EVs (indicated by I), when the model I BVP is solved. Along the path (indicated by II), some unspecified model II BVP is solved: in order to satisfy the right BC ($\hat{\mathbf{Q}} \cdot \mathbf{n} = 0$), an external vacuum layer of variable (unknown) thickness has to be added. Note that we do not actually solve this model II BVP, but we construct the solution path for the model I eigenvalues by invoking that extended problem. However, the argument presented shows the physical significance of the solution path: it is not just a way of solving the EVP, but it constitutes an intrinsic structure of the complex ω-plane for a large class of spectral problems. Accordingly, the condition $W_2 = 0$ for the solution paths delineates the areas of the complex ω-plane where solutions are obtained of the extension with the model III BVP, where an external device either injects energy into the system ($W_2 < 0$) or extracts it ($W_2 > 0$). Thus, in order to maintain the time dependence $\exp(-i\omega t)$ of the perturbations, the energy W_2 should be absorbed by the external system if it is positive and supplemented by the external system if it is negative. This direct identification of energy supplied by the plasma to the vacuum at $x = x_2$ and extracted by the external system at $x = x_v$ only holds on average, when the "sloshing energy" of the plasma–vacuum oscillations with period $\tau \equiv 2\pi/\sigma$ is averaged out in the energy conservation law for the complete system. This averaging procedure is precisely the reason why the *imaginary* component of W enters.

The external device could be a system of magnetic coils coupled to a passive or active external circuit, but it could also be a mechanical device moving the plasma boundary. One could also imagine that the vacuum layer is replaced by another plasma (model II* BVP), but one should then demand that this second plasma is static. The whole point of considering an external layer of variable thickness is to permit different physical solutions *along the same solution path*. Since that path depends on the average Doppler–Coriolis shift, one cannot move the boundary x_v into the flowing part of the plasma without affecting the solution path. Obviously, keeping the solution path fixed is crucial for the determination of the structure of the unstable part of the spectrum.

As we will see in Chapter 13 for separable 1D systems, a unique path $\mathcal{P}_u^L$ is found for the left solutions for every choice of the pair of Fourier mode numbers referring to the ignorable directions (k_y and k_z for the plane case, m and k for the cylindrical case). Similarly, a unique *but different* path $\mathcal{P}_u^R$ is found for the right solutions, and a whole bundle of paths is found for mixed BVPs. Each of those paths is unique for the specified BVP, the imaginary part of the energy $W_2 = 0$ on each of them, and they all cross at the eigenvalues. Since we restrict the discussion to the left BVP, we will refer to $\mathcal{P}_u^L$ as the *single* path for the unstable modes in the separable 1D case. Inevitably, for non-separable 2D and 3D cases (like toroidal equilibria), at least one of the Fourier mode numbers fails to be a good quantum number so that we will obtain many paths of unstable modes then. Also, for separable 1D systems, we will find that the path may split into separate sub-paths in many cases of physical interest. For simplicity, we will continue to refer to $\mathcal{P}_u^L$ as a single path of unstable modes, leaving the multiplicity of the sub-paths understood, and also suppressing the superscript ^{L} from now on.

12.3.2 Approach to eigenvalues

So far, we have only considered the first half of the problem of the search for eigenvalues in the complex ω-plane. The full problem consists of two parts, viz.:

(1) "horizontal" iteration, for fixed ν_u, to determine the points $\sigma_u(\nu_u)$ of the path $\mathcal{P}_u$;

(2) iteration along the path, to determine where the eigenvalues are located on $\mathcal{P}_u$.

With respect to the second part, we will again invoke arguments that are general enough that the explicit procedures for 1D systems, to be expanded in Chapter 13, can be generalized, in principle, to 2D and 3D systems. Note that, for real values of ω, only the second part is relevant.

To complete the second part of the program, we need to consider solutions that are "oscillatory" on the path $\mathcal{P}_u$, or $\mathcal{P}_s$ for real ω, so that, eventually, both BCs (12.120) can be satisfied. For the time being, we will not assume satisfaction of the right BC yet. Consider two such (left) solutions, $\boldsymbol{\xi}_\alpha$ and $\boldsymbol{\xi}_\beta$, of the spectral equation (12.119) for two different values, ω_α and ω_β, of ω. To investigate the approach to eigenvalues, we need to study the behavior of these solutions relative to their inner product and distance $|\omega_\alpha - \omega_\beta|$ along the solution path. We first construct two basic quadratic forms, involving these quantities, which govern the relative behavior of arbitrary (not necessarily on the path) solutions.

Consider the spectral equations for $\boldsymbol{\xi}_\alpha$ and $\boldsymbol{\xi}_\beta^*$:

$$\mathbf{G}(\boldsymbol{\xi}_\alpha) - 2\omega_\alpha\, U\boldsymbol{\xi}_\alpha + \rho\omega_\alpha^2\boldsymbol{\xi}_\alpha = 0\,, \tag{12.129}$$

$$\mathbf{G}(\boldsymbol{\xi}_\beta^*) - 2\omega_\beta^*\,(U\boldsymbol{\xi}_\beta)^* + \rho\omega_\beta^{*2}\boldsymbol{\xi}_\beta^* = 0\,. \tag{12.130}$$

We construct two quadratic forms by taking the scalar product of the first equation with $\boldsymbol{\xi}_\beta^*$, resp. with $\omega_\beta^*\boldsymbol{\xi}_\beta^*$, and of the second equation with $\boldsymbol{\xi}_\alpha$, resp. with $\omega_\alpha\boldsymbol{\xi}_\alpha$, integrating over volume, subtracting and exploiting the pre-self-adjointness property (12.83) for $\mathbf{G}$ and the self-adjointness property (12.98)(b) for U. This yields

$$(\omega_\beta^* - \omega_\alpha)\left[\int \boldsymbol{\xi}_\beta^* \cdot U\boldsymbol{\xi}_\alpha\, dV - \tfrac{1}{2}(\omega_\beta^* + \omega_\alpha)\int \rho\boldsymbol{\xi}_\beta^* \cdot \boldsymbol{\xi}_\alpha\, dV\right]$$
$$= \tfrac{1}{2}\int \left(\xi_\beta^*\Pi_\alpha - \xi_\alpha\Pi_\beta^*\right) dS_2 \overset{\text{(1D)}}{=} \tfrac{1}{2}\left[\xi_\beta^*\Pi_\alpha - \xi_\alpha\Pi_\beta^*\right](x_2)\,, \tag{12.131}$$

$$(\omega_\beta^* - \omega_\alpha)\left[\int \boldsymbol{\xi}_\beta^* \cdot \mathbf{G}(\boldsymbol{\xi}_\alpha)\, dV - \omega_\beta^*\omega_\alpha\int \rho\boldsymbol{\xi}_\beta^* \cdot \boldsymbol{\xi}_\alpha\, dV\right]$$
$$= \omega_\alpha\int \left(\xi_\beta^*\Pi_\alpha - \xi_\alpha\Pi_\beta^*\right) dS_2 \overset{\text{(1D)}}{=} \omega_\alpha\left[\xi_\beta^*\Pi_\alpha - \xi_\alpha\Pi_\beta^*\right](x_2)\,, \tag{12.132}$$

where S_2 is the right boundary surface. These expressions are completely general, they make no use of special properties of $\boldsymbol{\xi}_\alpha$ and $\boldsymbol{\xi}_\beta$ (except that they are left solutions of the respective spectral equations), not even of the fact that ω_α and ω_β are on the paths $\mathcal{P}_u$ or $\mathcal{P}_s$.

Before we apply the quadratic forms (12.131) and (12.132) to the proposed study of the relative "motion" of the solutions along the solution path, it is useful to note that they summarize almost all previously obtained relations on complex eigenvalues (model I solutions), solution paths (model II solutions) and solutions for arbitrary values of ω (model III solutions).

(a) For $\omega = \omega_\alpha = \omega_\beta$, *but not on the solution path,* the surface integral transforms into the expression (12.123) for $W - W^* \equiv 2\mathrm{i}W_2$, so that we recover the relations (12.122) that hold for σ and ν in the areas III of the complex ω-plane depicted in Fig. 12.6.

(b) For $\omega = \omega_\alpha = \omega_\beta$ *on the solution path,* the surface integral vanishes by definition, so that we recover the relations (12.112)(b) that hold for complex eigenvalues ($\nu \neq 0$) as well as for complex values of ω on the solution path. For real $\omega = \omega_\alpha = \omega_\beta$, the two quadratic forms collapse into a trivial identity, of course indicating that the relations (12.112)(a) for $\nu = 0$ can be obtained directly from one spectral equation.

(c) For *different eigenvalues* $\omega_\alpha \neq \omega_\beta$, the surface integral again vanishes because both BCs are satisfied for $\boldsymbol{\xi}_\alpha$, as well as for $\boldsymbol{\xi}_\beta$, so that we obtain two relations on the "matrix elements" of the operators $\rho^{-1}U$ and $\rho^{-1}\mathbf{G}$:

$$\langle \boldsymbol{\xi}_\beta, \rho^{-1}U\boldsymbol{\xi}_\alpha \rangle = \tfrac{1}{2}(\omega_\beta^* + \omega_\alpha)\langle \boldsymbol{\xi}_\beta, \boldsymbol{\xi}_\alpha \rangle\,, \tag{12.133}$$

$$\langle \boldsymbol{\xi}_\beta, \rho^{-1}\mathbf{G}(\boldsymbol{\xi}_\alpha) \rangle = \omega_\beta^* \omega_\alpha \langle \boldsymbol{\xi}_\beta, \boldsymbol{\xi}_\alpha \rangle\,, \tag{12.134}$$

These relations exhibit a fundamental difference between the spectral problems for static and stationary plasmas. For static plasmas, $U \equiv 0$, so that $\langle \boldsymbol{\xi}_\beta, \boldsymbol{\xi}_\alpha \rangle = 0$ for eigenfunctions. This is no longer the case for stationary plasmas, so that *the eigenfunctions are not orthogonal,* in general. The relations (12.133) and (12.134) were used by Barston [24] to study stable finite dimensional Lagrangian systems.

(d) For our present study of *different (non-eigen)values* $\omega_\alpha \neq \omega_\beta$ *on the solution path,* the full expressions (12.131) and (12.132) need to be retained since the surface integral does not vanish.

These items once more demonstrate that the spectral problem of stationary plasmas is governed by the two linear operators U and $\mathbf{G}$, that are both self-adjoint, but that do not fit into a linear but into a quadratic eigenvalue problem for $\boldsymbol{\xi}$.

Returning to our study of the behavior of two arbitrary solutions $\boldsymbol{\xi}_\alpha$ and $\boldsymbol{\xi}_\beta$ on the solution paths, we now assume that ω_α and ω_β are close, so that $\boldsymbol{\xi}_\alpha$ and $\boldsymbol{\xi}_\beta$ are close in norm, and consistently develop the quadratic forms (12.131) and (12.132) to first order in the distance $|\omega_\alpha - \omega_\beta|$. To that end, we define variables $\boldsymbol{\xi}$ and $\Pi(\boldsymbol{\xi})$, corresponding to the average value $\omega = \sigma + \mathrm{i}\nu$ on the solution path, and perturbations $\boldsymbol{\eta}$ and $\Psi \equiv \Pi(\boldsymbol{\eta})$, corresponding to the perturbation $\delta + \mathrm{i}\epsilon$ of the eigenvalue parameter:

$$\boldsymbol{\xi} \equiv \tfrac{1}{2}(\boldsymbol{\xi}_\alpha + \boldsymbol{\xi}_\beta)\,, \quad \Pi \equiv \tfrac{1}{2}(\Pi_\alpha + \Pi_\beta)\,, \quad \sigma + \mathrm{i}\nu \equiv \tfrac{1}{2}(\omega_\alpha + \omega_\beta),$$

$$\boldsymbol{\eta} \equiv \tfrac{1}{2}(\boldsymbol{\xi}_\alpha - \boldsymbol{\xi}_\beta)\,, \quad \Psi \equiv \tfrac{1}{2}(\Pi_\alpha - \Pi_\beta)\,, \quad \delta + \mathrm{i}\epsilon \equiv \tfrac{1}{2}(\omega_\alpha - \omega_\beta). \quad (12.135)$$

We also need the inverse relations:

$$\boldsymbol{\xi}_\alpha \equiv \boldsymbol{\xi} + \boldsymbol{\eta}\,, \quad \Pi_\alpha \equiv \Pi + \Psi\,, \quad \omega_\alpha \equiv \sigma + \delta + \mathrm{i}(\nu + \epsilon)\,,$$

$$\boldsymbol{\xi}_\beta \equiv \boldsymbol{\xi} - \boldsymbol{\eta}\,, \quad \Pi_\beta \equiv \Pi - \Psi\,, \quad \omega_\beta \equiv \sigma - \delta + \mathrm{i}(\nu - \epsilon)\,. \quad (12.136)$$

We now distinguish the two very different cases of stable waves ($\nu = 0$: path $\mathcal{P}_\mathrm{s}$) and instabilities ($\nu \neq 0$: path $\mathcal{P}_\mathrm{u}$).

Approach to real eigenvalues For real ω, it is evident from the general quadratic equation (12.121), relating I, V, W_1 and W_2, that the condition $W_2 = 0$ should also hold for the solution path $\mathcal{P}_\mathrm{s}$. Hence, we may exploit the quadratic forms (12.131) and (12.132) also for this case. Inserting the inverses (12.136), and expanding in orders of δ, the leading order vanishes and the first order yields

$$2\delta \int \boldsymbol{\xi}^* \cdot \Big[U\boldsymbol{\xi} - \rho\sigma\boldsymbol{\xi} \Big]\, dV \;\; = \tfrac{1}{2} \int (\eta^* \Pi - \boldsymbol{\xi}^* \Psi + \eta \Pi^* - \boldsymbol{\xi} \Psi^*)\, dS_2\,, \quad (12.137)$$

$$2\delta \int \boldsymbol{\xi}^* \cdot \Big[\mathbf{G}(\boldsymbol{\xi}) - \rho\sigma^2\boldsymbol{\xi} \Big]\, dV = \sigma \int (\eta^* \Pi - \boldsymbol{\xi}^* \Psi + \eta \Pi^* - \boldsymbol{\xi} \Psi^*)\, dS_2\,, \quad (12.138)$$

where the second expression is actually redundant since it just reproduces the quadratic (12.106), or (12.121). Transforming the first expression by reverting to the original variables $\boldsymbol{\xi}_\alpha$, $\boldsymbol{\xi}_\beta$, σ_α and σ_β, and neglecting higher order terms in the volume integral on the LHS (since it is multiplied by the small parameter δ), this yields the required result:

$$-\mathrm{Re} \int \big(\xi_\beta^* \Pi_\alpha - \xi_\alpha^* \Pi_\beta \big)\, dS_2 \overset{(1\mathrm{D})}{=} -\Big[\xi_\beta \Pi_\alpha - \xi_\alpha \Pi_\beta \Big](x_2)$$

$$= 2(\sigma_\beta - \sigma_\alpha)\Big[\sigma_\alpha - \overline{V}(\sigma_\alpha) \Big] \int \rho |\xi_\alpha|^2\, dV \quad (stable\ waves)\,. \quad (12.139)$$

The surface integral consists of quantities that are readily available in whatever numerical integration technique is exploited. Its reduction for 1D systems automatically produces an expression involving only real quantities since the components of $\boldsymbol{\xi}$ may be chosen to be real when ω is real (as in the static case). To leading order, the frequency factor on the RHS may be transformed into the difference of the squares of the two Doppler–Coriolis shifted frequencies. Thus, the generalization of the expressions (7.169) and (7.171) of Section 7.4.3 [1], representing the oscillation theorem for static plasmas derived by Goedbloed and Sakanaka [183], becomes obvious. We will exploit the equality (12.139) in Section 13.1.3 in the final proof of the oscillation theorem $\mathcal{R}$ for real eigenvalues.

Evidently, in order to conclude monotonicity of the frequency factor on the RHS of Eq. (12.139), we need to exclude the zeros of the Doppler–Coriolis shifted frequency since that factor changes sign there. In its simplest form, this involves consideration of the *marginal stability transition* $\sigma = \sigma_0$ (for the BVP chosen), which is the value of σ where the path $\mathcal{P}_u$ meets the real axis (i.e. the path $\mathcal{P}_s$) and the solution-averaged Doppler–Coriolis shifted frequency vanishes. In that case,

$$\sigma_\alpha - \overline{V}(\sigma_\alpha) = f(\sigma_\alpha)(\sigma_\alpha - \sigma_0)\,, \quad f(\sigma_\alpha) \neq 0\,,$$

$$\sigma_0 = \overline{V}[\boldsymbol{\xi}(\sigma_0)] \equiv \frac{\displaystyle\int \boldsymbol{\xi}^* \cdot \mathbf{U}\boldsymbol{\xi}\, dV}{\displaystyle\int \rho|\boldsymbol{\xi}|^2\, dV} \quad \Rightarrow \quad \sigma_0(\nu = 0)\,. \tag{12.140}$$

This quantity can only be computed if the integrals exist, i.e. if σ_0 falls outside the continuous spectra. Frequently, this is not the case, so that the concept of ϵ-stability (end of Section 12.2.4) would have to be exploited, typically with $\epsilon \ll |\sigma_0|$:

$$\sigma_0^\epsilon = \overline{V}[\boldsymbol{\xi}(\sigma_0^\epsilon + i\epsilon)] \quad \Rightarrow \quad \sigma_0^\epsilon(\nu = \epsilon)\,. \tag{12.141}$$

However, this number will not be needed in our final formulation of the oscillation theorem since the theorem will apply only to the discrete spectrum, excluding the continua. More serious is the possibility of multiple solutions to Eq. (12.140), that we will encounter in Section 13.4.1. With rotation, due to the Coriolis shift, the solution path typically intersects the real axis twice. Let us call those marginal transition points σ_0 and σ_1. (In the case where more than two of such points turn up, one should take the two outermost ones.) Assuming $\sigma_1 \leq \sigma_0$, the sign of the expression $\sigma_\alpha - \overline{V}(\sigma_\alpha)$ is not definite in the range $\sigma_1 \leq \sigma_\alpha \leq \sigma_0$. We leave it understood that, if this range overlaps with a continuum, one of the points σ_0 or σ_1 is to be replaced by the pertinent extremum of the continuum. We will call such a range the *Doppler–Coriolis indefinite range*. Consequently, for the frequencies

$$\left\{ \begin{array}{ll} \sigma = \sigma_0 & \textit{(single marginal stability transition)} \\[2mm] \sigma_1 \leq \sigma \leq \sigma_0 & \textit{(Doppler–Coriolis indefinite range)} \end{array} \right. \tag{12.142}$$

the sign of the RHS of Eq. (12.139) is not definite. Outside those frequencies, the sign is definite and easily determined, resulting in oscillation theorem $\mathcal{R}$ of Section 13.1.3.

Approach to complex eigenvalues The reduction of the quadratic forms for complex values of ω is significantly more complicated, as we will see. Except for the cross-products entering the quadratic forms (12.131) and (12.132), we also need the quadratic forms involving the products of $\boldsymbol{\xi}_\alpha$ and $\boldsymbol{\xi}_\beta$ with themselves:

$$\sigma_\alpha = \overline{V}_\alpha, \quad \text{or} \quad \overline{W}_{2\alpha} = 0, \quad \text{and} \quad |\omega_\alpha|^2 = -\overline{W}_{1\alpha}, \tag{12.143}$$

$$\sigma_\beta = \overline{V}_\beta, \quad \text{or} \quad \overline{W}_{2\beta} = 0, \quad \text{and} \quad |\omega_\beta|^2 = -\overline{W}_{1\beta}. \tag{12.144}$$

Like the cross-products, these relations are just consequences of the spectral equations (12.129) and (12.130), where the additional assumption of constraining to the path, i.e. reality of W, is made. We need to work out the consequences of this assumption for the quadratic forms (12.131) and (12.132) as well.

Inserting the inverse relations (12.136) into Eqs. (12.129) and (12.130), we obtain the basic equations for $\boldsymbol{\xi}$ and $\boldsymbol{\eta}$:

$$\mathbf{G}(\boldsymbol{\xi}) - 2\omega\,U\boldsymbol{\xi} + \rho\omega^2\boldsymbol{\xi} = 0, \tag{12.145}$$

$$\mathbf{G}(\boldsymbol{\eta}) - 2\omega\,U\boldsymbol{\eta} + \rho\omega^2\boldsymbol{\eta} = 2(\delta + i\epsilon)\Big[U\boldsymbol{\xi} - \rho\omega\boldsymbol{\xi}\Big]. \tag{12.146}$$

Next, inserting the inverse relations into Eqs. (12.131) and (12.132), canceling the zeroth order terms by using Eqs. (12.143) and (12.144) and keeping the linear terms in δ and ϵ only, we obtain the corresponding first order quadratic form equations:

$$X \equiv \int \boldsymbol{\eta}^* \cdot \Big[U\boldsymbol{\xi} - \rho\sigma\boldsymbol{\xi}\Big]\, dV = (\delta - i\epsilon)I - \frac{i}{2\nu}\mathcal{S}, \tag{12.147}$$

$$Y \equiv -\tfrac{1}{2}\int \boldsymbol{\eta}^* \cdot \Big[\mathbf{G}(\boldsymbol{\xi}) - \rho|\omega|^2\boldsymbol{\xi}\Big]\, dV = -\omega\Big[(\delta - i\epsilon)I - \frac{i}{2\nu}\mathcal{S}\Big], \tag{12.148}$$

where the surface term in the RHSs is defined by

$$\mathcal{S} \equiv \tfrac{1}{2}\int (\xi^*\Psi - \eta\Pi^*)\, dS_2 \stackrel{(1\mathrm{D})}{=} \tfrac{1}{2}\Big[\xi^*\Psi - \eta\Pi^*\Big](x_2). \tag{12.149}$$

Since $\omega X + Y = 0$, the expression (12.148) for the quadratic form Y is actually redundant. Consequently, considering a solution $\boldsymbol{\xi}$ for $\omega = \sigma + i\nu$ on the solution path $\mathcal{P}_\mathrm{u}$, the relative location of nearby solutions $\boldsymbol{\xi} \pm \boldsymbol{\eta}$ on $\mathcal{P}_\mathrm{u}$, where $\|\boldsymbol{\eta}\| \ll \|\boldsymbol{\xi}\|$, is determined by the quadratic form X. Its quantitative value is obtained by substituting the solution $\boldsymbol{\eta}$ of the inhomogeneous spectral equation (12.146), where the factor on the RHS has a magnitude $|\delta + i\epsilon| \ll |\sigma + i\nu|$.

If desired, the latter equation could be solved with the same numerical techniques as used to solve the leading order homogeneous spectral equation (12.145), or by exploiting Green's functions to integrate the RHS (like in Section 10.1 [1]). However, we do not need to pursue this route to the end, since we may determine the requisite behavior from the scaling properties of $\boldsymbol{\eta}$ and X with the distance $\tfrac{1}{2}|\omega_\alpha - \omega_\beta| \equiv |\delta + i\epsilon|$ along the solution path. To that end, we renormalize η and

X to become *leading order* quantities:

$$\widetilde{\boldsymbol{\eta}} \equiv \frac{1}{\delta + i\epsilon}\,\boldsymbol{\eta}\,, \qquad \mathbf{G}(\widetilde{\boldsymbol{\eta}}) - 2\omega\,U\widetilde{\boldsymbol{\eta}} + \rho\omega^2\widetilde{\boldsymbol{\eta}} = 2\Big[U\boldsymbol{\xi} - \rho\omega\boldsymbol{\xi}\Big]\,, \qquad (12.150)$$

$$\widetilde{X} \equiv \frac{1}{\delta - i\epsilon}\,X \equiv \int \widetilde{\boldsymbol{\eta}}^{*}\cdot\Big[U\boldsymbol{\xi} - \rho\sigma\boldsymbol{\xi}\Big]\,dV = I - \frac{\delta + i\epsilon}{\delta^2 + \epsilon^2}\,\frac{i}{2\nu}\,\mathcal{S}\,. \qquad (12.151)$$

We then find from the path equation $\overline{W}_{2\alpha} = 0$, or $\overline{W}_{2\beta} = 0$, that $\mathcal{S}$ is real and from Eq. (12.151) that it is proportional to the imaginary part of $\widetilde{X}$:

$$\mathcal{S} = -\frac{\delta^2 + \epsilon^2}{\delta}\cdot 2\nu\widetilde{X}_2(\sigma,\nu) \qquad (\delta \neq 0)\,. \qquad (12.152)$$

The LHS and RHS of this equation consist of balancing small terms, of the order $|\delta + i\epsilon|$. The LHS contains the perturbations of the boundary values that are computed in any numerical procedure of solving the one-sided BVP for the spectral equation. For $\nu \neq 0$ (complex ω), since $\widetilde{X}_2$ does not depend on δ and ϵ, the RHS is a monotonic function along the solution path, as long as δ does not vanish. In the latter case, *the path is locally vertical* and, since both δ and $\widetilde{X}_2$ vanish, one should exploit the real part of the complex function $\widetilde{X}$ there:

$$\widetilde{X}_1 = I - \frac{\epsilon}{\delta}\widetilde{X}_2 \quad \Rightarrow \quad \mathcal{S} = \epsilon\cdot 2\nu\Big[\widetilde{X}_1(\sigma,\nu) - I\Big] \qquad (\delta = 0)\,. \qquad (12.153)$$

Hence, the RHS remains a monotonic function along the solution path, also in points where the path is vertical.

The expressions (12.152) and (12.153) determine the approach to eigenvalues, where we have assumed that ω_α and ω_β are two different values on the solution path and that at least one of them is not an eigenvalue so that the equalities (12.133) and (12.134) do *not* hold. Instead, the differences between the LHS and RHS of those two quadratic forms become both proportional to $\widetilde{X}_2 \neq 0$ if $\sigma_\beta \neq \sigma_\alpha$, resp. to $\widetilde{X}_1 - I \neq 0$ if $\sigma_\beta = \sigma_\alpha$. (It is a useful exercise to check this statement.) Hence, we do not need to know the values of these quantities, but just require them not to change sign in between eigenvalues. Reverting to the original variables $\boldsymbol{\xi}_\alpha$, $\boldsymbol{\xi}_\beta$, ω_α and ω_β, the equalities (12.152) and (12.153) finally yield the desired relationship between the boundary values and the distance along the solution path:

$$4\mathcal{S} = \mathrm{Re}\int (\boldsymbol{\xi}_\beta^{*}\Pi_\alpha - \boldsymbol{\xi}_\alpha^{*}\Pi_\beta)\,dS_2$$

$$\stackrel{(1D)}{=} \Big[\xi_{\beta 1}\Pi_{\alpha 1} - \xi_{\alpha 1}\Pi_{\beta 1} + \xi_{\beta 2}\Pi_{\alpha 2} - \xi_{\alpha 2}\Pi_{\beta 2}\Big](x_2)$$

$$= \begin{cases} \dfrac{|\omega_\beta - \omega_\alpha|^2}{\sigma_\beta - \sigma_\alpha}\cdot 4\nu\widetilde{X}_2(\sigma,\nu) & (\sigma_\beta \neq \sigma_\alpha) \\[2mm] -(\nu_\beta - \nu_\alpha)\cdot 4\nu\Big[\widetilde{X}_1(\sigma,\nu) - \tfrac{1}{2}I\Big] & (\sigma_\beta = \sigma_\alpha) \end{cases} \quad \textit{(instabilities)}. \quad (12.154)$$

We will exploit these equalities in Section 13.2.4 in the final proof of the oscillation theorem C for complex eigenvalues.

In conclusion, we have now obtained the required quadratic expressions (12.139) for real values of ω (stable waves) and (12.154) for complex values of ω (instabilities), with obvious monotonicity properties of the RHSs along the respective solution paths $\mathcal{P}_s$ and $\mathcal{P}_u$. It remains to substitute actual solutions in the LHSs and to demonstrate how they may be used to iterate towards eigenvalues. This will be done in Chapter 13 for 1D systems. This will finally lead to a kind of generalization of the Sturm–Liouville type of oscillation theorems discussed in Volume [1], Sections 7.4.3 and 9.4.1, for static plasma slabs and cylinders.

For stationary plasma equilibria, we will see that the *real* eigenvalues can still be obtained by considering a real solution ξ_α on a sub-interval of (x_1, x_2) bounded by two consecutive zeros (in the manner pioneered by Newcomb [347] in his study of the stability of the diffuse linear pinch) and then ask the question whether the distance between the zeros of another real solution ξ_β is larger or smaller when σ_β is larger than σ_α. The expression (12.139) will provide the answer to that. (Incidentally, for 2D and 3D systems, the zeros become nodal curves and surfaces, requiring much more analysis, even for static plasmas. This is evident from Courant and Hilbert I [98], Section VI.6. Their Figs. 7 and 8 show how a multiplicity of nodal curves, dividing a 2D domain into many sub-domains, by perturbation becomes a single spiraling curve dividing the domain in only two sub-domains.)

However, for *complex* eigenvalues, this approach fails because the zeros of the real and imaginary parts of solutions ξ_α cannot be forced to coincide. Also, the "thought experiment" to consider the consecutive zeros of ξ_α as virtual wall positions, so that stabilization or destabilization intuitively corresponds to moving the wall inward or outward, does not work anymore because (as mentioned in Section 12.3.1) the solution path $\mathcal{P}_u$ changes when the plasma boundaries are moved. This fundamental problem has been solved here by extending the BVP with an external vacuum bounded by a genuine wall, where the position $x = x_v$ of that wall can be changed without affecting the solution path. One could consider x_v as the control parameter determining the distribution of the eigenvalues of the extended model II problem, if one desires to solve that, but here it is used to relate it to the value of $\xi_\alpha(x_2)$ which is a kind of measure for the "distance" to eigenvalues of the restricted model I problem for arbitrary solutions ξ_α on the solution path. We will see in Sections 13.1.4 and 13.2.4 how the expression (12.153) can be exploited to provide a monotonic approach to eigenvalues for this restricted problem.

We are now fully prepared to complete this study with the determination of the spectra of stationary plasmas for the explicit examples of gravitating plane plasma slabs (Sections 13.1 and 13.2) and cylindrical plasmas (Sections 13.3 and 13.4).

12.4 Literature and exercises

Notes on literature

Stability of stationary equilibria

- Frieman & Rotenberg, 'On the hydromagnetic stability of stationary equilibria' [147], outline the stationary counterpart to the widely used MHD stability theory of static plasmas [203, 35], extensively discussed in Volume [1], when flows are admitted. Surprisingly (certainly for plasma-astrophysical applications where incorporation of flow is imperative), this more general theory remains underdeveloped. The present chapters are an attempt to restore the balance.

- Chandrasekhar, *Hydrodynamic and Hydromagnetic Stability* [84], is one of the first systematic presentations of the subject with numerous carefully worked out explicit examples: it is a real pleasure to see the master at work!

Exercises

[12.1] *Tokamak physics versus astrophysics*

Laboratory and astrophysical plasmas appear to be very different forms of matter. In this book, the common features are stressed.

- Why and when can a plasma in a fusion device, like a tokamak, be described with the same theoretical model as a plasma in the Universe?
- Using the same theory, there are important differences between the two cases though, in particular with respect to plasma equilibrium. What are those differences?
- Also with respect to the waves and instabilities, there are important differences. What are they?

[12.2] *Displacement field*

Newcomb has derived an expression which relates the gradient operator at the perturbed position to that at the unperturbed position. Explain why this expression needs special attention in nonlinear MHD, in particular in numerical nonlinear MHD.

[12.3] *Incompressible plasma*

In the exercises of the following chapters, the incompressible limit will be exploited frequently. Here, you will investigate what kind of implications this limit has. First, obtain the equations for an incompressible plasma from the ones for a compressible plasma by taking the limit $\gamma \to \infty$.

- Show that $\nabla \cdot \mathbf{v} = 0$ in the incompressible limit.
- Also show that $\nabla^0 \cdot \mathbf{v}^0 = 0$.
- Show that the displacement field satisfies $\nabla^0 \cdot \boldsymbol{\xi} = 0$.
- What does this limit imply for the generalized force operator $\mathbf{G}(\boldsymbol{\xi})$? What are the advantages of taking the incompressible limit?

[12.4] *Flow and self-adjointness*

Show that the operator $U \equiv -i\rho \mathbf{v} \cdot \nabla$ is self-adjoint. Motivate for every step what kind of argument you have used. Start from $\langle \boldsymbol{\eta}, \rho^{-1} U \boldsymbol{\xi} \rangle$.

- Indicate the differences in self-adjointness between the operators U and $\mathbf{G}$. Which one is the stronger? (Hint: consider the BCs to be satisfied by $\boldsymbol{\xi}$ in both cases.)

[12.5] *Determination of the solution path*

In this exercise you will make a summary and flow diagram of how to compute the solution paths for a one-dimensional system. This system can be described either in Cartesian or in cylindrical coordinates.

- Suppose you have selected a growth rate ν_{u}. It looks like there are still infinitely many possibilities to choose a value for the real part σ_{u} of the frequency. This is not so, σ_{u} should be selected from a certain range. Determine that range.
- Once you have selected a ν_{u} and a σ_{u}, you need to solve a certain differential equation. Which differential equation is that and what are the boundary conditions?
- Which condition has to be satisfied for the selected ν_{u} and σ_{u} to be part of the solution path and, therefore, to be an appropriate guess for the eigenvalue ω. Why are the ν_{u} and σ_{u} selected just a guess and not already the eigenvalue ω?
- Make a flow diagram of the computation of the solution path.

13

Shear flow and rotation

13.1 Spectral theory of plane plasmas with shear flow

13.1.1 Gravito-MHD wave equation for plane plasma flow

We apply the general theory of Chapter 12 to explicitly construct the basic structure of the spectrum of waves and instabilities for representative sets of stationary 1D equilibria, viz. plane slabs in this section, and the next, and to cylindrical plasmas in Sections 13.3 and 13.4. These two classes of problems are the returning, first-principle, ones in any explicit analysis of the waves and instabilities of fluids and plasmas, as already exemplified for static plasmas in Chapters 7 and 9[1]. With respect to MHD spectral theory, the study of the mentioned 1D problems very much functions like the study of the spectrum of the hydrogen atom in quantum mechanics. It reveals the basic complexity of the system, that is to be understood before the more complicated multi-dimensional systems can be studied fruitfully. In stability theory of stationary plasmas, this is even more urgent than in quantum mechanics since the basic spectral equation is quadratic in the eigenvalue and contains two operators, rather than one.

To that end, consider the model of a plane magnetized and gravitating plasma slab that was introduced in Section 7.3.2, Fig. 7.9[1], but now extended with a *plane shear flow* field:

$$\mathbf{B} = B_y(x)\,\mathbf{e}_y + B_z(x)\,\mathbf{e}_z\,, \quad \mathbf{g} = -g\,\mathbf{e}_x\,, \quad \rho = \rho(x)\,, \quad p = p(x)\,,$$

$$\mathbf{v} = v_y(x)\mathbf{e}_y + v_z(x)\mathbf{e}_z\,. \tag{13.1}$$

The functions $\rho(x), p(x), B_y(x)$ and $B_z(x)$ should satisfy the equilibrium equation (12.28), $(p + \frac{1}{2}B^2)' = -\rho g$, but $v_y(x)$ and $v_z(x)$ are completely arbitrary. The slab is confined between two solid boundaries at $x = x_1$ and $x = x_2$.

In applications, it is usually important that the magnetic field has *magnetic shear*, so that one can assume that the magnitude of $\mathbf{B}(x)$ is fixed but its direction should

49

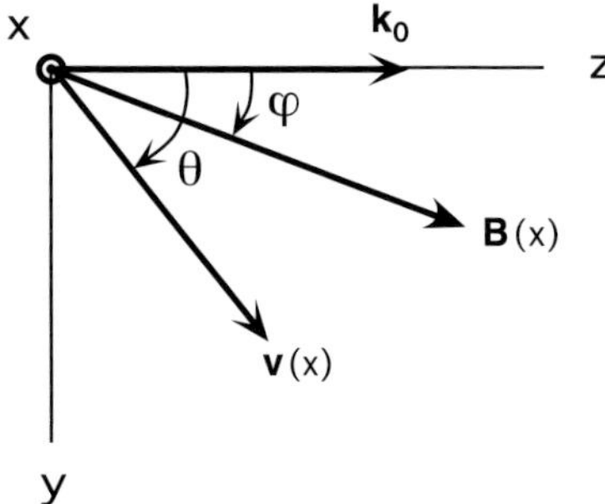

Fig. 13.1 Directions of the horizontal wave vector, magnetic field and velocity.

change with the vertical position x. On the other hand, *velocity shear* effects are usually represented by just assuming variation of the amplitude of a uni-directional flow field $\mathbf{v}(x)$. Without loss of generality, one can also choose the direction of the horizontal wave vector $\mathbf{k}_0$ to be along the z axis, making an angle θ with $\mathbf{v}$ and an angle $\varphi(x)$ with $\mathbf{B}$ (see Fig. 13.1). For the time being, we will not make these simplifying assumptions and just derive completely general expressions for the spectral equations of these equilibria.

In the reduction of the spectral equation for these equilibria, one of the two velocity-dependent terms in the expression (12.64) for the generalized force operator $\mathbf{G}$ vanishes since there are no centrifugal forces present, so that $\mathbf{v} \cdot \nabla \mathbf{v} = 0$, and the other one yields $-\rho(\mathbf{v} \cdot \nabla)^2 \boldsymbol{\xi}$, so that the eigenvalue problem (12.69), or rather (12.70), "nearly" simplifies to the old static one, with the same operator $\mathbf{F}$ but ω replaced by the Doppler shifted expression:

$$\mathbf{F}(\boldsymbol{\xi}) = -\rho(\omega - \rho^{-1}U)^2 \,\boldsymbol{\xi} \equiv -\rho(\omega + \mathrm{i}\mathbf{v} \cdot \nabla)^2 \,\boldsymbol{\xi}. \tag{13.2}$$

From the discussion in the previous chapter, it is evident that all the new physics associated with stationary plasma flows is hidden in this replacement.

As in Section 7.3.2 [1], we assume normal modes with plane wave dependence in the ignorable (y and z) directions,

$$\boldsymbol{\xi}(x, y, z, t) = \boldsymbol{\xi}(x; k_y, k_z) \, \mathrm{e}^{\mathrm{i}(k_y y + k_z z - \omega t)}, \tag{13.3}$$

where we recall that a distinctive hat for the Fourier amplitude was dropped for simplicity of the notation. For these Fourier normal modes (FNM), the operator U in Eq. (13.2) becomes a multiplication:

$$\omega + \mathrm{i}\mathbf{v} \cdot \nabla \overset{\mathrm{FNM}}{=} \widetilde{\omega}(x) \equiv \omega - \Omega_0(x), \qquad \Omega_0 \equiv \mathbf{k}_0 \cdot \mathbf{v}(x), \tag{13.4}$$

where $\mathbf{k}_0 \equiv k_y \mathbf{e}_y + k_z \mathbf{e}_z$ is the horizontal wave vector, $\Omega_0(x)$ is the *local Doppler shift* and $\widetilde{\omega}(x)$ is the *local Doppler shifted frequency* observed in a local frame

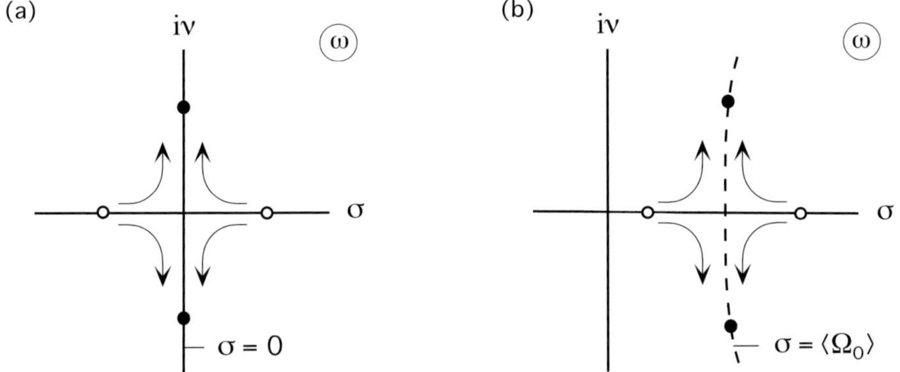

Fig. 13.2 Schematic transition to instability for (a) static and (b) stationary plasma.

co-moving with the plasma layer at the vertical position x. Hence, the eigenvalue problem becomes

$$\mathbf{F}(\boldsymbol{\xi}) = -\rho\tilde{\omega}^2\,\boldsymbol{\xi}\,, \tag{13.5}$$

subject to model I BCs on $\xi \equiv \xi_x$ at $x = x_1$ and $x = x_2$:

$$\left\{ \begin{array}{ll} \xi(x_1) = 0 & \textit{(left BC)} \\[2mm] \xi(x_2) = 0 & \textit{(right BC)} \end{array} \right. . \tag{13.6}$$

Since the equilibrium condition (12.28) is the same as for static plasmas, this implies that the equations derived in Section 7.3.2 [1] for the static plane slab remain valid for the stationary slab if just the replacement $\omega \rightarrow \tilde{\omega}(x)$ is made. Of course, the x-dependence of the local Doppler shift $\Omega_0(x)$ is the complicating factor. It opens up the possibility of new flow-driven (Kelvin–Helmholtz) instabilities.

Since $\tilde{\omega}$ depends on x through $\Omega_0(x)$, every discrete eigenvalue will be subject to a different solution-averaged Doppler shift $\langle \Omega_0(x) \rangle$ involving the solution ξ of the eigenvalue problem (13.5), (13.6) across the layer. This is the averaged Doppler–Coriolis shift $\overline{V}$ defined in Eq. (12.111), which now simplifies to

$$\overline{V} \equiv \langle \Omega_0(x) \rangle \equiv \frac{\displaystyle\int \rho\,\mathbf{k}_0 \cdot \mathbf{v}\,|\xi|^2\,dx}{\displaystyle\int \rho|\xi|^2\,dx}\,, \tag{13.7}$$

since there is no Coriolis contribution for plane flows. Inserting the left solution $\xi^{\mathrm{L}}(\sigma_{\mathrm{u}} + i\nu_{\mathrm{u}})$ of the one-sided BVP (13.5), (13.6)(a) into $\sigma_{\mathrm{u}} = \overline{V}$ yields the path $\mathcal{P}_{\mathrm{u}}^{\mathrm{L}}$ on which the unstable eigenvalues should be located. Schematically, this results in the difference between unstable modes in static and stationary equilibria as illustrated in Fig 13.2. When an equilibrium parameter is varied (recall from

Section 12.3 that this should be one that does not affect the solution path), two stable modes may approach each other along the real ω-axis, coalesce, and then become unstable. In the static case, this transition to instability is always through the marginal point $\omega = 0$. In the stationary case, this transition occurs at σ_0, or σ_0^ϵ (given by the solution of Eq. (12.140), or Eq. (12.141), with $U = \rho\Omega_0$), which is away from the origin in general so that *overstable* modes (propagating waves with an exponentially growing amplitude) appear. Hence, the unstable mode of Fig. 13.2(b) could have arisen because the static equilibrium of Fig. 13.2(a) is already unstable, so that the corresponding equilibrium with flow just moves that unstable eigenvalue off the imaginary axis into the complex plane. Alternatively, an overstable mode could also arise without a corresponding static instability as a result of a Kelvin–Helmholtz unstable flow profile. We will encounter examples of both in Section 13.1.2.

The spectral equation (13.5) is reduced by exploiting the *field line projection*,

$$\mathbf{e}_x \equiv \nabla x\,, \qquad \mathbf{e}_\perp \equiv (\mathbf{B}/B) \times \mathbf{e}_x\,, \qquad \mathbf{e}_\parallel \equiv (\mathbf{B}/B)\,,$$

$$\partial_x \equiv \mathbf{e}_x \cdot \nabla\,, \qquad k_\perp \equiv -i\mathbf{e}_\perp \cdot \nabla\,, \qquad k_\parallel \equiv -i\mathbf{e}_\parallel \cdot \nabla\,,$$

$$\xi \equiv \xi_x\,, \qquad \eta \equiv i\mathbf{e}_\perp \cdot \boldsymbol{\xi}\,, \qquad \zeta \equiv i\mathbf{e}_\parallel \cdot \boldsymbol{\xi}\,, \tag{13.8}$$

where, in contrast to the static case, *the variables ξ, η and ζ are no longer purely real in the stationary case* since $\omega = \sigma + i\nu$ is genuinely complex for instabilities. The resulting vectorial form of the *gravito-MHD wave equation for plane plasma flow* [455] is just Eq. (7.89) of Section 7.3.2 [1] with the replacement $\omega \rightarrow \widetilde{\omega}(x)$. Eliminating η and ζ yields the scalar ordinary differential equation (ODE) form of the gravito-MHD wave equation in terms of the normal displacement ξ,

$$\frac{d}{dx}\left(\frac{N}{D}\frac{d\xi}{dx}\right) + \left[A + \frac{B}{D} + \left(\frac{C}{D}\right)'\right]\xi = 0\,, \tag{13.9}$$

involving three regular coefficients with different powers of the acceleration of gravity (including the driving term $\rho' g$ of the Rayleigh–Taylor instability),

$$A(x;\widetilde{\omega}^2) \equiv \rho(\widetilde{\omega}^2 - \omega_A^2) + \rho' g\,,$$

$$B(x;\widetilde{\omega}^2) \equiv -k_0^2 \rho g^2 \,(\widetilde{\omega}^2 - \omega_A^2)\,,$$

$$C(x;\widetilde{\omega}^2) \equiv -\rho g\,\widetilde{\omega}^2(\widetilde{\omega}^2 - \omega_A^2)\,, \tag{13.10}$$

the numerator function N determining genuine (Alfvén and slow) singularities,

$$N(x;\widetilde{\omega}^2) \equiv (\gamma p + B^2)(\widetilde{\omega}^2 - \omega_A^2)(\widetilde{\omega}^2 - \omega_S^2)\,,$$

$$\omega_A^2(x) \equiv k_\parallel^2 \frac{B^2}{\rho}\,, \qquad \omega_S^2(x) \equiv k_\parallel^2 \frac{\gamma p B^2}{\rho(\gamma p + B^2)}\,, \tag{13.11}$$

and the denominator function D determining apparent (fast and slow turning point) singularities,

$$D(x; \tilde{\omega}^2) \equiv (\tilde{\omega}^2 - \omega_{\mathrm{f0}}^2)(\tilde{\omega}^2 - \omega_{\mathrm{s0}}^2) = \tilde{\omega}^4 - k_0^2(\gamma p + B^2)(\tilde{\omega}^2 - \omega_{\mathrm{S}}^2)/\rho \,,$$

$$\omega_{\mathrm{f0,s0}}^2(x) \equiv \tfrac{1}{2} k_0^2 \frac{\gamma p + B^2}{\rho} \left[1 \pm \left(1 - \frac{4 k_{\parallel}^2 \gamma p B^2}{k_0^2(\gamma p + B^2)^2} \right)^{1/2} \right]. \quad (13.12)$$

Equation (13.9), together with the BCs (13.6), uniquely determines the eigenvalues. The tangential components η and ζ follow from ξ by algebraic relations:

$$\eta = k_\perp \frac{(\gamma p + B^2)(\tilde{\omega}^2 - \omega_{\mathrm{S}}^2)\,\xi' - \rho g\, \tilde{\omega}^2\, \xi}{\rho D} \,,$$

$$\zeta = k_\parallel \frac{\gamma p (\tilde{\omega}^2 - \omega_{\mathrm{A}}^2)\,\xi' - g\,(\rho\tilde{\omega}^2 - k_0^2 B^2)\,\xi}{\rho D} \,. \quad (13.13)$$

The only formal difference with the equations for the static case is the replacement of ω^2 by $\tilde{\omega}^2(x; \omega)$.

For numerical integration, exploiting iterative solution of the two-sided BVP by means of a succession of one-sided (left or right) BVPs (see Section 13.1.4), the second order ODE (13.9) is suitably converted into an equivalent pair of first order ODEs for the function and its first derivative. Instead of ξ', it is more expedient to exploit the Eulerian perturbation Π of the total pressure, defined in Eq. (12.81),

$$\Pi = -(\gamma p + B^2)(\xi' + k_\perp \eta) - k_\parallel \gamma p\, \zeta + \rho g \xi = -\frac{1}{D}\,(N \xi' + C \xi)\,. \quad (13.14)$$

This yields

$$N \frac{d}{dx}\begin{pmatrix} \xi \\ \Pi \end{pmatrix} + \begin{pmatrix} C & D \\ E & -C \end{pmatrix}\begin{pmatrix} \xi \\ \Pi \end{pmatrix} = 0\,, \quad (13.15)$$

involving a new function

$$E(x; \tilde{\omega}^2) \equiv -[\rho(\tilde{\omega}^2 - \omega_{\mathrm{A}}^2) + \rho' g]N - \rho^2 g^2(\tilde{\omega}^2 - \omega_{\mathrm{A}}^2)^2$$

$$= -N\left(A + \frac{B}{D}\right) - \frac{C^2}{D} \quad (13.16)$$

connected with the functions of the second order formulation as indicated.

▷ **Apparent and spurious singularities** Recall from the analysis of the analogous problems for static plasma slabs (Sections 7.3.2 and 7.4.1 [1]) and cylinders (Section 9.2.1 [1]) that the formulation in terms of the second order ODE contains the apparent singularities $D = 0$, whereas the formulation in terms of first ODEs contains spurious singularities

$N = 0$. The apparent singularities gave rise to considerable confusion, as they were originally thought to be associated with additional continua by Grad [187], which was shown by Appert, Gruber and Vaclavik [10] to be incorrect by means of the formulation in terms of the first order system where these singularities simply did not turn up. The absence of $D = 0$ continua was further substantiated by Goedbloed [166] from the solution of the initial value problem (Section 10.2 [1]). In the second order formulation, the condition for the singularities $D = 0$ to be apparent, found by Greene [191], is

$$ NB + C^2 \sim D \,. \tag{13.17} $$

On the other hand, the first order formulation also has a defect, viz. the occurrence of a set of spurious singularities $N = 0$ on top of the genuine ones (since N multiplies both ξ' and Π' in Eq. (13.15): one factor too much). It was pointed out by Bondeson, Iacono and Bhattacharjee [56], for the analogous cylindrical problem with flow (see Section 13.3.1), that it is essential for the first order formulation that the determinant of the matrix in Eq. (13.15) is proportional to N,

$$ DE + C^2 \sim N \,. \tag{13.18} $$

This guarantees that the spurious $N = 0$ singularities can be eliminated again (leaving the genuine ones). It now emerges that *the problems of apparent and spurious singularities are complementary evils*: the condition (13.17) for the absence of $D = 0$ singularities and the condition (13.18) for the absence of spurious $N = 0$ singularities are just rearrangements of the same terms of the relation (13.16) connecting the two formulations. ◁

So far, we have stressed the formal similarities of the static and the stationary problems. We now need to study the differences. It is evident that the genuine ($N = 0$) and apparent ($D = 0$) singularities of the eigenvalue equation (13.9) have to be considered in detail in order to understand their role in the spectrum of stationary plasmas. Factoring these polynomials, e.g. the Alfvén component

$$ \widetilde{\omega}^2 - \omega_A^2 \equiv (\omega - \Omega_0)^2 - \omega_A^2 = (\omega - \Omega_A^+)(\omega - \Omega_A^-)\,, \quad \text{where } \Omega_0 \equiv \mathbf{k}_0 \cdot \mathbf{v}\,, $$

yields the new singular frequencies for equilibria with flow:

$$ N(x;\omega) = (\gamma p + B^2)(\omega - \Omega_A^+)(\omega - \Omega_A^-)(\omega - \Omega_S^+)(\omega - \Omega_S^-)\,, $$

$$ \Omega_A^\pm(x) \equiv \Omega_0(x) \pm \omega_A(x)\,, \qquad \Omega_S^\pm(x) \equiv \Omega_0(x) \pm \omega_S(x)\,, \tag{13.19} $$

and the new apparent singularities:

$$ D(x;\omega) = (\omega - \Omega_{f0}^+)(\omega - \Omega_{f0}^-)(\omega - \Omega_{s0}^+)(\omega - \Omega_{s0}^-)\,, $$

$$ \Omega_{f0}^\pm(x) \equiv \Omega_0(x) \pm \omega_{f0}(x)\,, \qquad \Omega_{s0}^\pm(x) \equiv \Omega_0(x) \pm \omega_{s0}(x)\,. \tag{13.20} $$

Here, the frequencies $\Omega_A^\pm$ and $\Omega_S^\pm$ indicate the *forward*[(+)] or *backward*[(−)] local Doppler-shifted Alfvén and slow frequencies in the *laboratory frame*. (Hence, the shift is $+\Omega_0$ here, whereas it is $-\Omega_0$ in Eq. (13.4) since that equation refers to the co-moving frame.) Since the analysis of the continuous spectra of static plasmas can be applied in completely the same way to stationary plasmas, the only change is that the static Alfvén continuum $\{\omega_A^2(x)\}$ is split into forward and backward Alfvén continua $\{\Omega_A^+(x)\}$ and $\{\Omega_A^-(x)\}$, and the static slow continuum $\{\omega_S^2(x)\}$ is split into forward and backward slow continua $\{\Omega_S^+(x)\}$ and $\{\Omega_S^-(x)\}$. Similarly, the collection of apparent singularities $\{\omega_{f0,s0}^2(x)\}$ is split into the sets of forward and backward turning point frequencies $\{\Omega_{f0,s0}^+(x)\}$ and $\{\Omega_{f0,s0}^-(x)\}$.

As in the static case, to complete the essential spectrum, consisting of continua and cluster points, one also needs to include the cluster points of the forward and backward fast magneto-sonic waves,

$$\Omega_F^\pm \equiv \pm\infty\,, \tag{13.21}$$

which formally also belong to the continuous spectrum. One easily proves from the definitions that the following ordering of the local frequencies (which are all real!) holds at each point $x_1 \le x \le x_2$ of the plasma:

$$\Omega_F^- \le \Omega_{f0}^- \le \Omega_A^- \le \Omega_{s0}^- \le \Omega_S^- \le \Omega_0 \le \Omega_S^+ \le \Omega_{s0}^+ \le \Omega_A^+ \le \Omega_{f0}^+ \le \Omega_F^+\,. \tag{13.22}$$

Computation of the collections of these frequencies over the interval $[x_1, x_2]$ should precede any computation of the spectrum since they determine its overall structure. One might object that, to determine global stability of gravitational instabilities (driven by the term $\rho'g$), one is not really interested in these "trouble makers". However, even in that case, one needs to know where these frequencies are located in order to properly deal with the marginal stability transition, i.e. the approach to the real ω-axis.

Fortunately, for the 1D equilibria under consideration, all of these frequencies are still (as in the static case) situated on the real ω-axis. We will study the continuous spectra in more detail in Section 13.1.3, where we also consider the hydrodynamic flow continuum $\{\Omega_0(x)\}$, or rather flow continua (since they are degenerate). These are not additional continua in MHD [174], but contained in the Alfvén and slow continua from which they emerge in the limit of vanishing magnetic field. Overlooking this limit has caused considerable confusion in the MHD literature.

13.1.2 Kelvin–Helmholtz instabilities in interface plasmas

As a non-trivial first example, consider the following extension of the interface problem discussed at the end of Chapter 6 of Volume [1], illustrated in Fig. 6.20. We now replace the bottom vacuum layer, that was considered there, by a second plasma with different flow and magnetic field parameters. In the resulting configuration, two homogeneous plane plasma layers with embedded constant magnetic fields $\mathbf{B}$ and $\hat{\mathbf{B}}$ and constant velocity fields $\mathbf{v}$ and $\hat{\mathbf{v}}$ are superposed, with a surface current $\mathbf{j}^*$ and a surface vorticity ω^* creating jumps in the magnitudes and directions of the magnetic fields and velocities at the interface. The jump of the velocity causes the plasma to be Kelvin–Helmholtz unstable. Since we also keep the vertical gravity field $\mathbf{g} = -g\mathbf{e}_x$, Rayleigh–Taylor instabilities will also be present. The change of direction of the magnetic field at the interface will cause stabilization by magnetic shear.

We summarize the assumptions on the various equilibrium quantities and recall (Section 12.2.1) that the equilibrium is essentially the same as for the static case.

– Upper layer ($0 < x \leq a$):

$$\rho = \text{const}\,, \quad \mathbf{v} = (0, v_y, v_z) = \text{const}\,, \quad \mathbf{B} = (0, B_y, B_z) = \text{const}\,,$$

$$p' = -\rho g \quad \Rightarrow \quad p = p_0 - \rho g x \quad (p_0 \geq \rho g a)\,. \tag{13.23}$$

– Lower layer ($-b \leq x < 0$):

$$\hat{\rho} = \text{const}\,, \quad \hat{\mathbf{v}} = (0, \hat{v}_y, \hat{v}_z) = \text{const}\,, \quad \hat{\mathbf{B}} = (0, \hat{B}_y, \hat{B}_z) = \text{const}\,,$$

$$\hat{p}' = -\hat{\rho} g \quad \Rightarrow \quad \hat{p} = \hat{p}_0 - \hat{\rho} g x\,. \tag{13.24}$$

– Jumps at the interface ($x = 0$):

$$p_0 + \tfrac{1}{2} B_0^2 = \hat{p}_0 + \tfrac{1}{2} \hat{B}_0^2 \qquad \text{(pressure balance)}\,, \tag{13.25}$$

$$\mathbf{j}^\star = \mathbf{n} \times [\![\mathbf{B}]\!] = \mathbf{e}_x \times (\mathbf{B} - \hat{\mathbf{B}}) \qquad \text{(surface current)}\,,$$

$$\boldsymbol{\omega}^\star = \mathbf{n} \times [\![\mathbf{v}]\!] = \mathbf{e}_x \times (\mathbf{v} - \hat{\mathbf{v}}) \qquad \text{(surface vorticity)}\,. \tag{13.26}$$

Recall from Section 4.5.2 [1] that the pressure balance equation (13.25) is a genuine boundary condition whereas the latter two equations are just implications of the jumps permitted by this equilibrium

We perform a normal mode analysis of this configuration, with Fourier harmonics in the ignorable directions: $\boldsymbol{\xi} \sim \exp\left[\mathrm{i}(k_y y + k_z z - \omega t)\right]$. Above, the example was called "simple" since the plasmas are taken incompressible and homogeneous so that the differential equations become trivial. The much harder part of the inhomogeneities and associated singularities will be considered in the following sections. In order to formulate a complete problem, we first present the general expression for the incompressible form of the wave equation and then specify to homogeneous plasmas. Taking the limit $\gamma \to \infty$ of Eq. (13.9) yields the general *wave equation for incompressible plasmas:*

$$\frac{d}{dx}\left[\rho(\tilde{\omega}^2 - \omega_{\mathrm{A}}^2)\frac{d\xi}{dx}\right] - k_0^2\left[\rho(\tilde{\omega}^2 - \omega_{\mathrm{A}}^2) + \rho' g\right]\xi = 0\,, \tag{13.27}$$

where $\tilde{\omega} \equiv \omega - \Omega_0 \equiv \omega - \mathbf{k}_0 \cdot \mathbf{v}$ is the Doppler shifted frequency and $\omega_{\mathrm{A}} \equiv$

$\mathbf{k}_0 \cdot \mathbf{B}/\sqrt{\rho_0}$ is the Alfvén frequency. For completeness, we also present the incompressible limit of the expressions (13.13) and (13.14) for the tangential variables η and ζ and the total pressure perturbation Π:

$$\eta \rightarrow -\frac{k_\perp}{k_0^2}\xi', \qquad \zeta \rightarrow -\frac{k_\parallel}{k_0^2}\xi', \qquad \Pi \rightarrow \frac{\rho}{k_0^2}(\tilde{\omega}^2 - \omega_A^2)\xi'. \qquad (13.28)$$

The incompressible counterpart of the system of first order ODEs (13.15) for ξ and Π can easily be constructed from these expressions.

In the present case, all equilibrium quantities are constant in the respective layers so that the ODEs simplify to equations with constant coefficients that can easily be solved. This yields the following solutions, for the variable ξ in the top layer:

$$\xi'' - k_0^2\xi = 0, \quad \text{with BC } \xi(a) = 0 \quad \Rightarrow \quad \xi = c\frac{\sinh\left[k_0(a - x)\right]}{\sinh\left(k_0 a\right)}, \qquad (13.29)$$

and for the variable $\hat{\xi}$ in the bottom layer:

$$\hat{\xi}'' - k_0^2\hat{\xi} = 0, \quad \text{with BC } \hat{\xi}(-b) = 0 \quad \Rightarrow \quad \hat{\xi} = \hat{c}\frac{\sinh\left[k_0(x + b)\right]}{\sinh\left(k_0 b\right)}. \qquad (13.30)$$

These eigenfunctions have the usual cusp-shaped form of *surface modes*. Obtaining them was trivial, but all intricacies of the analysis and the physics now reside in the boundary conditions at $x = 0$ that should determine the eigenvalues.

Since the bottom layer is a plasma rather than a vacuum, the relevant BCs connecting ξ and $\hat{\xi}$ at $x = 0$ are the interface conditions for model II* that were derived in Eqs. (6.144) and (6.147) [1]:

– *First interface condition* (continuity of the normal velocity):

$$[\![\mathbf{n} \cdot \boldsymbol{\xi}]\!] = 0 \quad \Rightarrow \quad \xi(0) = \hat{\xi}(0) = 0 \quad \Rightarrow \quad c = \hat{c}. \qquad (13.31)$$

– *Second interface condition* (pressure balance),

$$[\![\Pi + \mathbf{n} \cdot \boldsymbol{\xi}\,\mathbf{n} \cdot \nabla(p + \tfrac{1}{2}B^2)]\!] = 0, \quad \Pi \equiv -\gamma p\nabla \cdot \boldsymbol{\xi} - \boldsymbol{\xi} \cdot \nabla p + \mathbf{B} \cdot \mathbf{Q}. \qquad (13.32)$$

The latter equation needs to be reworked since $\gamma p\nabla \cdot \boldsymbol{\xi}$ is actually an undetermined quantity for incompressible plasmas ($\gamma \rightarrow \infty$, $\nabla \cdot \boldsymbol{\xi} \rightarrow 0$). We determine it by exploiting the incompressible expression (13.28) for Π. Inserting this expression into the second interface condition, exploiting the equilibrium conditions (13.23) and (13.24), and dividing by the first interface condition then yields

$$\left[\!\!\left[\frac{\rho}{k_0^2}(\tilde{\omega}^2 - \omega_A^2)\frac{\xi'}{\xi} - \rho g \right]\!\!\right] = 0, \qquad (13.33)$$

or

$$-\rho\left[(\omega - \Omega_0)^2 - \omega_{\mathrm{A}}^2\right]\coth(k_0 a) - k_0\rho g = \hat\rho\left[(\omega - \hat\Omega_0)^2 - \hat\omega_{\mathrm{A}}^2\right]\coth(k_0 b) - k_0\hat\rho g\,,$$
$$(13.34)$$

which is *the dispersion equation* for this configuration.

In the limit of small wavelength perturbations, $\coth(k_0 a) \approx \coth(k_0 b) \approx 1$ (walls effectively at ∞ and $-\infty$), the explicit solutions of this dispersion equation become:

$$\omega = \frac{\rho\Omega_0 + \hat\rho\hat\Omega_0}{\rho + \hat\rho} \pm \sqrt{-\frac{\rho\hat\rho(\Omega_0 - \hat\Omega_0)^2}{(\rho + \hat\rho)^2} + \frac{\rho\omega_{\mathrm{A}}^2 + \hat\rho\hat\omega_{\mathrm{A}}^2}{\rho + \hat\rho} - \frac{k_0(\rho - \hat\rho)g}{\rho + \hat\rho}}\,. \quad (13.35)$$

These solutions represent either two waves with real frequency or two modes with complex frequency (an instability and a damped wave), depending on the sign of the expression under the square root. When that expression is negative, an instability is obtained driven by a destabilizing *Kelvin–Helmholtz* contribution when $\Omega_0 \neq \hat\Omega_0$, a stabilizing *magnetic field line bending* contribution when $\omega_{\mathrm{A}}^2 \neq 0$ and/or $\hat\omega_{\mathrm{A}}^2 \neq 0$, and a destabilizing *Rayleigh–Taylor* contribution when $\rho > \hat\rho$ (or stabilizing when $\rho < \hat\rho$). Magnetic stabilization of both the Kelvin–Helmholtz and the Rayleigh–Taylor instability is obtained when

$$(\mathbf{k}_0 \cdot \mathbf{B})^2 + (\mathbf{k}_0 \cdot \hat{\mathbf{B}})^2 > \frac{\rho\hat\rho}{\rho + \hat\rho}\left[\mathbf{k}_0 \cdot (\mathbf{v} - \hat{\mathbf{v}})\right]^2 + k_0(\rho - \hat\rho)g\,. \quad (13.36)$$

Note that *magnetic shear* (different directions of $\mathbf{B}$ and $\hat{\mathbf{B}}$) is effective because it prevents vanishing of the magnetic terms for directions of the wave vector $\mathbf{k}_0$ perpendicular to the magnetic field. The expression for arbitrary wavelengths is easily derived from Eq. (13.34) by noting that every ρ occurs together with the term $\coth(k_0 a)$ and every $\hat\rho$ together with $\coth(k_0 b)$, except for the Rayleigh–Taylor term $-k_0(\rho - \hat\rho)g$. For long wavelength perturbations, $\coth(k_0 a) \approx (k_0 a)^{-1} \gg 1$ and $\coth(k_0 b) \approx (k_0 b)^{-1} \gg 1$ (walls effectively close), this leads to genuine competition between the three terms (all $\sim k_0$), so that stability depends on the precise choice of all those parameters.

These modes illustrate the generic behavior of instabilities shown in Fig. 13.2. Note that the dashed line of that figure becomes just a vertical straight line $\sigma = \langle\Omega_0\rangle = (\rho\Omega_0 + \hat\rho\hat\Omega_0)/(\rho + \hat\rho)$ for the present case. Summarizing, this sub-section demonstrates how equilibrium background flow may create new (Kelvin–Helmholtz) instabilities and affect existing (Rayleigh–Taylor) instabilities in magnetized plasmas. However, because of the simplifying assumption of homogeneous plasma layers, it hardly reveals the complexity of the problem for diffuse plasmas, as we will soon realize.

13.1.3 Continua and oscillation theorem $\mathcal{R}$ for real eigenvalues

To appreciate the complexity of the spectral problem when background flow is involved, consider the hydrodynamic (HD) case of a plane incompressible and inviscid, but inhomogeneous, fluid without gravity but with a horizontal flow velocity,

$$\mathbf{v} = v_y(x)\mathbf{e}_y + v_z(x)\mathbf{e}_z \,. \tag{13.37}$$

The Lagrangian time derivative of any one of the Eulerian perturbations ρ_1, $\mathbf{v}_1$, p_1 occurring in the HD counterpart of Eqs. (12.31)–(12.34) is given by

$$\left(\frac{Df}{Dt}\right)_1 \equiv \left(\frac{\partial f}{\partial t} + \mathbf{v}\cdot\nabla f\right)_1 \stackrel{\text{FNM}}{=} -i\widetilde{\omega}f_1 + f_0{}'v_{1x}\,, \qquad \widetilde{\omega} \equiv \omega - \Omega_0(x)\,, \tag{13.38}$$

where, as in Eq. (13.4), $\widetilde{\omega}(x)$ is the Doppler shifted frequency observed in a local frame co-moving with the fluid layer at position x. Here, the gradient operator parallel to the background velocity,

$$\rho^{-1}U \equiv -i\mathbf{v}\cdot\nabla \stackrel{\text{FNM}}{=} \mathbf{k_0}\cdot\mathbf{v} \equiv \Omega_0(x)\,, \tag{13.39}$$

gives rise to continuous spectra, the *flow continua*

$$\omega \in \{\Omega_{\mathrm{H}i} \equiv \Omega_0(x)|x_1 \leq x \leq x_2\}\,, \qquad (i = 1, 2)\,, \tag{13.40}$$

consisting of all local Doppler shifts, for which the Eulerian equations are singular. The multiplicity index i will be explained below.

Prior to the discovery of the flow continua by Case [77] in 1960 (see also Drazin and Reid [124], p. 149), the HD literature on the inviscid limit of the Navier–Stokes equations [310] was confused because the absence of a spectrum of discrete modes for the simplest flow profiles appeared to imply absence of stable oscillations. This paradox follows from the Eulerian representation in terms of the velocity perturbation $\mathbf{v}_1$, which we here relate to both the stream function χ that is usually exploited for incompressible HD problems and the Lagrangian fluid displacement $\boldsymbol{\xi}$, according to Eq. (12.40):

$$\mathbf{v}_1 = \mathbf{e}_z \times \nabla\chi = \left(\mathbf{v}_0\cdot\nabla + \frac{\partial}{\partial t}\right)\boldsymbol{\xi} = -i\widetilde{\omega}\boldsymbol{\xi} \;\Rightarrow\; \left\{ \begin{array}{l} v_{1x} = -ik_0\chi = -i\widetilde{\omega}\xi \\[2mm] v_{1y} = \chi' = -\widetilde{\omega}\eta \end{array} \right. . \tag{13.41}$$

For constant density ρ, these variables should satisfy the ODEs

$$\widetilde{\omega}\left(\frac{d^2\chi}{dx^2} - k_0^2\chi\right) - \widetilde{\omega}''\chi = 0\,, \quad \text{or} \quad \frac{d}{dx}\left(\widetilde{\omega}^2\frac{d\xi}{dx}\right) - k_0^2\widetilde{\omega}^2\xi = 0\,, \tag{13.42}$$

and be subject to the BCs $\chi(x_1) = \chi(x_2) = 0$, respectively $\xi(x_1) = \xi(x_2) = 0$. Assuming $\widetilde{\omega} \neq 0$ and choosing a linear velocity profile, $v_0 = a + bx$, so that $\widetilde{\omega}'' = 0$, these equations only possess solutions $\sim \exp(\pm k_0 x)$ that cannot satisfy the BCs.

Clearly, the assumption $\tilde{\omega} \neq 0$ should be dropped and one should consider singular modes and solve the initial value problem *à la* Landau [293] (see Section 2.3.3 [1]) in order to construct the stable response to an initial perturbation, which amounts to an integral representation over the continuous spectra (13.40). This is what Case [77] did, demonstrating that the initial value problem is well-posed for HD and that the continuum contributions damp out as t^{-1}.

For MHD, Hameiri [207] introduced the distinction between the Eulerian and Lagrangian descriptions, noting that the former admits solutions that are absent in the Lagrangian formulation. As pointed out at the end of Section 12.2.1, these additional solutions correspond to the *non-holonomic Eulerian entropy continua*

$$\omega \in \{\Omega_{\mathrm{E}} \equiv \Omega_0(x)\}, \tag{13.43}$$

for which the entropy perturbations $S_{\mathrm{E}1}$ cannot be expressed in terms of $\boldsymbol{\xi}$. Unfortunately, since the frequencies of these continua coincide with those of the flow continua (13.40) (which, on the contrary, are expressible in terms of $\boldsymbol{\xi}$!), the misunderstanding could arise that these continua are the same as the flow continua and, hence, that the flow continua also exist in MHD [416, 222, 40]. This confusion was eliminated only recently by Goedbloed *et al.* [174] by means of an initial value approach similar to that of Case, demonstrating that the plasma response due to the Alfvén and slow continua in MHD is completely analogous to the fluid response due to the flow continua $\Omega_{\mathrm{H}i}$ in HD. That there is no place in MHD for continua in addition to the *Alfvén and slow MHD continua* (13.19),

$$\omega \in \{\Omega_{\mathrm{A}}^{\pm} \equiv \Omega_0(x) \pm \omega_{\mathrm{A}}(x)\}, \quad \omega \in \{\Omega_{\mathrm{S}}^{\pm} \equiv \Omega_0(x) \pm \omega_{\mathrm{S}}(x)\}, \tag{13.44}$$

should be obvious from the fact that both of these expressions transform into the flow continua (13.40) in the limit of vanishing magnetic field. Also note that the spectral HD equation (13.42) is obtained in that limit from the incompressible spectral equation (13.27), which in turn derives from the full MHD wave equation (13.9). In other words: the HD flow continua (13.40) and the MHD Alfvén and slow continua (13.44) are "robust", i.e. they describe the global dynamics expressed in terms of holonomic variables, whereas the Eulerian entropy continua (13.43) are "flimsy", i.e. they are obtained from initial data that are incompatible with the ideal MHD constraints. Consequently, there is no coupling between the Eulerian entropy continuum modes and either the HD flow continua or the MHD Alfvén and slow continua, so that it is completely legitimate to restrict the analysis to the holonomic Lagrangian description in terms of $\boldsymbol{\xi}$, for MHD as well as for HD.

Extending the discussion now to compressible fluids and plasmas, the degeneracy of the Alfvén and slow continua is resolved and cluster spectra of fast waves with limiting frequencies $\Omega_{\mathrm{F}}^{\pm} \equiv \pm\infty$ and normal polarization ($\boldsymbol{\xi} \to \xi \mathbf{e}_x$) appear in

Table 13.1 *Essential spectra, with dominant variables, in HD and MHD.*

HD			MHD		
entropy: $\Omega_{\mathrm{E}} \equiv \{\Omega_0\}$	$S_{1\mathrm{E}}$	$(1\times)$	entropy: $\Omega_{\mathrm{E}} \equiv \{\Omega_0\}$	$S_{1\mathrm{E}}$	$(1\times)$
flow 1: $\Omega_{\mathrm{H1}} \equiv \{\Omega_0\}$	v_{1y}	$(1\times)$	Alfvén: $\Omega_{\mathrm{A}}^{\pm} \equiv \{\Omega_0 \pm \omega_{\mathrm{A}}\}$	η	$(2\times)$
flow 2: $\Omega_{\mathrm{H2}} \equiv \{\Omega_0\}$	v_{1z}	$(1\times)$	slow: $\Omega_{\mathrm{S}}^{\pm} \equiv \{\Omega_0 \pm \omega_{\mathrm{S}}\}$	ζ	$(2\times)$
sound: $\Omega_{\mathrm{P}}^{\pm} \equiv \pm\infty$	v_{1x}	$(2\times)$	fast: $\Omega_{\mathrm{F}}^{\pm} \equiv \pm\infty$	ξ	$(2\times)$

MHD, whereas analogous cluster spectra of sound waves with limiting frequencies $\Omega_{\mathrm{P}}^{\pm} \equiv \pm\infty$ and normal polarization ($\mathbf{v} \rightarrow v_{1x}\mathbf{e}_x$) appear in HD. Moreover, like the Alfvén and slow continua in MHD asymptotically characterize the two tangential (horizontal) degrees of freedom, those degrees of freedom are characterized in HD by the two flow continua Ω_{H1} and Ω_{H2}. Thus, the Lagrangian description involves four modes in HD and six in MHD, in the Eulerian description extended with the Eulerian entropy modes (see Table 13.1, where the different continua are associated with the physical variables that become dominant for those frequencies). This is how the continuous spectra along the real ω-axis asymptotically account for the different degrees of freedom of the linearized dynamics of stationary plasmas. (Note how the theme of counting variables, started in Sections 5.1 and 5.2 [1] for the nearly trivial case of waves in homogeneous fluids, keeps returning!)

Let us now consider the wider problem of how the continua "determine" the full spectral structure of waves and instabilities for stationary plasmas and fluids, i.e. we turn to the full spectral equation (13.9) for MHD and its $\beta \equiv 2p/B^2 \rightarrow \infty$ limit for compressible HD *with gravity*. From the analogous discussion for static plasmas in Section 7.4 [1], it is logical to try to involve an oscillation theorem linking the sequences of discrete modes to the extrema of the continua by means of some monotonicity property. However, the attempt to generalize the oscillation theorem of Section 7.4.3 [1] to stationary plasmas immediately runs into the problem that the eigenfunctions become complex, i.e. the triple (ξ, η, ζ) of the field line projection (13.8), which has been constructed so as to be real for static plasmas, now becomes intrinsically complex because the spectral variable $\widetilde{\omega}^2$ is no longer real. Hence, the solution of the spectral differential equation (13.9) for $\xi = \xi_1 + \mathrm{i}\xi_2$ would involve counting of nodes of both the real part ξ_1 and the imaginary part ξ_2 of the eigenfunctions. We may defer this tricky problem to the following section, where we will discuss the instabilities of stationary plasmas, since $\widetilde{\omega}^2$ does remain real for stable oscillations, $\nu = 0 \;\Rightarrow\; \widetilde{\omega} = \sigma - \Omega(x)$, so that ξ remains real. Hence, a partial answer to the above question may be obtained by considering how the continua determine the spectral structure for stable waves.

It is clear that the collection of Doppler shifts $\{\Omega_0(x)\}$, which in general is not part of the continuous spectrum in MHD, nevertheless significantly alters the spectrum of stable waves. In particular, notice that the marginal stability threshold σ_0, defined in Eq. (12.140), is necessarily situated inside $\{\Omega_0(x)\}$. Also, the new continua $\{\Omega_A^{\pm}(x)\}$ and $\{\Omega_S^{\pm}(x)\}$ are no longer symmetric with respect to the origin $\omega = 0$, and they now depend on x through two factors, viz. $\omega_{A,S}(x)$ and $\Omega_0(x)$. This, incidentally, opens up quite a number of new possibilities for resonant heating and for the existence of peculiar quasi-modes. To establish the connection between the continua and the structure of the stable part of the spectrum, we will prove the following **oscillation theorem $\mathcal{R}$ for stable waves** [174]:

The eigenvalues of stable oscillations of a stationary plasma slab [or cylinder] are monotonic (Sturmian or anti-Sturmian) in the number of nodes of the eigenfunction ξ [or χ] for real values of $\omega = \sigma$ outside the continua $\{\Omega_A^{\pm}(x)\}$ and $\{\Omega_S^{\pm}(x)\}$, outside the apparent singularity ranges $\{\Omega_{s0}^{\pm}(x)\}$ and $\{\Omega_{f0}^{\pm}(x)\}$, and outside the Doppler–Coriolis indefinite point σ_0 or range (σ_1, σ_0).

In square brackets, we have indicated that the theorem is also valid for cylindrical plasmas since the equations for $\chi \equiv r\xi_r$ have the same structure (see Section 13.3). To prove it, we exploit the quadratic form (12.139) involving an eigenfunction ξ_α, with eigenvalue σ_α, and a neighboring solution ξ_β of the one-sided BVP (12.119), (12.120)(a) satisfying the left BC, with parameter σ_β (which is *not* an eigenvalue). For the plane slab, we exploit the second order ODE (13.9) for ξ, or the system of first order ODEs (13.15) for ξ and Π, with BCs (13.6). (Corresponding equations for the cylinder may be found in Section 13.3.) Furthermore, we substitute the quartic polynomials (13.19) and (13.20) for N and D into the expression (13.14) for Π. The quadratic form (12.139) then reduces to the following lucid expression:

$$
-\xi_\beta(x_2)\Pi_\alpha(x_2) = \xi_\beta \left. \frac{N(\sigma_\alpha)}{D(\sigma_\alpha)}\, \xi_\alpha' \right|_{x=x_2}
$$

$$
\equiv \xi_\beta(\gamma p + B^2) \left. \frac{(\sigma_\alpha - \Omega_A^-)(\sigma_\alpha - \Omega_S^-)(\sigma_\alpha - \Omega_S^+)(\sigma_\alpha - \Omega_A^+)}{(\sigma_\alpha - \Omega_{f0}^-)(\sigma_\alpha - \Omega_{s0}^-)(\sigma_\alpha - \Omega_{s0}^+)(\sigma_\alpha - \Omega_{f0}^+)}\, \xi_\alpha' \right|_{x=x_2}
$$

$$
= 2(\sigma_\beta - \sigma_\alpha)\left[\sigma_\alpha - \overline{V}(\sigma_\alpha)\right] \int \rho |\xi_\alpha|^2\, dV \,. \tag{13.45}
$$

We now apply the usual Sturm–Liouville type of reasoning (see Section 7.4.3 [1]): If $\sigma_\beta > \sigma_\alpha$ and if $\xi_\alpha > 0$ on (x_1, x_2), so that $\xi_\alpha'(x_2) < 0$, the sign of $\xi_\beta(x_2)$ is necessarily negative (resp. positive), i.e. ξ_β oscillates faster (resp. slower) than ξ_α, if the sign of $N/[(\sigma_\alpha - \overline{V})D]$ is positive (resp. negative); QED.

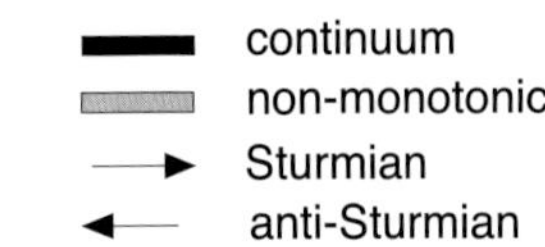

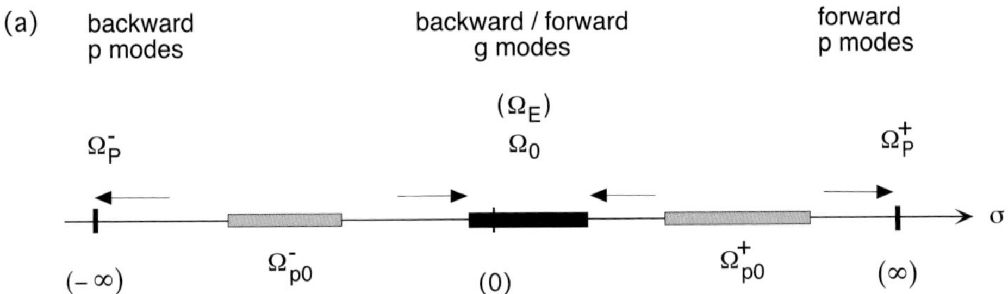

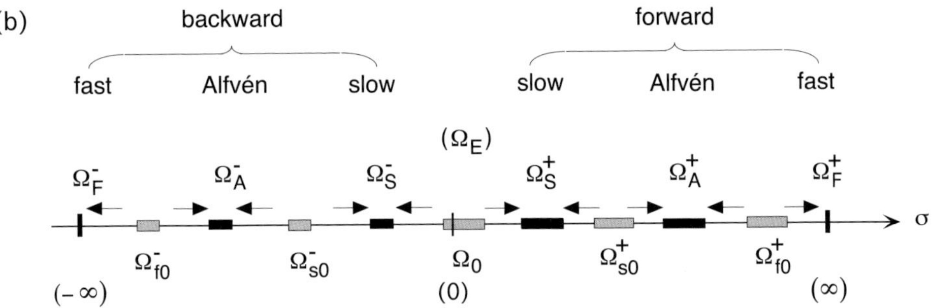

Fig. 13.3 Schematic spectrum of the stable oscillations of a stationary plane flow (a) in hydrodynamics and (b) in magnetohydrodynamics. Continua are labeled above, apparent singularities below the axis. Marginal transitions σ_0 (resp. σ_1) in Ω_0 are not indicated (the location of the origin $\sigma = 0$ inside Ω_0 is accidental).

Note that, in the above proof, the boundary points x_1 and x_2 of the plasma may be replaced by two subsequent zeros of the solution ξ_α, so that the theorem applies to all oscillatory parts of the solutions. The same arguments apply for right solutions of the one-sided BVP. Consequently, for $\sigma \notin \{\Omega_A^\pm\}$, $\{\Omega_S^\pm\}$, $\{\Omega_{s0}^\pm\}$, $\{\Omega_{f0}^\pm\}$, and $\neq \sigma_0$ (assuming a single marginal point), Sturmian or anti-Sturmian dependence of σ on the number of nodes of ξ is determined by the following expression:

$$\frac{N(\sigma)}{(\sigma - \sigma_0)D(\sigma)} \begin{cases} > 0 : & \textit{Sturmian} \\ \\ < 0 : & \textit{anti-Sturmian} \end{cases} . \tag{13.46}$$

This determines the spectral structure for real values of the eigenvalue parameter, as illustrated in Fig. 13.3.

In compressible HD, the Doppler shifts create the flow continua $\{\Omega_{Hi} \equiv \Omega_0\}$, with possible cluster spectra of forward and backward gravity-driven g modes at the ends [78, 128], and it also splits apart forward and backward pressure-driven p modes (sound waves). See Section 7.2 [1] for the corresponding static case, where the spectral equation (7.34) [1] provides the spectral equation for the stationary case through the transformation $\omega \to \widetilde{\omega}$. This also yields the expression for the apparent HD singularities $\Omega_{p0}^{\pm} \equiv \Omega_0 \pm k_0 c$, where c is the sound speed, separating the p and the g modes. In MHD (neglecting the Eulerian entropy modes), the Doppler shifts do not create continua but they also split apart the forward and backward parts of the spectrum. Thus, forward and backward continua $\Omega_S^{\pm}$, $\Omega_A^{\pm}$, $\Omega_F^{\pm}$ are separated by regions Ω_0, $\Omega_{s0}^{\pm}$, $\Omega_{f0}^{\pm}$ of non-monotonicity of the discrete spectrum which, otherwise, is either Sturmian ($\to$) or anti-Sturmian ($\leftarrow$) along the real ω-axis [455]. This is how the spectrum of a gravitating static equilibrium (depicted in Fig. 7.18 [1]) is changed by background flow.

The Sturmian and anti-Sturmian sub-spectra shown in Fig. 13.3 just indicate *possible* structures for sizeable background flow but small inhomogeneity, as may be obtained for any thin slice of plasma across the normal direction. This is so since the sequence of singular frequencies (13.22) is well-ordered for fixed position x, so that the collections of those frequencies may leave significant portions of the real ω-axis, but only if the inhomogeneities of the equilibrium are not too strong. One can stack slices like that to construct a (rather artificial) equilibrium exhibiting all sub-spectra. However, for realistic equilibria, the continua may fold over themselves, they may overlap, and they usually "swallow" large parts of the discrete sub-spectra. Except for the p-modes in HD and the fast modes in MHD, the continua even may not leave any space along the real axis for discrete modes. (Hence, the original embarrassment about complete absence of discrete modes in incompressible HD.)

In HD, possible ϵ-stability threshold(s) σ_0^{ϵ} are located immediately above the flow continua $\{\Omega_0\}$. In MHD (neglecting the Eulerian entropy modes), the Doppler range $\{\Omega_0\}$ does not correspond to continua but it also contains the stability threshold σ_0, or the range $\sigma_1 \leq \sigma \leq \sigma_0$, at least if it is not overlapped with the continua. Overlap of the Doppler range with continua is always present if the direction of the horizontal wave vector is somewhere (say at $x = x_0$) perpendicular to the magnetic field, so that $\omega_A(x_0) = 0$ and $\Omega_A^{\pm}(x_0) = \Omega_S^{\pm}(x_0) = \Omega_0(x_0)$. Hence, although HD and MHD spectral theories have many similarities, in particular that the Doppler range $\{\Omega_0\}$ delimits the strip of the complex ω plane where instabilities may occur for both, they differ in the important respect that this role of the Doppler range cannot be distinguished from that of the continua in HD, but it can in MHD. In this respect (of removing a degeneracy), MHD spectral theory is simpler than HD spectral theory. However, we will see what other ramifications occur.

13.1.4 Complex eigenvalues and the alternator

For instabilities of plasmas with background equilibrium flow, we have to address the fact that, due to the variable Doppler shift $\Omega_0(x)$, the eigenvalues $\omega = \sigma + i\nu$ and the eigenfunctions $\xi = \xi_1 + i\xi_2$ are irreducibly complex. Notice though that the x-dependence of the crucial function $\widetilde{\omega}(x)$ only resides in the real part:

$$\widetilde{\omega}(x) = \widetilde{\sigma}(x) + i\nu\,, \quad \widetilde{\sigma}(x) \equiv \sigma - \Omega_0(x)\,. \tag{13.47}$$

We will exploit the first order eigenvalue system (13.15), where we split all functions that occur into real and imaginary parts:

$$
\begin{pmatrix} \xi_1' \\ \xi_2' \\ \Pi_1' \\ \Pi_2' \end{pmatrix}
+
\begin{pmatrix}
\hat{C}_1 & -\hat{C}_2 & \hat{D}_1 & -\hat{D}_2 \\
\hat{C}_2 & \hat{C}_1 & \hat{D}_2 & \hat{D}_1 \\
\hat{E}_1 & -\hat{E}_2 & -\hat{C}_1 & \hat{C}_2 \\
\hat{E}_2 & \hat{E}_1 & -\hat{C}_2 & -\hat{C}_1
\end{pmatrix}
\begin{pmatrix} \xi_1 \\ \xi_2 \\ \Pi_1 \\ \Pi_2 \end{pmatrix}
= 0\,. \tag{13.48}
$$

The explicit expressions of the real and imaginary parts of the coefficients $\hat{C} \equiv C/N$, $\hat{D} \equiv D/N$ and $\hat{E} \equiv E/N$ are put in small print below.

▷ **Explicit expressions of the matrix elements** The elements of the first order system of ODEs (13.48) are given by the expressions

$$\hat{C}_1 \equiv (N_1 C_1 + N_2 C_2)/|N|^2\,, \quad \hat{C}_2 \equiv (N_1 C_2 - N_2 C_1)/|N|^2\,,$$

$$\hat{D}_1 \equiv (N_1 D_1 + N_2 D_2)/|N|^2\,, \quad \hat{D}_2 \equiv (N_1 D_2 - N_2 D_1)/|N|^2\,,$$

$$\hat{E}_1 \equiv (N_1 E_1 + N_2 E_2)/|N|^2\,, \quad \hat{E}_2 \equiv (N_1 E_2 - N_2 E_1)/|N|^2\,, \tag{13.49}$$

with the factors

$$N_1 = (\gamma p + B^2)\left[(R - \omega_{\rm A}^2)(R - \omega_{\rm S}^2) - I^2\right]\,, \quad N_2 = I(\gamma p + B^2)\left(2R - \omega_{\rm A}^2 - \omega_{\rm S}^2\right)\,,$$

$$D_1 = R^2 - I^2 - k_0^2\big((\gamma p + B^2)/\rho\big)(R - \omega_{\rm S}^2)\,, \quad D_2 = I\left[2R - k_0^2\big((\gamma p + B^2)/\rho\big)\right]\,,$$

$$C_1 = -\rho g\left[R(R - \omega_{\rm A}^2) - I^2\right]\,, \quad C_2 = -I\rho g\left(2R - \omega_{\rm A}^2\right)\,,$$

$$
\begin{aligned}
E_1 = {}&-\rho(\gamma p + B^2)\Big\{(R - \omega_{\rm A}^2 + \rho' g/\rho)\left[(R - \omega_{\rm A}^2)(R - \omega_{\rm S}^2) - I^2\right] \\
&- I^2(2R - \omega_{\rm A}^2 - \omega_{\rm S}^2)\Big\} - \rho^2 g^2\left[(R - \omega_{\rm A}^2)^2 - I^2\right]\,,
\end{aligned}
$$

$$
\begin{aligned}
E_2 = {}&-I\Big[\rho(\gamma p + B^2)\big\{(R - \omega_{\rm A}^2)(R - \omega_{\rm S}^2) - I^2 \\
&+ (R - \omega_{\rm A}^2 + \rho' g/\rho)(2R - \omega_{\rm A}^2 - \omega_{\rm S}^2)\big\} + 2\rho^2 g^2(R - \omega_{\rm A}^2)\Big]\,, \tag{13.50}
\end{aligned}
$$

where σ and ν only occur in the combinations

$$R(x; \sigma, \nu) \equiv \mathrm{Re}(\widetilde{\omega}^2) = \widetilde{\sigma}^2 - \nu^2\,, \quad I(x; \sigma, \nu) \equiv \mathrm{Im}(\widetilde{\omega}^2) = 2\widetilde{\sigma}\nu\,. \tag{13.51}$$

For *incompressible* plasmas ($\gamma p \to \infty$), the coefficients $\hat{C} \to 0$, $\hat{D} \to -k_0^2/[\rho(\widetilde{\omega}^2) - \omega_{\mathrm{A}}^2]$, $\hat{E} \to -\rho(\widetilde{\omega}^2 - \omega_{\mathrm{A}}^2) - \rho'g$, so that only the off-diagonal elements $\hat{D}_{1,2}$ and $\hat{E}_{1,2}$ survive:

$$\hat{D}_1 = -\frac{k_0^2(R - \omega_{\mathrm{A}}^2)}{\rho[(R - \omega_{\mathrm{A}}^2)^2 + I^2]}, \qquad \hat{D}_2 = \frac{k_0^2 I}{\rho[(R - \omega_{\mathrm{A}}^2)^2 + I^2]},$$

$$\hat{E}_1 = -\rho(R - \omega_{\mathrm{A}}^2) - \rho'g, \qquad \hat{E}_2 = -\rho I. \tag{13.52}$$

Note that, although the terms are significantly simpler, the complexity of the numerical procedure does not change by this transformation. ◁

The eigenvalues of instabilities are determined by solving the ODEs (13.48) subject to the BCs (12.120), which we now split into real and imaginary parts,

$$\text{BVP:} \quad \begin{cases} \xi_1(x_1) = \xi_2(x_1) = 0 & \textit{(left BCs)} \\[2mm] \xi_1(x_2) = \xi_2(x_2) = 0 & \textit{(right BCs)} \end{cases} . \tag{13.53}$$

Numerical solution of the ODEs for a given value of ω is straightforward if this two-sided BVP is turned into a one-sided BVP (according to the exposition of Section 12.3), by "shooting", either from the left or from the right. (Accurate numerical library subroutines exist to solve the one-sided BVP for a system of first order ODEs with any number of unknowns; see Section 7.5.1 [1] for the "shooting" method in the similar problem of static plasmas, where $\xi(x)$ and $\Pi(x)$ are real.) Hence, the above two-sided BVP will be turned into either a left one-sided BVP with solutions ξ^ℓ, or a right one with solutions ξ^r (lower case superscripts here since these solutions are not yet assumed to correspond to a solution path),

$$\text{left BVP:} \quad \xi_1^\ell(x_1) = \xi_2^\ell(x_1) = 0, \quad \Pi_1^\ell(x_1) = 1, \quad \Pi_2^\ell(x_1) = 0; \tag{13.54}$$

$$\text{right BVP:} \quad \xi_1^r(x_2) = \xi_2^r(x_2) = 0, \quad \Pi_1^r(x_2) = 1, \quad \Pi_2^r(x_2) = 0. \tag{13.55}$$

Here, we have chosen unit boundary conditions for $\Pi_1^\ell(x_1)$, or $\Pi_1^r(x_2)$, to initialize the derivatives of ξ^ℓ, or ξ^r. Note that this does not restrict the generality of the method since one can transform the solution found to correspond to any mix of initial derivatives by multiplying by a complex phase factor $\exp(\mathrm{i}\phi)$.

Suppose we shoot from the left. For arbitrary $\omega = \sigma + \mathrm{i}\nu$, we then solve the ODEs (13.48) by imposing the left BCs (13.54) on the complex functions ξ and Π. This provides a solution of the one-sided BVP that obviously does not qualify as an eigenfunction since, in general, it will not satisfy the right BCs (13.53)(b). One of those BCs can easily be satisfied, without changing the value of ω, by multiplying the solution found by a judicious choice of the mentioned phase factor:

$$\widetilde{\xi} = \mathrm{e}^{\mathrm{i}\phi}\xi, \quad \tan\phi \equiv -\xi_2(x_2)/\xi_1(x_2) \quad \Rightarrow \quad \widetilde{\xi}_2(x_2) = 0. \tag{13.56}$$

To satisfy the other BC, the value of ω needs to be changed iteratively such that the "error" $|\tilde{\xi}_1(x_2)|$ diminishes. If the iteration converges, eventually $\tilde{\xi}_1(x_2) = 0$ as well and a solution of the two-sided BVP with associated eigenvalue is obtained.

One might wish to exploit the phase factor to force the problem into the Sturm–Liouville type of studying the oscillatory behavior of the real components $\xi_1(x)$ and $\Pi_1(x)$, as was done in Section 13.1.3 for the real functions ξ and Π of stable oscillations. However, the freedom of choice of the arbitrary phase factor can be applied to any internal point $x \in [x_1, x_2]$, so that zeros of one of the components ξ_1 or ξ_2 can be created anywhere. This shows that the Sturm–Liouville type of counting nodes cannot be applied for complex solutions. We need to exploit another criterion for the approach to eigenvalues. Evidently, this will come from the exploitation of the solution paths $\mathcal{P}_{\mathrm{u}}^{\mathrm{L}}$ and $\mathcal{P}_{\mathrm{u}}^{\mathrm{R}}$, introduced in Section 12.3.

A concomitant advantage of not fixing the phase factor this way, but just letting it be determined by the condition (13.54) or (13.55), is the simplicity of the resulting boundary properties of the left and right solutions. The Wronskian of two arbitrary solutions ξ_α and ξ_β of the one-sided BVP for the same value of ω is constant,

$$\xi_\alpha(x)\Pi_\beta(x) - \xi_\beta(x)\Pi_\alpha(x) = \text{const}, \tag{13.57}$$

which is easily proved by differentiation with respect to x and substitution of the derivatives by the expressions of the system of first order ODEs (13.15). Application to ξ^r and ξ^ℓ at the end points x_1 and x_2 yields

$$\left[(\xi_1^r + i\xi_2^r)(\Pi_1^\ell + i\Pi_2^\ell)\right]_{x_1} = -\left[(\xi_1^\ell + i\xi_2^\ell)(\Pi_1^r + i\Pi_2^r)\right]_{x_2}, \tag{13.58}$$

which, upon application of the BC's (13.54) and (13.55) to $\Pi_{1,2}^\ell$ and $\Pi_{1,2}^r$, leads to the following useful relations between the boundary values of ξ^ℓ and ξ^r *on the opposite sides* (the end points) of the integration interval:

$$\xi_1^r(x_1) = -\xi_1^\ell(x_2), \qquad \xi_2^r(x_1) = -\xi_2^\ell(x_2). \tag{13.59}$$

This demonstrates the equivalence of shooting from the left and shooting from the right. We will exploit both though, doubling the computational effort, in order to create a third path of solutions, intermediate between the left and right paths, that has some attractive properties.

The solutions $\xi_1^\ell(x; \omega)$ and $\xi_1^r(x; \omega)$ of the respective BVPs may be obtained for any value of ω, but we now restrict those values to the solutions paths $\mathcal{P}_{\mathrm{u}}^{\mathrm{L}}$ and $\mathcal{P}_{\mathrm{u}}^{\mathrm{R}}$ defined by Eq. (12.127):

$$\begin{aligned}
\xi^\ell(x; \omega) &\rightarrow \xi^{\mathrm{L}}(x; \sigma_{\mathrm{u}}, \nu_{\mathrm{u}}), \quad \sigma_{\mathrm{u}} = \sigma_{\mathrm{u}}^{\mathrm{L}}(\nu_{\mathrm{u}}), \\
\xi^r(x; \omega) &\rightarrow \xi^{\mathrm{R}}(x; \sigma_{\mathrm{u}}, \nu_{\mathrm{u}}), \quad \sigma_{\mathrm{u}} = \sigma_{\mathrm{u}}^{\mathrm{R}}(\nu_{\mathrm{u}}).
\end{aligned} \tag{13.60}$$

Note that this implies that different values of ω apply for ξ^{L} and ξ^{R}, since they refer to different paths. Solving the nonlinear equation $\sigma_{\mathrm{u}} = \overline{V}[\xi^{\mathrm{L}}(\sigma_{\mathrm{u}}, \nu_{\mathrm{u}})]$ for $\mathcal{P}_{\mathrm{u}}^{\mathrm{L}}$, or

$\sigma_{\mathrm{u}} = \overline{V}[\boldsymbol{\xi}^{\mathrm{R}}(\sigma_{\mathrm{u}}, \nu_{\mathrm{u}})]$ for $\mathcal{P}_{\mathrm{u}}^{\mathrm{R}}$, involves the evaluation of the integral average of the operator U. According to the expression (12.128), this is equivalent to finding the zeros of the imaginary component $W_2^{\mathrm{L}}(\sigma_{\mathrm{u}}, \nu_{\mathrm{u}})$, or $W_2^{\mathrm{R}}(\sigma_{\mathrm{u}}, \nu_{\mathrm{u}})$, of the potential energy. For the numerical implementation, this expression is to be preferred over the integral since it does not require addition operations (with associated loss of accuracy), but just involves the boundary values of the solutions obtained. Not only that, but it also provides the quantity we are looking for, replacing the node counter of the real problem, viz. the *alternating ratio* R of the variables ξ and Π, or *alternator* for short,[1] which is real on the solution path:

$$W_2^{\mathrm{L}}(\sigma_{\mathrm{u}}, \nu_{\mathrm{u}}) = \tfrac{1}{2}\left[\xi_1^{\mathrm{L}}\Pi_2^{\mathrm{L}} - \xi_2^{\mathrm{L}}\Pi_1^{\mathrm{L}}\right]_{x_2} = 0$$

$$\Rightarrow \quad R^{\mathrm{L}} \equiv \frac{\xi^{\mathrm{L}}(x_2)}{\Pi^{\mathrm{L}}(x_2)} \overset{(\mathcal{P}_{\mathrm{u}}^{\mathrm{L}})}{=} \frac{\xi_1^{\mathrm{L}}(x_2)}{\Pi_1^{\mathrm{L}}(x_2)} \overset{(\mathcal{P}_{\mathrm{u}}^{\mathrm{L}})}{=} \frac{\xi_2^{\mathrm{L}}(x_2)}{\Pi_2^{\mathrm{L}}(x_2)}, \tag{13.61}$$

$$W_2^{\mathrm{R}}(\sigma_{\mathrm{u}}, \nu_{\mathrm{u}}) = -\tfrac{1}{2}\left[\xi_1^{\mathrm{R}}\Pi_2^{\mathrm{R}} - \xi_2^{\mathrm{R}}\Pi_1^{\mathrm{R}}\right]_{x_1} = 0$$

$$\Rightarrow \quad R^{\mathrm{R}} \equiv \frac{\xi^{\mathrm{R}}(x_1)}{\Pi^{\mathrm{R}}(x_1)} \overset{(\mathcal{P}_{\mathrm{u}}^{\mathrm{R}})}{=} \frac{\xi_1^{\mathrm{R}}(x_1)}{\Pi_1^{\mathrm{R}}(x_1)} \overset{(\mathcal{P}_{\mathrm{u}}^{\mathrm{R}})}{=} \frac{\xi_2^{\mathrm{R}}(x_1)}{\Pi_2^{\mathrm{R}}(x_1)}. \tag{13.62}$$

Both R^{L} and R^{R} exhibit tangent-like dependence on the imaginary part ν of the eigenvalue (see Fig. 13.6). Eigenvalues on the left, resp. right, solution path are approached for $R^{\mathrm{L}} \to 0$, resp. $R^{\mathrm{R}} \to 0$. Hence, the eigenvalue problem amounts to determining the zeros of R^{L} on the solution path $\mathcal{P}_{\mathrm{u}}^{\mathrm{L}}$, or the zeros of R^{R} on the solution path $\mathcal{P}_{\mathrm{u}}^{\mathrm{R}}$. Thus, the Sturm–Liouville analysis of counting internal nodes of real solutions is replaced by the study of the alternating ratio of boundary values of the complex solutions. Since numerical solution of systems of ODEs and determination of the zeros of real functions can be carried out with virtual unlimited accuracy, the study of the alternator along solution paths in the complex ω plane becomes the method of choice for the determination of the spectrum of stationary plasmas. In the next section, we will apply it to a variety of instabilities.

One may create a third path, intermediate between $\mathcal{P}_{\mathrm{u}}^{\mathrm{L}}$ and $\mathcal{P}_{\mathrm{u}}^{\mathrm{R}}$, that treats the two plasma boundaries on an equal footing. This is the *middle path* $\mathcal{P}_{\mathrm{u}}^{\mathrm{M}}$ obtained from the left and right solutions ξ^{ℓ} and ξ^{r} (now with lower case ℓ and r since they refer to values of ω on the middle path) by averaging the two average Doppler shifts:

$$\sigma_{\mathrm{u}} = \overline{V}[\boldsymbol{\xi}^{\mathrm{M}}(\sigma_{\mathrm{u}}, \nu_{\mathrm{u}})] \equiv \frac{\int\left(\boldsymbol{\xi}^{\ell *}\cdot U\boldsymbol{\xi}^{\ell} + \boldsymbol{\xi}^{r *}\cdot U\boldsymbol{\xi}^{r}\right)dV}{\int \rho\left(|\boldsymbol{\xi}^{\ell}|^2 + |\boldsymbol{\xi}^{r}|^2\right)dV} \quad \Rightarrow \quad \mathcal{P}_{\mathrm{u}}^{\mathrm{M}} \equiv \{\sigma_{\mathrm{u}}^{\mathrm{M}}(\nu_{\mathrm{u}})\}. \tag{13.63}$$

---

[1] Note the relationship to the *mechanical impedance* Z, defined in Eq. (10.99) of Section 10.5 [1] to describe the complex frequencies of leaky modes: $Z \equiv R^{-1}$.

One can show that the integral expression, in fact, corresponds to the average Doppler shift of the solution $\boldsymbol{\xi}^M \equiv \boldsymbol{\xi}^\ell \pm i\boldsymbol{\xi}^r$ with associated $\boldsymbol{\Pi}^M \equiv \boldsymbol{\Pi}^\ell \pm i\boldsymbol{\Pi}^r$, of some intermediate BVP (that need not be specified further), since the cross contributions of $\boldsymbol{\xi}^\ell$ and $\boldsymbol{\xi}^r$ to the integrals cancel. From these definitions and the relations (13.59), the boundary values of ξ^M on opposite sides of the integration interval are related by

$$\xi_1^M \equiv \xi_1^\ell \mp \xi_2^r \;\Rightarrow\; \xi_1^M(x_2) = \xi_1^\ell(x_2) = -\xi_1^r(x_1) = \mp\xi_2^M(x_1)\,,$$

$$\xi_2^M \equiv \xi_2^\ell \pm \xi_1^r \;\Rightarrow\; \xi_2^M(x_2) = \xi_2^\ell(x_2) = -\xi_2^r(x_1) = \pm\xi_1^M(x_1)\,, \qquad (13.64)$$

whereas the BCs (13.54) and (13.55) provide the boundary values of Π^M,

$$\Pi_1^M \equiv \Pi_1^\ell \mp \Pi_2^r \;\Rightarrow\; \Pi_1^M(x_2) = \Pi_1^\ell(x_2)\,, \qquad \Pi_1^M(x_1) = \mp\Pi_2^r(x_1) + 1\,,$$

$$\Pi_2^M \equiv \Pi_2^\ell \pm \Pi_1^r \;\Rightarrow\; \Pi_2^M(x_2) = \Pi_2^\ell(x_2) \pm 1\,, \quad \Pi_2^M(x_1) = \pm\Pi_1^r(x_1)\,. \quad (13.65)$$

Exploiting these boundary relations, the imaginary energy expression (12.123) for the path $\mathcal{P}_u^M$, in fact, reduces to the sum of left and right path contributions:

$$W_2^M(\sigma_u, \nu_u) = -\tfrac{1}{2}\Big[\xi_1^M \Pi_2^M - \xi_2^M \Pi_1^M\Big]_{x_1} + \tfrac{1}{2}\Big[\xi_1^M \Pi_2^M - \xi_2^M \Pi_1^M\Big]_{x_2}$$

$$= -\tfrac{1}{2}\Big[\xi_1^r \Pi_2^r - \xi_2^r \Pi_1^r\Big]_{x_1} + \tfrac{1}{2}\Big[\xi_1^\ell \Pi_2^\ell - \xi_2^\ell \Pi_1^\ell\Big]_{x_2} = W_2^r + W_2^\ell = 0\,.$$

$$(13.66)$$

The first line of this expression may be converted into

$$\Big[\xi_1^M(x_2)\mp\xi_2^M(x_1)\Big]\Big[\Pi_2^M(x_2)\mp\Pi_1^M(x_1)\Big] = \Big[\xi_2^M(x_2)\pm\xi_1^M(x_1)\Big]\Big[\Pi_1^M(x_2)\pm\Pi_2^M(x_1)\Big].$$

$$(13.67)$$

Hence, defining the symmetrized boundary displacement,

$$\hat{\xi}_1^M \equiv \tfrac{1}{2}[\xi_1^M(x_2) \mp \xi_2^M(x_1)] = \tfrac{1}{2}[\xi_1^\ell(x_2) - \xi_1^r(x_1)]\,,$$

$$\hat{\xi}_2^M \equiv \tfrac{1}{2}[\xi_2^M(x_2) \pm \xi_1^M(x_1)] = \tfrac{1}{2}[\xi_2^\ell(x_2) - \xi_2^r(x_1)]\,, \qquad (13.68)$$

and the average total pressure perturbation at the boundaries,

$$\hat{\Pi}_1^M \equiv \tfrac{1}{2}[\Pi_1^M(x_2) \pm \Pi_2^M(x_1)] = \tfrac{1}{2}[\Pi_1^\ell(x_2) + \Pi_1^r(x_1)]\,,$$

$$\hat{\Pi}_2^M \equiv \tfrac{1}{2}[\Pi_2^M(x_2) \mp \Pi_1^M(x_1)] = \tfrac{1}{2}[\Pi_2^\ell(x_2) + \Pi_2^r(x_1)]\,, \qquad (13.69)$$

the appropriate *alternator for the middle path* turns out to be

$$R^M \equiv \frac{\hat{\xi}^M}{\hat{\Pi}^M} \overset{(\mathcal{P}_u^M)}{=} \frac{\hat{\xi}_1^M}{\hat{\Pi}_1^M} \overset{(\mathcal{P}_u^M)}{=} \frac{\hat{\xi}_2^M}{\hat{\Pi}_2^M}\,. \qquad (13.70)$$

Note that the computation of all expressions for the middle path requires no operations in addition to the computation of the expressions for ξ^ℓ and ξ^r.

Based on the general theory expounded, the spectrum of flow-driven instabilities is now found by the following two, quantitatively very different, methods:

(1) *Contour plotting* In the following, as a visual aid, we first solve the eigenvalue problem graphically, in the full strip of the

$$
Doppler\ range: \begin{cases} \Omega_{0,\min} \leq \sigma \leq \Omega_{0,\max} \\[2mm] 0 \leq \epsilon \leq \nu \leq \nu_{\mathrm{ub}} \end{cases}, \tag{13.71}
$$

by contour plotting the following functions:

$$
W_2^{\mathrm{L/R/M}}(\sigma, \nu) = 0 \quad \text{\textit{(three solution paths)}},
$$

$$
\xi_{1\mathrm{e}}(\sigma, \nu) = 0 \quad \text{\textit{(real amplitude at the end point)}},
$$

$$
\xi_{2\mathrm{e}}(\sigma, \nu) = 0 \quad \text{\textit{(imaginary amplitude at the end point)}}. \tag{13.72}
$$

Here, ν_{ub} is an estimate of the upper bound of the growth rate as given by one of the integral expressions derived by Barston [24] and Hameiri [209], and $\xi_{1,2\mathrm{e}} \equiv \xi_{1,2}(x_{\mathrm{e}})$ refers to the end point of the integration interval. The first three curves produce the solution paths $\mathcal{P}_{\mathrm{u}}^{\mathrm{L}}$, $\mathcal{P}_{\mathrm{u}}^{\mathrm{R}}$ and $\mathcal{P}_{\mathrm{u}}^{\mathrm{M}}$, intersected at the eigenvalues by the latter two curves. Any pair of curves involving at least one of the latter two suffices to fix an eigenvalue, the intersection of the other curves at the same point can be taken as a check.

(2) *Eigenvalue search on solution paths* The complex eigenvalues are found by first fixing a series of one or more boxes, labeled by k, inside the Doppler range (13.71),

$$
k\ boxes: \begin{cases} \Omega_{0,\min} \leq \sigma_{\min}^{(k)} \leq \sigma_{\mathrm{u}} \leq \sigma_{\max}^{(k)} \leq \Omega_{0,\max} \\[2mm] 0 \leq \epsilon \leq \nu_{\min}^{(k)} \leq \nu_{\mathrm{u}} \leq \nu_{\max}^{(k)} \leq \nu_{\mathrm{ub}} \end{cases}, \tag{13.73}
$$

determining the fragments of the solution path inside these boxes and composing the full path from them,

$$
W_2^{\mathrm{L/R/M}}\!\left(\sigma_{\mathrm{u}}(\nu_{\mathrm{u}})\right) = 0 \quad \Rightarrow \quad \text{\textit{three paths:}}\ \mathcal{P}_{\mathrm{u}}^{\mathrm{L}}, \mathcal{P}_{\mathrm{u}}^{\mathrm{R}}, \mathcal{P}_{\mathrm{u}}^{\mathrm{M}}, \tag{13.74}
$$

and then computing the zeros of the three alternators on their respective solution paths,

$$
R^{\mathrm{L/R/M}}\!\left(\sigma_{\mathrm{u}}(\nu_{\mathrm{u}})\right) = 0 \quad \Rightarrow \quad \text{\textit{EVs:}}\ \{\sigma_n, \nu_n\},\ n = 1, 2, \ldots. \tag{13.75}
$$

The solutions of Eqs. (13.74) and (13.75) may be obtained by the very accurate root finding routines available (e.g., *Numerical Recipes* [385]), provided W_2 has opposite sign on the left and right borders (as is the case when the box is as large as the Doppler range itself) so that the paths intersect the bottom and top borders of the box. As will emerge from the examples of the following Section 13.2, the boxes should also be narrow enough to admit not more than one solution of Eq. (13.74).

Obviously, the first (2D) approach does not lead to very accurate results, and it is also computationally more expensive than the second (1D) method, since it blindly scans the complex ω-plane. However, it turns out to be very useful to sort out the relevant regions, i.e. the number and sizes of the boxes, for the search of eigenvalues by the second method.

Note that the three paths $\mathcal{P}_{\mathrm{u}}^{\mathrm{L}}$, $\mathcal{P}_{\mathrm{u}}^{\mathrm{R}}$, $\mathcal{P}_{\mathrm{u}}^{\mathrm{M}}$ themselves also cross at the eigenvalues, but that does not fix the eigenvalues uniquely since they also cross at intermediate points where $\hat{\Pi}_1 = \hat{\Pi}_2 = 0$, according to Eq. (13.66) (see, e.g., Fig. 13.5). Those points could be considered as the solutions of another eigenvalue problem where, for some or the other reason, one wishes to have the total pressure perturbation vanish at the boundary. Incidentally, this implies that the methods presented here can be applied to a wide variety of physical problems.

13.2 Case study: flow-driven instabilities in diffuse plasmas

We now possess the appropriate tools to explore the spectra of some representative stationary equilibria with respect to the effects of *gravity*, due to a non-constant density $\rho(x)$, of *velocity shear*, due to the variation of the magnitude of a uni-directional flow field $\mathbf{v}(x)$, and of *magnetic shear*, due to the varying direction of a constant amplitude magnetic field $\mathbf{B}(x)$ (see Fig. 13.1). We will choose parameters such that gravity-driven Rayleigh–Taylor instabilities and velocity-driven Kelvin–Helmholtz instabilities may occur simultaneously, but the current-driven magnetic shear will be limited to a stabilizing influence only. (Current-driven instabilities will be investigated in Chapter 14.) For the time being, we restrict the analysis to incompressible perturbations of plasmas described by the following quantities:

$$\mathbf{k}_0 = k_0\,\mathbf{e}_z\,, \qquad g = \text{const}\,, \qquad \rho(x) = \rho_0\,(1 - \delta x) \quad (x \in [0,1])\,,$$

$$\mathbf{v}(x) = v(x)(\sin\theta\,\mathbf{e}_y + \cos\theta\,\mathbf{e}_z)\,, \quad v(x) = v_0 + v_1(x - \tfrac{1}{2}) + v_2\sin\tau(x - \tfrac{1}{2})\,,$$

$$\mathbf{B}(x) = B_0\big[\sin\varphi(x)\mathbf{e}_y + \cos\varphi(x)\mathbf{e}_z\big]\,, \quad \varphi(x) = \varphi_0 + \alpha(x - \tfrac{1}{2})\,. \tag{13.76}$$

Note that the magnetic field is force-free, $\mathbf{j} = \nabla \times \mathbf{B} = \alpha\mathbf{B}$, where α corresponds to the magnetic shear, $\varphi' = \alpha$. As always, we exploit dimensionless variables, based on the three unit scales of length $a \equiv x_2 - x_1$, density ρ_0 and magnetic field strength B_0. Obviously, the profiles and the ten free parameters k_0, δ, g, θ, v_0, v_1, v_2, τ, φ_0, α can be extended to model any other desired distribution.

The equilibrium pressure may also be computed, $p(x) = p_0 - (x - \tfrac{1}{2}\delta x^2)g$, but it does not occur in the spectral equations for incompressible perturbations. For that case, the Alfvén and slow continua coincide so that the equilibria are characterized by the frequencies of the degenerate static continua (MHD) and of the Doppler range (MHD) or flow continua (HD),

$$\left.\begin{array}{c} \omega_{\mathrm{A/S}}(x) \equiv \dfrac{\mathbf{k}_0 \cdot \mathbf{B}}{\sqrt{\rho}} = k_0\,\dfrac{\cos\varphi(x)}{\sqrt{\rho(x)}} \\[2ex] \Omega_0(x) \equiv \mathbf{k}_0 \cdot \mathbf{v} = k_0\cos\theta\,v(x) \end{array}\right\} \;\Rightarrow\; \Omega_{\mathrm{A/S}}^{\pm}(x) \equiv \Omega_0(x) \pm \omega_{\mathrm{A/S}}(x)\,. \tag{13.77}$$

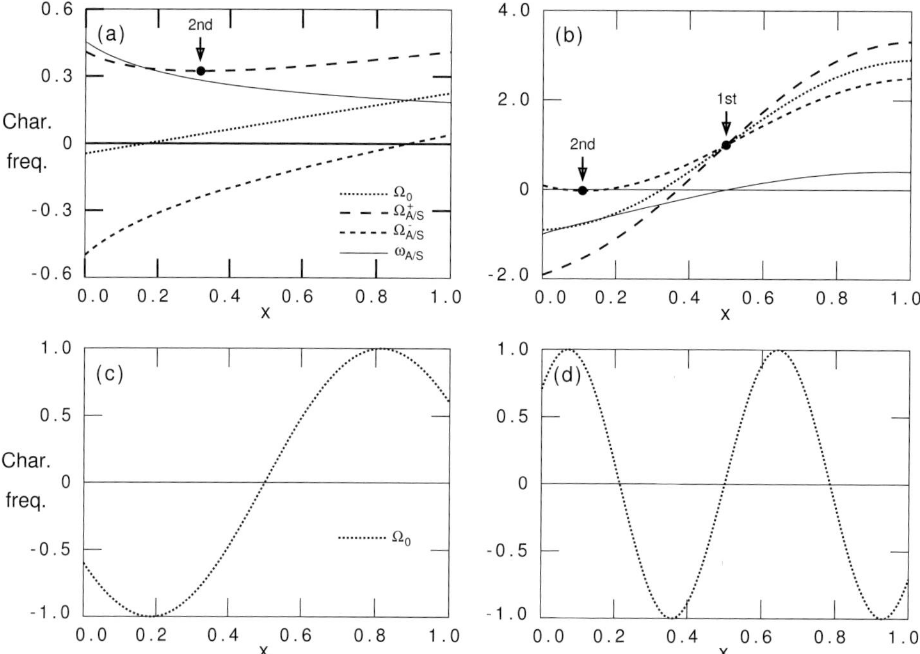

Fig. 13.4 Characteristic frequencies of four equilibria: (a) magnetized gravitating plasma without magnetic shear and with linear density and flow profiles (MHD1), (b) the same with magnetic shear and with sinusoidal flow profile, $\tau = 4$ (MHD2) [arrows indicate potential clustering from criteria with "2nd" or "1st" derivatives]; (c) fluid with sinusoidal flow profile, $\tau = 5$ (KH1), (d) the same, $\tau = 11$ (KH3).

These frequencies are plotted in Fig. 13.4 for two representative MHD examples and three HD ones (KH2 is omitted in the figure), which we here list, together with their parameter values, in the order in which they will be treated:

(a) *MHD1*: Rayleigh–Taylor instabilities of stationary plasma with uni-directional magnetic field and linear velocity profile (Figs. 13.4(a), 13.5–13.7),
$$k_0 = 1, \quad \delta = -5, \quad g = 15, \quad \varphi_0 = -0.35\pi, \quad \alpha = 0,$$
$$\theta = 0.35\pi, \quad v_0 = 0.2, \quad v_1 = 0.6, \quad v_2 = 0, \quad \tau = 0;$$

(b) *KH1/2/3*: Kelvin–Helmholtz instabilities of stationary fluid with sinusoidal velocity profile (Figs. 13.4(c),(d), 13.8–13.10),
$$k_0 = 1, \quad \delta = 0, \quad g = 0, \quad \theta = 0, \quad v_0 = v_1 = 0, \quad v_2 = 1, \quad \tau = 5/8/11;$$

(c) *MHD2*: Rayleigh–Taylor and Kelvin–Helmholtz instabilities of stationary plasma with magnetic shear and sinusoidal velocity profile (Figs. 13.4(b), 13.11–13.13),
$$k_0 = 1, \quad \delta = -5, \quad g = 100, \quad \varphi_0 = 0.5\pi, \quad \alpha = -\pi,$$
$$\theta = 0, \quad v_0 = 1, \quad v_1 = 2, \quad v_2 = 1, \quad \tau = 4.$$

We will refer to the different cases by the stated acronyms. In the MHD examples, we have freely chosen $\rho' > 0$ in order to study unstable "inverted atmospheres" (cf. Case [79] for a corresponding HD example), without addressing the question of what would cause such a density distribution (obviously, not gravity).

13.2.1 Rayleigh–Taylor instabilities of magnetized plasmas

In the first example (MHD1), the pure Rayleigh–Taylor instability mechanism is operating. The velocity field $\mathbf{v}$ and the magnetic field $\mathbf{B}$ are both taken unidirectional, making angles of $+0.35\pi$ and -0.35π with the horizontal wave vector. This yields forward continua that are well separated from both the Doppler range and the backward continua. This creates the "space" for an infinite sequence of gravitational (g) modes clustering at the minimum $x_0 \approx 0.318$ (which is also the edge) of the degenerate forward Alfvén/slow continua, where $(\Omega^+_{A/S})_{\min} \approx 0.323$ (indicated by "2nd" in Fig. 13.4(a)). The condition for clustering is derived from the spectral equation (13.27) for incompressible plasmas, in complete analogy to the derivation for static compressible plasmas presented in Section 7.4.4[1]. This yields clustering of the gravitational modes at the *real* frequency $\omega = \Omega^{\pm}_{A/S}(x_0)$,

$$\text{where } \Omega^{\pm'}_{A/S} = 0, \quad \text{if} \quad \begin{cases} \text{either} \quad k_0^2 \rho' g < \pm\tfrac{1}{4}\rho\,\omega_{A/S}(\Omega''_0 \pm \omega''_{A/S}) < 0, \\[2mm] \text{or} \quad k_0^2 \rho' g > \pm\tfrac{1}{4}\rho\,\omega_{A/S}(\Omega''_0 \pm \omega''_{A/S}) > 0. \end{cases}$$

$$\tag{13.78}$$

Of course, the $\pm$s in these expressions correspond to the different cases of forward and backward continua. The second condition is satisfied for the present example. The general conditions for clustering in stationary compressible plasmas (at the separate Alfvén and slow continua) have been derived by van der Holst *et al.* [455]. Note that one does *not* obtain the above criteria in the limit $\gamma \to \infty$ from those conditions (different factors $k_\perp^2$ and $k_\parallel^2$ occur) because the above ones are the result of the confluence of six of the eight $N = 0$ and $D = 0$ singularities of the general spectral equation (13.9) in the incompressible limit.

The cluster point $(\Omega^+_{A/S})_{\min}$ is *necessarily stable since it lies outside the Doppler range* $[-0.0454, 0.2270]$. Consequently, only a limited number ($n \leq 6$) of the cluster sequence is Rayleigh–Taylor unstable, as shown by the method 1 contours in Fig. 13.5. The three paths oscillate around the six eigenvalues (and the intermediate points mentioned above). As expected, the middle path provides the smoothest curve through them, so that it is the most attractive candidate for accurate computation of the eigenvalues. Along this path, the transition from unstable to stable occurs at $\sigma_0^M \approx 0.127$, i.e. far to the right of the edge $(\Omega^-_{A/S})_{\max} \approx 0.0417$ of the backward continua (indicated by the black piece of the σ-axis in the figure).

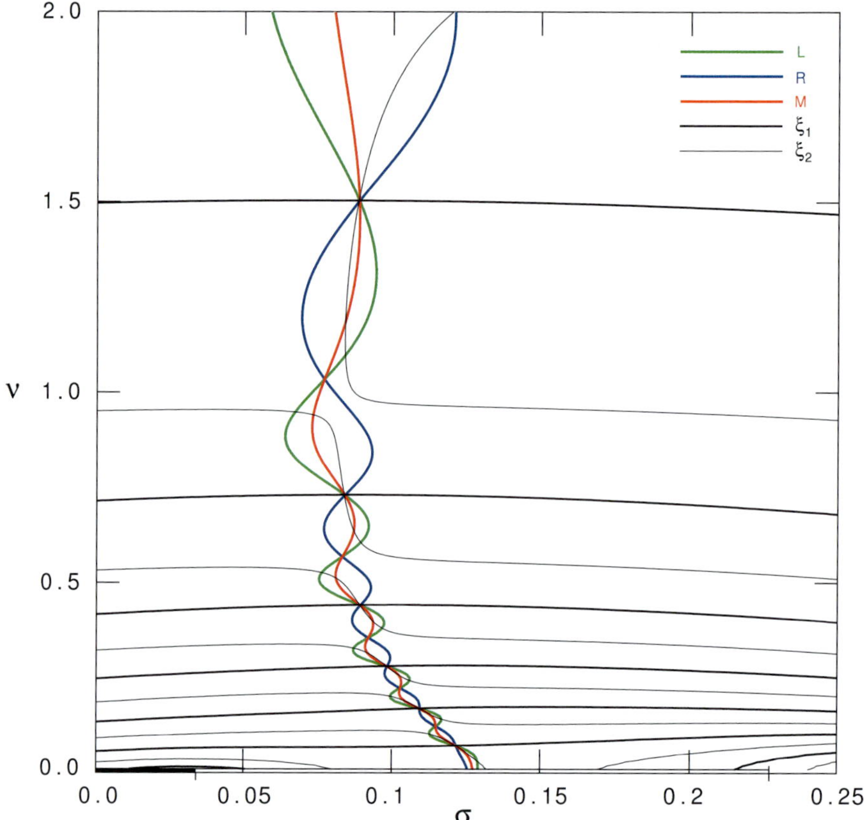

Fig. 13.5 Contours in the complex ω plane where the imaginary component of the energy vanishes, $W_2^{\mathrm{L/R/M}} = 0$, determining the left, right and middle paths, and contours of vanishing boundary values of the displacement vector, $\xi_{1\mathrm{e}} = 0$ and $\xi_{2\mathrm{e}} = 0$. Eigenvalues are located at the intersections of all curves (MHD1).

It should be stressed though that this transition does *not* correspond to a marginally stable state of this equilibrium since there is no eigenvalue at this point.

The alternator $R(\nu)$ along the middle path, obtained with method 2, is shown in Fig. 13.6(a). Whereas the real and imaginary components $\xi_{1\mathrm{e}}$ and $\xi_{2\mathrm{e}}$ of the displacement vector do not oscillate in step in the interior points of the plasma, the alternator keeps track of when they change sign simultaneously *at the plasma boundary,* so that $|\xi_\mathrm{e}| = 0$, which is necessary to have an eigenvalue. This can only happen for values of ω on the solution path. Because of the symmetry properties of the eigenvalue problem for the middle path, the absolute value of the total pressure perturbation also vanishes at points of that path, $|\Pi_\mathrm{e}| = 0$, as shown in Fig. 13.6(b). Of course, for a system of non-singular ODEs, those points cannot coincide with the former ones. In fact, the zeros of $|\xi_\mathrm{e}|$ and $|\Pi_\mathrm{e}|$ nicely alternate, as shown in

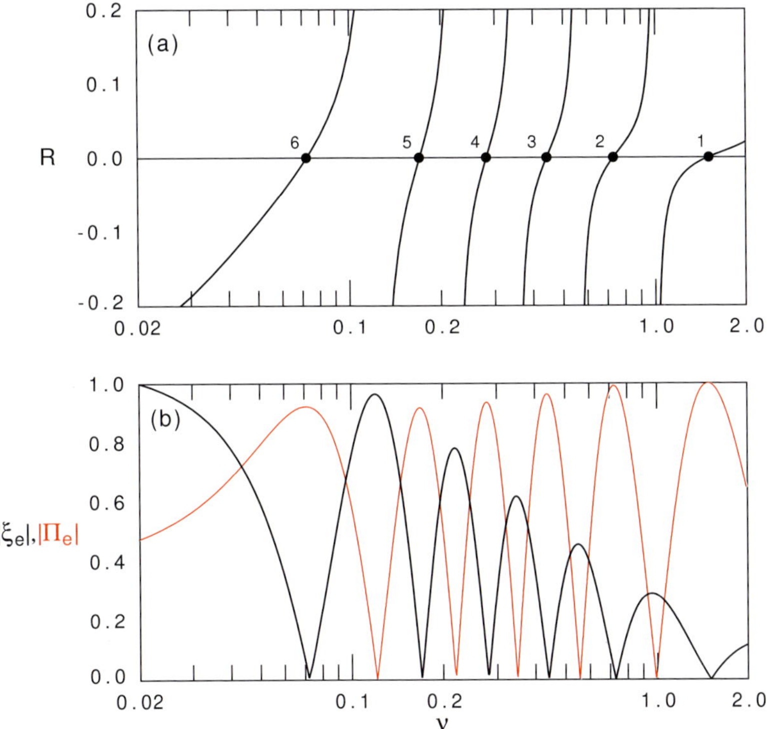

Fig. 13.6 (a) Alternator $R(\nu)$ and (b) absolute values $|\xi_e(\nu)|$ and $|\Pi_e(\nu)|$ along the middle path (MHD1). The eigenvalues occur for $R = 0$, where $|\xi_e| = 0$.

Fig. 13.6(b). Hence the appropriate name "alternator" for R. It turns out to be a perfect counter, where every new zero $|\Pi_e| = 0$ initiates a new branch of R, running from $+\infty$ to $-\infty$, so that the zeros $|\xi_e| = 0$ may be computed with very high precision.

The resulting full spectrum is shown in Fig. 13.7. The six unstable eigenvalues (and their damped complex conjugate counterparts) on the middle path join both the Sturmian sequence of stable g-modes, clustering at the mentioned minimum $(\Omega_{A/S}^{+})_{\min}$ of the forward continua, and another, isolated, stable g-mode on the anti-Sturmian part of the σ-axis, bounded by the edge of the backward continua. This is in perfect agreement with the oscillation theorem $\mathcal{R}$ of Section 13.1.3. The number of nodes n of the eigenfunctions $\xi(x)$, which are effectively real on the σ-axis, turns out to be a direct continuation (at $n = 7$ and $n = 7'$) of the label n of the unstable modes. Whereas along the real axis the number n counts the number of *internal nodes* of the real eigenfunctions, along the solution path $\mathcal{P}_u^M$ the label n of the alternator counts the number of changes of sign of both the real and the imaginary parts of the eigenfunctions ξ_{1e} or ξ_{2e} *at the boundary*. In

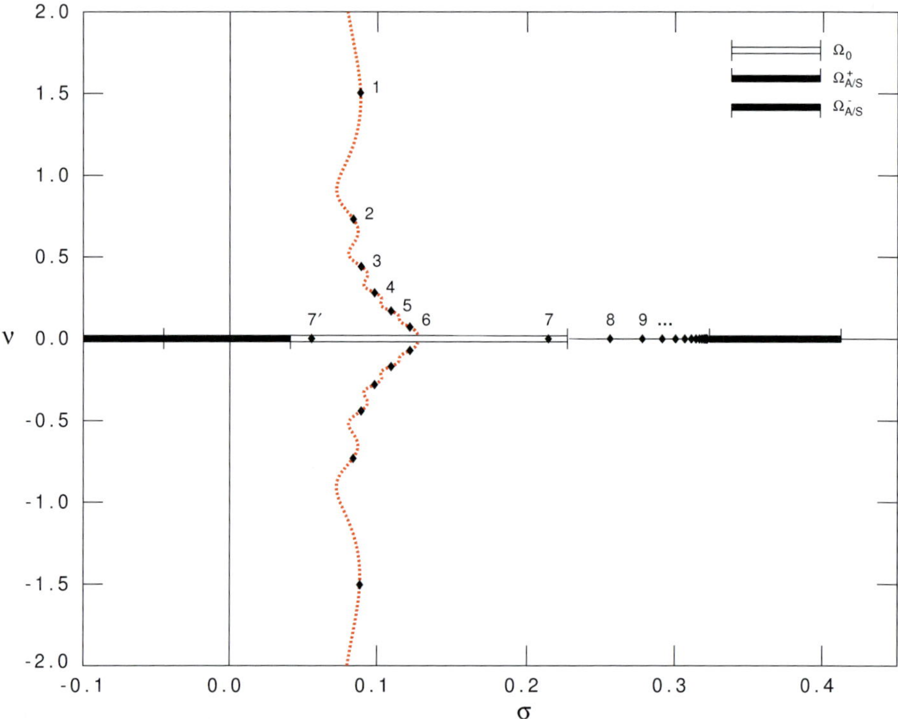

Fig. 13.7 Spectrum of discrete unstable modes, and their damped complex conjugates, on the middle path joins the stable oscillations on the real axis (MHD1). The Sturmian branch of the stable gravitational modes clusters towards the lower end point of the forward continua, the anti-Sturmian branch contains only one mode (indicated by $7'$).

this way, we have constructed a counter for complex eigenfunctions that does not require counting of the internal nodes of either the real or the imaginary part of the eigenfunctions. It appears that we have obtained a genuine generalization for complex eigenvalues of oscillation theorem $\mathcal{R}$ of Section 13.1.3. However, some serious hurdles need to be overcome yet.

13.2.2 Kelvin–Helmholtz instabilities of ordinary fluids

All flow-driven instabilities are complex but some are more complex, as we will see in this sub-section. In particular, the Kelvin–Helmholtz instability in the simplest configurations gives rise to surprisingly complex solution paths in the ω plane. We will first study them in HD and then in MHD.

We have encountered the Kelvin–Helmholtz instability in Section 13.1.2 when discussing instabilities of interface plasmas consisting of two layers with different densities, velocities, and magnetic fields. The resulting expression (13.34) for the growth rate of instabilities appears to be quite "general", but it could falsely be taken to suggest that the Kelvin–Helmholtz instability just requires a velocity gradient to drive it. In a streaming diffuse fluid or plasma, the conditions for instability are much more subtle than that, as could have been expected from the fact that there is no discrete spectrum of stable modes that could be studied perturbatively to find out about transitions to instability. In a sense, the Kelvin–Helmholtz instability is a perturbation of the flow continua $\Omega_{\mathrm{H}i}$, given by Eq. (13.40).

To appreciate the subtleties, let us return to a simple incompressible fluid layer with a uni-directional velocity of arbitrary amplitude, $\mathbf{v} = v(x)\mathbf{e}_z$. In either of the pertinent spectral equations (13.42) for the stream function χ, or for the fluid displacement ξ, the only expression that occurs (in addition to the factor k_0) is precisely the one that determines the flow continua,

$$\tilde{\omega}(x) \equiv \omega - \Omega_0(x) = \omega - k_0 v(x). \tag{13.79}$$

For a linearly increasing velocity profile, the spectral equation for χ immediately shows that the fluid is stable since the term $\tilde{\omega}'' = -\Omega_0'' = -k_0 v''$ vanishes so that only a continuous spectrum $\tilde{\omega} = 0$ remains (actually two, on top of each other, viz. the mentioned flow continua, as is evident from the spectral equation for ξ). Hence, it is necessary for instability to have a velocity profile with $v'' \neq 0$ over some region. This appears to imply that only local profile conditions matter, but we already made the parenthetical remark "in addition to the factor k_0". That factor cannot be scaled out of the problem since the length scale is fixed by the thickness $a \equiv x_2 - x_1$ of the fluid layer, determining the length scale of the inhomogeneities. This observation is crucial because the solutions of the spectral equation (13.42) become exponentially dominant for large k_0 (small wavelength), so that it is impossible to satisfy the boundary conditions then. Hence, in addition to the local criteria that are discussed below, the Kelvin–Helmholtz instability only occurs if k_0 is small enough, so that it is a *global, long wavelength, instability.*

The local conditions on the velocity profile are obtained as follows (see Drazin and Reid [124]). Depart from the spectral equation (13.42) for χ, multiply by χ^*, integrate over the plasma volume, transform the second derivative of χ by integration by parts and cancel the boundary terms by satisfying the BCs:

$$\chi'' - k_0^2\chi - \frac{\tilde{\omega}''}{\tilde{\omega}}\chi = 0 \quad \Rightarrow \quad -\int \frac{\tilde{\omega}''}{\tilde{\omega}}|\chi|^2\,dx = \int \left(|\chi'|^2 + k_0^2|\chi|^2\right)dx$$

$$\Rightarrow \quad \begin{cases} \text{Re:} \quad \displaystyle\int \frac{(\sigma - \Omega_0)\Omega_0''}{(\sigma - \Omega_0)^2 + \nu^2}\,dx = \int \left(|\chi'|^2 + k_0^2|\chi|^2\right)dx\,, \\[3mm] \text{Im:} \quad \displaystyle -\nu\int \frac{\Omega_0''}{(\sigma - \Omega_0)^2 + \nu^2}\,dx = 0\,. \end{cases} \tag{13.80}$$

From the imaginary part it follows that it is necessary for instability ($\nu \neq 0$) that the velocity profile should have an inflexion point, $v''(x_i) = 0$, somewhere in the flow (*Rayleigh's inflexion point theorem*). A sharper criterion follows by adding the two expressions after multiplication of the second one by $[\sigma - \Omega(x_i)]/\nu$:

$$-\int \frac{(\Omega_0 - \Omega_0(x_i))\Omega_0''}{(\sigma - \Omega_0)^2 + \nu^2}\, dx = \int \left(|\chi'|^2 + k_0^2|\chi|^2\right) dx > 0. \tag{13.81}$$

This implies that it is necessary for instability that the velocity profile should satisfy $[v - v(x_i)]v'' < 0$ somewhere in the flow (*Fjørtoft's theorem*).

In our second example (KH1/2/3), we have chosen the velocity profile to be sinusoidal with an inflexion point in the middle of the fluid, so that Fjørtoft's theorem is satisfied (nearly) everywhere:

$$v(x) = v_2 \sin \tau(x - \tfrac{1}{2}) \quad \Rightarrow \quad [v - v(x_i)]v'' = -\tau^2 v_2^2 \sin^2 \tau(x - \tfrac{1}{2}) \leq 0. \tag{13.82}$$

This does not imply that the fluid is unstable for every value of τ since we still have to consider the global boundary value problem, involving k_0. This nearly automatically implies numerical analysis, but we can obtain analytical stability thresholds for this example because of the highly symmetrical velocity profile chosen (Fig. 13.4(c) and (d)). As a result, these thresholds occur for $\omega = 0$, when $\widetilde{\omega}''/\widetilde{\omega} \to v''/v = -\tau^2$, so that marginal solutions are obtained for $\tau = \tau_n$:

$$\chi'' - (k_0^2 - \tau^2)\chi = 0 \quad \Rightarrow \quad \chi = A \sin n\pi x, \quad \tau_n \equiv \sqrt{k_0^2 + n^2\pi^2} \ (n = 1, 2, \cdots). \tag{13.83}$$

For our example, with $k_0 = 1$, this implies that instability sets in for $\tau > \tau_1 \approx 3.297$, whereas for $\tau > \tau_2 \approx 6.362$ and for $\tau > \tau_3 \approx 9.478$, etc., new unstable windows appear in the stability diagram, as depicted in Fig. 13.9. The three cases $\tau = 5$ (KH1), $\tau = 8$ (KH2) and $\tau = 11$ (KH3) have been chosen to illustrate how the Kelvin–Helmholtz instability gives rise to qualitatively different spectra in the different windows, as shown in the top frames of the figure.

Before we discuss those differences, we have to consider a striking new feature in our analysis of the instabilities of stationary fluids, as manifested by the contour plots of the solution paths of Figs. 13.8: for specified (left/right/middle) BVP, *the solution path no longer consists of a single curve, but it splits up!* Previously (Section 13.2.1), the solution path consisted of a single curve in the Doppler range running from $\nu = +\infty$ to $\nu = 0$ (and continuing to $\nu = -\infty$ for the complex conjugate part), so that a single point on the real σ-axis could be considered to represent the stability transition. With the Doppler range now corresponding to continuous spectra (the flow continua), this changes dramatically: there is still a continuous curve running from $\nu = +\infty$ to $\nu = \epsilon$ (and a corresponding one from $\nu = -\epsilon$ to $\nu = -\infty$), as there should be since the path equations (13.74) have at

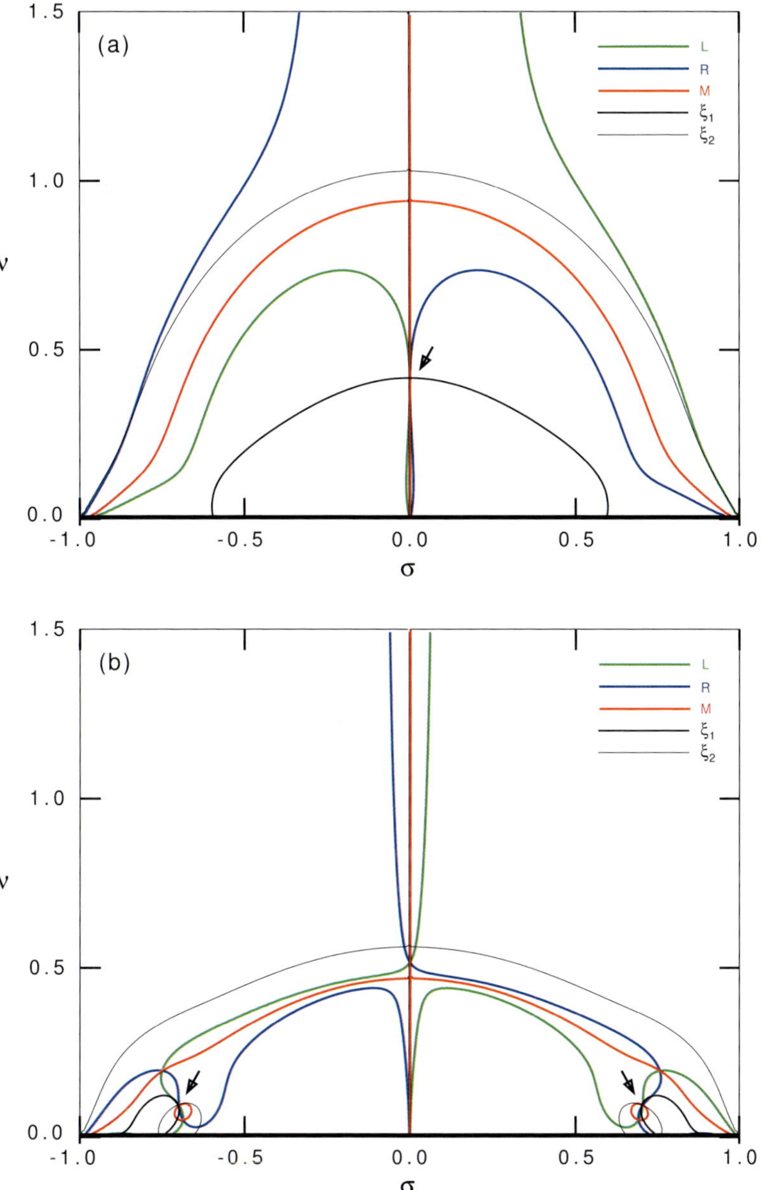

Fig. 13.8 Contours in the complex ω plane of $W_2^{\mathrm{L/R/M}} = 0$, determining the left, right and middle path, and of $\xi_{1\mathrm{e}} = 0$ and $\xi_{2\mathrm{e}} = 0$, determining the eigenvalues (indicated by the arrows), for fluid with sinusoidal flow profile, (a) $\tau = 5$ (KH1), (b) $\tau = 8$ (KH2).

least one solution for each value of ν, but there are now additional loops between points of the continuous spectrum.

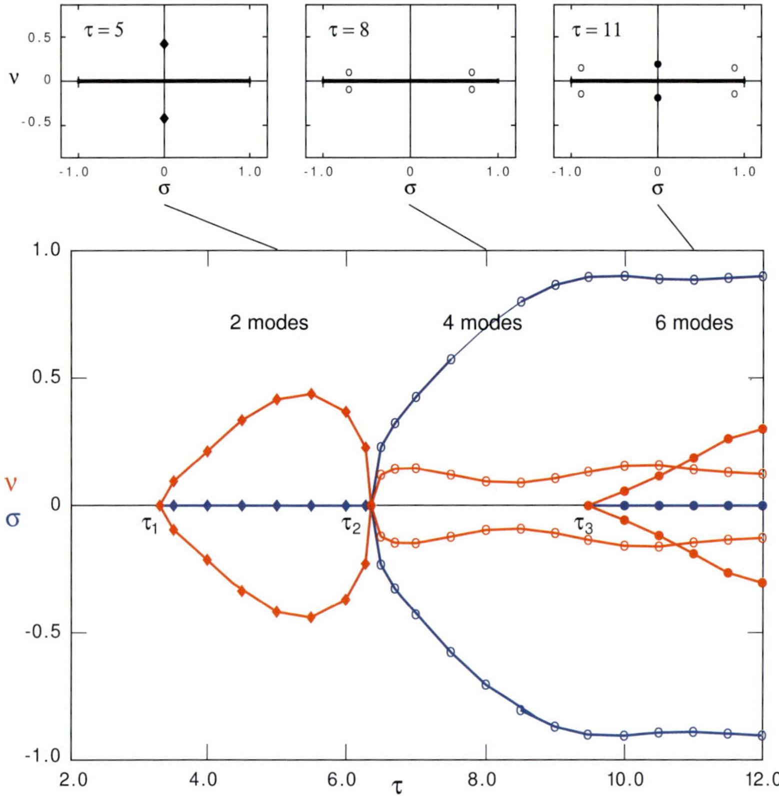

Fig. 13.9 Real (σ) and imaginary (ν) part of the eigenfrequency of the Kelvin–Helmholtz instability as a function of the parameter τ measuring the period of the sinusoidal flow profile. Every time τ increases beyond a critical $\tau_n \equiv [k_0^2 + (n\pi)^2]^{1/2}$, two more complex eigenvalues emerge from the origin $\omega = 0$. Insets above the figure show the spectra for $\tau = 5$ (KH1), $\tau = 8$ (KH2) and $\tau = 11$ (KH3).

In the example of Fig. 13.8(a) (KH1, $\tau_1 < \tau < \tau_2$), the eigenvalue (indicated by the arrow) is situated on the straight line $\sigma = 0$ through the origin for the middle solution, but it is situated on a loop between points of the continuum for the left and right solutions. This should not be taken to imply that the middle path represents a superior solution, since its symmetry is just a consequence of the symmetry of the first set of modes for these equilibria. When the value of τ is increased, as in Fig. 13.8(b) (KH2, $\tau_2 < \tau < \tau_3$), instead of the symmetric ($\sigma = 0$) pair of complex eigenvalues, four asymmetric ($\sigma \neq 0$) ones appear (see Fig. 13.9). In the contour plots, those eigenvalues are situated on wiggling loops of the left and right paths, whereas the middle path even ejects two small separate loops (not connected to the continuous spectrum) to host the eigenvalues.

Clearly, the occurrence of multiple solutions to the path equations (13.74) is a major complication for the systematic study of the spectra of stationary fluids and plasmas. This complication can not be avoided since the equation determining the path $\sigma_u(\nu_u)$ is a highly implicit and nonlinear equation, depending on the details of the equilibrium through the substitution of the solution $\xi(x; \sigma_u + i\nu_u)$ of the spectral equation. Moreover, the strange behavior of the solution paths is not a mathematical artifact of the method chosen to solve the spectral problem, but it represents the actual physics of the extended model II problem for a plasma with an adjacent vacuum layer of arbitrary thickness on the top (left solutions), on the bottom (right solutions), or both on top and bottom (middle solutions). Hence, we have to face this complication and study the topology of the solution paths.

Topology of the solution paths First of all, we need to introduce appropriate terminology for the occurrence of multiple solution paths. We will continue to use the term *path* for the whole system of *sub-paths* representing the solution of the extended model II problem, on some (not all!) of which the wanted eigenvalues of the restricted model I problem are found. *Sub-paths* will represent *disconnected* branches of the full solution path. For the example of the Kelvin–Helmholtz instability for $\tau = 8$, depicted in Fig. 13.8(b), it is obvious that the left solution path (in green) consists of, at least, two sub-paths, viz. (1) the single curve going down from $\sigma > 0, \nu_u = +\infty$ to the point $\sigma \approx -0.72, \nu = 0$ of the flow continuum, and (2) another single curve connecting the two points $\sigma = \nu = 0$ and $\sigma \approx 0.98, \nu = 0$ of the flow continuum. (We restrict the discussion to the upper, unstable, part of the ω plane.) Similarly, the right solution path (in blue) consists of two sub-paths, mirroring the left ones (replacing σ by $-\sigma$).

Next, what about the *fragments* of a sub-path, corresponding to the different boxes that were defined in Eq. (13.73) to guarantee single solutions of the path equations (13.74) inside the box? For example, the second sub-path of the left solution of Fig. 13.8(b) just discussed consists of four fragments, viz. the parts of the curve (a) going up (for $0 < \sigma < 0.11$), (b) going down ($0.11 < \sigma < 0.65$), (c) going up ($0.65 < \sigma < 0.76$), and (d) going down again ($0.76 < \sigma < 0.98$).

Recall why the boxes (13.73) were introduced in the algorithm for the eigenvalue search of Section 13.1.4. The path equation is solved for a range of values of ν_u by determining the zeros of $W_2(\sigma_u, \nu_u)$, which gives one or more solutions $\sigma_u = \sigma_u(\nu_u)$. In the case of multiple solutions, those solutions are to be associated with different sub-paths or with different fragments of the same sub-path, each inside its own box. Note that the distinction between these two cannot be determined from the local solution of the path equation, it depends on the overall topology of the full solution path, as revealed by the contour plots of $W_2(\sigma_u, \nu_u) = 0$. The fragments of a sub-path are separated by the extrema $\nu_u'(\sigma_u) = 0$ of the inverse relation $\nu_u = \nu_u(\sigma_u)$. Those extrema represent points on the path of minimum or maximum distance away from the continua. One might have expected that, at those points, some drastic change would occur, like change of the dependence of the alternator R on arc length from increasing to decreasing, or vice versa. However, anticipating the monotonicity of the alternator along the solution path (that will finally be proved in Section 13.2.4), the numerical results clearly point to the opposite: the alternator does not change its sense of monotonicity at these points, but it just remains monotonic as a function of arc length along the sub-path. In other words, the fragmentation demanded by the numerical algorithm is just a way of solving the problem, it does not reflect a change of monotonicity. Hence, after the problem has been solved, one should paste the fragments together to obtain a genuine sub-path: in the example of Fig 13.8(b), the four fragments labeled (a)–(d) constitute a single sub-path that should be considered as a whole.

From the numerical evidence collected so far (restricting the discussion to the upper part of the w plane), sub-paths are separated by the following special points:

– the point at infinity, $\nu_u = +\infty$, where all simple paths start,
– points $\nu_u = 0$ of the real axis, or of the continua, where all simple paths end,
– intersections of the path with itself.

After the fragments have been assembled into sub-paths, four kinds of sub-path may then be distinguished:

– a single curve starting at $\nu_u = +\infty$ and ending at a point $\nu_u = 0$ of the real axis, or of the continua (again, restricting the discussion to the upper part of the w-plane),
– curves (with $\nu_u \neq 0$) connecting two points $\nu_u = 0$ of the continua,
– separate closed loops,
– separate sub-paths resulting from breakup of one of the above three types of curves by an intersection of the path with itself.

The latter three kinds of sub-path may be absent. In that case, the first sub-path (which is always there) constitutes the full path.

Consequently, the middle path depicted in Fig. 13.8(b) consists of six sub-paths: (1) the ν-axis from $\nu = \infty$ to $\nu \approx 0.47$, (2) the ν-axis from $\nu \approx 0.47$ to $\nu = 0$, (3) the loop connecting the point $\sigma \approx -0.97$ of the flow continuum to the point $\sigma = 0$, $\nu \approx 0.47$ at the apex, (4) the loop connecting the latter apex to the point $\sigma \approx 0.97$ of the flow continuum, (5) an isolated loop containing the left eigenvalue ($\sigma \approx -0.7024$, $\nu \approx 0.0952$), (6) an isolated loop containing the right eigenvalue ($\sigma \approx 0.7024$, $\nu \approx 0.0952$). This example represents a very intricate limit because the path intersects with itself, causing the breakup into the first four sub-paths, and it splits off the two isolated loops containing the eigenvalues.

The intersection of the middle path with itself not only splits the degenerate sub-path along the ν-axis into the two sub-paths (1) and (2), and the loop connecting the two points $\sigma \approx \pm 0.97$ of the flow continuum into the sub-paths (3) and (4), but it also causes the monotonicity along these two curves to change sign at the intersection point (again contrary to what one might have expected!). Consequently, the composite curves (1)–(2) and (3)–(4) cannot be considered as sub-paths, but the composite curves (1)–(3) and (2)–(4), or alternatively the composites (1)–(4) and (3)–(2), might be considered as sub-paths in the sense we have defined (in that the alternator is monotonic along the full composite curve). Hence, the first four sub-paths of the middle path solution could be considered to represent only two (properly chosen) sub-paths. Significantly, those two composite sub-paths have the same monotonicity properties as the two left and right solution sub-paths.

Finally, again anticipating the monotonicity of the alternator along the solution path: the isolated loops (5) and (6) of the middle path not only contain a zero of $|\xi|$ (the eigenvalue) but also a zero of $|\Pi|$, so that the alternator R maintains its monotonicity as a function of arc length along each loop by changing sign twice (like the tangent over a period π).

We now return to the discussion of the calculation of the Kelvin–Helmholtz spectra of Figs. 13.8 and 13.9. The corresponding eigenfunctions are shown in Fig. 13.10. For maximum readability of the plots, they were renormalized as

$$\bar{\xi}_{1,2} \equiv \xi_{1,2}/a_0 \,, \quad \bar{\eta}_1 \equiv \eta_1/a_1 \,, \quad \bar{\eta}_2 \equiv \eta_2/a_2 \,, \quad \bar{\Pi}_1 \equiv \Pi_1/b_1 \,, \quad \bar{\Pi}_2 \equiv \Pi_2/b_2 \,,$$

$$a_0 \equiv (|\xi_1|, |\xi_2|)_{\max} \,, \quad a_1 \equiv (|\xi_1|, |\eta_1|)_{\max} \,, \quad a_2 \equiv (|\xi_2|, |\eta_2|)_{\max} \,,$$

$$b_1 \equiv |\Pi_1|_{\max} \,, \quad b_2 \equiv |\Pi_2|_{\max} \,, \tag{13.84}$$

with the scale factors $a_{01} \equiv a_0/a_1$, $a_{21} \equiv a_2/a_1$, b_1 and b_2 indicated. Thus, the relative proportion of the normal components ξ_1 and ξ_2 is maintained, but the much larger tangential components η_1 and η_2 are scaled down to fit the same frame.

The eigenfunctions of Fig. 13.10 clearly show the two aspects of the Kelvin–Helmholtz instability: they are global, not confined to a particular part of the plasma, but they do require the proximity of the flow continua in order to develop large, rather local, variations of the amplitude of the displacement ξ. The corresponding plots of the perturbed velocity $v_1 \sim \tilde{\omega}\xi$ (not shown) do not exhibit such large variations because of the smallness of the factors $\tilde{\sigma}$ and ν close to the continua. Note the qualitative difference between the symmetric/anti-symmetric eigenfunctions depicted in Figs. 13.10(a) and (c), for $\tau = 5$ (KH1) and $\tau = 11$ (KH3), and the asymmetric eigenfunctions depicted in Figs. 13.10(b) and (d), for $\tau = 8$ (KH2) and $\tau = 11$ (KH3). (Only the eigenfunctions for $\sigma \geq 0$ are shown.) The symmetric eigenfunctions correspond to purely exponentially growing modes, with "smoothed jumps" precisely at the positions where $\sigma = 0$ and $\Omega_0(x) = 0$, according to Fig. 13.4(a) and (b), and the asymmetric eigenfunctions correspond

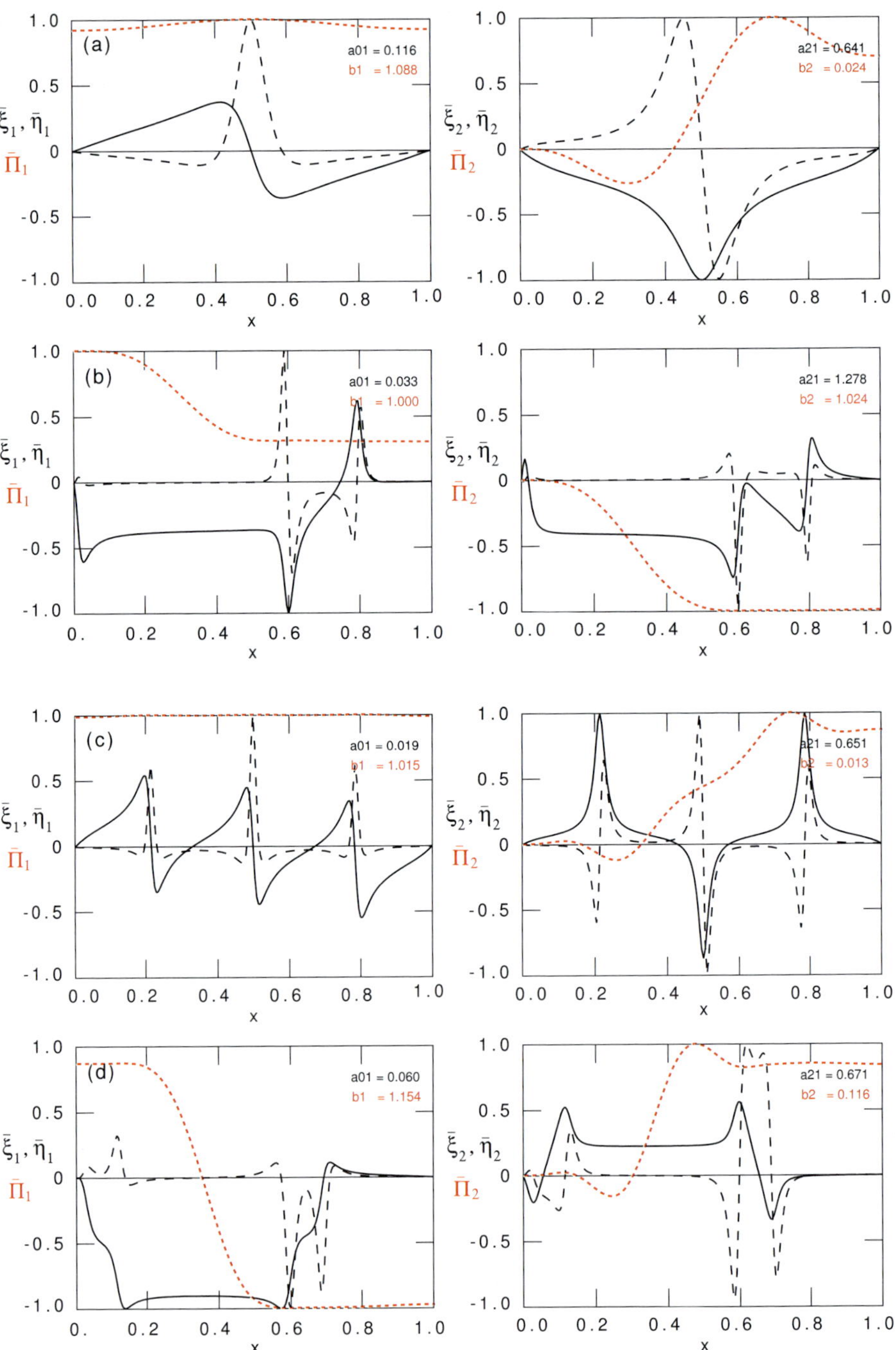

Fig. 13.10 Eigenfunctions of the Kelvin–Helmholtz instability for sinusoidal flow for (a) $\tau = 5$ (purely exponential mode, KH1), (b) $\tau = 8$ (overstability, KH2), (c) $\tau = 11$ (purely exponential mode, KH3) and (d) $\tau = 11$ (overstability, KH3).

to overstable modes, where "smoothed jumps" occur at the asymmetric positions $\sigma - \Omega_0(x) = 0$.

13.2.3 Gravito-MHD instabilities of stationary plasmas

We now superpose the Rayleigh–Taylor and Kelvin–Helmholtz mechanisms in a plasma with a sheared magnetic field. In the first example (MHD1, Fig. 13.4(a)), a minimum of the forward Alfvén/slow continua gave rise to Rayleigh–Taylor instabilities connected with a *stable* cluster sequence concentrated at the lower edge of those continua. We now choose the parameters of the equilibria (13.76) such that a minimum of the *backward* continua $\Omega^-_{\mathrm{A/S}}$ occurs inside the Doppler range (MHD2, Fig. 13.4(b)), so that the full cluster sequence of Rayleigh–Taylor modes becomes unstable. Since the backward continua are also embedded in the much larger range of the forward continua, a rather intricate spectral structure arises.

As in the HD examples, the flow profile is chosen sinusoidal so that Kelvin–Helmholtz instabilities appear as well. The minimum $(\Omega^-_{\mathrm{A/S}})_{\min} \approx -0.0242$ occurs at $x_0 \approx 0.112$ (indicated by "2nd" in Fig. 13.4(b)), so that the Rayleigh–Taylor modes are expected to be localized on the left part of the plasma interval, whereas Fjørtoft's theorem (13.81) suggests that the Kelvin–Helmholtz modes should be concentrated on the right. In fact, the paths of unstable modes shown in Figs 13.11 and 13.12, and the eigenfunctions shown in Fig 13.13, support these expectations. However, we first have to address some features that are less intuitive.

First, in the presence of gravity and density stratification, there are more appropriate criteria than the ones of Rayleigh and Fjørtoft. These involve the *Richardson number J*, which is a measure of the gravitational potential energy associated with the differences of the density with respect to the kinetic energy residing in the relative motions of the different fluid layers. By means of an estimate of the energy change due to the displacement of a fluid element (see Chandrasekhar [84], p. 491), or by means of the construction of a special quadratic form (see Drazin and Reid [124], p. 327), one can show that it is *sufficient for HD stability* that

$$J \equiv -\frac{\rho' g}{\rho v'^2} \geq \tfrac{1}{4}, \quad \text{or} \quad \rho' g + \tfrac{1}{4}\rho v'^2 \leq 0 \quad \text{(everywhere)}. \tag{13.85}$$

Vice versa, a necessary criterion for instability is that $J < \tfrac{1}{4}$ somewhere in the fluid. These criteria are due to Howard [233]. Hence, a fluid with a gravitationally stable density stratification ($\rho' \leq 0$) may still be unstable if the shear flow is large enough. For the present example, although the criterion (13.85) is violated everywhere, this does not automatically imply instability since Howard's criterion is not valid for magnetized plasmas, where magnetic shear may overcome both the gravitational and the shear flow instability drives.

Second, concerning MHD stability, except for the mentioned extremum $\Omega_{\mathrm{A/S}}^-$, one also notices an additional special point in Fig. 13.4(b) (indicated by "1st") that lies precisely in the middle of the plasma ($x = 0.5$), where $\omega_{\mathrm{A/S}} = 0$ and, consequently, the two continuum frequencies $\Omega_{\mathrm{A/S}}^+$ and $\Omega_{\mathrm{A/S}}^-$ degenerate into the Doppler frequency Ω_0. For this frequency, the spectral equation (13.27) for incompressible plasmas develops a quadratic singularity so that, again, a cluster sequence could appear there, of course governed by an entirely different cluster criterion than the one given by Eq. (13.78). Since this possibility is not realized for the present example, and since its discussion really interrupts the flow of the argument, we put it as a detour in small print here.

▷ **Cluster criteria and local gravitational interchanges** If a resonant surface $\mathbf{k}_0 \cdot \mathbf{B} = 0$ occurs at some point x_{res} in the plasma, i.e. $\omega_{\mathrm{A}}(x_{\mathrm{res}}) = 0$, the frequencies of the forward and backward continua coincide with the Doppler frequency. Hence, at the *real* frequency $\omega = \Omega_{\mathrm{A/S}}^+ = \Omega_{\mathrm{A/S}}^- = \Omega_0$, i.e. in the limit $\nu \to 0$, the singular factor of the spectral equation (13.27) may be expanded as

$$\tilde{\omega}^2 - \omega_{\mathrm{A}}^2 \equiv (\omega - \Omega_0)^2 - \omega_{\mathrm{A}}^2 \approx (\Omega_0'^2 - \omega_{\mathrm{A}}'^2)s^2 \,, \qquad s \equiv x - x_{\mathrm{res}} \,, \tag{13.86}$$

so that we obtain an equation similar to Eq. (7.197)[1], governing local gravitational interchange modes:

$$\frac{d}{ds}\left[s^2 (1 + \cdots) \frac{d\xi}{ds} \right] - q_0 (1 + \cdots)\xi = 0\,, \qquad q_0 \equiv \left(\frac{k_0^2 \rho' g}{\rho(\Omega_0'^2 - \omega_{\mathrm{A}}'^2)} \right)_{x_{\mathrm{res}}}. \tag{13.87}$$

The indices $\mu_{1,2} = -\frac{1}{2} \pm \sqrt{1 + 4q_0}$ are complex, associated with infinitely oscillatory behavior of the solutions at $x = x_{\mathrm{res}}$, when $1 + 4q_0 < 0$, so that clustering of stable or unstable modes occurs at $\omega = \Omega_{\mathrm{A/S}}^+(x_{\mathrm{res}}) = \Omega_{\mathrm{A/S}}^-(x_{\mathrm{res}}) = \Omega_0(x_{\mathrm{res}})$,

$$\text{where } \omega_{\mathrm{A}} = 0\,, \quad \text{if } \begin{cases} \text{either} & k_0^2 \rho' g < -\frac{1}{4}\rho(\Omega_0'^2 - \omega_{\mathrm{A}}'^2) < 0\,, \\[2mm] \text{or} & k_0^2 \rho' g > \frac{1}{4}\rho(\omega_{\mathrm{A}}'^2 - \Omega_0'^2) > 0\,. \end{cases} \tag{13.88}$$

The first condition applies to gravitationally stable plasmas ($\rho' < 0$) when shear flow dominates over magnetic shear, and the second one applies to gravitationally unstable plasmas ($\rho' > 0$) when magnetic shear dominates over shear flow. In the present example, the latter applies but the second part of the cluster condition is not satisfied, as is evident from the continuum profiles of Fig. 13.4(b) at the resonant point.

In the limits of HD ($\omega_{\mathrm{A}} \to 0$) and static MHD ($\Omega_0 \to 0$), the two mutually exclusive conditions (13.88) for clustering reduce to a single possibility for each case, with precisely the opposite relation to stability. Considering the limit to HD first, the cluster condition (13.88)(a) reduces to the inequality (13.85) (without the equality sign) so that it refers to clustering of stable modes. This condition was studied by Case in his second 1960 paper [78] for an exponentially decreasing density and a linear velocity profile, based on the analysis of the complex zeros of the solutions of the corresponding spectral equation by Dyson [128]. When the condition (13.85) is satisfied, in addition to the flow continua, the spectrum consists of a cluster sequence of stable g modes clustering towards the edge of the flow continua. However, it was also shown that this particular class of equilibria is stable for all values of J, demonstrating that Howard's criterion (13.85) is only sufficient,

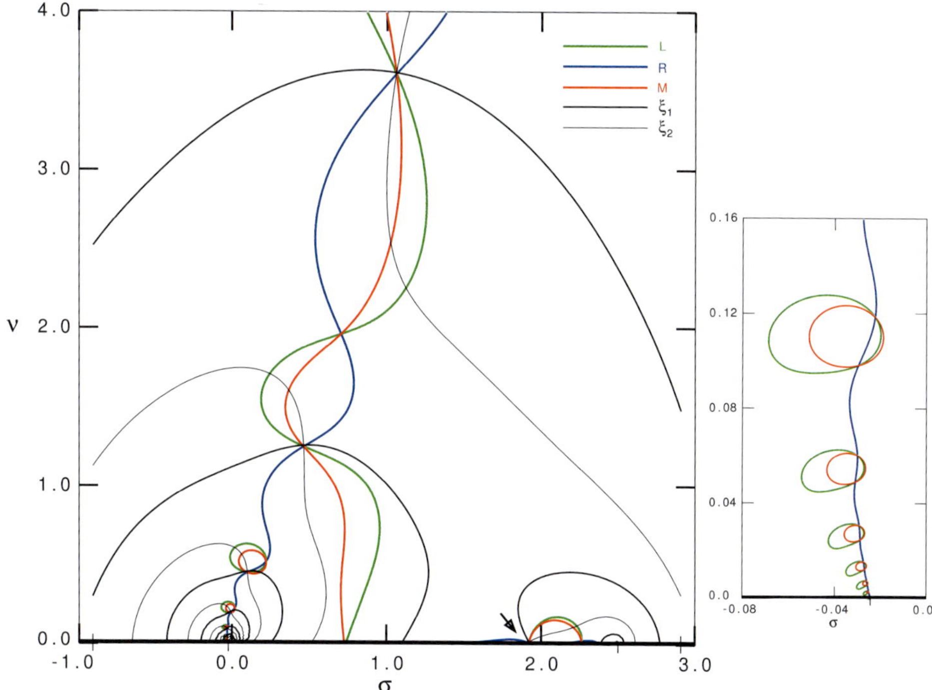

Fig. 13.11 Contours of $W_2^{\mathrm{L/R/M}} = 0$, corresponding to the left, right and middle path, and of $\xi_{1e} = 0$ and $\xi_{2e} = 0$, for a gravitating plasma with magnetic shear and sinusoidal flow profile (MHD2). A cluster spectrum of Raylcigh–Taylor instabilities is continuously connected by the right path, whereas the left and middle paths transform into a series of ever smaller closed loops at the eigenvalues (see inset). The arrow indicates an isolated flow-driven instability close to the continuum.

not necessary for stability. For completeness: a third paper in 1960 by Case [79] concerns the spectrum of static HD, where an unstable continuum is found, but this is due to the consideration of a gravitationally unstable inverted atmosphere *of infinite extent*.

Considering the limit to static MHD, the negation of the cluster condition (13.88)(b) yields the Schwarzschild–"Suydam" stability criterion (7.199)[1] (in the limit $\gamma \to \infty$):

$$\rho' g - \tfrac{1}{4} B^2 \varphi'^2 \leq 0 \qquad \text{(necessary for stability in static MHD)}. \qquad (13.89)$$

When satisfied, it predicts stability because the stabilizing magnetic shear is larger than the destabilizing gravitational drive. It appears logical to combine this stability criterion for static MHD with Howard's HD criterion (13.85), but there is no justification for that (there exists no MHD counterpart of Howard's lucky juggling with quadratic forms). All one gets from the cluster condition (13.88)(b) is a sufficient criterion for instability involving two conditions that both have to be satisfied:

$$\rho' g > \tfrac{1}{4} B^2 \varphi'^2 - \tfrac{1}{4} \rho v'^2 > 0 \qquad \text{(sufficient for *instability* in stationary MHD)}. \quad (13.90)$$

Since this criterion is not satisfied in our MHD2 example, this just implies that there are

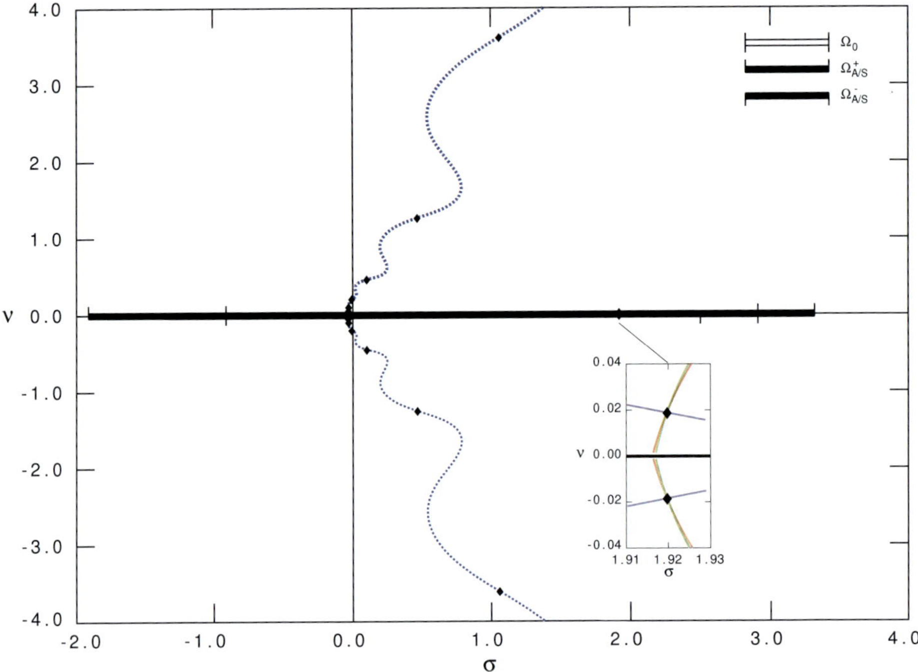

Fig. 13.12 Spectrum of discrete unstable modes (with right path) and stable continuum oscillations for a gravitating plasma with magnetic shear and sinusoidal flow profile (MHD2). Rayleigh–Taylor type instabilities cluster towards the internal extremum of the backward continua. An isolated Kelvin–Helmholtz instability, together with its complex conjugate, is shown by the inset.

no local instabilities clustering at the resonant frequency in the middle of the plasma. Of course, as demonstrated by our results, this does not exclude either the occurrence of local instabilities, clustering elsewhere, or the occurrence of global instabilities.

Finally, the generalization of the cluster criterion (13.88)(b) for compressible plasmas yields two separate criteria for Alfvén and slow resonances, which exhibit subtle crossover limits that were first studied for pressure-driven interchanges in cylindrical geometry by Hameiri [207] and Bondeson *et al.* [56]. The counterpart for gravitational interchanges in plane slab geometry is due to van der Holst *et al.* [455].   ◁

The solution paths for this example, shown in Figs. 13.11 and 13.12, exhibit the different stabilizing and destabilizing mechanisms at work. For large growth rates ($\nu > 2$), the three paths appear to be tuned towards the global gravitational interchanges associated with the resonant frequency $\omega = \Omega_{A/S}^{+} = \Omega_{A/S}^{-} = \Omega_0 = 1$ at $x = 0.5$. The eigenfunctions (not shown in Fig. 13.13) are completely global there. Thus, although local resonant instabilities do not occur in this example since the right inequality of the conditions (13.90) is not satisfied, the three terms of the left inequality do indicate the three mechanisms that produce global instability.

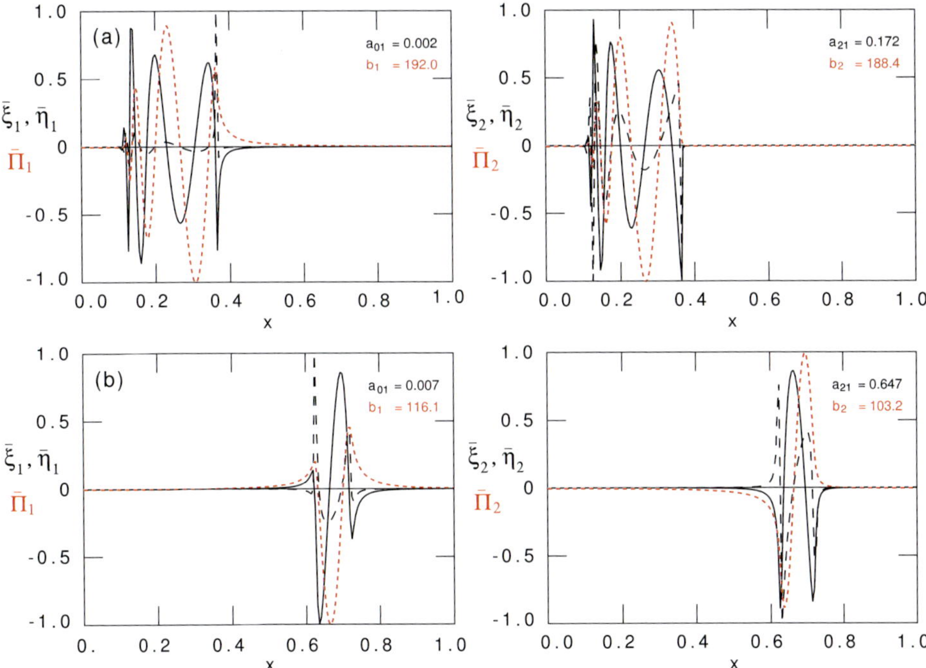

Fig. 13.13 Eigenfunctions for a gravitating plasma with magnetic shear and sinusoidal flow profile corresponding to the spectrum shown in Fig. 13.11 (MHD2): (a) one of the Rayleigh–Taylor modes ($n = 12$) of the cluster spectrum, (b) isolated Kelvin–Helmholtz instability. Renormalization is according to Eqs. (13.84).

For smaller values of the growth rate ν, with the approach of the continua, the violation of the local interchange conditions (13.78)(b) at the extremum of the backward continua permits an infinite sequence of local interchange instabilities clustering towards the edge of the backward continua, which are embedded in the forward continua. One of the corresponding eigenfunctions is shown in the top frame of Fig. 13.13. Notice, again, that the three terms of the instability conditions (13.78)(b) indicate the three mechanisms at work for this local instability which, although described by second instead of first derivatives of the equilibrium profiles, are the same as for the global interchanges. (Incidentally, the sensitive dependence on details of the equilibrium profiles is a recurring theme in MHD stability of plasmas. Progress on this topic, both in fusion research and astrophysics, crucially depends on progress of experimental and observational detection techniques to determine those profiles. Without that, the theoretical effort in MHD spectral theory becomes meaningless.)

The cluster sequence of local gravitational interchanges is presented quite differently by the three solution paths. In this case, the right path (associated with a finite amplitude $\xi_e \neq 0$ on the left) apparently permits a continuous dependence, while the left path and, consequently, also the middle path appear to be unable to represent the approach to the cluster point by a continuous sub-path. Instead, they eject a series of closed loops, with the eigenvalues on them, of ever smaller sizes when the cluster point is approached (see inset of Fig. 13.11). Recall that these small solution sub-paths actually represent complete cycles of the alternator R, doubly connecting the eigenvalue point, where $|\xi_e| = 0$ and $R = 0$, with the point where $|\Pi_e| = 0$, where R flips sign from $+\infty$ to $-\infty$. It may be of interest for future investigations (e.g. of the external excitation of instabilities in laboratory or astrophysical plasmas) to note that these closed loops actually represent "bubbles" of positive imaginary energy W_2 immersed in a "sea" of negative W_2, according to the schematic representation of Fig. 12.6.

The continuing single sub-paths of the left and middle path (associated with finite amplitude $\xi_e \neq 0$ on the right) carry on towards a point $\sigma \approx 0.74$ of the continua in between the gravitational cluster sequence on the left and an emerging Kelvin-Helmholtz instability on the right. In fact, the latter instability (represented by the inset of Fig. 13.12) has an eigenfunction that is localized on the right part of the plasma (Fig. 13.13(b)).

From the two eigenfunctions presented in Fig. 13.13 one cannot easily draw the above conclusions, viz. that the left one (shown on top) belongs to a gravitational interchange cluster sequence and that the right one (shown at the bottom) is actually a gravitationally modified Kelvin–Helmholtz instability. That conclusion was reached by some additional "numerical experiments". If one moves in the right boundary (e.g. by just considering the slice $0 \leq x \leq 0.8$), the growth rates of the most global modes decrease, but the cluster sequence of local modes remains concentrated at the same cluster point. However, if one moves in the left boundary (i.e. one removes a sizeable fraction of the gravitationally unstable part), the growth rate of the Kelvin–Helmholtz instability increases! Most convincingly, if one just switches off gravity (which one can do for these equilibria without affecting the continuum frequencies), the growth rate increases enormously (the eigenvalue moves from $\sigma \approx 1.92$, $\nu \approx 0.0187$ to $\sigma \approx 1.72$, $\nu \approx 0.168$) and the eigenfunction then precisely shows the kind of dependence of the Kelvin–Helmholtz instability exhibited in Fig. 13.10(a). In other words: the Kelvin–Helmholtz instability is suppressed by the gravitational instability in these equilibria.

In conclusion, all eigenvalues of the spectral problem may be determined by each of the three solution paths, but some may be easier to compute by one path (like the cluster sequence of gravitational interchanges which are continuously connected by the right path in this case) and some may be easier to compute by another path (like the Kelvin–Helmholtz instability by the left path). The best choice is usually suggested by the geometry of the contour plots of $W_2(\sigma, \nu) = 0$.

13.2.4 Oscillation theorem C for complex eigenvalues

In Section 12.3, we introduced the method of solution paths in the complex ω plane and derived the two quadratic forms (12.139) for stable waves and (12.154) for instabilities that govern the approach to eigenvalues. One might have expected that the details of actually finding the eigenvalues for 1D equilibria would follow by just substituting the pertinent quantities of the ODEs into those quadratic forms. In principle, this is what we have done in Section 13.1.3 for real eigenvalues, where we had to deal with the singularities of the continuous spectra. This complication is of the same kind as considered already for the case of static equilibria. However, for complex frequencies, the approach to eigenvalues is very different from that of real eigenvalues. In particular, conceptional obstacles had to be overcome of a completely different kind (not encountered in the static case) before we could arrive at the present stage where we may finally formulate the oscillation theorem that governs the distribution of eigenvalues in the complex ω plane.

We first needed to define a node counter for complex eigenfunctions that could replace the usual node counting of real functions of the standard Sturm–Liouville type of approach. This was done by, first, introducing the solution path approach in Section 12.3.1 and, next, introducing the alternator R in Section 13.1.4, based on considering the boundary values of the displacement ξ and of the total pressure perturbation Π for the open (model II) system rather than considering the internal nodes of ξ for the closed (model I) system, as illustrated by Fig. 12.6. Furthermore, as demonstrated by the in-depth case study of the instabilities of stationary plasmas of the present section, we had to develop proper numerical techniques and introduce appropriate terminology in order to compute the solutions paths and to deal with their breakup into separate sub-paths. This now being out of the way, we may finally conclude our quest.

Guided by the numerical results obtained, we now put everything together in the **Oscillation theorem C for instabilities**.

The complex eigenvalues of a stationary plasma slab [or cylinder] are situated on the solution path consisting of the solutions $\omega = \sigma_u(\lambda) + i\nu_u(\lambda)$ of the spectral differential equation for which the solution-averaged Doppler–Coriolis shifted frequency $\sigma_u - \overline{V}(\omega_u)$, or the energy transfer across the boundary W_2, vanishes. The alternator $R \equiv \xi_e/\Pi_e$ [or χ_e/Π_e] is a real and monotonic function of arc length λ along the solution path, or along each separate sub-path, in between the zeros of $|\Pi_e|$ separating the eigenvalues.

To prove the theorem, we reduce the surface integral occurring in the LHS of the quadratic form (12.154) for complex ω to an expression in terms of the increment of the alternator along the solution path. From the path equations (12.128), or (13.61), we obtain the following expressions for the values of the alternator at points $\omega = \omega_\alpha$ and $\omega = \omega_\beta$ of the same sub-path of the solution path:

$$W_{\alpha 2} = 0 \quad \Rightarrow \quad R_\alpha = \frac{\xi_{\alpha 1}}{\Pi_{\alpha 1}}(x_e) = \frac{\xi_{\alpha 2}}{\Pi_{\alpha 2}}(x_e),\tag{13.91}$$

$$W_{\beta 2} = 0 \quad \Rightarrow \quad R_\beta = \frac{\xi_{\beta 1}}{\Pi_{\beta 1}}(x_e) = \frac{\xi_{\beta 2}}{\Pi_{\beta 2}}(x_e).\tag{13.92}$$

Exploiting these expressions for evaluation of the surface integral (12.149) yields:

$$4\mathcal{S} \equiv \mathrm{Re}\left[\xi_\beta^* \Pi_\alpha - \xi_\alpha^* \Pi_\beta\right] = \xi_{\beta 1}\Pi_{\alpha 1} - \xi_{\alpha 1}\Pi_{\beta 1} + \xi_{\beta 2}\Pi_{\alpha 2} - \xi_{\alpha 2}\Pi_{\beta 2}$$

$$= \left(1 + \frac{\Pi_{\alpha 2}\Pi_{\beta 2}}{\Pi_{\alpha 1}\Pi_{\beta 1}}\right)(\xi_{\beta 1}\Pi_{\alpha 1} - \xi_{\alpha 1}\Pi_{\beta 1})$$

$$= (\Pi_{\alpha 1}\Pi_{\beta 1} + \Pi_{\alpha 2}\Pi_{\beta 2})(R_\beta - R_\alpha) \approx 2|\Pi_\alpha|^2(R_\beta - R_\alpha),\tag{13.93}$$

where we leave it understood that all quantities are evaluated at $x = x_e$. The last approximation holds when ω_α and ω_β are close, which is the basic assumption of the whole analysis. Hence, the quadratic form (12.154) reduces to

$$R_\beta - R_\alpha = \begin{cases} \dfrac{|\omega_\beta - \omega_\alpha|^2}{\sigma_\beta - \sigma_\alpha} \cdot \dfrac{2\nu_\alpha \widetilde{X}_2}{|\Pi_\alpha|^2} & (\sigma_\beta \neq \sigma_\alpha) \\[4mm] -(\nu_\beta - \nu_\alpha) \cdot \dfrac{2\nu_\alpha(\widetilde{X}_1 - I)}{|\Pi_\alpha|^2} & (\sigma_\beta = \sigma_\alpha) \end{cases}.\tag{13.94}$$

We have seen in Section 12.3.2 that $\widetilde{X}_2$ and $\widetilde{X}_1$ on the RHS are definite quantities of order unity. Consequently, the alternator R is a monotonic quantity in the arc length $\lambda \equiv |\omega_\beta - \omega_\alpha|$ along any sub-path in between the zeros of $|\Pi_e|$, so that the eigenvalues (where $|\xi_e| = 0$) should lie in between them, QED.

Note that, whereas the oscillation theorems for the stable waves of a static plasma (Section 7.4.3 [1]) and of the stable waves of a stationary plasma (Section 13.1.3) are related straightforwardly (as the limits $\Omega_0 \to 0$ and $\sigma_0 \to 0$), the corresponding oscillation theorems for the instabilities appear hardly related. Of course, this is due to the fact that the complex eigenvalues of stationary plasmas first require the construction of the solution sub-paths. Even after they have been obtained, the limit for the instabilities of static plasmas still represents a rather singular limit of the present oscillation theorem since the operator U vanishes then and the path is vertical, so that the quantity $\widetilde{X}_2/(\sigma_\beta - \sigma_\alpha)$ is undefined and should be replaced by $-(\widetilde{X}_1 - I)/(\nu_\beta - \nu_\alpha)$. A related complication has already been encountered for the middle path solutions of the Kelvin–Helmholtz instability presented in Fig. 13.8: also with a symmetric flow profile, the imaginary axis may become part of the solution path.

13.3 Spectral theory of rotating plasmas

13.3.1 MHD wave equation for cylindrical flow

The substantial modification of the Hain–Lüst equation for cylindrical plasmas with flow, in particular rotation, was first obtained by Hameiri [207, 209], and later studied in more detail by Bondeson *et al.* [56]. We here present the extension by Keppens *et al.* [255] with the effects of a gravitational field of cylindrical symmetry, which is applied in Section 13.4.2 to the thin slice approximation of the rotational instabilities of accretion discs.

The one-dimensional cylindrical equilibrium, described by the density $\rho(r)$, the pressure $p(r)$, the magnetic field components $B_\theta(r)$ and $B_z(r)$, and the velocity components $v_\theta(r)$ and $v_z(r)$, is restrained by just the single differential equation (12.30) in the radial coordinate r. For now, the gravitational potential $\Phi(r)$ is considered to be due to a singular mass distribution on the axis $r = 0$. In Section 13.4.2, we will consider a more appropriate limit for application to a genuine gravitational field, viz. that of a compact central mass surrounded by an accretion disk (replacing the line singularity by a point singularity).

Because of the cylinder symmetry, we consider Fourier components of the normal modes of the form

$$\boldsymbol{\xi}(r, \theta, z, t) = \left[\xi_r(r)\mathbf{e}_r + \xi_\theta(r)\mathbf{e}_\theta + \xi_z(r)\mathbf{e}_z \right] e^{i(m\theta + kz - \omega t)}, \tag{13.95}$$

but now (in contrast to the plane case of Sections 13.1 and 13.2) the two velocity-dependent terms of the generalized force operator $\mathbf{G}$, defined in Eq. (12.64), produce contributions that cannot be cast in the simple form (13.5). In particular, because the gradient operator not only acts on the vector components of $\boldsymbol{\xi}$ but also on the unit vectors, the effect of the operator $U \equiv -i\rho\mathbf{v} \cdot \nabla$ is to produce not only a local Doppler shift $\Omega_0(r)$, but also an additional local *Coriolis shift* proportional to the vector product of the rotation vector $(v_\theta/r)\mathbf{e}_z$ and $\boldsymbol{\xi}$:

$$U\boldsymbol{\xi} \equiv -i\rho\mathbf{v} \cdot \nabla\boldsymbol{\xi} = \rho\Omega_0\boldsymbol{\xi} - i\rho(v_\theta/r)\mathbf{e}_z \times \boldsymbol{\xi}, \tag{13.96}$$

$$\Omega_0(r) \equiv mv_\theta/r + kv_z. \tag{13.97}$$

Hence, in cylindrical geometry, the two velocity-dependent terms of $\mathbf{G}$ become

$$\nabla \cdot (\boldsymbol{\xi}\rho\mathbf{v} \cdot \nabla\mathbf{v}) = (\rho\mathbf{v} \cdot \nabla\mathbf{v})\nabla \cdot \boldsymbol{\xi} + \boldsymbol{\xi} \cdot \nabla(\rho\mathbf{v} \cdot \nabla\mathbf{v})$$

$$= -\left[(\rho v_\theta^2/r)\nabla \cdot \boldsymbol{\xi} + (\rho v_\theta^2/r)'\xi_r\right]\mathbf{e}_r - (\rho v_\theta^2/r^2)\xi_\theta\mathbf{e}_\theta, \tag{13.98}$$

$$-\nabla \cdot (\rho\mathbf{v}\mathbf{v} \cdot \nabla\boldsymbol{\xi}) = -\rho\mathbf{v} \cdot \nabla(\mathbf{v} \cdot \nabla\boldsymbol{\xi}) = U\rho^{-1}U\boldsymbol{\xi}$$

$$= \rho\Omega_0^2\boldsymbol{\xi} - 2i\rho\Omega_0(v_\theta/r)\mathbf{e}_z \times \boldsymbol{\xi} + \rho(v_\theta^2/r^2)(\xi_r\mathbf{e}_r + \xi_\theta\mathbf{e}_\theta). \tag{13.99}$$

They combine to yield the following form of the spectral equation for cylindrical plasmas given by Bondeson *et al.* [56]:

$$\mathbf{F}(\boldsymbol{\xi}) - \left[(\rho v_\theta^2/r)\nabla\cdot\boldsymbol{\xi} + (\rho v_\theta^2/r^2)'r\xi_r\right]\mathbf{e}_r$$

$$+ 2\mathrm{i}\rho\widetilde{\omega}(v_\theta/r)\mathbf{e}_z\times\boldsymbol{\xi} + \rho\widetilde{\omega}^2\boldsymbol{\xi} = 0\,, \tag{13.100}$$

where

$$\widetilde{\omega} \equiv \omega - \Omega_0 = \omega - mv_\theta/r - kv_z \tag{13.101}$$

is the local Doppler-shifted frequency (but not involving the local Coriolis shift!). In the absence of rotation, $v_\theta = 0$, the simple "quasi-static" form (13.5) is again obtained. Note that the operator $\mathbf{F}$, which is just an abbreviation of the terms appearing in the force operator (12.65) for static equilibria, is *not self-adjoint* for the present stationary case. This is evident by the appearance of the two asymmetric radial terms above which combine with $\mathbf{F}$ to yield the symmetric matrix representation below, but only after judicious application of the equilibrium equation.

▷ **Three-dimensional wave equation and reduction** We exploit again the field line projection (9.23)–(9.26)[1], with dimensionless "wave numbers" $g \equiv G/B$ and $f \equiv F/B$, where $G \equiv mB_z/r - kB_\theta$ and $F \equiv mB_\theta/r + kB_z$, and dimensionless field components $b_\theta \equiv B_\theta/B$ and $b_z \equiv B_z/B$. Defining

$$\mathbf{e}_r \equiv \nabla r\,, \qquad \mathbf{e}_\perp \equiv (\mathbf{B}/B)\times\mathbf{e}_r\,, \qquad \mathbf{e}_\parallel \equiv \mathbf{B}/B\,,$$

$$\partial_r \equiv \mathbf{e}_r\cdot\nabla\,, \qquad g \equiv -\mathrm{i}\mathbf{e}_\perp\cdot\nabla\,, \qquad f \equiv -\mathrm{i}\mathbf{e}_\parallel\cdot\nabla\,,$$

$$\xi \equiv \xi_r\,, \qquad \eta \equiv \mathrm{i}\mathbf{e}_\perp\cdot\boldsymbol{\xi}\,, \qquad \zeta \equiv \mathrm{i}\mathbf{e}_\parallel\cdot\boldsymbol{\xi}\,, \tag{13.102}$$

the spectral equation may be written in matrix form as

$$(\mathbf{H} - 2\rho\widetilde{\omega}\mathbf{C} + \rho\widetilde{\omega}^2\mathbf{I})\cdot\mathbf{X} = 0\,, \qquad \mathbf{X} \equiv (\xi,\eta,\zeta)^{\mathrm{T}}\,. \tag{13.103}$$

Here, $\mathbf{H} \equiv \mathbf{H}_A + \mathbf{H}_b$, where $\mathbf{H}_A$ is the previous matrix of the static case without gravity and flow, defined in the LHS of Eq. (9.28)[1] (to obtain it, watch out not to use the static equilibrium relation!) and $\mathbf{H}_b$ represents the effects of gravity and (partially) rotation, whereas $\mathbf{C}$ represents the (Coriolis) effects of rotation:

$$\mathbf{H}_A \equiv \begin{pmatrix} \dfrac{d}{dr}\dfrac{\gamma p + B^2}{r}\dfrac{d}{dr}r - f^2 B^2 - r\left(\dfrac{B_\theta^2}{r^2}\right)' & \dfrac{d}{dr}g(\gamma p + B^2) - \dfrac{2kB_\theta B}{r} & \dfrac{d}{dr}f\gamma p \\[2ex] -\dfrac{g(\gamma p + B^2)}{r}\dfrac{d}{dr}r - \dfrac{2kB_\theta B}{r} & -g^2(\gamma p + B^2) - f^2 B^2 & -fg\gamma p \\[2ex] -\dfrac{f\gamma p}{r}\dfrac{d}{dr}r & -fg\gamma p & -f^2\gamma p \end{pmatrix}\,,$$

$$\mathbf{H}_b \equiv -\begin{pmatrix} \rho r(\Phi'/r)' & g\Lambda & f\Lambda \\ g\Lambda & 0 & 0 \\ f\Lambda & 0 & 0 \end{pmatrix}\,, \qquad \mathbf{C} \equiv \dfrac{v_\theta}{r}\begin{pmatrix} 0 & b_z & b_\theta \\ b_z & 0 & 0 \\ b_\theta & 0 & 0 \end{pmatrix}\,. \tag{13.104}$$

The function

$$\Lambda(r) \equiv \rho(v_\theta^2/r - \Phi') \tag{13.105}$$

represents the deviation from static MHD equilibrium caused by gravity and rotation. Equivalently, according to the equilibrium relation (12.30), it may be considered as the deviation from pure Keplerian HD flow caused by pressure gradients and Lorentz forces. Elimination of the tangential components η and ζ,

$$\eta = \frac{G\widetilde{S}(r\xi)' - \left[2k\gamma pF(B_\theta F + \rho v_\theta \widetilde{\omega}) - 2kB_\theta B^2 \rho \widetilde{\omega}^2 - (2B_z \rho v_\theta \widetilde{\omega} + rG\Lambda)\rho\widetilde{\omega}^2\right]\xi}{rBD},$$

$$\zeta = \frac{\gamma pF\widetilde{A}(r\xi)' + \left[2k\gamma pG(B_\theta F + \rho v_\theta \widetilde{\omega}) + (2B_\theta \rho v_\theta \widetilde{\omega} + rF\Lambda)(\rho\widetilde{\omega}^2 - h^2 B^2)\right]\xi}{rBD},$$

$$\tag{13.106}$$

with $\widetilde{A}$, $\widetilde{S}$, D and h^2 defined in Eqs. (13.109) and (13.110), and substitution into the radial component of Eq. (13.103) leads to the spectral differential equation (13.107) for ξ. ◁

The eigenvalue problem can be cast in the form of a second order ordinary differential equation for the radial component $\chi \equiv r\xi$ of the plasma displacement,

$$\frac{d}{dr}\left(\frac{N}{D}\frac{d\chi}{dr}\right) + \left[A + \frac{B}{D} + \left(\frac{C}{D}\right)'\right]\chi = 0, \tag{13.107}$$

with coefficients N, D, A, B, C defined in (13.109), (13.110), (13.113)–(13.115), and subject to model I BCs (wall at $r = a$),

$$\chi(0) = 0, \qquad \chi(a) = 0. \tag{13.108}$$

Recall that the radial component ξ has been replaced by the variable χ just to be able to express the BCs on the origin in this way, rather than to have to distinguish between the finite $|m| = 1$ Fourier components of $\xi(0)$ and the vanishing ones.

With the addition of stationary flows to the equilibrium, the expressions with the eigenvalue ω transform into expressions involving the local Doppler shifted frequency $\widetilde{\omega}$. Together with the parallel gradient operator $F \equiv mB_\theta/r + kB_z$, they determine the *Alfvén and slow continuum singularities* $N = 0$, where

$$N \equiv \frac{1}{r}\widetilde{A}\widetilde{S} \equiv \frac{\rho^2(\gamma p + B^2)}{r}(\omega - \Omega_A^+)(\omega - \Omega_A^-)(\omega - \Omega_S^+)(\omega - \Omega_S^-),$$

$$\widetilde{A} \equiv \rho\widetilde{\omega}^2 - F^2, \qquad \widetilde{S} \equiv (\gamma p + B^2)\rho\widetilde{\omega}^2 - \gamma pF^2, \tag{13.109}$$

and the *apparent (fast and slow turning point) singularities* $D = 0$, where

$$D \equiv \rho^2\widetilde{\omega}^4 - h^2\widetilde{S} \equiv \rho^2(\omega - \Omega_{f0}^+)(\omega - \Omega_{f0}^-)(\omega - \Omega_{s0}^+)(\omega - \Omega_{s0}^-),$$

$$h^2 \equiv f^2 + g^2 \equiv m^2/r^2 + k^2. \tag{13.110}$$

The associated definitions of the *forward and backward Alfvén and slow continua*,

$$\Omega_A^\pm(r) \equiv \Omega_0(r) \pm \omega_A(r), \qquad \Omega_S^\pm(r) \equiv \Omega_0(r) \pm \omega_S(r),$$

$$\omega_A^2(r) \equiv \frac{F^2}{\rho}, \qquad \omega_S^2(r) \equiv \frac{\gamma p}{\gamma p + B^2}\frac{F^2}{\rho}, \tag{13.111}$$

and of the *forward and backward apparent fast and slow singularities,*

$$\Omega_{\text{f0}}^{\pm}(r) \equiv \Omega_0(r) \pm \omega_{\text{f0}}(r)\,, \qquad \Omega_{\text{s0}}^{\pm}(r) \equiv \Omega_0(r) \pm \omega_{\text{s0}}(r)\,, \tag{13.112}$$

$$\omega_{\text{f0,s0}}^2(x) \equiv \tfrac{1}{2}h^2\frac{\gamma p + B^2}{\rho}\left[1 \pm \sqrt{1 - \frac{4\gamma p F^2}{h^2(\gamma p + B^2)^2}}\right],$$

only differ from the plane slab definitions in that they contain an "azimuthal wave number" contribution m/r in the effective wave number h, which replaces k_0. With these definitions, and also supplementing the cluster frequency $\Omega_{\text{F}}^{\pm} \equiv \pm\infty$ of the fast magneto-sonic modes, the ordering (13.22) of the local frequencies of the genuine and apparent singularities remains valid without change. Notice though that $\Omega_{\text{f0}}^{\pm} \to \pm\infty$ for $r \to 0$ (another sign that it cannot be a genuine singularity).

Cylindrical flow does not change the formal form of the singular expressions, but the coefficients A, B and C are modified substantially:

$$A \equiv \frac{1}{r}\left[\widetilde{A} + r\left(\frac{B_\theta^2 - \rho v_\theta^2}{r^2}\right)' + \rho'\Phi'\right]\,, \tag{13.113}$$

$$\begin{aligned}
B \equiv \; & -\frac{4}{r^3}\Big\{(\rho\widetilde{\omega}^2 - k^2\gamma p)(B_\theta F + \rho v_\theta\widetilde{\omega})^2 \\
& + B_\theta^2\rho\widetilde{\omega}^2\Big[h^2(B_\theta^2 - \rho v_\theta^2) - 2(m/r)(B_\theta F + \rho v_\theta\widetilde{\omega})\Big] \\
& - r\Lambda\Big[h^2(B_\theta\widetilde{\omega} + F v_\theta)\rho\widetilde{\omega}B_\theta - (m/r)(B_\theta F + \rho v_\theta\widetilde{\omega})\rho\widetilde{\omega}^2\Big] + \tfrac{1}{4}r^2\Lambda^2 h^2\widetilde{A}\Big\}\,,
\end{aligned} \tag{13.114}$$

$$C \equiv \frac{2}{r^2}\Big\{(m/r)(B_\theta F + \rho v_\theta\widetilde{\omega})\widetilde{S} - \rho\widetilde{\omega}^2\Big[(B_\theta\widetilde{\omega} + F v_\theta)\rho\widetilde{\omega}B_\theta - \tfrac{1}{2}r\Lambda\widetilde{A}\Big]\Big\}\,. \tag{13.115}$$

Note that the function $A(r;\omega)$ is a quadratic polynomial in the eigenvalue parameter ω, whereas the functions N, D, B and C are quartic polynomials in ω.

The differential equation (13.107) is formally identical to the spectral wave equation (13.9) for the stationary plane slab, but also to the spectral equations for the static case presented in Volume [1], Chapter 7 for the plane slab and Chapter 9 for the cylindrical plasma (with slightly redefined coefficients for the latter, see below). Of course, this does not imply that the spectral problem for the stationary cylinder is just as simple as that for the static cylinder, but it does imply that the analytical tools of *solution path* and *oscillation theorems* governing the distribution of discrete eigenvalues on them, developed in Sections 12.3, 13.1 and 13.2, and applied there to shear flow of plane plasmas, can be applied [virtually without change] to cylindrical plasmas as well. This we have already indicated in the formulations of the two oscillation theorems of Section 13.1.3 and 13.2.4.

The corresponding formulation of two coupled first order differential equations for χ and the Eulerian total pressure variation $\Pi \equiv -(1/D)(N\chi' + C\chi)$ reads:

$$
N\frac{d}{dr}\begin{pmatrix}\chi \\ \Pi\end{pmatrix} + \begin{pmatrix}C & D \\ E & -C\end{pmatrix}\begin{pmatrix}\chi \\ \Pi\end{pmatrix} = 0\,,
\tag{13.116}
$$

where the new function

$$
E \equiv -N\left(A + \frac{B}{D}\right) - \frac{C^2}{D} \qquad \left[\Rightarrow NB + C^2 \sim D, \quad DE + C^2 \sim N\right]
\tag{13.117}
$$

$$
\equiv -\frac{N}{r}\left[\tilde{A} + r\left(\frac{B_\theta^2 - \rho v_\theta^2}{r^2}\right)' + \rho'\Phi'\right]
$$

$$
+ \frac{4}{r^4}\left\{(B_\theta F + \rho v_\theta\tilde\omega)^2\tilde{S} - \left[(B_\theta\tilde\omega + F v_\theta)\rho\tilde\omega B_\theta - \tfrac{1}{2}r\Lambda\tilde{A}\right]^2\right\}
\tag{13.118}
$$

is a sixth order polynomial in ω.

▷ **Redefined quantities** The functions appearing in the equivalent second and first order formulations above have been defined slightly differently from the old ones exploited in Section 9.2.1 [1]. The old quantities, indicated with the subscript $_\mathrm{o}$, are obtained from the present ones in the limit of static equilibrium and vanishing gravity as follows:

$$
D \to D_\mathrm{o}\,, \qquad N \to N_\mathrm{o}/r\,, \qquad B \to V_\mathrm{o}\,, \qquad E \to E_\mathrm{o}\,,
$$

$$
A \to U_\mathrm{o} + 2\left(B_\theta^2/r^2\right)'\,, \qquad C \to W_\mathrm{o} - 2\left(B_\theta^2/r^2\right)D_\mathrm{o} \equiv C_\mathrm{o}\,.
\tag{13.119}
$$

With the new definitions, the old function W_o has become redundant and the more concise form (13.117) for the relationship between A, B, C and E is obtained.   ◁

With the present definitions of the basic polynomials, the formulations in terms of the second order ODE (13.107) for $\chi \equiv r\xi$ and the system of first order ODEs (13.116) for χ and Π for the cylinder are identical to the corresponding ones, (13.9) and (13.15), for the plane slab. Moreover, the fundamental relationship (13.117) between the quantities of the second order ODE and the quantities of the first order ODEs is identical to the relation (13.16) for the analogous plane slab relations obtained in Section 13.1.1. Hence, the conclusion reached there on the two complementary relations (13.17) and (13.18) also applies here, i.e. *the apparent singularities of the second order representation and the redundant singularities of the first order representation are complementary evils*, also for the cylindrical case. The controversies on the different formulations of the spectral problem (related to the elimination of apparent singularities [187], [10], [166]) thus finally come to rest by observing that one representation is not superior to the other and that the relationship (13.117) is essential for both to remove their spurious singularities.

We have now presented all the basic equations for the analysis of the spectrum of stationary plasma flow in cylindrical geometry. One, physically very important, structural difference with the analysis for the plane slab is the fact that, in the path equation

$$\sigma - \overline{V}(\omega) = 0 \,, \tag{13.120}$$

the *solution averaged Doppler–Coriolis shift,* defined in Eq. (13.7) for the plane slab, now acquires a Coriolis contribution:

$$\overline{V} \equiv \langle \rho^{-1} U \rangle = \frac{\int \rho \big\{ \Omega_0 |\boldsymbol{\xi}|^2 + (v_\theta/r)\,[\xi^*(b_z\eta + b_\theta\zeta) + \xi(b_z\eta^* + b_\theta\zeta^*)] \big\}\, r\,dr}{\int \rho |\boldsymbol{\xi}|^2\, r\,dr} \,.$$

$$\tag{13.121}$$

Here, the intermediate substitution of the azimuthal component in terms of the field line projected components, $i\xi_\theta \equiv b_z\eta + b_\theta\zeta$, has been made to highlight the fact that the Coriolis shift, like the Doppler shift, is real for real frequencies. Of course, in the actual analysis, the further substitution (13.106) to expressions in terms of χ' and χ is made. The Coriolis shift has important consequences for the monotonicity of real eigenvalues and for the solution path of complex eigenvalues. For cylindrical plasmas, the solution path $\mathcal{P}_{\mathrm{u}}$, again obtained from Eq. (13.120), or $W_2 = 0$, is no longer confined to the strip $\{\Omega_0(r)\}$ in the complex ω plane. Moreover, this path will cross the real ω axis in more than one point, as will be illustrated in Section 13.4.1.

13.3.2 Local stability*

In the stability theory of static cylindrical plasmas, Suydam's criterion (9.118)[1] for *interchanges localized about rational magnetic field lines* plays an important role. Recall from the analysis of Section 9.4.1 [1] that this condition is obtained by expanding the marginal equation of motion ($\omega = 0$) about a singular point $r = r_\mathrm{s}$ where the tangential "wave vector" $\mathbf{h} \equiv (m/r)\mathbf{e}_\theta + k\mathbf{e}_z$ of the perturbations is perpendicular to the field lines:

$$F(r_\mathrm{s}) \equiv (k_\parallel B)_{r_\mathrm{s}} = [(k + \mu m)B_z]_{r_\mathrm{s}} = 0 \,, \quad \mu \equiv \frac{B_\theta}{rB_z}\left[\approx \frac{1}{qR_0}\ (\text{slender torus})\right].$$

$$\tag{13.122}$$

(The last expression in square brackets shows the approximate relationship to toroidal geometry by relating the inverse pitch μ of the helical magnetic field lines on a "periodic" cylinder of finite length $2\pi R_0$ to the toroidal safety factor q of an equivalent slender torus with large radius R_0, where $r/R_0 \ll 1$.) The physical significance of this resonance condition is that the magnetic field perturbations, associated with the dominant Alfvén wave dynamics, are minimized so that interchanges driven by smaller effects, like pressure gradients, can develop. The mathematical expression of this resonance is the confluence of four genuine and two apparent singularities of the spectral differential equation:

$$F = 0 \quad \Rightarrow \quad \omega^2 = \omega_A^2 = \omega_S^2 = \omega_{s0}^2 = 0 \qquad (\text{at } r = r_s). \qquad (13.123)$$

For these special points of the continuous spectrum, the stability problem reduces to a (necessary) local stability criterion in closed form.

For the analogous theory of stationary plasmas, we need to consider resonant modes ($F = 0$) that are marginal in the co-moving Doppler-shifted frame ($\widetilde{\omega} = 0$):

$$F = 0, \ \widetilde{\omega} = 0 \quad \Rightarrow \quad \omega = \Omega_0 = \Omega_A^+ = \Omega_A^- = \Omega_S^+ = \Omega_S^- = \Omega_{s0}^+ = \Omega_{s0}^-$$
$$(\text{at } r = r_s), \qquad (13.124)$$

so that, again, confluence of six singularities occurs. However, the tacit assumption that $\widetilde{\omega} = 0$ is a marginal point needs justification now. If this point were not part of the continuous spectrum, it would require proof that it is a solution of the path equation (13.120) with $\overline{V}(\sigma) = \Omega_0$. From the expression (13.121) for $\overline{V}$, it is obvious that this implies that the Coriolis shift contribution should vanish for these solutions. In fact, for the very localized solutions of the continuum modes, the Coriolis shift contribution of the integral (13.121) is much smaller than the Doppler shift contribution since the latter comes with the factor $|\xi|^2 \sim |\eta|^2 \sim |\chi'|^2$ in the integrand and the former with the factor $|\xi_\theta^* \xi| \sim |\chi' \chi| \ll |\chi'|^2$, using the reductions (13.106) of η and ζ to χ' and χ. Of course, none of the integrals exists in the continuous spectrum since χ' contains δ-function contributions. However, approaching the continuous spectrum from above by exploiting the concept of ϵ-stability, as expressed by Eq. (12.141), the integrals do exist and the smallness argument remains valid, as long as ϵ is small and does not vanish, so that the solution of the path equation becomes

$$\sigma_0^\epsilon \approx \Omega_0(r_s). \qquad (13.125)$$

Violation of the local stability criterion then implies that the underlying point of the continuous spectrum is a cluster point of unstable modes lying on this path.

Expansion of the coefficients of the spectral differential equation (13.107) about the singular point $r = r_s$ yields

$$\widetilde{\omega} \approx -\Omega_0' s, \quad F \approx F' s, \quad F' = -k B_z \mu'/\mu, \quad \text{where } s \equiv r - r_s, \qquad (13.126)$$

so that the functions N and D defined in Eqs. (13.109) and (13.110) become

$$N \equiv \frac{1}{r} \widetilde{A} \widetilde{S}, \qquad D \approx -(k^2 B^2/B_\theta^2) \widetilde{S}, \qquad (13.127)$$

$$\widetilde{A} \approx -(1 - \widetilde{M}^2) F'^2 s^2, \quad \widetilde{S} \approx -(\gamma p + B^2)(M_c^2 - \widetilde{M}^2) F'^2 s^2.$$

Here, the *trans-Alfvénic* factor involves the *shear Alfvén Mach number* $\widetilde{M}$ measuring the shear of the background velocity with respect to the shear of the Alfvén velocity, both taken along the parallel "wave vector" $\mathbf{k}_\| \equiv \mathbf{h}\,(m = -k/\mu)$, at the resonant surface:

$$\widetilde{M} \equiv -\frac{\widetilde{\omega}}{\omega_A} \approx \left[\frac{\sqrt{\rho}\,\Omega_0'}{F'}\right]_{r_s} \equiv \left[\frac{(\mathbf{k}_\| \cdot \mathbf{v})'}{(\mathbf{k}_\| \cdot \mathbf{v}_A)'}\right]_{r_s}, \tag{13.128}$$

whereas the *trans-slow* factor $M_c^2 - \widetilde{M}^2$ involves the *critical* (or *cusp*) value M_c of the shear Alfvén Mach number, where the shear of the background velocity becomes equal to the shear of the continuum slow wave speed at the resonant surface:

$$M_c \equiv \frac{\omega_S}{\omega_A} \equiv \sqrt{\frac{\gamma p}{\gamma p + B^2}}\,. \tag{13.129}$$

The rest of the analysis proceeds similarly to that of Section 9.4.1 [1].

In the expansion of N/D, the familiar magnetic shear term $(\mu'/\mu)^2$ is now multiplied by the trans-Alfvénic factor,

$$\frac{N}{D} \approx \alpha s^2\,, \qquad \alpha \equiv \frac{B_\theta^2 B_z^2}{r B^2}(1 - \widetilde{M}^2)\left(\frac{\mu'}{\mu}\right)^2\,, \tag{13.130}$$

and the spectral differential equation for $\omega \approx \Omega_0(r_s)$ reduces to

$$\alpha \frac{d}{ds}\left(s^2 \frac{d\chi}{ds}\right) - \beta\chi = 0\,, \qquad \beta \equiv -\left[A + \frac{B}{D} + \left(\frac{C}{D}\right)'\right]_{r_s}, \tag{13.131}$$

where the explicit form for β, involving the trans-Alfvénic as well as the trans-slow factor, can be read off from the resulting form of the local stability criterion (13.134) below. Close to the singularity $s = 0$, the solutions behave as

$$\chi = a s^{\nu_1} + b s^{\nu_2}\,, \qquad \nu_{1,2} = -\tfrac{1}{2} \pm \tfrac{1}{2}\sqrt{1 + 4\beta/\alpha}\,, \tag{13.132}$$

which gives rise to solutions with infinitely many oscillations as $s \to 0$ if the indices are complex. Hence, local stability demands the opposite:

$$1 + 4\beta/\alpha > 0\,. \tag{13.133}$$

Substitution of the coefficients α and β in this inequality produces the following, rather formidable, generalization of Suydam's criterion of a cylindrical plasma with background flow and a radial gravitational field:

$$\frac{1}{1 - \widetilde{M}^2}\left[p' - \frac{\sqrt{\rho}\,v_\theta B_z^2 \widetilde{M}}{B_\theta}\frac{\mu'}{\mu} + \frac{r^2 B^2}{2 B_\theta^2}\left(\frac{\rho v_\theta^2}{r^2}\right)' - \frac{r B^2}{2 B_\theta^2}\rho'\Phi' + \frac{2\rho v_\theta^2}{r} - \Lambda\right]$$

$$- \frac{2(\sqrt{\rho}\,v_\theta - B_\theta \widetilde{M})\left[\gamma p B_\theta(\sqrt{\rho}\,v_\theta - B_\theta \widetilde{M}) - r B^2 \widetilde{M}\Lambda\right]}{r B_\theta(\gamma p + B^2)(M_c^2 - \widetilde{M}^2)(1 - \widetilde{M}^2)}$$

$$- \frac{r B^2 \Lambda^2}{2 B_\theta^2(\gamma p + B^2)(M_c^2 - \widetilde{M}^2)} + \tfrac{1}{8}r B_z^2\left(\frac{\mu'}{\mu}\right)^2 > 0\,. \tag{13.134}$$

Recall that the gravitational terms are to be used only in an annular thin slice approximation of a disk with central compact object of mass M_* (see Fig. 13.16(a) below), producing a radial acceleration $\mathbf{g} = -g\mathbf{e}_r$, where $g = \Phi' \approx GM_*/r^2$. In that context, $\Lambda \equiv \rho(v_\theta^2/r - \Phi')$ represents deviations from Keplerian flow.

The local stability criterion (13.134), without the gravitational terms, was first obtained by Hameiri [209], and investigated numerically by Bondeson *et al.* [56]. By means of a boundary layer analysis [207, 56], violation of the criterion was shown to lead to a solution path (in our present terminology) consisting of two straight lines crossing at $\widetilde{\omega} = 0$, with a sequence of unstable discrete modes on each of them. The boundary layer analysis consists of asymptotic matching of the "external" solutions of Eq. (13.132) to the "inner", inertial, solutions of the spectral equation, where the approximations (13.127) for $\widetilde{A}$ and $\widetilde{S}$ are extended with the imaginary contributions of the growth rate ν. This yields an accurate approximation of the cluster frequencies for $\nu \to 0$, whereas the more global modes with larger growth rates need to be computed numerically. Obviously, this analysis breaks down for $\widetilde{M}^2 \to 1$ and $\widetilde{M}^2 \to M_c^2$. Those limits are associated with possible transfer of the cluster sequences [56] from the resonant point to a minimum or maximum of one of the continua (see text in small print below).

Without shear flow ($v_\theta' = v_z' = 0 \Rightarrow \widetilde{M} = 0$) and gravity ($\Phi' = 0$), only the first and last term of the criterion (13.134) survive and Suydam's criterion [427] is recovered. However, the simple picture of instability driven by a pressure gradient and stabilized by magnetic shear is no longer valid in the presence of shear flow. Many terms occur and their influence on stability or instability changes depending on the magnitude of $\widetilde{M}^2$, giving rise to the following shear flow regimes:

$$\left\{ \begin{array}{ll} 0 \;\le\; \widetilde{M}^2 < M_c^2 & \textit{(sub-slow)}, \\[2mm] M_c^2 \le \widetilde{M}^2 < 1 & \textit{(slow/sub-Alfvénic)}, \\[2mm] 1 \;\le\; \widetilde{M}^2 & \textit{(super-Alfvénic)}. \end{array} \right. \tag{13.135}$$

In the absence of rotation but with shear flow in the longitudinal direction, the pressure gradient of Suydam's criterion is divided by the trans-Alfvénic factor and an additional compressional term appears that is divided by both factors:

$$\frac{1}{1 - \widetilde{M}^2}p' - \frac{2B_\theta^2 M_c^2 \widetilde{M}^2}{r(M_c^2 - \widetilde{M}^2)(1 - \widetilde{M}^2)} + \tfrac{1}{8}rB_z^2\left(\frac{\mu'}{\mu}\right)^2 > 0. \tag{13.136}$$

This simplified criterion was exploited in the boundary layer analyses cited above. The flow regimes will return in a different guise in Chapter 18 (unstable toroidal Alfvén–slow continua) and Chapter 20 (transonic MHD flows and shocks).

In the limit of an annulus with $r \to \infty$, but finite Δr, a plane slab is obtained. In that limit, only the fourth term of the first line and the two terms of the last line of Eq. (13.134) survive. Dividing out a factor $rB^2/(2B_\theta^2)$, the criterion of van der Holst *et al.* [455] for stability of gravitational interchanges in stationary plasmas with plane shear flow is recovered:

$$ -\frac{1}{1-\widetilde{M}^2}\,\rho'g - \frac{M_c^2}{M_c^2-\widetilde{M}^2}\,\frac{\rho^2 g^2}{\gamma p} + \tfrac{1}{4}B^2\varphi'^2 > 0 \,. \tag{13.137} $$

This criterion exhibits obvious similarity with the cylindrical criterion (13.136), but the compressional term spoils the simple analogy between pressure-driven interchanges in cylinder and gravitational interchanges in plane slab geometry. Without flow, the interchange criterion (7.199)[1] for static plasmas is recovered.

▷ **Transfer of cluster sequences** In the incompressible limit ($\gamma p \to \infty$, $M_c^2 \to 1$) of the criterion (13.137), the trans-Alfvénic and trans-slow factors coincide and one recovers the criteria (13.88)(a) and (13.88)(b) [or (13.90)] (involving first derivatives) studied in Section 13.2.3. Recall that, for the example of the equilibrium of Fig. 13.4(b), no clustering occurs at the resonant point (indicated with "1st") but, instead, there is a sequence of unstable modes clustering at the extremum of the slow/Alfvén continuum $\Omega^-_{A/S}$ (indicated with "2nd") since the cluster condition (13.78)(b) (involving second derivatives) is violated there. This transfer of cluster sequences is demonstrated by the spectra and eigenfunctions of Figs. 13.11–13.13.

In the compressible case, for the plane as well as the cylindrical equilibrium, reconsidering the expansion (13.127), e.g. for the slow continua,

$$ \Omega^\pm_S \approx \Omega_0 + (\widetilde{M} \pm M_c)(F/\sqrt{\rho})s + \tfrac{1}{2}\Omega^{\pm\prime\prime}_S s^2 \,, \tag{13.138} $$

it is clear that, when $\widetilde{M} \to \pm M_c$, the given derivation of the cluster criterion (13.134) needs to be modified since the second derivatives become dominant then. In that case, the sequence of unstable modes at the resonant point may be transferred to an unstable sequence clustering at the minimum ($\Omega^{\pm\prime\prime}_S > 0$) or maximum ($\Omega^{\pm\prime\prime}_S < 0$) of the slow continua. For gravitational interchanges in plane slab geometry, the relevant cluster conditions, derived by van der Holst *et al.* [455], are generalizations (replacing ω''_S by $\Omega^{\pm\prime\prime}_S$) of the static conditions (7.180) and (7.181) of Volume [1]. For the analogous cylindrical case, this transfer is discussed at length for purely longitudinal flows by Bondeson *et al.* [56] and for rotations by Wang *et al.* [472]. ◁

13.3.3 *WKB approximation*

In the astrophysics literature on MHD instabilities in cylindrical models, e.g. of rotating stars [145] or accretion disks [127, 433, 276], frequent use is made of the WKB approximation for waves in inhomogeneous media. This leads to a so-called *local dispersion equation*. Let us apply standard WKB analysis to the second order differential equation (13.107) for χ, written in the form $(f\chi')' - g\chi = 0$. This involves solutions of the form

$$ \chi(r) = p(r)\,\exp\left[\mathrm{i}\int q(r)\,dr\right], \tag{13.139} $$

where BCs on χ lead to quantization of the definite form of the integral over q. The expressions $p(r)$ and $q(r)$ should be determined such that these solutions are correct to leading order in the inhomogeneity. Substitution into Eq. (13.107) gives

$$- fpq^2 - gp + \mathrm{i}(f'pq + 2fp'q + fpq') + fp'' + f'p' = 0. \tag{13.140}$$

Introducing a length scale L for the radial inhomogeneities ($\sim$ the width of the disk), where $(qL)^{-1} \ll 1$, and neglecting the second order terms $fp'' + f'p'$ compared to the first order ones, this yields the following expressions for the effective radial "wave number" q and the slowly varying amplitude p:

$$q \approx (-g/f)^{1/2}, \qquad p \approx (-fg)^{-1/4}. \tag{13.141}$$

Hence, the local dispersion equation reads:

$$q^2 \approx -\frac{g}{f} = \frac{D}{N}\left[A + \frac{B}{D} + \left(\frac{C}{D}\right)' \right]. \tag{13.142}$$

Here, first derivatives in the expansions for D, N, A, B and C are to be kept, but second derivatives should be dropped since terms of $\mathcal{O}(qL)^{-2}$ have been assumed to be negligible in the derivation above. Accordingly, the last term of equation (13.142) contributes, so that the local dispersion equation becomes a polynomial of tenth degree in ω.

▷ **Erroneous form of the local dispersion equation** If one naively inserts solutions of the form (13.139) for both $\chi(r)$ and $\Pi(r)$ into the system of first order ODEs (13.116), one obtains the dispersion equation

$$q_{\mathrm{wrong}}^2 \approx -\frac{1}{N^2}(C^2 + DE) \overset{(13.117)}{=} \frac{D}{N}\left(A + \frac{B}{D} \right), \tag{13.143}$$

which is evidently wrong since the last term of Eq. (13.142) is missing. This erroneous dispersion equation was actually presented in Eq. (9) of Keppens *et al.* [255], fortunately without implications for the explicit spectral results presented there since they were obtained by numerical integration of the exact system of differential equations. The correct form of the dispersion equation is obtained from the first order system by admitting a phase difference between the variables χ and Π:

$$\Pi(r) = [s(r) + \mathrm{i}t(r)] \, \exp\left[\mathrm{i} \int q(r)\,dr \right]. \tag{13.144}$$

One then has to determine the four quantities $p(r)$, $q(r)$, $s(r)$, $t(r)$ from the real and imaginary parts of the two equations (13.116) such that the expressions are valid to $\mathcal{O}(qL)^{-1}$. Of course, this again leads to the correct local dispersion equation (13.142). It demonstrates though that, for WKB analysis, the second order differential equation (13.107) is to be preferred over the first order system (13.116). ◁

It may come as a surprise that the correct dispersion equation is a polynomial of tenth degree in ω rather than a sixth degree polynomial, as would be the case if a homogeneous plasma permitting six discrete eigenvalues (forward and backward slow, Alfvén and fast waves) were just slightly perturbed. Apparently, it is impossible to "slightly perturb". The reason is, once more, that apparent singularities $D = 0$ lead to spurious roots due to the term $\sim D(C/D)'$ of the local dispersion equation (13.142). In general, this additional term is not negligible and, hence, necessary to get correct results from the local dispersion equation. However, it also introduces spurious roots for frequencies on one side of an apparent singularity $D(r; \omega) = 0$, viz. the side where $q^2 \to \infty$. At points in the radial domain corresponding to those frequencies, the function $g(r)$ jumps from $-\infty$ to $+\infty$ so that the assumptions underlying the WKB analysis are violated. These roots should be eliminated a posteriori. Hence, except that the local dispersion equation (13.142) provides useful estimates of growth rates (see e.g. Section 13.4.2 on the magneto-rotational instability), solution of the exact differential equation (13.107) is greatly to be preferred.

Whereas the approach of cluster points or edges of the continua $\{\Omega_S^\pm\}$, $\{\Omega_A^\pm\}$ and $\{\Omega_F^\pm\}$ implies that the local approximation becomes ever more valid, the approach of the edges of the apparent singularity ranges $\{\Omega_{s0}^\pm\}$ and $\{\Omega_{f0}^\pm\}$ implies precisely the opposite. Even in the absence of the mentioned derivative term, proper analysis of the latter two frequency ranges reveals that the WKB approximation is not valid there since the solutions are not oscillatory but evanescent (exhibiting turning point behavior). Actual computation of growth rates of instabilities for accretion disks from the local dispersion equation, involving matching of the WKB solutions to analytic approximations in the turning point regions, has been carried out by Blokland *et al.* [53].

13.4 Rotational instabilities

13.4.1 Rigid rotation of incompressible plasmas

In order to obtain some understanding of the Coriolis effects on the distribution of eigenvalues in the complex ω plane, we consider a special case that can be solved analytically, both with respect to the solution path and with respect to the eigenvalues on it. Since the coefficients of the ODEs (13.107) and (13.116) are rather formidable, making explicit analysis cumbersome, we first simplify them by taking the *incompressible limit of the spectral equation* in its second order form,

$$r\frac{d}{dr}\left[\frac{\rho\widetilde{\omega}^2 - F^2}{h^2 r}\frac{d\chi}{dr}\right] - \left[\rho\widetilde{\omega}^2 - F^2 + r\left(\frac{B_\theta^2 - \rho v_\theta^2}{r^2}\right)' + \rho'\Phi'\right.$$
$$\left. - \frac{4k^2(B_\theta F + \rho v_\theta \widetilde{\omega})^2}{h^2 r^2(\rho\widetilde{\omega}^2 - F^2)} - r\left(\frac{2m(B_\theta F + \rho v_\theta \widetilde{\omega})}{h^2 r^3}\right)'\right]\chi = 0\,, \quad (13.145)$$

or in its first order form,

$$\frac{d}{dr}\begin{pmatrix} \chi \\ \Pi \end{pmatrix} + \begin{pmatrix} \hat{C} & \hat{D} \\ \hat{E} & -\hat{C} \end{pmatrix}\begin{pmatrix} \chi \\ \Pi \end{pmatrix} = 0 , \tag{13.146}$$

$$\hat{C} \equiv \frac{C}{N} \equiv \frac{m\alpha}{r} , \qquad \hat{D} \equiv \frac{D}{N} \equiv -\frac{h^2 r}{\rho\widetilde{\omega}^2 - F^2} ,$$

$$\hat{E} \equiv \frac{E}{N} \equiv -\frac{1}{r}(1 - \alpha^2)(\rho\widetilde{\omega}^2 - F^2) - \left(\frac{B_\theta^2 - \rho v_\theta^2}{r^2}\right)' - \frac{1}{r}\rho'\Phi' .$$

Defining the function

$$\alpha(r) \equiv \frac{2(B_\theta F + \rho v_\theta \widetilde{\omega})}{r(\rho\widetilde{\omega}^2 - F^2)} , \tag{13.147}$$

the tangential components of $\boldsymbol{\xi}$ and the total pressure perturbation simplify to

$$\eta = -\frac{1}{h^2 r}(g\chi' - fk\alpha\chi) , \qquad \zeta = -\frac{1}{h^2 r}(f\chi' + gk\alpha\chi) , \tag{13.148}$$

$$\Pi = -\mathrm{i}\,\frac{\rho\widetilde{\omega}^2 - F^2}{k}\,\xi_z = \frac{\rho\widetilde{\omega}^2 - F^2}{h^2 r}\left(\chi' + \frac{m}{r}\alpha\chi\right) , \tag{13.149}$$

whereas the azimuthal component of $\boldsymbol{\xi}$ in the Coriolis part of the Doppler–Coriolis shift expression (13.121) for $\overline{V}$ becomes

$$\mathrm{i}\xi_\theta \equiv b_z\eta + b_\theta\zeta = -\frac{1}{h^2 r}\left(\frac{m}{r}\chi' - k^2\alpha\chi\right) . \tag{13.150}$$

These expressions suggest how the equilibrium should be chosen in order to get analytical solutions.

Neglecting gravity and choosing equilibria where both $B_\theta \sim r$ and $v_\theta \sim r$, so that two coefficients of the spectral equations (13.145) and (13.146) vanish and α becomes constant, the solutions can be written in terms of Bessel functions. This amounts to an extension of the static equilibrium with *helical magnetic field of constant pitch*, considered in Section 9.3.2 [1], to a *rigidly rotating equilibrium*:

$$B_\theta = \mu r , \quad B_z = B_0 , \quad \rho = \rho_0 , \quad v_\theta = \lambda r , \quad v_z = \text{const.} \tag{13.151}$$

(We here indicate the rotation frequency with λ, rather than the more usual Ω, to avoid confusion with the Doppler shifted frequencies for which we already exploited that symbol.) We assume the plasma to be confined within a cylinder of radius $r = a$. Of course, we exploit scale independence by setting $B_0 = 1$, $\rho_0 = 1$ and $a = 1$. In the absence of gravity, the equilibrium condition (12.30) demands that the pressure profile (not appearing in the spectral equation though) is parabolic:

$$p = p_0 - (\mu^2 - \tfrac{1}{2}\lambda^2)r^2 . \tag{13.152}$$

For these equilibria, the spectral equation (13.145) reduces to

$$ r \frac{d}{dr} \left[\frac{1}{h^2 r} \frac{d\chi}{dr} \right] - \left[1 - \left(\alpha^2 - \frac{2m\alpha}{h^2 r^2} \right) \frac{k^2}{h^2} \right] \chi = 0 , \qquad (13.153) $$

where the parameter

$$ \alpha \equiv \frac{2(\mu F + \lambda \widetilde{\omega})}{\widetilde{\omega}^2 - F^2} , \quad \text{with} \ \ F \, (\equiv \omega_{\mathrm{A}}) \equiv m\mu + k , \qquad (13.154) $$

is now constant. (We here exploit a more appropriate definition, $\alpha \equiv \alpha_\mathrm{o}/m$, where the corresponding parameter (9.97) of Section 9.3.2 [1] is indicated by α_o.) Note that all equilibrium dependence is now represented by this parameter.

Defining a kind of modified wave number

$$ k^* \equiv k\sqrt{1 - \alpha^2} \equiv i\ell^* , \qquad \ell^* \equiv k\sqrt{\alpha^2 - 1} , \qquad (13.155) $$

the solutions of Eq. (13.153) are expressions in terms of modified Bessel functions with argument $k^* r$ when $\alpha^2 < 1$, or ordinary Bessel functions with argument $\ell^* r$ when $\alpha^2 > 1$. (We have encountered such solutions in Section 9.3 [1], with $\alpha = 0$, for perturbations of a vacuum field in Eq. (9.57) and, with different definitions of α, for a θ-pinch in Eq. (9.76) and for a constant-pitch magnetic field in Eq. (9.98).) Suppressing the arbitrary amplitude factor, these solutions read:

$$ \chi = \begin{cases} k^* r \mathrm{I}'_{|m|}(k^* r) - m\alpha \mathrm{I}_{|m|}(k^* r) & (0 \le \alpha^2 < 1) \\[2mm] [|m| + \frac{1}{2}(1 \pm 1)h^2 r^2] r^{|m|} & (\alpha = \pm \mathrm{sg}(m)) \\[2mm] \ell^* r \mathrm{J}'_{|m|}(\ell^* r) - m\alpha \mathrm{J}_{|m|}(\ell^* r) & (\alpha^2 > 1) \end{cases} \qquad (13.156) $$

(The reader may wish to obtain these solutions, which here "dropped from the air", by deriving the equivalent second order ODE for Π, which is just a Bessel equation, so that $\Pi \sim \mathrm{I}_{|m|}(k^* r) \sim \mathrm{J}_{|m|}(\ell^* r)$, and expressing χ in terms of Π' and Π.) These solutions satisfy the BC on the origin, $\chi(0) = 0$. The complete BVP involves the BC on the wall as well, obviously requiring oscillatory Bessel functions:

$$ \chi(1) = \ell^* \mathrm{J}'_{|m|}(\ell^*) - m\alpha \mathrm{J}_{|m|}(\ell^*) = 0 \quad \Rightarrow \quad \ell^* = \ell_i^* \ \ (i = 1, 2, \ldots) . \qquad (13.157) $$

This yields a discrete set of solutions labeled by the index i. Through the relationships (13.155) between ℓ^* and α and (13.154) between α and ω, this determines the *discrete eigenvalues*. The spectral equation (13.153) is singular for $\alpha^2 \to \infty$, when the constant factor $\widetilde{\omega}^2 - F^2$ vanishes. This yields the *forward and backward degenerate Alfvén–slow continua*, collapsed into the two points

$$ \Omega_{\mathrm{A/S}}^{\pm} \equiv \Omega_0 \pm \omega_{\mathrm{A}} , \quad \Omega_0 = m\lambda + k v_z , \quad \omega_{\mathrm{A}} \, (\equiv F) = m\mu + k . \qquad (13.158) $$

These frequencies act as the limiting frequencies of the discrete spectrum, determined by the BVP (13.157), which clusters both from above and from below so that *four* discrete sub-spectra are obtained. Hence, the label i should be supplemented with some other discrete labels to obtain four infinite sequences.

The additional labels, distinguishing the sub-spectra, are obtained by first solving the quadratic equation (13.154) for $\omega(\alpha)$ and then connecting α to ℓ_i^* by means of the quadratic equation (13.155):

$$\omega = \omega_{a/s,i}^{s_1,s_2} \equiv \Omega_0 + \lambda/\alpha_i + s_1\sqrt{F^2 + 2(\mu/\alpha_i)F + (\lambda/\alpha_i)^2}\,, \quad s_1 = \pm 1\,,$$

$$\alpha = \alpha(\ell_i^*) \equiv -s_2\sqrt{1 + (\ell_i^*/k)^2}\,, \quad s_2 = \pm 1\,. \tag{13.159}$$

Consistent with our previous notation, the Doppler shifted frequencies are indicated with upper case symbol Ω and subscripts $_{A/S}$ for the continua, but with lower case symbol ω and subscripts $_{a/s}$ for the discrete eigenvalues. With the additional labels s_1 and s_2, this yields four discrete sub-spectra along the real ω axis, if the discriminant in the expression (13.159)(a) for ω is positive, viz. forward or backward ($s_1 = 1$ or -1) and Sturmian or anti-Sturmian ($s_2 = 1$ or -1) Alfvén/slow modes. If the discriminant becomes negative, the forward Sturmian and backward anti-Sturmian branches meet and then bifurcate into branches with overstable modes (labeled $^{0+}$) and damped modes (labeled $^{0-}$) on it. This yields the following schematic structure for the spectrum of modes:

$$\left[\omega_{a/s,i}^{-+}\,,\ \Omega_{A/S}^-\,,\ \omega_{a/s,i}^{--}\right]\,,\ \begin{bmatrix}\omega_{a/s,i}^{0+}\\\omega_{a/s,i}^{0-}\end{bmatrix}\,,\ \left[\omega_{a/s,i}^{++}\,,\ \Omega_{A/S}^+\,,\ \omega_{a/s,i}^{+-}\right].$$

$$\textit{(backward)} \qquad \textit{(complex)} \qquad \textit{(forward)} \qquad (13.160)$$

This is illustrated in Fig. 13.14 for a particular, unstable, case ($\lambda < \mu$). However, anticipating the discussion below, it is to be noted that the association of the label s_1 with forward or backward modes and of the label s_2 with Sturmian or anti-Sturmian modes only holds asymptotically, in the approach of the cluster points $\Omega_{A/S}^+$ and $\Omega_{A/S}^-$.

The top frame of Fig. 13.14 shows the spectrum for the static case. It consists of four discrete sub-spectra, clustering at the degenerate Alfvén/slow continuum frequencies $\pm\omega_{A/S}$, and located partially along the real axis (stable modes) and partially along the imaginary axis (exponentially growing and damped modes). The bottom frame shows how this spectrum is modified by rotation. This produces a *fixed Doppler shift* $\Omega_0 = m\lambda + kv_z$ (the dashed vertical line) as well as a *varying Coriolis shift,* deforming the locus of unstable modes into the circular path

$$\omega = \hat{\sigma} + \hat{\tau}e^{i\phi} \quad (0 \le \phi < 2\pi)\,. \tag{13.161}$$

The explicit expressions for $\hat{\sigma}$ and $\hat{\tau}$ are given below. Increasing the rotation rate $|\lambda|$, the radius of this circular path shrinks, until it finally vanishes at one

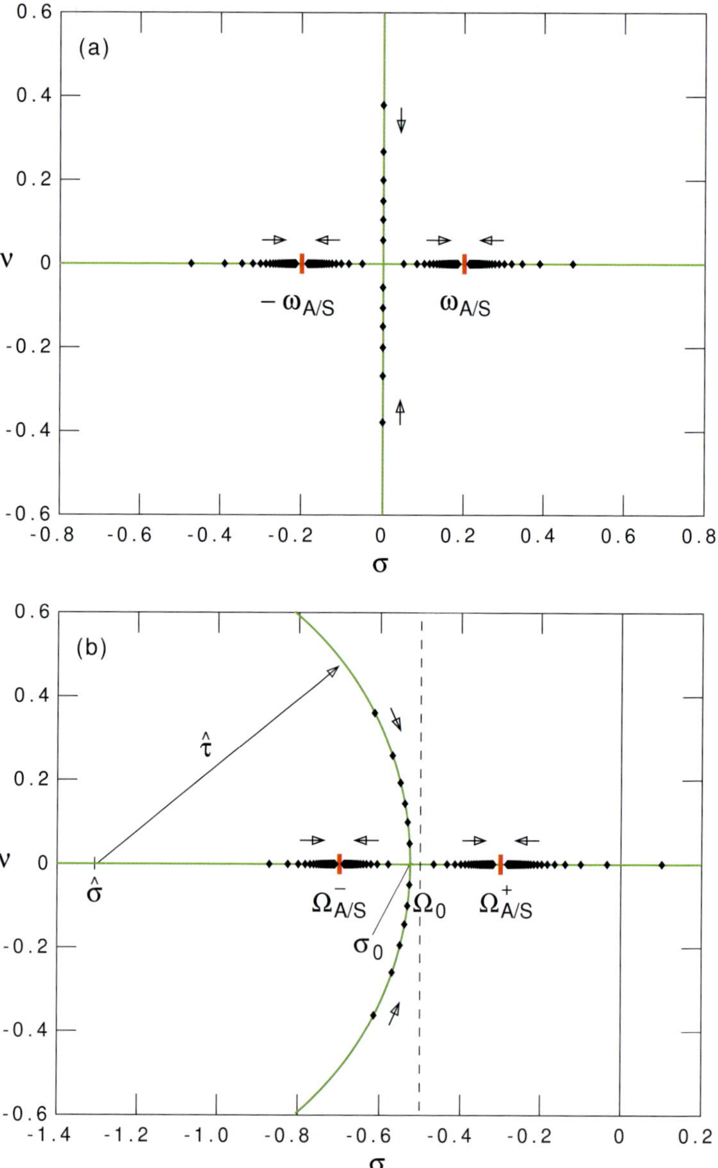

Fig. 13.14 Spectrum of incompressible cylindrical plasma with magnetic field of constant pitch, $\mu \equiv B_\theta/r = 1$, for mode numbers $m = -2$ and $k = 2.2$: (a) static case, $\lambda \equiv v_\theta/r = 0$; (b) rigidly rotating plasma, $\lambda = 0.25$, $v_z = 0$. Left solution paths are green, arrows indicate the sense of monotonicity of the four sequences of discrete modes clustering at the degenerate Alfvén /slow continua.

of the degenerate continua $\Omega^{\pm}_{\text{A/S}}$ when $|\lambda| = |\mu|$. (Notice that the Doppler shift remains, but the Coriolis shift vanishes there; in Chapter 18, we will encounter continua where this is *not* the case.) For faster rotation, $|\lambda| > |\mu|$, the plasma is stable. Hence, for this particular class of equilibria, rotation exclusively stabilizes.

Let us analyze how the solution obtained is related to the general method of solution paths and oscillation theorems developed in Sections 12.3, 13.1 and 13.2. In the present case, the solution path $\mathcal{P}^{\mathrm{L}}$ of left solutions is obtained by dropping the right BC (13.157) so that ℓ^*, or rather α, becomes a continuous parameter:

$$\alpha = -s_2\sqrt{1 + (\ell^*/k)^2}\,, \quad s_2 = \pm 1\,. \tag{13.162}$$

The solution path is then simply given by the solution of the quadratic equation

$$\omega = \Omega_0 + \lambda/\alpha + s_1\sqrt{F^2 + 2(\mu/\alpha)F + (\lambda/\alpha)^2}\,, \quad s_1 = \pm 1\,. \tag{13.163}$$

This path is along the real ω axis if the discriminant is positive. When it vanishes, the two real sub-paths meet and turn into the complex plane. This happens at two frequencies:

$$\sigma = \sigma_{0,1} = \overline{V}(\nu = 0) = \hat{\sigma} \pm \hat{\tau}$$

$$= \Omega_0 + \lambda/\alpha_{0,1} = \Omega_0 - F\frac{\mu}{\lambda}\left(1 \mp \sqrt{1 - \lambda^2/\mu^2}\right). \tag{13.164}$$

When increasing the counting parameter ℓ^*, the subscript $_0$ indicates the location on the sub-path $\mathcal{P}_{\mathrm{s}}^{\mathrm{L}}$ where the modes become unstable and the subscript $_1$ indicates the location on the sub-path $\mathcal{P}_{\mathrm{u}}^{\mathrm{L}}$ where they become stable again. Hence, σ_0 and σ_1 indicate the *two marginal stability transitions* (see Fig. 13.14, the point σ_1 lies outside the frame). Instability amounts to determining whether any of the eigenvalues have crossed the marginal point σ_0, but not yet σ_1.

For the range of α in between α_0 and α_1, the discriminant is negative and the real and imaginary parts of Eq. (13.163) provide the expressions for $\sigma_{\mathrm{u}}(\alpha)$ and $\nu_{\mathrm{u}}(\alpha)$ of the sub-path $\mathcal{P}_{\mathrm{u}}^{\mathrm{L}}$ of unstable solutions. This yields the mentioned circle of radius $\hat{\tau}$ around the point $\sigma = \hat{\sigma}$, defined by

$$\hat{\sigma} \equiv \tfrac{1}{2}(\sigma_0 + \sigma_1) = \Omega_0 - \frac{\mu}{\lambda}F\,, \quad \hat{\tau} \equiv \tfrac{1}{2}|\sigma_0 - \sigma_1| = \left|F\frac{\mu}{\lambda}\sqrt{1 - \lambda^2/\mu^2}\right|. \tag{13.165}$$

Of course, this is the solution path which, in general, is obtained from the path equation (13.120), as one might verify by substituting the Bessel function solutions (13.156) into the expression (13.121) for $\overline{V}$ and evaluating the integrals. In this manner, by varying the continuous parameter α, one obtains the solution paths $\mathcal{P}_{\mathrm{s}}^{\mathrm{L}}$ and $\mathcal{P}_{\mathrm{u}}^{\mathrm{L}}$ and, with minor additional effort, the discrete eigenvalues on them by determining the zeros of the alternator

$$R(\alpha) \equiv \frac{\xi(1)}{\Pi(1)} \sim \frac{\ell^* J'_{|m|}(\ell^*)}{J_{|m|}(\ell^*)} - m\alpha\,. \tag{13.166}$$

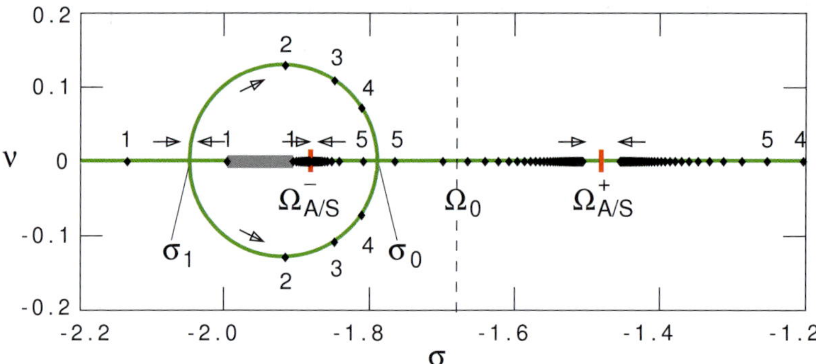

Fig. 13.15 Spectrum of rigidly rotating incompressible cylindrical plasma with magnetic field of constant pitch, $\mu = 1$, $\lambda = 0.84$, $v_z = 0$, $m = -2$ and $k = 2.2$. The first five discrete modes of each sequence are numbered (outside frame: three eigenvalues $\sigma \approx -0.686, -0.990, -1.127$ of forward anti-Sturmian sequence). The grey strip indicates non-monotonicity in the Doppler–Coriolis range (σ_1, σ_0).

In this case, this is nothing else but rephrasing of the BVP (13.157). However, recall that the generic procedure does not depend on being able to reduce the solution to known special functions but just requires a numerical algorithm for solving the general spectral differential equation, where extension with compressibility and arbitrary distributions of magnetic field and rotation frequency is just a matter of defining appropriate function routines for the coefficients.

With respect to our oscillation theorem $\mathcal{R}$, the two points σ_0 and σ_1 mark the *Doppler–Coriolis indefinite range* $\sigma_1 \leq \sigma \leq \sigma_0$, introduced in Eq. (12.142) of Section 12.3.2, where the sign of $\sigma - \overline{V}(\nu = 0)$ becomes indefinite. This implies possible non-monotonicity of the real eigenvalues in that range, as illustrated in Fig. 13.15. Compared to the equilibrium of Fig. 13.14, the rotation rate λ is increased so that the eigenvalues have spread over the full circle and one of them has "intruded" the space of the backward Sturmian sequence $\omega_{a/s}^{-+}$ (which is squeezed into the narrow range to the left of the cluster point). Consequently, since two discrete modes, both labeled with a 1 (on opposite sides of the grey strip in the figure), are situated on the same sub-path, the discrete spectrum cannot be monotonic in the number of nodes of the eigenfunction there. This is in agreement with the exclusion of the Doppler–Coriolis indefinite range (σ_1, σ_0) in oscillation theorem $\mathcal{R}$ of Section 13.1.3. Also notice that the "intruder" is a result of the merger of the two stable sequences $\omega_{a/s}^{--}$ and $\omega_{a/s}^{++}$ at σ_0 and subsequent merger of the resulting unstable sequences $\omega_{a/s}^{0+}$ and $\omega_{a/s}^{0-}$ at σ_1, so that the superscripts s_1 and s_2 of the modes cannot simply refer to backward/forward and Sturmian/anti-Sturmian anymore.

The spectrum of Fig. 13.15 also illustrates the role of the two marginal points for the formulation of stability or instability criteria. To guarantee the existence of an unstable range of eigenvalues ℓ_i^*, i.e. a range of the associated parameter $\alpha(\ell_i^*)$ defined by Eq. (13.159) in between the two values $\alpha_0 = \lambda/\tilde{\sigma}_0$ and $\alpha_1 = \lambda/\tilde{\sigma}_1$ given by Eq. (13.164), the following criterion on the magnitude of the parallel wave vector $k_\parallel \equiv F \equiv m\mu + k$ should be satisfied:

$$\frac{|\mu|(1 - \sqrt{1 - \lambda^2/\mu^2})}{\sqrt{1 + (\ell_i^*/k)^2}} < |m\mu + k| < \frac{|\mu|(1 + \sqrt{1 - \lambda^2/\mu^2})}{\sqrt{1 + (\ell_i^*/k)^2}} . \tag{13.167}$$

Given the eigenvalues for the lowest modes of the present equilibrium ($\ell_1^* \approx 4.267$, $\ell_2^* \approx 7.557$, $\ell_3^* \approx 10.766$, $\ell_4^* \approx 13.946$, $\ell_5^* \approx 17.113$), one easily checks that only ℓ_2^*, ℓ_3^* and ℓ_4^* are unstable, in agreement with Fig. 13.15. One obtains the maximum unstable range for instabilities *localized about a particular rational surface,* i.e. by assuming large values of the mode numbers, $|m| \to \infty$ and $|k| \to \infty$ but such that $|m\mu + k|$ remains finite, so that the denominators $\to 1$:

$$|\mu|(1 - \sqrt{1 - \lambda^2/\mu^2}) < |m\mu + k| < |\mu|(1 + \sqrt{1 - \lambda^2/\mu^2}). \tag{13.168}$$

For static (non-rotating) plasmas, this condition reduces to the usual instability range of *quasi-interchange modes* on the two sides of $k_\parallel = 0$, with marginal stability at $k_\parallel = 0$ (see Fig. 9.15 [1]). With rotation ($\lambda \neq 0$), the latter wave vector direction becomes stable, and the unstable range even completely disappears when the rotation is fast enough: *rigid rotation stabilizes quasi-interchange instabilities if* $|\lambda| > |\mu|$. In terms of the so-called poloidal Alfvén Mach number, this stability condition requires the rotation speed to be *super-Alfvénic:*

$$M_{\mathrm{A,p}}^2 \equiv v_\theta^2/v_{\mathrm{A}\theta}^2 = \lambda^2/\mu^2 < 1. \tag{13.169}$$

This condition is strictly valid for incompressible plasmas only.

Discrete modes on circular solution paths were first found by Spies [416] for interchange modes in a rigidly rotating θ-pinch plasma with a density profile. Hameiri [208] extended this analysis to admit shear flow stabilization of the centrifugal analogy of the Rayleigh–Taylor instability. Analytical and numerical results on spectra of rigidly rotating plasmas with a constant pitch magnetic field, also permitting compressibility so that the Alfvén/slow degeneracy is lifted, have been derived by Nijboer *et al.* [352].

Of course, the examples of Kelvin–Helmholtz instabilities of Section 13.2 will prevent us from drawing oversimplified conclusions with respect to the stability, and particularly with respect to the geometry of the solution paths in the complex ω plane, for arbitrary velocity distributions (with velocity shear) from the very special equilibrium treated in this section. Whereas the example does serve to highlight the important effects of the Coriolis shift on the spectrum of rotating plasmas, it is actually a rather devious, possibly misleading, representation of the theory of spectra of stationary plasmas. The point is that a *real* representation of the Bessel function solutions (13.156) suffices here, since the zeros given by Eq. (13.157) are real, whereas the path on which these solutions are situated is obtained by solving the quadratic equation (13.154) for ω, again assuming *real* α. In other words: the example is so simple that obtaining the solution path as well as the eigenvalues is nearly trivial. Without the examples of Section 13.2.2 on the breakup of solution paths by shear flow, one might be seduced to draw false conclusions from this example. Nevertheless, it is useful as an illustration of how rotation changes the geometry of solution paths through the Coriolis effect. For example, contrary to the spectra of plane shear flow equilibria of Section 13.2, where at least one solution sub-path off the real axis is always present (though there may not be unstable modes on it), with rotation even this may not be the case. In the present example, for $|\lambda| \geq |\mu|$, the plasma is not only stable, but solution paths away from the real axis do not exist: Doppler shifts and Coriolis shifts are entirely different physical phenomena.

13.4.2 *Magneto-rotational instability: local analysis*

As an application of the rather formidable spectral problem for cylindrical plasmas with flow outlined in Section 13.3.1, we will consider the stability problem of an accretion disk around a compact object. This object may be a young stellar object (mass $M_* \sim M_\odot$) or an active galactic nucleus (massive black hole with $M_* \sim 10^9 M_\odot$). For this incredible range of objects, the same fundamental problem arises, viz. how can accretion on such objects occur at all on a reasonable time scale? Without dissipation this would be impossible, because a disk (that would already be there for whatever reason) would conserve angular momentum and, hence, rotate forever without change. Some form of viscosity is needed to facilitate the transfer of angular momentum to larger distances. But, with viscosity there is still a problem since the ordinary molecular viscosity coefficient is much too small to produce accretion of the correct order of magnitude. A turbulent increase of this coefficient might give the right answer. For such a process, small-scale instabilities are needed. However, no hydrodynamic instabilities were found for these disks. It is generally assumed that the resolution of this problem involves the *magneto-rotational instability (MRI)*, as first suggested by Balbus and Hawley [18]. The instability itself was already described thirty years earlier by Velikhov [467] and Chandrasekhar [83], of course, without application to accretion disks. We here present a derivation of the instability conditions to demonstrate that the analysis of Section 13.3.1 is actually applicable to genuine astrophysical objects.

We simplify the schematic axi-symmetric (2D) representation of an accretion disk, shown in Fig. 4.1 [1], even further by *neglecting vertical equilibrium varia-*

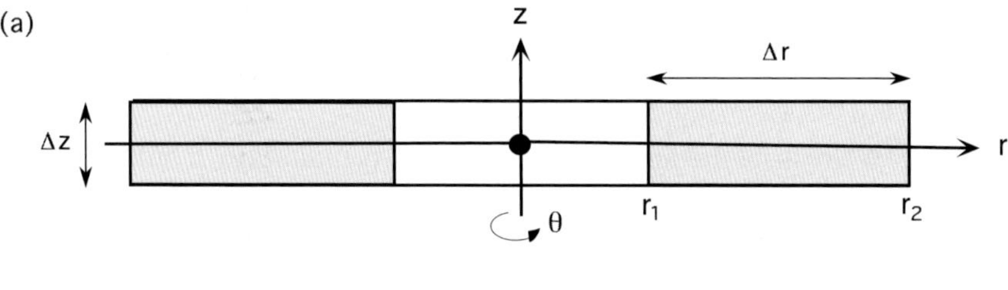

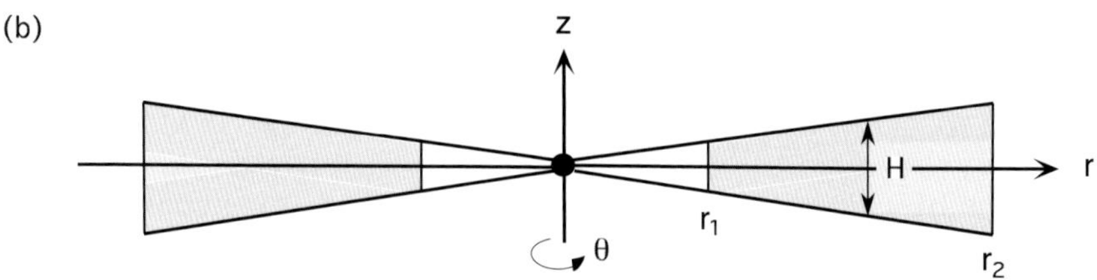

Fig. 13.16 Schematic geometries for accretion disk: (a) cylindrical slice model; (b) quasi two-dimensional model.

tions so that the annular cylindrical (1D) thin slice model of Fig. 13.16(a) is obtained. (One might object that is not a disk at all anymore, but this is how plasma-astrophysicists grapple with the problem of turbulent transport in these objects.)

We specify the gravitational potential to be due to a compact object of mass M_* at the origin,

$$\Phi = -\frac{GM_*}{\sqrt{r^2 + z^2}} \approx -\frac{GM_*}{r}, \tag{13.170}$$

where the latter, cylindrical, approximation is appropriate for *short wavelengths fitting the disk in the vertical direction:*

$$k\,\Delta z \gg 1. \tag{13.171}$$

We wish to investigate the instabilities in the limit of *small magnetic fields:*

$$\beta \equiv \frac{2p}{B^2} \gg 1. \tag{13.172}$$

This approximation justifies the simplification of the spectral equation (13.107) to the *incompressible limit* (13.145). It is also consistent with incompressibility to assume a *constant density* so that the explicit gravitational term $\rho'\Phi'$ disappears from the spectral equation (of course, not the implicit dependence through the equilibrium equation (12.30) which the functions ρ, p, B_θ, B_z and v_θ still have to satisfy). Next, we simplify the magnetic configuration to the most simple form consistent with a rotating magnetized disk by choosing a *purely vertical and constant mag-*

netic field and a *purely azimuthal velocity field:*

$$B_\theta = 0 , \quad v_z = 0 \quad \Rightarrow \quad \omega_A = kB_z/\sqrt{\rho} = \text{const} , \quad \Omega_0 = mv_\theta/r . \quad (13.173)$$

Furthermore, we restrict the analysis to *vertical wave numbers k* only:

$$m = 0 \quad \Rightarrow \quad \Omega_0 = 0 , \text{ so that } \tilde{\omega} = \omega . \quad (13.174)$$

The spectral problem then simplifies to the solution of

$$(\omega^2 - \omega_A^2) r \frac{d}{dr}\left(\frac{1}{r}\frac{d\chi}{dr}\right) - k^2\left[\omega^2 - \omega_A^2 - r\left(\frac{v_\theta^2}{r^2}\right)' - \frac{4\omega^2 v_\theta^2/r^2}{\omega^2 - \omega_A^2}\right]\chi = 0 , \quad (13.175)$$

subject to the BCs $\chi(r_1) = \chi(r_2) = 0$. As always, these conditions do not imply that no dynamics occurs outside this range, but just that we wish to minimize the consequences of our ignorance about it by considering modes that are localized within this range. With all these approximations, nearly everything is constant, except the angular rotation frequency $\lambda(r) \equiv v_\theta(r)/r$.

To concord with the astrophysical literature, we replace the symbol λ by the standard notation for the *angular frequency* $\Omega(r)$ (not be confused with $\Omega_0(r)$ now) and a derived quantity called the *epicyclic frequency* $\kappa(r)$:

$$\Omega\,[\equiv \lambda] \equiv \frac{v_\theta}{r} , \qquad \kappa^2 \equiv \frac{1}{r^3}(r^4\Omega^2)' = 2r\Omega\Omega' + 4\Omega^2 . \quad (13.176)$$

The latter quantity is a measure of how much the specific angular momentum $\hat{L}(r)$ of the disk deviates from a constant distribution:

$$\hat{L} \equiv L/\rho \equiv rv_\theta \equiv r^2\Omega \quad \Rightarrow \quad \hat{L}' = 0 , \text{ if } \kappa^2 = 0 . \quad (13.177)$$

In terms of these quantities, the spectral equation may be written as

$$r\frac{d}{dr}\left(\frac{1}{r}\frac{d\chi}{dr}\right) - k^2\left[1 - \frac{\kappa^2(r)}{\omega^2 - \omega_A^2} - \frac{4\omega_A^2\Omega^2(r)}{(\omega^2 - \omega_A^2)^2}\right]\chi = 0 . \quad (13.178)$$

This equation exposes the important difference between the stability of a hydrodynamic disk ($\omega_A^2 = 0$) and the stability of a magnetohydrodynamic disk ($\omega_A^2 \neq 0$), even when the magnetic field is arbitrarily small ($\omega_A^2 \to 0$).

In the hydrodynamic limit, one easily derives *Rayleigh's circulation criterion* (see below, or Drazin and Reid [124], p. 69), stating that the fluid is stable to axisymmetric disturbances ($m = 0$) when

$$\kappa^2 \geq 0 \quad \text{everywhere} , \quad (13.179)$$

which is why κ^2 is called Rayleigh's discriminant in hydrodynamics. Neglecting the influence of the pressure gradient on the equilibrium relation (12.30), this criterion is satisfied for *Keplerian rotation*,

$$\frac{\rho v_{\theta,\mathrm{K}}^2}{r} = \rho\Phi' = \rho\frac{GM_*}{r^2} \;\Rightarrow\; \Omega_{\mathrm{K}}^2 = \frac{GM_*}{r^3} \;\Rightarrow\; \kappa_{\mathrm{K}}^2 = \frac{GM_*}{r^3} > 0\,, \quad (13.180)$$

so that such equilibria are hydrodynamically stable under rather wide assumptions. This explains why interest shifted to magnetohydrodynamic instabilities to explain the turbulent increase of the dissipation processes in accretion disks.

(The restriction to incompressible plasmas and axi-symmetric modes is actually made to highlight the merits of the MRIs. It implies exclusion of quite a number of instabilities, e.g. the hydrodynamic one found by Papaloizou and Pringle [361] (see Pringle and King [388], Chapter 14), which is a global $m \neq 0$ overstability operating in compressible thin disks as well as thick tori for $\kappa^2 = 0$.)

To obtain a stability criterion for the magnetohydrodynamic case, we first have to construct the solution path $\mathcal{P}_{\mathrm{u}}$ from the path equation (13.120), with the expression (13.121) for the Doppler–Coriolis shift. Under the present assumptions, $m = 0$, $v_z = 0$, $B_\theta = 0$ and incompressibility, the Doppler shift Ω_0 vanishes, whereas the auxiliary function α defined in Eq. (13.147) and the tangential expressions (13.148) reduce to

$$\alpha(r) = \frac{2\Omega(r)\omega}{\omega^2 - \omega_{\mathrm{A}}^2}\,, \qquad \mathrm{i}\xi_\theta = \eta = \frac{\alpha}{r}\chi\,, \qquad \mathrm{i}\xi_z = \zeta = -\frac{1}{kr}\chi'\,, \qquad (13.181)$$

so that the remaining Coriolis shift becomes

$$\overline{V} = 4\mathrm{Re}\left(\frac{\omega}{\omega^2 - \omega_{\mathrm{A}}^2}\right)\frac{\displaystyle\int \Omega(r)^2 \,|\xi|^2 r\,dr}{\displaystyle\int (|\xi|^2 + |\eta|^2 + |\zeta|^2) r\,dr}\,. \qquad (13.182)$$

A trivial solution (the imaginary ω axis) of the path equation (13.120) for $\mathcal{P}_{\mathrm{u}}$ is found immediately by inspection:

$$\sigma = \overline{V}(\omega = \mathrm{i}\nu) = 0\,. \qquad (13.183)$$

It is evident from the symmetry of the perturbations and the equilibrium that no other solutions, with $\sigma \neq 0$, can exist. Hence, in a problem with a genuine rotation profile $\Omega(r)$, we have managed to produce solution paths $\mathcal{P}_{\mathrm{s}}$ and $\mathcal{P}_{\mathrm{u}}$ (the real and imaginary ω axes) that are identical to those of a static equilibrium! Needless to say that this lucky circumstance will get lost for more general perturbations and equilibria, with $m \neq 0$ and $B_\theta \neq 0$ (see Section 13.4.3).

Next, we note that the construction of the eigenvalues on the solution paths requires the solution of the spectral differential equation (13.178) which, like in the static case, only depends on the squared eigenvalue parameter, ω^2, which we now know to be real. Hence, instability or stability depends on whether or not there are eigenvalues $\omega^2 < 0$. We construct a quadratic form by multiplying the differential equation by χ, integrating over the plasma interval $r_1 \leq r \leq r_2$, integrating by parts and canceling the boundary term by applying the BCs:

$$(f\chi')' - g\chi = 0 \quad \Rightarrow \quad \int_{r_1}^{r_2} (f\chi'^2 + g\chi^2) r\, dr = 0. \tag{13.184}$$

Here, f and g indicate the coefficients of Eq. (13.178), but it is clear that the argument applies to any second order differential equation with real coefficients. (We have encountered it many times before in Volume [1].) Clearly, to obtain eigenfunctions (i.e. oscillatory solutions satisfying the BCs) for real values of ω^2, the ratio of these coefficients should be negative,

$$\frac{g}{f}(r;\omega^2) \equiv k^2 \left[1 - \frac{\kappa^2(r)}{\omega^2 - \omega_A^2} - \frac{4\omega_A^2 \Omega^2(r)}{(\omega^2 - \omega_A^2)^2} \right] < 0, \tag{13.185}$$

over at least some sub-interval $r_1 \leq r_A < r < r_b \leq r_2$. From our oscillation theorems $\mathcal{R}$ (Section 13.1.3) and $\mathcal{C}$ (Section 13.2.4), which reduce to the ones for the static case (Section 9.4.1 [1]) under the present approximations, stability or instability is determined by the absence or presence of oscillatory solutions of the marginal equation of motion, obtained from Eq. (13.178) in the limit $\omega \to 0$. Hence, the criterion for the absence of magneto-rotational instabilities becomes

$$\omega_A^2 + \kappa^2 - 4\Omega^2 \geq 0 \quad \text{everywhere}, \tag{13.186}$$

significantly different from the criterion (13.179) for absence of HD instabilities. Inserting the definition (13.176) for κ^2, the MHD criterion is violated for Keplerian disks *when the magnetic field is sufficiently small* ($\omega_A^2 \to 0$),

$$\kappa_K^2 - 4\Omega_K^2 \equiv r(\Omega_K^2)' \overset{(13.180)}{=} -3\Omega_K^2 < 0. \tag{13.187}$$

Hence, in contrast to hydrodynamic disks, disks with a small magnetic field are always unstable with respect to the magneto-rotational instability [18]. This is the reason for the popularity of research of MRIs in accretion disks.

Two problems remain to be addressed.

(a) Why does the MHD condition (13.186) not transform into the HD condition (13.179) in the limit $\omega_A^2 \to 0$?

(b) Can one actually find parameters for the disk and a range of unstable wave numbers k that satisfy all mentioned smallness assumptions imposed on the solutions?

Both problems are most effectively solved by means of the consideration of localized modes of the type introduced in Section 13.3.3.

The apparent contradiction between pure hydrodynamics ($\omega_A^2 = 0$) and the hydrodynamic limit of MHD ($\omega_A^2 \to 0$) is resolved when the growth rate of instabilities is taken into account. This is best illustrated by considering modes that are sufficiently *localized radially* to exploit the WKB solution (13.139) with

$$q\,\widetilde{\Delta r} \gg 1 \qquad (\,\widetilde{\Delta r} \sim r_1 \ll \Delta r \equiv r_2 - r_1\,).$$

(13.188)

The restriction to $\widetilde{\Delta r}$ instead of Δr is necessary here since the equilibrium, i.e. $\Phi(r)$, has an $\mathcal{O}(1)$ variation over a much narrower range than the width of the disk. With the ratio g/f of the coefficients of the approximate spectral equation (13.178) as given by Eq. (13.185), the *local dispersion equation* (13.142) reduces to

$$q^2 = -\frac{g}{f} \;\Rightarrow\; (k^2+q^2)(\omega^2-\omega_\mathrm{A}^2)^2 - k^2\kappa^2(\omega^2-\omega_\mathrm{A}^2) - 4k^2\omega_\mathrm{A}^2\Omega^2 = 0\,,$$

(13.189)

giving two solutions,

$$\omega^2 = \omega_\mathrm{A}^2 + \frac{1}{2}\frac{k^2}{k^2+q^2}\kappa^2 \pm \sqrt{\left(\frac{1}{2}\frac{k^2}{k^2+q^2}\kappa^2\right)^2 + 4\frac{k^2}{k^2+q^2}\omega_\mathrm{A}^2\Omega^2}\,.$$

(13.190)

This yields the following approximations under various assumptions.

(1) As always, the asymptotic spectral structure is obtained from the *most local modes* ($q^2 \gg k^2 \gg 1$) forming cluster spectra immediately above and below the degenerate Alfvén/slow continua $\Omega^{\pm}_{\mathrm{A/S}} = \pm\omega_\mathrm{A}$,

$$\omega^2 \approx \omega_\mathrm{A}^2 \pm 2(k/q)\omega_\mathrm{A}\Omega \quad (\textit{continua and cluster spectra})\,.$$

(13.191)

(2) The expressions for the *most global modes* ($q^2 \ll k^2$) yield the instability criteria,

$$\omega^2 \approx \omega_\mathrm{A}^2 + \tfrac{1}{2}\kappa^2 \pm \tfrac{1}{2}\sqrt{\kappa^4 + 16\omega_\mathrm{A}^2\Omega^2} \;\Rightarrow\; \begin{cases} \kappa^2 < 0\,, \quad \text{for } \omega_\mathrm{A}^2 = 0 \quad (\textit{HD}) \\[2mm] \kappa^2 - 4\Omega^2 < -\omega_\mathrm{A}^2 \quad\quad (\textit{MHD}) \end{cases}.$$

(13.192)

(3) *Small magnetic field strengths* ($\omega_\mathrm{A}^2 \ll \kappa^2 \sim \Omega^2$) yield the two relevant kinds of modes,

$$\omega^2 \approx \begin{cases} \dfrac{k^2}{k^2+q^2}\kappa^2 + (4\Omega^2/\kappa^2 + 1)\,\omega_\mathrm{A}^2 \quad (\textit{epicyclic modes}) \\[4mm] -(4\Omega^2/\kappa^2 - 1)\,\omega_\mathrm{A}^2 \quad\quad\quad\quad\quad (\textit{MRIs}) \end{cases}.$$

(13.193)

In the limit of vanishing magnetic field ($\omega_\mathrm{A}^2 \to 0$), the upper solution is unstable if Rayleigh's stability criterion (13.179) is violated, i.e. for rotation profiles $\Omega(r)$ that fall off faster than r^{-2}. For those profiles, the lower solution is stable. However, for rotation falling off with a lower power, like Keplerian rotation $\Omega \sim r^{-3/2}$, the upper solution is stable and the lower solution (corresponding to MRI) is unstable, but with a vanishing growth rate. Consequently, for Keplerian rotation profiles, the MRI dominates as long as ω_A^2 is finite, but sufficiently small.

Consequently, the maximum growth rate for MRIs is obtained from the bottom expression (13.193)(b), which is valid for the most global modes ($q^2 \ll k^2$). For increasing values of q, a whole sequence of MRIs is obtained, crossing the marginal point $\omega = 0$ and bifurcating there, and finally joining the $-$ pair of cluster spectra (13.191) for $q \to \infty$. In that limit, the epicyclic modes (13.193)(a) transform into the $+$ pair. This structure is illustrated in Section 13.4.3 (Fig. 13.17).

With respect to the order of magnitude of the different physical parameters for which the MRI will be operating in the disk, we first utilize scale-independence by setting three parameters equal to 1. This time, since we want to study the limits of small and vanishing magnetic field strength, it is not expedient to exploit B_z for that purpose. Instead, we normalize frequencies with respect to the angular rotation frequency at $r = r_1$ and eliminate dimensions by fixing the following three parameters:

$$r_1 \equiv 1, \quad \rho \equiv 1, \quad \Omega(r_1) \equiv 1 \qquad \textit{(scale independence)}. \qquad (13.194)$$

The geometry of a thin and wide accretion disk is then represented by a small parameter ϵ and a large parameter Δ:

$$\epsilon \equiv \Delta z \ll 1 \ll \Delta \equiv \Delta r \equiv r_2 - r_1 \qquad \textit{(thin and wide disk)}. \qquad (13.195)$$

The longitudinal wave number k and the radial "wave number" q should be chosen such that the conditions (13.171) and (13.188) are satisfied, e.g.

$$k \sim \epsilon^{-2} \gg \epsilon^{-1} \sim q \gg 1 \qquad \textit{(local disturbances)}. \qquad (13.196)$$

We can now estimate the order of magnitude of the magnetic field B_z, i.e. of the Alfvén frequency ω_A, required to give significant growth of the MRI during one rotation period (the condition (13.186) excludes growth faster than that):

$$\omega_{\mathrm{MRI}} \sim \omega_A \equiv k B_z \sim \Omega(1) \equiv 1 \quad \Rightarrow \quad B_z \sim k^{-1} \ll \epsilon \ll 1. \qquad (13.197)$$

Assuming $k \sim \epsilon^{-2}$, this implies the very small order of magnitude of the magnetic field, $B_z \sim \epsilon^2 \Omega(1)$ (in dimensionless units), *above which the MRI switches off.* Even with the successes of modeling turbulent processes in accretion disks by means of the magneto-rotational instability (see Balbus and Hawley [19], Stone *et al.* [422]), this still begs the question whether nature actually makes use of these processes. Eventually (in a very distant future), this question can only be answered by means of extremely high-resolution observations, unequivocally establishing the magnetic signature of the modes involved.

13.4.3 Magneto-rotational instability: numerical solutions

Since the window in parameter space of the MRI is so narrow, it is logical that intensive research was conducted on extended models of accretion disks where the mentioned restrictive conditions were dropped one by one. We here discuss one particular model, by Keppens *et al.* [255], since it illustrates the complexity of the spectrum of waves and instabilities of rotating equilibria with "numerically exact" solutions of the full spectral equation (13.107) for a cylinder with a relevant choice of the radial distribution of the equilibrium variables for accretion disks. This brings in dependence on *compressibility* (i.e. lifting of the degeneracy of Alfvén and slow sub-spectra and appearance of the fast sub-spectrum), a *toroidal magnetic field* ($B_\theta \neq 0$) and *toroidal mode numbers* $m \neq 0$. (Note that "toroidal" here refers to the azimuthal component, in contrast to cylindrical models for tokamaks where the same angle is always called "poloidal" angle, simulating a torus by a periodic cylinder.) The results presented were obtained with a numerical code (LEDAFLOW [353]) dating from before the development of the concepts of solution path and alternator, but clearly illustrating their central importance.

The cylindrical equilibrium equation (12.30) is satisfied for disks with a power dependence of the radial variable r:

$$v_\theta = v_1 r^{-\frac{1}{2}}, \quad \rho = r^e, \quad p = p_1 r^{e-1}, \quad B_\theta = B_{\theta 1} r^{\frac{1}{2}(e-1)}, \quad B_z = B_{z1} r^{\frac{1}{2}(e-1)}, \tag{13.198}$$

where approximate Keplerian rotation is assumed. The normalization is chosen slightly differently from Eq. (13.194), viz. $GM_* \equiv 1$, but $v_1 \equiv v_\theta(1) \equiv \Omega(1) \approx 1$, where the exact expression for v_1 is given in Eq. (13.201) below. In the present context of an *ideal* stationary equilibrium, the exponent e is completely arbitrary. It will be fixed to the value $e = -3/2$, corresponding to a particular class of self-similar *dissipative* stationary disk solutions, given by Spruit *et al.* [417], with radial accretion velocity v_r ($\ll v_\theta$, see below) and associated outward angular momentum transport due to turbulent viscosity, modeled with the parameter α introduced by Shakura and Sunyaev [410]. Related to their celebrated scaling, we generalize the cylindrical model of Fig. 13.16(a) somewhat, to the quasi-2D model of Fig. 13.16(b), by estimating the pressure from the vertical component of the two-dimensional equilibrium equation by balancing the pressure gradient and gravitational acceleration in the z-direction (see Frank, King and Raine [139], p. 87):

$$\frac{1}{\rho}\frac{\partial p}{\partial z} = -\frac{\partial \Phi}{\partial z} \approx -\frac{z}{r^3} \quad \Rightarrow \quad p \approx p_1 r^{-5/2}\left(1 - \frac{z^2}{2H^2}\right), \quad H \approx \sqrt{p_1}\, r . \tag{13.199}$$

Here, H is the scale height of the disk, thus assumed to be determined by thermal effects only. This very crude estimate of the two-dimensional structure of the equilibrium permits us to relate the magnitude of the pressure to the inverse aspect ratio of the disk:

$$\sqrt{p_1} = \epsilon \equiv \frac{H}{r} . \tag{13.200}$$

Note that these considerations on dissipation and geometry only fix the orders of magnitude of the *ideal one-dimensional* equilibrium that will be investigated with respect to the MRI and its generalizations. It thus becomes a four-parameter family in terms of ϵ, the pitch μ_1 of the magnetic field lines at $r = r_1$, the ratio β between kinetic and magnetic pressure, and the width $\Delta \equiv r_2 - r_1$ of the disk:

$$v_\theta = v_1 r^{-\frac{1}{2}}, \quad \rho = r^{-\frac{3}{2}}, \quad p = \epsilon^2 r^{-\frac{5}{2}}, \quad B_\theta = \mu_1 B_{z1} r^{-\frac{5}{4}}, \quad B_z = B_{z1} r^{-\frac{5}{4}},$$

$$v_1 = \left[1 - 2.5\epsilon^2 \left(1 + \frac{1}{\beta} \frac{1 + 0.2\mu_1^2}{1 + \mu_1^2} \right) \right]^{1/2},$$

$$\beta \equiv 2p/B^2, \quad \mu_1 \equiv B_{\theta 1}/B_{z1}, \quad B_{z1} = \epsilon \Big/ \sqrt{\tfrac{1}{2}\beta(1 + \mu_1^2)}. \tag{13.201}$$

This completely determines the equilibrium for the spectral study.

Two side remarks are in order: first, the inverse aspect ratio ϵ of the disk and the iso-thermal sound speed for these equilibria are related by

$$c_s \equiv \sqrt{p/\rho} = \epsilon/\sqrt{r} \quad \Rightarrow \quad \epsilon = c_s/v_\theta, \tag{13.202}$$

showing that the rotation speeds involved are necessarily supersonic ($v_\theta \gg c_s$). Second, the frequently cited estimate of Shakura and Sunyaev [410] of the kinematic viscosity due to turbulent eddies with sizes not exceeding H exhibits a simple scaling in terms of the dimensionless parameter α (estimated to have a magnitude of order 0.1–1), that also justifies the neglect of the radial inflow velocity:

$$\nu = \alpha c_s H = \alpha \epsilon^2 \sqrt{r} \quad \Rightarrow \quad v_r \approx -\frac{3\nu}{2r} = -\tfrac{3}{2}\alpha \epsilon^2 r^{-\frac{1}{2}} \ll v_\theta = r^{-\frac{1}{2}}. \tag{13.203}$$

Simple and attractive but, obviously, to be taken with a grain of salt. After all, the magnitude of α is not a constant of nature but should be the eventual outcome of the MRI calculations.

Figure 13.17 shows the complete MHD spectrum of *axi-symmetric modes* for an equilibrium (13.201) with parameters $\epsilon = 0.1$, $\mu_1 = 1$, $\beta = 2000$, $\Delta = 9$, perturbed with wave numbers $m = 0$ and $k = 70$. The different discrete modes, corresponding to different numbers of radial nodes (i.e. of the wave number q in the local approximation of Section 13.4.2), exhibit the monotonicity of the alternator R required by the oscillation theorems $\mathcal{R}$ and $\mathcal{C}$. Due to the choice $k \gg 1$, they are very densely distributed over the central part of the imaginary ω axis and smoothly join the sequence of stable epicyclic modes along the real axis, eventually clustering at the continua. Because so many discrete modes are calculated, the distinction between continua and discrete modes is somewhat blurred in the figure. In conclusion, the slow and Alfvén modes interact to produce both the MRIs and the stable epicyclic modes, and it supports the general expectation that the many local MRIs may provide the required source of turbulent increase of the viscosity to yield accretion flows of the correct order of magnitude.

The numerical spectrum of the full range of local and global modes of Fig. 13.17

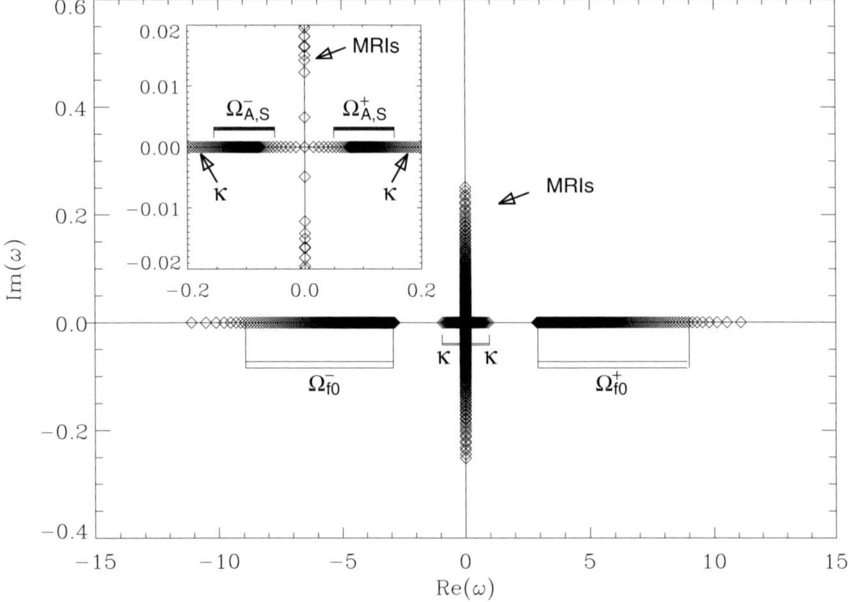

Fig. 13.17 MHD spectrum for weakly magnetized accretion disk for axi-symmetric ($m = 0$) modes. Magneto-rotational instabilities (MRIs) and stable epicyclic modes (κ) are associated with cluster spectra of interacting Alfvén and slow modes about the overlapping forward and backward continua $\{\Omega^+_{A,S}\}$ and $\{\Omega^-_{A,S}\}$; fast modes starting from the turning point ranges $\{\Omega^+_{f0}\}$ and $\{\Omega^-_{f0}\}$ cluster towards $\Omega^\pm_F \equiv \pm\infty$; $\epsilon = 0.1$, $\mu_1 = 1$, $\beta = 2000$, $\Delta = 9$, $k = 70$. (Adapted from Keppens *et al.* [255].)

agrees qualitatively with the local analysis of Section 13.4.2, but also quantitatively if the following considerations on mode localization are taken into account. The approximation (13.193)(b) for the growth rate of global MRIs ($q^2 \ll k^2$) yields

$$\nu_{\text{MRI}} \approx \sqrt{4\Omega^2/\kappa^2 - 1}\,\omega_A = \sqrt{3}\,\omega_A = kB_{z1}\sqrt{3/\rho} \approx 0.271/\sqrt{r}. \qquad (13.204)$$

This agrees with the maximum growth rate of Fig. 13.17 for modes localized at the inside ($r = 1$), whereas modes localized on the outside ($r - \Delta = 9$) have much reduced growth rates since the associated value of q is much larger. Actually, the more localized modes have to cross the origin for some value of $q \gg 1$ since they eventually have to join onto the cluster spectra (13.191) at the continua $\Omega^\pm_A = \pm\omega_A$ and $\Omega^\pm_S = \pm\omega_S$ for $q \to \infty$. The degeneracy of the incompressible Alfvén/slow continua $\Omega^\pm_{A/S}$ is now lifted but, because $\beta = 2000$, it is clear that this is just a tiny effect, much smaller than the width of the continua produced by the r dependence, so that the two pairs of continua still largely overlap.

Along the real axis, except for the mentioned cluster spectra, there is also the

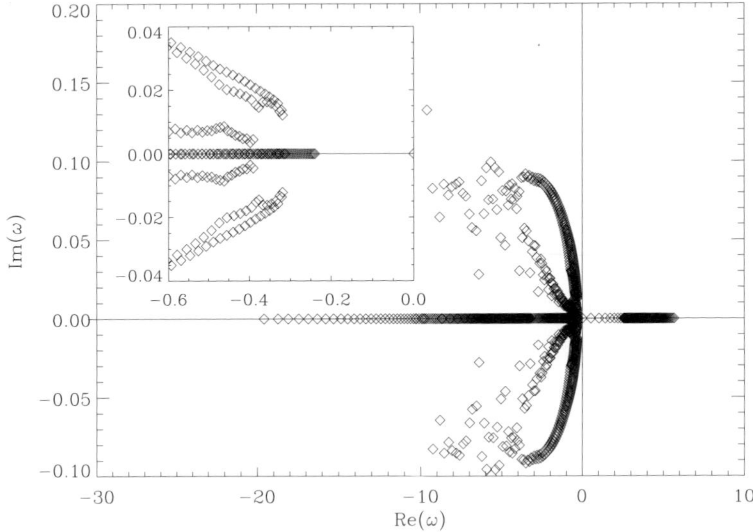

Fig. 13.18 MHD spectrum for weakly magnetized accretion disk for non-axi-symmetric perturbations. Doppler shifted MRIs and epicyclic modes are associated with cluster spectra of interacting Alfvén and slow modes about the overlapping continua, dominated by a large Doppler shift: $\Omega_{A,S}^{+} \approx \Omega_{A,S}^{-} \approx \Omega_0$, extending from $\mathrm{Re}\,\omega = -10$ to -0.37; fast modes and wide turning point ranges $\{\Omega_{f0}^{-}\}$ and $\{\Omega_{f0}^{+}\}$ clutter the real axis; $\epsilon = 0.1$, $\mu_1 = 1$, $\beta = 2000$, $\Delta = 9$, $m = -10$, $k = 70$. (Adapted from Keppens *et al.* [255].)

sequence (13.193)(a) of stable epicyclic modes (indicated by κ in the figure),

$$\omega_\kappa \approx \pm\kappa = \pm 1/\sqrt{r}\,, \tag{13.205}$$

with a maximum value $|\omega_\kappa| \approx 1$. In addition, because of the finite value of β, the fast modes, that are situated at $\pm\infty$ for incompressible plasmas, appear to come much closer to the slow and Alfvén sub-spectra than one might have expected. This is due to the power dependence of the sound speed, so that the fast modes have frequencies

$$\omega_f \approx \pm\sqrt{(k^2 + q^2)\gamma p/\rho} \approx \sqrt{\gamma}\,k\epsilon/\sqrt{r} = 9.04/\sqrt{r}\,, \tag{13.206}$$

with a minimum value of $|\omega_f| \approx 3$ corresponding to localization at the outer edge. These modes are spread over a wide range, containing the non-monotonicity ranges $\Omega_{f0}^{+} = \omega_{f0}$ and $\Omega_{f0}^{-} = -\omega_{f0}$, where $\omega_{f0} \equiv \omega_f(q = 0)$, indicated in the figure and clustering towards the two limiting values $\Omega_F^{\pm} \equiv \pm\infty$.

Extending the analysis to *non-axi-symmetric modes* (Fig. 13.18), with azimuthal mode number $m = -10$, the spectrum is dominated by the large Doppler shift, resulting in a Doppler range $\{\Omega_0\} = [-10, -0.37]$ for the parameters chosen. The

MRIs are now shifted away from the imaginary ω axis, but they remain situated in the mentioned Doppler strip since the Coriolis shift is negligible for this particular case. This is in agreement with the theory developed in this chapter. The growth rates of the instabilities have the same order of magnitude as for the axi-symmetric modes, but they are reduced by about a factor 2.

The Doppler shift $\Omega_0 = m/(r\sqrt{r})$ is largest for the inner edge of the disk and smallest for the outer edge. Since the fastest growing (global) modes are concentrated on the inside and the most local modes are concentrated on the outside, the Doppler shift is largest for the fastest growing instabilities and remains small for the near marginal ones, in agreement with the figure. Even though $\beta = 2000$ is large, the fast magneto-sonic modes and the associated turning point ranges $\{\Omega_{f0}^-\} = [-19.04, -3.38]$ and $\{\Omega_{f0}^+\} = [-0.96, 2.64]$ come close to the continua, and $\{\Omega_{f0}^-\}$ even overlaps with them, because of the large extension of the disk. Hence, identification of the modes along the real axis in terms of slow, Alfvén and fast becomes meaningless: except at the edges of the continua, they are all of mixed type.

Several extensions have been made for cylindrical plasmas with $\beta \sim 1$ (so-called "equipartition"). According to the disk expansion (13.200) represented in Fig. 13.16(b), this really implies a *fat disk* so that incorporation of toroidal effects becomes essential. In Chapter 18, we show that, due to coupling of the Alfvén and slow modes, the continua themselves may become unstable in that case so that magnetically dominated instabilities become even more important then.

13.5 Literature and exercises

Notes on literature

Hydrodynamic and magnetohydrodynamic stability

- Chandrasekhar, *Hydrodynamic and Hydromagnetic Stability* [84], discusses a great variety of topics in fluid dynamic stability, like thermal instability, effects of shear flow, rotation, gravity and stability of jets and cylinders. Magnetic field effects enter at many places, e.g. in Chapter IX in the discussion of the magneto-rotational instability, shortly earlier discovered by Velikhov [467] and independently by Chandrasekhar [83].

- Drazin & Reid, *Hydrodynamic Stability* [124], is the classical compendium on the subject of linear hydrodynamic stability, treating both inviscid and viscous flows with respect to thermal convection, shear flow and rotation, and containing a short introduction of some topics in nonlinear stability.

- Hameiri, 'Spectral estimates, stability conditions, and the rotating screw-pinch' [209], presents the first systematic study of the spectrum of stationary cylindrical plasmas.

- Bondeson, Iacono & Bhattacharjee, 'Local magnetohydrodynamic instabilities of cylindrical plasma with sheared equilibrium flows' [56], extend Hameiri's results with a numerical analysis of local instabilities at resonant magnetic surfaces.

Astrophysical flows

- Balbus & Hawley, 'Instability, turbulence, and enhanced transport in accretion disks' [19], point out the significance of the magneto-rotational instability as a possible mechanism for turbulent enhancement of angular momentum transport in accretion disks about compact objects.
- Pringle & King, *Astrophysical Flows* [388], is a basic textbook on the fluid dynamical processes relevant to astrophysics, covering wave propagation, shocks, spherical flows, stellar oscillations, and instabilities driven by magnetic fields, thermal conduction, gravity, shear flow and rotation.

Exercises

[13.1] *Convection in a plasma slab*

In this exercise we investigate the convective instability of a plasma slab. Assume that the slab has a constant flow and magnetic field.

- Derive the spectral equation in the incompressible limit.
- Derive the local dispersion equation using a solution of the form $\xi \sim \mathrm{e}^{\mathrm{i}qx}$, assuming that $q\Delta x \gg 1$.
- What are the solutions of this dispersion equation?
- What is the role of the magnetic field?

[13.2] *Gravito-acoustic waves in a stationary fluid*

Similarly to Section 7.2.3 [1], we are going to investigate gravito-acoustic waves, i.e. neglect the magnetic field, but we do take flow into account.

- Derive an equation for the pressure assuming that the density is exponentially decaying, $\rho = \rho_0 \mathrm{e}^{-\alpha x}$, and that the gravity is constant. Show that the sound speed c and the decay parameter α are constant.
- Writing

$$(\tilde{\omega}^2 - k_0^2 c^2)\left[\tilde{\omega}^2/(\tilde{\omega}^2 - k_0^2 c^2)\right]' = -k_0^2 \lambda(x,\omega),$$

  find the expression for $\lambda(x,\omega)$.
- Derive from the general spectral equation a second order differential equation for the displacement ξ. Write this equation in the form $\xi'' + f(x,\omega)\xi' + g(x,\omega)\xi = 0$.
- Looking at this differential equation, one might think to try solutions of the form $\xi \sim C \exp[(a \pm \mathrm{i}q)x]$. Explain, why this kind of solutions cannot be used in general.
- Assume that the velocity is constant. Explain, why the form mentioned in the previous question can now be used. Show that both $f(x,\omega)$ and $g(x,\omega)$ are real.
- Derive the dispersion equation from the spectral equation.
- Determine the eigenfrequencies from the dispersion equation.

[13.3] *Magneto-rotational instability and convection*

In Section 13.4.2, the magneto-rotational instability is discussed assuming constant density. Here, we drop this assumption, which allows us to study convective instabilities.

- Derive the spectral equation in the incompressible limit. Assume a constant magnetic field in the vertical direction, and a purely azimuthal velocity field. Consider only axi-symmetric perturbations. (Hint: use the function $\Lambda(r)$ defined in Eq. (13.105).)
- Derive the local dispersion equation from the spectral equation, assuming oscillatory solutions of the form $\chi \sim \exp(\mathrm{i}qr)$ with $q\Delta r \gg 1$.
- Determine the solutions of the dispersion equation.
- Derive a stability criterion from them and explain the role of the magnetic field.

[13.4] *Magneto-rotational instability and non-zero azimuthal magnetic field*

In the previous exercise, you have investigated the magneto-rotational instability in combination with convection in the absence of an azimuthal magnetic field. Here, you will look at the case of a non-vanishing azimuthal magnetic field.

- Convince yourself that the azimuthal magnetic field cannot be constant.
- Repeat the first two questions of the previous exercise, including an azimuthal magnetic field. Find a local dispersion equation of the form $\omega^4 + a_2\omega^2 + a_1\omega + a_0 = 0$, and determine the coefficients a_2, a_1 and a_0.
- The solutions of this dispersion equation cannot be obtained analytically. You have to find them numerically. To do so, an equilibrium has to be specified. Work out the coefficients for the equilibrium specified in Eq. (13.201) for an accretion disk.
- Compute the solutions of the dispersion equation making use of Laguerre's method (see *Numerical Recipes* [385]; in IDL use FZ_ROOTS, in Matlab use ROOTS1). Use the following parameters for the equilibrium: $\epsilon = 0.1$ (thin disk), $\mu_1 = 1$ (inclusion of azimuthal magnetic field), $\beta = 1000$ (weakly magnetized plasma), and $GM_* = 1$. As a starting value for Laguerre's method use the complex number $\widehat{\omega} = \mathrm{i}$. What do you notice? What happens if you include an azimuthal magnetic field?
- Make a plot of the growth rate and oscillation frequency as a function of the vertical wavenumber k. What is the minimum value for the wavenumber k? What do you conclude from this plot?

[13.5] *WKB analysis*

In the previous exercises, we have assumed $q\Delta x \gg 1$, or $q\Delta r \gg 1$, depending on whether we describe the plasma in Cartesian or in cylindrical coordinates. This assumption is part of the WKB analysis of the spectral equation. Here, you are going to perform the WKB analysis for plasma in Cartesian coordinates. (For cylindrical coordinates, this can be done in a similar fashion.)

- To perform a proper WKB analysis, write the displacement ξ in the following form:

$$\xi(x, y, z) = p(x)\exp\left[\mathrm{i}\int_{x_0}^{x} q(s)ds + \mathrm{i}(k_y y + k_z z)\right],$$

  where $p(x)$ and $q(s)$ are the amplitude and "wavenumber" in the x-direction, respectively. Calculate the first and second derivative of ξ with respect to x.
- The spectral equation (13.9) can be written as

$$\frac{d}{dx}\left(f(x)\frac{d\xi}{dx}\right) - g(x)\xi = 0.$$

  Work this out by inserting the expressions found in the previous question.
- Derive an expression for the amplitude $p(x)$ in terms of $f(x)$ and $q(x)$.

- Now apply the WKB approximation. This implies that $(qL)^{-1} \ll 1$, where $L \sim d/dx$ is the length-scale of the variation of the background. Show that the spectral equation of the previous question reduces to an algebraic equation. This is the local dispersion equation. What is the approximate expression for the amplitude $p(x)$?

- Take a closer look at the expression for $g(x)$, especially the last term. Find an argument that justifies neglecting this term compared to the other ones. (Hint: try to identify a term proportional to $\widetilde{\omega}^2 - \omega_A^2$ in the expression for D.)

- Derive the local dispersion equation by inserting the expressions for $f(x)$ and $g(x)$. Show that this dispersion equation is a third order polynomial in $\widetilde{\omega}^2$ and compute all coefficients of it.

14

Resistive plasma dynamics

14.1 Plasmas with dissipation

14.1.1 Conservative versus dissipative dynamical systems

We have already come across the enormous difference between conservative (ideal) MHD and dissipative (resistive, viscous, etc.) MHD in Volume [1], Chapter 4. This difference runs through all of classical dynamics of discrete and continuous media. It involves quite different physical assumptions and corresponding different mathematical solution techniques. An instructive example is spectral theory (Volume [1], Chapter 6) which is classical, consistent and misleadingly beautiful for ideal MHD, but full of unresolved problems in resistive MHD. The classical part concerns self-adjoint linear operators in Hilbert space, analogous to quantum mechanics, and stability analysis by means of an energy principle. When dissipation is important, precisely these two "sledge hammers" are missing in the dynamical systems workshop. Even the definition of what is an important, i.e. physically dominant, contribution to the dynamics deserves extreme care. This is best illustrated by the general description of the dynamics of ordinary fluids which is fundamentally different for ideal fluids, characterised by an infinite Reynolds number, and viscous fluids, characterised by a finite Reynolds number. This is even so for extremely large Reynolds numbers, in a certain sense irrespective of how large this number is. Viscous boundary layers always arise in real fluids. This qualitative difference between ideal and dissipative dynamics, with the occurrence of boundary layers, also applies to MHD when resistivity is introduced. This gives rise to internal resistive boundary layers, facilitating new modes of instability, as we will see in Section 14.2. The physical cause of these instabilities is the loss of conservation of magnetic flux, leading to reconnection of magnetic field lines (Section 14.4). The implications for the structure of the resistive spectrum are only partly understood (Section 14.3). Resistive MHD is only one of a number of extensions of the ideal MHD model which come under the name of "extended MHD" (Section 14.4.4).

14.1.2 Stability of force-free magnetic fields: a trap

As a preliminary to the study of resistive instabilities, to be undertaken in Section 14.2, let us first investigate the ideal MHD stability of about the simplest system that deserves analysis, viz. a plane current-carrying plasma slab. The reader is warned in advance that an educational trap has been laid in this section.

The current will be chosen such that the magnetic field has a constant magnitude but its direction varies. The simplest case to treat is a so-called force-free magnetic field with a constant ratio α between the current and the magnetic field:

$$\mathbf{j} = \nabla \times \mathbf{B} = \alpha \mathbf{B}, \quad \alpha = \text{const}, \tag{14.1}$$

or, in components,

$$j_y = -B'_z = \alpha B_y, \qquad j_z = B'_y = \alpha B_z.$$

This equation can easily be integrated:

$$\mathbf{B} = B_0 [\sin \varphi(x) \mathbf{e}_y + \cos \varphi(x) \mathbf{e}_z], \quad \varphi(x) = \alpha x, \quad B_0 = \text{const}, \tag{14.2}$$

representing a field with a uniformly varying direction (Fig. 14.1). The plasma is considered to be confined between two perfectly conducting plates at $x = x_1$ and $x = x_2$. Hence, the parameter αa, where $a \equiv x_2 - x_1$, is a measure for the total current through the plasma. In the absence of gravity, equilibrium would still permit a finite, but constant, pressure. However, we will neglect pressure altogether by considering a zero-β plasma. We wish to investigate the stability of this configuration.

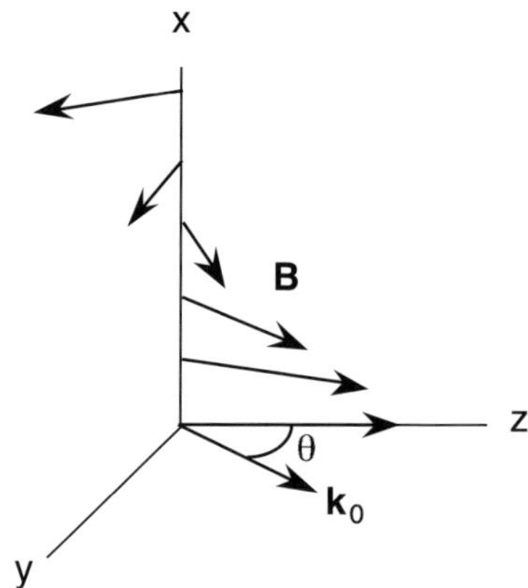

Fig. 14.1 Plane force-free field.

(a) An interesting stability result ...

Stability may be investigated by means of the energy principle, i.e. the study of the sign of the energy $W[\boldsymbol{\xi}]$ of the perturbations $\boldsymbol{\xi}$, exploiting the explicit expression (6.85) of Volume [1] for the fluid energy. As in Section 7.3.2 [1], we decompose the displacement vector $\boldsymbol{\xi}(\mathbf{r})$ in Fourier components, according to Eq. (7.78), and we study the stability of the separate modes. For the present problem, the expression for the fluid energy W simplifies to

$$W = \tfrac{1}{2} \int \left(|\mathbf{Q}|^2 + \alpha \mathbf{B} \cdot \boldsymbol{\xi}^* \times \mathbf{Q} \right) dx\,, \tag{14.3}$$

where we have normalised W with respect to the area in the $y - z$ plane. Following the textbook of Schmidt [402], p. 141, we may simplify the algebra by using the vector potential $\mathbf{A}$,

$$\mathbf{Q} = \nabla \times \mathbf{A}\,, \qquad \mathbf{A} \equiv \boldsymbol{\xi} \times \mathbf{B}\,, \tag{14.4}$$

so that

$$W = \tfrac{1}{2} \int \left[|\nabla \times \mathbf{A}|^2 - \alpha \mathbf{A}^* \cdot \nabla \times \mathbf{A} \right] dx\,. \tag{14.5}$$

According to Section 6.4.4 [1], we may now minimize W subject to some convenient normalization, for which we choose the helicity (see Section 4.3.4 [1]) of the perturbations:

$$K \equiv \tfrac{1}{2} \int \mathbf{A}^* \cdot \nabla \times \mathbf{A}\, dx = \text{const}\,. \tag{14.6}$$

The proper way to minimize W subject to such a constraint is to minimize another quadratic form, viz.

$$\widetilde{W} \equiv W + \lambda K = \tfrac{1}{2} \int \left[|\nabla \times \mathbf{A}|^2 - (\alpha - \lambda)\mathbf{A}^* \cdot \nabla \times \mathbf{A} \right] dx\,, \tag{14.7}$$

where the constraint is absorbed by means of an undetermined Lagrange multiplier λ that is to be determined together with $\mathbf{A}$. Since

$$\nabla \cdot \left[\mathbf{A}^* \times (\nabla \times \mathbf{A}) \right] \overset{(A.12)}{=} \nabla \times \mathbf{A}^* \cdot \nabla \times \mathbf{A} - \mathbf{A}^* \cdot \nabla \times \nabla \times \mathbf{A}\,, \tag{14.8}$$

we may integrate the expression for $\widetilde{W}$ by parts:

$$\widetilde{W} = \tfrac{1}{2} \left[\mathbf{A}^* \times (\nabla \times \mathbf{A}) \cdot \mathbf{n} \right]_{x_1}^{x_2} + \tfrac{1}{2} \int \mathbf{A}^* \cdot \left[\nabla \times \nabla \times \mathbf{A} - (\alpha - \lambda)\nabla \times \mathbf{A} \right] dx\,. \tag{14.9}$$

The boundary term vanishes by virtue of the boundary conditions $\mathbf{B} \cdot \mathbf{n} = 0$ and $\boldsymbol{\xi}^* \cdot \mathbf{n} = 0$ applied to $\mathbf{A} \equiv \boldsymbol{\xi} \times \mathbf{B}$. Consequently, for arbitrary $\mathbf{A}^*$, the quadratic form $\widetilde{W}$ is minimized by solutions of the Euler–Lagrange equation

$$\nabla \times \nabla \times \mathbf{A} - (\alpha - \lambda)\nabla \times \mathbf{A} = 0\,. \tag{14.10}$$

This may be written as another force-free field equation for the perturbations:

$$\nabla \times \mathbf{Q} = \tilde{\alpha}\,\mathbf{Q}\,, \qquad \tilde{\alpha} \equiv \alpha - \lambda\,. \tag{14.11}$$

Equation (14.11) is an eigenvalue equation, where $\tilde{\alpha}$ (and, hence, λ) is determined by imposing the boundary condition $\mathbf{n}\cdot\mathbf{Q} = 0$ at $x = x_1$ and $x = x_2$. Inserting such a solution into the expression (14.9) for $\widetilde{W}$ gives $\widetilde{W} = 0$, so that

$$W \;=\; \widetilde{W} - \lambda K = (\tilde{\alpha} - \alpha)\,\tfrac{1}{2}\int \mathbf{A}^* \cdot \nabla \times \mathbf{A}\,dx$$

$$\stackrel{(14.10)}{=} \frac{\tilde{\alpha} - \alpha}{\tilde{\alpha}}\,\tfrac{1}{2}\int \mathbf{A}^* \cdot \nabla \times \nabla \times \mathbf{A}\,dx$$

$$\stackrel{(14.8)}{=} \frac{\tilde{\alpha} - \alpha}{\tilde{\alpha}}\,\tfrac{1}{2}\int |\nabla \times \mathbf{A}|^2\,dx \stackrel{(14.4)}{=} \frac{\tilde{\alpha} - \alpha}{\tilde{\alpha}}\,\tfrac{1}{2}\int |\mathbf{Q}|^2\,dx\,. \tag{14.12}$$

Hence, $W < 0$ and the system appears to be unstable if the equation (14.11) has an eigenvalue $\tilde{\alpha}$ such that

$$0 < \tilde{\alpha} < \alpha\,. \tag{14.13}$$

It remains to determine the eigenvalue $\tilde{\alpha}$.

We now have to study the Euler equation (14.11) in detail to find out whether the condition (14.13) can be satisfied for the slab model. In this model, the Euler equation may be reduced to an ordinary second order differential equation for the normal component of $\mathbf{Q}$, which may be solved analytically so that we find an explicit stability criterion. To that end, write Eq. (14.11) in components,

$$ik_y Q_z - ik_z Q_y = \tilde{\alpha} Q_x\,,$$

$$ik_z Q_x - Q_z' = \tilde{\alpha} Q_y\,,$$

$$Q_y' - ik_y Q_x = \tilde{\alpha} Q_z\,. \tag{14.14}$$

Note that only two of these three equations are independent since $\mathbf{Q}$ is a magnetic field perturbation, so that

$$\nabla \cdot \mathbf{Q} = Q_x' + ik_y Q_y + ik_z Q_z = 0\,. \tag{14.15}$$

Hence, two variables suffice, for which we choose the normal components of the magnetic field and the current:

$$Q \equiv -iQ_x\,, \qquad R \equiv i(\nabla \times \mathbf{Q})_x = k_z Q_y - k_y Q_z\,. \tag{14.16}$$

Substituting the reverse expressions,

$$Q_x = iQ\,, \qquad Q_y = -(k_y Q' - k_z R)/k_0^2\,, \qquad Q_z = -(k_z Q' + k_y R)/k_0^2\,,$$

$$\tag{14.17}$$

into Eq. (14.14)(a) produces a simple relationship between R and Q,

$$R = -\tilde{\alpha}Q, \qquad (14.18)$$

whereas the two other components (14.14)(b),(c) just reduce to the same, second order, differential equation for Q:

$$Q'' + (\tilde{\alpha}^2 - k_0^2)Q = 0, \quad k_0^2 \equiv k_y^2 + k_z^2. \qquad (14.19)$$

This is the Euler equation we were looking for.

The solution of Eq. (14.19) which vanishes for $x = x_1$ and $x = x_2$ reads:

$$Q = C \sin \sqrt{\tilde{\alpha}^2 - k_0^2}\, x, \qquad (14.20)$$

where

$$\sqrt{\tilde{\alpha}^2 - k_0^2} = n\pi/a, \quad a \equiv x_2 - x_1.$$

Hence, the instability criterion (14.13) is fulfilled for

$$\tilde{\alpha}^2 = k_0^2 + \frac{n^2\pi^2}{a^2} < \alpha^2, \qquad \text{or} \quad (k_0/\alpha)^2 + (n\pi/(\alpha a))^2 < 1. \qquad (14.21)$$

This gives an unstable region in the $k_0/\alpha - \alpha a$ plane, as sketched in Fig. 14.2(a). Moving to the right in the shaded area subsequently $n = 1, n = 2, \ldots$ become unstable. The marginal modes (for which $\tilde{\alpha} = \alpha$) are distinguished by the number of nodes, $n - 1$, of Q on the interval (x_1, x_2) (Fig. 14.2(b)). Notice that in the long wavelength limit, $k_0 = 0$, every time αa increases with πa, i.e. every time the magnetic field has changed its direction by $180°$, a mode with one more node becomes unstable. This appears to be a perfectly reasonable result: the analysis indicates the existence of *long wavelength instabilities that are driven by the current,* which has to surpass a certain critical value given by $\alpha a = \pi$.

(b) ... which, however, does not make sense ...

Let us double-check the result obtained by rederiving it from a formulation in terms of the displacement $\boldsymbol{\xi}$ rather than the magnetic field perturbation $\mathbf{Q}$. It is then expedient to exploit *the field line projection* for $\boldsymbol{\xi}$, which was also introduced in Section 7.3.2 [1]:

$$\boldsymbol{\xi} = \xi\mathbf{e}_x - \mathrm{i}\eta\mathbf{e}_\perp - \mathrm{i}\zeta\mathbf{e}_\parallel, \qquad (14.22)$$

where $\mathbf{e}_\perp$ and $\mathbf{e}_\parallel$ were defined in Eq. (7.79) and the relationship with the Cartesian components was given in Eq. (7.83) [1]. Using this projection, when computing derivatives, recall that one should take care of the fact that these unit vectors are x-dependent: $\partial_x \mathbf{e}_\perp = -\alpha\mathbf{e}_\parallel$, $\partial_x \mathbf{e}_\parallel = -\alpha\mathbf{e}_\perp$. (This, incidentally, shows that α is proportional to the variation with height of the directional angle φ of the magnetic

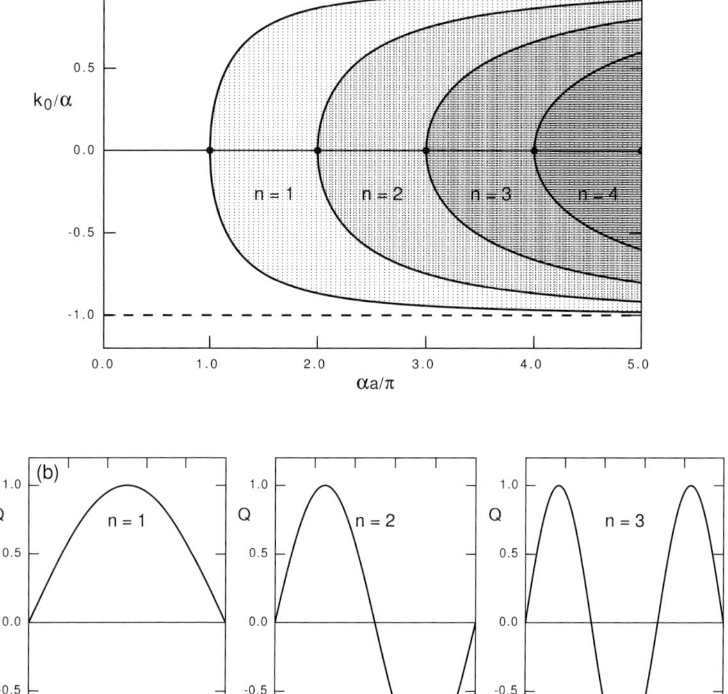

Fig. 14.2 Plane force-free field: (a) "stability" diagram; (b) marginal modes.

field shown in Fig. 7.9[1]: $\alpha = \varphi'$.) Next, express the components of $\mathbf{Q} = \nabla \times (\boldsymbol{\xi} \times \mathbf{B})$ in the components ξ, η and ζ of the displacement vector, where we notice that the last component does not appear:

$$Q_x = \mathrm{i}F\xi \,,$$

$$Q_y = -(B_y\xi)' + k_z B\eta = -B_y\xi' - \alpha B_z\xi + k_z B\eta \,,$$

$$Q_z = -(B_z\xi)' - k_y B\eta = -B_z\xi' + \alpha B_y\xi - k_y B\eta \,, \tag{14.23}$$

so that

$$Q \equiv -\mathrm{i}Q_x = F\xi \,,$$

$$R \equiv k_z Q_y - k_y Q_z = G\xi' - \alpha F\xi + k_0^2 B\eta \,, \tag{14.24}$$

where

$$F \equiv k_y B_y + k_z B_z \quad \text{and} \quad G \equiv k_y B_z - k_z B_y \tag{14.25}$$

are the two projections of the horizontal wave vector having the properties

$$F' = \alpha G \quad \text{and} \quad G' = -\alpha F. \tag{14.26}$$

Using Eqs. (14.23), it is now straightforward to express the potential energy (14.3) in terms of ξ and η:

$$\begin{aligned}
W &= \tfrac{1}{2} \int_{x_1}^{x_2} \Big[|Q_x|^2 + |Q_y|^2 + |Q_z|^2 - \alpha \xi_x^*(B_y Q_z - B_z Q_y) \\
&\qquad\qquad - \alpha(B_z \xi_y^* - B_y \xi_z^*) Q_x \Big] \, dx \\
&= \tfrac{1}{2} \int_{x_1}^{x_2} \Big[F^2 \xi^2 + (\alpha B \xi - F \eta)^2 + (B \xi' + G \eta)^2 \\
&\qquad\qquad - \alpha^2 B^2 \xi^2 + 2\alpha B F \xi \eta \Big] \, dx \\
&= \tfrac{1}{2} \int_{x_1}^{x_2} \Big[F^2 (\xi^2 + \eta^2) + (B \xi' + G \eta)^2 \Big] \, dx > 0.
\end{aligned} \tag{14.27}$$

Hence, the potential energy of the perturbations is positive definite so that we conclude that *the slab is trivially stable!* See Ref. [175].

We may obtain the minimizing perturbations by rearranging terms:

$$W = \tfrac{1}{2} \int_{x_1}^{x_2} \Big[F^2 (\xi'^2 / k_0^2 + \xi^2) + (k_0 B \eta + G \xi' / k_0)^2 \Big] \, dx$$

so that W is minimized for perturbations that satisfy

$$k_0 B \eta + G \xi' / k_0 = 0 \tag{14.28}$$

and

$$(F^2 \xi')' - k_0^2 F^2 \xi = 0. \tag{14.29}$$

One easily checks that the latter equation corresponds to Eq. (14.19) with $Q = F\xi$ for $\tilde{\alpha} = \alpha$: the minimising equations are equivalent. There is no mistake in the algebra!

(c) ... and why.

To see what went wrong let us plot the eigenfunctions ξ corresponding to the eigenfunctions Q shown in Fig. 14.2(b). Writing $F = k_0 B_0 \cos(\alpha x - \theta)$, with θ as defined in Fig. 14.1, we find:

$$\xi = \frac{Q}{F} = \frac{1}{k_0 B_0} \frac{\sin(n \pi x / a)}{\cos(\alpha x - \theta)}, \tag{14.30}$$

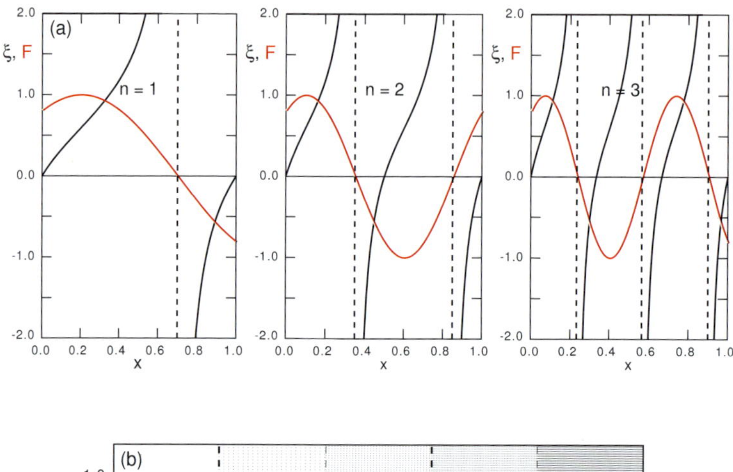

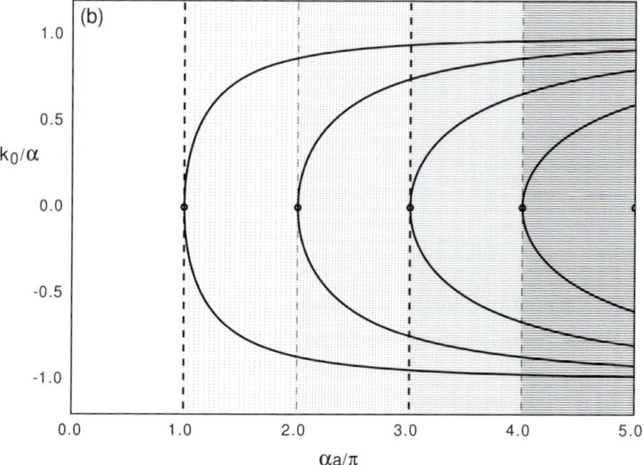

Fig. 14.3 Plane force-free field: (a) marginal modes in terms of ξ ; (b) singularities $F = 0$ of ξ in the shaded areas $\alpha a \geq n\pi$ $(n = 1, 2, \ldots)$.

as plotted in Fig. 14.3(a). Hence, if a solution Q exists such that W as given in Eq. (14.12) is negative, $\alpha a > \pi$ and ξ develops a singularity (Fig. 14.3(b)). For every zero that is added in Q, at least one zero is added to the function F because F oscillates faster than or at least as fast as Q. It is clear that these singularities are of such a nature that the norm

$$\|\boldsymbol{\xi}\|^2 = \int (\xi^2 + \eta^2 + \zeta^2)\, \rho\, dx = \int \left[\xi^2 + G^2 \xi'^2 / (k_0^4 B^2) \right] \rho\, dx \to \infty \, ,$$

where η from Eq. (14.28) and $\zeta = 0$ have been substituted. Hence, *the trial functions Q used in deriving the stability criterion (14.21) do not correspond to physically permissible displacements $\boldsymbol{\xi}$.*

Fortunately, there is still a use for the nice stability diagrams obtained. Observe that apparently a reservoir of energy is available that could drive instabilities if the associated displacement $\boldsymbol{\xi}$ only were realizable. Such is the case if we allow a small amount of resistivity in the system so that the relation $Q = F\xi$ of ideal MHD has to be replaced by one that has extra terms proportional to the resistivity. These terms limit the amplitude of the displacement $\boldsymbol{\xi}$ at the singularity (and, therefore, also the perturbed current that is flowing there). As a result, the unstable energy reservoir is tapped and resistive instabilities develop. Such instabilities are called *tearing modes*. We will investigate these modes in detail in the next section and prove that the stability diagram 14.2(a) applies for those modes.

In cylindrical geometry, the picture becomes more complicated yet. Then, ideal MHD instabilities of force-free fields also develop. This is a subtle effect, due to the additional magnetic curvature, where the magnetic field variable Q may just oscillate a little faster than the function F in certain regions of the $k_0/\alpha - \alpha a$ parameter plane. This was shown by Voslamber and Callebaut [470] by a careful analysis taking proper care of the singularities. The corresponding calculation of the growth rates and eigenfunctions was carried out by Goedbloed and Hagebeuk [177] (see Figs. 9.17 and 9.18 [1]).

14.2 Resistive instabilities

14.2.1 Basic equations

We now present the *resistive normal mode analysis* of the plane slab. The starting point is the nonlinear resistive MHD equations as given in Volume [1], Chapter 2, Eqs. (2.126)–(2.129), which we here summarise for the convenience of the reader:

$$\frac{\partial \rho}{\partial t} = -\nabla \cdot (\rho \mathbf{v}) , \tag{14.31}$$

$$\rho \left(\frac{\partial \mathbf{v}}{\partial t} + \mathbf{v} \cdot \nabla \mathbf{v} \right) = -\nabla p + \rho \mathbf{g} + \mathbf{j} \times \mathbf{B} , \qquad \mathbf{j} = \nabla \times \mathbf{B} , \tag{14.32}$$

$$\frac{\partial p}{\partial t} = -\mathbf{v} \cdot \nabla p - \gamma p \nabla \cdot \mathbf{v} + (\gamma - 1) \eta |\mathbf{j}|^2 , \tag{14.33}$$

$$\frac{\partial \mathbf{B}}{\partial t} = -\nabla \times \mathbf{E} = \nabla \times (\mathbf{v} \times \mathbf{B}) - \nabla \times (\eta \mathbf{j}) . \tag{14.34}$$

Note that resistivity enters through the Ohmic dissipation term in the pressure equation and through the resistive diffusion in the flux equation. In particular, the latter effect is responsible for substantial modifications of the stability analysis.

We will linearise Eqs. (14.31)–(14.34) for small amplitude perturbations about a static equilibrium. Strictly speaking, the assumption of static equilibrium is not justified since resistivity causes the magnetic field to decay. However, since *the magnetic Reynolds number R_m is very large* for situations of interest, this is a very slow process operating on a time scale of the order of $R_\mathrm{m} \cdot \tau_\mathrm{A}$, where τ_A is the characteristic Alfvén time for ideal MHD phenomena. The resistive modes considered in this section will turn out to exponentiate on a much faster time scale, proportional to a broken power of the magnetic Reynolds number, so that the background equilibrium may be considered static. (We here follow standard practice and "write magnetic Reynolds number", although "Lundquist number" would be the more appropriate terminology since the background flow velocity is neglected so that the equilibrium is properly characterized by the value of the Alfvén velocity; see Volume [1], Section 4.4.1.)

Considering a plasma slab with background equilibrium quantities depending on the transverse coordinate x only, we make the usual Ansatz

$$f(\mathbf{r}, t) = f_0(x) + f_1(x)\mathrm{e}^{\mathrm{i}(k_y y + k_z z - \omega t)} , \tag{14.35}$$

where $f_0(x)$ refers to equilibrium quantities and $f_1(x)$ to perturbations. The equilibrium is described by the variables ρ_0, p_0 and $\mathbf{B}_0$, where we will suppress the subscript 0 for convenience, and the perturbations are described by the variables $\delta \equiv \rho_1$, $\mathbf{v} \equiv \mathbf{v}_1$, $\pi \equiv p_1$ and $\mathbf{Q} \equiv \mathbf{B}_1$, so that there is no need for the subscript 1 either.

Assuming a constant resistivity η, the linearised evolution equations read:

$$\frac{\partial \delta}{\partial t} = -\nabla \cdot (\rho \mathbf{v}) , \tag{14.36}$$

$$\rho \frac{\partial \mathbf{v}}{\partial t} = -\nabla \pi + \delta \mathbf{g} - \mathbf{B} \times (\nabla \times \mathbf{Q}) + (\nabla \times \mathbf{B}) \times \mathbf{Q} , \tag{14.37}$$

$$\frac{\partial \pi}{\partial t} = -\mathbf{v} \cdot \nabla p - \gamma p \nabla \cdot \mathbf{v} + 2(\gamma - 1)\eta \, \nabla \times \mathbf{B} \cdot \nabla \times \mathbf{Q} , \tag{14.38}$$

$$\frac{\partial \mathbf{Q}}{\partial t} = \nabla \times (\mathbf{v} \times \mathbf{B}) + \eta \nabla^2 \mathbf{Q} . \tag{14.39}$$

The resistive terms spoil the possibility of integrating the equations for δ, π and $\mathbf{Q}$ to get expressions in terms of the displacement vector $\boldsymbol{\xi}$ alone, as could be done in ideal MHD. We can still exploit the latter variable, but it will not be possible to eliminate the magnetic field perturbation $\mathbf{Q}$. Thus, the new feature of resistive MHD is the distinction between fluid flow, described by $\boldsymbol{\xi}$, and magnetic field evolution, described by $\mathbf{Q}$, since *magnetic field and fluid do not necessarily move together anymore.*

We now introduce *a projection* based on *the direction of inhomogeneity* (x) and *the two directions in the horizontal plane* defined with respect to the horizontal wave vector $\mathbf{k}_0 = (0, k_y, k_z)$:

$$u \equiv v_x ,$$

$$v \equiv (\nabla \times \mathbf{v})_x = -\mathrm{i}(k_z v_y - k_y v_z) ,$$

$$w \equiv \nabla \cdot \mathbf{v} - v_x{}' = \mathrm{i}(k_y v_y + k_z v_z),$$

$$Q \equiv -\mathrm{i}Q_x,$$

$$R \equiv \mathrm{i}(\nabla \times \mathbf{Q})_x = \mathrm{i}\, j_{1x} = k_z Q_y - k_y Q_z. \tag{14.40}$$

Here, u is the normal velocity, v is the normal vorticity, w is the horizontal part of the compressibility, Q is the normal magnetic field perturbation and R is the perturbed normal current. In terms of these physical variables the eigenvalue problem becomes

$$-\mathrm{i}\omega\,\delta = -(\rho u)' - \rho w,$$

$$-\mathrm{i}\rho\omega\,u = -\pi' - g\delta + k_0^{-2}(FQ' + GR)' - FQ,$$

$$-\mathrm{i}\rho\omega\,v = -G'Q + FR,$$

$$-\mathrm{i}\rho\omega\,w = k_0^2\pi - F'Q - GR,$$

$$-\mathrm{i}\omega\,\pi = -p'u - \gamma p(u' + w) - 2(\gamma - 1)\eta\,k_0^{-2}\left[F'(Q'' - k_0^2 Q) + G'R'\right],$$

$$-\mathrm{i}\omega\,Q = Fu + \eta(Q'' - k_0^2 Q),$$

$$-\mathrm{i}\omega\,R = (Gu)' - Fv + Gw + \eta(R'' - k_0^2 R), \tag{14.41}$$

where $k_0^2 \equiv k_y^2 + k_z^2$, and F and G are the projections of the horizontal wave vector $\mathbf{k}_0$ onto the magnetic field:

$$G \equiv \mathbf{e}_x \cdot (\mathbf{k}_0 \times \mathbf{B}) = k_y B_z - k_z B_y, \qquad F \equiv \mathbf{k}_0 \cdot \mathbf{B} = k_y B_y + k_z B_z. \tag{14.42}$$

Recall that the equilibrium is inhomogeneous through the quantities $\rho(x)$, $p(x)$, $B_y(x)$, $B_z(x)$, so that F and G also depend on x. As usual, the prime denotes differentiation with respect to x. The system (14.41) is suitable for numerical integration where the main difficulty is the presence of a complex eigenvalue spectrum.

Alternatively, one could introduce the components of the displacement vector $\boldsymbol{\xi}$ again,

$$u \equiv -\mathrm{i}\omega\xi, \qquad v \equiv -\mathrm{i}\omega\sigma, \qquad w \equiv -\mathrm{i}\omega\tau, \tag{14.43}$$

and eliminate the variables δ, σ, τ and π to obtain a sixth order system of three coupled second order differential equations for ξ, Q and R. The ideal MHD second order differential equation (7.91)[1] for ξ would be contained as the limiting case $\eta \to 0$. We will not pursue this line here since it would involve too many terms in an exposition that is already complicated enough if the mere essentials are presented. This we intend to do.

An important simplification results from *the assumption of incompressibility*. This is justified for the kind of resistive modes we will study, as can be checked after the solutions have been obtained. The incompressible limit is formally obtained from Eqs. (14.41) by taking the limits $\gamma \to \infty$ and $\nabla \cdot \mathbf{v} \to 0$ simultaneously in such a way that the product $\gamma p \nabla \cdot \mathbf{v}$ and, hence, π remains finite but undetermined. Consequently, equation (14.41)(e) for π should be dropped (so that the complicated Ohmic dissipation term in square brackets also disappears from the problem) and replaced by the constraint of incompressibility, $\nabla \cdot \mathbf{v} = 0$. The latter implies that $w = -u'$, so that the variable w is known in terms of u, and equation (14.41)(d) for w can then be used to determine π. Furthermore, the variables δ, v, and w may be expressed in terms of $\xi \equiv u/(-i\omega)$, Q and R so that we obtain the following sixth order sytem:

$$\eta \left[(\rho\omega^2 \xi')' - k_0^2 (\rho\omega^2 + \rho' g)\, \xi + F'' Q \right] + i\omega F \, (Q - F\xi) = 0 \,,$$

$$\eta \, (Q'' - k_0^2 Q) + i\omega \, (Q - F\xi) = 0 \,,$$

$$\eta \, (R'' - k_0^2 R) + i\omega \, (R - G'\xi) - \frac{iF}{\rho\omega} \, (FR - G'Q) = 0 \,. \tag{14.44}$$

A pleasant surprise is that the variable R does not occur in the first pair of equations, so that we may drop the last equation and restrict the study to the fourth order system for the variables ξ and Q alone.

Obtaining the incompressible ideal MHD equations from these equations by taking the limit $\eta \to 0$ is tricky. First, we have to expand Eq. (14.44)(b) to first order,

$$Q = F\xi + \frac{i\eta}{\omega}(Q'' - k_0^2 Q) \approx F\xi + \frac{i\eta}{\omega}\left[(F\xi)'' - k_0^2 F\xi \right], \tag{14.45}$$

and then insert the result in Eq. (14.44)(a):

$$\left[(\rho\omega^2 - F^2)\, \xi' \right]' - k_0^2(\rho\omega^2 - F^2 + \rho' g)\xi = 0 \,, \qquad Q = F\xi \,. \tag{14.46}$$

This agrees with the ideal MHD Eq. (7.91) of Volume [1], in the incompressible limit $\gamma p \to \infty$, where field and fluid move together again. Our task now is to analyse what happens when these equations are replaced by Eqs. (14.44) for the resistive evolution and what is the role of the ideal MHD limit (14.46).

14.2.2 Tearing modes

We will now study the so-called *tearing modes*, which result in breaking and rejoining of the magnetic field lines. Starting from the incompressible resistive MHD equations (14.44), the following assumptions are appropriate:

– the analysis is restricted to eigenvalues corresponding to purely exponential instability so that a real and positive eigenvalue parameter can be defined,

$$\lambda \equiv -i\omega > 0 \,; \tag{14.47}$$

– it is assumed that the density $\rho = \mathrm{const}$, which eliminates ideal MHD gravitational instabilities, since $\rho' g = 0$, so that the slab is stable in the ideal MHD description.

The modes are then described by the resistive MHD equations in the following form:

$$\eta \left[\lambda^2(\xi'' - k_0^2\xi) - (F''/\rho)\, Q \right] + \lambda(F/\rho)\,(Q - F\xi) = 0\,,$$

$$\eta\,(Q'' - k_0^2 Q) - \lambda\,(Q - F\xi) = 0\,. \tag{14.48}$$

Note that all terms in these equations are real now.

In a problem like this it is imperative to enumerate the degrees of freedom by defining *dimensionless parameters*. One can then make various assumptions on the smallness of those parameters to exploit them in *asymptotic expansions*. In the present problem, the thickness of the slab a, the density ρ and the magnitude of magnetic field at the mid plane B_0 are taken as units of length, mass and time, exploiting the definition of the Alfvén velocity, $v_{\mathrm{A}} \equiv B_0/\sqrt{\rho}$, so that $\tau_{\mathrm{A}} \equiv a/v_{\mathrm{A}}$. This implies that a, ρ and B_0 should not be counted as free parameters since they simply fix the dimensions and then disappear from the problem, only to return in the end when actual dimensional numbers need to be computed for comparison with observed quantities.

Our first assumption on parameters is that the wavelength in the horizontal plane is comparable to the transverse size a of the plasma:

$$k_0 a \sim 1\,. \tag{14.49}$$

This expresses the fact that tearing modes should be considered as *large-scale macroscopic MHD modes* which do involve a small-scale resistive effect operating in the normal (x) direction, as we will see, but it does not require localization or small wavelengths in the transverse (y, z) directions. Next, as already mentioned, we will *exploit the magnetic Reynolds number*[1] *as an ordering parameter*:

$$(R_{\mathrm{m}})^{-1} \equiv \eta/(\mu_0 a v_{\mathrm{A}}) \ll 1\,. \tag{14.50}$$

The equilibrium decays on a diffusion time scale τ_{D} that is much longer than the characteristic Alfvén time τ_{A} for ideal MHD. We wish to study resistive modes that exponentiate much faster than the resistive diffusion time, but much slower than the ideal MHD times:

$$(\tau_{\mathrm{D}})^{-1} \equiv (R_{\mathrm{m}})^{-1} v_{\mathrm{A}}/a \ll \lambda \ll v_{\mathrm{A}}/a \equiv (\tau_{\mathrm{A}})^{-1}\,. \tag{14.51}$$

---

[1] For once, we have explicitly written the constant μ_0 in the definition of R_{m} to refresh our memory, although it is suppressed again in most of the rest of this chapter.

This is possible if we can find modes with a growth rate λ that scales as a broken power of the magnetic Reynolds number: $\lambda \sim (R_{\mathrm{m}})^{-\nu} v_{\mathrm{A}}/a$, where $0 < \nu < 1$. This will turn out to be the case. Since R_{m} is huge, this provides enough parameter space for our asymptotic analysis.

With this ordering, for small η, the resistive equations (14.48) automatically would lead to the ideal MHD equations,

$$\left[(\lambda^2 + F^2/\rho)\, \xi' \right]' - k_0^2(\lambda^2 + F^2/\rho)\, \xi = 0\,, \qquad Q = F\xi\,, \qquad (14.52)$$

since the expansion (14.45) yields

$$Q \approx F\xi + (\eta/\lambda)\left[(F\xi)'' - k_0^2 F\xi \right] \approx \left[1 + \mathcal{O}(\lambda\tau_{\mathrm{D}})^{-1} \right] F\xi \approx F\xi\,. \qquad (14.53)$$

Here, the resistive correction of order $(\lambda\tau_{\mathrm{D}})^{-1} \equiv \eta/(\lambda a^2)$ is negligible according to the left part of the approximation (14.51) *if ξ is assumed to have $\mathcal{O}(1)$ variations only*. However, we know from our previous analysis of Section 14.1 that the assumption of finite variation on ξ is not justified when ideal MHD singularities $F = 0$ occur. Then (choosing the origin of the x-coordinate to coincide with the singularity), $\xi_{\mathrm{ideal}} \sim 1/x \to \infty$ while the magnetic field variable Q remains finite. Hence, the resistive terms in Eq. (14.53) cannot be neglected in a small layer surrounding the ideal MHD singularity where they limit the amplitude of ξ and the related current density perturbation. Outside this layer, ideal MHD is appropriate. Consequently, the problem may be analyzed by distinguishing three regions, viz. *two outer ideal MHD regions* where F is not small, and *one inner resistive layer* surrounding the point $F = 0$. The solutions of the three regions have to be matched to each other so that there should exist overlap regions where the resistive as well as the ideal solutions are valid.

The singularity $F = 0$ can occur anywhere on the interval $x_1 \leq x \leq x_2$, but we will position it at $x = x_0 = 0$ for simplicity. This involves both a simplification of notation (expansions in x rather than $x - x_0$) and of some part of the analysis (which distinguishes between modes that are even or odd with respect to $x = 0$). Of course, these restrictions can easily be relaxed. Our choice of the zero point does not limit the physics of the problem in any way, since we are free to shift the x-axis to simplify the algebra.

(a) Outer ideal MHD regions

The outer ideal MHD regions cover most of the real axis from $x = x_1 = -\tfrac{1}{2}a$ to $x = x_2 = \tfrac{1}{2}a$, where $F \neq 0$, except for a small region around $x = 0$ where F is small and resistive effects dominate. Obviously, the size of the resistive region is determined by the resistivity, but in a way that has to be determined yet. In the outer ideal MHD regions we may simplify the equations (14.52) even further by noting that

$$\frac{\lambda^2}{F^2/\rho} \sim (\lambda a/v_{\mathrm{A}})^2 \ll 1 \,,$$

according to the estimates (14.49) and the right part of the inequalities (14.51). Hence, the *marginal* ($\lambda = 0$) ideal MHD equations are appropriate in the outer layer:

$$(F^2\xi')' - k_0^2 F^2 \, \xi = 0 \,, \qquad Q = F\xi \,. \tag{14.54}$$

This implies that the evolution of tearing modes is so slow that inertia is negligible on the ideal MHD time scale. The associated boundary conditions are

$$\xi(x_1) = \xi(x_2) = 0 \,. \tag{14.55}$$

In terms of Q, the basic differential equation (14.54) transforms into

$$Q'' - (k_0^2 + F''/F)\, Q = 0 \,. \tag{14.56}$$

If there is no ideal MHD singularity ($F \neq 0$ everywhere) Eq. (14.54) has no solutions satisfying the boundary conditions (14.55), as we have seen in Section 14.1, and hence there are no resistive instabilities in that case. To get resistive instabilities we need a point $F = 0$ so that the ideal MHD solution ξ blows up and Q' is discontinuous at that point (dashed lines in Fig. 14.4).

Let us formally solve Eq. (14.54) in the two outer regions $(x_1, -\epsilon)$ and (ϵ, x_2), where matching to the resistive layer solutions should take place at $x = \pm\epsilon$. The main idea is that the singularity $x = 0$ is eliminated, but that the ideal MHD solutions are valid close enough to this point to permit a singular expansion. Since $F \approx F_0' x$ for $|x| \sim \epsilon$, two solutions are obtained, a large one, ξ_l, and a small one, ξ_s, which may be continued up to the outer boundaries $x = x_{1,2}$. The solutions in the two outer regions are distinguished by a superscript $+$ for $x > \epsilon$ and $-$ for $x < -\epsilon$. Consequently, the general solutions in the two outer regions may be written as

$$\xi^\pm(x) = A^\pm \xi_l^\pm(x) + B^\pm \xi_s^\pm(x) \,, \tag{14.57}$$

where

$$\xi_l = \frac{1}{x}\left[1 + \tfrac{1}{2}(k_0 x)^2 + \cdots \right] \,, \qquad \xi_s = 1 - \tfrac{1}{4}(k_0 x)^2 + \cdots \,. \tag{14.58}$$

The two ratios B^+/A^+ and B^-/A^- between the small and large contributions are determined by the boundary conditions (14.55). Whereas close to the singularity the large contributions dominate, for matching to the resistive inner layer solutions the small contributions are essential as well. Matching will require a prescribed mix of large and small solutions, as we will see. This is expressed most

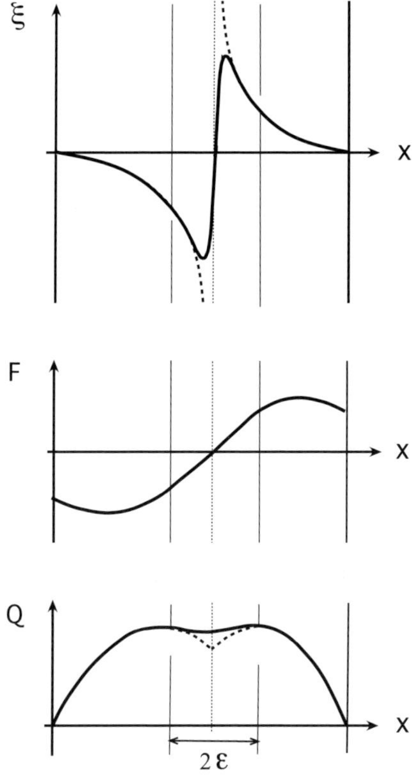

Fig. 14.4 Resistive (drawn) and ideal (dashed) MHD solutions at the ideal MHD
singularity $F = 0$.

clearly by the logarithmic derivatives of the magnetic field perturbation which are
completely determined by the solutions in the ideal regions:

$$\frac{1}{Q}\frac{dQ}{dx}\bigg|^{\text{outer}}_{x=\epsilon\downarrow 0} \approx \frac{B^+}{A^+} = -\frac{\xi^+_1(x_2)}{\xi^+_s(x_2)}, \qquad \frac{1}{Q}\frac{dQ}{dx}\bigg|^{\text{outer}}_{x=-\epsilon\uparrow 0} \approx \frac{B^-}{A^-} = -\frac{\xi^-_1(x_1)}{\xi^-_s(x_1)}.$$

$$(14.59)$$

These two quantities are all that the outer ideal MHD regions can provide and it
will turn out to be sufficient.

Notice that the two logarithmic derivatives (14.59) are not equal in the limit
$\epsilon \to 0$, so that a jump is obtained. This jump is permitted in ideal MHD and it corresponds to the occurrence of a *surface current perturbation*, which is extremely
stabilising. Neglecting this contribution is the physical cause of the erroneous instability results obtained in Section 14.1. In resistive MHD, the jump is smoothed
through the resistive layer solutions, which will be computed next, and the instability results expressed by Fig. 14.2 will turn out to become applicable then.

(b) Inner resistive layer

The resistive layer surrounds the ideal MHD singularity $F = 0$, which we have positioned at $x = 0$. In this region, the function F may be approximated as

$$F \equiv \mathbf{k}_0 \cdot \mathbf{B} \approx F_0' \, x \,. \tag{14.60}$$

We will call the width of the resistive layer δ. Obviously,

$$\delta \ll a \,, \tag{14.61}$$

because of the small value of the resistivity, and it is logical to expect it to scale with a negative power of the magnetic Reynolds number. One of our aims is to determine this power. Hence, we are in the peculiar situation of having to prescribe boundary conditions at an unknown location. We will see that this problem is solved by following the example of *boundary layer analysis* of fluid mechanics.

Note that matching to the outer ideal MHD solutions is considered to take place at $x = \pm\epsilon$ so that we formally need to impose

$$\delta \ll \epsilon \ll a \,. \tag{14.62}$$

This can easily be arranged again by means of some broken power of the magnetic Reynolds number, which we could choose after the scaling of δ with R_{m} has been determined. The precise magnitude of ϵ should not matter, though, as long as the approximation (14.60) is valid there.

In the resistive layer, the differential equations (14.48) hold. They may be further simplified on the basis of the approximations (14.49), (14.51) and (14.61). To that end, the orders of magnitude of the different terms are compared:

$$\left| \frac{k_0^2 \, \xi}{\xi''} \right| \sim \delta^2/a^2 \ll 1 \,, \qquad \left| \frac{\eta(F''/\rho)\,Q}{\lambda(F/\rho)\,Q} \right| \sim \left| \frac{\eta k_0^2 \, Q}{\lambda Q} \right| \sim \frac{(R_{\mathrm{m}})^{-1} v_{\mathrm{A}}/a}{\lambda} \ll 1 \,.$$

Consequently, the resistive layer equations may be simplified to

$$\eta \lambda \xi'' - (F^2/\rho)\,\xi = -(F/\rho)\,Q \,,$$

$$\eta Q'' - \lambda Q = -\lambda F \xi \,. \tag{14.63}$$

It is to be noted that this significant simplification now only holds in the resistive layer since the approximate ideal MHD equations (14.54) are no longer contained as a limiting case (which was the case for the original pairs of equations (14.48) and (14.52)). Consequently, we have lost the option of (numerically) integrating the resistive equations all the way to the plasma boundaries at $x = x_{1,2}$, i.e. we are now forced to consider the matching problem.

(c) Scaling and matching

Before we do this, we finally perform the promised scaling of the resistive layer equations (14.63) exploiting three of the four dimensionless parameters that occur in this problem, viz.

$$\lambda a/v_{\mathrm{A}}\,, \qquad R_{\mathrm{m}} \equiv a v_{\mathrm{A}}/\eta\,, \qquad H \equiv F_0' a/(k_0 B_0)\,, \qquad K \equiv k_0 a\,. \qquad (14.64)$$

Note that K appears as a separate parameter in the outer ideal MHD equations (14.54). By means of these parameters all variables may be rescaled:

$$s \equiv \left[R_{\mathrm{m}} K H\right]^{1/3} (x/a)\,, \qquad\qquad \bar{\lambda} \equiv \left[R_{\mathrm{m}}/(K^2 H^2)\right]^{1/3} (\lambda a/v_{\mathrm{A}})\,,$$

$$\Phi \equiv \left[R_{\mathrm{m}}/(K^2 H^2)\right]^{-1/3} (\xi/a)\,, \qquad \Psi \equiv Q/B_0\,. \qquad (14.65)$$

The explicit dependence on the parameters R_{m}, K and H then disappears and only the scaled eigenvalue $\bar{\lambda}$ appears in the resistive equations:

$$\bar{\lambda}\,\frac{d^2\Phi}{ds^2} - s^2\,\Phi = - s\,\Psi\,,$$

$$\frac{d^2\Psi}{ds^2} - \bar{\lambda}\,\Psi = -\bar{\lambda}\,s\,\Phi\,, \quad \text{or} \quad \frac{d^2\Psi}{ds^2} = -\bar{\lambda}^2 \frac{1}{s}\,\frac{d^2\Phi}{ds^2}\,. \qquad (14.66)$$

The problem has now been reduced to the solution of a fourth order system of ODEs with only one parameter ($\bar{\lambda}$). This system has to be solved on an interval in terms of the stretched coordinate s which corresponds to a very narrow region in physical space. Also, notice that the tendency of the original variable ξ to blow up in the limit of small resistivity has been absorbed in the normalization of the scaled variable Φ.

▷ **Fourth order differential equation** The equations (14.66) may be written explicitly as a fourth order equation in terms of Φ alone:

$$\bar{\lambda} s^2 \frac{d^4\Phi}{ds^4} - 2\bar{\lambda} s \frac{d^3\Phi}{ds^3} + (2\bar{\lambda} - \bar{\lambda}^2 s^2 - s^4)\frac{d^2\Phi}{ds^2} - 2s^3 \frac{d\Phi}{ds} = 0\,. \qquad (14.67)$$

Notice that the lowest order term with Φ is missing so that one of the four solutions is a constant. Since this simplification is not essential (for other resistive modes it does not occur, Section 14.2.3), we will not base our analysis on it. Equation (14.67) may be investigated with respect to singular behavior by means of the usual series expansion methods [243], [33]. One discovers that the point $s = 0$ has become a regular point (of course, this was the whole point of replacing the ideal MHD equations by the resistive ones) and the points $s = \pm\infty$ are irregular singular points. The latter permit a singular expansion which is to be matched asymptotically to the outer ideal MHD solutions. It is useful to consider the three formal expansions in the resistive region (about $s = 0$, $s = \infty$ and $s = -\infty$) to clarify the behavior of the solutions and the kind of boundary conditions needed. Once this has been established, Eq. (14.67) can be solved by whatever method is convenient (numerical or approximate analytical).

By straightforward substitution, we find the following expansions about $s = 0$:

$$\Phi = \Phi^0(s) = \sum_{i=1}^{4} C_i^0 \Phi_i^0, \qquad \Psi = \Psi^0(s) = \sum_{i=1}^{4} C_i^0 \Psi_i^0,$$

$$\Phi_1^0 = 1, \qquad\qquad \Psi_1^0 = s,$$

$$\Phi_2^0 = s + \frac{1}{20\bar{\lambda}} s^5 + \cdots, \qquad \Psi_2^0 = -\frac{1}{12}\bar{\lambda} s^4 + \cdots,$$

$$\Phi_3^0 = s^3 + \frac{3\bar{\lambda}}{20} s^5 + \cdots, \qquad \Psi_3^0 = -6\bar{\lambda} + \cdots,$$

$$\Phi_4^0 = s^4 + \frac{\bar{\lambda}}{15} s^6 + \cdots, \qquad \Psi_4^0 = -12\bar{\lambda} s + \cdots. \qquad (14.68)$$

As expected, all solutions are regular near $s = 0$, and they are either odd or even, consistent with the reflection symmetry about $s = 0$ of the resistive equations (14.66) or (14.67).

The behavior at the outer edges is considerably more complicated, as shown by the following expansions for $s \to \infty$ and $s \to -\infty$ (indicated by the superscripts $+$ and $-$):

$$\Phi = \Phi^{\pm}(s) = \sum_{i=1}^{4} C_i^{\pm} \Phi_i^{\pm}, \qquad\qquad \Psi = \Psi^{\pm}(s) = \sum_{i=1}^{4} C_i^{\pm} \Psi_i^{\pm},$$

$$\Phi_1^{\pm} = 1, \qquad\qquad \Psi_1^{\pm} = s\,\Phi_1^{\pm},$$

$$\Phi_2^{\pm} = \frac{1}{s}\left(1 - \frac{\bar{\lambda}^2}{3s^2} + \cdots\right), \qquad \Psi_2^{\pm} \approx s\,\Phi_2^{\pm},$$

$$\Phi_3^{\pm} = e^{s^2/(2\sqrt{\bar{\lambda}})}\frac{1}{s}|s|^{\frac{1}{2}+\frac{1}{2}\bar{\lambda}\sqrt{\bar{\lambda}}}\left(1 + \frac{\alpha^{\pm}}{s^2} + \cdots\right), \qquad \Psi_3^{\pm} \approx -\frac{\bar{\lambda}^2}{s}\Phi_3^{\pm},$$

$$\Phi_4^{\pm} = e^{-s^2/(2\sqrt{\bar{\lambda}})}|s|^{-\frac{1}{2}-\frac{1}{2}\bar{\lambda}\sqrt{\bar{\lambda}}}\left(1 + \frac{\beta^{\pm}}{s^2} + \cdots\right), \qquad \Psi_4^{\pm} \approx -\frac{\bar{\lambda}^2}{s}\Phi_4^{\pm}. \qquad (14.69)$$

The coefficients $\alpha^{\pm}$ and $\beta^{\pm}$ are known, in principle, but they are not important. Clearly, at the outer edges of the resistive layer, there are two pairs of solutions, $\Phi_1^{\pm}$ and $\Phi_2^{\pm}$, which exhibit the ideal MHD relationship $\Psi \approx s\Phi$, and two pairs, $\Phi_3^{\pm}$ and $\Phi_4^{\pm}$, which do not. Consequently, the first two solutions can be used to match to the outer ideal MHD solutions, and the latter two should not play a role. For the solutions $\Phi_4^{\pm}$ this is automatically the case since they are exponentially small there. The two solutions $\Phi_3^{\pm}$, though, are exponentially unbounded and should be excluded by imposing the boundary conditions of *regularity*, so that $C_3^- = C_3^+ = 0$. What remains is the other boundary conditions of *matching of the logarithmic derivatives of the magnetic field perturbation* to the outer ideal MHD expressions given in Eq. (14.59). One easily checks that these expressions just involve the two solutions $\Phi_1^{\pm}$ and $\Phi_2^{\pm}$: $(\Psi'/\Psi)^- = C_1^-/C_2^-$ and $(\Psi'/\Psi)^+ = C_1^+/C_2^+$. Notice that the two exponentially small solutions $\Phi_4^{\pm}$ have not been excluded; they just do not contribute to the boundary conditions directly. Hence, everything now counts correctly: we have two regularity conditions and two conditions on the logarithmic derivatives, i.e. four boundary conditions on a fourth order ODE.

Of course, the three expansions (14.68) and (14.69) all refer to one and the same solution of the fourth order ODE (14.67), i.e. we may write

$$\Phi = \sum_{i=1}^{4} C_i^- \Phi_i^- = \sum_{i=1}^{4} C_i^0 \Phi_i^0 = \sum_{i=1}^{4} C_i^+ \Phi_i^+. \qquad (14.70)$$

Therefore, the three sets of fundamental solutions $\{\Phi_i^-\}$, $\{\Phi_i^0\}$ and $\{\Phi_i^+\}$ possess a linear relationship to each other which could be found explicitly, if desired, by analytic continuation or by straightforward numerical integration of Eq. (14.67). Consequently, the three sets of constants $\{C_i^-\}$, $\{C_i^0\}$ and $\{C_i^+\}$ are linearly dependent as well:

$$C_i^- = \sum_{j=1}^{4} \gamma_{i,j}\, C_j^0, \qquad C_i^0 = \sum_{j=1}^{4} \delta_{i,j}\, C_j^+, \tag{14.71}$$

where the coefficients $\gamma_{i,j}$ and $\delta_{i,j}$ are then known, in principle. The point is that one set of four constants is sufficient to fix the solution, but these four can not coincide with any one of the three separate sets since boundary conditions at different locations are involved. For example, from the present analysis, it should be clear that these four constants could be C_4^-, C_1^-/C_2^-, C_4^+ and C_1^+/C_2^+. $\qquad\qquad\triangleleft$

In conclusion, appropriate boundary conditions to be imposed on the resistive layer equations (14.66) are:

$$\Phi \Big|_{\substack{\text{inner}\\ s\to-\infty}} \text{regular}\,, \qquad \frac{1}{\Psi}\frac{d\Psi}{ds}\Big|_{\substack{\text{inner}\\ s\to-\infty}} \approx \left(\frac{x}{s}\right)\frac{1}{Q}\frac{dQ}{dx}\Big|_{\substack{\text{outer}\\ x\uparrow 0}}\,,$$

$$\Phi \Big|_{\substack{\text{inner}\\ s\to\infty}} \text{regular}\,, \qquad \frac{1}{\Psi}\frac{d\Psi}{ds}\Big|_{\substack{\text{inner}\\ s\to\infty}} \approx \left(\frac{x}{s}\right)\frac{1}{Q}\frac{dQ}{dx}\Big|_{\substack{\text{outer}\\ x\downarrow 0}}\,, \tag{14.72}$$

which translate into the following conditions on the constants:

$$C_3^- = 0\,, \qquad C_1^-/C_2^- = B^-/A^-\,,$$

$$C_3^+ = 0\,, \qquad C_1^+/C_2^+ = B^+/A^+\,. \tag{14.73}$$

Alternatively, since we have arranged the singularity to be located in the middle of the interval, we can restrict the analysis to even and odd modes about the midplane. In particular, we can study modes that are odd in Φ and even in Ψ since they are the first ones to become unstable. The appropriate boundary conditions for this system read:

$$\Phi(0) = 0\,, \qquad\qquad \Psi'(0) = 0\,,$$

$$\Phi \Big|_{\substack{\text{inner}\\ s\to\infty}} \text{regular}\,, \qquad \frac{1}{\Psi}\frac{d\Psi}{ds}\Big|_{\substack{\text{inner}\\ s\to\infty}} \approx \left(\frac{x}{s}\right)\frac{1}{Q}\frac{dQ}{dx}\Big|_{\substack{\text{outer}\\ x\downarrow 0}}\,, \tag{14.74}$$

which translates into the following conditions on the constants:

$$C_1^0 = 0\,, \qquad C_4^0 = 0\,,$$

$$C_3^+ = 0\,, \qquad C_1^+/C_2^+ = B^+/A^+\,. \tag{14.75}$$

Notice that the factor (x/s) just represents the different normalizations of the inner and outer regions according to Eq. (14.65).

We could now solve Eqs. (14.66). Unfortunately, this would have to be done numerically, since the system is still not elementary enough to allow for analytical expressions. Therefore, one additional step will be taken to obtain closed answers justifying the scaling assumptions.

(d) Approximate solution

From a study of the numerical results, Furth, Killeen and Rosenbluth in their classical paper on the resistive instabilities [152] proposed the so-called *"constant Ψ"* *approximation* to obtain explicit solutions. The idea is that the function Φ is the one which misbehaves in ideal MHD and, consequently, exhibits large variations in the limit of small resistivity, but the magnetic field variable Ψ only exhibits moderate variations. Its behavior only counts to produce the correct numerical magnitude of the derivative to connect to the outer solutions through the boundary conditions (14.72). If Φ were known, that part of the information could be obtained with high accuracy by just integrating the second form of Eq. (14.66)(b). On the other hand, the correct behavior of Φ can be obtained from Eq. (14.66)(a) without much influence of the magnetic field contribution on the right hand side: just putting $\Psi \approx \Psi_0 = \text{const}$ will do. This approximation turns out to produce results that are quite close to the numerical values. Once it has been made, the rest of the analysis is straightforward.

In the "constant Ψ" approximation, the variables can be scaled once more:

$$\bar{s} \equiv \bar{\lambda}^{-1/4}\, s\,, \qquad \bar{\Phi} \equiv \bar{\lambda}^{1/4}\, \Phi/\Psi_0\,, \qquad \bar{\Psi} \equiv \Psi/\Psi_0\,, \tag{14.76}$$

so that the eigenvalue $\bar{\lambda}$ is also eliminated from the ODE for Φ, and the equations (14.66) transform into

$$\frac{d^2\bar{\Phi}}{d\bar{s}^2} - \bar{s}^2\, \bar{\Phi} = -\bar{s}\, \bar{\Psi} \approx -\bar{s}\,,$$

$$\frac{d^2\bar{\Psi}}{d\bar{s}^2} = -\bar{\lambda}^{3/2}\, \frac{1}{\bar{s}}\, \frac{d^2\bar{\Phi}}{d\bar{s}^2}\,. \tag{14.77}$$

In other words: we obtain an inhomogeneous second order ODE for $\bar{\Phi}$, whereas the logarithmic derivative of $\bar{\Psi}$, i.e. $\bar{\Psi}'$, may be obtained by integrating Eq. (14.77)(b) once. We will now exploit the symmetry of the problem, i.e. we assume $\bar{\Phi}$ to be odd, by using the boundary conditions (14.74) which, for the present variables, become

$$\bar{\Phi}(0) = 0\,, \qquad\qquad \bar{\Psi}'(0) = 0\,,$$

$$\bar{\Phi}\Big|_{\bar{s}\to\infty} \text{regular}\,, \qquad \bar{\Psi}'\Big|_{\bar{s}\to\infty} \approx \text{const} \left[= (x/\bar{s})\, Q'/Q \Big|_{x\downarrow 0} \right]\,, \tag{14.78}$$

where the constant is just the logarithmic derivative of the magnetic field of the outer ideal MHD solutions.

The solution of the ODE (14.77)(a) for $\bar{\Phi}$, subject to the boundary conditions (14.78), can be represented in terms of a definite integral over an auxiliary variable u:

$$\bar{\Phi} = \tfrac{1}{2}\bar{s}\int_0^1 (1-u^2)^{-1/4}\, e^{-\frac{1}{2}\bar{s}^2 u}\, du\,. \tag{14.79}$$

One easily checks the correctness of this expression by substitution into the differential equation, whereas the behavior for small and large $\bar{s}$ demonstrates satisfaction of the boundary conditions:

$$\bar{\Phi}(\bar{s}) \approx \frac{C}{2\sqrt{\pi}}\,\bar{s}\,, \qquad \bar{\Psi}'(\bar{s}) \approx \bar{\lambda}^{3/2}\bar{s} \qquad (\bar{s}\ll 1)\,,$$

$$\bar{\Phi}(\bar{s}) \approx \frac{1}{\bar{s}}\,, \qquad \bar{\Psi}'(\bar{s}) \approx \tfrac{1}{2}C\bar{\lambda}^{3/2} \qquad (\bar{s}\gg 1)\,, \tag{14.80}$$

where the constant

$$C = \sqrt{2\pi}\int_0^1 \sqrt{u}\,(1-u^2)^{-1/4}\,du = 2\pi\Gamma(\tfrac{3}{4})/\Gamma(\tfrac{1}{4}) = 2.1236\ldots \tag{14.81}$$

results from an integration over the auxiliary variable.

The actual identification of $\bar{\Psi}'\big|_{\bar{s}\to\infty}$ with the logarithmic derivative of the outer ideal MHD solutions, according to the boundary condition (14.78)(b), provides the dispersion equation for the computation of the growth rate of the tearing modes. We will rewrite this condition in terms of the contributions of both outer boundaries:

$$\frac{d\bar{\Psi}}{d\bar{s}}\bigg|_{\substack{\text{inner}\\ \bar{s}\to\infty}} - \frac{d\bar{\Psi}}{d\bar{s}}\bigg|_{\substack{\text{inner}\\ \bar{s}\to-\infty}} = -\bar{\lambda}^{3/2}\int_{-\infty}^{\infty} d\bar{s}\,\frac{1}{\bar{s}}\frac{d^2\bar{\Phi}}{d\bar{s}^2} = C\,\bar{\lambda}^{3/2} = \left(\frac{x/a}{\bar{s}}\right)\Delta'\,, \tag{14.82}$$

where Δ' is the jump of the logarithmic derivative of the magnetic field perturbation of the outside ideal MHD solution,

$$\Delta' \equiv \frac{a}{Q}\left(\frac{dQ}{dx}\bigg|_{\substack{\text{outer}\\ x\downarrow 0}} - \frac{dQ}{dx}\bigg|_{\substack{\text{outer}\\ x\uparrow 0}}\right), \tag{14.83}$$

which is determined by the solutions of Eq. (14.56).

Substituting the scaling factors from the definitions (14.65) and (14.76) back into Eq. (14.82) then results in the explicit expression for *the growth rate of the tearing mode*:

$$\lambda = R_{\mathrm{m}}^{-3/5}(KH)^{2/5}\left(\frac{\Delta'}{C}\right)^{4/5} v_{\mathrm{A}}/a\,, \tag{14.84}$$

which justifies our original assumption of broken powers of the magnetic Reynolds number. An estimate of *the resistive layer width* is obtained from the relation $s\sim 1$, which gives:

$$\delta \sim R_{\mathrm{m}}^{-2/5}(KH)^{-2/5}\left(\frac{\Delta'}{C}\right)^{1/5} a\,, \tag{14.85}$$

which also conforms to our assumptions.

The tearing mode requires Δ' to be positive so that the outer ideal MHD solutions determine its stability. It should be noted that the growth rate λ of Eq. (14.84) depends on the dimensionless parameters K and H both through the explicit power $(KH)^{2/5}$ and through the implicit dependence $\Delta' = \Delta'(H, K)$. This may be illustrated by the force free magnetic field of Section 14.1, where the value of Δ' can be computed easily. In that case (taking $\theta = \pi/2$, so that the singularity occurs at $x = 0$ and $F = k_0 B_0 \sin \alpha x$), $Q^{\pm} = \mp D \sin \sqrt{\alpha^2 - k_0^2} \, (x \mp \tfrac{1}{2}a)$, so that

$$
\begin{aligned}
\Delta' &= -2a\sqrt{\alpha^2 - k_0^2} \, \cot\left(\tfrac{1}{2}a\sqrt{\alpha^2 - k_0^2}\right) \\
&= -2\sqrt{H^2 - K^2} \, \cot\left(\tfrac{1}{2}\sqrt{H^2 - K^2}\right).
\end{aligned}
\tag{14.86}
$$

Its value is positive, as required for tearing instability, when $(\alpha a)^2 - (k_0 a)^2 \equiv H^2 - K^2 > (n\pi)^2$. This agrees with the stability diagrams of Section 14.1 so that this analysis has finally received its proper re-interpretation now in the context of resistive MHD: *the plane force-free field is unstable with respect to long wavelength tearing instabilities*. The expressions (14.84) and (14.86) clearly show that tearing modes are long wavelength instabilities ($K \sim 1$ and $|K| < |H|$) driven by *the current*, as expressed by the parameter H.

14.2.3 Resistive interchange modes

A completely different kind of resistive mode is the *resistive gravitational interchange mode*. The driving force of this instability may be expressed by the dimensionless parameter

$$
G \equiv \rho' g a^2 / B_0^2 \,,
\tag{14.87}
$$

which was neglected when transforming Eq. (14.44) to Eq. (14.48). We will not enter the detailed analysis of this mode, but just give some of the final results in order to illustrate the wide range of possible outcomes of a boundary layer analysis.

First of all, it should be remarked that the stability criterion for the corresponding ideal MHD gravitational interchanges is given by Eq. (7.199) in the incompressible limit ($\gamma p \to \infty$):

$$
\rho' g \leq \tfrac{1}{4} B^2 \varphi'^2 \,, \quad \text{i.e.} \quad G \leq \tfrac{1}{4} H^2 \,.
\tag{14.88}
$$

This criterion is completely analogous to the well-known Suydam criterion where a pressure gradient driven instability is balanced by the shear of the magnetic field lines. Similarly, the gravitational interchanges are driven by *a density gradient* (heavy fluid on top of a lighter fluid), but possibly stabilised by *the shear* term H.

Hence, the same parameter H enters as in the tearing mode analysis, but with an entirely different result: tearing instabilities are driven by the current, interchanges are stabilised by shear which is also created by the current.

However, *stabilisation by shear of the magnetic field lines is another ideal MHD property that is lost when resistivity is introduced.* The stability criterion for resistive interchange modes simply becomes

$$G \leq 0, \tag{14.89}$$

i.e. only lighter fluid on top of heavier fluid allowed! By a similar boundary layer analysis as for the tearing modes, the growth rate is found to be given by

$$\lambda \sim R_{\mathrm{m}}^{-1/3} \left(\frac{KG}{H} \right)^{2/3} v_{\mathrm{A}}/a, \tag{14.90}$$

whereas the resistive layer width, in this case, is given by

$$\delta \sim R_{\mathrm{m}}^{-1/3} \left(\frac{G}{K^2 H^4} \right)^{1/6} a. \tag{14.91}$$

Hence, resistive interchanges are no longer stabilised by magnetic shear (the parameter H), but their growth rate is strongly diminished with the $-1/3$ power of the magnetic Reynolds number according to Eq. (14.90).

14.3 Resistive spectrum

14.3.1 Resistive wall mode

The basic mechanism of resistive instability due to magnetic flux reconnection is also operating in an entirely different location than inside the plasma, viz. in the conducting wall that is supposed to stabilize the external MHD instabilities. In particular, if operating parameters of fusion experiments are pushed up to obtain higher values of beta, current-driven external instabilities frequently interfere, and they are usually stabilized by the presence of a conducting wall. Apart from the fact that a conducting wall close to a plasma is not very desirable in a fusion reactor, the conductivity itself turns out to be a problem.

Consider an external kink mode which is stabilized by a conducting wall, but which would be unstable without that wall. For simplicity, we analyze this mode for a flat current distribution in the "straight tokamak" approximation, appropriate for low-beta plasmas. From the expression of the growth rate given by Eq. (9.99) of Volume [1], we find for a wall position normalized to the plasma radius, $w \equiv b/a$, and mode numbers $n = -1$ and $m > 0$:

$$\bar{\omega}^2 \equiv \frac{\omega^2}{\epsilon^2 \tau_{\mathrm{A}}^2} = \frac{2(q-m)\left[q - (m - 1 + w^{-2|m|})\right]}{q^2(1 - w^{-2|m|})}. \tag{14.92}$$

Here, ϵ is the inverse aspect ratio of the slender tokamak torus that is represented by a periodic straight cylinder of length $2\pi R_0$ and plasma radius a, and τ_A is the characteristic Alfvén time across the plasma:

$$\epsilon \equiv \frac{a}{R_0} \ll 1 \,, \qquad \tau_A \equiv \frac{v_A}{a} = \frac{B_0}{a\sqrt{\mu_0 \rho_0}} \,. \tag{14.93}$$

Considering the external $m = 2$ kink mode with a wall close to the plasma, only a narrow instability range $1 + w^{-4} < q < 2$ remains and the mode is stable in the wide range $1 < q < 1 + w^{-4}$, but violently unstable when the wall is taken away. Let us consider what happens in that range when the wall is not taken away, but the finite resistivity of it is taken into account.

The expression (14.92) for the ideal MHD growth rate is found by solving the ODE for the radial displacement ξ, obtained from the Hain–Lüst equation in the low-beta "straight tokamak" approximation,

$$\left\{\frac{r}{m^2}\left[\bar{\omega}^2 - (n + m/q)^2\right](r\xi)'\right\}' - \left[\bar{\omega}^2 - (n + m/q)^2\right]\xi = 0 \quad (0 \le r \le 1), \tag{14.94}$$

and the ODE for the radial perturbation $\hat{Q}_i$ of the magnetic field in the vacuum,

$$\left[\frac{r}{m^2}(r\hat{Q}_i)'\right]' - \hat{Q}_i = 0 \qquad (1 \le r \le w), \tag{14.95}$$

joining them by applying the BC at the plasma–vacuum interface,

$$\frac{\bar{\omega}^2 - (n + m/q)^2}{m^2(n + m/q)}\frac{(r\xi)'}{\xi} + \frac{2}{mq} = -\frac{n + m/q}{m^2}\frac{(r\hat{Q}_i)'}{\hat{Q}_i} \qquad (r = 1), \tag{14.96}$$

and applying either one of the following BCs on the outside:

$$\hat{Q}_i = 0 \,, \quad \text{for } r = \begin{cases} w & \textit{(infinitely conducting wall)} \\[2mm] \infty & \textit{(no wall)} \end{cases}. \tag{14.97}$$

For constant current density in the plasma, the safety factor $q(r) = \text{const}$, so that the factor in square brackets in the ODE (14.94) becomes constant, and we obtain the following simple solutions for ξ and $r\hat{Q}_i$:

$$r\xi = A_1 r^{|m|} \,, \qquad r\hat{Q}_i = A_2(r^{|m|} + Cr^{-|m|}) \,. \tag{14.98}$$

Substitution into the BC (14.96) yields the dispersion equation (14.92).

The BC (14.96) at the plasma surface is obtained by dividing the two plasma–vacuum interface conditions derived in Section 6.6.1 [1]. Note that possible problems with a vanishing denominator are absent here since $\xi(r = 1) \neq 0$ by the definition of an external mode. The obvious advantage of this procedure is that the amplitudes A_1 and A_2 disappear from the problem, so that only the factor C remains. If desired, one could assume a more realistic current profile and solve the ODE (14.94), but the equation for the determination of growth rate would be implicit. This will change the magnitude of the growth rate of the resistive wall mode, but not the formal scaling of the expression that we will derive.

The problem of the loss of wall stabilization of external kink modes when the resistivity of the wall is taken into account was first discussed by Pfirsch and Tasso [371]. They proved that an MHD unstable configuration cannot be stabilized by the introduction of resistive walls, i.e. with respect to stability it does not make a difference whether or not a resistive wall is present. Of course, the crucial question then becomes whether the growth rate of the ensuing *resistive wall mode* (RWM) is compatible with plasma confinement on the characteristic time scale needed for nuclear fusion. This problem was addressed by Goedbloed *et al.* [182], who solved the dispersion equation of resistive wall modes for a high-beta screw pinch (a toroidal device differing from a tokamak by the presence of a stabilizing layer of force-free currents in a tenuous outer plasma replacing the "vacuum"). The outcome was positive for the high-beta device, considered promising at the time, but rather marginal for tokamaks. In this section, we will follow that paper but replace the Bessel functions, needed to represent perturbations of high-beta pinches, by the much simpler representation in terms of powers of r, appropriate for the low-beta tokamak approximation. This changes the numerical magnitude of the growth rates, but not the qualitative spectral picture presented in Fig. 14.5.

Obviously, the difference between a perfectly conducting wall and a resistive wall resides in the BC that is applied at the wall. Instead of the usual conducting wall BC, $\mathbf{n} \cdot \hat{\mathbf{Q}} = 0$, which guarantees that no flux leaves the volume enclosed by the wall, we should now apply the following conditions:

$$\left.\begin{aligned} [\![\mathbf{n} \cdot \hat{\mathbf{Q}}]\!] &= 0 \\[2ex] [\![\mathbf{n} \cdot \nabla \hat{\mathbf{Q}}]\!] &= -\mathrm{i}(\mu_0 \omega / \eta^*)\mathbf{n} \cdot \hat{\mathbf{Q}} \end{aligned}\right\} \quad (\text{at } r = w)\,, \qquad (14.99)$$

$$\mathbf{n} \cdot \hat{\mathbf{Q}} \to 0 \qquad\qquad (\text{for } r \to \infty)\,. \qquad (14.100)$$

The resistive wall condition (14.99)(b) is found by integrating the equation for the magnetic field perturbation across a thin wall of thickness δ and resistivity η, and taking the limits $\delta \to 0$ and $\eta \to 0$ such that the surface resistivity $\eta^* \equiv \eta/\delta$ remains finite (in the same way as the procedure described in Section 4.5 [1] to obtain jump conditions for plasma discontinuities).

It is expedient to introduce an abbreviation for the characteristic time scale associated with resistive diffusion of the magnetic field perturbation through the wall:

$$\tau_\mathrm{D} \equiv \mu_0 b / \eta^*\,. \qquad (14.101)$$

The pertinent resistive wall BC, replacing the original BC (14.97), is obtained by dividing the two expressions (14.99) and inserting τ_{D}:

$$\frac{(r\hat{Q}_{\mathrm{i}})'}{\hat{Q}_{\mathrm{i}}} = \frac{(r\hat{Q}_{\mathrm{e}})'}{\hat{Q}_{\mathrm{e}}} + \mathrm{i}\tau_{\mathrm{D}}\omega \qquad (\text{at } r = w). \qquad (14.102)$$

Here, the exterior solution $\hat{Q}_{\mathrm{e}}$ should obey the BC (14.100), i.e.

$$\hat{Q}_{\mathrm{e}} \to 0 \qquad (r \to \infty), \qquad (14.103)$$

so that

$$r\hat{Q}_{\mathrm{e}} = A_3 r^{-|m|}, \qquad (14.104)$$

The full eigenvalue problem now becomes to solve the ODEs (14.94) and (14.95) for $\xi(r)$ and $\hat{Q}_{\mathrm{i}}$, connect them by applying the plasma–vacuum BC (14.96), solve the ODE analogous to (14.95) for $\hat{Q}_{\mathrm{e}}$, subject it to the BC (14.103) at infinity, and finally connect $\hat{Q}_{\mathrm{i}}$ and $\hat{Q}_{\mathrm{e}}$ by applying the resistive wall BC (14.102).

It remains to substitute the expressions (14.98) and (14.104) for ξ, $\hat{Q}_{\mathrm{i}}$ and $\hat{Q}_{\mathrm{e}}$ into the BCs (14.96) and (14.102). This gives

$$\frac{\bar{\omega}^2 - (n + m/q)^2}{|m|(n + m/q)} + \frac{2}{mq} = -\frac{n + m/q}{|m|}\frac{1 - C}{1 + C} \qquad (r = 1), \qquad (14.105)$$

$$|m|\frac{1 - Cw^{-2|m|}}{1 + Cw^{-2|m|}} = -|m| + \mathrm{i}\tau_{\mathrm{D}}\omega \qquad (\text{at } r = w), \qquad (14.106)$$

which, upon elimination of C, yields the *dispe rsion equation for resistive wall modes:*

$$\bar{\omega}^2 - (n + m/q)^2 + 2(\mathrm{sgn}(m)/q)(n + m/q)$$

$$= (n + m/q)^2 \frac{2|m| - (1 + w^{-2|m|})\mathrm{i}\tau_{\mathrm{D}}\omega}{2|m| - (1 - w^{-2|m|})\mathrm{i}\tau_{\mathrm{D}}\omega}, \qquad (14.107)$$

or, in expanded cubic form,

$$\mathrm{i}\tau_{\mathrm{D}}\omega\left(\frac{\mathrm{i}\omega}{\epsilon\tau_{\mathrm{A}}}\right)^2 - D\left(\frac{\mathrm{i}\omega}{\epsilon\tau_{\mathrm{A}}}\right)^2 + E(\mathrm{i}\tau_{\mathrm{D}}\omega) - F = 0,$$

$$D \equiv \frac{2|m|}{1 - w^{-2|m|}}, \qquad E \equiv \frac{2(n + m/q)}{1 - w^{-2|m|}}\left[n + m/q - \frac{\mathrm{sgn}(m)}{q}(1 - w^{-2|m|})\right],$$

$$F \equiv \frac{4|m|(n + m/q)}{1 - w^{-2|m|}}\left[n + m/q - \frac{\mathrm{sgn}(m)}{q}\right]. \qquad (14.108)$$

From this expression, when the resistivity of the wall is increased, from $\eta^* = 0$ to $\eta^* = \infty$, the spectrum of eigenvalues of the external waves and instabilities qualitatively changes as indicated in Fig. 14.5:

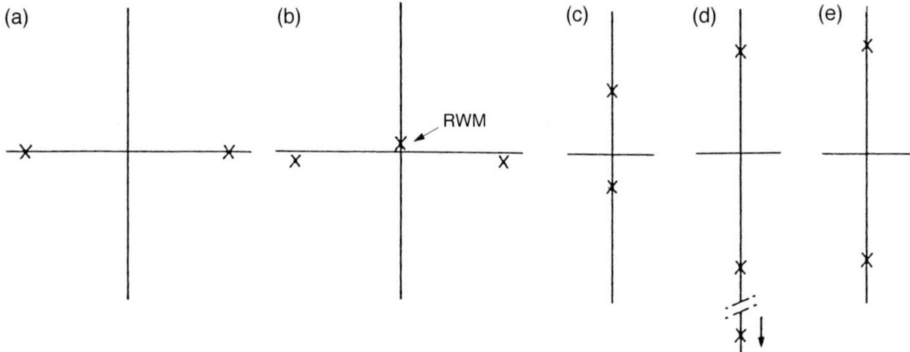

Fig. 14.5 Transition of wall-stabilized external kink modes to unstable and damped kink modes by increasing the resistivity of the wall: (a) for $\eta^* = 0$, the two kink modes are wall stabilized, (b) for small η^*, the kink modes become damped and a purely growing resistive wall mode pops out of the origin, (c) for $\eta^* = \eta^*_{\rm crit}$, the damped kink modes coalesce and become purely imaginary, (d) for $\eta^* > \eta^*_{\rm crit}$, one of the damped modes moves off to $-{\rm i}\infty$, (e) for $\eta^* = \infty$, the unstable kink mode and its complex conjugate is obtained. (From Goedbloed *et al.* [182].)

(a) for vanishing resistivity of the wall, $\eta^* \sim 1/\tau_{\rm D} = 0$, from the first and third term, we obtain the two stable kink oscillations described by Eq. (14.92):

$$\bar{\omega}^2 = E \equiv \bar{\omega}^2_{\rm w} > 0 \,; \tag{14.109}$$

(b) for small, but non-vanishing, resistivity, the two stable kink modes become resistively damped, and a new mode, the resistive wall mode, pops out of the origin:

$$\omega_{1,2} \approx \pm \epsilon \tau_{\rm A} E - \tfrac{1}{2}({\rm i}/\tau_{\rm D})(D - F/E) \,, \qquad {\rm i}\tau_{\rm D}\omega_3 \approx F/E \,; \tag{14.110}$$

(c) increasing the resistivity still further, the real part of the stable oscillations vanishes for a certain critical value $\eta^* = \eta^*_{\rm crit}$, so that a degenerate stable oscillation together with a rapidly growing resistive wall mode is obtained;

(d) for very large η^*, the two degenerate modes split apart, one moves off to $-{\rm i}\infty$ and the other one becomes the complex conjugate partner of the unstable kink mode in the absence of a wall;

(e) finally, for $\eta^* \to \infty$, from the second and fourth term, we obtain the unstable kink mode and its conjugate partner described by Eq. (14.92) in the limit $w \to \infty$:

$$\bar{\omega}^2 = F/D \equiv \bar{\omega}^2_\infty < 0 \,. \tag{14.111}$$

Clearly, the crucial expression above is the imaginary ω_3 given by Eq. (14.110), which is the *growth rate of the resistive wall mode* for small resistivity:

$$\omega_{\rm rwm} \approx -\frac{\rm i}{\tau_{\rm D}}\frac{F}{E} = -\frac{\rm i}{\tau_{\rm D}}\frac{2|m|}{1 - w^{-2|m|}}\frac{\bar{\omega}^2_\infty}{\bar{\omega}^2_{\rm w}} \,. \tag{14.112}$$

Because of our starting assumption, the two factors $\bar{\omega}_\infty^2$ and $\bar{\omega}_w^2$ have opposite sign so that the mode is unstable. Hence, with respect to stability, a resistive wall cannot change a mode from unstable to stable, but it does decrease the growth rate in proportion to the conductivity of the wall.

It is now just of matter of inserting numbers to find out whether the growth rate can be made small enough to be of no concern for fusion. Of course, for that purpose, realistic current profiles and a proper description of the toroidal geometry need to be taken into account. Several ways of eliminating the resistive wall mode for future fusion devices have been investigated. We just mention the consideration of RF stabilization [176], plasma rotation [57, 39, 199], and active feedback control [138, 137] to slow down the modes. The fact that there is a lot of ongoing research on this topic shows that the answer to the question is far from comforting.

14.3.2 Spectrum of homogeneous plasma

We will now consider the influence of resistivity of the plasma itself, starting with the stable part of the spectrum, in particular of the Alfvén waves. As we will see, similar spectral structures as in the resistive wall mode occur. However, in this case, there is no formal justification of the neglect of the resistive decay of the background equilibrium since the frequencies of the waves will be modified by a damping that may operate on a time scale comparable to that of resistive diffusion. Nevertheless, we will follow standard practice and neglect this effect, pending an overall theory that incorporates both resistive diffusion of the background and resistive decay of the waves.

First, let us study the case when the background equilibrium is completely homogeneous. In that case, Eqs. (14.44) for incompressible plasmas transform into

$$\eta\, \rho\omega^2 (\xi'' - k_0^2 \xi) - \mathrm{i}\omega F\,(Q - F\xi) = 0\,,$$

$$\eta\,(Q'' - k_0^2 Q) + \mathrm{i}\omega\,(Q - F\xi) = 0\,. \tag{14.113}$$

In ideal MHD, this reduces to

$$(\omega^2 - \omega_A^2)(\xi'' - k_0^2 \xi) = 0\,, \tag{14.114}$$

so that $\pm\omega_A$ represents a continuum of two infinitely degenerate eigenvalues.

Now, we may also assume harmonic dependence in the x-direction, so that $d/dx = \mathrm{i}k_x$. Introducing the resistive parameter

$$\tilde{\eta} \equiv \frac{\eta k^2}{2\omega_A}\,, \qquad k^2 \equiv k_x^2 + k_y^2 + k_z^2\,. \tag{14.115}$$

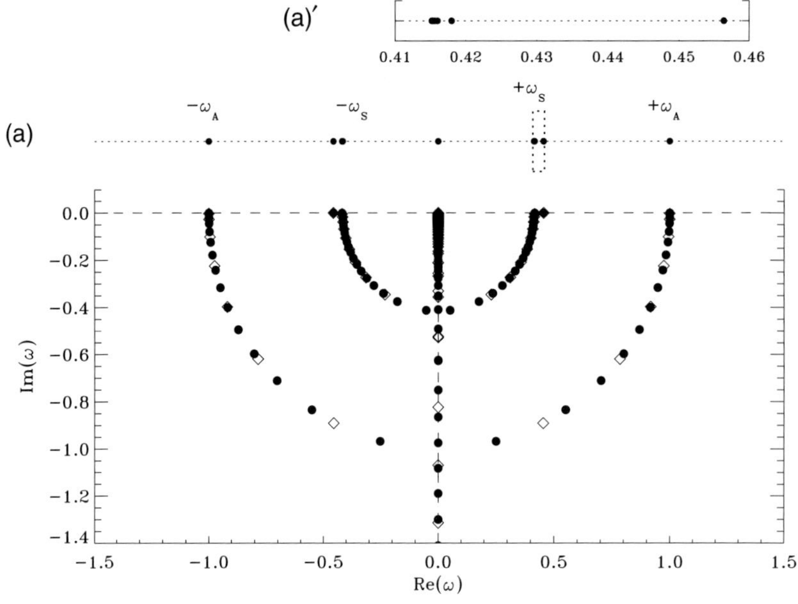

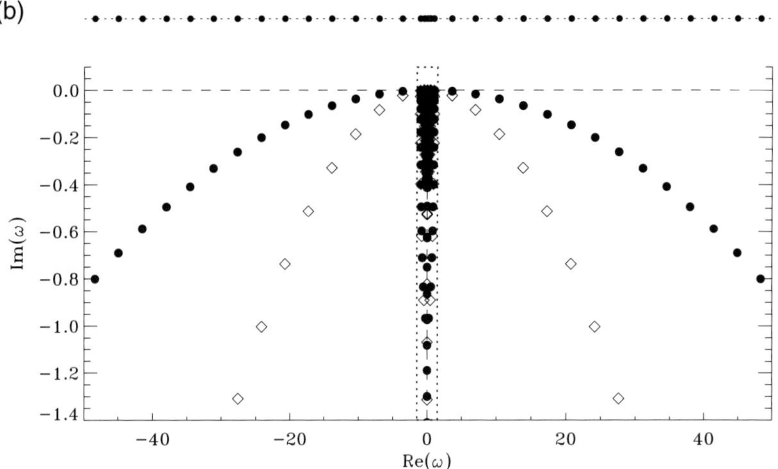

Fig. 14.6 Linear ideal versus resistive MHD spectra for a uniform, compressible plasma. The ideal (purely stable) results are indicated above the two panels, which combine results for $\eta = 0.001$ (filled circles) and $\eta = 0.005$ (diamonds). Note the difference in the horizontal axis (oscillation frequency): the bottom panel (b) shows most clearly the fast mode sequence, while the top panel (a) zooms in on Alfvén and slow modes. Dashed boxes indicate the range of a zoomed view above. The insert (a)′ illustrates the anti-Sturmian sequence of ideal MHD slow modes.

the dispersion equation may be written as

$$\omega^2 + 2i\widetilde{\eta}\,\omega_A \omega - \omega_A^2 = 0\,, \tag{14.116}$$

with the solution

$$\omega = \omega_A\left(\pm\sqrt{1 - \widetilde{\eta}^2} - i\widetilde{\eta}\right). \tag{14.117}$$

For varying values of $\widetilde{\eta}$, the modes are lying on a semi-circle in the lower half plane, starting from the real axis at $\omega = \pm\omega_A$ ($\widetilde{\eta} = 0$), intersecting the negative imaginary axis at $\omega = -i\omega_A$ ($\widetilde{\eta} = 1$), and then moving along the negative imaginary axis to $\omega = 0$ and $\omega = -i\infty$ ($\widetilde{\eta} \to \infty$).

The analysis above concentrated on the Alfvén modes for a homogeneous, incompressible plasma, and showed analytically that resistivity introduces specific curves (semi-circles) in the damped frequency half-plane on which the modes reside. The spacing of the resistive Alfvén eigenmodes on these curves is influenced by the resistivity in accord with Eq. (14.115), but the radius of the semi-circle is independent of the resistivity η.

In Fig. 14.6, we show the effect of finite resistivity on the entire MHD spectrum consisting of fast, Alfvén and slow modes of a homogeneous, compressible plasma. Using numerical techniques discussed more extensively in Chapter 15, an overview of all modes with mode numbers $k_y = 0$, $k_z = 1$, for a uniform layer of density $\rho = 1$, magnetic field $\mathbf{B} = (0,0,1)$ and $\beta = 0.25$ (with $\gamma = 5/3$) is computed easily. The figure combines the spectrum obtained under ideal MHD $\eta = 0$ conditions, together with its modification when the resistivity parameter has a constant value $\eta = 0.001$ or $\eta = 0.005$. Note that, in ideal MHD, the homogeneous plasma slab has an infinite Sturmian sequence of fast modes accumulating to infinity, a degenerate Alfvén frequency ω_A and an anti-Sturmian sequence of slow modes accumulating to the (constant) slow frequency ω_S. The latter is shown in more detail in the top insert of the figure. This behavior was explained previously in Chapter 7 [1] and also shown in Fig. 7.8. When resistivity is included, the bottom part of Fig. 14.6 shows that the fast mode sequence shifts from purely oscillatory to damped, where the resistivity *does influence the precise location of the damped fast mode sequence* in the complex plane. As expected, the higher the mode frequency in the Sturmian sequence, the more it is damped. The top panel demonstrates clearly that both the Alfvén and the slow modes relocate to semi-circles in the damped half-plane (and the negative imaginary axis), and these semi-circle locations remain uninfluenced by resistivity. However, the spacing of the individual resistive Alfvén or slow modes on these curves is affected by resistivity changes. This complements and agrees with the above analytical results for the resistive Alfvén modes in the incompressible case.

14.3.3 Spectrum of inhomogeneous plasma

For inhomogeneous plasmas, the resistive spectrum should be some kind of modification of the continuous spectrum of ideal MHD. Resistivity changes the order of the system so that the singularities due to the vanishing of the coefficients in front of the highest derivatives disappear. Hence, one should expect the continua to split up into discrete modes. This is indeed what happens. However, it happens in a rather unexpected fashion, which is already suggested by the case of the homogeneous plasma discussed in the previous section. There, the continuum frequencies, which were infinitely degenerate in ideal MHD, relocated to sequences of discrete resistive eigenmodes on semi-circles and the negative imaginary axis in the stable half-plane. As a mathematical analysis for the resistive spectrum of an inhomogeneous plasma becomes rather formidable, we here restrict ourselves to discussing exemplary spectra computed numerically. This has the advantage that no simplifying assumptions on, e.g., incompressibility, or non-overlapping continua have to be made.

A relevant example revisits the force-free equilibrium from Eq. (14.2) analyzed in Section 14.1. While most of the analysis presented there used incompressible conditions, the main conclusions on stability against resistive tearing modes as governed by criteria in Eq. (14.21) or Eq. (14.86) can be expected to hold in the compressible case as well. When considering a plasma of constant density $\rho = 1$, $\beta = 0.15$ and magnetic field $\mathbf{B} = (0, \sin(\alpha x), \cos(\alpha x))$ on $x \in [-0.5, 0.5]$ with specific heat ratio $\gamma = 5/3$, the tearing mode stability criterion is violated for the choice $k_y = 0.49$, $k_z = 0$, when $\alpha = 4.73884$. For this combination of wave numbers and equilibrium parameters, the Alfvén and slow continua all go through zero at $x = 0$, and contain internal extrema. These overlapping continuum ranges are shown in the top panel of Fig. 14.7. The bottom panel shows the resistive MHD spectrum for a resistivity of $\eta = 0.0001$. Only the Alfvén and slow modes are seen in this frequency range, and one recognizes the following.

- The ideal, stable continuum ranges are replaced by sequences of discrete resistive eigenmodes which lie on specific curves in the stable half-plane. Repeating the computation for differing values of η influences the spacing of the modes on these curves, but not the curves themselves.

- For the case of the overlapping continua shown in the figure, the curves still vaguely resemble semi-circles, as for the homogeneous case in Fig. 14.6, but complicated by branches which split off in almost triangular patterns. The tips of the triangles approach the stable real axis at oscillation frequencies that correspond to an internal extremum, an end value (at $x = \pm 0.5$) or a zero value of the ideal continua.

- While the ideal continua are thus dramatically transformed into an intricate structure of discrete modes in the stable half-plane, the unstable tearing mode is still prominently

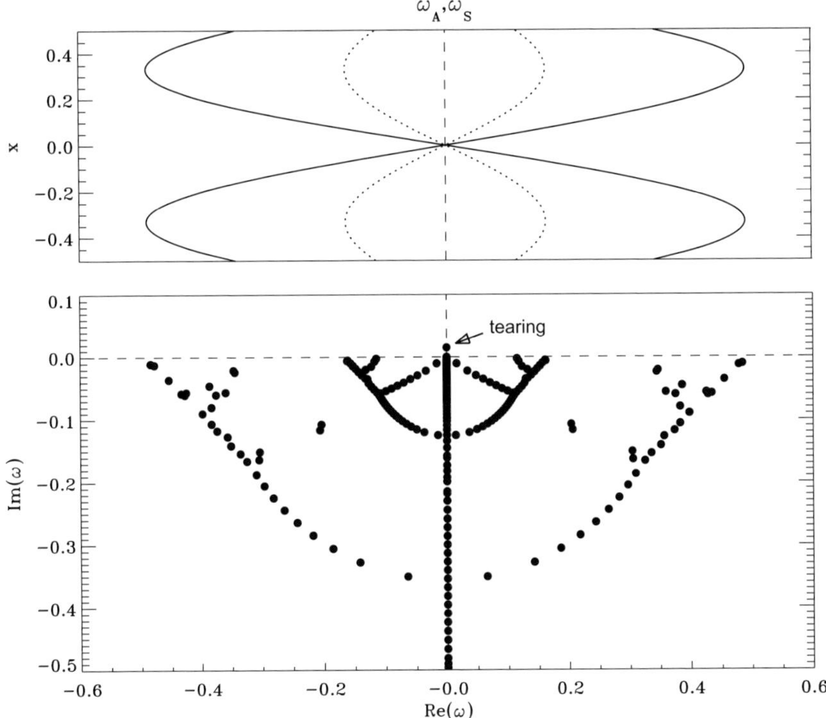

Fig. 14.7 Typical Alfvén and slow magneto-sonic parts of the resistive MHD spectrum for a force-free slab. Note the (isolated) unstable tearing mode. The ideal continuous spectra are indicated above (oscillation frequency versus x-location in the slab), and show the distinct connection between their end point ($x = \pm 0.5$) and internal extremal values with the curves on which resistive modes are found.

present with its growth rate in agreement with the scaling given by Eq. (14.84) for varying resistivity. This latter point is demonstrated in Fig. 14.8, where the theoretical predictions are compared with numerically obtained growth rates.

The latter conclusion underlines the fact that instabilities form the most robust part of the MHD spectrum (when going from ideal to resistive conditions), and that the concepts of σ-stability (Chapter 6) for static plasmas and ϵ-stability (Chapter 12) for stationary plasmas can be expected to remain of practical use.

In addition to the unstable tearing mode and the resistive discrete eigenmodes on curves that are the counterpart of the ideal continua, Fig. 14.7 also shows some more isolated modes which appear within the triangular sections of the curves. A better example is discussed in Chapter 15, Fig. 15.11, where (half of) the resistive sub-spectrum of Alfvén and slow modes is shown (rotated over 90 degrees)

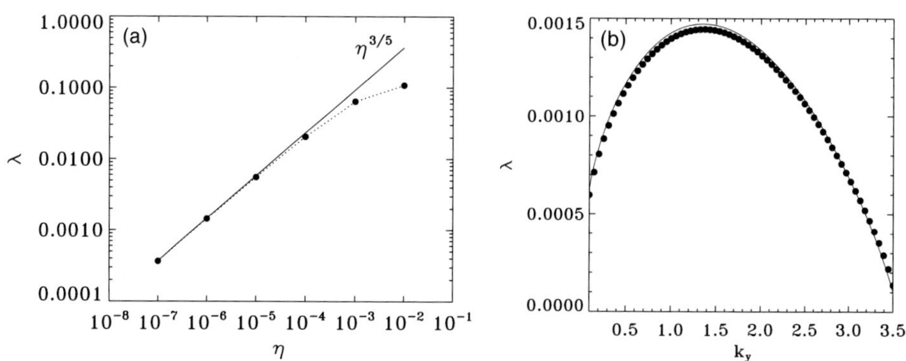

Fig. 14.8 Linear resistive MHD results for the tearing mode in a force-free slab: (a) growth rate as a function of η for $k_y = 1.5$ and all other parameters as in Fig. 14.7; (b) predicted versus computed growth rate variation at fixed $\eta = 10^{-6}$.

for a cylindrical plasma equilibrium and mode numbers with non-overlapping, well-separated ideal Alfvén and slow continuum ranges. The isolated mode in the triangular section of the Alfvén sub-spectrum in Fig. 15.11 has been identified as an ideal quasi-mode, and these locally damped (due to local resonance with the continuum frequencies) global modes form the most robust part of the *stable* frequency spectrum. They represent the natural oscillation modes of the system and explain how perfect coupling between a driver and its excited plasma loops can be achieved (see Chapter 11 [1]). As mentioned there, these (ideal) quasi-modes demonstrate a damping rate which becomes *independent of resistivity* for large Reynolds numbers (which distinguishes them from the resistive modes on the curves). A recent study [464] has investigated their intricate connection with the resistive eigenmodes on the curves, by varying the governing equilibrium parameters.

Finally, we may wonder how the combined ingredients of resistivity *and equilibrium flow* manifest themselves in this tearing-unstable, force-free slab. For the equilibrium from Figs. 14.7–14.8, we can conclude that in the static case, maximal growth occurs for modes with $k_y = 1.5$. To investigate the effect of equilibrium flow on this mode, we now add a linear flow profile $v_y(x) = 0.15\,x$, leading to Doppler shifted frequency ranges $\Omega_{\mathrm{A,S}}^{\pm}$. With this choice of flow profile, the spectrum still contains the tearing instability, and the non-trivial effect of flow on the linear MHD spectrum is illustrated in Fig. 14.9. The discrete modes seen in the damped half-plane are again found on curves, which still show a clear link with the extremal and edge values of the Doppler shifted continua. The computational result also shows the purely oscillatory Eulerian entropy continuum modes, as well as evidence for more curves appearing in the frequency range of the local Doppler

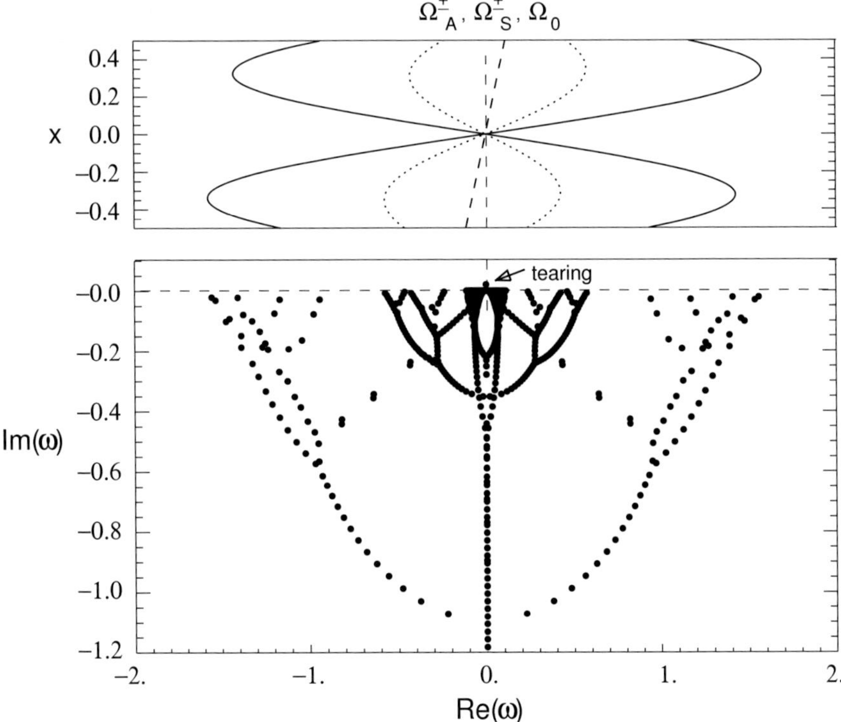

Fig. 14.9 Typical Alfvén and slow magneto-sonic parts of the resistive MHD spectrum for a force-free slab with a linear flow profile. Note the (isolated) unstable tearing mode. The (Doppler-shifted) ideal continuous spectra are indicated above together with the local Doppler shift, and show the distinct connection between their end point ($x = \pm 0.5$) and internal extremal values with the curves on which resistive modes are found.

shift $\Omega_0(x)$. It should be emphasized that the combined effects of equilibrium flow, magnetic shear and finite resistivity on the full MHD spectrum are not fully understood yet, and future studies using MHD spectroscopy for these configurations are called for. By varying the equilibrium flow profile, one can then investigate subtle effects on global, unstable modes: the tearing mode in this force-free field configuration can either be suppressed or rendered more unstable, depending on the precise flow strength and profile.

In closing this section, it is appropriate to mention that extensive research has been devoted in hydrodynamics to the similar problem of the effect of viscosity on the spectrum of waves and instabilities. In local WKB solutions of the Orr–Sommerfeld equation, an important role is played by the Stokes lines in the complex ω plane (see, e.g. Drazin and Reid [124], Section 27.3). Eigenvalues associated with wildly oscillating eigenfunctions are found on particular curves, like

those of Figs. 14.6, 14.7, 14.9. These techniques have been transferred to magnetohydrodynamics (see, e.g., Pao & Kerner [360], Riedel [391] and Lifschitz [308],
Section 7.15). However, in contrast to the resistive instabilities, like tearing modes,
the role of the stable, damped, part of the spectrum remains rather unclear. Some
relatively recent research ties this up to the concept of *pseudo-spectrum* (see Borba
et al. [61]), developed by Trefethen [446] for hydrodynamics. It is beyond the
scope of this book to describe these developments in detail.

14.4 Reconnection

We have thus far concentrated on how resistivity modifies the linear MHD spectrum, pointing out the existence of various unstable tearing-type modes, involving
reconnection. However, reconnection plays an important role in nearly all dynamic
phenomena in space and laboratory plasmas, where it manifests itself in intrinsically nonlinear evolutions as well. Magnetic reconnection rightfully deserves fundamental study on its own, a fact reflected in excellent modern textbooks on the
subject (e.g. by Priest and Forbes [387] and by Biskamp [46]). In what follows, we
restrict ourselves to discussing its role in the temporal evolution of a very simple,
planar configuration, the so-called Harris sheet. This configuration has played a
central role in several collaborative modeling "challenges", such as the "Geospace
Environmental Modeling (GEM) Magnetic Reconnection Challenge" [42] and the
"Newton Challenge" [43]. In both these challenges, the nonlinear evolution of the
2D system was computed and compared with a large variety of codes: ranging from
conventional resistive MHD, over various extended MHD models, to fully particle
based, kinetic treatments. In Section 14.4.4, we briefly summarize the important
findings of this multi-code approach, but first discuss what can be concluded from
standard resistive MHD modeling alone.

14.4.1 Reconnection in 2D Harris sheet

(a) Linear stability properties The Harris sheet configuration is a planar MHD
equilibrium with a horizontal magnetic field given by

$$B_x(y) = B_0 \tanh(y/\lambda_B) \,. \tag{14.118}$$

The corresponding current layer thus has a thickness measured by $2\lambda_B$. One further
assumes a uniform temperature $T = T_0$ and a density profile given by

$$\rho(y) = \rho_0 \cosh^{-2}(y/\lambda_B) + \rho_\infty \,. \tag{14.119}$$

When exploiting a scaling where $B_0 = 1$, $\rho_0 = 1$ and $2\lambda_B = 1$, all velocities
become normalized to the Alfvén speed, and time is measured in Alfvén crossing

times of the initial current sheet width. From pressure balance, one then finds $T_0 = 0.5$, and the reference equilibrium configuration takes $\rho_\infty/\rho_0 = 0.2$ and a ratio of specific heats fixed at $\gamma = 5/3$. The problem is then fully specified when supplemented with the domain size and the employed boundary conditions. The standard case takes $-L_x/2 < x < +L_x/2$ and $-L_y/2 < y < +L_y/2$ with $L_x = 25.6$ and $L_y = 12.8$. The simulations use periodic boundary conditions horizontally, while at $y = \pm L_y/2$ perfectly conducting, impenetrable walls are usually assumed. In the GEM challenge, this initial equilibrium is perturbed by an additional magnetic perturbation specified as

$$B_{1x}(x,y) = -\psi_1 \frac{\pi}{L_y} \cos\left(\frac{2\pi x}{L_x}\right) \sin\left(\frac{\pi y}{L_y}\right),$$

$$B_{1y}(x,y) = +\psi_1 \frac{2\pi}{L_x} \sin\left(\frac{2\pi x}{L_x}\right) \cos\left(\frac{\pi y}{L_y}\right). \tag{14.120}$$

In the original study, one adopts an amplitude $\psi_1 = 0.1$, which is deliberately taken so large that nonlinearity dominates instantly. When we restrict the discussion to the case of uniform resistivity, the only remaining parameter is then the value of the magnetic Reynolds number, as quantified by its dimensionless inverse η. In what follows, we will first point out the linear stability properties for this choice of parameters, to make a direct link with our foregoing discussion of the tearing mode. Then, we will discuss its nonlinear evolution: not unexpectedly, a decisive parameter will turn out to be the value of the resistivity η, as it can lead to radically different reconnection scenarios.

We first present in Fig. 14.10 the numerically determined eigenfunctions for the (most unstable) linear tearing mode of this reference configuration, when taking $\eta = 10^{-4}$. The linear mode is assumed to be purely two-dimensional, and adopts a Fourier dependence $\exp(ik_x x)$ with $k_x = 2\pi/L_x$, so that its horizontal wavelength exactly fits the chosen box size. The growth rate λ of the exponential growth $\exp(\lambda t)$ for this mode has the same dependence on the resistivity η and on the wave number k_x, as shown by Fig. 14.8. (Note that the direction of inhomogeneity is now taken as the y-direction, in contrast to the discussion in the previous sections.) As can be deduced from these plots, the tearing mode shows the characteristic magnetic field perturbations discussed earlier, with the component $B_{1y}(y)$ to be compared with Q as sketched in Fig. 14.4. The density eigenfunction clearly shows that compressibility effects can no longer be neglected in this case, while the tearing mode also manifests a distinct temperature variation across the resistive inner layer. Varying the growth rate versus wave number at constant $\eta = 10^{-4}$ shows that this mode attains its maximal growth (like the tearing mode of Fig. 14.8) at a somewhat shorter wavelength than that preferred by the horizontal

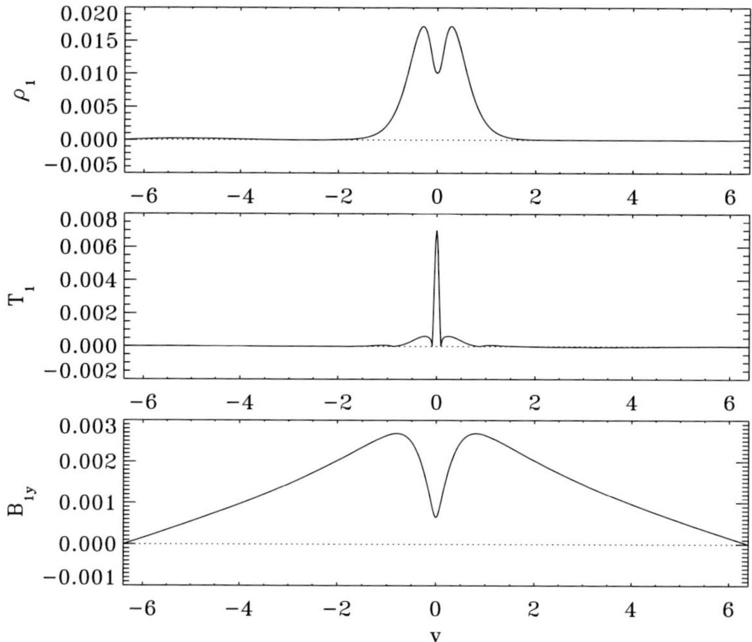

Fig. 14.10 Linear resistive MHD results for the Harris sheet equilibrium exploited in the GEM challenge.

box size $L_x = 25.6$. Finally, the variation of the growth rate with resistivity confirms the analytic scaling behavior $\lambda \sim \eta^{3/5}$, although convincingly only for values of the corresponding magnetic Reynolds number beyond 10^4. It must be stressed that linear MHD codes do possess the required accuracy to compute eigenmodes at realistic Reynolds numbers of order 10^9 or more. As explained in Chapter 15, this requires the use of a sufficiently accurate numerical representation, if needed combined with grid accumulation, and aptly chosen generalized eigenvalue solvers. In contrast, for the nonlinear simulations discussed next, direct numerical simulations that achieve magnetic Reynolds numbers of 10^4 are still at the limit of many of the high resolution methods exploited to date (such as the shock-capturing, finite volume methods described in more detail in Chapter 19).

(b) Nonlinear evolution Although the reference Harris sheet equilibrium exploited for the nonlinear simulations is tearing-unstable, the large amplitude perturbation used in Eq. (14.120) imposes an initial magnetic island perturbation with a width comparable to the current layer. This is mainly motivated by the GEM challenge's aim to compare nonlinear reconnection rates found from both fluid and kinetic models. For purely uniform resistivity, the original resistive MHD simulations

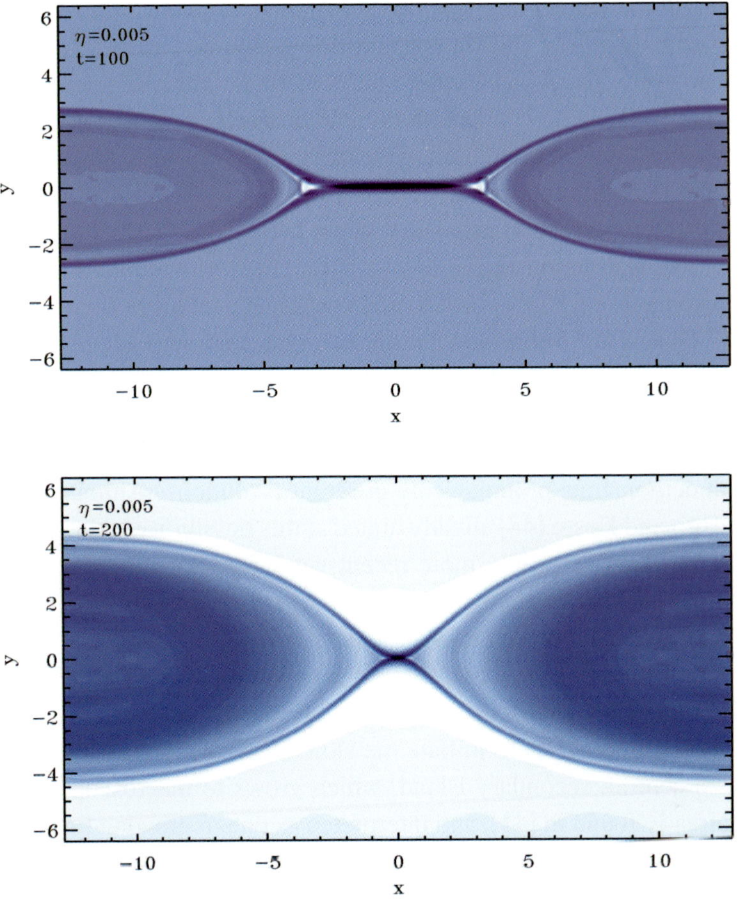

Fig. 14.11 Nonlinear evolution of the Harris sheet equilibrium, perturbed in accord with the GEM challenge, for uniform resistivity with $\eta = 0.005$. Shown is the out-of-plane current distribution at $t = 100$ and $t = 200$.

from Birn and Hesse [44] already demonstrated pronounced differences in the evolution when varying η from 0.005 down to 0.001. Their fairly low resolution simulations followed the current sheet dynamics for several hundreds of Alfvén crossing times. At their highest resistivity value $\eta = 0.005$, the initial current sheet collapses centrally to form an elongated, narrow dissipative layer connecting the two halves of the magnetic island located at the periodic sides. This island gradually grows in size, at a slow rate set by the near-steady reconnection occurring through the Y-shaped endpoints of the elongated current layer. This rate can be computed analytically from a consideration of the stationary ($\partial/\partial t = 0$) resistive MHD equations across a diffusion region of macroscopic length $2L$ and width $2l$, e.g. observing that mass conservation demands $Lv_i = lv_o$, where v_i and v_o denote the (vertical) inflow and (lateral) outflow velocities, respectively. This was originally done by Sweet and further analyzed by Parker [363], and the resulting so-called Sweet–Parker reconnection layer is characterized by a very slow inflow speed $v_i \sim \sqrt{\eta}v_A$, an Alfvénic outflow velocity $v_o \sim v_A$, and a current sheet width related to the length L and resistivity value by $l \approx \sqrt{\eta}L$. Fig. 14.11 shows the evolution of the perturbed Harris sheet when taking $\eta = 0.005$, at two consecutive times. The simulation here employs automated grid-adaptivity to achieve an effective resolution of 1920×1920, and essentially recovers the original results from [44]. During the entire simulation, a single dominant island grows in size, as mediated by the near-steady reconnection occurring across the central current layer.

When the magnetic Reynolds number is increased, the same Harris sheet configuration can demonstrate pronouncedly different evolutions. Although the original results by Birn and Hesse [44] already hinted at this possibility by including a lower $\eta = 0.001$ case as well, only more recent work by Lapenta [296] has convincingly demonstrated the complexity attainable by (visco-)resistive MHD evolutions at the larger, physically more realistic, magnetic Reynolds numbers. Figure 14.12 presents the evolution of the identical configuration as simulated in Fig. 14.11, for $\eta = 0.001$ (a fivefold increase in the magnetic Reynolds number). Now, the central current sheet still collapses to initiate the slow reconnection process, but in addition forms a central, secondary island, which grows to macroscopic dimensions. This was already found in [44], and interpreted as due to growing linearly unstable eigenmodes of smaller wavelength. (This secondary tearing is not seen at smaller magnetic Reynolds numbers as the initial current then dissipates much faster.) This secondary island eventually undergoes a sudden merger with the larger island structure, an effect attributable to the so-called coalescence instability mediated by the mutual attraction of parallel current filaments. Continuing the simulation further shows the renewed appearance of a smaller island structure in the central current layer, which later on is again seen to coalesce. This process already indicates the transition to a new, highly non-steady, reconnection regime that is characterized by sudden tearing-type disruptions of the current layer.

As first pointed out by Lapenta [296], even lower resistivity values where $\eta = 10^{-4}$ show a dramatic changeover from the original Sweet–Parker reconnection regime to one where the collapsed current layer spontaneously disrupts chaotically in repeated island chains. The resulting reconnection rate in the self-feeding, turbulent reconnection phase is dramatically increased beyond the slow Sweet–Parker rate. Figure 14.13 again shows a representative snapshot of the evolution, this time characterized by almost randomly appearing, small island chains, which again show sudden mergers with the larger islands due to the coalescence instabil-

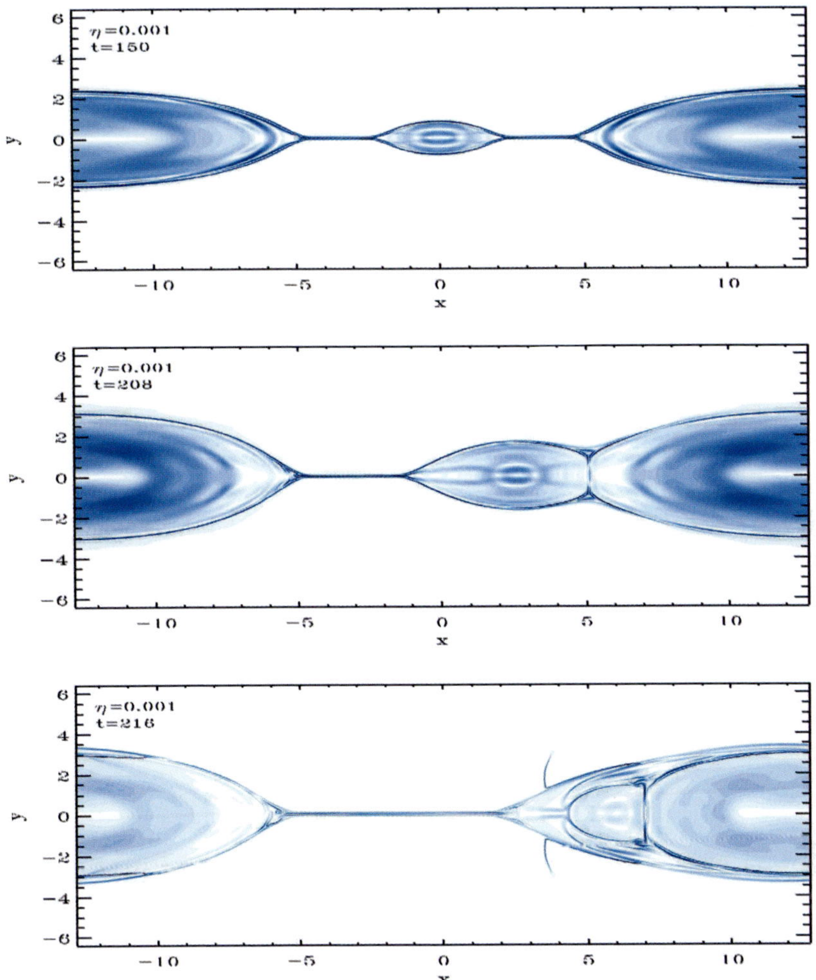

Fig. 14.12 Nonlinear evolution of the Harris sheet equilibrium, perturbed in accord with the GEM challenge, for uniform resistivity with $\eta = 0.001$. Shown are schlieren plots of the density at consecutive times.

ity. Hc attributes the onset of the resulting fast reconnection regime to the formation of closed circulation patterns centered on the multiple reconnection sites, self-feeding at X-type points. In Fig. 14.13, the sites where this fast reconnection occurs demonstrate a clear resemblance to that obtained in the other well-known, stationary Petschek reconnection model, involving slow MHD shocks and co-spatial current sheets found in an X-type configuration. In the next subsection, this stationary Petschek reconnection process is briefly discussed. The self-feeding, turbulent reconnection found at sufficiently low values of η is highly non-stationary,

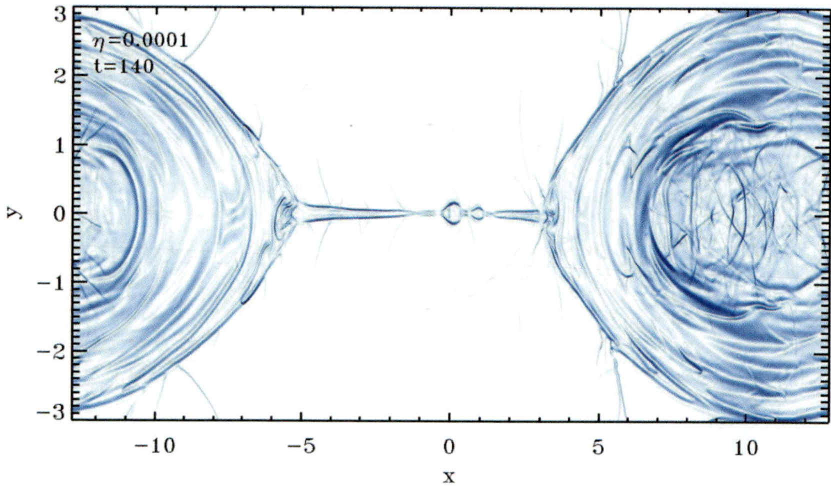

Fig. 14.13 Nonlinear evolution of the Harris sheet equilibrium, perturbed in accord with the GEM challenge, for uniform resistivity with $\eta = 0.0001$. Shown is a schlieren plot of the density at $t = 140$.

but physically very relevant since it occurs on Alfvénic time scales, and by nature distributes the reconnection over a macroscopic region.

14.4.2 Petschek reconnection

The slow reconnection rate obtained in the stationary Sweet–Parker regime is in stark contrast with that found in violent solar flares, or the one realized in several laboratory reconnection experiments. To overcome this shortcoming of the modeling based on stationary resistive MHD, Petschek [370] proposed a viable alternative involving the formation of slow shock fronts across which most of the energy conversion takes places. The central diffusion region is in the Petschek model reduced in size as compared to the elongated current layer associated with the Sweet–Parker regime, and four standing shock waves emerge from it in an overall X-type configuration.

Tóth *et al.* [443] applied their resistive MHD, Versatile Advection Code (see Section 19.3.2) to study Petschek-type reconnection of magnetic field lines. Figure 14.14 shows two snapshots of one of their simulations. Initially, there is a Harris sheet equilibrium with $v_x = v_y = B_x = 0$ and $B_y = \tanh(x/L)$ with L the width of the initial current layer. This initial state spontaneously transits to a configuration containing a localized dissipative layer (left picture) by imposing a spatially non-uniform, anomalous resistivity centered about the origin. A pair of slow mode shocks propagate away from the reconnection layer, eventually form-

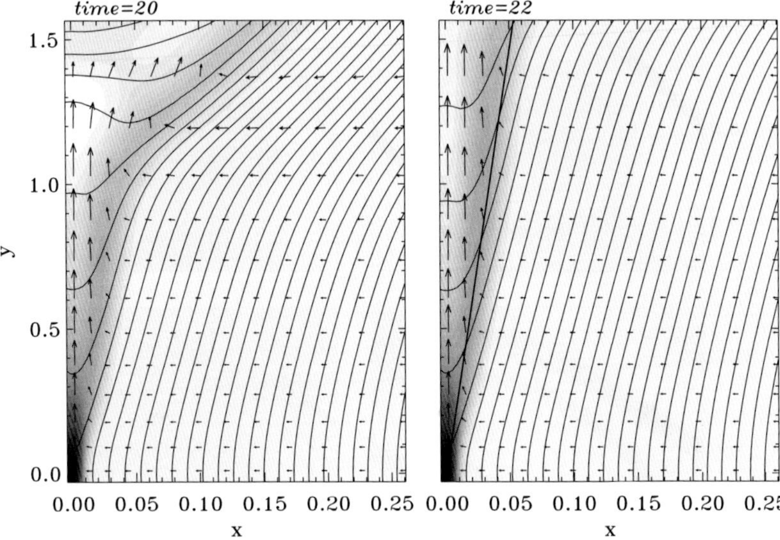

Fig. 14.14 Magnetic field lines, velocity field (arrows), and current density j_z (grey-scale) for two snapshots of a magnetic reconnection problem which evolves to a stationary Petschek solution. Only the upper right quarter of the X-type structure is shown because of the symmetry. (From Tóth *et al.* [443].)

ing a standing X configuration. The material crossing those shocks accelerates to Alfvénic velocities and gets heated in the process. The figure shows how the configuration eventually evolves to a steady state Petschek reconnection regime. While most computations assume non-uniform resistivity, it has been shown that at nearly uniform resistivity a stationary configuration can be obtained numerically [28]. At the same time, this classical stationary Petschek regime is just one realization of the various fast reconnection mechanisms known to date. For a discussion of these, we refer the interested reader to Priest and Forbes [387] and Biskamp [46].

14.4.3 Kelvin–Helmholtz induced tearing instabilities

The Harris sheet configuration discussed thus far clearly demonstrates the surprising complexity for planar resistive MHD evolutions. With various tearing-type instabilities already present for static equilibria, one can anticipate even more complex scenarios for stationary equilibria (the topic of the previous Chapters 12 and 13) containing initial current sheets. Since the linear resistive MHD spectrum of planar, stationary configurations is still actively being researched, we end our discussion of resistive MHD reconnection with an example of tearing-type disruptions induced by pure Kelvin–Helmholtz instabilities. In Keppens *et al.* [267], a

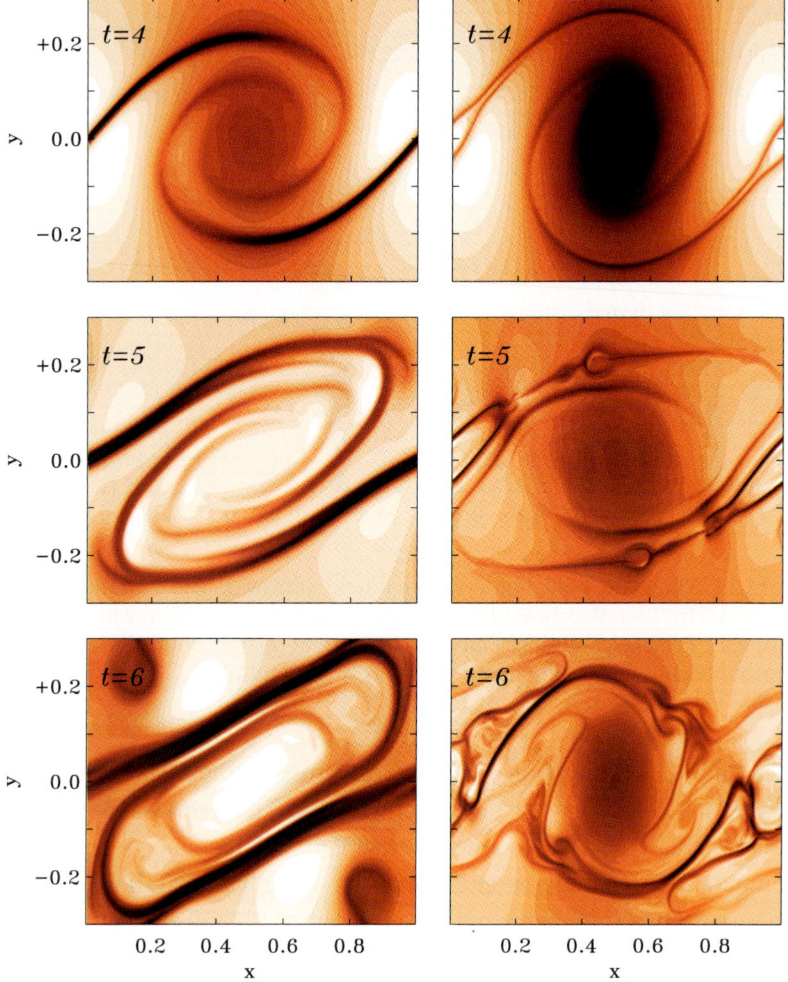

Fig. 14.15 Evolution of the density for a Kelvin–Helmholtz unstable shear flow for an initially uniform (left) versus a reversed (right) flow-aligned magnetic field. In the presence of an initial narrow current layer (right panels), the Kelvin–Helmholtz mode triggers tearing-type instabilities at the vortex periphery. (From [267]).

numerical survey of a planar shear flow $v_x = v_0 \tanh(y/a)$, embedded in uniform plasma conditions, was performed. At the high initial plasma beta $\beta = 120.2$, the configuration is known to be Kelvin–Helmholtz unstable for subsonic flow conditions $v_0 < c = \sqrt{\gamma p_0/\rho_0}$, with a weak stabilizing influence on the growth rate by the tension of the flow-aligned magnetic field. The authors compared the linear Kelvin–Helmholtz growth rates, as well as the nonlinear evolution of the system, for a uni-directional as well as a "reversed" magnetic field at $t = 0$. The reversed case discontinuously changes the direction of the magnetic field at the

$y = 0$ mid-plane, and represents the limit of a co-spatial magnetic Harris sheet $B_x = B_0 \tanh(y/b)$ for the limit $b \to 0$. At sufficiently high resolution, combined with low values of the uniform resistivity, the reversed case demonstrated the distinct possibility of sudden tearing-type events. Small-scale magnetic islands can appear suddenly in the current sheets that are amplified by the Kelvin–Helmholtz mode development. The vortical flow accompanying the Kelvin–Helmholtz mode warps the initial field in spiral patterns at the vortex periphery, and a case which contrasts the evolution of a uniform versus a reversed field configuration is shown in Fig. 14.15. The density "islands" coincide with magnetic islands, and their sudden appearance allows a rapid transition to magneto-turbulent flow conditions.

This example shows that the nonlinear dynamics of flowing, current-carrying, plasmas is a topic that deserves further study within a purely resistive MHD context. Even their linear stability properties have not been charted to the amount of detail obtained for static equilibria. Moreover, reconnection in more complicated 3D configurations is even more intricate, with a staggering increase in topological possibilities for reconnecting anti-parallel field lines. Chapter 8 of Priest and Forbes [387] provides a readable account of its possible manifestations in 3D.

14.4.4 Extended MHD and reconnection

As mentioned earlier on, the 2D Harris sheet configuration has been used as a benchmark configuration in both the GEM and the Newton challenge projects to identify the essential physics required to properly model collisionless magnetic reconnection. Whereas the GEM challenge adds the fairly large perturbation given by Eq. (14.120) to bring the system to a nonlinear reconnection regime, the Newton challenge differs in the way this nonlinear regime gets accessed. Rather than imposing a perturbation in the current sheet at $t = 0$, it gradually moves the field lines at $y = \pm L_y/2$ inwards at a prescribed velocity. This inflow is at most 10% of the Alfvén velocity and decays beyond about 60 Alfvén crossings, so it gently forces the central reconnection. Both collaborative challenges used a large variety of models to simulate the evolution, with codes ranging from resistive MHD to full particle (kinetic) treatments. The findings of both these efforts were rather similar, and the reconnection rates obtained from different models are shown in Fig. 14.16. We can summarize their results as follows.

- The reconnection rate found in conventional resistive MHD simulations was significantly smaller than that in any simulation which allows a minimal decoupling of the electron and ion dynamics. However, the resistivity values adopted (and those effectively reached by the numerics) for uniform resistivity models were typically higher than $\eta = 0.001$, so that the eventual transition to the chaotic, fast reconnection regime as shown in Fig. 14.13 was not appreciated fully.

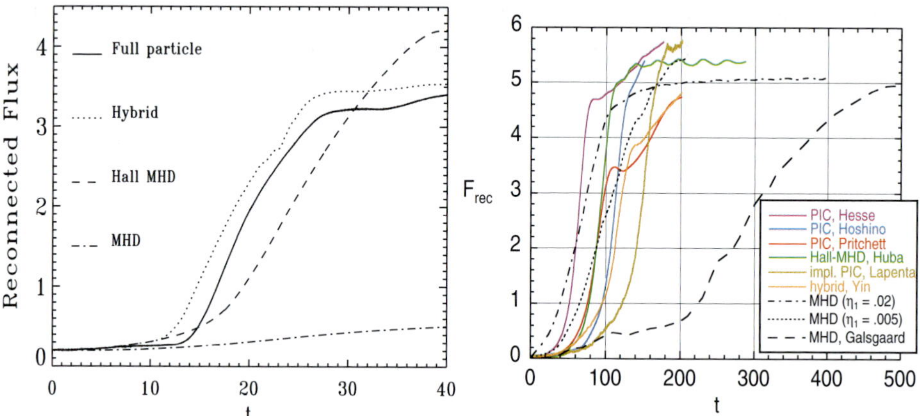

Fig. 14.16 Reconnection rates as obtained for the GEM challenge [42] (left panel) and Newton challenge [43] (right panel) simulations.

- All treatments that consider at least the effect of the Hall term in a generalized Ohm's law show that reconnection occurs fast (with Alfvénic inflow rates). This reconnection rate is surprisingly similar between Hall-MHD, hybrid, or full particle models. The latter could indicate that reconnection is insensitive to the details of the electron dynamics and the dissipation mechanism.

- In order to raise the reconnection rate from resistive MHD computations to the faster rate found from the more advanced models, one can adopt anomalously raised, localized, resistivity prescriptions.

These conclusions from both 2D magnetic reconnection studies then call for suitable extensions of the standard MHD model. The simplest of such models is that of Hall-MHD, which can be regarded as a straightforward generalization of the single fluid MHD description. It includes the effect of the Hall current in the generalized Ohm's law, and brings in whistler type waves, which have a faster phase speed at shorter wavelengths or higher frequencies. The Hall-MHD model thereby enables a faster reconnection than obtained in pure resistive MHD. Extended MHD models of increasing physical complexity can be rigorously derived along the lines given in Chapter 3 [1], where the derivation of the MHD equations from kinetic theory was discussed. In so doing, one can gradually relax the inherent assumptions of the single fluid MHD description, allowing a descent in the length and time scale hierarchy. Recalling these hierarchies from Eq. (3.143) [1], we have

$$\lambda_{\mathrm{MHD}} \equiv |\nabla|^{-1} \quad \sim a \quad \gg R_{\mathrm{i}} \quad \left[\gg \delta_{\mathrm{e}} \gg R_{\mathrm{e}} \gg \lambda_{\mathrm{D}} \right],$$

$$\tau_{\mathrm{MHD}} \equiv |\partial/\partial t|^{-1} \sim a/v_{\mathrm{A}} \gg \Omega_{\mathrm{i}}^{-1} \left[\gg \omega_{\mathrm{p}}^{-1} \sim \Omega_{\mathrm{e}}^{-1} \right], \qquad (14.121)$$

where all symbols have their usual meaning: a and v_A are the typical MHD length scale and Alfvén velocity, R_i and R_e denote the ion and electron gyro-radii, δ_e is the electron skin depth, λ_D the Debye length, Ω_i and Ω_e denote the ion and electron gyro-frequencies, while the intermediate plasma frequency is ω_p. One must then similarly invoke a kind of maximal ordering (Section 3.4.1 [1]) to derive various extended MHD descriptions, valid up to a pre-chosen length and time scale. Obviously, each extended model must adopt a suitable closure strategy and somehow parameterize the effect of the higher order moments of the distribution function, i.e. enter the realm of modern transport theory. Since the computational effort and techniques can also differ significantly from one extended MHD model to the next (with particle in cell or PIC treatments usually prevailing at the kinetic levels, while finite volume or finite difference methods reign at fluid-like levels), this is also the domain of many advanced coding efforts. Ultimately, one would like to develop adaptive physics solvers, where not only the employed (grid or particle) resolution changes dynamically, but also the used model differs on different parts of the computational domain.

▷ **Generalized Ohm's law and Hall-MHD** As a concrete example of an extended MHD model, we here "derive" the Hall-MHD model and later on discuss how it modifies the familiar Alfvén wave dynamics. Our starting point will be an intermediate set of single fluid equations, Eqs. (3.135), (3.148)–(3.150) [1], rigorously derived from kinetic theory and repeated here for convenience:

$$\frac{\partial \rho}{\partial t} + \nabla \cdot (\rho \mathbf{v}) = 0, \tag{14.122}$$

$$\rho \frac{\partial \mathbf{v}}{\partial t} + \rho \mathbf{v} \cdot \nabla \mathbf{v} + \nabla p - \tau \mathbf{E} - \mathbf{j} \times \mathbf{B} = 0, \tag{14.123}$$

$$-\mu \left(\frac{m_i}{Ze}\right)^2 \frac{1}{\rho} \left[\frac{\partial \mathbf{j}}{\partial t} + \nabla \cdot (\mathbf{jv} + \mathbf{vj}) \right] - \frac{m_i}{Ze} \frac{1}{\rho} \left[(1-\mu)\mathbf{j} \times \mathbf{B} - \frac{Z-\mu}{Z+1} \nabla p \right]$$
$$+ \mathbf{E} + \mathbf{v} \times \mathbf{B} = \eta \mathbf{j}, \tag{14.124}$$

$$\frac{\partial p}{\partial t} + \mathbf{v} \cdot \nabla p + \gamma p \nabla \cdot \mathbf{v} = (\gamma - 1) \eta |\mathbf{j}|^2. \tag{14.125}$$

We recall that the first equation (14.122) denotes total mass conservation where $\rho = n_e m_e + n_i m_i$, the second (14.123) is the total momentum equation where $\rho \mathbf{v} = n_e m_e \mathbf{u}_e + n_i m_i \mathbf{u}_i$, obtained as a mass-weighted combination of both electron and ion momentum equations, the third Eq. (14.124) is the generalized Ohm's law, while Eq. (14.125) is the heat balance equation for the total pressure $p = p_e + p_i$. In the latter, we already neglected pressure anisotropies (no ion/electron viscosities) so that we are left with a scalar pressure. We also implicitly assumed an interest in time scales beyond the temperature equilibration time scale (identical temperature for ions and electrons), adopted quasi-neutrality such that $n_e = Z n_i$ (hence the charge density vanishes $\tau = 0$), while the ion–electron momentum transfer due to collisions has been quantified by means of a scalar resistivity η. All other symbols have their usual meaning, with the ratio $\mu = Z m_e / m_i$ of mass over charge appearing as a small parameter. Note that electron and ion pressures are to a good approximation given by $p_e = Zp/(1+Z)$ and $p_i = p/(1+Z)$. The generalized Ohm's law

Eq. (14.124) was obtained as a charge-weighted average of the ion and electron momentum equations and it can be used to consistently extend the ideal or resistive MHD model. For these standard MHD descriptions, the electric field is calculated from the simple form $\mathbf{E} + \mathbf{v} \times \mathbf{B} = \eta\mathbf{j}$. This is then used in combination with the pre-Maxwell equations

$$\mathbf{j} = \mu_0^{-1}\nabla \times \mathbf{B}, \qquad \frac{\partial \mathbf{B}}{\partial t} + \nabla \times \mathbf{E} = 0, \qquad \nabla \cdot \mathbf{B} = 0. \tag{14.126}$$

When we exploit the smallness of the mass over charge ratio μ, we can alternatively write

$$\begin{aligned}
\mathbf{E} &= -\left(\mathbf{v} - \frac{\mathbf{j}}{en_e}\right) \times \mathbf{B} + \eta\mathbf{j} - \frac{\nabla p_e}{en_e} + \frac{m_e}{e^2 n_e}\left[\frac{\partial \mathbf{j}}{\partial t} + \nabla \cdot (\mathbf{j}\mathbf{v} + \mathbf{v}\mathbf{j})\right] \\
&= -\mathbf{u}_e \times \mathbf{B} + \eta\mathbf{j} - \frac{\nabla p_e}{en_e} + \frac{m_e}{e^2 n_e}\left[\frac{\partial \mathbf{j}}{\partial t} + \nabla \cdot (\mathbf{j}\mathbf{v} + \mathbf{v}\mathbf{j})\right].
\end{aligned} \tag{14.127}$$

The Hall current $(en_e)^{-1}\mathbf{j} \times \mathbf{B}$ contribution to the electric field breaks the frozen-in condition of the ideal MHD induction equation, in a collisionless manner. Hence, even when only the first term in Eq. (14.127) is taken along, field lines are no longer forced to follow the flow due to decoupled ion–electron dynamics, in a way which is separate from the ion–electron collisional effects encoded in the scalar resistivity. One can incorporate various effects beyond the resistive MHD model, by computing the electric field from this generalized Ohm's law and inserting the result in Eq. (14.126)(b). The extra terms in Eq. (14.127) can be omitted when electron pressure is not important ($p_e = 0$) and electron inertia can be neglected. In practice, one may also find Hall-MHD models where anisotropic pressure effects are incorporated, or one partially accounts for electron inertia, noting that this generalized Ohm's law is to be seen as a form of the electron momentum equation. ◁

▷ **Hall-MHD and ion whistler waves** The most basic Hall-MHD model merely takes the first term in Eq. (14.127) along, and then notes that the remaining equations contain essentially ion dynamics information where $\rho = n_i m_i$ while $n_e = Z n_i$, and the smallness of μ allows to write $\mathbf{v} = \mathbf{u}_i$ and $\mathbf{u}_e = \mathbf{v} - \mathbf{j}/en_e$. When assuming negligible resistivity, $\eta = 0$, the main modification from the ideal MHD description can be appreciated immediately from analyzing the linear wave modes of a static uniform plasma. When we linearize the governing equations about a cold plasma state $p_0 = 0$ with density ρ_0, and assume a uniform background field $\mathbf{B}_0 = B_0 \mathbf{e}_z$, the dispersion relation can be computed in the usual fashion. When we restrict the analysis to plane waves $e^{i(\mathbf{k}\cdot\mathbf{r}-\omega t)}$ with a wave vector purely parallel to the magnetic field $\mathbf{k} = k_\parallel \mathbf{e}_z$, one finds

$$\left(\omega^2 - \omega_A^2\right)^2 = \left(\omega_A^4/\Omega_i^2\right)\omega^2. \tag{14.128}$$

In this equation, the Alfvén frequency is the usual $\omega_A^2 = k_\parallel^2 v_A^2$, while the ion gyro-frequency $\Omega_i = ZeB_0/m_i$. Note that, formally, the limit $(\Omega_i)^{-1} \to 0$ boils down to omitting the right hand side term, and yields the ideal MHD result with both fast and Alfvén waves at the Alfvén frequency. The modification induced by Hall-MHD is the fact that both waves are now dispersive, i.e. their phase speed $v_{ph} = \omega/k_\parallel$ depends on the wave number. Note that the long wavelength limit yields the ideal MHD Alfvén frequency, while at short wavelengths one finds $v_{ph} \sim k_\parallel v_A^2/\Omega_i$. The shortest wavelengths thus travel fastest (which means trouble for computational approaches), or one may also conclude that $v_{ph} \sim \sqrt{\omega}$. The highest frequencies arrive first and the wave thus has a descending pitch on arrival, hence the name "whistler". ◁

14.5 Literature and exercises

Notes on literature

Resistive instabilities

– The threat of resistive instabilities to plasma confinement for fusion was realized very early. Two seminal papers on the subject already appeared in 1963: 'Finite resistivity instabilities of a sheet pinch' by Furth, Killeen & Rosenbluth [152] and 'On the stability of hydromagnetic systems with dissipation' by Coppi [92].

– Most textbooks on plasma physics have chapters on resistive instabilities, e.g. *Theory of Toroidally Confined Plasmas* (Chapter 5) by White [483]; *Plasma Physics* (Chapter 17) by Sturrock [426]; *Introduction to Plasma Physics* (Chapter 20) by Goldston & Rutherford [185]; *Fundamentals of Plasma Physics* (Chapter 12) by Bellan [32].

Resistive spectrum

– The subject of MHD spectral theory of dissipative plasmas has not developed yet to the textbook level. A start can be found, though, in Section 7.15 of *Magnetohydrodynamics and Spectral Theory* by Lifschitz [308].

– The subject of resistive MHD instabilities has been investigated by too many authors to even attempt to give a representative sample. However, fundamental analytical studies by Pao & Kerner [360], by Riedel [391], and by Borba *et al.* [61], and systematic numerical studies by Kerner *et al.* on cylindrical plasmas [274], quasi-modes [378], and large-scale computing for tokamaks [272], and similar studies by McMillan *et al.* [329] come close to the viewpoint of the present book.

Reconnection

– The subject of reconnection is central in the exposition of the sequence of the three books *Nonlinear Magnetohydrodynamics* [45], *Magnetic Reconnection in Plasmas* [46], and *Magnetohydrodynamic Turbulence* [47] by Biskamp.

– *Magnetic Reconnection; MHD Theory and Applications* by Priest & Forbes [387] provides the fundamentals of the subject, including reconnection in 3D, and gives many applications to solar and astrophysical plasmas, finishing with cosmic particle acceleration.

Exercises

[14.1] *Ideal MHD versus resistive MHD*

The difference between ideal and resistive MHD is that one takes resistivity into account (where η is usually assumed to be small).

– What is the main consequence of this?

– Why can we not use the same strategy for analyzing the stability of resistive plasmas as for ideal plasmas?

[14.2] *Tearing modes*

In this chapter, a detailed analysis of tearing modes is presented.

– Explain why those modes do not appear in ideal plasmas.

– What essential condition is needed to make sure that a tearing mode may appear?

- The analysis of tearing modes is carried out in three regions. Specify for each region what kind of assumptions has been used and which equations have to be solved.

[14.3] *Resistive spectrum of a homogeneous incompressible plasma*

The resistive spectrum of a homogeneous incompressible plasma is discussed in Section 14.3. In this exercise, you derive the equations presented in that section and you make a plot of the resistive spectrum. Consider a plasma slab between two perfect conducting plates at a distance L.

- Start from the resistive equations (14.44) and reduce them for a homogeneous plasma.
- In a homogeneous plasma, the displacement field component ξ can be written as $\xi(x, y, z; t) = \hat{\xi}(k_x, k_y, k_z; \omega) \exp[i(k_x x + k_y y + k_z z - \omega t)]$. Explain why you can make this assumption. What are the boundary conditions on the walls?
- Derive the resistive equations using the assumption of the previous question.
- Write the derived equations in matrix form and derive the dispersion equation.
- Solve this dispersion equation.
- Look at the solutions of the dispersion equation and note that there is one special solution. Which one is it and why is it special?
- Plot both solutions of the dispersion equation in the complex ω-plane using your favorite plotting program. What kind of scaling can be used to plot the solutions?

[14.4] *Resistive spectrum of a homogeneous compressible plasma*

Figure 14.6 shows the resistive spectrum of a homogeneous compressible plasma. This spectrum has been created with one of the numerical methods discussed in Chapter 15. The dispersion equation can also be derived as in the previous exercise.

- Start from the resistive equations (14.41) and reduce them for a homogeneous plasma representing perturbations $\xi(x, y, z; t)$ as in Exercise [14.3].
- Write the equations in matrix form.
- Show that the resulting dispersion equation can be written as

$$\omega \left[\omega^2 - k_\parallel^2 b^2 + i\eta k^2 \omega \right] \left[\omega^4 - k^2 (b^2 + c^2)\omega^2 + k^2 k_\parallel^2 b^2 c^2 \right. $$
$$\left. + i\eta k^2 \omega(\omega^2 - k^2 c^2) \right] = 0 \,,$$

where $b^2 \equiv B^2/\rho$ and $c^2 \equiv \gamma p/\rho$ are the squares of the Alfvén and the sound speed.
- Show that this dispersion equation reduces to Eq. (14.116) in the incompressible limit.
- Take the limit of vanishing resistivity and discuss the resulting solutions.
- Reproduce the plots of Fig. 14.6 using your favorite plotting program. Make sure that you can use Laguerre's method (see Numerical recipes [385], in IDL use FZ_ROOTS, in Matlab use ROOTS1) to compute the roots of the fourth order polynomial. What kind of boundary conditions should be used?

[14.5] *Ion whistler waves*

Derive the dispersion relation for the ion whistler waves for an ideal Hall-MHD description of a cold uniform plasma. Complete the discussion of the wave modes with a categorization of the marginal modes at $\omega = 0$. (You know that there will be three more waves in addition to the four given by Eq. (14.128), so be careful to distinguish spurious modes from physically meaningful degenerate solutions!)

15

Computational linear MHD

Computational magnetohydrodynamics is a very active research field due to the increasing demand for quantitative results for realistic magnetic configurations on the one hand and the availability of ever more computer power on the other [373]. Many MHD phenomena can not be described by analytical methods in all of their complexity although simplified analytical models have led to indispensable insight into the fundamental physics of various magnetohydrodynamic processes. The intricate geometry of present tokamaks, for instance, forces theory to resort to computer simulations as the mathematics is not fully tractable anymore. The fast increase of computer speed and memory allows simulations with ever more "physics" in the equations and taking into account the full 3D geometrical effects.

While the governing ideal MHD equations form a set of *nonlinear, hyperbolic, partial differential equations*, we already encountered many magneto-fluid phenomena which are adequately modeled by means of the linearized MHD equations. In this chapter, we concentrate mostly on computational approaches for *linear* MHD problems, and introduce several basic numerical concepts and techniques along the way. We give a brief overview of the most frequently encountered spatial discretizations to translate any problem expressed as a (set of) differential equation(s) into a discrete linear algebraic problem, and discuss commonly used strategies for solving the resulting linear systems and generalized eigenvalue problems. Representative applications cover MHD spectroscopic computations for diagnosing eigenoscillations and stability of given, possibly pre-computed, MHD equilibria, as well as steady-state and time dependent solutions to externally driven MHD configurations. Both ideal and non-ideal linear MHD problems are encountered. We focus on robust and widely used methods and review several indicative results in order to demonstrate their typical use and power.

In Section 15.1, the different spatial discretization techniques that are most common in linear MHD are discussed and illustrated by means of a generic *model steady-state problem*. The application of these techniques to solve representa-

177

tive linear MHD boundary value problems is illustrated in Section 15.2. In Section 15.3, we discuss the linear algebraic methods used to solve the linear systems obtained after spatial discretization. The direct and iterative linear system solvers are applied in Section 15.4 to solve some illustrative linear MHD *initial value problems*. Finally, in Section 15.5, we sum up the criteria to select a numerical method for solving a specific linear MHD problem.

15.1 Spatial discretization techniques

Even after linearization, the MHD equations remain a fairly complicated set of, generally time-dependent, partial differential equations. To introduce the basic concepts connected to spatial discretization techniques, we consider a generic one-dimensional *model problem*. This is related to the *Sturm–Liouville equation*, which is a second-order linear differential equation of the form

$$ -\frac{d}{dx}\left[p(x)\frac{du(x)}{dx}\right] + q(x)u(x) = \lambda\, w(x)u(x)\,, \tag{15.1} $$

where the coefficient functions $p(x)$ and $q(x)$ and the weight function $w(x)$ are given and real, and λ is one of the eigenvalues to be computed, together with the associated unknown solution $u(x)$. In general, the boundary conditions originate from specific physical problems and they are usually chosen to guarantee reality of the eigenvalues. The self-adjoint form of the equation is closely related to the quadratic forms that may be constructed from it, and that are used to prove that reality. A central topic is the construction of suitable function spaces of orthogonal eigenfunctions. One of the issuing attractions is the so-called *Sturmian monotonicity property* of the eigenvalues λ_i in terms of the number of nodes of those eigenfunctions $u_i(x)$.

This classical problem occurs in many parts of physics, in particular it has been instrumental in unraveling the spectra of elementary quantum mechanical systems. Also, in the study of the eigenvalues of the MHD spectrum of one-dimensional equilibria, like a gravitating slab or cylindrical plasma, a second order differential equation occurs, as we have seen in Chapters 7 and 9 of Volume [1], which is not in Sturm–Liouville form, but, nevertheless exhibits similar monotonicity properties, like Sturmian or anti-Sturmian behavior of the eigenvalues. Although the coefficients of the differential equation have a much more involved dependence on the eigenvalue parameter, similarly powerful oscillation theorems could be formulated and proven. They could even be extended to stationary configurations, with complex eigenvalues, as discussed in the Chapters 12 and 13 of the present volume. This shows that the basic theory of the Sturm–Liouville equation, and its generalizations, have wide applicability.

In the present chapter, the different discretization methods will be illustrated for the related *inhomogeneous* problem

$$-\frac{d}{dx}\left[p(x)\frac{du(x)}{dx}\right] + q(x)u(x) = f(x)\,, \tag{15.2}$$

on the finite closed interval $[0, 1]$, again with the functions $p(x)$, $q(x)$ and $f(x)$ given and real, but without an eigenvalue parameter for the time being. The solution $u(x)$ will be subjected to the following boundary conditions. At $x = 0$ the function value is assumed to be known,

$$u(0) = 0\,, \tag{15.3}$$

which represents a boundary condition of the "Dirichlet" type, and at $x = 1$ we will impose

$$\alpha u(1) + \beta\frac{du}{dx}(1) = F\,, \tag{15.4}$$

which is a more general, inhomogeneous, boundary condition involving both u and du/dx. The quantities α, β and F are given constants. Notice that the function f in the RHS of Eq. (15.2) depends on x, which makes the problem non-trivial, even when $p(x)$ and $q(x)$ are constant functions.

Physically, the above problem might correspond to the displacement of a string or a magnetic field line with variable density that is fixed at one end ($x = 0$) and neither fixed nor completely free at the other end (e.g. connected to a spring), or to the distribution of the temperature of a gas or a plasma which is kept at a fixed temperature at one end ($x = 0$) and heated at the other end ($x = 1$). Note that Eq. (15.2) and the boundary conditions (15.3) and (15.4) represent a *steady-state problem*, and not an initial value problem. In more spatial dimensions, it will correspond to an elliptic boundary-value problem, e.g. governed by Laplace's equation. Initial value problems will be discussed in Section 15.4.

The differential form of Eq. (15.2) inherently assumes that (first and) second order derivatives of $u(x)$ exist on the domain $[0, 1]$. Associated with the differential form, the function $u(x)$ must obey the *integral form*

$$\int_0^1 \left[-\frac{d}{dx}\left(p(x)\frac{du(x)}{dx}\right) + q(x)u(x) - f(x)\right] dx = 0\,. \tag{15.5}$$

For our model problem, this reduces to

$$\left[-p(x)\frac{du(x)}{dx}\right]_0^1 + \int_0^1 [q(x)u(x) - f(x)]\,dx = 0\,, \tag{15.6}$$

thus requiring less regularity of the sought function $u(x)$.

15.1.1 Basic concepts for discrete representations

To solve the two-point boundary-value problem posed by Eq. (15.2) and the boundary conditions (15.3) and (15.4) numerically, we will need to make a choice for the numerical representation of the unknown function $u(x)$. When the functions $p'(x)/p(x)$ and $q(x)/p(x)$ are analytic in all $x \in [0,1]$, the sought exact solution $u(x)$ is mathematically well-behaved and defined on the entire continuous interval $[0,1]$. However, a computational approach always needs to represent $u(x)$ *discretely*, i.e. involving a *finite* number of unknowns. We will indicate the discrete representation for $u(x)$ involving N unknowns with $\hat{u}^N(x)$. As will be explained in following sections, depending on the chosen representation, these unknowns may directly relate to function values $u(x_i)$ at a finite set of particular pre-chosen points $x_i \in [0,1]$, with $i = 1,\ldots,N$ (note that in $x_0 = 0$ we have $u(0) = 0$ due to boundary condition (15.3)), or to the average values of $u(x)$ in N sub-intervals of $[0,1]$, or even to more general expansion coefficients used in finite function series representations (e.g. a Fourier series). Either way, the discretization function $\hat{u}^N(x)$ is an approximation of $u(x)$.

In order to quantify how good this approximation is, we need to define a norm so that we can measure the "distance" between the two functions. Several choices are possible here. For example, for quadratically integrable functions on a domain $[a,b]$, the L^2-norm quantifying the distance between functions f and g is

$$\|f - g\|_2 \equiv \left(\int_a^b [f(x) - g(x)]^2 \, dx \big/ (b - a) \right)^{1/2}. \tag{15.7}$$

This L^2-norm is especially useful for linear problems because Fourier analysis is often used for these problems, and Parseval's relation says that a function and its Fourier transform have the same L^2-norm. Frequently encountered norms, written for a finite dimensional vector $\mathbf{u} = \{u_i\}$, with $i = 1,\ldots,N$, are the L^p-norm and the L^∞-norm defined as

$$\|\mathbf{u}\|_p \equiv \left(\frac{1}{N} \sum_{i=1}^N |u_i|^p \right)^{1/p} \quad \text{and} \quad \|\mathbf{u}\|_\infty \equiv \max_i |u_i|, \tag{15.8}$$

respectively, where $p = 1$ and $p = 2$ are the most popular L^1-norm and L^2-norm in their discrete form. The concept of "convergence" of a numerical solution is then always connected to a certain norm. The series $\{\hat{u}^N(x)\}_{N\to\infty}$ *converges* to the function $u(x)$ *in the L^2-norm* when

$$\lim_{N\to\infty} \|u(x) - \hat{u}^N(x)\|_2 = 0. \tag{15.9}$$

Note that this involves the *global truncation error*, i.e. the difference between the exact solution $u(x)$ and the approximation $\hat{u}^N(x)$ over the entire domain. Although the global truncation error is important, it will typically be easier to quantify the so-called *local truncation error* E_i between a local function value (or derivative) such as $u(x_i) \equiv u_i$ (or $du/dx(x_i) \equiv u'_i$) and the local approximation $\hat{u}^N(x_i)$, i.e. $E_i = u_i - \hat{u}^N(x_i)$. Knowledge of the local truncation error will often enable an estimate of the maximal global error.

Another important cause of errors in numerical solutions is the fact that computers calculate with a finite number of decimals while real numbers usually have infinitely many decimals. This yields a *round-off error* R_i:

$$R_i = \hat{u}^N(x_i) - U_i, \tag{15.10}$$

where U_i denotes the actually calculated value with the given method. The absolute value of the *total error* made in calculating u_i is given by

$$|u_i - U_i| = |u_i - \hat{u}^N(x_i) + \hat{u}^N(x_i) - U_i|$$

$$\leq |u_i - \hat{u}^N(x_i)| + |\hat{u}^N(x_i) - U_i| \leq |E_i| + |R_i|. \tag{15.11}$$

The total error on the function value u_i is thus limited by the sum of the absolute values of the local truncation error and the round-off error. In the context of spatio-temporal problems, it will therefore be important to use a method where truncation errors do not grow unbounded, as discussed further in this chapter.

Another frequently encountered measure of the quality of the approximation is the *residual*, which is the equation to be solved as evaluated for the approximation. For our model Eq. (15.2) this means

$$\hat{r}^N(x) \equiv -\frac{d}{dx}\left[p(x)\frac{d\hat{u}^N}{dx}\right] + q(x)\hat{u}^N - f(x). \tag{15.12}$$

This residual $\hat{r}^N(x)$ can be quantified even when the exact solution is not available. A numerical scheme is said to be *consistent* when this residual vanishes in the limit $N \to \infty$ for all x, i.e. when the approximation $\hat{u}^N(x)$ converges for all x to the exact solution $u(x)$.

Other criteria that are useful in evaluating and selecting numerical schemes are *efficiency*, *accuracy* and *stability*. The latter two criteria are related to the requirement that the deviation of the computed values from the exact solution of the differential equation is small. The criterion *accuracy* is concerned with the global and local truncation errors and the round-off errors discussed above. Usually, round-off errors may be ignored in comparison with the truncation errors if the scheme is *stable*. *Numerical stability* is concerned with error propagation. As a matter of fact, even if truncation and round-off errors are small, a numerical scheme will be of little value for time-dependent problems when the small errors grow rapidly in time. In Section 15.4 we will provide a quantitative measure for both accuracy and stability by means of a linear dispersion analysis of some of the numerical schemes discussed in this chapter. Last but not least, the criterion *efficiency* is concerned with optimizing the costs, in terms of CPU time and computer memory, for a given quality of the simulation model.

15.1.2 Finite difference methods

The finite difference approximation is the most widely used computational method. In the finite difference method (FDM), the continuous domain of the independent variable x is replaced by a finite number of discrete points, the *grid points* or *mesh points*. The grid points can be equidistant, as in Fig. 15.1, where we subdivided the domain $[0, 1]$ into N intervals of the same width (introducing $N + 1$ grid points), but they can also be accumulated in places where a higher resolution is required. All dependent functions $g(x)$ on the continuous domain are then approximated by their local values $\{g_i\}$ on the mesh, $g_i \equiv g(x_i)$ with $i = 0, \ldots, N$.

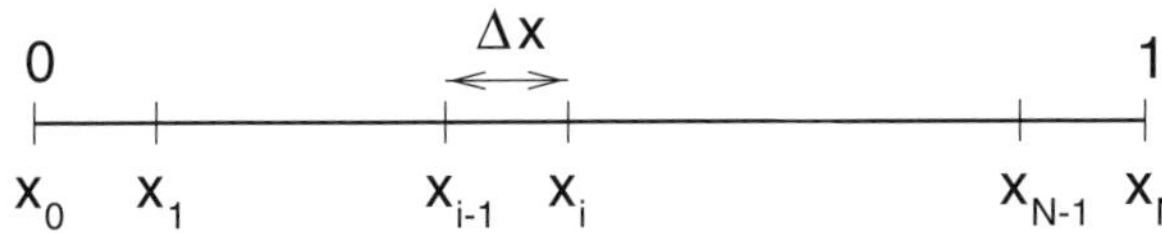

Fig. 15.1 Discrete representation of the domain $[0, 1]$ by $N + 1$ equidistant grid points: $x_i = i\Delta x$ with $\Delta x = 1/N$.

The representation of the local derivatives is based on truncated Taylor series expansions. For example, when u_i denotes the value of the variable u in grid point x_i, the value of u in grid point x_{i+1} is given by

$$u_{i+1} = u_i + u_i'\Delta x + \tfrac{1}{2}u_i''(\Delta x)^2 + \mathcal{O}(\Delta x)^3 . \qquad (15.13)$$

Hence, up to *first-order accuracy* in $\Delta x \equiv x_{i+1} - x_i$, we have

$$u_{i+1} = u_i + u_i'\Delta x . \qquad (15.14)$$

Solving Eq. (15.14) for the first order derivative u_i' in x_i yields

$$u_i' = (u_{i+1} - u_i)/\Delta x + \mathcal{O}(\Delta x) , \qquad (15.15)$$

which is a *first-order forward difference* expression for the derivative. Similarly, a Taylor series expansion for u_{i-1} about u_i yields a *first-order backward difference* expression, and by subtracting the two Taylor series expansions, one gets

$$u_i' = \tfrac{1}{2}(u_{i+1} - u_{i-1})/\Delta x + \mathcal{O}(\Delta x)^2 , \qquad (15.16)$$

which is a *second-order central difference* expression for the first derivative.

▷ **Exercise** Show that the expression

$$u'_i = \left[\tfrac{1}{2}\alpha\,(u_{i+1} - u_{i-1}) + \tfrac{1}{4}(1 - \alpha)\,(u_{i+2} - u_{i-2})\right]/\Delta x$$

yields a fourth-order approximation when $\alpha = 4/3$. ◁

Upon substitution of Eq. (15.16) into Eq. (15.13) one obtains a *second-order* accurate expression for the second derivative u''_i of u in x_i, as the third order term in the Taylor series in Eq. (15.13) cancels out:

$$u''_i = (u_{i+1} - 2u_i + u_{i-1})/(\Delta x)^2 + \mathcal{O}(\Delta x)^2\,. \tag{15.17}$$

Expressions for higher derivatives of u can be found in a similar way. Hence, finite difference approximations are simple to derive and easy to code on regular meshes.

▷ **Exercise** Show that the formula given by Eq. (15.17) can be interpreted as the forward difference of the first derivative, when backward differences are used to discretely evaluate these first derivatives. ◁

The FDM applied to Eq. (15.2) replaces the differential equation by its finite difference representation in all mesh points. For all interior mesh points x_i with $i \neq 0$ or $i \neq N$, use of the central difference formulas (15.16)–(15.17) leads obviously to a representation with a second order local truncation error. For this model problem, the only complication to guarantee the same second order accuracy for the full solution on the grid is a corresponding second order treatment of the boundary points $x_0 = 0$ and $x_N = 1$. The boundary condition (15.3) is easily imposed: it corresponds directly to $u_0 = 0$. However, condition (15.4) involves the first order derivative in $x_N = 1$, so we need to derive a second order backward difference expression for use in this right boundary condition.

While it is possible to do this by suitably combining Taylor expansions for u_i, u_{i-1} and u_{i-2} as above, an alternative and equivalent method to derive such expressions is based on local polynomial fits. In the case at hand, where we seek a second order backward difference expression for u'_i, we envision a locally linear dependence for u' which corresponds to a quadratic formula for u. Hence, for grid point $x_N = 1$, we write locally $x = x_N + y$, and assume that $u(x)$ can be expressed by a second-order polynomial, namely:

$$u(x_N + y) = a_0 + a_1 y + a_2 y^2 \quad \Rightarrow \quad u'(x_N + y) = a_1 + 2a_2 y\,. \tag{15.18}$$

Applying this expression to the last three grid points x_N, $x_{N-1}\,(= x_N - \Delta x)$, and $x_{N-2}\,(= x_N - 2\Delta x)$ yields:

$$\begin{aligned}
u_N &= a_0\,, \\
u_{N-1} &= a_0 - a_1\Delta x + a_2(\Delta x)^2\,, \\
u_{N-2} &= a_0 - a_1(2\Delta x) + a_2(2\Delta x)^2\,.
\end{aligned} \tag{15.19}$$

Solving these three equations for the wanted $u'(x_N) = a_1$ gives the expression:

$$u'_N = a_1 = \tfrac{1}{2}\left(3u_N - 4u_{N-1} + u_{N-2}\right)/\Delta x + \mathcal{O}(\Delta x)^2 \,. \tag{15.20}$$

This formula can be used to impose the boundary condition (15.4) in a second order treatment of the model problem. Note that the second order central difference discretization of our model problem described by Eq. (15.2) leads to an algebraic system with a tridiagonal coefficient matrix. Once solved, the representation $\{u_i\}$ is clearly an incomplete description of $u(x)$ for $x \in [0, 1]$, but the function $u(x)$ can be approximated at any point of the interval by, e.g., using a second order, linear interpolation between adjacent grid points. Thus, when x lies in the interval $[x_i, x_{i+1}]$ we have:

$$u(x) \approx \hat{u}(x) = \left(\frac{x - x_i}{x_{i+1} - x_i}\right) u_{i+1} + \left(\frac{x_{i+1} - x}{x_{i+1} - x_i}\right) u_i \,. \tag{15.21}$$

$\triangleright$ **Exercise** Derive a second-order *forward* difference expression for u'_N using the polynomial fit method to find

$$u'_N = \tfrac{1}{2}(-u_{N+2} + 4u_{N+1} - 3u_N)/\Delta x + \mathcal{O}(\Delta x)^2 \,. \tag{15.22}$$

Also, derive a second-order one-sided difference expression for the second derivative u''_i, which requires a polynomial for $u(x_i + y)$ which is cubic in y, and find

$$u''_i = (-u_{i+3} + 4u_{i+2} - 5u_{i+1} + 2u_i)/(\Delta x)^2 + \mathcal{O}(\Delta x)^2 \,. \tag{15.23}$$

Note that this second order formula involves four grid points. The number of grid points needed to evaluate derivatives discretely is referred to as the "stencil" of the method. The stencil of formula (15.23) thus includes grid point x_i and three grid points to its right. $\triangleleft$

For use in multi-dimensional problems (as in most MHD applications), finite difference expressions for partial derivatives in any spatial direction are given by similar expressions. Expressions for mixed derivatives can easily be derived, e.g. by considering that

$$\frac{\partial^2 u}{\partial x \partial y} = \frac{\partial}{\partial x}\left(\frac{\partial u}{\partial y}\right) \,.$$

If we write the occurring x-derivative as a central difference of the y-derivatives according to Eq. (15.16), and also use this central difference expression for the latter y-derivatives, we derive a second order accurate expression for the mixed derivative as:

$$\left(\frac{\partial^2 u}{\partial x \partial y}\right)_{i,j} = \frac{u_{i+1,j+1} + u_{i-1,j-1} - u_{i-1,j+1} - u_{i+1,j-1}}{4\Delta x \Delta y} + \mathcal{O}((\Delta x)^2, (\Delta y)^2) \,.$$

Generalizations to higher order Achieving higher than second order accuracy with the FDM is conceptually simple, but normally comes at the price of handling wider stencils. This also means that boundary treatments need particular attention to achieve the same overall order of accuracy. Fourth order central finite difference formulas for the first and second derivative involve up to five grid points, so that adjacent to boundaries both one-sided and semi-one-sided formulas are needed. These formulas can again most directly be obtained by means of the local polynomial fit method. Representative $\mathcal{O}(\Delta x)^4$ formulas are

$$u_i' = \tfrac{1}{12}(-u_{i+2} + 8u_{i+1} - 8u_{i-1} + u_{i-2})/\Delta x\,,$$

$$u_i' = \tfrac{1}{12}(u_{i+3} - 6u_{i+2} + 18u_{i+1} - 10u_i - 3u_{i-1})/\Delta x\,,$$

$$u_i' = \tfrac{1}{12}(-3u_{i+4} + 16u_{i+3} - 36u_{i+2} + 48u_{i+1} - 25u_i)/\Delta x\,,$$

$$u_i'' = \tfrac{1}{12}(-u_{i+2} + 16u_{i+1} - 30u_i + 16u_{i-1} - u_{i-2})/(\Delta x)^2\,,$$

$$u_i'' = \tfrac{1}{12}(u_{i+4} - 6u_{i+3} + 14u_{i+2} - 4u_{i+1} - 15u_i + 10u_{i-1})/(\Delta x)^2\,,$$

$$u_i'' = \tfrac{1}{12}(-10u_{i+5} + 61u_{i+4} - 156u_{i+3} + 214u_{i+2} - 154u_{i+1} + 45u_i)/(\Delta x)^2\,.$$

$$(15.24)$$

▷ **Exercise** Derive these formulas and also derive the equivalent formulas at the other boundary of the computational domain, i.e. at $x = 1$. Your result should be consistent with replacing indices $i \pm * \leftrightarrow i \mp *$ and $\Delta x \leftrightarrow -\Delta x$. ◁

An alternative means to achieve higher order accuracy within the finite difference framework, while maintaining a compact stencil, is to exploit so-called compact or implicit FD schemes. The basic idea is to exploit the local values for the derivatives u_i' and u_i'' as additional unknowns, and complement the equations to solve with implicit formulas linking these additional unknowns to the local function values in discrete expressions with the desired accuracy, as e.g. in

$$\tfrac{1}{2}(u_i' + u_{i+1}') = (u_{i+1} - u_i)/\Delta x + \mathcal{O}(\Delta x)^2, \qquad (15.25)$$

which is easily found from combining the central second order formula (15.16) for u_i' with the backwards formula (15.20) for u_{i+1}'. Fourth order expressions are

$$\tfrac{1}{6}(u_{i+1}' + 4u_i' + u_{i-1}') = \tfrac{1}{2}(u_{i+1} - u_{i-1})/\Delta x + \mathcal{O}(\Delta x)^4,$$

$$\tfrac{1}{12}(u_{i+1}'' + 10u_i'' + u_{i-1}'') = (u_{i+1} - 2u_i + u_{i-1})/(\Delta x)^2 + \mathcal{O}(\Delta x)^4. \quad (15.26)$$

The above examples yield, for the second order relation (15.25) and for the fourth order relations (15.26), a bidiagonal and two tridiagonal systems in the derivatives, respectively, to be solved together with the discretized equation. Since these systems effectively couple all grid points, no explicit relation can be written for u_i' in terms of neighboring function values u_i alone, hence their implicit nature.

A more detailed description of FDM concepts can e.g. be found in Hirsch [226], Chapter 4. The most attractive feature of finite differences is that it can be very easy to implement them. A negative point is the quality of the approximation between grid points, which is poor in the FDM. This problem can be solved by considering a finite element approach which focuses more on the approximation to the solution of the differential equation rather than on the equation itself.

15.1.3 Finite element method

In the *finite element method* (FEM), the dependent variables are approximated by a finite set of local piecewise polynomials. A good introduction to the finite element method can be found in Strang and Fix [424]. Consider again the problem posed by Eq. (15.2) and the boundary conditions (15.3) and (15.4). The domain $D = [0, 1]$ can be divided into a finite number of equally (as in Fig. 15.1) or unequally sized sub-domains $[x_{i-1}, x_i]$ for $i = 1, \ldots, N$ which are this time called the "elements" connecting the "nodes" x_i. The solution $u(x)$ is then approximated by a linear combination of *basis functions* which are local piecewise polynomials on the sub-intervals $[x_{i-1}, x_{i+1}]$ and taken to vanish outside these finite intervals, hence the name "finite" elements. The basis functions are in turn constructed from element *shape functions* of a local coordinate ξ, typically defined on the standard interval of the local coordinate $\xi \in [-1, 1]$. There are two linear finite element shape functions, corresponding to the two degrees of freedom in a linear profile, fixed by requiring a unit value at one node and a zero value at the other node:

$$N_1(\xi) = \tfrac{1}{2}(1 - \xi) \quad \text{and} \quad N_2(\xi) = \tfrac{1}{2}(1 + \xi). \tag{15.27}$$

These shape functions are combined into the "tent" functions $h_i(x)$ depicted in Fig. 15.2, which form the linear basis functions to approximate:

$$u(x) \approx \hat{u}(x) = \sum_{i=0}^{N} u_i h_i(x). \tag{15.28}$$

The approximate solution $\hat{u}$ can be interpreted as belonging to the function space spanned by the linear finite element basis functions h_i given by

$$h_i(x) \equiv \begin{cases} \dfrac{x - x_{i-1}}{x_i - x_{i-1}} & \text{for } x_{i-1} \leq x \leq x_i, \\[2mm] \dfrac{x_{i+1} - x}{x_{i+1} - x_i} & \text{for } x_i \leq x \leq x_{i+1}, \\[2mm] 0 & \text{elsewhere.} \end{cases} \tag{15.29}$$

Approximations to derivatives are obtained by differentiating Eq. (15.28). Note that these linear elements are such that the coefficients in the expansion (15.28) directly relate to local grid point values u_i. Obviously, other interpolation schemes are possible, based on other shape functions. For instance, the approximation $\hat{u}$ can be obtained by considering piecewise constant shape functions. In this case, the FDM can be considered as a special case of the FEM approach. Other choices include piecewise quadratic and/or cubic shape functions, to be discussed below.

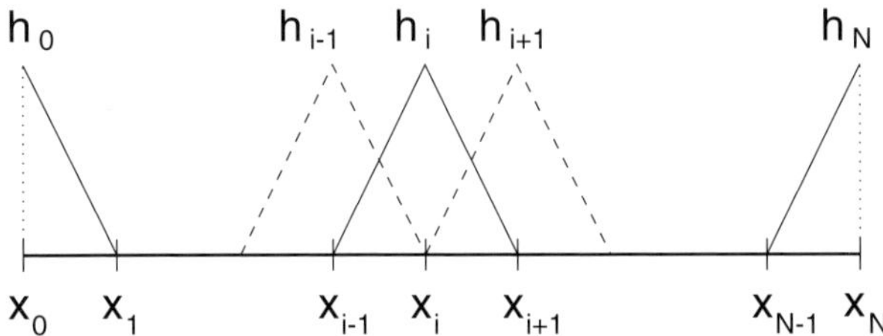

Fig. 15.2 Linear finite elements.

In any case, a linear system of algebraic equations for the coefficients $\{u_i\}$ is obtained by requiring that $N + 1$ weighted integrals of the *residual* given in Eq. (15.12), i.e. the equation to solve evaluated for the approximation $\hat{u}$, are zero. For the model problem defined on the domain $D = [0, 1]$, the $N + 1$ weighted integrals of the residual are

$$\int_0^1 w_l \left[\left(-\frac{d}{dx} p(x) \frac{d}{dx} + q(x) \right) \hat{u}(x) - f(x) \right] dx = 0, \qquad l = 0, 1, \ldots N.$$
$$(15.30)$$

The still unspecified *weight functions* are denoted by w_l. In general, the function space consisting of all linear combinations of the basis functions $\{h_i; i = 0, 1, \ldots N\}$ is finite. Hence, the function $\hat{u}(x)$ defined by Eq. (15.28) may not be equal to the exact solution $u(x)$ in all points of the interval $[0, 1]$. Convergence requires completeness of the function space of the basis functions *and* of the space of the weight functions, i.e. as N increases, the approximation (15.28) must become better and the residual must vanish in the limit $N \to \infty$. The form (15.30) is called the *weighted residual formulation* and, with a certain choice for the weight functions, this formulation yields an approximate solution for each finite N.

In the finite element method, the weight functions are often chosen to be equal to the basis functions themselves, i.e. $w_l = h_l$, such that the residual $r(x)$ is made orthogonal, in the sense of an inner product defined by the integral $\int_0^1 h_l(x)\, r(x)\, dx$, to the function space of the basis functions. The method is then called the *Galerkin method*. For the model problem (15.2) this yields

$$\int_0^1 h_l \left[\left(-\frac{d}{dx} p(x) \frac{d}{dx} + q(x) \right) \hat{u}(x) - f(x) \right] dx = 0, \quad l = 0, 1, \ldots N, \quad (15.31)$$

where $\hat{u}$ is given by Eq. (15.28). Hence, we have a linear algebraic system of $N+1$ equations for the $N+1$ unknowns $\{u_i\}_{i=0,1,\dots N}$.

The integrals occurring in the Galerkin method can often be simplified by performing integrations by parts on the integrands with the highest order derivatives. For problem (15.2) we could proceed with the Galerkin method (15.31) as follows:

$$-\left[h_l\,p(x)\frac{d\hat{u}}{dx}\right]_0^1 + \int_0^1 \frac{dh_l}{dx}\,p(x)\frac{d\hat{u}}{dx}\,dx + \int_0^1 h_l(x)[q(x)\hat{u} - f(x)]\,dx = 0\,,$$

$$l = 0,1,\dots N\,. \quad (15.32)$$

The integration by parts results in a different formulation of the problem, which allows solutions which are less "smooth" in the sense that they have to be continuously differentiable to a lower degree. Indeed, due to the integration by parts the derivatives that occur in the expression are of a lower order. Therefore, this formulation is called the *weak formulation* of the problem. For the weak form (15.32), this means that functions with discontinuous first derivative are allowed, whereas in the original differential form (15.2) the first order derivatives have to be continuous and differentiable. In this sense, the weak formulation is closer to the integral form of the equations, and this is of interest to numerically represent more irregular, physically realizable, solutions.

The boundary terms generated by the integrations by parts can often be used to impose the boundary conditions. Boundary conditions that can enter the equations in this way are called *natural boundary conditions*. For example, for our model problem (15.2) this yields for the boundary condition (15.4), with linear elements:

$$-\left[h_l p(x)\frac{d\hat{u}}{dx}\right]_0^1 = \delta_{lN}h_N(1)p(1)\left(\frac{\alpha}{\beta}h_N(1)u_N - \frac{F}{\beta}\right)\,, \quad l = 0,1,\dots N\,,$$

$$(15.33)$$

since h_N is the only basis function different from zero in $x = x_N = 1$ and h_0 has to be left out to satisfy boundary condition (15.3), as we will see below. The first term in the RHS of (15.33) yields a contribution to the coefficient matrix of the system, called the "stiffness matrix" in FEM terminology, while the second term defines a "source" term. The other boundary conditions, such as condition (15.3) of our model problem, have to be imposed explicitly and are called *essential boundary conditions*. This is done by limiting the space of basis functions to those basis functions that satisfy these boundary conditions. For boundary condition (15.3) of our model problem this means that all basis functions that are non-zero in $x = 0$ have to be left out, i.e. h_0 has to be left out in the case of linear elements.

▷ **Exercise** Consider model problem (15.2) with $p(x) \equiv 1$ and $f(x) \equiv 0$ and show that the Galerkin method in combination with linear finite elements reduces to the standard second order finite difference scheme when the trapezoidal rule $\int_a^b f(x)dx \approx \frac{1}{2}(b-a)[f(a)+f(b)]$ is used and partial integration is performed on the term with the second derivative. ◁

Construction of the system matrix For the model problem (15.2), the system of equations for the coefficients $\{u_i\}_{i=0}^{N}$ is obtained upon substitution of expansion (15.28) for $\hat{u}$ in the "weak" form (15.32):

$$-h_l \delta_{lN} p \frac{d\hat{u}}{dx}\bigg|_1 + \sum_{i=0}^{N}\left[\int_0^1 \left(\frac{dh_l}{dx}\frac{dh_i}{dx}p + h_l q h_i\right)dx\right]u_i = \sum_{i=0}^{N}\left[\int_0^1 h_l f dx\right],$$
$$l = 0, 1, \ldots N, \quad (15.34)$$

where the boundary condition (15.4) still has to be substituted in the first (surface) term. For simple functions $p(x)$, $q(x)$ and $f(x)$, the integrals can be calculated by hand, and many are actually vanishing due to the chosen *finite* elements. In general, however, the coefficient functions of the differential equation(s) can be complicated functions of x and the finite elements themselves can be taken as higher order polynomials too. The integrals are then calculated by numerical integration or "quadrature" formulas, which themselves need to reach a particular order of accuracy to guarantee the overall accuracy of the solution. Gaussian quadrature formulas for integration on the standard interval $\xi \in [-1, 1]$ involve the judicious choice of n weights w_i and the locations of n integration points ξ_i such that all polynomials of degree $2n - 1$ are evaluated exactly by the discrete formula

$$\int_{-1}^1 f(\xi)d\xi = \sum_{i=1}^{n} w_i f(\xi_i). \quad (15.35)$$

Table 15.1 gives the positions of the evaluation points ξ_i and the weights w_i for this Gaussian integration up to $n = 4$.

Table 15.1 *Weights and evaluation points in double precision accuracy for n-point Gaussian quadrature. (From Abramowitz and Stegun [3].)*

n	weights w_i	evaluation points ξ_i
1	2	0
2	1	$\pm 1/\sqrt{3}$
3	0.888 888 888 888 889	0
	0.555 555 555 555 556	$\pm 0.774 596 669 241 483$
4	0.652 145 154 862 546	$\pm 0.339 981 043 584 856$
	0.347 854 845 137 454	$\pm 0.861 136 311 594 053$

The most practical way to construct the coefficient matrix in an actual implementation is then by performing a loop over the elements $[x_{i-1}, x_i]$ instead of over the intervals $[x_{i-1}, x_{i+1}]$. This seems more complicated at first sight but it turns

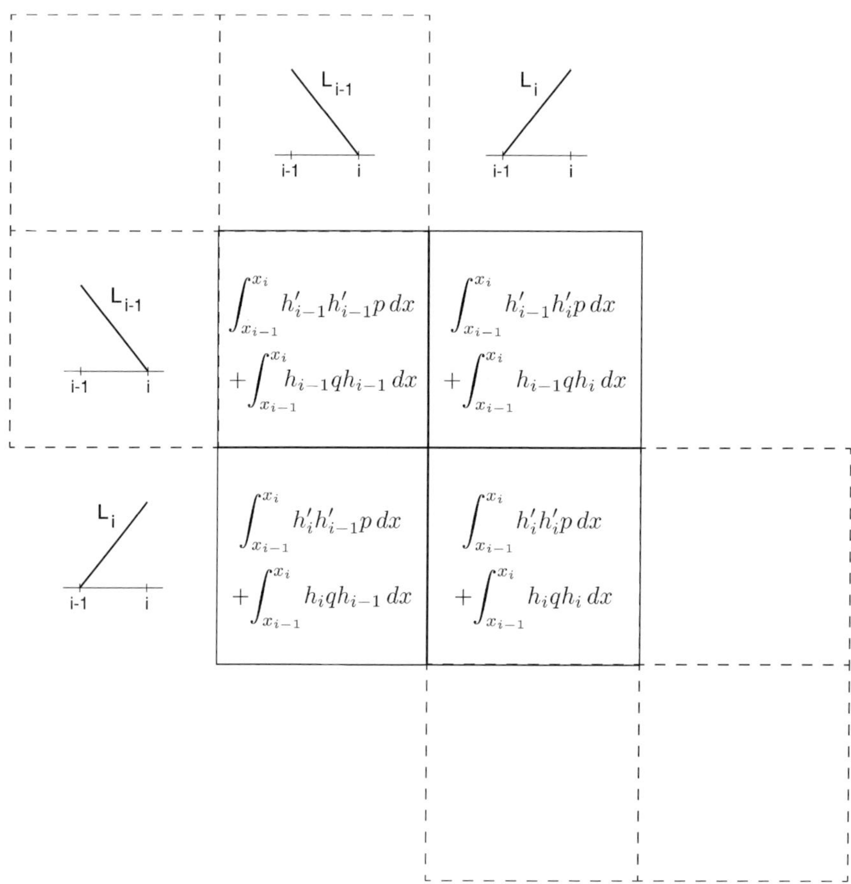

Fig. 15.3 The sub-matrix resulting from integrations over the element $[x_{i-1}, x_i]$ with linear finite elements. The position of the sub-matrices resulting from the neighboring elements $[x_{i-2}, x_{i-1}]$ and $[x_i, x_{i+1}]$ are indicated with dashed boxes. Notice that each diagonal entry gets contributions from two neighboring elements.

out to be the easiest way to program the coefficient matrix, simply because the integration can be done in *exactly the same way* (i.e. with the same subroutine) for all linear element shape functions which constitute the basis functions. In the interval $[x_{i-1}, x_i]$ *only* the linear basis functions h_{i-1} and h_i are non-zero. This yields four combinations ($h_{i-1} h_{i-1}$, $h_{i-1} h_i$, $h_i h_{i-1}$ and $h_i h_i$) that are all four computed in the same step (Fig. 15.3). The contribution of the neighboring elements is added in the next step of the loop. As a result, each row of the coefficient matrix, except for the first one and the last one, are computed in two subsequent steps.

Generalizations to higher order So far we have only discussed linear finite elements. With linear elements the FEM is equivalent to the FDM (see one of the

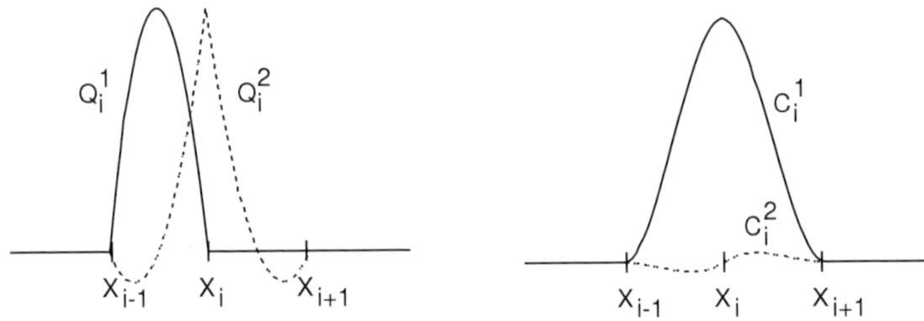

Fig. 15.4 Quadratic (left) and cubic "Hermite" (right) finite elements.

exercises above) so that there seem to be no advantages that justify the use of this more complex method. However, a local polynomial fit of a function $u(x)$ becomes more accurate when higher order polynomials are used. Hence, it is quite natural to consider higher order finite elements. Piecewise quadratic elements yield a better approximation. One can e.g. add the midpoint $x_{i-\frac{1}{2}}$ (or $\xi = 0$ on the standard interval $\xi \in [-1, 1]$) as an extra node internal to the element. Three quadratic element shape functions can be defined to attain a unit value in one node and zero at both other nodes, namely

$$N_1(\xi) = -\tfrac{1}{2}\xi(1-\xi), \quad N_2(\xi) = 1 - \xi^2, \quad N_3(\xi) = \tfrac{1}{2}\xi(1+\xi). \quad (15.36)$$

These can be combined into two basis functions for the quadratic finite elements:

$$Q_i^1(x) \equiv \begin{cases} \dfrac{4(x - x_{i-1})(x_i - x)}{(x_i - x_{i-1})^2} & \text{for } x_{i-1} \le x \le x_i, \\ 0 & \text{elsewhere,} \end{cases} \quad (15.37)$$

and

$$Q_i^2(x) \equiv \begin{cases} \dfrac{(2x - x_i - x_{i-1})(x - x_{i-1})}{(x_i - x_{i-1})^2} & \text{for } x_{i-1} \le x \le x_i, \\ \dfrac{(2x - x_{i+1} - x_i)(x - x_{i+1})}{(x_{i+1} - x_i)^2} & \text{for } x_i \le x \le x_{i+1}, \\ 0 & \text{elsewhere.} \end{cases} \quad (15.38)$$

The expansion used to approximate $u(x)$ with quadratic elements is written as

$$u(x) \approx \hat{u}(x) = \sum_{i=0}^{N} \left[Q_i^1(x)u_{i1} + Q_i^2(x)u_{i2} \right]. \quad (15.39)$$

Note that the expansion coefficients now correspond to local function values $u_i = u_{i2}$ and some non-trivial relation between u_{i1}, u_{i2} and the local approximations to derivatives u_i' (in fact, u_{i1} is the approximated function value at the midpoint $(x_{i-1} + x_i)/2$). The derivatives of these quadratic elements, however, are not continuous at the nodes (see the schematic representation of Fig. 15.4).

With two cubic elements per interval one can also make the derivatives continuous at the nodes and approximate both the original differential equation and its derivative. In this way a higher accuracy can be achieved (fourth-order for the cubic elements [424]) and derivatives of higher order can be approximated *without abandoning the local nature of the elements, so that the system matrix remains compact.* This should be compared with the higher order expressions for the FDM method, which typically require wider stencils and, hence, more computer storage. The approximation of $u(x)$ with cubic Hermite elements is written as

$$u(x) \approx \hat{u}(x) = \sum_{i=0}^{N} \left[C_i^1(x)u_{i1} + C_i^2(x)u_{i2} \right]. \tag{15.40}$$

The basis functions for cubic Hermite elements $C_i^1(x)$ and $C_i^2(x)$ are defined by

$$C_i^1(x) \equiv \begin{cases} \left(\dfrac{x - x_{i-1}}{x_i - x_{i-1}}\right)^2 \left(3 - 2\dfrac{x - x_{i-1}}{x_i - x_{i-1}}\right) & \text{for } x_{i-1} \leq x \leq x_i, \\ \left(\dfrac{x_{i+1} - x}{x_{i+1} - x_i}\right)^2 \left(3 - 2\dfrac{x_{i+1} - x}{x_{i+1} - x_i}\right) & \text{for } x_i \leq x \leq x_{i+1}, \\ 0 & \text{elsewhere,} \end{cases} \tag{15.41}$$

and

$$C_i^2(x) \equiv \begin{cases} (x - x_i)\left(\dfrac{x - x_{i-1}}{x_i - x_{i-1}}\right)^2 & \text{for } x_{i-1} \leq x \leq x_i, \\ (x - x_i)\left(\dfrac{x_{i+1} - x}{x_{i+1} - x_i}\right)^2 & \text{for } x_i \leq x \leq x_{i+1}, \\ 0 & \text{elsewhere.} \end{cases} \tag{15.42}$$

Now, the expansion coefficients in (15.40) directly correspond to local function values $u_i = u_{i1}$ and derivatives $u_i' = u_{i2}$. These Hermite cubic elements are built up from the four shape functions

$$N_1(\xi) = \tfrac{1}{4}(1 - \xi)^2(2 + \xi), \qquad N_2(\xi) = \tfrac{1}{4}(1 - \xi)^2(1 + \xi),$$

$$N_3(\xi) = \tfrac{1}{4}(1 + \xi)^2(2 - \xi), \qquad N_4(\xi) = -\tfrac{1}{4}(1 + \xi)^2(1 - \xi). \tag{15.43}$$

The four degrees of freedom in a cubic formulation have then been fixed by requiring a double zero at one node, and either a unit value and zero derivative *or* a zero value and unit derivative at the other node. The four shape functions combine to the two basis functions C_i^1 and C_i^2 per node and yield 4×4 combinations in each element $[x_{i-1}, x_i]$ so that the "stiffness" matrix gets a block–tridiagonal structure with sub-blocks of size 2×2 (see Fig. 15.5). Thus, with this better approximation for our 1D model problem, we obtain a system of $2(N + 1)$ equations with a "sparse" banded coefficient matrix: only the diagonal and the first three upper and lower diagonals contain non-zero entries.

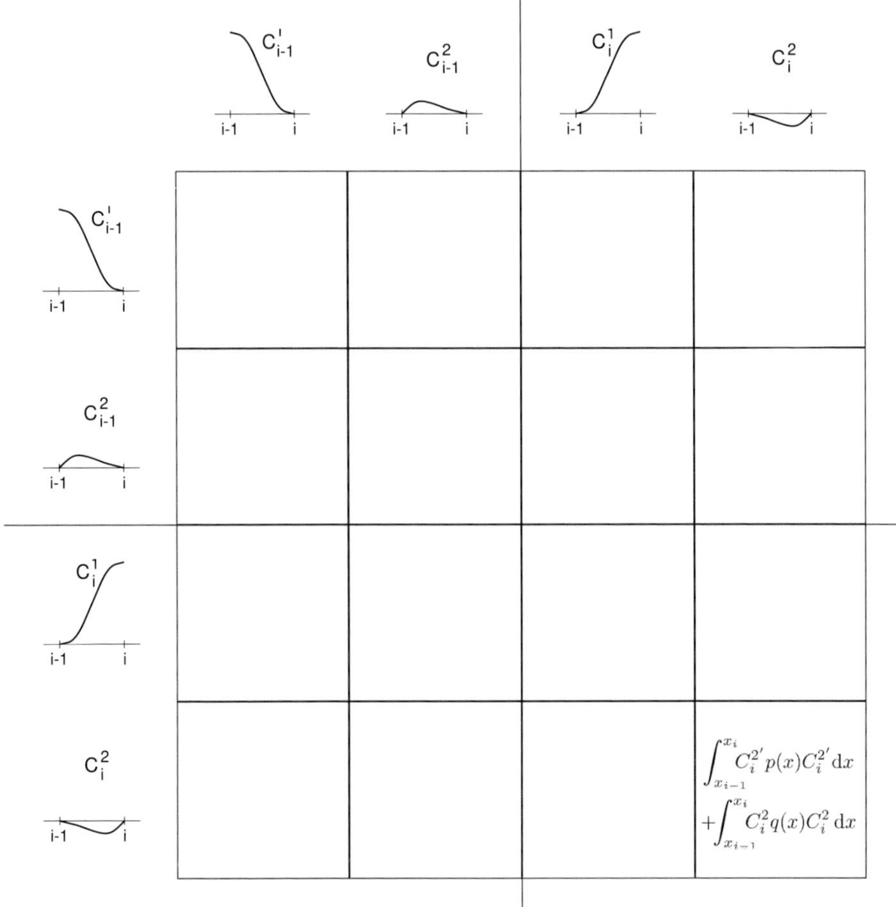

Fig. 15.5 The sub-matrix resulting from the integrations over the subinterval $[x_{i-1}, x_i]$ with cubic "Hermite" finite elements.

Multi-dimensional techniques and MHD applications The finite element technique can be generalized to more than one spatial dimension. In two dimensions, linear elements become "pyramids" (generalized "tent" functions) e.g. as shown schematically on a triangular grid in Fig. 15.6. Moreover, the "elements" can have curved sides which yields a more accurate representation of domains with curved boundaries (cf. Section 16.3.3 where two-dimensional equilibria determined with isoparametric mapping are discussed). Therefore, the finite element method is very popular as it combines high accuracy with high flexibility. As a matter of fact, the superiority of the FEM with respect to the FDM becomes more apparent the more irregular the grid is and the more curved the boundaries are. In MHD, where we have to deal with systems of equations and both scalar and vector quantities, it be-

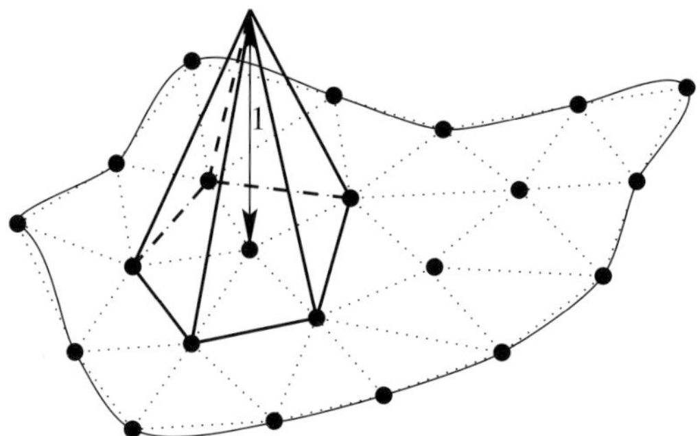

Fig. 15.6 Example of a two-dimensional linear finite element ("tent function") on an irregular triangular grid as discretization of an irregular domain.

comes possible to use a different kind (linear, quadratic, ...) of element for each unknown function. As we will see in Section 15.1.5, this hybrid choice will be necessary for the three components of vector quantities, in order to avoid "spectral pollution" in the computations of linear eigenmodes [9].

Finite elements are typically used for discretization in the direction normal to the flux surfaces in many linear and nonlinear MHD codes. For 1D slab or cylindrical equilibrium configurations, the spatial dependence of the linear perturbations about the equilibrium state in the symmetry directions is handled trivially by selecting one Fourier mode pair at the time. The non-trivial normal direction is treated with the FEM method in 1D MHD spectral codes such as LEDA (large-scale eigenvalue solver for the dissipative Alfvén spectrum) [274] for stability analysis of 1D slab or cylindrical plasma models, or its extension LEDAFLOW to stationary 1D equilibria with external gravity, suitable for MHD spectroscopic studies of stratified atmospheres, astrophysical jets or 1D accretion disk models [353]. Also, the same technique returns for diagnosing the linear dynamics of 2D MHD equilibria, with translational or axi-symmetric invariance. Linear MHD codes in this category that exploit FEM for treating the direction normal to the flux surfaces are e.g. POLLUX (program on line-tied loops under excitation) [205] for the study of line-tied coronal loop configurations; MARS (magnetohydrodynamic resistive spectrum) [313], NOVA (non-variational code) [88] and CASTOR (complex Alfvén spectrum of toroidal plasmas) [272] for both ideal and non-ideal MHD spectral analysis of tokamak plasmas. Examples of earlier tokamak spectral codes based on the ideal MHD variational formulation are ERATO [198] and PEST (Princeton equilibrium, stability and transport package) [194]. The FEM method is also heavily used in nonlinear MHD computations, in particular to accurately compute the 2D MHD equilibria themselves. Examples include tokamak equilibrium codes like CHEASE (cubic Hermite element axi-symmetric static equilibrium) [320, 321] and HELENA [238], or more general axi-symmetric stationary MHD equilibrium solvers such as FINESSE (finite element solver for stationary equilibria) [29]. In these MHD equilibrium solvers, one typically uses 2D finite elements for handling both the normal and poloidal direction.

Comparing finite differences and finite elements The finite difference and finite element methods are compared in Table 15.2 and Fig. 15.7. The details of this test case are explained in the Exercise section at the end of this chapter and, in particular, in Exercise 15.9. It involves the solution of the problem

$$- \left(1 + x^2\right) u''(x) - 2x\, u'(x) + 2\, u(x) = f(x) \tag{15.44}$$

on the interval $[0, 1]$ with boundary conditions $u(0) = u(1) = 0$. For the case shown, the true solution is $u(x) = x^4(1 - x)^4$, which actually then defines the function $f(x)$, as explained in the Exercises.

Table 15.2 *Infinity norm* $\|\mathbf{u}_{\text{computed}} - \mathbf{u}_{\text{true}}\|$, *where* $\mathbf{u}_{\text{true}}$ *is the vector of true values at the* $M - 1$ *mesh points of the error at the mesh points for various methods and numbers* M *of interior mesh points.*

Method	$M = 9$	$M = 99$	$M = 999$
first order finite difference	2.5108×10^{-4}	1.5125×10^{-5}	1.4149×10^{-6}
second order finite difference	1.0269×10^{-4}	1.0702×10^{-6}	1.0707×10^{-8}
linear finite elements	1.0384×10^{-5}	1.0265×10^{-7}	1.0265×10^{-9}
quadratic finite elements	7.0052×10^{-6}	1.6224×10^{-9}	1.9604×10^{-13}

Note the observed convergence rate r as we increase the resolution: if the error drops by a factor of 10^r, when the number of grid points is increased by a factor of 10, the observed convergence rate is r. From Table 15.2, it is clear that the convergence rates for this smooth test function are close to the theoretical convergence rates $r = 1$ and $r = 2$ for the first and second order finite difference implementations, respectively, and for the finite element implementation based on linear finite elements (for which $r = 2$). Notice that the quadratic finite element approximation (which actually exploits a combination of linear and quadratic elements) has $r = 4$, which is better than the $r = 3$ we might expect. This is called super-convergence and happens because we only measured the error at the mesh points, whereas the $r = 3$ result was for the average value of the error over the entire interval.

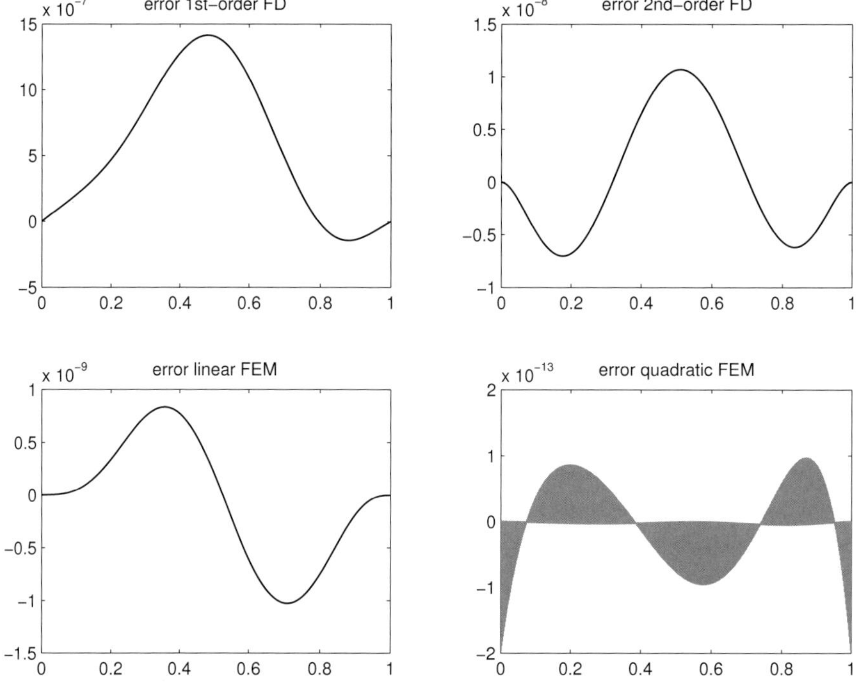

Fig. 15.7 The error for various methods and for 999 interior mesh points.

15.1.4 Spectral methods

Since the FDM essentially works with local function values and the FEM with local polynomial representations, both the FDM and the FEM are very useful for approaching very localized solutions. This is of great practical value for MHD applications, and one can optimally benefit from this in combination with irregular grids, using accumulation of the mesh points in the region(s) where the solution is expected to have large gradients. Often, however, the solution is not localized at all in one or more spatial directions. Also, the solution may be periodic in one or more spatial directions, e.g. in the toroidal and poloidal directions in a tokamak. In such cases the *spectral method* provides a valuable alternative. The spectral method is in some sense very similar to the finite element method. The important difference is that the shape functions used to approximate the solution are now "global" functions. These are functions that are non-zero on the whole domain, i.e. the same domain as the differential operator itself. There are several possibilities, and the most convenient choices involve "orthonormal" functions, such as Legendre and Chebyshev polynomials on $[-1, +1]$ and $\{\sin nx\}_{n=1}^{\infty}$ on the interval $[0, \pi]$.

Fourier harmonics are the most well-known choice for the global expansion functions. They are used for linear MHD problems with periodic boundary conditions. Every truncation of the Fourier series of the solution yields an approximation. Hence, for a function $u(x)$ on a finite domain $[0, L]$, we may write

$$u(x) \approx \hat{u}(x) = \sum_{k=0}^{N-1} \hat{u}_k e^{\mathrm{i}\,k2\pi x/L}. \tag{15.45}$$

Since every term used in the RHS expansion is periodic, $\hat{u}(x = 0) = \hat{u}(x = L)$. In the case when the exact solution does not obey this relation (e.g. in our model problem (15.2) when $\beta \neq 0$ or $F \neq 0$ in the RHS boundary condition (15.4)), other global functions should be used, or, when the equation is linear, one can split the problem and first solve the homogeneous equation with inhomogeneous BCs and then the inhomogeneous equation with homogeneous BCs. The same periodicity is then typically implied for any function appearing in the equation to be solved, like the functions $p(x)$ and $q(x)$ in Eq. (15.2). The procedure to obtain a linear system for the N complex-valued expansion coefficients $\hat{u}_k$ closely follows the procedure explained for the FEMs. In the Fourier representation, derivatives with respect to x turn into multiplications with factors $\mathrm{i}\,k\,2\pi/L$, as a result of differentiating the basis functions $e^{\mathrm{i}\,k2\pi x/L}$ and replacing the equations by their projections onto the basis functions (as in the Galerkin method and the weak form). In the complex notation used in (15.45), one has to multiply with the complex conjugates of the shape functions, i.e. $e^{-\mathrm{i}\,l2\pi x/L}$, and then use the orthogonality relation

$$\frac{1}{L} \int_0^L e^{\mathrm{i}\,(k-l)2\pi x/L}\, dx = \delta_{kl} \tag{15.46}$$

to obtain the weak Galerkin form of the equations.

For our model problem (15.2), the additional functions like q are then written as $q(x) = \sum_{m=0}^{N-1} \hat{q}_m \exp(\mathrm{i}m2\pi x/L)$, and likewise for p and f. This in turn yields contributions of the corresponding terms to the kth row of the coefficient matrix, e.g. the second LHS term in Eq. (15.2):

$$\text{row } k \Rightarrow \frac{1}{L} \int_0^L e^{-\mathrm{i}k2\pi x/L} \sum_{m=0}^{N-1} \sum_{l=0}^{N-1} \hat{q}_m \hat{u}_l e^{\mathrm{i}\,(m+l)2\pi x/L}\, dx = \sum_{l=0}^{N-1} \hat{q}_{k-l} \hat{u}_l, \tag{15.47}$$

due to the orthogonality relation (15.46) of the Fourier modes, making the integral vanish except when $m = k - l$. Notice that, due to the x-dependence of the coefficient function $q(x)$, there still is a summation in Eq. (15.47), which means that all the $\hat{u}_l$-components couple. This summation is the discrete convolution of both Fourier series, and its presence makes the coefficient matrix resulting from the weak formulation a *full* matrix, in contrast to the sparse coefficient matrices in the FDM and FEM.

For periodic functions with all their derivatives periodic as well, the truncated Fourier series can be shown to yield exponential convergence, i.e. convergence faster than any algebraic power (N^{-p}). This so-called spectral or infinite-order accuracy makes the spectral approach the method of choice for periodic problems. Its advantage lies in its tendency to make minimal or no phase errors. For non-linear incompressible flow simulations focusing on transitions to turbulent flows it is unrivaled in accuracy.

Fourier expansion coefficients and local function values   In the expansion (15.45), it seems that we now have N complex valued unknown Fourier coefficients, while in the discussion of the FDM and the FEM we always considered only N real-valued degrees of freedom. At the same time, the expansion is meant to approximate the real function $u(x)$ on the finite domain $[0, L]$. The relation between the Fourier coefficients and local function values u_j on an equidistant grid $x_j = j\Delta x$ for $j = 0, \ldots, N$ with fixed grid spacing $\Delta x = L/N$ (as shown in Fig. 15.1 for $L = 1$) is then as follows. First, the periodicity implies $u_0 = u_N$, so that we have only N possibly different function values u_j to determine. Introducing the Nth root of unity $w = \mathrm{e}^{\mathrm{i}\,2\pi/N}$, local function values u_j form the discrete Fourier transform (DFT) of the Fourier coefficients $u_j = \sum_{k=0}^{N-1} \hat{u}_k w^{jk}$. Reversely, the expansion coefficients themselves are the inverse DFT of the local function values:

$$\hat{u}_k = \frac{1}{N} \sum_{j=0}^{N-1} u_j \left(w^{-1}\right)^{jk} . \tag{15.48}$$

Since we also have $w^m = w^{m+nN}$ for all (positive or negative) integer values m, n, several symmetry properties between the N complex Fourier components can be obtained from the knowledge that the function $u(x)$ we are approximating is real, so that $u_j = u_j^*$. Indeed, it is straightforward to deduce, using the orthogonality relation (15.46), that

$$\hat{u}_0 = \hat{u}_0^* = \frac{1}{N} \sum_{j=0}^{N-1} u_j , \tag{15.49}$$

relating the first (real) Fourier coefficient to the arithmetic mean of the local function values. Similarly, one finds a symmetry about $N/2$ as

$$\hat{u}_k = \hat{u}_{N-k}^* , \tag{15.50}$$

meaning that indeed only N real numbers are to be determined, consistent with the equivalent number of real local function values u_j, with $j = 0, \ldots, N-1$. Finally, the same observation is true between positive and negative indices, namely

$$\hat{u}_k = \hat{u}_{-k}^* , \tag{15.51}$$

which is the reason for restricting the sum to positive indices only in (15.45). Note that these relations also play a role for evaluating the discrete convolution term in (15.47), which we can write as

$$\sum_{l=0}^{N-1} \hat{f}_{k-l}\hat{u}_l \;\Rightarrow\; \sum_{l=0}^{k} \hat{f}_{k-l}\hat{u}_l + \sum_{l=k+1}^{N-1} \hat{f}^*_{l-k}\hat{u}_l \,. \tag{15.52}$$

Finally, once the Fourier coefficients are determined as solutions of the resulting linear system, the local function values u_j are most efficiently computed using a fast Fourier transform algorithm, which computes the N DFTs in order $N \log_2 N$ operations. This relies on the following split in even and odd terms of the DFT formula $u_j = \sum_{k=0}^{N-1} \hat{u}_k \mathrm{e}^{\mathrm{i}\,2\pi\, jk/N}$ for the special case where $N = 2^s$:

$$u_j = \sum_{m=0}^{N/2-1} \hat{u}_{2m}\mathrm{e}^{\mathrm{i}\,2\pi\, jm/(N/2)} + \mathrm{e}^{\mathrm{i}\,2\pi\, j/N}\sum_{m=0}^{N/2-1} \hat{u}_{2m+1}\mathrm{e}^{\mathrm{i}\,2\pi\, jm/(N/2)} \,. \tag{15.53}$$

Each occurring sum is itself recognized as a DFT, yielding a recursive algorithm of order $N \log_2 N$.

The periodicity of the Fourier modes can be a disadvantage when the solution is *not periodic*. The obtained approximation is then poor at the boundaries, where "Gibbs" phenomena occur: the Fourier series of a piecewise continuously differentiable periodic function behaves peculiarly as a result of the fact that the nth partial sum of the Fourier series has large oscillations near the jump, and often the maximum of the partial sum is higher than that of the function itself (cf. Section 19.1.3). For non-periodic problems on a finite domain $[-1, 1]$, one can take the orthonormal *Legendre polynomials* as global expansion functions. These Legendre polynomials are given by

$$P_n(x) = \frac{1}{2^n}\sum_{l=0}^{n/2}(-1)^l \frac{(2n-2l)!}{l!\,(n-l)!\,(n-2l)!}x^{n-2l}\,, \quad n = 0, 1, 2, \ldots \tag{15.54}$$

These functions obey the orthogonality relation

$$\int_{-1}^{1} P_m(x)\,P_n(x)\,dx = \frac{2}{2n+1}\delta_{mn}\,, \tag{15.55}$$

and any function $u(x)$ on the domain $[-1, 1]$ can be approximated by truncating its Legendre series given by

$$u(x) = \sum_{k=0}^{\infty} P_k(x)\frac{2k+1}{2}\int_{-1}^{1} u(x')\,P_k(x')\,dx' \equiv \sum_{k=0}^{\infty} \tilde{u}_k P_k(x)\,. \tag{15.56}$$

For any arbitrary continuous function $u(x)$ (or even quadratically integrable u, under additional assumptions, see e.g. Wang and Guo [473]), the completeness of the set of Legendre polynomials guarantees that this series converges to $u(x)$ for any point $x \in [-1, 1]$. As for the Fourier series, equations for the expansion coefficients $\tilde{u}_k$ for any finite approximation $\hat{u}(x) = \sum_{k=0}^{N-1} \tilde{u}_k P_k(x)$ are derived from the weak form, using Eq. (15.55).

In a completely analogous fashion, one could use *Chebyshev polynomials* instead, which are defined as

$$
\begin{cases}
T_0(x) = 1 , \\
T_n(x) = \frac{n}{2} \sum_{l=0}^{n/2} (-1)^l \frac{(n-l-1)!}{l!\,(n-2l)!} (2x)^{n-2l} , \quad n = 1, 2, \ldots,
\end{cases}
\tag{15.57}
$$

with corresponding orthogonality relations (involving a weight function):

$$
\int_{-1}^{1} T_m(x)\, T_n(x)\, \frac{1}{\sqrt{1-x^2}}\, dx =
\begin{cases}
0 & n \neq m , \\
\pi/2 & n = m > 0 , \\
\pi & n = m = 0 .
\end{cases}
\tag{15.58}
$$

These Chebyshev polynomials obey the recursion formula $T_{n+1}(x) = 2x T_n(x) - T_{n-1}(x)$ for $n \geq 1$, with $T_0(x) = 1$ and $T_1(x) = x$. These polynomials are, e.g., used in the SPECLS (spectral compressible linear stability) code for linear HD and MHD stability computations. This Chebyshev collocation code was first developed for compressible hydrodynamic linear eigenvalue problems for one-dimensional stationary planar shear flows by Macaraeg *et al.* [322], and subsequently extended to the compressible MHD case by Dahlburg and Einaudi [102].

Non-Galerkin spectral approaches In the description of the spectral methods, we only mentioned the Galerkin approach where the weight functions in the weighted residual (15.30) are taken identical to the global expansion functions. In this spectral Galerkin approach, essential boundary conditions exclude those expansion functions not obeying the boundary conditions from the set of expansion functions used, similar to what we discussed for dealing with essential boundary conditions in the FEM Galerkin method. There are two other popular variants of the spectral methods, namely the *collocation approach* and the *tau approach*. The spectral collocation technique uses as weight functions in (15.30) the delta function $\delta(x - x_i)$ in a suitably chosen set of collocation points x_i. Hence, the collocation approach uses again local function values, and the global expansion functions are used only to evaluate local derivatives. In the spectral tau method, one can handle e.g. non-periodic boundary conditions too, viz. by allowing also weight functions which do not individually satisfy the boundary conditions. In essence, the boundary conditions themselves then need to be expressed in a weighted residual form as well. A thorough discussion of spectral methods is the textbook by Canuto *et al.* [75].

15.1.5 Mixed representations

Different discretization techniques are often combined in MHD calculations, as we need to deal with up to three spatial dimensions and eight scalar fields (two vectors and two scalars). If we consider linear perturbations about an MHD equilibrium, the equilibrium introduces preferred directions inside and normal to its magnetic flux surfaces. The direction normal to the magnetic flux surfaces is then usually discretized by a finite difference or a finite element method because the solutions are often localized due to singular or nearly singular behavior in this direction. In addition, these two methods have the advantage that they allow mesh accumulation in boundary layers or other regions where the solutions vary rapidly, e.g. the resonant layers created by resonant heating (cf. Chapter 11 [1]). On the other hand, for the two periodic coordinates in a tokamak, spectral methods are widely used. Below, we make some additional observations about certain discretization combinations for linear MHD computations. In particular, we discuss some exemplary combinations of FDM and FEM discretizations with (pseudo-)spectral methods and their consequences for the discrete spatial representations used for the three components of the occurring vector quantities (velocity and magnetic field).

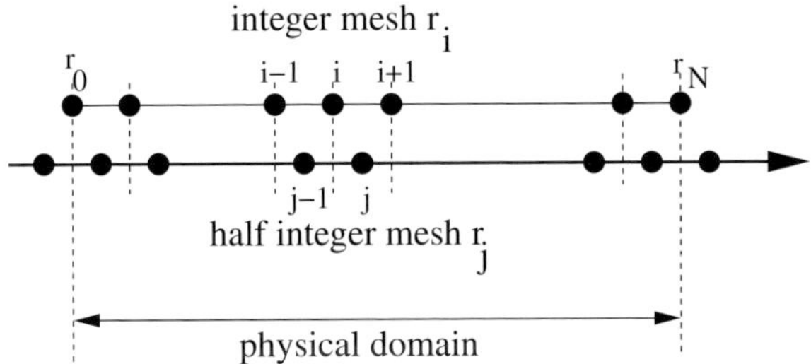

Fig. 15.8 An example of a staggered mesh, consisting of an equidistant integer mesh r_i, and a half-integer mesh $r_j = r_{i+\frac{1}{2}}$.

Assuming for simplicity a cylindrical coordinate system for a generic periodic tubular flux configuration, Fourier representations in the directions about and along the "loop" or "flux rope" axis can be exploited for all quantities. The radial dependence may then be handled by a finite difference approach. As explained in what follows, solenoidal magnetic fields can be ensured on the numerical level by exploiting a *staggered mesh*, essentially using different grid point locations for different vector components. In the simplest case of an equidistant base grid $r_i = i\Delta r$ for $i = 0, 1, \ldots, N$, where N intervals divide the loop radius $r_N = N\Delta r$, one uses as an additional radial grid the half-integer locations $r_j \equiv r_{i+\frac{1}{2}} = (r_i + r_{i+1})/2$, as shown in Fig. 15.8. The main advantage is that the first order forward difference formula for u_i' given by Eq. (15.16) can be interpreted as a second order central difference formula for $u'_j = u'_{i+\frac{1}{2}}$. This second order formula now involves only u_i and u_{i+1}. To handle boundary conditions on this half-integer mesh, one extends this mesh with *ghost cells* appearing *outside* the domain $[0, r_N]$ (corresponding to $r_{-\frac{1}{2}}$ and $r_{N+\frac{1}{2}}$). For the three components of the magnetic field vector, one can then choose to represent the radial field component B_r on the integer mesh, while the B_θ and B_z components are linked to the half-integer mesh. In combination with the Fourier representation used in their (θ, z) dependence, the following discrete formula for $\nabla \cdot \mathbf{B} = 0$ is obtained for each Fourier mode pair (m, n) from $\exp[\mathrm{i}\,(m\theta + 2\pi n/L)]$ (where L is the loop length) on the half-integer mesh:

$$(\nabla \cdot \hat{\mathbf{B}})_j = 0 = \frac{1}{r_j \Delta r}\left(r_{i+1}\hat{B}_{r,i+1} - r_i \hat{B}_{r,i}\right) + \frac{\mathrm{i}m}{r_j}\hat{B}_{\theta,j} + \frac{\mathrm{i}2\pi n}{L}\hat{B}_{z,j}\,. \quad (15.59)$$

This can be seen as a bi-diagonal system for the Fourier coefficients $\hat{B}_{r,i}$, once $\hat{B}_{\theta,j}$ and $\hat{B}_{z,j}$ are known. This procedure to determine the radial magnetic field component can then replace the discretized radial component of the induction equation. Additionally, one can guarantee the analytical identity $\nabla \cdot (\nabla \times \mathbf{B}) = 0$ in its discrete equivalent by taking

$$(\nabla \times \hat{\mathbf{B}})_{r,i} = \frac{\mathrm{i}m}{2r_i}(\hat{B}_{z,j} + \hat{B}_{z,j-1}) - \frac{\mathrm{i}\pi n}{L}(\hat{B}_{\theta,j} + \hat{B}_{\theta,j-1})\,,$$

$$(\nabla \times \hat{\mathbf{B}})_{\theta,j} = \frac{\mathrm{i}\pi n}{L}(\hat{B}_{r,i+1} + \hat{B}_{r,i}) - \frac{1}{2\Delta r}(\hat{B}_{z,j+1} - \hat{B}_{z,j-1})\,,$$

$$(\nabla \times \hat{\mathbf{B}})_{z,j} = \frac{1}{2r_j \Delta r}\left[r_{i+1}(\hat{B}_{\theta,j+1} + \hat{B}_{\theta,j}) - r_i(\hat{B}_{\theta,j} + \hat{B}_{\theta,j-1})\right]$$
$$- \frac{\mathrm{i}m}{2r_j}(\hat{B}_{r,i+1} + \hat{B}_{r,i})\,. \quad (15.60)$$

This is obviously desirable physically as well, and boils down to using simple second order linear interpolation between the two staggered grids.

Another, somewhat related, issue is relevant for all MHD eigenvalue computations. For detailed numerical diagnosis of the stability of a particular MHD equilibrium, it is of course desirable that the numerical eigenvalues converge to correct physical eigenvalues when the exploited number of grid points is increased. However, both finite difference and finite element discretizations of eigenvalue problems may exhibit "spectral pollution", where unphysical, "polluting", eigenvalues belong to either strongly distorted, or simply spuriously introduced branches of the discrete equivalent of the dispersion relation. Briefly put, avoiding spectral pollution entails the physical requirement that the chosen discretization for the eigenfunctions should be able to satisfy constraints like

$$\nabla \cdot \mathbf{v} = 0\,, \quad (15.61)$$

$$\nabla \cdot \mathbf{B} = 0\,, \quad (15.62)$$

in every point, i.e. also in the entire continuous interval between subsequent mesh points. In the case of a FE representation, it is easily seen how this necessitates the use of elements of a different order for the three different velocity components (and similarly for the magnetic vector field). Suppose a cylindrical coordinate system is used and the radial variation is represented by FE while Fourier modes handle the periodic directions. The radial derivative on the radial velocity in the constraint (15.61) forces the use of finite elements for this component of one order higher than for the other velocity components. This combination is then termed a "conforming" FE discretization. Of course, the MHD equations do not impose these constraints, but some eigenmodes "choose" to satisfy them. Hence, to get uniform convergence of the entire computed spectrum, i.e. for every eigenvalue, one should be able to satisfy constraints like Eqs. (15.61) and (15.62). An analysis of "spectral pollution" occurring in simple model eigenvalue problems for both FD and FE discretizations can be found in Ref. [314], where remedies involve the use of staggered representations for FD schemes, or the use of a "finite hybrid element method" [196]. The latter involves the choice of different order FE representations for vector components and their derivatives as occurring in the corresponding weak forms. The precise mixture of FE orders must be such that each term in the weak form can have the same functional dependence, e.g. mixing piecewise linear and piecewise constant elements to get an overall piecewise constant dependence. More on "spectral pollution" can be found in the papers [9, 390], and an extensive discussion is given in the book [197], treating finite element methods as used in linear MHD spectral solvers (in particular, ERATO).

For linear MHD eigenvalue computations, ideal spectral solvers can exploit the variational formulation in terms of the Lagrangian displacement vector $\boldsymbol{\xi}$. Avoiding spectral pollution in FE methodology then entails the use of different order elements for its components to ensure one can discretely obey $\nabla \cdot \boldsymbol{\xi} = 0$. Examples of such codes for tokamak spectroscopic studies are PEST and ERATO. The PEST code Fourier analyzes the poloidal angle variation, and uses a conforming linear and piecewise constant FE mixture in the direction normal to the flux surfaces. ERATO uses 2D hybrid FE discretizations in the poloidal cross-section, and exploits the finite hybrid element approach again mixing constant and linear finite shape functions. Also in cases where the three components of the linearized velocity field and magnetic field vector (or the vector potential) are handled with a FEM representation (in at least one spatial direction), the use of mixed FEM representations has become a standard practice. For example, the MARS code uses Fourier modes in the poloidal direction and mixes constant and linear finite elements in the radial flux coordinate. CASTOR again handles the poloidal angle in Fourier representation, and uses conforming quadratic and cubic FE representations in the flux coordinate direction.

15.2 Linear MHD: boundary value problems

We now specify the discussion to the linearized MHD equations. We will explain how a choice of spatial discretization for all occurring variables, together with an assumed time-dependence, turns the mathematical problem into a boundary value problem. First, steady state calculations of externally driven dissipative plasmas are discussed. Next, the corresponding eigenvalue problems are solved as an alternative and complementary approach. In both cases, a linear system of equations containing a large number of unknowns is obtained. Solution strategies for this linear algebraic problem are discussed separately in Section 15.3. A third approach, viz. the determination of the full time-accurate solution of the initial value problem, will be treated in Section 15.4 as this involves also the discretization of time.

15.2.1 Linearized MHD equations

Linearized MHD studies the dynamic response of a plasma, initially in equilibrium, to "small" perturbations. The equations are linearized around the equilibrium, which is usually assumed to be static. This latter assumption is not essential for the numerical methods discussed, and we have encountered examples of linear MHD phenomena for stationary equilibria as well. For the static equilibrium case, when we also include the effect of finite resistivity on the linear response, the linearized resistive MHD equations can be written in the following dimensionless form, where the indices 0 refer to equilibrium quantities and the indices 1 refer to perturbed quantities:

$$\frac{\partial \rho_1}{\partial t} = -\nabla \cdot (\rho_0 \mathbf{v}_1), \tag{15.63}$$

$$\rho_0 \frac{\partial \mathbf{v}_1}{\partial t} = -\nabla(\rho_0 T_1 + \rho_1 T_0) + (\nabla \times \mathbf{B}_0) \times (\nabla \times \mathbf{A}_1)$$
$$- \mathbf{B}_0 \times (\nabla \times \nabla \times \mathbf{A}_1), \tag{15.64}$$

$$\rho_0 \frac{\partial T_1}{\partial t} = -\rho_0 \mathbf{v}_1 \cdot \nabla T_0 - (\gamma - 1)\rho_0 T_0 \nabla \cdot \mathbf{v}_1$$
$$+ 2\eta(\gamma - 1)(\nabla \times \mathbf{B}_0) \cdot (\nabla \times \nabla \times \mathbf{A}_1), \tag{15.65}$$

$$\frac{\partial \mathbf{A}_1}{\partial t} = -\mathbf{B}_0 \times \mathbf{v}_1 - \eta \nabla \times \nabla \times \mathbf{A}_1. \tag{15.66}$$

The temperature T is used instead of the pressure $p = \rho T$, and the vector potential $\mathbf{A}_1$ has been used instead of $\mathbf{B}_1$ itself: $\mathbf{B}_1 = \nabla \times \mathbf{A}_1$. As a result, the magnetic field is guaranteed to be "divergence-free", i.e. $\nabla \cdot \mathbf{B}_1$ is always zero.[1] Gravity

---

[1] While $\nabla \cdot \nabla \times \mathbf{A}_1 = 0$ is obviously true analytically, in a numerical approach it requires that also the discrete equivalent of the divergence and curl operators behave in this fashion! Remark further that the vector potential

is ignored for now, as it does not influence the numerics significantly (at least for external gravitational fields). Dissipative effects, however, do affect the choice of discretization method for both space and time dependencies. Electrical resistivity is taken into account to illustrate how. In one-dimensional configurations, i.e. configurations in which the equilibrium quantities depend on one spatial coordinate only, the equilibrium force balance $\nabla p_0 = \mathbf{j}_0 \times \mathbf{B}_0$ corresponds to an ordinary differential equation which can be solved easily. We have seen examples of this in Chapters 7 and 9 [1] and here we will assume that the equilibrium is known.

In three spatial dimensions the system (15.63)–(15.66) consists of eight partial differential equations for eight unknowns ($\mathbf{v}_1$ and $\mathbf{A}_1$ each have three components). Interchanging LHS and RHS, the system (15.63)–(15.66) can be written symbolically in the form

$$\mathbf{L}\mathbf{u} = \mathbf{R}\frac{\partial \mathbf{u}}{\partial t}, \tag{15.67}$$

when we introduce the state vector $\mathbf{u}$. In cylindrical coordinates, this vector can be taken simply as

$$\mathbf{u}^{\mathrm{T}} = (\rho_1,\ v_{1r},\ v_{1\theta},\ v_{1z},\ T_1,\ A_{1r},\ A_{1\theta},\ A_{1z}). \tag{15.68}$$

However, other possibilities for $\mathbf{u}$ may e.g. exploit projections of the velocity perturbation on the three orthogonal directions of the local field line triad. In any case, the operators $\mathbf{L}$ and $\mathbf{R}$ contain equilibrium quantities and spatial differential operators. With the state vector as in Eq. (15.68), the $\mathbf{R}$ operator is diagonal and can be read off from the LHS of the system (15.63)–(15.66). The derivative with respect to time is written explicitly in the form (15.67) because it plays a *key role* in the distinction between *three types of problem*. This relates to the three possibilities of having (i) a prescribed time-dependence (as e.g. in *steady, externally driven* problems); (ii) an exponential or oscillatory time-dependence, with growth rates or frequencies to be computed as physically realizable eigenfrequencies for the given equilibrium (i.e. *eigenvalue* problems); or (iii) an unknown time-dependence, to be determined along with the spatial dependence of the solution. In many cases, these three possibilities correspond to *three different numerical approaches to the same problem* which yield *complementary information* (see also Section 6.3.1 [1], but notice that we discuss them here in different order). Since cases (i) and (ii) can be interpreted as boundary value problems, where in essence only the spatial dependence of linear perturbations is left to determine, we will discuss them first in the following sections.

$\mathbf{A}_1$ is only determined up to an arbitrary gauge $\nabla\Phi$. This freedom can be used to set the potential Φ (or a specific component of $\mathbf{A}_1$) equal to zero.

15.2.2 *Steady solutions to linearly driven problems*

Many dynamical systems evolve to a steady state and one is often not interested in the temporal evolution but just in the eventual steady state itself. For parameter studies of the efficiency of the plasma–driver coupling, time dependent computer simulations are much too CPU intensive. The steady-state approach yields the stationary state at once ("in one time step" so to speak). This makes very extensive parameter studies possible, at the price of skipping almost all information on the time scales of the heating process. We already encountered an analytically tractable example in Section 11.1 [1] of a periodically driven dissipative system where the steady-state was of particular interest. There we analyzed a semi-infinite plasma slab adjacent to a vacuum in which an external antenna (surface current) induces periodic perturbations which are resonantly absorbed. We assumed that the system had evolved to a steady state in which all physical quantities oscillate harmonically with the frequency ω_d imposed by the external source. Representative numerical results of the steady-state approach have been encountered previously as well: steady-state quantifications of solar p-mode absorption by sunspots taken from [190] were given in Figs. 11.15 and 11.16 [1]. The parameter scans shown in those figures were performed numerically, using the generic approach explained below. An example is the parametric study of the efficiency of the plasma–driver coupling in MHD wave heating schemes for solar coronal loops by means of resonant absorption. In the same spirit, impedance scans of the response of a tokamak plasma to an external "antenna" current can be performed in a systematic fashion. In the context of MHD spectroscopy for laboratory plasmas, such computer simulated impedance scans must be compared to measured ones, in turn yielding information on the internal profiles of the equilibrium quantities.

In computational linear MHD, the steady state of a problem involving a driving frequency ω_d can easily be determined by assuming it has been reached already. In practice this is done by imposing the time behavior $\sim \exp(-i\omega_d t)$ in the equations, i.e. by *replacing the time derivative by a multiplication with* $-i\omega_d$. After discretization of all spatial dependencies of all eight components of the state vector $\mathbf{u}$ in Eq. (15.68), using any of the discrete representations introduced in Section 15.1, we need to apply the appropriate method to translate the set of linear PDEs (15.67) into a linear system for all the expansion coefficients. Indicating the corresponding vector of unknowns with $\mathbf{x}$, the system (15.67) then reduces to

$$(\mathbf{A} + i\omega_d \mathbf{B})\mathbf{x} = \mathbf{f}, \qquad (15.69)$$

where $\mathbf{A}$ and $\mathbf{B}$ are now algebraic matrices. The RHS vector $\mathbf{f}$ results from imposing the boundary conditions related to the external driving source (see later).

As a concrete example, we consider a combination of model configurations II and III of Section 4.6 [1]: a static cylindrical equilibrium plasma in which the equilibrium quantities only depend on the radial coordinate r, surrounded by a vacuum and a perfectly conducting wall at $r = r_{\rm w}$. The plasma is perturbed by a periodic (AC) current in a helical coil in the vacuum region (at $r = r_{\rm a}$, with $r_{\rm p} < r_{\rm a} < r_{\rm w}$). This is the cylindrical equivalent of the slab configuration studied analytically in Chapter 11 [1] (apart from the conducting wall). The linear dynamics of this system is described by Eqs. (15.63)–(15.66) with appropriate boundary conditions. With a mono-periodic driving current, the dissipative system will reach a steady state after a finite time. In order to determine this steady state, we assume all perturbed quantities have a time behavior of the form $\exp(-i\omega_{\rm d}t)$, with $\omega_{\rm d}$ the frequency of the "antenna". In the late 1970s and early 1980s, this configuration was often considered with periodic boundary conditions in the longitudinal z-direction as a first approximation of a tokamak. We will describe the system in the common cylindrical coordinates. Due to the one-dimensional nature of the equilibrium, the θ and z-dependence of the perturbed quantities is handled trivially in a Fourier fashion, with in this case only one Fourier mode to consider for each of these coordinates, hence $e^{i(m\theta+kz)}$. In the radial direction, two kinds of finite element are used to avoid spectral pollution. Here, this is accomplished by choosing finite elements of one order higher for the (suitably scaled) components v_{1r}, $A_{1\theta}$ and A_{1z} than for the other ones since these components appear differentiated in the r-direction in discrete expressions for $\nabla \cdot \mathbf{v}_1$ and $\nabla \times \mathbf{A}_1$. For their scaled counterparts $\bar{v}_1$, $\bar{A}_2$ and $\bar{A}_3$ cubic Hermite finite elements are used, while $\bar{\rho}$, $\bar{v}_2$, $\bar{v}_3$, $\bar{T}$ and $\bar{A}_1$ are approximated by quadratic finite elements. As explained in Section 15.1.3, we then typically use two finite elements per grid point in the expansion, making the total number of unknowns $2 \times 8 \times N$, with N the number of radial grid points. In the Galerkin procedure, the discretization (15.39) or (15.40) is inserted in the system (15.63)–(15.66), and the weak form of this system is obtained by multiplying it with each of the $2 \times 8 \times N$ finite elements and integrating over the plasma volume [424]. The application of the Galerkin procedure then yields the algebraic system (15.69). The matrices appearing in (15.69) are block-tridiagonal matrices with sub-blocks of dimension 16×16. Furthermore, $\mathbf{A}$ is a non-Hermitian matrix, and $\mathbf{B}$ is a positive-definite matrix.[2]

In order to obtain a well-defined problem with a unique solution, boundary conditions need to be specified. As in the similar slab problem studied in Chapter 11 [1], the vacuum solution can be determined analytically. In cylindrical geometry, however, the vacuum solution is obtained in terms of modified Bessel functions.

---

[2] A complex square matrix $\mathbf{A} = (a_{ij})$ is called Hermitian when it is equal to its conjugate transpose, i.e. $\mathbf{A}^{\rm H} = (a_{ji}^*) = \mathbf{A}$. For a real matrix, this reduces to a symmetric matrix. A complex matrix $\mathbf{A}$ is positive definite when, for all non-zero complex column vectors $\mathbf{x}$, we have $\mathbf{x}^{\rm H}\mathbf{A}\mathbf{x} > 0$.

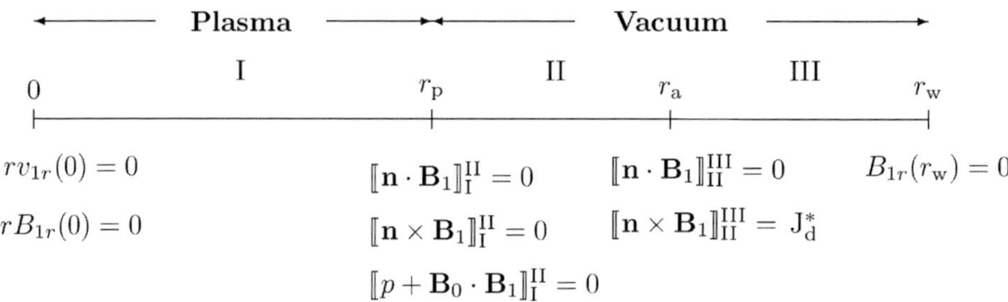

Fig. 15.9 Boundary conditions needed to obtain a unique solution of the steady state of a periodic cylindrical plasma that is driven externally by a current $\mathbf{J}_{\mathrm{d}}^{*}$ in an external helical coil at $r = r_{\mathrm{a}}$.

The system (15.63)–(15.66) is of order six in r (due to the dissipative terms) while the modified Bessel equation describing the vacuum solutions in each of the regions $r_{\mathrm{p}} \leq r \leq r_{\mathrm{a}}$ and $r_{\mathrm{a}} \leq r \leq r_{\mathrm{w}}$ is of order two. Hence, ten boundary conditions are needed for a unique solution. These boundary conditions are summarized in Fig. 15.9. The two regularity conditions at the magnetic axis are imposed by dropping the finite elements that do not satisfy them in the space of shape functions. Hence, these boundary conditions are treated as *essential* boundary conditions (cf. Section 15.1.3). The boundary conditions at the plasma–vacuum interface become *natural* [239] in a very similar way to that in the model problem considered in Section 15.1: they are imposed by exploiting the surface terms that arise upon integrating by parts, i.e. by formulating the problem in the weak form. This eventually yields a linear system of the form

$$(\mathbf{A}' + \mathrm{i}\omega_{\mathrm{d}} \, \mathbf{B})\mathbf{x} = \mathbf{f}', \tag{15.70}$$

with $\mathbf{A}'$ and $\mathbf{f}'$ a slight modification of $\mathbf{A}$ and $\mathbf{f}$, respectively, both resulting from imposing the natural boundary conditions. As explained above, the coefficient matrix $\mathbf{A}' + \mathrm{i}\omega_{\mathrm{d}} \, \mathbf{B}$ is of order $16N$. While this can become quite large, it is a sparse block tridiagonal matrix, which can be stored in band storage mode, saving a lot of computer memory.

Fig. 15.10, taken from [376], shows the result of a numerically obtained parameter scan for the 1D driven plasma configuration from Fig. 15.9. As a function of the driving frequency (indicated with ω_{p} in the figure), the fractional absorption f_{a} measures the ratio of the Ohmically dissipated energy to the total energy emitted by the antenna. The equilibrium in the plasma region was characterized by a parabolic axial current density, a constant axial magnetic field component and uniform density. For a magnetic Reynolds number of $R_{\mathrm{m}} = 10^{8}$, and mode numbers of the helical antenna surface current $(m, n) = (2, 1)$ in an $\exp(\mathrm{i}m\theta + \mathrm{i}nkz)$ Fourier dependence, the scan revealed the existence of an optimal driving frequency with

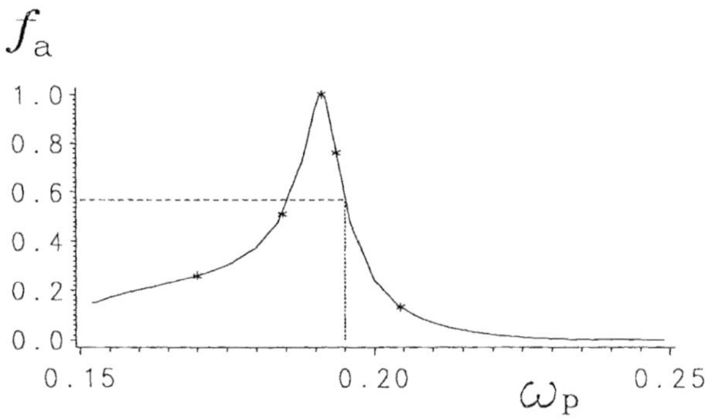

Fig. 15.10 Fractional absorption f_a versus driving frequency ω_p of the driven plasma—vacuum–antenna–wall system of Fig. 15.9. This steady-state computation identified the possibility for perfect coupling (at driving frequency $\omega_p = 0.1911$), where all energy supplied is converted into heat. (From Poedts *et al.* [376].)

perfect coupling, i.e. $f_a = 1$, or 100%. The possibility to explore numerically in a systematic fashion the effect of equilibrium parameters, perturbation parameters (driving frequency and its mode numbers) and model parameters such as the resistivity has helped us to realize how the presence of resistively damped, collective "quasi"-modes located in the frequency range of the Alfvén continuum can play a prominent role in the energetics of driven magnetic loop configurations.

15.2.3 MHD eigenvalue problems

In the normal mode approach, more information on the linear dynamics is obtained by computing the spectrum of all eigenoscillations. Imposing an a priori unknown time dependence of the form $\exp(\lambda t)$, with $\lambda \equiv -i\omega$, results, in a similar fashion as explained for the driven case (Eq. (15.69)), in the discrete equivalent of Eq. (15.67) for the general eigenvalue problem:

$$(\mathbf{A} - \lambda \mathbf{B})\mathbf{x} = 0. \tag{15.71}$$

The possibly complex eigenvalue λ now needs to be determined as part of the solution procedure and $\mathbf{x}$ corresponds to the discrete representation of the eigenvector. Quite a variety of algorithms exists for the solution of such large-scale eigenvalue problems [268]. In Section 15.3 below, we briefly discuss the QR algorithm, inverse vector iteration, and the more modern Krylov-subspace based Jacobi–Davidson technique.

This eigenvalue formulation is an efficient approach to determine growth rates of instabilities. In addition, it allows one to explore the stable part of the spectrum and to study e.g. continuum damping of quasi-modes such as encountered in Fig. 10.16

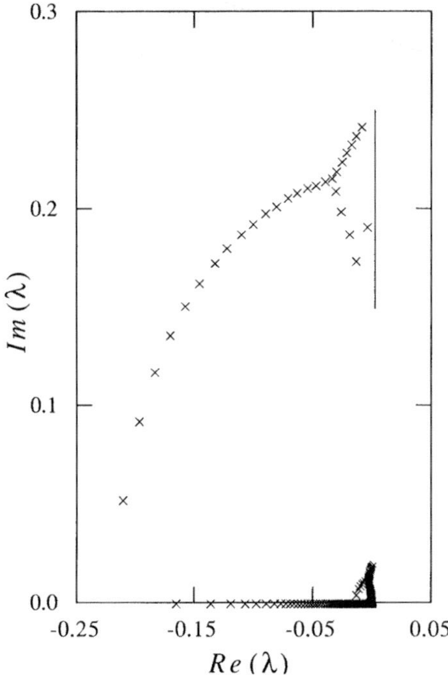

Fig. 15.11 Typical Alfvén and slow magneto-sonic parts of the resistive MHD spectrum, with a quasi-mode in the triangle; $\lambda \equiv -i\omega$. (From Poedts and Kerner [378].)

of Chapter 10 [1]. Also in the previous volume, an example for a cylindrical plasma showing all three (slow, Alfvén and fast) mode types and the possibility of unstable interchanges is depicted in Fig. 9.11 [1]. Note that when we use the same spatial discretization technique as for the corresponding driven problem, e.g. a FEM and its weak Galerkin formulation of the resistive MHD eigenvalue problem, this leads to a complex non-Hermitian matrix eigenvalue problem (15.71) with the matrices **A** and **B** of course *exactly the same* as considered before in the driven problem. It is of interest to note that in the non-ideal MHD eigenvalue solvers based on these FEM discretizations, realistically low values of the resistivity coefficient η can be handled accurately, confirming predictions of growth rate dependence as fractional powers of η for analytically tractable unstable modes.

In Fig. 15.11 part of the resistive MHD spectrum (including a so-called ideal MHD quasi mode) is shown for $\eta = 5 \times 10^{-5}$, as calculated by Poedts and Kerner [378]. These authors determined the full resistive MHD spectrum of a periodic cylindrical plasma column with the QR method, which will be discussed below, using 51 grid points in the radial direction. The static equilibrium considered in this calculation consisted of a finite length "periodic cylinder" with aspect ratio 10, limited by a perfectly conducting wall and with a constant plasma density and a parabolic current density profile. The weakly damped discrete resistive eigenmodes lie on typical curves in the complex plane. The resistive Alfvén eigenmodes lie on the upper λ-shaped curve with a bifurcation point where two branches approach the ends of the ideal Alfvén continuum. The ideal Alfvén continuum, ranging from $\lambda = (0, 0.15)$ to $\lambda = (0, 0.25)$ in this case, is also indicated on the plot. Decreasing the value of the plasma resistivity results in an upward shift of the resistive eigenmodes along the same curves so that the density of the eigenmodes on these curves increases. The weakly damped eigenmode with frequency situated in the triangle formed by the ideal Alfvén continuum and the two legs of the λ-shaped curve with resistive Alfvén eigenmodes in the complex plane, corresponds to an "ideal quasi mode". The oscillatory part of the frequency of this ideal quasi-mode is precisely 0.191, consistent with the result in Fig. 15.10. This weakly damped global mode manifests itself as the natural oscillation of the plasma and explains the temporal evolution of the driven system, as discussed in Chapter 11 [1]. The complex frequencies in the lower part of Fig. 15.11, near the origin, belong to the slow magneto-sonic sub-spectrum which does not couple to the Alfvén sub-spectrum in first order. Notice the discrepancy in time scales between the Alfvén modes and the slow magneto-sonic modes. The fast magneto-sonic modes have much higher frequencies and are not shown in the plot.

15.2.4 Extended MHD examples

The significant advantage of a numerical approach over an analytical one for computing the linear eigenmodes of an ideal MHD equilibrium configuration is the relative ease of including more physical effects in the description of the linear dynamics. To illustrate this observation, we discuss three examples of eigenvalue computations of essentially 1D MHD equilibria which use a different spatial discretization method for the non-trivial direction.

(a) Magneto-thermal instabilities for solar coronal arcades Our first example is taken from Van der Linden, Goossens and Hood [460], containing an MHD eigenvalue computation for a solar coronal "arcade" configuration. The 1D equilibrium taken for the "arcade" is rather rudimentary and is actually a pure z-pinch field with $\mathbf{B}_0 = (r/(1 + r^2))B_c\mathbf{e}_\theta$ and with a constant temperature throughout (isothermal). The "arcade" is then the top half of the cylinder when it is oriented such that its symmetry axis is the magnetic neutral line of the overarching arcade field lines. However, the eigenvalue computation includes various non-adiabatic effects of importance in the solar coronal environment, in particular optically thin radiative losses and a parameterized coronal heating prescription. This implies that both a source and sink of internal energy is included, appearing as a term $-(\gamma - 1)\rho\mathcal{L}$

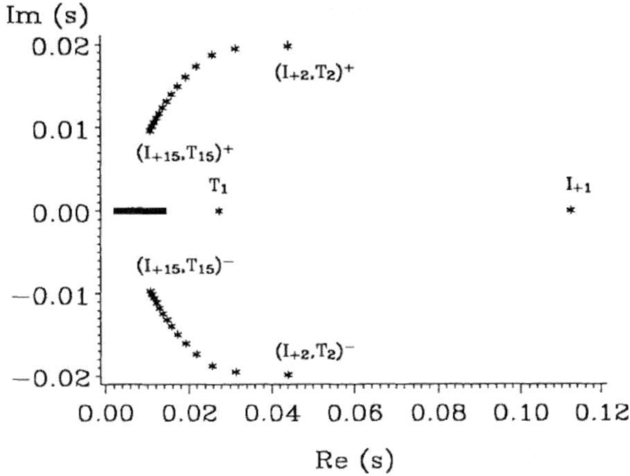

Fig. 15.12 Unstable part of the MHD spectrum for $m = 0$ and $k = 1$ perturbations of a z-pinch "solar arcade model". Discrete thermal instabilities labeled with T_n (where n is the number of radial nodes of the eigenfunction) coalesce from a certain n onwards with unstable pinching modes I_n to form overstable magneto-thermal mode pairs; $s \equiv -i\omega$. (From Van der Linden *et al.* [460].)

on the RHS of the governing nonlinear equation for the temperature evolution:

$$\rho \frac{\partial T}{\partial t} = -\rho \mathbf{v} \cdot \nabla T - (\gamma - 1)p\nabla \cdot \mathbf{v} - (\gamma - 1)\rho \mathcal{L} + (\gamma - 1)\nabla \cdot (\boldsymbol{\kappa} \cdot \nabla T) . \quad (15.72)$$

The function $\mathcal{L}(\rho, T)$ represents the energy gain–loss per unit mass, due to non-adiabatic processes *other than those due to thermal conduction*. The thermal conduction itself will be anisotropic in strongly magnetized plasmas, hence the appearance of the tensorial conductivity coefficient $\boldsymbol{\kappa}$.

For the internal energy losses due to optically thin radiation, a dependency of the form $\sim \rho^2 T^\alpha$ is appropriate, with the precise proportionality coefficient as well as the fractional temperature dependence α varying for different temperature ranges. Writing $\rho \mathcal{L} = \chi \rho^2 T^\alpha - \rho h$, in which h denotes the unknown "coronal heating" per unit mass, the equilibrium itself is then assumed to obey $\rho_0 \mathcal{L}(\rho_0, T_0) = \nabla \cdot (\boldsymbol{\kappa}_0 \cdot \nabla T_0)$. In Ref. [460], the z-pinch equilibrium was used to compare the analytic predictions obtained for unstable magnetic and *thermal condensation modes* using WKB techniques to those found by means of a numerical eigenvalue solution. The numerical analysis used the same (mixed order) FEM discretization discussed in Section 15.2.2 for the radial variation of the eigenfunctions, with a fixed Fourier dependence $\exp i(m\theta + kz)$. Boundary conditions impose a regularity condition at the axis, while a fixed perfectly conducting wall was adopted at the cylinder radius for simplicity.

For $m = 0$ eigenmodes, which are axi-symmetric, $\mathbf{k} \cdot \mathbf{B}_0 = 0$ and the influence of purely field-aligned thermal conduction can be dropped. Both the slow and Alfvén continuum collapse onto the origin for $\mathbf{k} \cdot \mathbf{B}_0 = 0$, but an infinite sequence of unstable discrete $m = 0$ magnetic pinching modes with real growth rates λ can accumulate to the marginal frequency $\lambda = 0$. In the presence of non-adiabatic effects, another continuous range of purely exponentially growing (or damped) modes exists known as the *thermal continuum*. This thermal continuum range can be derived analytically for 1D planar or cylindrical equilibria [459]. It represents an additional continuous range in the eigenfrequency plane corresponding to purely exponentially growing or damped perturbations, in contrast to the wave-like Alfvén and slow continua. In the presence of anisotropic thermal conduction, the neglect of cross-field thermal conduction is needed to maintain a continuous range, as any finite $\kappa_\perp$ will replace the continuum with a dense set of discrete modes [458], in a way reminiscent of what happens to the Alfvén and slow continua when going from ideal to resistive computations. This highlights that this thermal continuum is due to the fact that each field line can cool or heat independently of the adjoining ones when perpendicular thermal conduction is absent. Incorporating a heat-loss function and only field-aligned thermal conduction (and no resistivity) introduces the thermal continuum, leaves the familiar Alfvén continuum unmodified from the ideal case, and renders the slow continuum into a continuous range of complex overstable or damped wave modes influenced by non-adiabatic effects. In addition to the continuum range for the thermal condensation modes, an additional sequence of discrete thermal modes may appear at its edges.

Figure 15.12 shows the computed unstable part of the MHD spectrum for axisymmetric modes with axial wave number $k = 1$ under equilibrium parameters where (i) the entire thermal continuum is situated in the unstable half-plane (seen as solid line segments on the $\mathrm{Im}(s) = 0$ axis, where $s \equiv \lambda \equiv -\mathrm{i}\omega$); (ii) the infinite sequence of discrete unstable magnetic pinching modes is seen to start as indicated by the mode labeled with I_{+1}; (iii) discrete thermal modes with fundamental mode labeled with T_1 exist beyond the thermal continuum range. The numerical analysis shows that from a certain overtone onwards, the two discrete sequences merge in coalesced overstable magneto-thermal mode pairs (with complex conjugate eigenvalues λ and λ^*). This is a particular example of a computational MHD eigenmode analysis where an analytic treatment alone of these intricate mode couplings is challenging but almost necessarily incomplete. In further computational studies [458] of more realistic helical equilibrium fields of coronal loops the effects of anisotropic thermal conduction as well as finite resistivity [244] have been included. The inclusion of a finite (but small) cross-field conduction replaces the thermal continuum with a dense set of discrete thermal eigenmodes, and introduces very localized rapidly varying density eigenfunctions for those modes with eigenvalues located in the original continuum range. These could be responsible for the formation of fine structure during the thermal condensation process leading to solar prominences. In tokamak plasmas, the thermal instability is believed to cause the multifaceted asymmetric radiation from the edge (MARFE) phenomenon, where low charge state impurity radiation gives rise to a condensation instability that is primarily located at the high field side of the tokamak plasma edge (hence it is asymmetric in the poloidal cross-section, but toroidally symmetric).

▷ **Thermal condensation modes** Thermal instability can be encountered as a result of the precise thermodynamic dependence of the energy gain–loss function in Eq. (15.72). The linearized version of this equation will contain the partial derivatives $\partial \mathcal{L}/\partial \rho|_T$ taken at constant temperature and $\partial \mathcal{L}/\partial T|_\rho$ taken at constant density. Non-gravitational unstable thermal condensation modes can be predicted to form even in a static unmagnetized medium of constant density and temperature, if the isobaric criterion

$$\left.\frac{\partial \mathcal{L}}{\partial T}\right|_p = \left.\frac{\partial \mathcal{L}}{\partial T}\right|_\rho - \frac{\rho}{T}\left.\frac{\partial \mathcal{L}}{\partial \rho}\right|_T < 0 \tag{15.73}$$

is encountered. The ideal gas law $p \sim \rho T$ has been assumed. A runaway condensation can result when a local cooling causes a net increase of the radiative energy losses, resulting in a further cooling triggering yet more radiative losses and hence an unstable situation. This condensation mode is an exponentially growing mode, and is the marginal entropy mode discussed in Section 5.2.2 [1] driven unstable due to non-adiabatic influences, as already analysed in detail in a seminal paper by Field [134], following on an earlier suggestive work by Parker [362]. In the same purely hydrodynamic case, the sound waves, which are purely oscillatory wave modes in adiabatic situations, can be driven overstable if the isentropic instability criterion

$$\left.\frac{\partial \mathcal{L}}{\partial T}\right|_S = \left.\frac{\partial \mathcal{L}}{\partial T}\right|_\rho + \frac{\rho^2}{p(\gamma - 1)}\left.\frac{\partial \mathcal{L}}{\partial \rho}\right|_T < 0 \tag{15.74}$$

is met. In the presence of isotropic thermal conduction, these criteria are modified with a stabilizing influence of thermal conduction on both the purely exponential condensation mode and the overstable sound waves for short wavelengths. ◁

(b) Shear flow instabilities in current sheet configurations A second example of MHD eigenvalue computations is taken from Dahlburg and Einaudi [102]. There, the linear eigenmodes of a 1D current-vortex sheet are computed. The planar current-vortex sheet is a *stationary* equilibrium configuration in ideal MHD, characterized by a uniform density $\rho_0 = 1$ and pressure $p_0 = 1/(\gamma M^2)$, but with a shear flow profile given by $\mathbf{v}_0 = \tanh y \, \mathbf{e}_x$ and a co-spatial current sheet with $\mathbf{B}_0 = A \tanh y \, \mathbf{e}_x + A \operatorname{sech} y \, \mathbf{e}_z$. The dimensionless parameters M and A denote the sonic Mach number and the Alfvén Mach number, respectively. The magnetic field configuration is a force-free field where the magnetic field rotates about the y-axis in an out-of-plane fashion near $y = 0$ to become aligned with the anti-parallel planar flow in both half spaces at $y = \pm\infty$.

The linearized MHD equations now contain many additional terms due to the equilibrium velocity profile, and in Ref. [102] the effects due to (isotropic) thermal

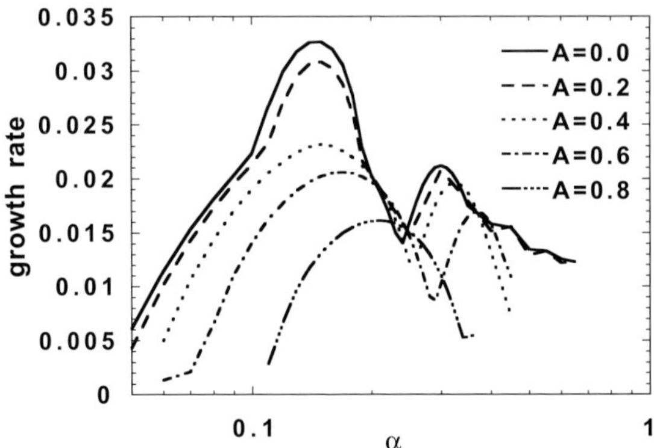

Fig. 15.13 Growth rates of overstable, oscillatory, weakly evanescent eigenmodes for varying streamwise wave number α of a planar current-vortex sheet. The stationary equilibrium considered is characterized by a sonic Mach number $M = 1$ and Alfvén Mach number A. (From Dahlburg and Einaudi [102].)

conduction, finite magnetic resistivity and viscous stresses are also taken along. Both the streamwise x-direction and the spanwise z-direction are handled in a single Fourier mode pair for the perturbations $\exp[\mathrm{i}(\alpha x + \beta z)]$, while the spatial variation of the eigenfunctions in the cross stream y-direction is discretized using Chebyshev polynomials in a spectral collocation formalism. The SPECLS (spectral compressible linear stability) code used was originally developed for pure hydrodynamic eigenvalue computations by Macaraeg *et al.* [322] and modified in this work to handle MHD configurations. The boundary conditions in the cross stream directions represent free slip conditions: setting the gradients of all fluctuating quantities to zero at fixed distances $y = \pm L$. This allows one to compute eigenmodes which are only weakly evanescent and oscillatory in their asymptotic $y \to \pm\infty$ behavior. Eventually, the obtained large generalized eigenvalue problem is solved with a QZ algorithm.

The numerical approach then investigated parametrically the unstable part of the MHD spectrum for flow and magnetically dominated regimes. Depending on their relative strength, surface type ordinary Kelvin–Helmholtz modes can be found (for $M < 1$ and small A), but also a class of overstable, oscillatory, weakly evanescent modes occur when $M \approx 1$ and $A < 1$. Figure 15.13 shows the growth rate of these latter modes for $M = 1$ and varying A, over a range of streamwise wave numbers α. Clearly, extensive numerical parameter studies are an essential means to categorize the many intricate relationships between shear-flow driven Kelvin–Helmholtz, resistive tearing modes, and all other ideal and non-ideal instabilities for these and similar MHD equilibrium configurations with co-spatial velocity and magnetic shear.

(c) Line-tied MHD spectrum Our examples of MHD eigenvalue computations thus far considered 1D equilibrium configurations (infinite slab or cylinder). The spatial dependence of the eigenfunctions in the ignorable directions is then handled by a single Fourier mode pair. However, virtually all astrophysical and laboratory MHD equilibrium states are at least two-dimensional in nature. For example, coronal loops are anchored (line-tied) at both ends in the dense photospheric environment and can show considerable curvature depending on their loop length to radius ratio. Even when neglecting their curvature and treating only their internal variation with cylinder radius, the line-tying boundary conditions render the MHD stability and eigenoscillation problem intrinsically 2D. For axi-symmetric tokamak plasmas, only the toroidal direction for the linear perturbations can be treated with a single Fourier mode at a time. In both cases, one needs to choose appropriate discretization combinations for the two non-trivial spatial directions in the corresponding MHD spectroscopic analysis.

In the case of line-tied loops, several authors [468, 106, 461] considered the 2D eigenvalue problem for axi-symmetric cylindrical models. The static MHD equilibrium is then still only constrained by the 1D radial equilibrium condition

$$\frac{d}{dr}\left(p_0 + \tfrac{1}{2}B_0^2\right) = -\frac{B_{\theta 0}^2}{r} \,. \tag{15.75}$$

However, the eigenvalue problem takes account of a finite loop length and considers eigenoscillations which obey line-tying boundary conditions. Line-tying typically imposes rigid wall conditions for the velocity perturbations

$$\mathbf{v}_1 = 0 \ \text{ at } \ z = \pm L \,, \tag{15.76}$$

for a loop of length $2L$. The poloidal θ direction can be handled trivially as $\exp(\mathrm{i}\,m\theta)$, but the line-tying will introduce coupling between all axial Fourier modes $\exp(\mathrm{i}\,n\pi z/L)$. In the case when Eqs. (15.63)–(15.66) are considered, all linear quantities are written as

$$f_1(r,\theta,z) \rightarrow \sum_{n=-\infty}^{n=+\infty} (-1)^n \hat{f}_{nm}(r)\mathrm{e}^{\mathrm{i}(n\pi z/L + m\theta)} \,, \tag{15.77}$$

and all θ derivatives are, as usual, replaced by multiplications with factors $\mathrm{i}m$. However, both first and second order derivatives with respect to z have to be handled carefully: not all perturbed quantities or their z-derivatives have equal values at $z = \pm L$ and hence the periodic continuation of them will not behave continuously at $z = \pm L$. Partial integration of the (multiplied) equations by $\int_{-L}^{+L} dz \, \exp(-\mathrm{i}\,k\pi z/L)$ introduces unknown "surface terms", with jumps across $z = \pm L$, in addition to the usual Fourier substitutions of first and second derivatives. One then uses a truncation of the series representation (15.77) up to N

terms. The $8N$ equations for the different Fourier modes $\exp(\mathrm{i}\,n\pi z/L)$ following from (15.63)–(15.66) are then coupled through these unknown surface terms. The procedure to remove these unknown surface terms subtracts the Nth equation for each linear variable from the $N-1$ previous equations, and replaces those equations of the Nth set containing them by the rigid wall line-tying conditions, which read in the form (15.77) as

$$\sum_{n=-\infty}^{n=+\infty} \hat{v}_{r1;nm}(r) = 0\,. \tag{15.78}$$

Truncating this and similar relations for the other velocity components to N terms yields a finite set of coupled ordinary differential equations for the radial variation of the eigenfunctions, supplemented with boundary conditions on axis and at the loop radius. Velli *et al.* [468, 106] combined this axial representation of the eigenfunctions using a truncated Fourier series with a finite difference treatment of the radial direction, while Van der Linden *et al.* [461] combined it with the FEM representation discussed above. Both used it to quantify the stabilizing influence of line-tying and finite loop lengths on selected kink $m = \pm 1$ modes. In such cases, one may suffice with a reasonably small number (up to 10) of axial Fourier modes to get a converged eigenvalue, when these axial modes are chosen to correspond in wavelength to the most unstable modes for the equivalent infinite length cylinder.

15.3 Linear algebraic methods

After discretization, all computational linear MHD applications for determining steady solutions to driven problems eventually lead to a large linear algebraic system. Similarly, determining the MHD spectrum of a particular equilibrium configuration mathematically boils down to the numerical solution of a generalized eigenvalue problem. We now briefly discuss some of the most suitable linear algebraic methods for these purposes. In practice, known properties of the coefficient matrix (symmetry or Hermitian property, sparseness pattern, …) will influence the optimal algorithm choice for solving the linear system or eigenvalue problem. For more details on linear algebraic methods we refer to the textbook on iterative Krylov methods by van der Vorst [463].

15.3.1 Direct and iterative linear system solvers

The numerical solution to the linearly driven magnetized plasma loop discussed in Section 15.2.2 led to a linear system given by the matrix equation (15.70). In principle, this matrix problem can be solved directly: one then LU-factorizes the occurring coefficient matrix and subsequently solves the system by means of standard available linear algebra library routines (e.g. LAPACK). In the LU decomposition, the coefficient matrix is written as the product of a lower triangular matrix $\mathbf{L}$ and an upper triangular matrix $\mathbf{U}$ so that:

$$(\mathbf{A}' + \mathrm{i}\omega_{\mathrm{d}}\,\mathbf{B})\mathbf{x} = (\mathbf{LU})\mathbf{x} = \mathbf{L}\,\underbrace{(\mathbf{Ux})}_{\mathbf{y}} = \mathbf{f}', \tag{15.79}$$

where the diagonal elements of the matrix $\mathbf{L}$ are all equal to one. The LU decomposition allows a fast solution of the system in two trivial steps: first the system $\mathbf{Ly} = \mathbf{f}'$ is solved by forward substitution; next the system $\mathbf{Ux} = \mathbf{y}$ is solved by backward substitution. The LU decomposition of a matrix is obtained by successive standard Gaussian transformations, and its computational cost scales as order n^3 for a full $n \times n$ matrix. For banded sparse matrices, this cost may decrease when a good direct solver is used.

For very large linear systems, for which direct methods may become unfeasible and computationally intensive, various iterative techniques have become standard practice. A modern, practical, guide to iterative methods for linear systems is found in Ref. [23]. An important class of iterative methods can be interpreted as *projection* methods. Projection methods seek an approximate solution of the linear system $\mathbf{Ax} = \mathbf{b}$ at iteration step k in a subspace $\mathbf{x}_0 + K_k$, with $\mathbf{x}_0$ an initial guess, by imposing the Petrov–Galerkin condition: $\mathbf{b} - \mathbf{Ax}_k \perp L_k$. Alternative choices of L_k and K_k correspond to different methods. In what follows, we discuss the basic idea of two particularly powerful iterative methods, namely the generalized minimal residual (GMRES) method [401], and the bi-conjugate gradient stabilized (Bi-CGSTAB) algorithm [462]. Both of these methods are suitable for dealing with non-symmetric (real) or non-Hermitian (complex) matrices.

GMRES iterates on the solution of the linear system $\mathbf{Ax} = \mathbf{b}$ as follows. For an initial guess $\mathbf{x}_0$, the residual is indicated as $\mathbf{r}_0 = \mathbf{b} - \mathbf{Ax}_0$. The Krylov subspace of order k associated with this residual $\mathbf{r}_0$ and the matrix $\mathbf{A}$ is by definition the vector space spanned by the vector series $\{\mathbf{r}_0, \mathbf{Ar}_0, \mathbf{A}^2\mathbf{r}_0, \dots, \mathbf{A}^{k-1}\mathbf{r}_0\}$. Indicating this Krylov subspace with

$$K^k(\mathbf{A}, \mathbf{r}_0) \equiv \mathrm{span}\left\{\mathbf{r}_0, \mathbf{Ar}_0, \mathbf{A}^2\mathbf{r}_0, \dots, \mathbf{A}^{k-1}\mathbf{r}_0\right\}, \tag{15.80}$$

successive approximations $\mathbf{x}_k$ are obtained by minimizing the L^2-norm of the residual $\|\mathbf{b} - \mathbf{Ax}_k\|_2$ over the Krylov subspace $K^k(\mathbf{A}, \mathbf{r}_0)$. GMRES is then a projection method with the spaces $K_k = K^k(\mathbf{A}, \mathbf{r}_0)$ and $L_k = \mathbf{A}\,K^k(\mathbf{A}, \mathbf{r}_0)$. In this way, the solution $\mathbf{A}^{-1}\mathbf{b}$ is approximated by a matrix polynomial of order $k-1$: $\mathbf{x}_k = \mathbf{x}_0 + p_{k-1}(\mathbf{A})\mathbf{r}_0 \approx \mathbf{A}^{-1}\mathbf{b}$. In practice, the approximation $\mathbf{x}_k$ is only computed at the end of the entire iterative procedure. Each iteration merely constructs an orthonormal basis for the successive Krylov subspaces $K^k(\mathbf{A}, \mathbf{r}_0)$ by means of the following Arnoldi method [15]. Taking $\mathbf{v}_1 = \mathbf{r}_0/\|\mathbf{r}_0\|_2$, one computes

$$
\begin{aligned}
&\text{for } i = 1, 2, \ldots \text{ do} \\
&\quad \mathbf{v}_{i+1} = \mathbf{A}\mathbf{v}_i \\
&\quad \text{for } j = 1, 2, \ldots, i \text{ do} \\
&\qquad \mathbf{v}_{i+1} = \mathbf{v}_{i+1} - (\mathbf{v}_j^{\mathrm{T}} \mathbf{v}_{i+1}) \mathbf{v}_j \\
&\quad \text{end} \\
&\quad \mathbf{v}_{i+1} = \mathbf{v}_{i+1}/\|\mathbf{v}_{i+1}\|_2 \\
&\text{end}
\end{aligned}
\qquad .
$$

In each iteration, we thus need to evaluate one matrix–vector product, and $i + 1$ inner products between two vectors. Storage requirements also increase by one vector per iteration. This GMRES method demonstrates a fast convergence whenever the matrix eigenvalues are reasonably close together but away from zero, and the eigenvectors are nearly orthogonal. In the case when this is not true, one needs to design a suitable pre-conditioner, i.e. a matrix $\mathbf{P}$ approximating $\mathbf{A}$ in some way and easily invertible, such that the transformed system $\mathbf{P}^{-1}\mathbf{A}\mathbf{x} = \mathbf{P}^{-1}\mathbf{b}$ has a coefficient matrix $\mathbf{P}^{-1}\mathbf{A}$ with the desired property. The GMRES scheme is often implemented using a suitable restart strategy, which limits the maximal number of vectors used in the Krylov sequence $\{\mathbf{A}^k\mathbf{r}_0\}$.

The Bi-CGSTAB approach for solving $\mathbf{A}\mathbf{x} = \mathbf{b}$ works with two sequences of vectors (or "search directions") to update the iterate $\mathbf{x}_k$. Starting with guess $\mathbf{x}_0$ and residual $\mathbf{r}_0 = \mathbf{b} - \mathbf{A}\mathbf{x}_0$, the first search directions could be taken as

$$
\mathbf{p}_1 = \mathbf{r}_0, \qquad \mathbf{s}_1 = \mathbf{r}_0 - \frac{\mathbf{r}_0^{\mathrm{T}} \mathbf{r}_0}{\mathbf{r}_0^{\mathrm{T}} (\mathbf{A}\mathbf{p}_1)} \mathbf{A}\mathbf{p}_1. \tag{15.81}
$$

The basic iteration then updates the kth solution iterate as

$$
\mathbf{x}_k = \mathbf{x}_{k-1} + \frac{\mathbf{r}_0^{\mathrm{T}} \mathbf{r}_{k-1}}{\mathbf{r}_0^{\mathrm{T}} (\mathbf{A}\mathbf{p}_k)} \mathbf{p}_k + \frac{(\mathbf{A}\mathbf{s}_k)^{\mathrm{T}} \mathbf{s}_k}{(\mathbf{A}\mathbf{s}_k)^{\mathrm{T}} (\mathbf{A}\mathbf{s}_k)} \mathbf{s}_k. \tag{15.82}
$$

Denoting the residual with $\mathbf{r}_k = \mathbf{b} - \mathbf{A}\mathbf{x}_k$, the search directions are updated as

$$
\mathbf{p}_{k+1} = \mathbf{r}_k + \frac{\mathbf{r}_0^{\mathrm{T}} \mathbf{r}_k}{\mathbf{r}_0^{\mathrm{T}} (\mathbf{A}\mathbf{p}_k)} \frac{(\mathbf{A}\mathbf{s}_k)^{\mathrm{T}} (\mathbf{A}\mathbf{s}_k)}{(\mathbf{A}\mathbf{s}_k)^{\mathrm{T}} \mathbf{s}_k} \left(\mathbf{p}_k - \frac{(\mathbf{A}\mathbf{s}_k)^{\mathrm{T}} \mathbf{s}_k}{(\mathbf{A}\mathbf{s}_k)^{\mathrm{T}} (\mathbf{A}\mathbf{s}_k)} \mathbf{A}\mathbf{p}_k \right),
$$

$$
\mathbf{s}_{k+1} = \mathbf{r}_k - \frac{\mathbf{r}_0^{\mathrm{T}} \mathbf{r}_k}{\mathbf{r}_0^{\mathrm{T}} (\mathbf{A}\mathbf{p}_{k+1})} \mathbf{A}\mathbf{p}_{k+1}. \tag{15.83}
$$

Since the residual can also be written as

$$
\mathbf{r}_k = \mathbf{s}_k - \frac{(\mathbf{A}\mathbf{s}_k)^{\mathrm{T}} \mathbf{s}_k}{(\mathbf{A}\mathbf{s}_k)^{\mathrm{T}} (\mathbf{A}\mathbf{s}_k)} \mathbf{A}\mathbf{s}_k, \tag{15.84}
$$

the Bi-CGSTAB scheme requires per iteration step the evaluation of two matrix vector products $\mathbf{A}\mathbf{s}_k$ and $\mathbf{A}\mathbf{p}_k$ as well as four inner products $\mathbf{r}_0^\mathrm{T}\mathbf{r}_k$, $(\mathbf{A}\mathbf{s}_k)^\mathrm{T}(\mathbf{A}\mathbf{s}_k)$, $(\mathbf{A}\mathbf{s}_k)^\mathrm{T}\mathbf{s}_k$ and $\mathbf{r}_0^\mathrm{T}(\mathbf{A}\mathbf{p}_k)$. Most variants again augment this basic scheme with a suitable pre-conditioning.

15.3.2 Eigenvalue solvers: the QR algorithm

The column vector $\mathbf{x}$ is called a right *eigenvector* of an $n \times n$ matrix $\mathbf{A}$ with corresponding *eigenvalue* λ if

$$\mathbf{A}\mathbf{x} = \lambda\,\mathbf{x}. \tag{15.85}$$

A generalized eigenvalue problem involves two square matrices $\mathbf{A}$ and $\mathbf{B}$, and is of the form

$$\mathbf{A}\mathbf{x} = \lambda\,\mathbf{B}\mathbf{x}. \tag{15.86}$$

When $\mathbf{B}$ can be inverted[3] this becomes a standard eigenvalue problem for matrix $\mathbf{C} = \mathbf{B}^{-1}\mathbf{A}$. In principle, the eigenvalues of Eq. (15.85) can be found as the roots of the *characteristic equation*

$$|\mathbf{A} - \lambda\,\mathbf{I}_n| = 0, \tag{15.87}$$

so that there are always n eigenvalues, which can be degenerate. As a direct consequence of the definition, the eigenvectors are determined only up to a scale factor and the eigenvalues can be shifted:

$$(\mathbf{A} + \tau\mathbf{I}_n)\mathbf{x} = (\lambda + \tau)\,\mathbf{x}, \tag{15.88}$$

so that the eigenvectors are insensitive to shifts in the eigenvalues. Left eigenvectors are defined in a similar way, with a column vector $\mathbf{y}$ being a left eigenvector when $\mathbf{y}^\mathrm{T}\mathbf{A} = \mathbf{y}^\mathrm{T}\lambda$. These left eigenvectors have the *same eigenvalues* as right eigenvectors since, from the definition $\mathbf{A}\mathbf{x} = \lambda\mathbf{x}$, we find immediately that $\mathbf{x}^\mathrm{T}\mathbf{A}^\mathrm{T} = \mathbf{x}^\mathrm{T}\lambda$ and $|\mathbf{A}| = |\mathbf{A}^\mathrm{T}|$. When $\mathbf{X}_\mathrm{R}$ denotes the matrix with the right eigenvectors in its columns and $\mathbf{X}_\mathrm{L}$ denotes the matrix with the left eigenvectors in its rows, it can be shown that, with a proper normalization of the eigenvectors, $\mathbf{X}_\mathrm{L} = \mathbf{X}_\mathrm{R}^{-1}$ and

$$\mathbf{X}_\mathrm{R}^{-1}\mathbf{A}\mathbf{X}_\mathrm{R} = \mathrm{diag}(\lambda_1, \ldots, \lambda_n), \tag{15.89}$$

where the RHS denotes a diagonal matrix with the eigenvalues of $\mathbf{A}$ as elements on the diagonal. This is a special case of a *"similarity transform"*. Such similarity transforms do not affect the eigenvalues. As a matter of fact, with a transformation matrix $\mathbf{Z}$ we get: $|\mathbf{Z}^{-1}\mathbf{A}\mathbf{Z} - \lambda\mathbf{I}_n| = |\mathbf{Z}^{-1}||\mathbf{A} - \lambda\mathbf{I}_n||\mathbf{Z}| = |\mathbf{A} - \lambda\mathbf{I}_n|$. The strategy of

---

[3] For generalized eigenvalue problems, the QZ variant of QR avoids the inversion of $\mathbf{B}$.

many eigenvalue solvers therefore typically involves two steps: (1) reducing $\mathbf{A}$ to a simpler form by similarity transforms $\mathbf{A} \to \mathbf{P}_1^{-1}\mathbf{A}\mathbf{P}_1 \to \mathbf{P}_2^{-1}\mathbf{P}_1^{-1}\mathbf{A}\mathbf{P}_1\mathbf{P}_2 \to \ldots$, and (2) starting an iterative procedure. A thorough review of methods suitable for solving the complex generalized eigenvalue problem (15.86) can be found in Ref. [268].

The QR method in particular is based on a decomposition of matrix $\mathbf{A} = \mathbf{QR}$, with $\mathbf{R}$ an upper triangular matrix and $\mathbf{Q}$ an orthogonal matrix ($\mathbf{Q}^{\mathrm{T}}\mathbf{Q} = \mathbf{I}_n$). The QR algorithm consists of a succession of orthogonal transformations:

$$\mathbf{A} = \mathbf{Q}_0\mathbf{R}_0 \Rightarrow \mathbf{T}_0 \equiv \underbrace{\mathbf{R}_0\mathbf{Q}_0}_{\mathbf{Q}_0^{\mathrm{T}}\mathbf{A}\mathbf{Q}_0} = \mathbf{Q}_1\mathbf{R}_1 \Rightarrow \mathbf{T}_1 \equiv \mathbf{R}_1\mathbf{Q}_1 = \mathbf{Q}_2\mathbf{R}_2. \tag{15.90}$$

This yields in the ith iteration step:

$$\mathbf{T}_i \equiv \mathbf{R}_i\mathbf{Q}_i = \mathbf{Q}_{i+1}\mathbf{R}_{i+1} = \mathbf{Q}_i^{\mathrm{T}} \ldots \mathbf{Q}_0^{\mathrm{T}}\mathbf{A}\mathbf{Q}_0 \ldots \mathbf{Q}_i. \tag{15.91}$$

It can be shown that, when the eigenvalues λ_i are all different, $\mathbf{T}_i$ converges to an upper triangular matrix with the eigenvalues λ_i on its diagonal. When λ_k is p times degenerate, on the other hand, $\mathbf{T}_i$ evolves to an upper triangular matrix except for a diagonal block of order p (corresponding to the degenerate eigenvalue λ_k). The LAPACK routine HQR exploits this algorithm and determines all eigenvalues of the matrix $\mathbf{A}$. The method is very reliable and stable. However, the QR algorithm does not preserve the band structure of the matrices and the resulting memory requirements restrict the spatial resolution considerably. Moreover, the number of operations is proportional to n^3 for a $n \times n$ matrix. As a result, only "small" problems can be solved with this algorithm.

15.3.3 Inverse iteration for eigenvalues and eigenvectors

The QR algorithm can give a fair impression of the distribution of the eigenvalues across the complex plane, but to get information on the associated eigenvectors the products of all the transformation matrices must be kept, which is expensive. Inverse iteration allows one to achieve a more accurate numerical proxy for a single eigenvalue as well as its eigenvector, and at the same time this iteration is able to handle larger dimensions. The basic idea is that if $\mathbf{b}_0$ is a random vector and τ_0 an approximation of an eigenvalue λ_k, the vector $\mathbf{y}$ satisfying

$$(\mathbf{A} - \tau_0\mathbf{I}_n)\mathbf{y} = \mathbf{b}_0 \tag{15.92}$$

is an approximation of the eigenvector corresponding to λ_k. One can then start an iteration procedure and replace $\mathbf{b}_0$ by $\mathbf{y}$ to get a new $\mathbf{y}$ which is even closer to the eigenvector. This can be seen by writing both $\mathbf{y}$ and $\mathbf{b}_0$ in terms of all eigenvectors

$\{\mathbf{x}_i\}$ with $i = 1, \ldots, n$ of $\mathbf{A}$: $\mathbf{y} = \sum_i \alpha_i \mathbf{x}_i$ and $\mathbf{b}_0 = \sum_i \beta_i \mathbf{x}_i$. Substituting this in Eq. (15.92) we get

$$\alpha_i = \frac{\beta_i}{\lambda_i - \tau_0} \quad \text{and} \quad \mathbf{y}_i = \sum_i \frac{\beta_i}{\lambda_i - \tau_0} \mathbf{x}_i, \tag{15.93}$$

so that when τ_0 is close to λ_k and β_k is not accidentally too small, we have $\mathbf{y} \approx \mathbf{x}_k$ (remember that eigenvectors are only determined up to a scale factor).

It is also not difficult to understand how one obtains an improved approximation to λ_k. Suppose that in the ith iteration step we have $(\mathbf{A} - \tau_i \mathbf{I})\mathbf{y} = \mathbf{b}_i$ with a normalized RHS, $\mathbf{b}_i^{\mathrm{T}} \mathbf{b}_i = 1$. Using the fact that eigenvalues can be shifted, we can write from the definition $\mathbf{A}\mathbf{x}_k = \lambda_k \mathbf{x}_k$ that

$$(\mathbf{A} - \tau_i \mathbf{I}_n)\mathbf{x}_k = (\lambda_k - \tau_i)\mathbf{x}_k. \tag{15.94}$$

Since $\mathbf{y}$ is closer to $\mathbf{x}_k$ than $\mathbf{b}_i$, as shown above, we can replace $\mathbf{x}_k$ by $\mathbf{y}$ in Eq. (15.94) provided that λ_k is also replaced by the improved approximation τ_{i+1}. The value for τ_{i+1} then results from multiplying this expression by $\mathbf{b}_i^{\mathrm{T}}$:

$$\tau_{i+1} = \tau_i + \frac{1}{\mathbf{b}_i^{\mathrm{T}} \mathbf{y}}. \tag{15.95}$$

This inverse iteration preserves the band structure of the matrix and, hence, allows much higher resolutions. However, the CPU time consumption is also proportional to n^3 and the algorithm finds only one eigenvalue at a time, viz. the one closest to the "initial guess" τ_0. Moreover, the inverse iteration is less reliable in the sense that no convergence does not guarantee that there is no eigenvalue to be found in the neighborhood of the initial guess, which is particularly embarrassing when determining stability thresholds.

15.3.4 Jacobi–Davidson method

As an example of a more recently developed, very powerful, algorithm to compute a number of eigenvalues and their associated eigenvectors in the vicinity of a target value in the complex eigenvalue plane, we present the Jacobi–Davidson algorithm developed by Sleijpen and van der Vorst [412], applied to the generalized eigenvalue problem from Eq. (15.86). This is the general form obtained after discretization of the linearized MHD equations, where the coefficient matrix $\mathbf{A}$ is typically a complex-valued, non-Hermitian matrix, while $\mathbf{B}$ is self-adjoint and positive definite ($\mathbf{B}^{\mathrm{H}} = \mathbf{B}$ and $\mathbf{x}^{\mathrm{H}} \mathbf{B} \mathbf{x} > 0$ for any non-zero complex vector $\mathbf{x}$). The eigenvalues λ then correspond to eigenfrequencies ω of the perturbations about an MHD equilibrium configuration via $\lambda = -\mathrm{i}\omega$. As noted in earlier chapters, these eigenfrequencies can be located anywhere in the complex plane whenever the background equilibrium is in motion, or when incorporating non-ideal effects into the governing equations.

To obtain preferentially those eigenvalues in the vicinity of a specified target value σ, we can subtract $\sigma\mathbf{B}\mathbf{x}$ from both sides of Eq. (15.86) and obtain

$$(\mathbf{A} - \sigma\mathbf{B})\,\mathbf{x} = (\lambda - \sigma)\mathbf{B}\mathbf{x} \;\Rightarrow\; (\mathbf{A} - \sigma\mathbf{B})^{-1}\,\mathbf{B}\mathbf{x} = \frac{1}{\lambda - \sigma}\mathbf{x} \;\Rightarrow\; \mathbf{Q}\mathbf{x} = \mu\mathbf{x},$$
$$(15.96)$$

where we introduced the shifted eigenvalue $\mu \equiv 1/(\lambda - \sigma)$. This shift-and-invert strategy is generally applicable, at the cost of computing the matrix inverse of $(\mathbf{A} - \sigma\mathbf{B})$. For the block-tridiagonal matrices encountered in many of the linear MHD codes, one can affordably compute this inverse by performing its complete LU decomposition (to be done only once for each target σ) and thus take $\mathbf{Q} = (\mathbf{LU})^{-1}\mathbf{B}$. (The inverse of an upper triangular matrix is easily computed and remains upper triangular, and similarly for a lower triangular $\mathbf{L}$.) The following description of the Jacobi–Davidson algorithm [412] for computing several of the largest eigenvalues and their associated eigenvectors for the system (15.96) is adopted from [354].

The Jacobi–Davidson method iterates on an eigenvalue–eigenvector pair $(\mathbf{u}_k, \theta_k)$ by repeated (approximate) solves of linear systems of the same (large) dimension n as the eigenvalue problem under study. The large linear system is known as the "correction equation" and its solution vector $\mathbf{z}$ is taken orthogonal to the current eigenvector approximation, i.e. $\mathbf{u}_k^{\mathrm{H}}\mathbf{z} = 0$. The solution vector of the linear system is subsequently used to obtain an improved approximation to the eigenvalue–eigenvector combination in a step which computes a much smaller eigenvalue problem of order k exactly by direct methods, e.g. using the QR method and consecutive linear system solves. These two building blocks of the Jacobi–Davidson algorithm can be summarized as follows.

Given a proxy $\mathbf{u}_k$ to an eigenvector of Eq. (15.96) which is normalized such that $\|\mathbf{u}_k\|_2^2 = \mathbf{u}_k^{\mathrm{H}}\mathbf{u}_k = 1$, its eigenvalue can be approximated by $\theta_k = \mathbf{u}_k^{\mathrm{H}}\mathbf{Q}\mathbf{u}_k$. Defining the matrix $\mathbf{P} = \mathbf{u}_k\mathbf{u}_k^{\mathrm{H}}$, the multiplication of any vector with this $\mathbf{P}$ projects this vector on the sub-space spanned by the vector $\mathbf{u}_k$. In particular, $\mathbf{P}\mathbf{u}_k = \mathbf{u}_k$. Likewise, the matrix $\mathbf{I}_n - \mathbf{P}$ with $\mathbf{I}_n$ the identity matrix of order n projects a vector on the orthogonal complement of $\mathrm{span}\{\mathbf{u}_k\}$. The idea for obtaining a better guess for the eigenvector $\mathbf{x}$ from $\mathbf{Q}\mathbf{x} = \mu\mathbf{x}$ is then to write $\mathbf{x} = \mathbf{u}_k + \mathbf{z}$, where the correction vector $\mathbf{z}$ is taken orthogonal to $\mathbf{u}_k$, i.e. $\mathbf{u}_k^{\mathrm{H}}\mathbf{z} = 0$, and belongs to this orthogonal complement. This implies $\mathbf{P}\mathbf{z} = 0$. If we introduce the restriction operator $\mathbf{Q}_P$ of the matrix operator $\mathbf{Q}$ to the orthogonal complement by means of

$$\mathbf{Q}_P \equiv (\mathbf{I}_n - \mathbf{P})\mathbf{Q}(\mathbf{I}_n - \mathbf{P}), \qquad (15.97)$$

we know that $\mathbf{Q}_P\mathbf{u}_k = 0$ and we can rearrange (15.97) to $\mathbf{Q} = \mathbf{Q}_P + \mathbf{QP} + \mathbf{PQ} -$

PQP. Introducing the residual $\mathbf{r} = \mathbf{Q}\mathbf{u}_k - \theta_k\mathbf{u}_k$ (which is orthogonal to $\mathbf{u}_k$ since $\mathbf{u}_k^{\mathrm{H}}\mathbf{r} = 0$), we can rework the eigenvalue problem $\mathbf{Q}(\mathbf{u}_k + \mathbf{z}) = \mu(\mathbf{u}_k + \mathbf{z})$ to the equation

$$\left(\mathbf{Q}_P - \mu\mathbf{I}_n\right)\mathbf{z} = -\mathbf{r} + \left(\mu - \theta_k - \mathbf{u}_k^{\mathrm{H}}\mathbf{Q}\mathbf{z}\right)\mathbf{u}_k, \tag{15.98}$$

where we used the equality $\mathbf{P}\mathbf{Q}\mathbf{z} = (\mathbf{u}_k^{\mathrm{H}}\mathbf{Q}\mathbf{z})\mathbf{u}_k$. The unknown eigenvalue μ is then replaced by its approximation θ_k in the left hand side of Eq. (15.98), effectively replacing $\mathbf{z}$ with an approximation $\mathbf{z}_k$. Multiplying the equation with $\mathbf{u}_k^{\mathrm{H}}$ yields that $\mu = \theta_k + \mathbf{u}_k^{\mathrm{H}}\mathbf{Q}\mathbf{z}$ so we can retain only the residual contribution in the right hand side. The governing correction equation for $\mathbf{z}_k$ then is written as

$$\left(\mathbf{I}_n - \mathbf{P}\right)\left(\mathbf{Q} - \theta_k\mathbf{I}_n\right)\left(\mathbf{I}_n - \mathbf{P}\right)\mathbf{z}_k = -\mathbf{r}, \tag{15.99}$$

where $\mathbf{u}_k^{\mathrm{H}}\mathbf{z}_k = 0$. This is a large linear system, which is actually only solved approximately (consistent with the fact that $\mathbf{z}_k$ only approximates the actual $\mathbf{z}$ obeying Eq. (15.98)). This can be done by exploiting only a few iterations of a Krylov subspace iterative method such as GMRES.

The approximate solution of the correction equation (15.99) is then made orthogonal by a Gram–Schmidt procedure to all previous search directions $\mathbf{v}_k$ collected from the previous solves. This adds a new search direction $\mathbf{v}_{k+1}$ to use in the computation of the eigenvalue–eigenvector pairs $(\mathbf{u}_k, \theta_k)$. This part of the algorithm proceeds by noting that, at each step, the approximation $\mathbf{u}_k$ is a linear combination of the k search directions $\mathbf{v}_j$ with $j = 1, \ldots, k$. When writing them in an $n \times k$ matrix $\mathbf{V}_k$ as columns, we thus have $\mathbf{u}_k = \mathbf{V}_k\mathbf{s}$ with $\mathbf{s}$ a vector of length k. Due to the orthonormalization process in constructing the search directions, we have $\mathbf{V}_k^{\mathrm{H}}\mathbf{V}_k = \mathbf{I}_k$.

The residual vector $\mathbf{r} = \mathbf{Q}\mathbf{V}_k\mathbf{s} - \theta_k\mathbf{V}_k\mathbf{s}$ is then made orthogonal to the k search directions, so by multiplying with $\mathbf{V}_k^{\mathrm{H}}$ we get the $k \times k$ "projected" eigenvalue problem

$$\mathbf{V}_k^{\mathrm{H}}\mathbf{Q}\mathbf{V}_k\mathbf{s} = \theta_k\mathbf{s}. \tag{15.100}$$

This small eigenvalue problem is solved by a direct method, e.g. with QR all k eigenvalues are obtained, and their associated eigenvectors $\mathbf{s}$ can be computed. The eigenvector approximation is then $\mathbf{u}_k = \mathbf{V}_k\mathbf{s}$ for a particular set $(\theta_k, \mathbf{s})$ which solves system (15.100). In a suitable combination with the above scheme for extending the number of search directions by solving the correction equation (15.99), the algorithm can then compute a series of eigenvalue–eigenvector pairs of the original problem (15.96) up to the maximal dimension one uses for the projected problem (15.100). This can deliver a few tens of eigenvalues and eigenvectors in the vicinity of a target at once, and by scanning the target across the complex plane, one may obtain an accurate numerical approximation of the entire spectrum.

15.4 Linear MHD: initial value problems

The computation of the steady solution to a linearly excited, driven, MHD configuration discussed in Section 15.2.2, or the computation of its spectrum of eigenoscillations and instabilities presented in Sections 15.2.3 and 15.2.4, involved the solution of a boundary value problem and a suitable choice for the spatial discretization. In the efficient calculation of the steady state of a driven dissipative configuration, we directly assumed the asymptotically expected temporal periodicity of the driver. We then necessarily lost all information on the time scales to reach this steady state and had no information on transient phenomena that may occur in the initial driving phase. The study of this transient phase requires the accurate determination of the temporal evolution of the system, i.e. the integration of the MHD equations in time. In what follows, we first introduce several basic concepts associated with spatio-temporal discretizations, and then discuss specific strategies for and examples of initial value problems in linear MHD.

15.4.1 Temporal discretizations: explicit methods

Similar to our presentation of the various spatial discretization methods, we introduce some basic terminology concerning temporal discretization techniques by first considering a suitable model problem. It should be clear how these can subsequently be applied to the more complicated linearized MHD equations. Following e.g. LeVeque [302], it is instructive to focus on a seemingly trivial problem at first, namely the numerical solution to the linear advection equation. This equation considers a scalar variable u depending on one spatial coordinate x and on time t which obeys the following prototype hyperbolic equation:

$$\frac{\partial u}{\partial t} = -v\,\frac{\partial u}{\partial x}\,, \tag{15.101}$$

with v a given constant. The above equation is the simplest hyperbolic equation containing an "advection" term in the RHS. The equation just states that u is carried along (advected) without change of form in its arbitrary initial shape $u_0(x, t = 0)$ and the solution is a wave propagating in the positive x-direction (when $v > 0$):

$$u(x, t) = u_0(x - vt)\,. \tag{15.102}$$

Since we know the exact analytic solution at any time, the intricacies of numerical methods can easily be quantified. Moreover, Eq. (15.101) can be written in the form of a conservation law, viz.

$$\frac{\partial u}{\partial t} + \frac{\partial F}{\partial x} = 0\,, \tag{15.103}$$

showing that u is a conserved quantity with $F = vu$ its associated flux.

The numerical integration of Eq. (15.101) involves a discretization in both space and time. When we discretize the time coordinate in discrete time levels $t^n = n\Delta t$, the temporal treatment can be explicit, implicit, or "semi-implicit". In *explicit time integration*, values of quantities at a new time level t^{n+1} are computed from explicitly available information on time level t^n. Implicit or semi-implicit schemes will also involve the evaluation of the unknowns at the new time level, and will be discussed in Section 15.4.3. An explicit method for Eq. (15.101) could replace the spatial derivative with the second-order central difference (15.16) evaluated at t^n and the time derivative with the first-order accurate expression (15.15) to get

$$\frac{u_i^{n+1} - u_i^n}{\Delta t} = -v\,\frac{u_{i+1}^n - u_{i-1}^n}{2\,\Delta x}\,, \tag{15.104}$$

or, rearranging the terms,

$$u_i^{n+1} = u_i^n - v\,\frac{\Delta t}{\Delta x}\,\frac{u_{i+1}^n - u_{i-1}^n}{2}\,. \tag{15.105}$$

As before, the subscript refers to the grid point and the superscript to the time step, such that $u_i^n = u(i\Delta x, n\Delta t)$. The above scheme is called the *forward Euler scheme* or Euler's FTCS scheme (forward time central space), but we will shortly explain why this method is utterly useless. Formally, it is second-order accurate in space and first-order accurate in time and its "stencil" in the $x - t$-plane is shown in Fig. 15.14(a). The difference equation (15.105) is a *consistent* representation of the model PDE (15.101): the local truncation errors vanish in the limit $\Delta x \to 0$ and $\Delta t \to 0$.

In order to be of any practical value, a spatio-temporal discretization should not only be consistent, but also obey numerical stability constraints. Computers always round off numbers and a scheme solving the differential equation (15.101) is *stable* only if these round-off errors shrink or at least do not grow (accumulate) during the time progression. The growth or decay of these round-off errors in a numerical method can usually be evaluated by the *von Neumann method*. This is a *local* stability analysis, i.e. it is assumed that the coefficients of the difference equations vary slowly so they can be considered constant in both space and time. The round-off error $\epsilon(x, t)$ can then be Fourier analyzed in space and time and each Fourier term

$$\epsilon_k(x, t) = \hat{\epsilon}_k e^{\lambda t} e^{ikx} \tag{15.106}$$

can be studied separately. (Notice the similarity between stability analysis of a numerical scheme and stability analysis of a magneto-hydrostatic equilibrium.). The condition for a numerically stable scheme then reduces to

$$\left|\frac{\epsilon_k^{n+1}}{\epsilon_k^n}\right| = |e^{\lambda\Delta t}| \le 1\,, \qquad \text{for every } k\,. \tag{15.107}$$

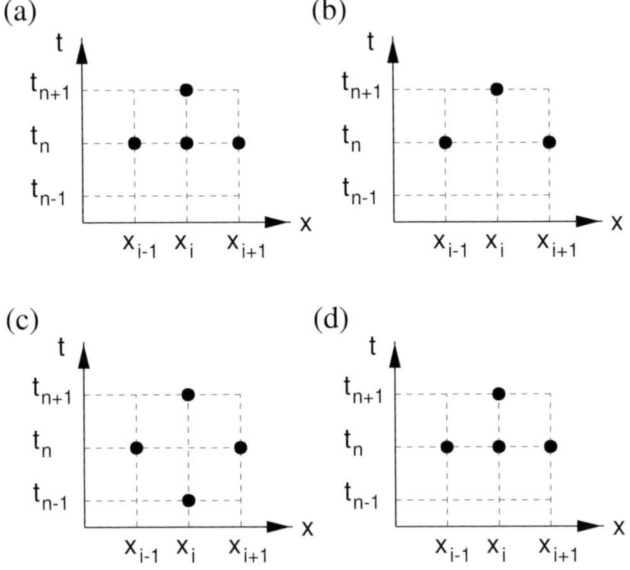

Fig. 15.14 Stencils in the x–t plane for (a) Euler's FTCS scheme, (b) Lax–Friedrichs scheme, (c) leapfrog scheme, (d) one-step Lax–Wendroff scheme.

When we analyze the numerical stability of the difference equation (15.105), we must admit that the numerical solution constitutes at best the exact solution, say E, plus the round-off error ϵ. Since the exact solution E satisfies the equation by definition, we need to insert Eq. (15.106) into Eq. (15.105) and we get, after dividing by ϵ_k^n,

$$e^{\lambda\Delta t} = 1 - \frac{v\Delta t}{\Delta x}\frac{e^{ik\Delta x} - e^{-ik\Delta x}}{2} = 1 - i\frac{v\Delta t}{\Delta x}\sin(k\Delta x). \tag{15.108}$$

We are thus led to conclude that this scheme is *unconditionally unstable* since

$$\left|\frac{\epsilon_k^{n+1}}{\epsilon_k^n}\right| > 1, \qquad \text{for every } k, \tag{15.109}$$

which makes the Euler FTCS scheme fairly useless. Note that if v were a function of x and t (or even u, which would bring in nonlinearity), it would still be treated as constant in the von Neumann method since this performs a *local* analysis. Nevertheless, this method is easy to apply and generally yields correct answers.

▷ **Exercise** Analyze a scheme similar to Eq. (15.105), viz. Euler's FTFS method (forward time forward space). How does the discretization of our model PDE look like in this scheme? Show that this method is also unconditionally unstable for $v < 0$. Use a physical argument to explain why the scheme is unstable then. ◁

One may alter the discrete formula (15.105) in several ways to resolve this numerical stability problem. Generally speaking, there are three types of solution. First, one can restore stability by the addition of "numerical diffusion" to damp the numerical (non-physical) instability. A second "cure" involves the use of a discretization with the same space-time symmetry as the original PDE. A third way out is the use of an *implicit* scheme. We start with examples of the first two strategies, while the third cure is discussed in Section 15.4.3. Upon replacing u_i^n in Eq. (15.105) by an average value of u^n between x_{i-1} and x_{i+1}, we get

$$u_i^{n+1} = \frac{u_{i+1}^n + u_{i-1}^n}{2} - v\frac{\Delta t}{\Delta x}\frac{u_{i+1}^n - u_{i-1}^n}{2}, \tag{15.110}$$

which is called the *Lax–Friedrichs scheme* (or Lax scheme). The scheme (15.110) is also a consistent, explicit scheme with a simple stencil shown in Fig. 15.14(b). Rearranging the terms in Eq. (15.110) yields

$$\frac{u_i^{n+1} - u_i^n}{\Delta t} = -v\frac{u_{i+1}^n - u_{i-1}^n}{2\Delta x} + \frac{(\Delta x)^2}{2\Delta t}\frac{u_{i+1}^n - 2u_i^n + u_{i-1}^n}{(\Delta x)^2}, \tag{15.111}$$

which can be recognized as a discretization of

$$\frac{\partial u}{\partial t} = -v\frac{\partial u}{\partial x} + \frac{(\Delta x)^2}{2\Delta t}\frac{\partial^2 u}{\partial x^2}. \tag{15.112}$$

In other words, we have added a diffusion term to the original equation which introduces "numerical dissipation" or "numerical viscosity". Hence, this is an example of the first cure to the stability problem. As a matter of fact, the von Neumann stability analysis for the Lax–Friedrichs scheme (15.110) yields

$$e^{\lambda\Delta t} = \cos\left(k\Delta x\right) - \mathrm{i}\frac{v\Delta t}{\Delta x}\sin\left(k\Delta x\right). \tag{15.113}$$

The resulting condition for stability reads:

$$C \equiv \frac{|v|\Delta t}{\Delta x} \leq 1, \tag{15.114}$$

which is a limitation of the time step Δt for a given resolution Δx. The condition (15.114) is called the *Courant–Friedrichs–Lewy* condition (CFL) after Courant, Friedrichs and Lewy who derived it in one of the first papers on finite difference methods in 1928 [96], and C is called the *Courant number*. The CFL condition is only a necessary condition for stability, not sufficient. Its physical meaning is that the explicit time step has to be smaller than the time required for the fastest wave in the system to propagate from one grid point to the next. Stated differently, "the domain of dependence of the differential equation should be contained in the domain of dependence of the discretized equations" [226, 477]. If the time step is too big, the physical domain of dependency is larger than the stencil dependency provided by the scheme and this lack of information leads to an instability. The physical domain of dependence is bounded by the physical characteristics in the $x - t$ plane corresponding to the fastest accessible signal speed. For our model problem the physical characteristics are given by $dx/dt = v$, i.e. they are straight parallel lines (see Fig. 15.15).

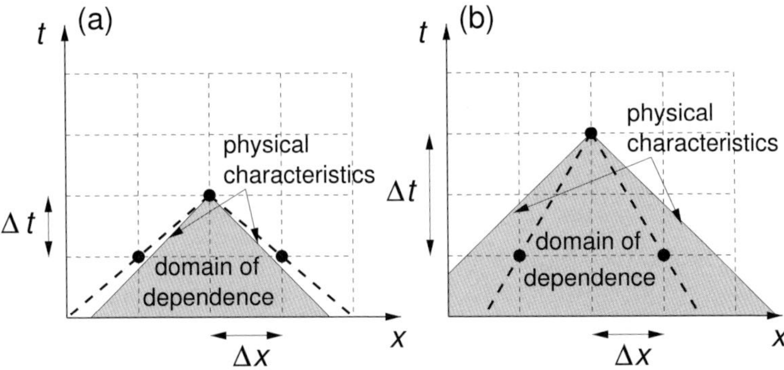

Fig. 15.15 Physical meaning of the CFL condition. (a) If Δt is small enough, the "physical" domain of dependence lies within the "numerical" domain of dependence. (b) If not, the scheme is not able to provide all the information needed to determine the solution in the next time step and becomes unstable. In the model problem, there is only one characteristic through each point, so that the situation sketched here applies to a second order wave equation with two characteristics $dx/dt = \pm v$.

Let us now use the "flux notation", i.e. the model equation (15.103) considered in "conservative form". For our model equation the "flux" is $F = vu$. This conservative form makes generalizations to the conservation laws of ideal MHD more obvious and ensures that the numerical schemes discussed below are directly applicable to the finite volume method often used for hyperbolic nonlinear problems, to be discussed in Section 19.1.4. Above, the unstable forward Euler scheme has been "stabilized" by adding numerical dissipation. Now, we discuss an example of a second "cure", viz. mimicking the symmetry of the PDE in the discretized equation. Using a central discretization for both x and t, e.g., one gets the "leapfrog" scheme:

$$u_i^{n+1} = u_i^{n-1} - \frac{\Delta t}{\Delta x}\left(F_{i+1}^n - F_{i-1}^n\right), \tag{15.115}$$

with F_i^n the "numerical" flux function, i.e. $F_i^n \equiv vu_i^n$. This scheme is *second-order* in both time and space. The time levels in the t-derivative "leapfrog" over the time levels in the x-derivative, as can be seen in Fig. 15.14(c). Notice that this scheme requires storage of u^n *and* u^{n-1} to determine u^{n+1}.

Another method which is also second-order in time and very extensively used for MHD (and CFD) calculations, is the *Lax–Wendroff scheme*. The Lax–Wendroff scheme is based on a truncated Taylor series expansion around $u(x, t)$:

$$u(x, t + \Delta t) \approx u(x, t) + \Delta t \, \frac{\partial u}{\partial t}(x, t) + \tfrac{1}{2}(\Delta t)^2 \, \frac{\partial^2 u}{\partial t^2}(x, t) \,. \qquad (15.116)$$

By using Eq. (15.101), the time-derivatives in this truncated series can be replaced by derivatives with respect to x. This yields the second-order accurate Lax–Wendroff scheme:

$$u_i^{n+1} = u_i^n - \tfrac{1}{2}\frac{\Delta t}{\Delta x} \, v \, (u_{i+1}^n - u_{i-1}^n) + \tfrac{1}{2}\frac{(\Delta t)^2}{(\Delta x)^2} \, v^2 \, (u_{i+1}^n - 2u_i^n + u_{i-1}^n) \,. \quad (15.117)$$

Clearly, this is another example of the first cure to the numerical instability problem: the LHS term and the first two RHS terms are identical, as in the forward Euler scheme, while the third term in the RHS adds numerical dissipation. The scheme is explicit and, hence, only conditionally stable. Its stencil is the same as for the Euler FTCS scheme and displayed in Fig. 15.14(d).

▷ **Exercise** Show that for our model problem the leapfrog scheme (15.115) requires the CFL condition to be satisfied for numerical stability. Notice that you obtain a quadratic equation in $\exp(\lambda \Delta t)$ now due to the occurrence of three time levels in the scheme. Further, show that for our model problem the Lax–Wendroff scheme (15.117) again requires the CFL condition to be satisfied for numerical stability. ◁

Example MHD application: p-mode interactions with sunspots An example linear MHD application where a Lax–Wendroff type scheme is used for the temporal integration of the linearized MHD equations is taken from Cally and Bogdan [74]. These authors simulated f- and p-mode interactions with a stratified "sunspot" in a purely planar approximation. The spot was represented by a mere slab of vertical field of sunspot strength, but took into account the realistically strong gravitational stratification of the plasma. Both internal "spot" plasma and the external unmagnetized medium had a polytropic stratification (taking account of the offsets required to ensure horizontal pressure balance) and the background density resulted from vertical hydrostatic equilibrium. Using a finite difference spatial discretization of the linearized MHD equations in conservation form, the evolution of the driven problem was simulated. The driver consisted of prescribing the horizontal velocity field at one side of the slab from the exact eigenfunction of the non-magnetic polytrope fitting the horizontal domain size. A staggered grid was employed to ensure numerical conservation. These simulations demonstrated convincingly that impinging f- or p-modes are partially converted to slow magneto-atmospheric gravity waves within the stratified magnetic slab. In turn, this leads to a clear deficit in the amplitude of the emerging, non-magnetic modes exiting the "spot" on the far side of the driver. As seen in Fig. 15.16, showing the horizontal velocity field when an incident p- mode is driving the slab from the left, the incoming acoustic mode

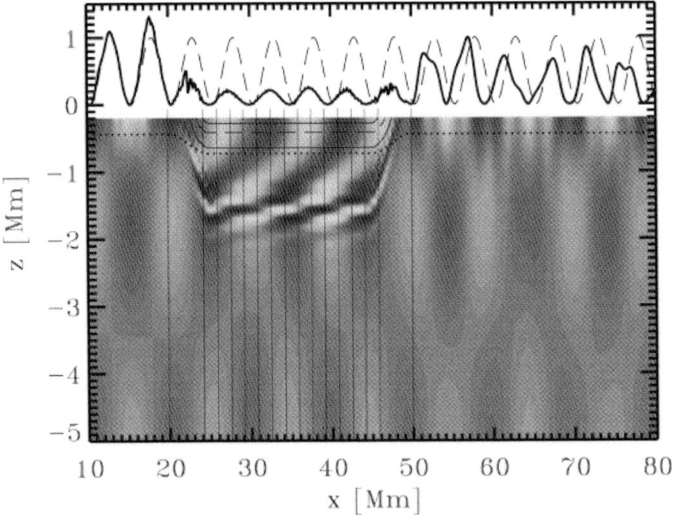

Fig. 15.16 A snapshot of the temporal evolution of a p-mode impinging on a stratified sunspot. The grey scale of the horizontal velocity shows that part of the incoming acoustic power (driver is at left) gets converted into downward propagating
magneto-acoustic slow modes in the magnetized slab representing the sunspot. The
top solid line indicates the square of the vertical velocity along a temperature iso-
therm marked by the dotted line. If the spot is removed, this velocity profile would
follow the dashed line shown. (From Cally and Bogdan [74].)

gets coupled at the fore boundary of the magnetic slab (between $x = 20$ Mm and
$x = 50$ Mm), at depths where $\beta \approx 1$, to downward propagating magneto-acoustic
slow or s-modes. These propagate away from this conversion layer along the vertical field lines, and form a very natural explanation for the observationally detected
p-mode power absorption in sunspot and plague regions. Indeed, at the far side,
the exciting p-mode re-emerges but is clearly reduced in amplitude and also mixed
with an f-mode of the same frequency but shorter horizontal wavelength. In more
realistically stratified sunspot models, the additional effects of resonant absorption
could act as an extra sink of incoming acoustic power, as already explained in
Chapter 11 [1].

Semi-discretization I All the schemes discussed so far are presented as if space
and time are discretized simultaneously. In practice, one may use the "method of
lines" also known as *semi-discretization* for treating a system of PDEs numerically.
This implies that one first only discretizes in space, turning the problem into a set
of ODEs in time. The latter can be solved by one of the numerous methods available for integrating ODEs, e.g. a second or fourth order Runge–Kutta scheme or a
predictor–corrector approach. Obviously, semi-discretization is very useful when

higher-order (> 2) accuracy is needed in time. The approach is also powerful because any spatial discretization method (FD, FE, spectral, or the finite volume method) of any accuracy can then be coupled to the ODE solver for the time discretization. For explicit schemes such as those discussed before, the CFL stability constraint on the time step will typically require the use of the same order of accuracy for treating the then separate temporal and spatial discretizations. This is no longer the case when dealing with implicit treatments, where as we will see the stability constraint will be lifted.

Runge–Kutta methods The application of the semi-discretization method on the model problem given by Eq. (15.101) yields an initial value problem determined by an ODE of the form

$$\frac{du}{dt} = f(t, u), \tag{15.118}$$

(where $f(t, u)$ denotes an expression depending on t and u) and an initial condition

$$u(t^0) = u^0. \tag{15.119}$$

When f and $\partial f / \partial u$ are continuous on an (open) rectangle in the (t, u)-plane containing the point (t^0, u^0) it can be proven that there exists a unique solution $u = \phi(t)$ of this problem in an interval around t^0. When Eq. (15.118) is nonlinear the solution of this problem is not trivial and often needs to be solved numerically, e.g. by the forward Euler scheme discussed in Section 15.4.1 or the backward Euler scheme discussed in Section 15.4.3. Both these schemes, however, are only first order accurate in time. There exists a class of schemes we now call the *Runge–Kutta methods* that enables much faster convergence in time. The Runge–Kutta formulas contain weighted averages of the value of $f(t, u)$ in different points in the time interval $[t^n, t^{n+1}]$. The second order formula uses a trial step at the mid-point of the interval to cancel out lower order error terms. It is given by

$$u^{n+1} = u^n + \Delta t\, k^{n2} + \mathcal{O}(\Delta t)^3, \tag{15.120}$$

with

$$k^{n1} = f(t^n, u^n), \qquad k^{n2} = f(t^n + \tfrac{1}{2}\Delta t, u^n + \tfrac{1}{2}\Delta t k^{n1}). \tag{15.121}$$

This two-step method is also known as the *predictor–corrector method*, where k^{n2} corresponds to evaluating the function f at time $t^{n+\frac{1}{2}}$. However, the method originally developed by Runge and Kutta is now called the classic fourth order four-step Runge–Kutta method, in short *the* Runge–Kutta method. It is given by

$$u^{n+1} = u^n + \Delta t\, \tfrac{1}{6}(k^{n1} + 2k^{n2} + 2k^{n3} + k^{n4}) + \mathcal{O}(\Delta t)^5, \tag{15.122}$$

where

$$k^{n1} = f(t^n, u^n), \qquad\qquad k^{n2} = f(t^n + \tfrac{1}{2}\Delta t, u^n + \tfrac{1}{2}\Delta t k^{n1}),$$

$$k^{n3} = f(t^n + \tfrac{1}{2}\Delta t, u^n + \tfrac{1}{2}\Delta t k^{n2}), \quad k^{n4} = f(t^n + \Delta t, u^n + \Delta t k^{n3}). \quad (15.123)$$

It is not difficult to show that the scheme (15.122) differs from the Taylor expansion of the exact solution by terms that are proportional to $(\Delta t)^5$, though the derivation is rather lengthy. On a finite time interval, the global truncation error is smaller than a constant times $(\Delta t)^4$. The scheme is thus *fourth order accurate* (for the time derivative) in the step size Δt. In other words, the order of consistency of this method is equal to four, and there are four intermediate steps necessary, viz. the determination of $k^{n1}, \ldots, k^{n4}$. The Runge–Kutta method thus converges three orders of magnitude faster than the Euler method with respect to the time step. Yet, it is relatively easy to implement and sufficiently accurate to tackle most problems efficiently. This is also true for so-called *adaptive* Runge–Kutta methods which take variable step sizes Δt where needed.

15.4.2 Disparateness of MHD time scales

In MHD, the CFL time step limitation for explicit codes is very severe due to the disparate times scales associated with the various MHD wave types. In practice, this makes explicit schemes useless for the computation of resistive instabilities, as it would involve too many discrete time steps and CPU time. In particular, the fast wave sub-spectrum is responsible for this problem since it accumulates at infinity. Let us analyze the consequences of the disparate time scales for simulating the linear (or nonlinear) dynamical behavior of a tokamak plasma. In tokamak geometry, the time scale associated with the compressional fast magneto-sonic wave is measured by the transit time of the fast magneto-sonic wave over the small plasma radius a, $\tau_{\text{fast}} \equiv a/v_{\text{f}}$, with v_{f} the phase velocity of the fast magneto-sonic wave. For the low-β plasmas in tokamaks (where β is the ratio of the plasma pressure and the magnetic pressure: $\beta \equiv 2\mu_0 p/B^2$), the fast magneto-sonic speed is nearly equal to the Alfvén speed, $v_{\text{f}} \approx v_{\text{A}}$, and, hence, $\tau_{\text{fast}} \approx a/v_{\text{A}}$. Shear Alfvén waves, on the other hand, propagate mainly along the magnetic field lines. Hence, the time scale related to these waves is measured by the transit time of the shear Alfvén wave over the *length* $2\pi R_0$ of the torus: $\tau_{\text{Alfv}} \equiv 2\pi R_0/v_{\text{A}}$, where R_0 is the major radius of the tokamak. In large aspect ratio $R_0/a \gg 1$ tokamaks (the argument also holds for large aspect ratio solar coronal loops), the time scales τ_{fast} and τ_{Alfv} differ substantially.

When resistive instabilities and Ohmic dissipation are studied, there is a third time scale of interest, viz. the time scale of resistive diffusion of magnetic fields,

$\tau_{\text{diff}} \equiv \mu_0 a^2/\eta$. The ratio of the compressional time scale to the time scale of resistive diffusion is equal to the magnetic Reynolds number R_{m}. In tokamak plasmas, R_{m} is typically 10^6–10^8 and in the large aspect ratio solar coronal loops R_{m} may reach values up to 10^{10}. Clearly, the resistive diffusion time scale is much longer than both τ_{fast} and τ_{Alfv} so that we have

$$\tau_{\text{fast}} \ll \tau_{\text{Alfv}} \ll \tau_{\text{diff}} \, . \tag{15.124}$$

Consequently, the study of wave problems in tokamaks or solar coronal loops where dissipation is important leads inevitably to "stiff" equations in which the dependent variables change on two or more very different scales. Stiff problems make explicit methods inefficient to use, or may even render them unstable, and the cure to solve this problem is the use of implicit methods.

15.4.3 Temporal discretizations: implicit methods

Consider again Euler's FTCS scheme. By evaluating the spatial derivative in this scheme in the $(n+1)$th time step instead of the nth, one obtains the "BTCS Euler scheme" (backward in time now):

$$u_i^{n+1} = u_i^n - v \, \frac{\Delta t}{\Delta x} \, \frac{u_{i+1}^{n+1} - u_{i-1}^{n+1}}{2} \, . \tag{15.125}$$

Notice that now u_i^{n+1} cannot be expressed in terms of known values at time index n anymore. It also depends on unknown values at time index $n+1$, viz. u_{i+1}^{n+1} and u_{i-1}^{n+1}. The scheme is therefore called *implicit*. For (15.125), this results in a tridiagonal system for u_i^{n+1}, $i = 1, \ldots N$. A von Neumann stability analysis of the scheme (15.125) yields

$$\mathrm{e}^{-\lambda \Delta t} = 1 + iv \frac{\Delta t}{\Delta x} \sin(k\Delta x) \, , \tag{15.126}$$

so that

$$|\mathrm{e}^{\lambda \Delta t}| = \frac{1}{|1 + iC \sin(k\Delta x)|} < 1 \qquad \text{for all } k \, . \tag{15.127}$$

Hence, this scheme is *unconditionally stable*, and any large time step Δt is in principle allowed. However, for wave problems, the time step must be much smaller than the wave period of interest to avoid numerical damping. This is so since the implicit treatment improves the stability of the scheme, but *not* the accuracy. Larger time steps introduce larger truncation errors. The stencil of scheme (15.125) is given in Fig. 15.17(a) where the three dots at time level $n+1$ imply the tridiagonal system solve. This introduces more computational work per time step, so the use of an implicit scheme only pays off when large enough time steps can be taken to make the total amount of work to reach a fixed future time decrease. For wave problems, this may often not be the case.

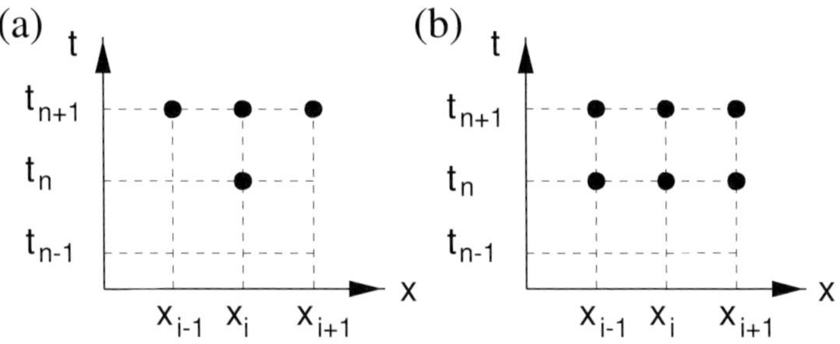

Fig. 15.17 Stencils of (a) BTCS Euler scheme and (b) Crank–Nicolson method.

The implicit backward Euler scheme (15.125) is only first-order in time. Upon writing the spatial differences in the right hand side of the explicit unstable forward Euler scheme (15.105) in terms of averages between the nth and the $(n+1)$th time step, one gets

$$u_i^{n+1} = u_i^n - \frac{1}{4} v \frac{\Delta t}{\Delta x} \left(u_{i+1}^{n+1} + u_{i+1}^n - u_{i-1}^{n+1} - u_{i-1}^n \right), \qquad (15.128)$$

which is called the *Crank–Nicolson method*. This scheme is second-order accurate in both time and space and again requires a tridiagonal system to be solved in each time step. Its stencil is shown in Fig. 15.17(b). The Crank–Nicolson method is widely used in CFD and computational MHD.

▷ **Exercise** Show that the scheme (15.128) is unconditionally stable. ◁

▷ **Semi-implicit schemes** The term "semi-implicit" for temporal discretization techniques is used for several schemes which are neither explicit nor fully implicit. Different options can be, e.g., to treat some variables explicitly and others implicitly, or to update only specific terms in the equations implicitly, or to relax the CFL condition somewhat by treating only the fastest sub-spectrum of waves implicitly. The latter alternative has been applied in particular for nonlinear MHD with a lot of success [214]. When the fast magneto-sonic waves are handled in an implicit way, the time step is limited by the CFL condition on the Alfvén waves. This is much less restrictive in tokamaks because of Eq. (15.124). ◁

Semi-discretization II Returning to the semi-discretization method, we can consider the quite general model problem involving a vector **u** of unknowns. These unknowns are introduced by the spatial discretization (FDM, FEM, (pseudo-)spectral, etc.) and one obtains an ordinary differential equation for **u**:

$$\frac{d\mathbf{u}}{dt} = \mathbf{f}(\mathbf{u}), \qquad (15.129)$$

where $\mathbf{f}$ could even be a nonlinear function of $\mathbf{u}$. Discretize this ODE in time by the following scheme:

$$\mathbf{u}^{n+1} = \mathbf{u}^n + \Delta t \left[\alpha \mathbf{f}(\mathbf{u}^{n+1}) + (1 - \alpha)\mathbf{f}(\mathbf{u}^n) \right], \tag{15.130}$$

with parameter α. For $\alpha = 1/2$ the above implicit scheme is called the *trapezoidal method*, while for $\alpha = 1$ the scheme reduces to the implicit "backward Euler" scheme. For $\alpha = 0$ we get the explicit forward Euler method. The scheme is only second-order accurate for $\alpha = 1/2$. As a matter of fact, with a central formula for the flux the trapezoidal method is exactly the same as the Crank–Nicolson method discussed above for the model problem (15.101). For a nonlinear $\mathbf{f}$, the above scheme involves the (iterative) solution of a nonlinear system at each time step. The system can be simplified by linearizing $\mathbf{f}(\mathbf{u}^{n+1})$, i.e. by replacing this term by

$$\mathbf{f}(\mathbf{u}^{n+1}) \approx \mathbf{f}(\mathbf{u}^n) + \frac{\partial \mathbf{f}^n}{\partial \mathbf{u}} (\mathbf{u}^{n+1} - \mathbf{u}^n), \tag{15.131}$$

where the matrix $\dfrac{\partial \mathbf{f}^n}{\partial \mathbf{u}}$ is called the *"Jacobian matrix"* of $\mathbf{f}$. With this linearization (15.131), the scheme (15.130) reduces to

$$\mathbf{u}^{n+1} = \mathbf{u}^n + \Delta t \alpha \frac{\partial \mathbf{f}^n}{\partial \mathbf{u}} (\mathbf{u}^{n+1} - \mathbf{u}^n) + \Delta t \mathbf{f}(\mathbf{u}^n)$$

$$\Rightarrow \quad \left[I - \Delta t \alpha \frac{\partial \mathbf{f}^n}{\partial \mathbf{u}} \right] \delta \mathbf{u} = \Delta t \mathbf{f}(\mathbf{u}^n), \tag{15.132}$$

which is a linear system for the unknowns $\delta \mathbf{u} \equiv \mathbf{u}^{n+1} - \mathbf{u}^n$. Notice that this equation is equivalent to the first step of a Newton iteration on Eq. (15.130), which may need to be expanded to a full Newton iteration in the case of nonlinear problems.

15.4.4 Applications: linear MHD evolutions

If we follow the semi-discretization approach for the linearized MHD equations (15.63)–(15.66), we obtain a system of ODEs in time. In a manner analogous to the discussion of the driven problem in Section 15.2.2, but now keeping the unknown temporal behavior, we can write

$$\mathbf{A}\mathbf{x} = \mathbf{B}\frac{d\mathbf{x}}{dt} + \mathbf{f}, \tag{15.133}$$

where $\mathbf{f}$ again results from a possible external driving source or initial perturbation. For linear (and even for nonlinear) stability calculations, the solution of this system as a function of time will automatically yield the expected growth at early times of the most unstable mode out of a random initial vector $\mathbf{f}$. Moreover, this initial value approach also enables us to simulate the nonlinear evolution of instabilities.

We discuss two example linear MHD applications where a different implicit time discretization was used to overcome the CFL time step constraint, thus enabling us to obtain quantitative insight into long-term, resistively controlled, phenomena.

(a) Determining growth rates of linear resistive instabilities Specifying the discussion to cylindrical geometry and linear MHD, we can combine, as in the steady-state approach, the finite-element discretization for the radial direction and the spectral discretization for the other two spatial directions. The Galerkin method then yields system (15.133) (without $\mathbf{f}$ for a non-driven case) with the matrices $\mathbf{A}$ and $\mathbf{B}$ exactly the same as before. If we perform a forward time discretization, we get the integration formula

$$\mathbf{x}^{n+1} = (\Delta t\, \mathbf{B}^{-1}\mathbf{A} + \mathbf{I})\mathbf{x}^n\,. \tag{15.134}$$

Numerical stability limits the time step to

$$\Delta t \leq 2/\lambda_{\mathrm{max}}\,, \tag{15.135}$$

where λ_{max} is the largest eigenvalue of the system $\mathbf{A}\mathbf{x} = \lambda\,\mathbf{B}\mathbf{x}$. This CFL condition is often not acceptable in MHD because λ_{max} tends to infinity in the limit of infinite resolution in the direction normal to the magnetic flux surfaces. The generalized trapezoidal method (15.130) then gives the following algorithm for the time advance:

$$[-\mathbf{B} + \alpha\,\Delta t\,\mathbf{A}]\mathbf{x}^{n+1} = -[\mathbf{B} + (1-\alpha)\,\Delta t\,\mathbf{A}]\mathbf{x}^n\,, \tag{15.136}$$

which is unconditionally stable for $\alpha \geq 0.5$.

Kerner, Jakoby and Lerbinger [273] applied this technique to analyze resistive waves (which requires an appropriate initialization), and current- and pressure-driven resistive instabilities for cylindrical equilibrium configurations. They typically used $\alpha = 0.52$ in order to introduce some numerical damping to damp out the fast modes. An example is shown in Fig. 15.18, where the growth rate of a resistive instability is quantified for such a cylindrical "tokamak-like" equilibrium configuration with a peaked current density and constant toroidal field. They considered the class of profiles

$$j_z = j_0(1 - r^2/a^2)^\nu, \qquad B_z = 1, \qquad \rho = 1,$$

with a the radius of the cylinder. For this class, the ratio of the safety factor on the surface and on the axis $q(a)/q(0) = \nu + 1$. They set $\nu = 1$ and adjusted the constant j_0 to vary $q(0)$, since j_0 is connected with $q(0)$ by $j_0 = 2k/q(0)$, where $k = 2\pi/L$ defines the periodicity length of the cylinder (simulating a tokamak with large aspect ratio). The wall is placed directly at the plasma surface (cf. model I in Section 4.6.1 [1]). This equilibrium is known to be unstable against the $m = 1$

tearing mode (see Section 14.2.2) if the $q = 1$ surface is located inside the plasma. This instability is avoided by making sure that $q > 1$ over the whole plasma radius. The $m = 2$ tearing mode is then the most dangerous instability. The growth rate of the most unstable mode is plotted versus $q(a)$ in Fig. 15.18. When the wall is placed directly at the surface, the $m = 2$ tearing mode is unstable for $2.20 \leq q(a) \leq 4.00$. The growth rates were obtained by studying the time evolution of an initial random starting vector with a time step $\Delta t = 400$ for the strong instabilities near $nq(a) = 3$, and $\Delta t = 1000$ for the weaker instabilities around $nq(a) = 2.1$. For $nq(a) \leq 2.3$ the instability changes from a current-driven into a pressure-driven mode.

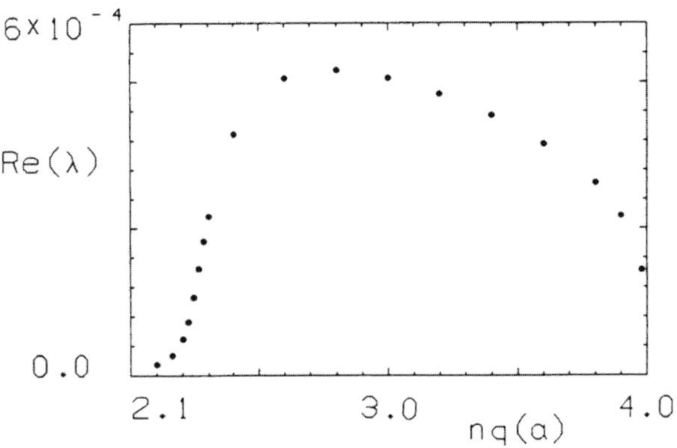

Fig. 15.18 Growth rates of resistive instabilities versus total current, parameterized by $q(a)$, showing the transition (at $nq = 2.3$) from pressure driven to current driven instabilities. (From Kerner *et al.* [273].)

When an external driving source excites the plasma, the implementation of the boundary conditions through the boundary terms in the weak form of the equations yields a "force" term $\mathbf{f}$ and a small change to the matrix $\mathbf{A} \rightarrow \mathbf{A}'$, as discussed in Section 15.2.2. Hence, we now retain $\mathbf{f}$ in (15.133) and the algorithm for the time advance is modified to

$$\underbrace{\left[-\mathbf{B} + \alpha\,\Delta t\,\mathbf{A}'\right]}_{\equiv\,\hat{\mathbf{A}}}\mathbf{x}^{n+1} = \underbrace{-\left[\mathbf{B} + (1 - \alpha)\,\Delta t\,\mathbf{A}'\right]}_{\equiv\,\hat{\mathbf{B}}}\mathbf{x}^n + \underbrace{\Delta t[(1 - \alpha)\mathbf{f}^n + \alpha\mathbf{f}^{n+1}]}_{\equiv\,\hat{\mathbf{f}}(t)}.$$

$$(15.137)$$

This equation for $\mathbf{x}^{n+1}$ is of the same form as Eq. (15.136) with the exception that the RHS is now *time dependent* through the driving term $\hat{\mathbf{f}}(t)$. However, the matrix operations are straightforward and preserve the block-tridiagonal structure. Hence, in each time step the above system (15.137) can be solved in the same way as the steady state problem (Section 15.2.2). Notice that the Δt is constant so that the matrices $\hat{\mathbf{A}}$ and $\hat{\mathbf{B}}$ are constant in time. This means that the CPU time consuming

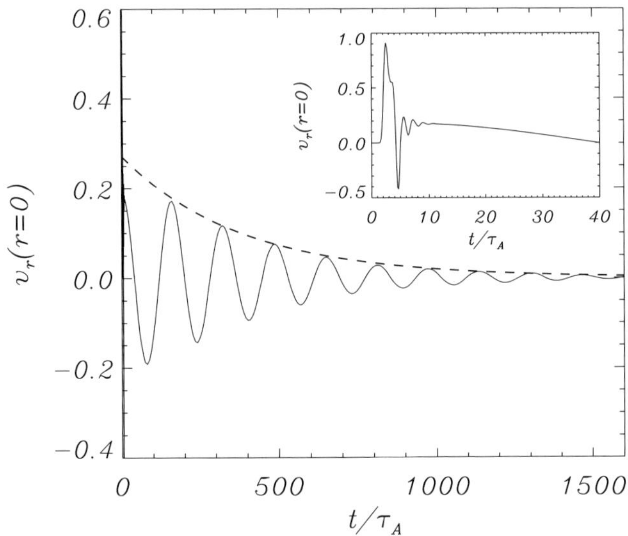

Fig. 15.19 The temporal evolution of the radial velocity at the loop axis, for a suddenly perturbed coronal loop. The inset shows the initial leaky transient, while the main plot shows the longer-period attenuated oscillation identified with a resonantly damped kink mode. (From Terradas *et al.* [434].)

LU factorization needs to be carried out only once. The solution process itself, however, must be carried out several 10^4 times since the time step Δt is limited by the period of the external driving source. As concrete examples of this approach, see the results on time scales and transient phenomena involving resonant absorption and wave heating effects discussed in Chapter 11 [1]. For example, Figs. 11.8 and 11.9 [1] have been obtained with the above semi-discretization method.

(b) Damped coronal loop oscillations A representative example of a linear MHD time evolution concentrates on solar coronal loop dynamics. The advent of modern high-resolution, high cadence, views on the highly structured coronal plasma by space missions such as the transition region and coronal explorer (TRACE, see e.g. the image in Fig. 8.10 [1]), has identified many MHD wave modes in coronal loops. Routinely, transverse oscillations are detected in these loops, e.g. in connection with a solar flare disruption of a neighboring arcade. Detailed information on the oscillation amplitudes, periods and damping times has been collected. Although the observed loop displacements are large, linear MHD theory can still be applicable since the associated velocities are small with respect to the Alfvén speed in the low-β coronal conditions. Terradas *et al.* [434] solved the linearized resistive MHD equations for a cylindrical coronal loop model exploiting a zero-β limit. In this limit, slow modes are eliminated and the governing equations need only retain radial and azimuthal velocity perturbations together with the linear field $\mathbf{B}_1$. The set of equations was semi-discretized, and the spatial discretization employed a finite element treatment of the radial direction, and a Fourier handling in the ignorable directions of the equilibrium. A single Fourier mode pair $(m, k_z) = (1, \pi/L)$ was chosen to isolate kink displacements of a line-tied loop of length L. The equilibrium taken considered a uniform axial magnetic field $\mathbf{B}_0 = B_0 \mathbf{e}_z$, and a smoothly varying density profile connecting the higher internal loop density ρ_i to the external coronal value ρ_e. The temporal discretization used the second order Crank–Nicolson approach, and simulated the long-term evolution of the response of the loop to a sudden radial velocity perturbation impinging on it. In Fig. 15.19, the typical temporal behavior of the radial velocity at the loop axis is shown: several phases characterized by totally different time scales are evident. The authors identified the first series of strongly attenuated short-period oscillations seen in the inset of Fig. 15.19 with a leaky eigenmode of the loop configuration. As the perturbation was selected to deposit a considerable amount of energy in the loop, this first phase corresponds to the resulting excitation of fast waves carrying energy radially off to infinity. The damped longer period oscillation eventually dominating the loop dynamics was then found to be in exact agreement with the presence of a resonantly damped quasi-mode. This global kink eigenoscillation is coupled to torsional Alfvén motions in a thin resonant layer due to the density variation. The strongly localized resistive dissipation quickly leads to the observed amplitude decay. By also computing the eigenmodes of the system, a one-to-one correspondence with the damped quasi-mode was established. This time-dependent analysis of loop oscillations convincingly showed the combined interpretative power of both eigenvalue and initial value approaches.

15.5 Concluding remarks

Computational MHD is essential for present-day modeling of fundamental plasma processes. We briefly discussed basic concepts of spatial and temporal discretization techniques, and gave various applications in computational linear MHD. Applications considered steady state as well as normal mode codes, together with time evolution codes for simulating linear wave dynamics. Numerical MHD simulations become more realistic as increasing computing power allows the consideration of higher Reynolds numbers (introducing smaller length scales and longer time scales), more complex geometries (curvature, stratification, X-points), and more "physics" (background flow, thermal conduction, etc.). This requires ever more efficient algorithms for dealing with large linear systems and eigenvalue problems, which exploit the power of massively parallel computer systems. We conclude this chapter by pointing out some overall guidelines for selecting suitable numerical methods. The choice of the numerical method for a specific problem should always consider the following criteria.

- *Consistency*: the approximation should converge to the real solution in the limit of vanishing Δt and Δx.
- *Numerical stability*: the round-off errors should not grow (accumulate).
- *Accuracy*: the approximation should be of a sufficiently high order in Δx and Δt or in the dimensionless parameters

$$\delta = \left| \frac{\Delta x}{u} \frac{\partial u}{\partial x} \right| \quad \text{and} \quad C = v \frac{\Delta t}{\Delta x}.$$

- *Efficiency*: the CPU time consumption and computer memory requirements should be optimized for a given accuracy.
- *Monotonicity*: a (locally) monotone solution at time t should remain monotonic at $t + \Delta t$, i.e. at all later times.

The last criterion is of particular relevance when dealing with nonlinear, shock-dominated problems. In Chapter 19, we continue the discussion of numerical methods, with a particular emphasis on nonlinear MHD computations.

15.6 Literature and exercises

Notes on literature

Finite differences and finite elements

- A good introduction to the basic theory and the practical implementation of the finite element method is given by William G. Strang and George J. Fix in *An Analysis of the Finite Element Method* [424]. Some sections are dated but the book is well-written, easy reading and contains many ideas on the behavior of finite elements which are interesting from both a mathematical and a practical point of view.
- A good introduction to the theory of finite difference and finite element methods is given by Stig Larsson and Vidar Thomée, *Partial Differential Equations with Numerical Methods* [297], Chapter 2, Sections 4.1 and 5.1.
- Dianne P. O'Leary's *Scientific Computing with Case Studies* [356] is a practical guide to the numerical solution of linear and nonlinear differential equations, optimization problems and eigenvalue problems. The exercises below are inspired by the homework assignment "Finite Differences and Finite Elements" of Chapter 23. With her consent, we adopted the assignment to the notation used by us.

Linear algebraic methods

- A good insight into the construction of iterative methods for the solution of linear systems with a large number of unknowns is provided by H. A. van der Vorst in *Iterative Krylov Methods for Large Linear Systems* [463]. Van der Vorst first discusses the main concepts and then explains how they lead to several efficient solvers such as CG, GMRES and Bi-CGSTAB. He also explains the main ideas behind the use of pre-conditioners and how they are constructed.

Exercises

[15.1] *Finite difference method*

We consider the simple problem

$$-[p(x)u'(x)]' + q(x)u(x) = f(x), \quad \text{for } x \in [0, 1],$$

where the functions $p(x)$, $q(x)$ and $f(x)$ are given and $u(0) = u(1) = 0$. We will assume that $0 < p(x) \le p_0$ and $q(x) \le 0$ for $x \in [0, 1]$.

- First rewrite the derivative as $-p(x)u''(x) - p'(x)u'(x)$. Then, use the first-order backward difference method (the equivalent of Eq. (15.15) of Section 15.1.2) for $u'(x)$ and the formula (15.17) for the second-order derivative $u''(x)$. Note that $p'(x)$ will be computed analytically so that no approximation to it is needed.

- Choose mesh points $x_j = jh$, where $h = 1/(M - 1)$ for some integer M, and solve for $u_j \approx u(x_j)$ for $j = 1, \ldots, M - 2$ to obtain an equation for each unknown by substituting the finite difference approximations for u'' and u' in the differential equation and then evaluating the equation at $x = x_j$.

- Let $M = 6$ and write the four finite difference equations at $x = 0.2, 0.4, 0.6$ and 0.8.

Notice that the matrix you constructed has non-zeros on only three bands around the main diagonal; all other elements are zero. The full matrix requires $(M - 2)^2$ storage locations, but, if we are careful, we can instead store all of the data in $\mathcal{O}(M)$ locations by agreeing to store only the non-zero elements, along with their row and column indices. This is a standard technique for storing sparse matrices, those whose elements are mostly zero.

[15.2] *Finite difference method: implementation*

Let us now see how this finite difference method is implemented.

- The Matlab function `finitediff1.m`, found on the website of Ref. [356], implements the finite difference method for our equation. The inputs are the parameter M and the functions p, q and f that define the equation. Each of these functions takes a vector of points as input and returns a vector of function values. (The function p also returns a second vector of values of p'.) The outputs of `finitediff1.m` are a vector ucomp of computed estimates of u at the mesh points xmesh, along with the matrix A and the right hand side g from which ucomp was computed, so that A ucomp = g.

- Add documentation to the function `finitediff1.m` so that a user could easily use it, understand the method, and modify the function if necessary.

[15.3] *Finite difference method: improvement*

There is a mismatch in `finitediff1.m` between our approximation to u'', which is second order in h, and our approximation to u', which is only first order. We can compute a better solution, for the same cost, by using a second order (central difference) approximation to u', so next we will make this change to our function.

- Define a central difference approximation to the first derivative by

$$u'(x) = [u(x + h) - u(x - h)]/(2h) + O(h^2).$$

- Modify the function of Problem 15.2 to produce a function `finitediff2.m` that uses this approximation in place of the first order approximation.

[15.4] *Finite element method: Galerkin approach*

For the finite element method, we keep the differential equation of Exercise 15.1.

– Apply the Galerkin approach and use integration by parts to obtain

$$\int_0^1 p(x) u_h'(x)\, \phi_h'(x) + q(x) u_h(x)\, \phi_h(x)\, dx = \int_0^1 f(x)\, \phi_h(x)\, dx,$$

for all functions $\phi_h \in S_h$, a subspace of H_0^1 (all functions that satisfy the boundary conditions and have a first derivative), with the integral of $(\phi_h'(x))^2$ on $[0, 1]$ finite.

[15.5] *Finite element method: preparation for implementation*

Choose for S_h the set of piecewise linear elements that are continuous and linear on each interval $[jh, (j+1)h]$, $j = 0, \ldots, M - 2$, where $h = 1/(M - 1)$. Use as basis of S_h, the set of hat functions ϕ_j, $j = 1. \ldots, M - 2$ defined by Eq. (15.29). These functions are designed to satisfy $\phi_j(x_j) = 1$ and $\phi_j(x_k) = 0$ if $j \neq k$.

– Put the unknowns u_j in a vector **u** and write the resulting system of equations as **Au** = **g**, where the (j, k) entry in **A** is $a(\phi_j, \phi_k)$ and the jth entry in **g** is (f, ϕ_j).

[15.6] *Finite element method: implementation (linear elements)*

Write a function `fe_linear.m` that has the same inputs and outputs as `finitediff1.m` but computes the finite element approximation to the solution using piecewise linear elements. Remember to store **A** as a sparse matrix.

[15.7] *Finite element method: implementation (higher order elements)*

Better accuracy can be achieved if we use higher order elements; for example, piecewise quadratic elements would produce a result within $\mathcal{O}(h^3)$ for smooth data. A convenient basis for this set of elements is the piecewise linear basis plus $M - 1$ quadratic functions ψ_j that are zero outside $[x_{j-1}, x_j]$ and satisfy

$$\psi_j(x_j) = 0, \quad \psi_j(x_{j-1}) = 0, \quad \psi_j(x_{j-1} + h/2) = 1 \quad \text{(for } j = 1, \ldots, M - 1).$$

– Write a function `fe_quadratic.m` that has the same inputs and outputs as the function `finitediff1.m` but computes the finite element approximation to the solution using piecewise quadratic elements. In order to keep the number of unknowns comparable to the number in the previous functions, let the number of intervals be $m = M/2$. If you order the basis elements as $\psi_1, \phi_1 \ldots, \psi_{m-1}, \phi_{m-1}, \psi_m$ then the matrix **A** will have five non-zero bands around the main diagonal.

– Compute one additional output `uval` which is the finite element approximation to the solution at the $m - 1$ interior mesh points and the m midpoints of each interval, where the $2m - 1$ equally spaced points are ordered smallest to largest. (In our previous methods, this was equal to `ucomp`, but now the values at the mid-points of the intervals are a linear combination of the linear and quadratic elements.)

[15.8] *Comparison of finite element methods versus finite difference methods*

Now we have four solution algorithms, so we define a set of functions for experimentation:

$$u_1(x) = x(1 - x)\, e^x,$$

$$u_2(x) = \begin{cases} u_1(x) \\ x(1 - x)\, e^{2/3} \end{cases}, \qquad u_3(x) = \begin{cases} u_1(x) & (x \leq 2/3) \\ x(1 - x) & (x > 2/3) \end{cases},$$

$$p_1(x) = 1, \qquad p_2(x) = 1 + x^2,$$

$$p_3(x) = \begin{cases} p_2(x) \\ (x - 1/3) + 10/9 \end{cases}, \qquad p_4(x) = \begin{cases} p_2(x) & (x \leq 1/3) \\ 1 + 2x^2 & (x > 1/3) \end{cases},$$

$$q_1(x) = 0, \qquad q_2(x) = 2, \qquad q_3(x) = 2x.$$

The function $f(x)$ is the one obtained under the specified true solutions mentioned below. Use your four algorithms to solve seven problems:

- p_1 with q_j ($j = 1, 2, 3$) and true solution u_1.
- p_j ($j = 2, 3$) with q_1 and true solution u_1.
- p_1 with q_1 with true solution u_j ($j = 2, 3$).

Compute three approximations for each algorithm and each problem, choose for the number of unknowns 9, 99 and 999. For each approximation, print $\|\mathbf{u}_{\text{computed}} - \mathbf{u}_{\text{true}}\|$ where $\mathbf{u}_{\text{true}}$ is the vector of true values at the $M - 1$ mesh points. Discuss the results.

- How easy is it to program each of the four methods? Estimate how much work Matlab does to form and solve the linear systems. (The work to solve the tridiagonal systems should be about $5M$ multiplications, for the 5-diagonal systems $11M$ multiplications, so you just need to estimate the work in forming each system.)
- For each problem, note the observed convergence rate r: if the error drops by a factor of 10^r when M is increased by a factor of 10, the observed convergence rate is r.
- Explain any deviations from the theoretical convergence rate: $r = 1$ and $r = 2$ for the two finite difference implementations, $r = 2$ and $r = 3$ for the finite element ones.

[15.9] *Test problem comparison of FEM versus FDM*

Use your four algorithms to solve the test problems that yielded Fig. 15.7 and Table 15.2, viz. p_2 with q_2 and true solution $u = x^4(1 - x)^4$. Compute three approximations for each algorithm with 9, 99 and 999 for the number of unknowns. For each approximation, print $\|\mathbf{u}_{\text{computed}} - \mathbf{u}_{\text{true}}\|$, where $\mathbf{u}_{\text{true}}$ is the vector of true values at the $M - 1$ mesh points.

[15.10] *Wave modes in unmagnetized, uniform, radiating gases*

Compute the dispersion relation of linear perturbations about a static (non-magnetized) homogeneous gas of constant density ρ_0, pressure p_0 and temperature T_0, which can lose internal energy by means of radiative losses.

- Assuming a radiative loss function $\mathcal{L}(\rho, T)$, which vanishes for the equilibrium conditions $\mathcal{L}(\rho_0, T_0) = 0$, obtain this dispersion relation by linearizing the set of (dimensionless) equations given by

$$\frac{\partial \rho}{\partial t} + \nabla \cdot (\rho \mathbf{v}) = 0, \qquad \rho\left(\frac{\partial}{\partial t} + \mathbf{v} \cdot \nabla\right)\mathbf{v} + \nabla p = 0,$$

$$\left(\frac{\partial}{\partial t} + \mathbf{v} \cdot \nabla\right)p + \gamma p \nabla \cdot \mathbf{v} + (\gamma - 1)\rho\mathcal{L}(\rho, T) = 0, \qquad p = \rho T.$$

- Assume for the linear perturbations ρ_1, $\mathbf{v}_1$, p_1 and T_1 the usual $\exp \mathrm{i}(\mathbf{k} \cdot \mathbf{r} - \omega t)$ dependence, with wave vector $\mathbf{k}$ and eigenfrequency ω. (Note: since the precise dependence of $\mathcal{L}$ on thermodynamic variables ρ and T is further unspecified, linearization will introduce partial derivatives $\mathcal{L}_\rho \equiv \partial\mathcal{L}/\partial\rho|_T$ and $\mathcal{L}_T \equiv \partial\mathcal{L}/\partial T|_\rho$.)
- Renaming $\lambda \equiv -\mathrm{i}\omega$, discuss the essential cubic dispersion relation in λ in terms of its physical implications: what changes from the case when radiative losses are absent?
- Assuming that the radiative terms are small and the cubic in λ has then one real root and a pair of complex conjugate roots, can you derive an instability criterion for the real root from the constant term in the cubic?
- Discuss the basic physical mechanism to form condensations in a uniform radiating gas in the absence of gravity through this "thermal instability".

Part IV

Toroidal plasmas

16

Static equilibrium of toroidal plasmas

16.1 Axi-symmetric equilibrium

16.1.1 Equilibrium in tokamaks

The aim of the theory of plasma equilibrium in any configuration is to *determine the global magnetic confinement topology* and the physical characteristics of the underlying basic equilibrium state. For most fusion applications, to first approximation, this state is assumed to be static, i.e. the background plasma velocity and the time derivative of the other variables vanish, $\mathbf{v} = 0$ and $\partial/\partial t\{\rho, p, \mathbf{B}\} = 0$. A superficial impression might be that this must correspond to the most boring example of plasma behavior, viz. total absence of dynamics: a corresponding fluid dynamics problem hardly exists. Of course, the reason for our interest in this state is the prospect of obtaining clean, abundant, and cheap energy from controlled thermonuclear fusion reactions. At present, the most promising candidate to reach this goal is the tokamak configuration, in which the assumption of static equilibrium is satisfied to a rather high degree of precision.

The MHD equations for static equilibrium are about the best satisfied plasma equations we know. If a plasma is sitting at rest, it is hard to imagine it satisfying any other conditions than the following ones:

$$\mathbf{j} \times \mathbf{B} = \nabla p \qquad \textit{(pressure balance)}, \qquad (16.1)$$

$$\mathbf{j} = \nabla \times \mathbf{B} \qquad \textit{(Ampère's law)}, \qquad (16.2)$$

$$\nabla \cdot \mathbf{B} = 0 \qquad \textit{(basic law of magnetic flux)}. \qquad (16.3)$$

In fact, if the equation of pressure balance is not satisfied, the plasma is immediately accelerated to huge velocities and there is no way to prevent it from being smashed into the wall and causing severe damage to the equipment.

In Section 1.2.3 of Volume [1], we have schematically summarized the history of magnetic plasma confinement experiments aimed at the eventual construction of a thermonuclear reactor. After the declassification of the subject in 1958, some ten years of rather unsuccessful research of high-beta plasmas in z-pinch and θ-pinch configurations followed, exhibiting dramatic demonstration of lack of equilibrium, instabilities and end losses, all operating on the typical MHD time scale of microseconds. It was entirely unclear how time scales of these fusion experiments could ever be extended to seconds or minutes, let alone to steady state operation. All this changed by the announcement of progress by Soviet scientists with an entirely different confinement scheme, called *tokamak*, at the IAEA Novosibirsk conference of 1968 [16]. (See the review paper on tokamaks by Artsimovich [17] and Section 5.2 of Braams and Stott [66] on the history of this period.) Soon after this, tokamaks were constructed in many countries, and the next thirty years witnessed steady increase of the triple product $n\tau_\mathrm{E}\tilde{T}$, of density, energy confinement time and temperature in these devices, with a factor of about 10^6 nearing the required value for ignition of $5 \times 10^{21}\,\mathrm{m}^{-3}\,\mathrm{s\,keV}$ by the end of the twentieth century (see Fig. 1.1.1 of Wesson [481]). Based on this phenomenally successful upgrading, the International Tokamak Experimental Reactor (ITER) is presently being constructed in Cadarache, France.

Whereas the simple schemes of θ- and z-pinch easily produced the temperatures needed for thermonuclear ignition by shock heating, they fell short by a factor of at least a million with respect to the required confinement times. Crudely speaking, the tokamak configuration cures the main problems of the z-pinch (its instability due to the curvature of the magnetic field B_θ) and of the θ-pinch (its end losses) by combining them into a single configuration (see Fig. 1.4 [1]). With respect to the θ-pinch end losses, those are simply eliminated by closing the plasma column onto itself by means of a "toroidal" confinement chamber and the current-driven instabilities of the z-pinch are eliminated by keeping the toroidal plasma current I_φ below the Kruskal–Shafranov limit (2.160) [1] so that the resulting "poloidal" magnetic field B_p (corresponding to B_θ in cylinder geometry) remains much smaller than the dominant toroidal magnetic field B_φ (corresponding to B_z). Hence, the magnetic field structure is now helical and the confinement geometry is toroidal, rather than cylindrical. Most important, the plasma β ($\equiv 2p/B^2$) is significantly decreased and the induction of currents is slow, so that shock heating has to be abandoned and replaced by other heating methods (Ohmic, neutral beams, ion and electron cyclotron resonance heating and, eventually, heating by the fusion produced α particles themselves.). Although the tokamak configuration thus eliminates the mentioned insurmountable problems of θ-pinches and z-pinches, quite a variety of lesser equilibrium and stability problems have to be addressed still. These will be the subject of the present and the next chapter.

Concerning equilibrium in tokamaks, returning to the misleadingly simple looking equations (16.1)–(16.3), the very first remark to be made is that these equations are actually *nonlinear* partial differential equations, as follows immediately by substituting Ampère's law (16.2) into the pressure balance equation (16.1). We have encountered the central importance of the basic law (16.3) for magnetic flux many times in Volume [1]. Hence, the solution of the combined equations

$$(\nabla \times \mathbf{B}) \times \mathbf{B} = \nabla p, \qquad \text{or} \quad -\nabla(p + \tfrac{1}{2}B^2) + \mathbf{B} \cdot \nabla \mathbf{B} = 0, \qquad (16.4)$$

$$\text{and} \quad \nabla \cdot \mathbf{B} = 0, \qquad (16.5)$$

with the single boundary condition

$$\mathbf{n} \cdot \mathbf{B} = 0 \quad \textit{(plasma boundary)}, \qquad (16.6)$$

for plasmas without circular cylinder symmetry (2D: tokamaks, 3D: stellarators) constitutes a nonlinear problem with many intriguing geometrical features that turns out not to be a boring problem at all!

From the original equations (16.1) and (16.2), it follows that the magnetic field and the current density are orthogonal to the pressure gradient:

$$\mathbf{B} \cdot \nabla p = 0, \qquad \mathbf{j} \cdot \nabla p = 0. \qquad (16.7)$$

Hence, the *magnetic surfaces* spanned by the magnetic field lines and the current density lines are also constant pressure surfaces. If none of the quantities ∇p, $\mathbf{B}$ and $\mathbf{j}$ vanishes anywhere in the plasma volume, these surfaces form an infinite continuous sequence of nested magnetic toroids around a single closed curve, called the *magnetic axis* [287]. Consequently, the magnetic surfaces may be labeled by a single variable Ψ, so that $p = p(\Psi)$. We will soon see that the most effective choice for that variable is the poloidal magnetic flux.

For tokamaks, the equilibria still have the additional property of *axi-symmetry* ($\partial/\partial\varphi = 0$), which is of great help in explicit calculations. Fig. 16.1 shows a schematic representation of the magnetic configuration. Because of the axi-symmetry of the configuration, it is expedient to exploit special cylindrical coordinates R, Z, φ, where R is the distance from the symmetry axis, Z is the vertical coordinate and φ is the (ignorable) toroidal angle (see Appendix A.2.4). Although the toroidal magnetic field component is much larger than the poloidal component in tokamaks, the latter one is usually considered as the primary quantity since it effectively describes the magnetic geometry of the tokamak. This is done by defining the toroidal and poloidal magnetic fluxes through the respective surfaces S_Φ and S_Ψ (indicated by the shaded areas in Fig. 16.1):

$$\Phi \equiv \frac{1}{2\pi} \int B_\varphi \, dS_\Phi, \qquad \Psi \equiv \frac{1}{2\pi} \int B_\mathrm{p} \, dS_\Psi, \qquad (16.8)$$

where the normalization factor $1/(2\pi)$ is introduced to simplify the relationship between the polodial flux and the poloidal magnetic field; see Eq. (16.84) below.

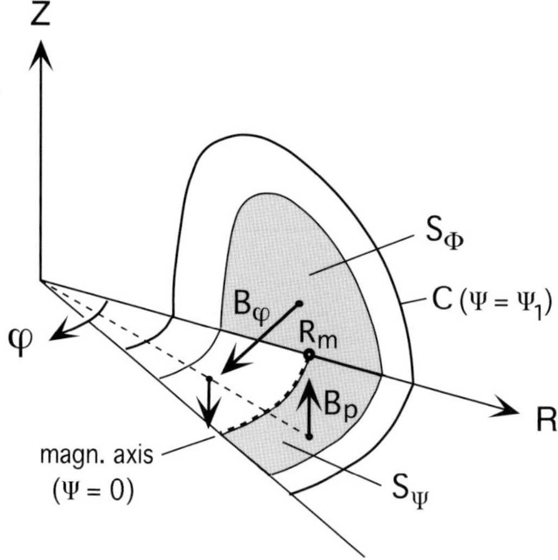

Fig. 16.1 Toroidal and poloidal magnetic field components B_φ and B_{p} in a tokamak and cross-sections S_Φ and S_Ψ for the corresponding magnetic fluxes Φ and Ψ within a magnetic surface lying between the magnetic axis and the plasma boundary, indicated by its cross-section (the curve C) with the poloidal plane $\varphi = 0$. Of course, the surface S_Φ also includes the reflected part $Z \leq 0$ of the shaded area and it can be taken at any angle φ. Similarly, the surface S_Ψ covers the full range $0 \leq \varphi \leq 2\pi$ and it can be taken between the magnetic axis and any circle $R = \mathrm{const}$, $Z = \mathrm{const}$ lying in the magnetic surface.

In Fig. 16.1, the poloidal flux surface S_Ψ is chosen in the horizontal plane $Z = 0$, containing the magnetic axis (the dashed circle $R = R_{\mathrm{m}}$), so that the poloidal field is pointing up for $R > R_{\mathrm{m}}$ and down for $R < R_{\mathrm{m}}$. From the flux tube concept (see Section 4.2.1 [1]), it is obvious though that the outer boundary of S_Ψ may be any horizontal circle lying in the magnetic surface. Hence, the infinitely many nested magnetic flux surfaces are most effectively labeled with the value of Ψ, which thus plays the role of a "radial" coordinate, running from $\Psi = 0$ at the magnetic axis to $\Psi = \Psi_1$ at the plasma boundary, where Ψ_1 is the total poloidal flux confined within the plasma. The toroidal magnetic flux is then considered as a function of the poloidal flux, $\Phi = \Phi(\Psi)$.

Once this is established, the helicity of the magnetic field lines is expressed as the derivative of the toroidal flux with respect to the poloidal flux. This quantity is known as the *safety factor*, or *inverse rotational transform*, of the field lines:

$$q(\Psi) \equiv \frac{d\Phi}{d\Psi}. \tag{16.9}$$

This function, which runs from $q_0 \equiv q(0)$ on the magnetic axis to $q_1 \equiv q(\Psi_1)$ on the plasma boundary, is the most important physical variable in the stability analysis of tokamaks. An alternative expression for the safety factor may be obtained from Eq. (16.9) by considering the poloidal flux $d\Psi = RB_{\mathrm{p}}\,dx$ through an infinitesimal annulus between two flux surfaces separated by a local distance dx and the corresponding toroidal flux $d\Phi = (2\pi)^{-1}\oint(B_\varphi dx)d\ell$, where $\oint d\ell$ is the line integral along the closed boundary curve of the toroidal flux surface S_Φ (see Wesson [481], Section 3.4). This yields

$$q(\Psi) = \frac{1}{2\pi}\oint_\Psi \frac{B_\varphi}{RB_{\mathrm{p}}}\,d\ell\,, \qquad (16.10)$$

which now expresses the safety factor as some average of the ratio B_φ/B_{p}, i.e. of the tangent of the field line.

Before we discuss the implications of the details of the Ψ-dependence of the safety factor, it is useful to stress the importance of the global properties of the equilibrium, which are mainly determined by two physical parameters and one geometrical parameter. Concerning the latter, the geometry is mainly controlled by *the inverse aspect ratio of the torus,*

$$\epsilon \equiv a/R_0\,, \qquad (16.11)$$

where a is the half width of the plasma column and R_0 is the distance of the center of that column to the symmetry axis. For simplicity, we will usually (except in Sections 16.1.4 and 16.2.3) assume that the plasma boundary coincides with a "wall" surrounding it so that a and R_0 are just geometrical properties of that wall. Moreover, we will assume up–down symmetry of the equilibrium with respect to the horizontal plane $Z = 0$. None of these assumptions is essential for the argument, they just simplify the analysis. (In particular, divertors for impurity control usually destroy up–down symmetry.) The physical parameters are the value of the *safety factor at the edge of the plasma,*

$$q_1 \equiv q(\Psi_1)\,, \qquad (16.12)$$

and the value of the *average kinetic pressure compared to the magnetic pressure,*

$$\beta \equiv \frac{2\langle p\rangle}{\langle B_\varphi^2\rangle}\,, \qquad \langle f\rangle \equiv \int f dV\,. \qquad (16.13)$$

We will return to these determining parameters in Section 16.1.4 when we discuss global confinement in tokamaks, i.e. equilibrium "in the large". Of course, all the local dependence of the plasma variables has to be determined as well. This can only be done if the magnetic geometry is known in detail, i.e. if the nonlinear partial differential equations (16.1)–(16.3) have been solved. This will concern us in Section 16.2, but we will first define the relevant properties of the magnetic field lines and coordinates describing them.

16.1.2 Magnetic field geometry

In order to appreciate the meaning of the average in the definition (16.10) of the safety factor, we need to consider the magnetic field lines themselves. Introducing the infinitesimal tangent vector ds_{fl} to them, the equation for the field lines may be written as

$$\mathbf{B} \times ds_{\mathrm{fl}} = 0 \,. \tag{16.14}$$

To work out the consequences of this equation, it is expedient to exploit coordinates based on the poloidal flux Ψ and the toroidal angle φ, rather than the cylindrical coordinates (R, Z, φ), which do not "see" the geometry of the magnetic field. The third coordinate must be a poloidal angle indicating the position on the flux surface for given φ. That coordinate, θ, could be a polar angle but, since any function of θ that increases monotonically from 0 to 2π is again an acceptable poloidal co-ordinate, it is convenient to exploit the arbitrariness to construct coordinates with simplifying properties. Two such coordinate systems are frequently exploited, viz. *orthogonal flux coordinates*, where the gradient of the poloidal angle is chosen orthogonal to $\nabla\Psi$ and $\nabla\varphi$, and *straight field line coordinates*, where the poloidal angle is chosen such that the field lines are straightened out in the representation of the tangential plane. The details of these coordinates are put in small print since they are a detour of the present exposition, but the reader who is not familiar with these constructions is advised *not* to skip this part since it is central to understanding of the geometry of the equilibrium needed in this chapter and the next.

▷ **Flux coordinates** The construction of these coordinates involves a task that recurs in equilibrium calculations, viz. the determination of the connection between the ordinary geometrical coordinates and the coordinates based on the flux:

$$\big[(X,Y,Z) \;\Rightarrow\;\big] \quad (R,Z,\varphi) \quad \Rightarrow \quad (\Psi,\theta,\varphi)\,. \tag{16.15}$$

This is actually an entirely non-trivial problem since it involves, first, the solution of the Grad–Shafranov equation (see Section 16.2), providing $\Psi = \Psi(R, Z)$, and, next, the construction of $\theta = \theta(R, Z)$ from the condition that describes the property desired of the poloidal angle. For now, we simply assume that these problems have been solved (we return to them in Sections 16.2 and 16.3), so that these two functions are known.

For *orthogonal flux coordinates* (Ψ, χ, φ), for which $\nabla\chi \cdot \nabla\Psi = 0$, we can then construct the line element,

$$(ds)^2 = h_1^2 (d\Psi)^2 + h_2^2 (d\chi)^2 + h_3^2 (d\varphi)^2 \,, \tag{16.16}$$

the volume element and the Jacobian,

$$dV = J_\circ\, d\Psi\, d\chi\, d\varphi \,, \qquad J_\circ \equiv (\nabla\Psi \times \nabla\chi \cdot \nabla\varphi)^{-1} = h_1 h_2 h_3 \,, \tag{16.17}$$

the gradient operator,

$$\nabla = \mathbf{e}_\Psi \frac{1}{h_1} \frac{\partial}{\partial \Psi} + \mathbf{e}_\chi \frac{1}{h_2} \frac{\partial}{\partial \chi} + \mathbf{e}_\varphi \frac{1}{h_3} \frac{\partial}{\partial \varphi} \,, \tag{16.18}$$

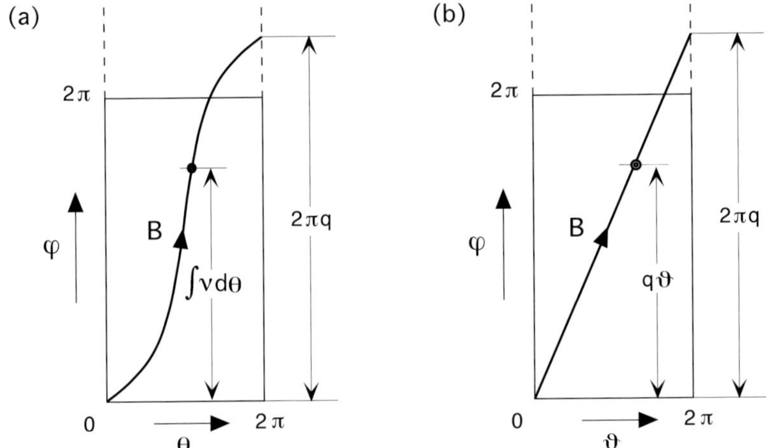

Fig. 16.2 Field line and safety factor q in plane tangential to a magnetic surface for (a) arbitrary flux coordinates (Ψ, θ, φ); (b) straight field line coordinates $(\Psi, \vartheta, \varphi)$.

and all other vector expressions (see Appendix A.2.7), which involve the three scale factors

$$ h_1 \equiv \frac{1}{|\nabla\Psi|} = \frac{1}{RB_p}, \quad h_3 \equiv \frac{1}{|\nabla\varphi|} = R \;\Rightarrow\; h_2 \equiv \frac{1}{|\nabla\chi|} = \frac{J_0}{h_1 h_3} = J_0 B_p. \quad (16.19) $$

This provides all expressions needed to account for the geometry of magnetic field lines in orthogonal flux coordinates (assuming the coordinate connections (16.15) are known). These coordinates are frequently used in analytical calculations, but not in numerical ones since they provide poor angular resolution in plasmas with elongated cross-sections.

For *straight field line coordinates* $(\Psi, \vartheta, \varphi)$, we determine the function $\vartheta(\theta)$ by straightening out the field lines in the tangential (ϑ, φ) plane (see Fig. 16.2) for each flux surface $\Psi = \text{const}$. This coordinate system has the obvious advantage of a simple representation of the field lines, which is extremely important in stability calculations. However, it is non-orthogonal so that its use requires the knowledge of the four elements of the metric tensor (see Appendix A.3.1):

$$ g_{11} \equiv \left|\frac{\partial \mathbf{r}}{\partial \Psi}\right|^2, \quad g_{12} \equiv \frac{\partial \mathbf{r}}{\partial \Psi} \cdot \frac{\partial \mathbf{r}}{\partial \vartheta}, \quad g_{22} \equiv \left|\frac{\partial \mathbf{r}}{\partial \vartheta}\right|^2, \quad g_{33} \equiv \left|\frac{\partial \mathbf{r}}{\partial \varphi}\right|^2 = R^2. \quad (16.20) $$

We just provide the expressions for the volume element and the Jacobian,

$$ dV = \mathcal{J}\,d\Psi d\vartheta d\varphi, \quad \mathcal{J} \equiv (\nabla\Psi \times \nabla\vartheta \cdot \nabla\varphi)^{-1} = \sqrt{[g_{11}g_{22} - (g_{12})^2]\,g_{33}}, \quad (16.21) $$

and leave explicit determination of the metric elements for later.   ◁

In the orthogonal flux coordinate system (Ψ, χ, φ), the magnetic field and the infinitesimal tangent vector are expressed as

$$ \mathbf{B} = (0, B_p, B_\varphi), \quad ds_{fl} = (0, J_0 B_p d\chi, R d\varphi), \quad (16.22) $$

so that Eq. (16.14) yields the local direction ν of the field lines in the representation of the plane tangential to the magnetic surfaces:

$$B_{\mathrm{p}}(R\,d\varphi - J_{\mathrm{o}}B_\varphi\,d\chi) = 0 \quad \Rightarrow \quad \nu \equiv \left.\frac{d\varphi}{d\chi}\right|_{\mathrm{fl}} = \frac{J_{\mathrm{o}}B_\varphi}{R}. \tag{16.23}$$

Hence, since $d\ell = J_{\mathrm{o}}B_{\mathrm{p}}\,d\chi$, the safety factor q just represents the *progression of the field line over the toroidal angle after one full poloidal revolution*:

$$q(\Psi) = \frac{1}{2\pi}\oint_\Psi \frac{B_\varphi}{RB_{\mathrm{p}}}\,d\ell = \frac{1}{2\pi}\oint_\Psi \nu(\Psi,\chi)\,d\chi. \tag{16.24}$$

This is illustrated in Fig. 16.2(a) for arbitrary flux coordinates (Ψ,θ,φ) on a magnetic surface with $q(\Psi) > 1$.

Actually, inside a magnetic surface, orthogonality of $\nabla\Psi$ and $\nabla\chi$ is not of much help. It is much more expedient to exploit the arbitrariness of the poloidal coordinate to construct a coordinate ϑ for which the field lines are straight, as shown in Fig. 16.2(b). In the definition of q, the line element along the poloidal circumference of a flux surface differs for the different coordinate systems since their Jacobians are different. In particular,

$$d\ell = JB_{\mathrm{p}}\,d\theta = J_{\mathrm{o}}B_{\mathrm{p}}\,d\chi = \mathcal{J}B_{\mathrm{p}}\,d\vartheta \tag{16.25}$$

for arbitrary flux coordinates (Ψ,θ,φ), with Jacobian J; orthogonal flux coordinates (Ψ,χ,φ), with Jacobian J_{o}; and straight field line coordinates, with Jacobian $\mathcal{J}$, respectively. Obviously, the last expression is the best choice since the local direction of the field lines then coincides with the global direction, as expressed by the relationship between the safety factor and the Jacobian that does not depend on the position on the flux surface:

$$q(\Psi) = \frac{\mathcal{J}B_\varphi}{R} \quad \textit{(straight field line coordinates)}. \tag{16.26}$$

Running ahead of our presentation (see Chapter 17), these coordinates are most appropriate to describe local stability since that basically involves Alfvén waves traveling along the field lines in a curved magnetic geometry.

An important concept in this context is that of *rational magnetic field lines*, situated on *a rational magnetic surface*. The latter is a surface where the field lines close upon themselves after M revolutions the short way (poloidally) and N revolutions the long way (toroidally) around the torus (see Fig. 16.3, for $M = 2$ and $N = 3$), so that the safety factor is a rational number there:

$$q_{\mathrm{rat}} = \frac{N}{M} \quad \left(= \frac{\text{number of toroidal revolutions}}{\text{number of poloidal revolutions}}\right). \tag{16.27}$$

If q is irrational, the field line does not close onto itself but covers the magnetic surface ergodically. The importance of rational field lines can be seen from the expression of magnetic perturbations on a rational magnetic surface. Expressing the

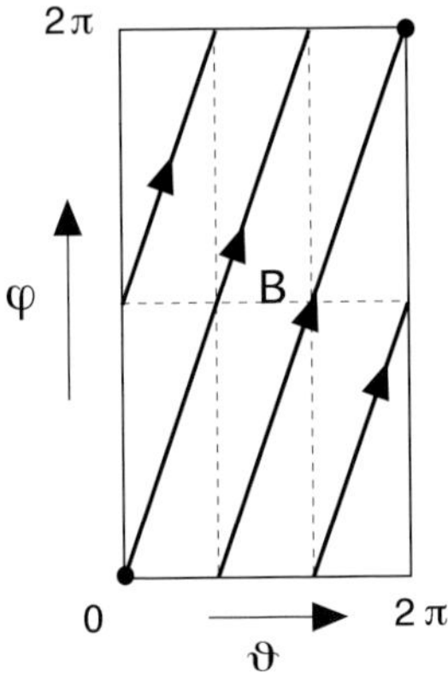

Fig. 16.3 Rational field line ($q = 3/2$).

doubly periodic perturbations in the form of an expansion in Fourier components of the plasma displacement normal to the magnetic surfaces,

$$\xi(\Psi, \vartheta, \varphi) = \sum_m \sum_n \xi_{mn}(\Psi)\, e^{i(m\vartheta + n\varphi)} , \tag{16.28}$$

where m is the poloidal mode number and n is the toroidal mode number, the main field line bending perturbation is roughly (neglecting curvature contributions) proportional to the parallel gradient operator acting on ξ:

$$\mathbf{B} \cdot \nabla \xi \sim \mathbf{k}_0 \cdot \mathbf{B}\, \xi \sim (m + nq)\, \xi . \tag{16.29}$$

Since field line bending is associated with a large increase of the potential energy, this term must vanish, or be small, for almost all instabilities that occur in tokamaks (not for astrophysical plasmas, recall the discussion of Section 12.1.2). The expression vanishes when the "wave vector" $\mathbf{k}_0$ is perpendicular to the magnetic field, i.e. when the wave fronts of the perturbation on the magnetic surface are parallel to the magnetic field lines so that they are minimally bent. According to Eq. (16.29), this happens when the ratio of mode numbers is rational, which only happens on a rational surface where *the field lines are resonant with the perturbation:*

$$q_{\text{rat}} = -\frac{m}{n} \quad \left(= -\frac{\text{poloidal mode number}}{\text{toroidal mode number}} \right) . \tag{16.30}$$

(Note that the adjectives "poloidal" and "toroidal" have switched position here.) Consequently, most local MHD stability analysis (Suydam's criterion (9.145)[1], the Mercier criterion and ballooning mode theory, see Chapter 17) centers about the resonant surfaces and field lines. (Therefore, if such resonances are forbidden because of different longitudinal boundary conditions, like line-tying of solar coronal flux tubes at the photospheric boundary, stability theory is completely changed and configurations are encountered which are much more stable; see Goedbloed and Halberstadt [178].)

16.1.3 Cylindrical limits

With the introduction of straight field line coordinates, the magnetic geometry has been reduced to a representation resembling that of a cylindrical plasma with circular cross-section. This is helpful because much intuition on equilibrium and stability comes from circular cylinder theory. Hence, let us consider the limit of a slender torus to a "periodic" straight cylinder, the so-called *straight tokamak* approximation, already introduced in Sections 9.1.1 and 9.4.4 [1]. We will use this limit to discuss the connection of the rather subtle effects of the safety factor and flux function distributions to the robust ones of the total current flowing in the plasma and the global pressure balance requirements.

Recall from Section 9.1.1 [1] that in a straight plasma cylinder with a circular cross-section, because of the symmetries $\partial/\partial\theta = 0$ and $\partial/\partial z = 0$, the solution of the equilibrium equations (16.1)–(16.3) is a function of the radius r alone. Moreover, $B_r = 0$ and $j_r = 0$, whereas the density profile $\rho(r)$ is completely arbitrary in static equilibria. The remaining equilibrium functions $p(r)$, $B_\theta(r)$, $B_z(r)$ have to satisfy just one differential equation, viz.

$$\frac{d}{dr}\left(p + \tfrac{1}{2}B^2\right) = -\frac{B_\theta^2}{r}, \tag{16.31}$$

so that two of those three functions can be chosen arbitrarily. The components of the current density are then determined by Ampère's law,

$$j_\theta = -\frac{dB_z}{dr}, \qquad j_z = \frac{1}{r}\frac{d}{dr}(rB_\theta), \tag{16.32}$$

which completes the description of cylindrical equilibria. In conclusion, in addition to $\rho(r)$, two of the three functions $p(r)$, $B_\theta(r)$, $B_z(r)$ can be freely chosen. This corresponds to the experimental freedom to create different magnetic confinement configurations, like those of a θ-pinch or a z-pinch (Figs. 16.4(a) and (b)). The same freedom is present in toroidal equilibria, where it needs to be represented very carefully in order to enable comparison of experimental data with theoretical assumptions on the equilibrium (see Section 16.3).

As discussed in Section 16.1.1 on the history of the different approaches to plasma confinement for fusion, the tokamak configuration may be considered as the combination of a θ-pinch with a much smaller pressure (and hence a smaller dip in the longitudinal field) and a z-pinch with a smaller plasma current (Fig.16.4(c)). This gives rise to a finite safety factor with a radially increasing profile, typically (but not necessarily) increasing monotonically from $q_0 \sim 0.8$ to $q_1 \sim 2.5$.

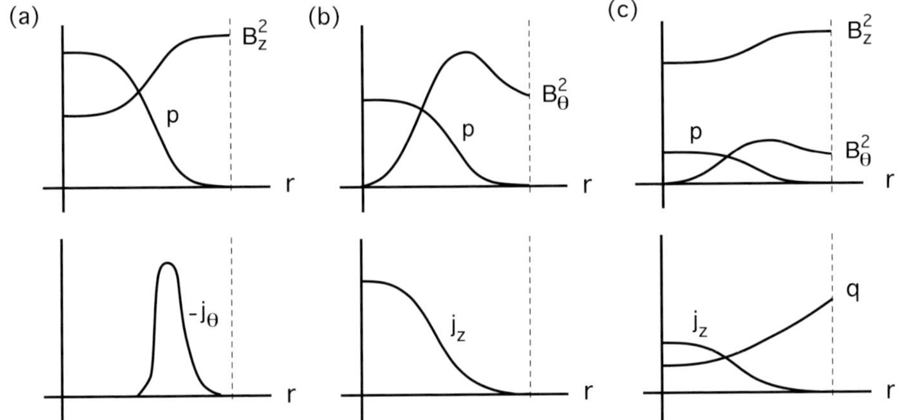

Fig. 16.4 Schematic cylindrical equilibrium profiles for (a) θ-pinch, (b) z-pinch, (c) "straight tokamak" with periodicity over $L_z \equiv 2\pi R_0$.

To define the safety factor q in straight circular cylinder geometry, one first needs to consider the original toroidal problem for a slender torus with *major radius* R_0 and *minor radius* $r = a$ (Fig. 16.5), so that the *inverse aspect ratio*, defined in Eq. (16.11), is small:

$$\epsilon \equiv a/R_0 \ll 1 . \tag{16.33}$$

In the limit $\epsilon \to 0$, this torus is represented as a "periodic cylinder" of length L_z and radius a, where the cylindrical coordinate z is related to the toroidal angle φ and the periodicity length L_z according to

$$z = [\varphi/(2\pi)]\,L_z , \qquad L_z \equiv 2\pi R_0 . \tag{16.34}$$

The nested magnetic surfaces become nested periodic cylinders of radius $r \leq a$. On each of those cylinders, the safety factor is easily visualized by unrolling that cylinder, which results in the straight field line representation of Figs. 16.2(b) and 16.3. In the limit $\epsilon \to 0$, the two magnetic fluxes defined in Eq. (16.8) become

$$\Phi_{\text{cyl}} = \int_0^r B_z r\,dr , \qquad \Psi_{\text{cyl}} = R_0 \int_0^r B_\theta\,dr , \tag{16.35}$$

so that any of the definitions (16.9), (16.10) or (16.24) of the safety factor yields

$$q_{\text{cyl}}(r) = \frac{d\Phi}{d\Psi} = \frac{1}{2\pi} \oint \frac{d\varphi}{d\theta}\bigg|_{\text{fl}} d\theta = \frac{r B_z}{R_0 B_\theta} . \tag{16.36}$$

The value of the safety factor at the edge of the plasma, defined in Eq. (16.12) is

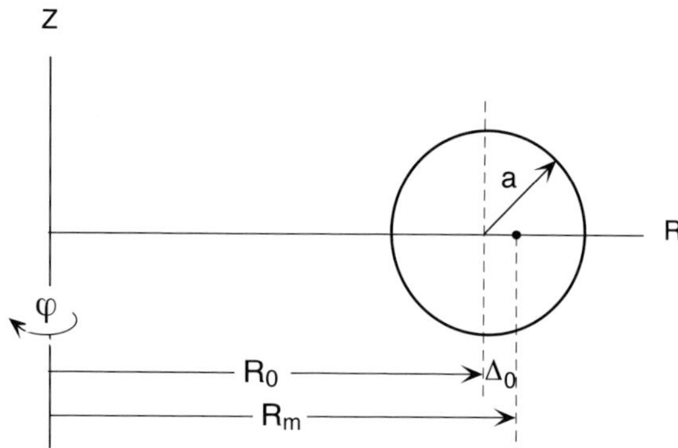

Fig. 16.5 Large aspect ratio torus with Shafranov shift Δ_0 of the magnetic axis.

then a *direct measure for the total toroidal current flowing in the plasma:*

$$q_{\text{cyl}}(a) = \frac{aB_z(a)}{R_0 B_\theta(a)} = \frac{2\pi a^2 B_z(a)}{R_0 I_z} \equiv (2\pi\epsilon a B_0)/I_z \,, \tag{16.37}$$

where the constant value $B_0 \equiv B_z(a)$ in the last expression anticipates the dimensional scaling that will be made in Section 16.3.1.

The importance of extracting the magnitude of the safety factor at the plasma boundary, $q_{\text{cyl}}(a)$, from the detailed radial distribution $q_{\text{cyl}}(r)$ is to stress the very different physical consequences of the toroidal current and the field line geometry. In particular, as already noted in Section 2.4.3 [1], when this difference is not appreciated, it leads to confusion on the cause of the $m = 1$ external kink mode instability. In a periodic cylinder with circular cross-section, the condition for stability of the external kink mode is the celebrated *Kruskal–Shafranov limit:*

$$q_{\text{cyl}}(a) > 1 \,, \quad \text{or} \quad I_z < 2\pi\epsilon a B_0 \,, \tag{16.38}$$

already encountered in Eqs. (2.160) and (9.91). This condition just depends on the value of $q_{\text{cyl}}(a)$, not on the details of $q_{\text{cyl}}(r)$, i.e. the current distribution, as clearly illustrated in Wesson's stability diagram [480], reproduced in Fig. 9.19 [1]. If this condition is violated, the result is just as bad as lack of equilibrium: the plasma is smashed onto the wall on a time scale of microseconds. Hence, the Kruskal–Shafranov limit is an essential limit on the parameter regime of tokamaks (the very reason of the necessity to decrease the toroidal current of a z-pinch to that of a tokamak, illustrated in Figs. 16.4(b) and(c)). The confusion arises because Eq. (16.38) appears to suggest, as frequently stated in the literature, that external kink mode instability occurs because the field lines close onto themselves after one revolution the short way and one revolution the long way around the torus.

However, this is just an unfortunate consequence of degeneracy of the circular cross-section cylinder, which disappears for genuine toroidal equilibria with non-circular cross-section. (To avoid further confusion: field line topology does play a dominant role in the stability of *internal* $m = 1$ kink modes, requiring $q_{\mathrm{cyl}}(0) > 1$. However, this is a much softer condition, as demonstrated by the fact that tokamaks usually operate in a regime where this condition is violated.)

The other global parameter describing the overall equilibrium of tokamaks is the *ratio of the average plasma pressure confined to the average pressure of the confining toroidal magnetic field*, defined in Eq. (16.13):

$$\beta \equiv \frac{2 \int p \, dV}{\int B_\varphi^2 \, dV} \approx \frac{2 \int p \, dS}{\int B_z^2 \, dS} \approx \frac{2}{B_0^2} \langle p \rangle, \qquad \langle p \rangle \equiv \frac{\int p \, dS}{\int dS}. \tag{16.39}$$

Here, the volume integral is over the total plasma volume and the surface integral is over the poloidal cross-section of it. The last two approximations come from the straight cylinder approximation, whereas the first one is due to the general orders of magnitude in a tokamak:

$$p \sim B_{\mathrm{p}}^2 \left(\approx B_\theta^2 \right) \ll B_\varphi^2 \left(\approx B_z^2 \right) \quad \Rightarrow \quad \beta \sim \epsilon^2 \ll 1, \tag{16.40}$$

as illustrated in Fig. 16.4(c). The importance of β is that it is a measure of the thermonuclear power obtained for a given magnetic field strength (Wesson [481], Section 3.5; Freidberg [141], Section 5.5.6). Because of the estimate (16.40), the total pressure is dominated by the contribution of the magnetic field:

$$P \equiv p + \tfrac{1}{2} B^2 \approx \tfrac{1}{2} B^2. \tag{16.41}$$

One gets an impression of the magnitude of this pressure by just inserting typical values of tokamaks, e.g. for $B = 5 \, \mathrm{T}$, we already found in Section 12.1.1 that

$$P \approx 10^7 \, \mathrm{N/m^2} = 1000 \, \mathrm{metric \ tons} = 100 \, \mathrm{atm} \,! \tag{16.42}$$

Clearly, magnetic pressures are huge and need to be balanced carefully. It is to be noted though that the dominant contribution of the above pressure is exerted on the external coils of the tokamak which, therefore, need a very strong supporting structure. The internal pressures, even though smaller by an order of magnitude, are not negligible either (e.g., for $\beta = 5\%$ the plasma pressure $p \approx 5 \, \mathrm{atm}$). Moreover, since β is a figure of merit for future fusion reactors, there is an urgent need to try to increase it. This implies that some part of the huge magnetic pressure estimated above should ultimately be shifted towards the plasma interior, that would then resemble more the θ-pinch configuration of Fig. 16.4(a). However, we will see in Section 16.1.4 that, within the tokamak confinement scheme, the best one can do eventually is to obtain values of $\beta \sim \epsilon$.

This upper estimate of the order of magnitude of β implies that the toroidal geometry is essential for the proper description of overall pressure balance, so that the cylindrical approximation fails. This is most clearly illustrated by one of the effects of toroidal pressure balance, viz. the outward shift of the magnetic surfaces relative to those of a circular cross-section cylinder: the *Shafranov shift* (Fig. 16.5). In the next section, we will see that the order of magnitude of this shift depends on the order of magnitude of β, as expressed by the following two expansion schemes:

$$q_1 \sim 1 \,, \quad \beta \sim \epsilon^2 \ll 1 \quad \Rightarrow \quad \Delta_0 \sim \epsilon \qquad (\textit{low } \beta \textit{ tokamak}) \,, \qquad (16.43)$$

$$q_1 \sim 1 \,, \quad \beta \sim \epsilon \ll 1 \quad \Rightarrow \quad \Delta_0 \sim 1 \qquad (\textit{high } \beta \textit{ tokamak}) \,. \qquad (16.44)$$

In the first expansion scheme, the zeroth order is the straight circular cylinder and toroidal effects enter as subsequent higher order corrections. In the second one, the cylindrical approximation fails since the equilibrium is essentially two-dimensional. Hence, we now have to turn to a proper toroidal description of tokamak equilibrium.

To summarize: except for the freedom of choice of the equilibrium profiles, two other properties of toroidal configurations can be anticipated in the context of straight cylinder theory. These have to do with the total toroidal current flowing in the plasma, i.e. the global magnetic field line geometry expressed by the edge safety factor, and with some aspects of the bulk forces associated with the magnetic pressure. However, a proper description of the magnetic pressure and of the kinetic pressure effects in the plasma requires a genuine toroidal theory.

16.1.4 Global confinement and parameters

Why can one not obtain the desirable high values of β by just bending the cylindrical θ-pinch column of Fig. 16.4(a) into a torus, i.e. by closing the configuration onto itself? To answer that question, we need to consider the complete equilibrium problem, both of *internal pressure balance* inside the plasma and of the position control of the plasma column as a whole by means of magnetic fields produced by currents in *external coils* (Shafranov [409], p. 124; Miyamoto [334], Section 7.7, and [335], Section 6.3; Wesson [481], Section 3.1; Freidberg [141], Section 11.7). In a tokamak, the large magnetic field B_z of the θ-pinch becomes the main toroidal magnetic field component B_φ. That component is primarily produced by poloidal currents I_p^{ex} in the external toroidal field coils, whereas the internal poloidal plasma current density $\mathbf{j}_p$ produces a relatively small deviation from that externally produced "vacuum magnetic field" distribution. Neglecting the poloidal plasma currents for the time being, the toroidal magnetic field is obtained by integrating Ampère's law (16.2) along a circle of radius R in the mid-plane ($Z = 0$), enclosing

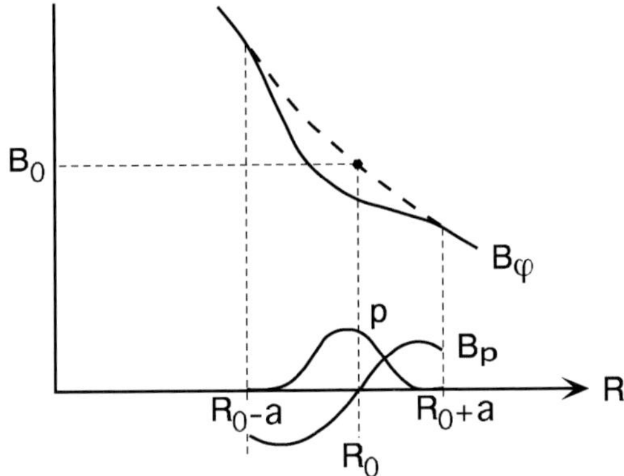

Fig. 16.6 Schematic tokamak equilibrium at the mid plane ($Z = 0$).

a surface S_R, and applying Gauss' law (A.14):

$$I_{\mathrm{p}}^{\mathrm{ex}} \approx \int \mathbf{j} \cdot \mathbf{n}\, dS_R = \oint \mathbf{B} \cdot d\mathbf{l}_R = 2\pi R B_\varphi \;\Rightarrow\; B_\varphi \approx R_0 B_0 / R. \tag{16.45}$$

Here, B_0 is the value of the external toroidal magnetic field on the plasma boundary at $R = R_0$, see Fig. 16.6. (For the scaling to be made later, in Section 16.3.1, it is important to note that this parameter is exact, independent of the approximation made.) The $1/R$ dependence of the toroidal magnetic field implies that the plasma column cannot be in equilibrium with this field since it produces a magnetic pressure that is much larger on the inside, the high field side ($R = R_0 - a$), than on the outside, the low field side ($R = R_0 + a$). The result is an outward force in the R-direction, which is larger than the inward force due to magnetic tension, as we will see below. To ensure equilibrium, counter measures have to be taken.

The counter measures taken are, first, induction of a (secondary) *toroidal plasma current* I_φ by means of coupling, due to the change in time of the poloidal magnetic flux through the central hole of the torus, to the toroidal current I_φ^{ex} in a set of (primary) windings on the outer legs of a transformer (see Fig. 1.5 of Volume [1]). This contributes the z-pinch part (Fig. 16.4(b)) to the tokamak confinement scheme, with the advantage of enhanced flexibility due to the additional parameter I_φ^{ex}. An obvious disadvantage is that tokamak operation now becomes limited to the time scale of resistive decay of the plasma current. This is the main reason for continued interest in the stellarator approach to plasma confinement, where such currents are not needed. (Eventual steady state operation of tokamaks has become feasible, though, through the kinetic effects of current drive by radiofrequency heating and the possible production of a toroidal bootstrap current by pressure gradients; see the review papers by Fisch [136] and Boozer [60].)

The induction of the toroidal plasma current appears to have an adverse effect on the equilibrium, viz. the production of an additional outward force, called the *hoop force*. This follows from the basic fact of electrodynamics that a current-carrying ring tends to increase its size in order to reduce the magnetic field strength for given magnetic flux trapped inside the ring. The hoop force can be obtained from the expression for the magnetic energy of the poloidal field of a thin current-carrying ring ($a \ll R$) with circular cross-section (Shafranov [409], p. 122):

$$ W_{\rm p} = \tfrac{1}{2} \int B_{\rm p}^2 \, dV = \tfrac{1}{2} L_\varphi I_\varphi^2 \,, \quad L_\varphi = R[\,\ln(8R/a) + \tfrac{1}{2}\ell_{\rm i} - 2\,] \,, \qquad (16.46) $$

where L_φ is the self-inductance of the ring. The internal contribution to the self-inductance, $\ell_{\rm i}$, defined below, is a positive quantity of order unity which depends on the distribution of the current (Landau and Lifschitz [295], p. 124; recall that the factor 4π of the Gaussian system of units is to be replaced by the factor μ_0 of the mks system of units, which is consistently dropped here). The hoop force is then given by:

$$ F_{\rm h} = -\frac{\partial W_{\rm p}}{\partial R}\bigg|_{L_\varphi I_\varphi} = \tfrac{1}{2} I_\varphi^2 \frac{\partial L_\varphi}{\partial R} \approx \tfrac{1}{2} I_\varphi^2 \,[\,\ln(8R_0/a) + \tfrac{1}{2}\ell_{\rm i} - 1\,] > 0\,, \quad (16.47) $$

i.e. it points outward, QED.

The second, crucial, counter measure is the creation of a homogeneous *vertical magnetic field* (in the Z-direction) by means of a toroidal current $I_{\varphi,\rm vf}^{\rm ex}$ in a set of external vertical field coils (Fig. 16.7). Of course, this counter measure is to be taken together with the first one, since a homogeneous vertical magnetic field has no effect on the toroidal θ-pinch (non-)equilibrium part of the tokamak, but it does have an effect on the toroidal z-pinch part by interacting with the toroidal plasma current. This is easily seen from the direction of the poloidal magnetic field vectors of Fig. 16.1: the external vertical field will increase the magnitude of the poloidal field on the outside, but decrease it on the inside. The resulting magnetic pressure will be inward. This inward force can be estimated from the Lorentz force of the vertical field on a wire carrying the toroidal plasma current I_φ:

$$ F_{\rm vf} \approx -2\pi R_0 I_\varphi B_{\rm v} \,. \qquad (16.48) $$

It is now simply a matter of turning the knob on the current $I_{\varphi,\rm vf}^{\rm ex}$ in the vertical field coils to keep the plasma column at the desired equilibrium position.

By means of the mentioned three external current parameters $I_{\rm p}^{\rm ex}$ (controlling the toroidal magnetic field B_φ), $I_\varphi^{\rm ex}$ (controlling the poloidal magnetic field $B_{\rm p}$), and $I_{\varphi,\rm vf}^{\rm ex}$ (controlling the external vertical magnetic field $B_{\rm v}$), the tokamak configuration obtains the necessary flexibility which has produced the impressive increase in performance towards controlled fusion described in Section 16.1.1. It remains to be shown how the value of β is determined by these parameters. This requires

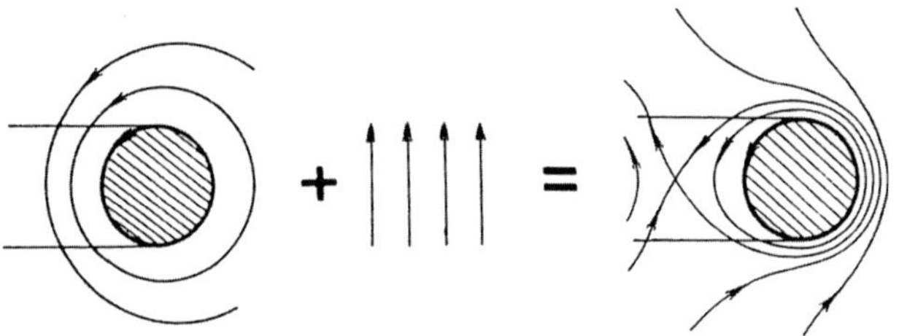

Fig. 16.7 Adding an external vertical field to ensure equilibrium in a tokamak. (From Mukhovatov and Shafranov [342].)

the consideration of the volume averaged effects of the gradients of the plasma pressure and of the toroidal field pressure, giving rise to forces F_p and F_{B_φ}.

Following Freidberg [141], Section 11.7.7, these forces are approximated by computing the different weighting due to the larger volume on the outside than on the inside. Exploiting polar coordinates r, θ, with the origin in the center of the plasma, $R = R_0$, $Z = 0$, i.e. neglecting the Shafranov shift, so that r and θ effectively become "cylindrical" coordinates, this only involves toroidal corrections of the volume element,

$$dV = 2\pi R r \, dr \, d\theta \approx 2\pi R_0 [1 + (r/R_0) \cos\theta] r \, dr \, d\theta \,, \tag{16.49}$$

coupled to the $\cos\theta$ variations of the outward unit vector $\mathbf{e}_R$,

$$\mathbf{e}_R = \cos\theta \, \mathbf{e}_r - \sin\theta \, \mathbf{e}_\theta \,, \tag{16.50}$$

so that the volume integrals just involve averaging of the $\cos^2\theta$ contributions:

$$-\int \mathbf{e}_R \cdot \nabla f \, dV \approx -2\pi R_0 \int_0^a \int_0^{2\pi} [1 + (r/R_0)\cos\theta] \cos\theta \, f' r \, dr \, d\theta$$
$$= -2\pi^2 \int_0^a f' r^2 \, dr = 2\pi^2 \left[2\int_0^a f r \, dr - a^2 f(a) \right] \equiv 2\pi^2 a^2 \left[\langle f \rangle - f(a) \right]. \tag{16.51}$$

Hence, the final volume averages just reduce to integration of the leading order, "cylindrical", contributions of the variables over the radius r:

$$\langle f \rangle \equiv \frac{1}{V} \int f \, dV \approx \frac{2}{a^2} \int_0^a f r \, dr \,. \tag{16.52}$$

All this will be justified rigorously in Section 16.2.2 from a large aspect ratio expansion of the Grad–Shafranov equation in the "shifted circle" approximation.

In effect, the leading order outward force due to pressure gradients becomes

$$F_p \equiv -\int \mathbf{e}_R \cdot \nabla p \, dV \approx 2\pi^2 a^2 \langle p \rangle \,. \tag{16.53}$$

For the calculation of the outward force due to the toroidal magnetic field, the $1/R$ dependence of B_φ, indicated by Eq. (16.45), should be accounted for:

$$B_\varphi(r,\theta) \approx (R_0/R)B_{\varphi0}(r) \approx [1 - (r/R_0)\cos\theta]B_{\varphi0}(r),\qquad(16.54)$$

where we now also include internal structure due to poloidal currents, indicated by the "cylindrical" factor $B_{\varphi0}(r)$. As a result of the $1/R^2$ dependence of the toroidal magnetic field pressure, the integrand of Eq. (16.51) flips sign, so that the associated outward force becomes

$$F_{B_\varphi} \equiv -\int \mathbf{e}_R \cdot \left[\nabla(\tfrac{1}{2}B_\varphi^2) + B_\varphi^2/R\right] dV \approx -\int \mathbf{e}_R \cdot \nabla(\tfrac{1}{2}B_\varphi^2)\, dV$$

$$\approx -\int \frac{R_0^2}{R^2}\, \mathbf{e}_R \cdot \nabla(\tfrac{1}{2}B_{\varphi0}^2)\, dV \approx -2\pi^2 a^2\left[\langle\tfrac{1}{2}B_{\varphi0}^2\rangle - \tfrac{1}{2}B_{\varphi0}^2(a)\right].(16.55)$$

The poloidal field also has a $1/R$ dependence, $B_{\mathrm{p}}(r,\theta) \approx (R_0/R)B_{\mathrm{p}0}(r)$, producing an outward force that has already been accounted for by the hoop force (16.47). Of course, the three functions $B_{\mathrm{p}0}(r)$, $p(r)$ and $B_{\varphi0}(r)$ have to satisfy the cylindrical equilibrium equation (16.31), which becomes

$$(p + \tfrac{1}{2}B_{\varphi0}^2)' = -\frac{B_{\mathrm{p}0}}{r}(rB_{\mathrm{p}0})'.\qquad(16.56)$$

It is now expedient to convert the toroidal field force (16.55) to an expression in terms of the plasma pressure and the poloidal field. This is done by applying the operator $\int dr\, r^2$ to Eq. (16.56) and integrating by parts, giving

$$-\langle\tfrac{1}{2}B_{\varphi0}^2\rangle + \tfrac{1}{2}B_{\varphi0}^2(a) = \langle p\rangle - \tfrac{1}{2}B_{\mathrm{p}0}^2(a).\qquad(16.57)$$

Hence

$$F_p + F_{B_\varphi} = 2\pi^2 a^2[2\langle p\rangle - \tfrac{1}{2}B_{\mathrm{p}0}^2(a)] = \tfrac{1}{2}I_\varphi^2(\beta_{\mathrm{p}} - \tfrac{1}{2}),\qquad(16.58)$$

where the new parameter

$$\beta_{\mathrm{p}} \equiv \frac{2\langle p\rangle}{B_{\mathrm{p}0}^2(a)} = \frac{8\pi^2 a^2}{I_\varphi^2}\langle p\rangle\qquad(16.59)$$

represents the average plasma pressure compared to the magnetic pressure of the poloidal field at the plasma boundary, i.e. measured in terms of the square of the total toroidal current flowing in the plasma.

Adding up the four forces yields global equilibrium,

$$F_{\mathrm{tot}} \equiv F_p + F_{B_\varphi} + F_{\mathrm{h}} + F_{\mathrm{vf}}\qquad(16.60)$$

$$= \tfrac{1}{2}I_\varphi^2\left[\beta_{\mathrm{p}} + \ln(8R_0/a) + \tfrac{1}{2}\ell_{\mathrm{i}} - \tfrac{3}{2} - \frac{4\pi R_0}{I_\varphi}B_{\mathrm{v}}\right] = 0,\qquad(16.61)$$

provided the vertical field has the proper magnitude

$$B_{\mathrm{v}} = \frac{I_\varphi}{4\pi R_0}\left[\beta_{\mathrm{p}} + \ln(8R_0/a) + \tfrac{1}{2}\ell_{\mathrm{i}} - \tfrac{3}{2}\right].$$ (16.62)

Here, the parameter

$$\ell_{\mathrm{i}} \equiv \frac{\langle B_{\mathrm{p}0}^2\rangle}{B_{\mathrm{p}0}^2(a)}$$ (16.63)

represents the average poloidal magnetic field pressure, which is a function of the radial distribution of the toroidal current density.

We have now encountered the *main parameters governing global equilibrium* in a tokamak with circular plasma cross-section. Taking the external "engineering" parameters $I_{\mathrm{p}}^{\mathrm{ex}}$, I_φ^{ex} and $I_{\varphi,\mathrm{vf}}^{\mathrm{ex}}$, respectively producing the toroidal, the poloidal and the vertical magnetic field, for granted, and zooming in onto the resultant equilibrium characteristics of the plasma column itself, we may distinguish

(1) the *trivial scaling parameters* a (plasma radius) and B_0 (external magnetic field at the center of the plasma, $R = R_0$) that fix the size of the plasma and the overall magnetic field strength: just to be used to create dimensionless quantities according to our usual scale-independence argument;

(2) the *geometry parameter* $\epsilon \equiv a/R_0$ that fixes the aspect radius of the plasma torus, and possible elongation parameters of the plasma cross-section;

(3) the *global confinement parameters* β_{p} and $B_{\mathrm{p}0}(a) \sim I_\varphi$, fixing the average amount of plasma pressure confined and the total toroidal current flowing in the plasma;

(4) *distribution parameters* describing the details of the pressure $p(r)$ and the toroidal current density $j_{\varphi 0}(r)$, or the poloidal field $B_{\mathrm{p}0}(r)$, for given global parameters. (The equilibrium condition (16.56), later to be replaced by the exact conditions from the Grad–Shafranov equation, then fixes the third profile, that of the toroidal field B_φ.)

We will consider these scaling considerations in more detail in Section 16.3.1, but item (3) needs to be elaborated here since it still involves the dimensional poloidal magnetic field and toroidal current variables that appear to be rather different from the safety factor q, that might be expected at this point.

In fact, an obvious way to create a dimensionless parameter representing the poloidal field at the plasma edge, $B_{\mathrm{p}0}(a)$, or the toroidal plasma current, I_φ, is to relate them to the safety factor (16.10), evaluated at the plasma edge:

$$\left.\begin{aligned}q_1 &= \frac{1}{2\pi}\oint_{\Psi_1}\frac{B_\varphi}{RB_{\mathrm{p}}}\,d\ell\\[2mm]I_\varphi &\equiv \iint \nabla\times\mathbf{B}\cdot\mathbf{e}_\varphi\,dS = \oint B_{\mathrm{p}}\,d\ell\end{aligned}\right\} \;\Rightarrow\; q_{1,\mathrm{cyl}} = \frac{aB_0}{R_0 B_{\mathrm{p}0}(a)} = \frac{2\pi a^2 B_0}{R_0 I_\varphi},$$ (16.64)

as already suggested in Eq. (16.37). Since elongating the plasma cross-section has become an important method of increasing the admissible value of β in tokamaks, with respect to both equilibrium and stability, let us consider the corresponding expression for a straight cylinder with elliptical cross-section [299]:

$$q_{1,\mathrm{cyl}} = \frac{\pi(a^2 + b^2)B_0}{R_0 I_\varphi}, \tag{16.65}$$

where a and b define the plasma cross-section through the points $R = R_0 \pm a$ and $Z = \pm b$. It would appear that β_p and $q_{1,\mathrm{cyl}}$ become the most appropriate global dimensionless parameters to describe equilibrium in a tokamak.

A serious objection needs to be raised, though, against the tacit identification of the robust effects of the toroidal current, expressed by the cylindrical parameter $q_{1,\mathrm{cyl}}$, and the subtle toroidal effects of field line geometry, expressed by the parameter $q(\Psi_1)$. For example, in tokamaks with a divertor, the line integral (16.64)(a) for the edge safety factor q_1 blows up due to the x-point at the plasma boundary, where $B_\mathrm{p} = 0$, whereas the line integral (16.64)(b) for I_φ just stays finite. Similarly, the ultimate limit of tokamak equilibrium by means of a vertical field, just discussed, implies that the x-point depicted in Fig. 16.7 intrudes into the plasma, again implying that $q_1 \to \infty$, whereas I_φ stays finite. Hence, we introduce a parameter that is formally identical to $q_{1,\mathrm{cyl}}$, but actually only measures the magnitude of the toroidal current, viz. the *modified safety factor* q^*:

$$q^* \equiv \frac{L^2 B_0}{2\pi R_0 I_\varphi}, \qquad L \equiv \oint d\ell \approx \pi\sqrt{2(a^2 + b^2)}. \tag{16.66}$$

This definition arose in stability studies of skin current high-β tokamaks [120], but turns out to be most adequate for our present general discussion of tokamak equilibrium. (We here drop the later modification [153, 179] of the definition of q^* and return to the earlier one since it conserves the mentioned relation with $q_{1,\mathrm{cyl}}$ for non-circular cross-sections.)

It is to be noted that the definition (16.66) is an exact one for toroidal geometry: it only involves the trivial item (1) scaling parameters a and B_0 and the item (2) geometry parameters b and R_0, that may be considered to be *external parameters,* i.e. they can be accurately prescribed (e.g. by the size of the limiters) and they do not require determination through plasma diagnostics. In contrast, the safety factor, both on edge and at the magnetic axis, though extremely important for local plasma dynamics, is only indirectly known by means of various diagnostics (usually delivering no better than 10% accuracy, up till now).

We now also need to replace the quasi-cylindrical definition (16.59) for β_p by an exact toroidal definition, for which we choose [153]

$$\beta_\mathrm{p} \equiv \frac{8\pi S\langle p\rangle}{I_\varphi^2}, \qquad S \equiv \iint dx\, dy \approx \pi ab, \tag{16.67}$$

where, again, an accurate external geometry parameter appears, viz. the surface area S of the plasma cross-section. Finally, we redefine the basic confinement parameter β by comparing the average plasma pressure with the magnetic pressure of the external toroidal field, i.e. again in terms of an external parameter:

$$\beta \equiv \frac{2\langle p \rangle}{B_0^2} . \tag{16.68}$$

With the global equilibrium parameters β_{p} and q^* thus defined, the parameter β, though crucial as a measure for the performance of fusion reactors, just becomes a secondary quantity:

$$\frac{\beta}{\epsilon} = C \, \frac{\epsilon \beta_{\mathrm{p}}}{q^{*2}} , \qquad C \equiv \frac{L^4}{16\pi^3 a^2 S} \approx \begin{cases} 1 & \text{(circle)} \\[2mm] \dfrac{(1 + b^2/a^2)^2}{4b/a} & \text{(ellipse)} \end{cases} . \tag{16.69}$$

Here, the constant C just depends on the cross-sectional shape of the plasma, i.e. it may be considered as an external parameter that is accurately known. With this parameterization, the two orderings introduced in Eqs. (16.43) and (16.44) of Section 16.1.3 can be expressed more appropriately as

$$q^* \sim 1, \qquad \beta/\epsilon \sim \epsilon \beta_{\mathrm{p}} \sim \begin{cases} \epsilon & \text{(low-beta tokamak)} \\[2mm] 1 & \text{(high-beta tokamak)} \end{cases}, \tag{16.70}$$

where the high-beta tokamak regime is the one where the limiting values of the parameters are found.

We have purposely replaced the usual definition of β, involving the average of the toroidal magnetic field pressure $\langle \frac{1}{2} B_\varphi^2 \rangle$, given in Eq. (16.39), by the above one which just involves the external magnetic field pressure $\frac{1}{2} B_0^2$. Also, the volume averages of plasma variables, defined in Eq. (16.52), have been replaced with cross-sectional averages. This way, the influence of poorly known internal distributions of plasma parameters is restricted to the necessary minimum, viz. to the pressure distribution entering $\langle p \rangle$. Of course, once an accurate equilibrium solver is constructed, all desired derived quantities can be computed "exactly", *necessarily making bold assumptions on those poorly known distributions though*, but this is just a numerical detail, not be confused with the basic scaling of plasma equilibrium variables.

The relation (16.69) summarizes the main constituents in the optimization procedure of tokamaks with respect to equilibrium and stability. Roughly speaking, as illustrated in Fig. 16.8, the value of the parameter $\epsilon \beta_{\mathrm{p}}$ is limited from above by equilibrium considerations, and the value of q^* is limited from below by stability conditions. With respect to equilibrium, we have already indicated that there is a

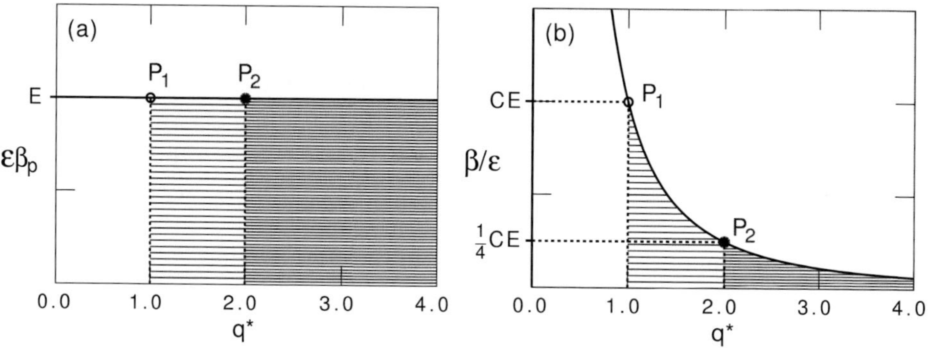

Fig. 16.8 Schematic equilibrium–stability diagram. (a) An x point intruding into the plasma imposes a limit $\epsilon\beta_{\mathrm{p}} \leq E$ to the equilibrium, external $m = 1$ and $m = 2$ instabilities impose the limits $q^* \geq 1$, respectively $q^* \geq 2$, to the stability. (b) This translates into the limits $\beta/\epsilon \leq CE$ (point P_1), resp. $\beta/\epsilon \leq \frac{1}{4}CE$ (point P_2).

limit due to the fact that the vertical magnetic field produces an x-point that will hit the plasma, and thus destroy the coherence of the nesting of the magnetic surfaces, if $\epsilon\beta_{\mathrm{p}}$ is increased beyond a certain value. Let us call that value E. For circular plasma cross-section, this number is easily estimated from the expression for the poloidal magnetic field B_{p}, which involves a parameter Λ that describes the poloidal variation (see Eq. (16.109) of the following section):

$$E \equiv \epsilon\beta_{\mathrm{p,max}} = \epsilon\Lambda_{\mathrm{max}} \approx 1 \,. \tag{16.71}$$

From the relation (16.69), it is clear that to obtain a maximum value of β, one wishes to push the toroidal current to a maximum, i.e. to choose the value of q^* as small as possible. Obviously, external kink mode stability will ultimately limit that value to $q^* = 1$, which is the Kruskal–Shafranov limit. However (see Wesson's stability diagram, reproduced in Fig. 9.19 of Volume [1]), the $m = 2$ external kink mode presents a much severer condition since it increases the limit to $q^* = 2$. Consequently, a rough estimate of the combined equilibrium–stability limit on β is given by

$$\beta_{\mathrm{max}} = C\,\frac{\epsilon^2\beta_{\mathrm{p,max}}}{q^{*2}} = \tfrac{1}{4}CE\,\epsilon \,. \tag{16.72}$$

For a circular plasma cross-section, this implies that β should not exceed $\frac{1}{4}\epsilon$, which is less than 0.1 for usual aspect ratios. It is clear that the increase of this value by cross–sectional shaping is most desirable, but one should also take the ordinary engineering wisdom into account to stay away from ultimate operating boundaries with a margin of a least 10%.

In principle, these severe equilibrium and stability limits can easily be overcome by enclosing the plasma with a conducting wall. With perfect conductivity, the value of β_p can even be increased without a limit since the poloidal flux trapped between the plasma and the wall will always prevent the plasma from actually hitting the wall. Similarly, a conducting wall that is close enough will stabilize any external kink mode with $m \geq 2$. However, proximity of a conducting wall is prohibited in a fusion reactor. More fundamentally, even when close, a conducting wall is no longer a cure for improving either equilibrium or stability limitations since the perfect conductivity required is simply no longer there on fusion time scales: the poloidal flux required for equilibrium leaks away and the resistive wall mode grows unimpeded on those time scales (see Section 14.3.1). Although Fig. 16.8 gives a good impression of the order of magnitude of the limits on β, it is clearly an over-simplification: toroidal effects deform the straight lines of external kink stability limits at higher β. Most important, in toroidal geometry, new pressure-driven instabilities appear that restrict the operating windows even further.

16.2 Grad–Shafranov equation

16.2.1 Derivation of the Grad–Shafranov equation

The previous section presented the qualitative features and global parameterization of plasma equilibrium in a torus. To proceed further, to the construction of general axi-symmetric equilibria, which is presently possible virtually without any real limitations of the numerical precision, we need to derive the central partial differential equation (PDE) that describes the spatial dependence of the poloidal magnetic flux. This equation is generally known as the *Grad–Shafranov equation*, after the authors that published it in the 1950s [188, 406], although Lüst and Schlüter [319] also published it independently at the same time.

We will exploit again the special cylindrical coordinates R, Z, φ, where φ is the toroidal angle, R is the distance to the symmetry axis and Z is the vertical coordinate, so that R and Z are just Cartesian coordinates in the poloidal plane. Note that the order of these cylindrical coordinates is not the usual one. It is chosen to provide the most logical connection with the different toroidal coordinates (Appendices A.2.4–A.2.7). In particular, notice that the replacements (A.62) to convert vector operators from the usual r, θ, z coordinates to the present R, Z, φ coordinates involves a minus sign: $d\theta = -d\varphi$.

The derivation of the Grad–Shafranov equation involves the following steps.

(a) From the divergence equation (16.3) and Ampère's law (16.2), the poloidal field and the poloidal current are derivable from *stream functions* $\Psi(R, Z)$ and $I(R, Z)$:

$$\nabla \cdot \mathbf{B} = \frac{1}{R}\frac{\partial(RB_R)}{\partial R} + \frac{\partial B_Z}{\partial Z} = 0 \;\Rightarrow\; B_R = -\frac{1}{R}\frac{\partial \Psi}{\partial Z}, \quad B_Z = \frac{1}{R}\frac{\partial \Psi}{\partial R}, \qquad (16.73)$$

$$\nabla \cdot \mathbf{j} = \frac{1}{R}\frac{\partial(Rj_R)}{\partial R} + \frac{\partial j_Z}{\partial Z} = 0 \;\Rightarrow\; j_R = \frac{1}{R}\frac{\partial I}{\partial Z}, \quad j_Z = -\frac{1}{R}\frac{\partial I}{\partial R}. \quad (16.74)$$

Of course, Ψ is the poloidal magnetic flux (normalized by dividing by a factor 2π). Alternatively, solving $\nabla \cdot \mathbf{B} = 0$ by means of $\mathbf{B} = \nabla \times \mathbf{A}$, the poloidal flux turns out to be related to the toroidal component of the vector potential:

$$\Psi = -RA_\varphi. \qquad (16.75)$$

(b) The toroidal and poloidal components of Ampère's law (16.2) provide associated expressions for the toroidal current density j_φ and the poloidal current stream function I:

$$Rj_\varphi = R\left(\frac{\partial B_Z}{\partial R} - \frac{\partial B_R}{\partial Z}\right) = R\frac{\partial}{\partial R}\left(\frac{1}{R}\frac{\partial \Psi}{\partial R}\right) + \frac{\partial^2 \Psi}{\partial Z^2} \equiv \Delta^*\Psi, \qquad (16.76)$$

$$j_R = \frac{\partial B_\varphi}{\partial Z}, \quad j_Z = -\frac{1}{R}\frac{\partial(RB_\varphi)}{\partial R} \;\Rightarrow\; I \equiv RB_\varphi. \qquad (16.77)$$

Here, the special symbol Δ^* indicates a Laplacian-like operator where the order of the factors R and $1/R$ is reversed with respect to the ordinary Laplacian.

(c) Finally, the toroidal and poloidal components of the pressure balance equation (16.1) imply that the stream function I of the poloidal current and the pressure p are *flux functions* (i.e. functions of the flux Ψ) which are related to the toroidal current j_φ:

$$\frac{\partial p}{\partial \varphi} = j_R B_Z - j_Z B_R = \frac{1}{R^2}\left(\frac{\partial I}{\partial Z}\frac{\partial \Psi}{\partial R} - \frac{\partial I}{\partial R}\frac{\partial \Psi}{\partial Z}\right) = 0 \;\Rightarrow\; I \equiv I(\Psi), \quad (16.78)$$

$$\left.\begin{aligned}
\frac{\partial p}{\partial R} &= j_Z B_\varphi - j_\varphi B_Z = \left(-\frac{II'}{R^2} - \frac{j_\varphi}{R}\right)\frac{\partial \Psi}{\partial R} \\[4pt]
\frac{\partial p}{\partial Z} &= j_\varphi B_R - j_R B_\varphi = \left(-\frac{II'}{R^2} - \frac{j_\varphi}{R}\right)\frac{\partial \Psi}{\partial Z}
\end{aligned}\right\} \;\Rightarrow\; p = p(\Psi), \qquad (16.79)$$

$$p' = -\frac{II'}{R^2} - \frac{j_\varphi}{R}. \qquad (16.80)$$

Here and in the following, the prime indicates differentiation with respect to Ψ.

Summarizing, from Eqs. (16.76) and (16.80) it follows that the equilibrium is described by an elliptic nonlinear PDE, *the Grad–Shafranov equation*, for the poloidal flux $\Psi = \Psi(R, Z)$:

$$\left[\Delta^*\Psi \equiv R^2 \nabla \cdot \left(\frac{1}{R^2}\nabla\Psi\right) \equiv \Delta\Psi - \frac{2}{R}\frac{\partial \Psi}{\partial R} \equiv\right]$$

$$R\frac{\partial}{\partial R}\left(\frac{1}{R}\frac{\partial \Psi}{\partial R}\right) + \frac{\partial^2 \Psi}{\partial Z^2} = -II' - R^2 p' \;\left[= Rj_\varphi\right], \qquad (16.81)$$

which has to satisfy the boundary condition

$$\Psi = \Psi_1 = \text{const} \quad \text{(on the plasma cross-section)}. \tag{16.82}$$

A considerable complication is that this plasma cross-section, in general, is also unknown since it represents the interface between the plasma and the external vacuum region which is determined by another nonlinear problem, viz. the *external free-boundary problem* with given currents in external coils. Here, we will assume that this problem is solved separately so that the cross-sectional shape is known. (Alternatively, one could assume that the plasma is surrounded by a closely fitting external wall of the desired shape, or, as we will do in the next sections, one could prescribe a desired shape of the plasma cross-section and compute what external field would produce it: an ill-posed, though physically very relevant, problem.) From Eqs. (16.77), (16.78), (16.79) it follows that the RHS of the Grad–Shafranov equation (16.81) contains *two completely arbitrary flux functions,*

$$I \equiv R B_\varphi = I(\Psi), \quad \text{and} \quad p = p(\Psi). \tag{16.83}$$

This arbitrariness is the toroidal counterpart of the freedom to specify two variables in cylindrical equilibria, that we frequently encountered. Specifying the functions $I = I(\Psi)$ and $p = p(\Psi)$, everything else will be determined from the solution of the Grad–Shafranov equation. For example, *the poloidal field and current* are determined from Eqs. (16.73) and (16.74):

$$\mathbf{B}_\text{p} = \frac{1}{R}\,\mathbf{e}_\varphi \times \nabla\Psi\,, \qquad \mathbf{j}_\text{p} = -I'\,\mathbf{B}_\text{p}\,. \tag{16.84}$$

This completes the derivation of the Grad–Shafranov equation.

16.2.2 Large aspect ratio expansion: internal solution

As a first application that requires the solution of the Grad–Shafranov equation, we revisit the global equilibrium problem of Section 16.1.4. We now justify all the rather ad hoc approximations that were made there and derive precise expressions for all the quantities, both inside the plasma (this section) and in the outer vacuum region that produces the necessary vertical field (next section).

The poloidal flux Ψ of the internal plasma region is determined by the full Grad–Shafranov equation (16.81). To solve it, we anticipate that, for a circular outer cross-section of the plasma, the cross-sections of the magnetic surfaces inside are approximately shifted circles, where the shift Δ varies from Δ_0 at the magnetic axis to 0 at the plasma surface. We assume this shift to be small with respect to the plasma radius: $\Delta/a \sim \epsilon \ll 1$. Of course, these assumptions are to be justified by the solutions that will be obtained. Our main task is to determine the magnitude of

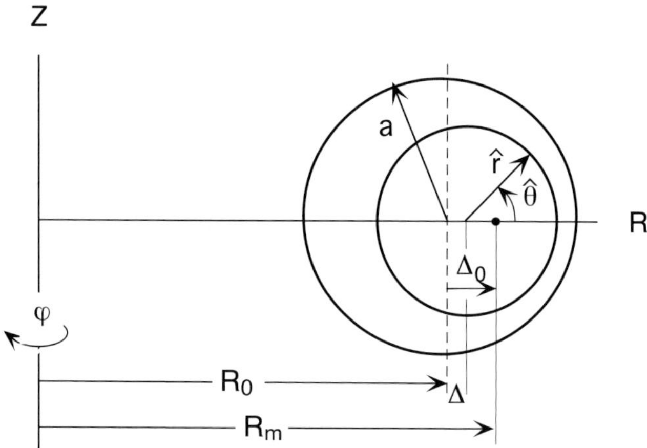

Fig. 16.9 "Shafranov" shifted circle approximation of the flux surfaces: non-orthogonal coordinates $\hat{r}, \hat{\theta}$ are defined with respect to the centers of the flux surfaces, which are shifted by an amount $\Delta(\hat{r})$, to be computed, with respect to the center of the plasma ($R = R_0$); the magnetic axis ($R = R_{\mathrm{m}}$) is indicated by a dot.

Δ for the different magnetic surfaces. For the solution of this problem, we exploit the *non-orthogonal shifted circle coordinate system* $\hat{r}, \hat{\theta}, \varphi$ (Fig. 16.9) based on the "Shafranov" shift Δ:

$$R = R_0 + \hat{r}\cos\hat{\theta} + \Delta(\hat{r}), \qquad Z = \hat{r}\sin\hat{\theta}. \tag{16.85}$$

Note that the three terms of R are of the order ϵ^{-1}, 1 and ϵ, respectively. Actually, our definition of $\Delta(\hat{r})$ differs from the shift $\Delta_{\mathrm{Shafr}}(\hat{r}) \equiv \Delta_0 - \Delta(\hat{r})$ defined by Shafranov in that we measure it with respect to the center of the plasma ($R = R_0$), whereas Shafranov defined it with respect to the magnetic axis ($R = R_{\mathrm{m}}$). This modification is not meant to be original, but to implement the strategy discussed in Section 16.1.4 to define parameters, as much as possible, in terms of precisely known external quantities.

The covariant metric elements of the shifted circle coordinates are given by

$$\hat{g}_{11} = 1 + 2\Delta'\cos\hat{\theta} + \Delta'^2 \approx 1 + 2\Delta'\cos\hat{\theta}, \qquad \hat{g}_{12} = -\hat{r}\Delta'\sin\hat{\theta},$$

$$\hat{g}_{22} = \hat{r}^2, \qquad \hat{g}_{33} = R^2 \approx R_0^2\left[1 + 2(\hat{r}/R_0)\cos\hat{\theta}\right],$$

$$\hat{J} = \sqrt{\det(\hat{g}_{ij})} = R\hat{r}(1 + \Delta'\cos\hat{\theta}) \approx R_0\hat{r}[1 + (\hat{r}/R_0 + \Delta')\cos\hat{\theta}], \tag{16.86}$$

where each first expression is exact and each second expression is correct to first order in ϵ. We also need the contravariant components of the metric tensor:

$$\hat{g}^{11} = (R^2/J^2)\hat{g}_{22}, \qquad \hat{g}^{12} = -(R^2/J^2)\hat{g}_{12},$$

$$\hat{g}^{22} = (R^2/J^2)\hat{g}_{11}\,, \qquad \hat{g}^{33} = 1/R^2\,. \tag{16.87}$$

To first order in the inverse aspect ratio, the Grad–Shafranov equation becomes

$$\Delta^*\Psi = \frac{R^2}{\hat{J}}\partial_i\left(\frac{\hat{J}}{R^2}\hat{g}^{ij}\partial_j\Psi\right)$$

$$\approx (1 - 2\Delta'\cos\hat\theta)\frac{1}{\hat{r}}\frac{\partial}{\partial\hat{r}}\left(\hat{r}\frac{\partial\Psi}{\partial\hat{r}}\right) - (1/R_0 - \Delta'/\hat{r} + \Delta'')\cos\hat\theta\frac{\partial\Psi}{\partial\hat{r}} + \frac{1}{\hat{r}^2}\frac{\partial^2\Psi}{\partial\hat\theta^2}$$

$$= -I\frac{dI}{d\Psi} - R^2\frac{dp}{d\Psi} \approx -\frac{d}{d\Psi}\left(\tfrac{1}{2}I^2 + R_0^2 p\right) - 2R_0\hat{r}\frac{dp}{d\Psi}\cos\hat\theta\,, \tag{16.88}$$

where the arbitrary functions $I(\Psi)$ and $p(\Psi)$ are yet to be specified. Here, the terms with $\partial\Psi/\partial\hat\theta$ and $\partial^2\Psi/\partial\hat{r}\partial\hat\theta$ have been neglected since first order multipliers turn them into second order quantities.

We now apply the low-β tokamak approximation introduced in Section 16.1.3. To get dimensionless quantities that one can compare, we exploit a characteristic length scale and magnetic field strength, for which we choose the plasma radius a and the "vacuum" magnetic field strength B_0 at $R = R_0$, i.e. the trivial scaling parameters of Section 16.1.4. This yields the following orders of magnitude for the main quantities:

$$\hat{r}/a \sim 1\,, \quad B_\varphi/B_0 \sim 1\,, \quad \beta \sim p/B_0^2 \sim B_{\rm p}^2/B_0^2 \sim \epsilon^2\,. \tag{16.89}$$

The merit of the shifted circle approximation is that, in the usual expansion

$$\Psi(\hat{r}, \hat\theta) = \Psi_0(\hat{r}) + \Psi_1(\hat{r})\cos\hat\theta + \cdots\,, \tag{16.90}$$

the first order flux Ψ_1 vanishes identically since the angular distortion of the magnetic surfaces, which it represents, is already accounted for by the coordinates. Hence, satisfaction of the BC (16.82) is automatic in this order.

Before we continue with the solution of the Grad–Shafranov equation, it is useful to digress on certain technicalities of the use of non-orthogonal coordinates. The details are given in Appendix A.3, but we here illustrate some of the subtleties by working out the expression (16.84) for the poloidal field. Denoting the basis vectors of the non-orthogonal coordinates $\hat{r}$, $\hat\theta$, φ as

$$\mathbf{a}^1 \equiv \nabla\hat{r}\,, \qquad \mathbf{a}^2 \equiv \nabla\hat\theta\,, \qquad \mathbf{a}^3 \equiv \nabla\varphi\,, \tag{16.91}$$

the gradient of the poloidal flux is written as

$$\nabla\Psi = (\partial_i\Psi)\,\mathbf{a}^i = \frac{\partial\Psi}{\partial\hat{r}}\mathbf{a}^1 + \frac{\partial\Psi}{\partial\hat\theta}\mathbf{a}^2\,, \tag{16.92}$$

so that the poloidal field becomes

$$\mathbf{B}_{\rm p} = \frac{1}{R}\mathbf{e}_\varphi \times \nabla\Psi = \nabla\varphi \times \nabla\Psi = \mathbf{a}^3 \times \left(\frac{\partial\Psi}{\partial\hat{r}}\mathbf{a}^1 + \frac{\partial\Psi}{\partial\hat\theta}\mathbf{a}^2\right)$$

$$\overset{(A.69)}{=} \frac{1}{\hat{J}} \frac{\partial \Psi}{\partial \hat{r}} \mathbf{a}_2 - \frac{1}{\hat{J}} \frac{\partial \Psi}{\partial \hat{\theta}} \mathbf{a}_1 \approx \frac{1}{\hat{J}} \frac{\partial \Psi}{\partial \hat{r}} \mathbf{a}_2 . \tag{16.93}$$

This yields the angular dependence of the poloidal magnetic field amplitude,

$$B_{\mathrm{p}} = |B_{\mathrm{p}}| = \frac{\sqrt{\mathbf{a}_2 \cdot \mathbf{a}_2}}{\hat{J}} \frac{\partial \Psi}{\partial \hat{r}} \equiv \frac{\sqrt{g_{22}}}{\hat{J}} \frac{\partial \Psi}{\partial \hat{r}} \approx \frac{1}{R_0} \left[1 - (\hat{r}/R_0 + \Delta') \cos \hat{\theta} \right] \frac{\partial \Psi}{\partial \hat{r}} , \tag{16.94}$$

which plays a central role in these sections.

We now work out the expanded Grad–Shafranov equation (16.88) in the low-β tokamak ordering. This yields the following *leading order contribution:*

$$\frac{1}{\hat{r}} \frac{d}{d\hat{r}} \left(\hat{r} \frac{d\Psi_0}{d\hat{r}} \right) = -\frac{d}{d\Psi_0} \left(\tfrac{1}{2} I^2 + R_0^2 p \right) . \tag{16.95}$$

From the leading order part of the poloidal field expression (16.94), derivatives with respect to Ψ_0 may now be converted into derivatives with respect to $\hat{r}$:

$$B_{\mathrm{p}0} = \frac{1}{R_0} \frac{d\Psi_0}{d\hat{r}} \quad \Rightarrow \quad \frac{d}{d\Psi_0} = \frac{1}{R_0 B_{\mathrm{p}0}} \frac{d}{d\hat{r}} . \tag{16.96}$$

Hence, to significant order, the poloidal field expression (16.94) becomes

$$B_{\mathrm{p}}(\hat{r}, \hat{\theta}) \approx [1 - (\hat{r}/R_0 + \Delta') \cos \hat{\theta}] B_{\mathrm{p}0}(\hat{r}) , \tag{16.97}$$

where $B_{\mathrm{p}0}(\hat{r})$ is free so far. Moreover, to get a balance between the LHS and RHS of Eq. (16.95), the function I^2 must be constant to leading and first order, so that

$$I \equiv RB_{\varphi} \approx (R_0 + \hat{r} \cos \hat{\theta} + \Delta)(B_{\varphi}^{(0)} + B_{\varphi}^{(1)} + B_{\varphi}^{(2)})$$

$$\approx R_0 B_{\varphi}^{(0)} + R_0 B_{\varphi}^{(1)} + \hat{r} \cos \hat{\theta} B_{\varphi}^{(0)} + R_0 B_{\varphi}^{(2)} + \hat{r} \cos \hat{\theta} B_{\varphi}^{(1)} + \Delta B_{\varphi}^{(0)}$$

$$\Rightarrow B_{\varphi}^{(0)} = \mathrm{const} , \qquad B_{\varphi}^{(1)} = -(\hat{r}/R_0) \cos \hat{\theta} B_{\varphi}^{(0)} . \tag{16.98}$$

Hence, to significant order, the expression for the toroidal field becomes

$$B_{\varphi} \approx [1 - (\hat{r}/R_0) \cos \hat{\theta}] B_{\varphi}^{(0)} + B_{\varphi}^{(2)} \approx [1 - (\hat{r}/R_0) \cos \hat{\theta}](B_{\varphi}^{(0)} + B_{\varphi}^{(2)})$$

$$\Rightarrow B_{\varphi}(\hat{r}, \hat{\theta}) \approx [1 - (\hat{r}/R_0) \cos \hat{\theta}] B_{\varphi 0}(\hat{r}) , \tag{16.99}$$

where we have lumped together the zeroth and second order into the "cylindrical" function $B_{\varphi 0}(\hat{r})$, which is also free so far. Consequently, the leading order equilibrium relation (16.95) reduces to the cylindrical equilibrium relation (16.56),

$$\frac{d}{d\hat{r}} (p + \tfrac{1}{2} B_{\varphi 0}^2) + \frac{B_{\mathrm{p}0}}{\hat{r}} \frac{d}{d\hat{r}} (\hat{r} B_{\mathrm{p}0}) = 0 , \tag{16.100}$$

where the variables p, $B_{\mathrm{p}0}$ and $B_{\varphi 0}$ should be considered as functions of $\hat{r}$, *two of which may be chosen freely,* as in the analogous cylindrical equilibrium problem. An expedient choice is $p(\hat{r})$ and $B_{\mathrm{p}0}(\hat{r})$, or the related current density $j_{\varphi 0}(\hat{r})$.

The next, and final, order of the expanded Grad–Shafranov equation (16.88) is the *first order contribution*:

$$-2\Delta'\frac{d^2\Psi_0}{d\hat{r}^2} - (\Delta'' + \Delta'/\hat{r} + 1/R_0)\frac{d\Psi_0}{d\hat{r}} = -2R_0\hat{r}\frac{dp}{d\Psi_0}\,. \qquad (16.101)$$

Converting $d\Psi_0/dr$ into $R_0 B_{\mathrm{p}0}$ in this relation yields the crucial inhomogeneous differential equation for the determination of $\Delta(\hat{r})$:

$$\frac{d}{d\hat{r}}\left(\hat{r}B_{\mathrm{p}0}^2\frac{d}{d\hat{r}}\Delta\right) = (\hat{r}/R_0)(2\hat{r}p' - B_{\mathrm{p}0}^2)\,. \qquad (16.102)$$

For given $p = p(\hat{r})$ and $B_{\mathrm{p}0}^2(\hat{r})$, the two integrations required are straightforward, so that the problem may be considered solved. We first construct the first integral of Eq. (16.102) and transform it by integration by parts:

$$\begin{aligned}
\Delta'(\hat{r}) &= (1/R_0)\cdot\frac{1}{\hat{r}B_{\mathrm{p}0}^2(\hat{r})}\int_0^{\hat{r}}(2\hat{r}p' - B_{\mathrm{p}0}^2)\,\hat{r}d\hat{r} \\
&= (1/R_0)\cdot\frac{1}{\hat{r}B_{\mathrm{p}0}^2(\hat{r})}\left[2\hat{r}^2 p(\hat{r}) - \int_0^{\hat{r}}(4p + B_{\mathrm{p}0}^2)\,\hat{r}d\hat{r}\right], \quad (16.103)
\end{aligned}$$

and then integrate the result to produce the shift itself, satisfying the BC at the plasma boundary, viz. $\Delta(a) = 0$:

$$\Delta(\hat{r}) = -\int_{\hat{r}}^a\Delta'(\hat{r})\,d\hat{r} \quad\Rightarrow\quad \Delta_0 \equiv \Delta(0) = -\int_0^a\Delta'(\hat{r})\,d\hat{r}\,. \qquad (16.104)$$

This finally gives the unknown displacement Δ_0 of the magnetic axis. Incidentally, notice that, only now, after obtaining the solution characterized by $\Psi_0 = \Psi_0(\hat{r})$ and $\Delta = \Delta(\hat{r})$, the coordinate system has become explicitly known. This is the general *a posteriori* feature of exploiting magnetic flux based coordinates.

It remains to extract the global physical characteristics from these solutions. To that end, we average over the toroidal plasma volume,

$$dV = \hat{J}\,d\hat{r}\,d\hat{\theta}\,d\varphi \approx 4\pi^2 R_0\,\hat{r}\,d\hat{r} \quad\Rightarrow\quad V \approx 2\pi^2 a^2 R_0\,, \qquad (16.105)$$

so that the rough approximation (16.52) for the average of leading order quantities gets a precise meaning in terms of the shifted circle coordinate $\hat{r}$:

$$\langle f\rangle \equiv \frac{1}{V}\int f(\hat{r},\hat{\theta})\,dV \approx \frac{2}{a^2}\int_0^a f_0(\hat{r})\,\hat{r}\,d\hat{r}\,. \qquad (16.106)$$

With this understanding, the definitions (16.59) and (16.63) of Section 16.1.4 for the *poloidal beta*, β_{p}, and of the *internal inductance*, ℓ_{i}, of the plasma, now also get a precise meaning. Their normalization with respect to the average poloidal field at the plasma boundary demonstrates the central importance of the *total toroidal current flowing in the plasma,*

$$B_{\mathrm{p0}}(a) = \frac{I_\varphi}{2\pi a} \,, \tag{16.107}$$

and, hence, of the modified safety factor q^* defined in Eq. (16.66).

With the mentioned definitions of the two global parameters β_{p} and ℓ_{i}, the solution (16.103) for Δ' at the plasma boundary becomes

$$\Delta'(a) = -\epsilon(\beta_{\mathrm{p}} + \tfrac{1}{2}\ell_{\mathrm{i}}) \,, \tag{16.108}$$

where the edge pressure is assumed to vanish, $p(a) = 0$. From Eq. (16.97), this yields the required expression for the poloidal field at the plasma boundary:

$$B_{\mathrm{p}}(a, \theta) = \frac{I_\varphi}{2\pi a} \left(1 + \epsilon \Lambda \cos \theta \right) ,$$

$$\Lambda \equiv -1 - \Delta'(a)/\epsilon = \beta_{\mathrm{p}} + \tfrac{1}{2}\ell_{\mathrm{i}} - 1 \,. \tag{16.109}$$

We have replaced $\hat{\theta}$ by the ordinary polar angle θ because $\Delta(a) = 0$ implies that the Shafranov coordinates coincide with the ordinary polar coordinates at the plasma boundary: $\hat{r} = r = a \Rightarrow \hat{\theta} = \theta$. The poloidal dependence of $B_{\mathrm{p}}(a, \theta)$, expressed through Λ, summarizes the main physical properties of the internal solution Ψ, which will be *applied as a BC on the external solution* Ψ^{ex} in the next section.

Extrapolating the expression (16.109) to the high-beta tokamak regime, which is strictly speaking invalid here since we have assumed the low-beta tokamak ordering, but which nevertheless catches the essential physics, a limit on the equilibrium appears when the vertical field has to be increased so much that the poloidal field vanishes on the inside plasma boundary, $B_{\mathrm{p}}(a, \pi) = 0$:

$$\epsilon \Lambda \approx \epsilon \beta_{\mathrm{p}} = 1 \,. \tag{16.110}$$

This yields the limiting equilibrium value E of Eq. (16.71).

Finally, it is of interest to notice that β_{p} also determines the *overall* radial dependence of the "cylindrical" part $B_{\varphi 0}^2(\hat{r})$ of the toroidal magnetic field pressure [481]. This follows by applying a similar reasoning as led to Eq. (16.57):

$$\int_0^a \frac{dB_{\varphi 0}^2}{d\hat{r}} \, \hat{r}^2 \, d\hat{r} = -2 \int_0^a j_{\mathrm{p0}} B_{\varphi 0} \hat{r}^2 \, d\hat{r} = a^2 \left(B_0^2 - \langle B_{\varphi 0}^2 \rangle \right) = \frac{I_\varphi^2}{4\pi^2} (\beta_{\mathrm{p}} - 1) \,. \tag{16.111}$$

Clearly, the value $\beta_{\mathrm{p}} = 1$ separates *paramagnetic equilibria* with the "wrong" direction of the poloidal current (viz. increasing the toroidal field in the plasma with respect to that in the vacuum) from *diamagnetic equilibria* where the poloidal current is in the proper direction (viz. θ-pinch like) to facilitate higher values of the plasma pressure confined:

$$\begin{cases} j_{\mathrm{p}0} B_{\varphi 0} > 0 \quad \textit{(outward Lorentz force)} : \quad \beta_{\mathrm{p}} < 1\,, \\ j_{\mathrm{p}0} B_{\varphi 0} < 0 \quad \textit{(inward Lorentz force)} : \quad \beta_{\mathrm{p}} > 1\,. \end{cases} \tag{16.112}$$

For the latter case, on average, the plasma pressure "digs a hole" in the toroidal magnetic field pressure, as shown in Fig. 16.6.

▷ **Explicit solutions for Wesson profiles** It is instructive to substitute in the expressions derived the simple model distributions for the plasma pressure and the toroidal current density introduced by Wesson [481], Section 3.7, viz.

$$p = p_0(1 - \bar{r}^2)\,, \quad j_{\varphi 0} = j_0(1 - \bar{r}^2)^\nu\,, \quad \text{where} \quad \bar{r} \equiv \hat{r}/a\,. \tag{16.113}$$

Recall that this class of current profiles was extensively exploited to study the stability of "straight tokamaks" (see Section 9.4.4 and Fig. 9.19 [1]). In particular, recall that the overall magnetic shear, expressed by the ratio of the cylindrical safety factors at the plasma surface and the magnetic axis, is $q_1/q_0 = \nu + 1$. For $\nu = 1$ (parabolic pressure and current profile), the explicit expressions of the "cylindrical" quantities become

$$B_{\mathrm{p}0}(\bar{r}) = \frac{I_\varphi}{2\pi a} \cdot \bar{r}(2 - \bar{r}^2)\,, \qquad I_\varphi = \tfrac{1}{2}\pi a^2 j_0\,,$$

$$\beta_{\mathrm{p}} = \left(\frac{2\pi a}{I_\varphi}\right)^2 p_0\,, \qquad \ell_{\mathrm{i}} = \tfrac{11}{12} \approx 0.917\,, \tag{16.114}$$

and the final integrations (16.103) and (16.104) of the Shafranov shift yield

$$\Delta'(\hat{r}) \equiv \frac{1}{a}\frac{d}{d\bar{r}}\Delta(\bar{r}) = -\epsilon \cdot \frac{\bar{r}(\beta_{\mathrm{p}} + 1 - \tfrac{2}{3}\bar{r}^2 + \tfrac{1}{8}\bar{r}^4)}{(2 - \bar{r}^2)^2}$$

$$\Rightarrow \quad \Delta'(a) = -\epsilon(\beta_{\mathrm{p}} + \tfrac{11}{24})\,, \tag{16.115}$$

$$\Delta(\bar{r}) = \tfrac{1}{4}\epsilon a \cdot \left[\frac{(2\beta_{\mathrm{p}} + \tfrac{5}{6} - \tfrac{1}{4}\bar{r}^2)(1 - \bar{r}^2)}{2 - \bar{r}^2} + \tfrac{1}{3}\ln(2 - \bar{r}^2) \right]$$

$$\Rightarrow \quad \Delta_0 \equiv \Delta(0) = \tfrac{1}{4}\epsilon a(\beta_{\mathrm{p}} + \tfrac{5}{12} + \tfrac{1}{3}\ln 2)\,. \tag{16.116}$$

The first relation reproduces Eq. (16.108) for the derivative of the shift at the plasma surface. The second one yields the shift of the magnetic axis, in agreement with Fig. 3.7.2 of [481], where graphs of Δ_0 for varying ν (i.e. ℓ_{i}) are given. Explicit calculation of the toroidal field is left as an exercise for the reader.   ◁

16.2.3 Large aspect ratio expansion: external solution

In this section, we determine the vertical magnetic field needed for equilibrium, extending the procedure first presented by Shafranov [408], applied by Greene *et al.* [192] to a tokamak compression experiment, and extensively discussed by

Miyamoto [334], [335](a). This requires solution of the external "vacuum" magnetic field equations

$$\nabla \times \mathbf{B}^{ex} = 0, \qquad \nabla \cdot \mathbf{B}^{ex} = 0 \tag{16.117}$$

$$\Rightarrow \quad \mathbf{B}_p^{ex} = \frac{1}{R}\mathbf{e}_\varphi \times \nabla \Psi^{ex}, \qquad \Delta^* \Psi^{ex} = 0, \tag{16.118}$$

so that the poloidal flux Ψ^{ex} of the external region is determined by the Grad–Shafranov equation (16.81) with vanishing RHS, i.e. vanishing toroidal current. The source of the external field, viz. the distribution of the toroidal current $I_{\varphi,vf}^{ex}$ in the vertical field coil(s) (Section 16.1.4), is not considered here but, instead, a BC is imposed on Ψ^{ex} that yields a homogeneous vertical field $\mathbf{B}_v$ at large distances from the plasma. This is sufficient to determine the amplitude B_v. Of course, for the design of external coils for plasma control or the interpretation of magnetic diagnostics, the actual current distribution can not be ignored [7, 290, 13].

In the absence of skin currents on the plasma boundary, the pertinent BCs for this problem follow from prescribing the dependence (16.109) of B_p at the plasma boundary and the dependence of Ψ^{ex} far away (at "infinity") from the plasma:

$$\left.\begin{array}{l} \Psi^{ex}(a,\theta) = \text{const} \\[2ex] B_p^{ex}(a,\theta) = \dfrac{I_\varphi}{2\pi a}(1 + \epsilon\Lambda\cos\theta) \end{array}\right\} \quad (\textit{at the plasma boundary}), \tag{16.119}$$

$$\Psi^{ex} = \Psi_{pl}^{ex} + \Psi_{vf}^{ex} \;\rightarrow\; C + \Psi_{pl,\infty}^{ex} + \tfrac{1}{2}B_v R^2 \quad (\textit{at "infinity"}), \tag{16.120}$$

where $\Lambda \equiv \beta_p + \tfrac{1}{2}\ell_i - 1$, the flux $\Psi_{pl,\infty}^{ex}$ corresponds to the far field due to the plasma current and Ψ_{vf}^{ex} is the flux of the vertical field.

It should be noticed that the boundary value problem (BVP) of solving the PDE (16.118) with BCs (16.119)–(16.120), suffers from two defects.

(1) Disregarding the BC (16.120), the remaining BVP is already sufficient to completely determine the solution Ψ^{ex}. However, it is *ill-posed*: instead of solving the elliptic problem starting from Dirichlet conditions on the two boundaries, the problem is solved by posing Cauchy conditions on the internal boundary and integrating outward. This is not forbidding by itself (many problems in science and engineering are ill-posed and effective numerical procedures exist to solve them), but it does have peculiar consequences, as we will see.

(2) The additional BC (16.120) really makes the problem *over-determined*, which implies that we will have to relax it in some way. We will discuss how this may be done when we have obtained the solution of the restricted BVP (16.118)–(16.119).

To solve the "Grad–Shafranov" equation (16.118) with the BC (16.119), we exploit *orthogonal toroidal coordinates* μ, η, φ, using the notation of Morse and

Feshbach [341], p. 1301. They have the following relationship to the cylindrical coordinates R, Z, φ and resultant scale factors of the differential operators:

$$R = R_{\rm c}\,\frac{\sinh\mu}{D}\,,\quad Z = R_{\rm c}\,\frac{\sin\eta}{D}\,,\qquad D \equiv \cosh\mu - \cos\eta\,,$$

$$h_1 = h_2 = \frac{R_{\rm c}}{D}\,,\quad h_3 = R\,,\qquad J = h_1 h_2 h_3\,. \tag{16.121}$$

The full range of these coordinates is indicated by the following scheme:

$$\mu = \left\{ \begin{array}{lll} 0 : & R = 0\,,\ \text{or}\ Z = \pm\infty\,,\ \text{or}\ R = \infty & \textit{(infinity)} \\[4pt] \mu_1 : & R = R_0 + a\cos\theta\,,\quad Z = a\sin\theta & \textit{(plasma boundary)} \\[4pt] \infty : & R = R_{\rm c}\,,\quad Z = 0 & \textit{(concentration point)} \end{array} \right. \tag{16.122}$$

The "radial" coordinate curves $\mu = \text{const}$ of this system are circles with centers $R = R_{\rm c}\coth\mu$ and radii $r = R_{\rm c}/\sinh\mu$, fitted to the circular plasma boundary that is indicated by the value $\mu = \mu_1$, so that

$$R_0 = R_{\rm c}\coth\mu_1\,,\quad a = R_{\rm c}/\sinh\mu_1 \tag{16.123}$$

$$\Rightarrow\quad \epsilon^{-1} \equiv R_0/a \equiv \cosh\mu_1 \approx \tfrac{1}{2}e^{\mu_1} \gg 1\,,\qquad R_{\rm c} = R_0\sqrt{1 - \epsilon^2}\,.$$

These coordinates are not very practical inside the plasma since the concentration point $(R_{\rm c}, 0)$ is shifted inward with respect to the plasma center $(R_0, 0)$, whereas the magnetic axis is shifted outward. We exploit them only in the outer range,

$$0 \le \mu \le \mu_1 \equiv \cosh^{-1}(R_0/a) \approx \ln(2R_0/a)\,, \tag{16.124}$$

where μ_1 need not be very large for $e^{-\mu_1} \approx \tfrac{1}{2}\epsilon$ to be very small.

In terms of these coordinates, the vacuum Grad–Shafranov equation becomes

$$\Delta^* \Psi^{\rm ex} \equiv \frac{R^2}{J}\left[\frac{\partial}{\partial\mu}\left(\frac{1}{R}\frac{\partial\Psi^{\rm ex}}{\partial\mu} \right) + \frac{\partial}{\partial\eta}\left(\frac{1}{R}\frac{\partial\Psi^{\rm ex}}{\partial\eta} \right) \right] = 0\,. \tag{16.125}$$

This equation may be solved by the transformation

$$\Psi^{\rm ex}(\mu, \eta) = \frac{G(\mu, \eta)}{\sqrt{D(\mu, \eta)}} = \frac{1}{\sqrt{D(\mu, \eta)}} \sum_{m=-\infty}^{\infty} \hat{G}_m(\mu)\cos m\eta\,, \tag{16.126}$$

where the Fourier harmonics $\hat{G}_m$ of the function $G(\mu, \eta)$ satisfy the ODEs

$$\sinh\mu\,\frac{d}{d\mu}\left(\frac{1}{\sinh\mu}\frac{d\hat{G}_m}{d\mu} \right) - (m^2 - \tfrac{1}{4})\hat{G}_m = 0\,, \tag{16.127}$$

having derivatives of the two kinds of Legendre functions as solutions. Hence, the general up–down symmetric solution of Eq. (16.125) may be written as

$$\Psi^{\mathrm{ex}} = \frac{1}{\sqrt{D(\mu,\eta)}} \sum_{m=0}^{\infty} (2 - \delta_{m0})\Big[a_m S_m(\mu) + b_m T_m(\mu)\Big] \cos m\eta\,,$$

$$\begin{cases} S_m(\mu) \equiv \sinh\mu\, \dfrac{d}{d\mu} P_{m-\frac{1}{2}}(\cosh\mu) \equiv \sinh\mu\, P^1_{m-\frac{1}{2}}(\cosh\mu)\,, \\[2mm] T_m(\mu) \equiv \sinh\mu\, \dfrac{d}{d\mu} Q_{m-\frac{1}{2}}(\cosh\mu) \equiv \sinh\mu\, Q^1_{m-\frac{1}{2}}(\cosh\mu)\,, \end{cases} \qquad (16.128)$$

where $P_{m-\frac{1}{2}}$ and $Q_{m-\frac{1}{2}}$ are the zero order toroidal harmonics (or Legendre functions of the first and second kind), and $P^1_{m-\frac{1}{2}}$ and $Q^1_{m-\frac{1}{2}}$ are the *first order toroidal harmonics*. (The higher order toroidal harmonics may be exploited to calculate 3D perturbations with toroidal dependence $\exp(in\varphi)$, like external kink modes.) The functions S_m and T_m are related to the Fock functions f_m and g_m, exploited by Shafranov [407, 408], through $S_m \equiv (m^2 - \frac{1}{4})f_m$ and $T_m \equiv (m^2 - \frac{1}{4})\pi g_m$. They satisfy the Wronskian identity $S_m T'_m - T_m S'_m = (m^2 - \frac{1}{4})\sinh\mu$.

▷ **Scalar potential** Just for completeness: the vacuum field equations (16.117) may also be solved by means of the scalar potential, $\mathbf{B}^{\mathrm{ex}} = \nabla\Phi^{\mathrm{ex}}$, satisfying $\Delta\Phi^{\mathrm{ex}} = 0$; see Biermann *et al.* [41]. The latter is solved by the transformation $\Phi = \sqrt{D}F$, where the Fourier harmonics $\hat{F}_m$ of $F(\mu,\eta)$ satisfy Legendre's equation proper, so that the solution

$$\Phi^{\mathrm{ex}}(\mu,\eta) = \sqrt{D(\mu,\eta)} \sum_{m=-\infty}^{\infty} \Big[c_m P_{m-\frac{1}{2}}(\cosh\mu) + d_m Q_{m-\frac{1}{2}}(\cosh\mu)\Big] \cos m\eta \quad (16.129)$$

involves the Legendre functions themselves; see [341], p. 1303.   ◁

Close to the plasma boundary (large μ), the exact solution (16.128) for $\Psi^{\mathrm{ex}}(\mu,\eta)$ may be approximated by the leading terms in an expansion in powers of $\mathrm{e}^{-\mu}$:

$$[D(\mu,\eta)]^{-1/2} \approx \sqrt{2}\,\mathrm{e}^{-\frac{1}{2}\mu}(1 + \mathrm{e}^{-\mu}\cos\eta)\,, \qquad (16.130)$$

$$\begin{cases} S_0(\mu) \approx -\dfrac{1}{2\pi}(\mu + \ln 4 - 2)\mathrm{e}^{\frac{1}{2}\mu}\,, \quad S_m(\mu) \approx \dfrac{(m-1)!}{2\Gamma(m-\frac{1}{2})\sqrt{\pi}}\,\mathrm{e}^{(m+\frac{1}{2})\mu} \\[1mm] \hspace{6.5cm} (m \geq 1)\,, \\[2mm] T_m(\mu) \approx -\dfrac{\Gamma(m+\frac{3}{2})\sqrt{\pi}}{2m!}\,\mathrm{e}^{-(m-\frac{1}{2})\mu} \qquad (\text{all } m)\,. \end{cases} \qquad (16.131)$$

For the full hypergeometric expressions, see Abramowitz and Stegun [3], p. 332, or Morse and Feshbach [341], pp. 1302 and 1329. To assist the reader who wishes to reproduce the algebra, in the rest of this section, we indicate by ≪...≫ the different stumbling blocks from errors and confusing notations encountered in the literature. ≪ *In the latter reference, note that the functions P and Q are defined with extra powers of i and -1 respectively, that the expression for Q on p. 1302 is correct, but the one on p. 1329 is incorrect, and that the Fourier coefficients of the expansion for $1/\sqrt{D}$ should be a factor 2 larger for $m \neq 0$ than those given on pp. 1304 and 1330.* ≫

Since the low-β tokamak approximation of the internal solution contains only two Fourier harmonics, to solve the restricted BVP (16.118)–(16.119) with these approximate expressions, we only need to substitute the boundary values of the $m = 0$ and $m = 1$ harmonics into the external solutions Ψ^{ex} and B_{p}^{ex}:

$$\Psi^{\text{ex}}(\mu_1, \eta) \approx -\frac{1}{\pi\sqrt{2}}\bigg\{ (\ln(8/\epsilon) - 2)a_0 + \tfrac{1}{2}\pi^2 b_0$$

$$+ \tfrac{1}{2}\epsilon\Big[(\ln(8/\epsilon) - 2)a_0 - 8\epsilon^{-2}a_1 + \tfrac{1}{2}\pi^2(b_0 + 3b_1) \Big] \cos\eta \bigg\} = \text{const}\,, \tag{16.132}$$

$$B_{\text{p}}^{\text{ex}}(\mu_1, \eta) = -\left(\frac{1}{h_1 R}\frac{\partial \Psi^{\text{ex}}}{\partial \mu} \right)_{\mu_1} \approx \frac{\epsilon^{-1}}{\pi\sqrt{2}R_0^2}\bigg\{ a_0 - \tfrac{1}{2}\epsilon\Big[(\ln(8/\epsilon) + 1)a_0$$

$$+ 8\epsilon^{-2}a_1 + \tfrac{1}{2}\pi^2(b_0 + 3b_1) \Big] \cos\eta \bigg\} = \frac{I_\varphi}{2\pi a}(1 + \epsilon\Lambda\cos\eta)\,. \tag{16.133}$$

In the last equality we used $\theta \approx \eta$, obtained from the coordinate relationship $R(\mu, \eta) = R_0 + r\cos\theta$, $Z(\mu, \eta) = r\sin\theta$. To the order required, this yields

$$\cos\eta \approx \cos\theta\,, \qquad \mu - \mu_1 \approx \ln(a/r) - \tfrac{1}{2}\epsilon(a/r - r/a)\cos\theta\,. \tag{16.134}$$

The transformed BCs (16.132) and (16.133) determine three of the four constants,

$$a_0 = \frac{1}{\sqrt{2}}R_0 I_\varphi\,, \qquad a_1 = -\tfrac{1}{8}\epsilon^2(\Lambda + \tfrac{3}{2})a_0\,,$$

$$b_0 + 3b_1 = -\frac{2}{\pi^2}[\Lambda + \ln(8/\epsilon) - \tfrac{1}{2}]a_0\,, \tag{16.135}$$

so that the solution close to the plasma boundary is represented by

$$\Psi^{\text{ex}}(\mu, \eta) \approx -\frac{R_0 I_\varphi}{2\pi}\bigg\{ \mu - \mu_1 + \ln(8/\epsilon) - 2 + \tfrac{1}{2}\pi^2 b_0/a_0$$

$$+ \Big[\mu - \mu_1 - (\Lambda + \tfrac{3}{2})(1 - e^{2(\mu - \mu_1)}) \Big] e^{-\mu}\cos\theta \bigg\}\,, \tag{16.136}$$

or

$$\Psi^{\text{ex}}(r, \theta) \approx -\frac{R_0 I_\varphi}{2\pi}\bigg\{ \ln(8R_0/r) - 2 + \tfrac{1}{2}\pi^2 b_0/a_0$$

$$- \tfrac{1}{2}\Big[\ln(r/a) + (\Lambda + \tfrac{1}{2})(1 - a^2/r^2) \Big] (r/R_0)\cos\theta \bigg\}\,. \tag{16.137}$$

Apart from the arbitrary constant b_0/a_0, the solution is now completely determined in terms of a truncated Fourier expansion around the plasma boundary.

We now have to consider whether it is possible at all to satisfy the additional "BC" (16.120) for the vertical field and, if so, to determine its amplitude B_{v}. To that end, we first extract the plasma self field $\Psi^{\text{ex}}_{\text{pl},\infty}$ from the solution obtained and then evaluate whether the remainder conforms to the vertical flux $\Psi^{\text{ex}}_{\text{pl,v}} = \tfrac{1}{2}B_{\text{v}}R^2$. (For simplicity, we exploit the arbitrariness of b_0/a_0 to set $C = 0$.)

Far away from the plasma, the actual current distribution in the plasma can be ignored so that the plasma self field is related to the exact expression of the vector potential for a ring current I_φ at $R = R_0$ (see Jackson [247], Section 5.5):

$$\Psi^{\rm ex}_{\rm pl,\infty} = -RA_\varphi = -R_0 I_\varphi \cdot \sqrt{R/R_0}\, \frac{(2-k^2)K(k^2) - 2E(k^2)}{2\pi k}\,,$$

$$k^2 \equiv \frac{4R/R_0}{(1+R/R_0)^2 + (Z/R_0)^2}\,. \tag{16.138}$$

Converting these expressions into μ, η coordinates, the square root $\sqrt{R/R_0} \approx \sqrt{\sinh\mu/D}$, the parameter k becomes a function of μ alone, $k^2 \equiv 1 - {\rm e}^{-2\mu}$, and the quotient with the elliptic integrals brings in the function $-S_0(\mu)/\sqrt{2\sinh\mu}$, where $S_0(\mu) = -\frac{1}{2}\cosh\mu\, P_{-1/2}(\cosh\mu) + \frac{1}{2} P_{1/2}(\cosh\mu)$. Hence,

$$\Psi^{\rm ex}_{\rm pl,\infty} = R_0 I_\varphi\, \frac{S_0}{\sqrt{2D}} = a_0\, \frac{S_0(\mu)}{\sqrt{D(\mu,\eta)}}\,. \tag{16.139}$$

This expression for the plasma far field not only justifies the neglect of $a_1 S_1(\mu)$, since $a_1 \sim \epsilon^2$ according to Eq. (16.135), but of all terms $a_m S_m(\mu)$ for $m \geq 1$.

≪ *In the derivation of these expressions, a serious source of confusion had to be identified: in contrast to Jackson [247] and others [189], we considered it more logical to indicate the argument of the elliptic integrals as it appears in the integrals, viz. as k^2 and not as k, i.e. to follow the convention of Abramowitz and Stegun [3], Chapter 17. However, in their expressions for the Legendre functions in terms of elliptic integrals [3], Section 8.13, they inconsistently exploit the wrong argument k.* ≫

It remains to express the vertical field in μ, η coordinates and to convert it into a series of the functions $T_m(\mu)$. To that end, we exploit a general identity (see, e.g., B. Braams [65]) relating those functions to the Fourier coefficients of $\sqrt{2D}$:

$$\pi\sqrt{2D(\mu,\eta)} = \sum_{m=0}^{\infty} \frac{2-\delta_{m0}}{m^2 - \frac{1}{4}}\, T_m(\mu)\cos m\eta \tag{16.140}$$

$$\Rightarrow\quad T_m(\mu) = (m^2 - \tfrac{1}{4}) \int_0^{\pi} \sqrt{2D(\mu,\eta')}\,\cos m\eta'\, d\eta'\,. \tag{16.141}$$

By differentiating this identity twice, and exploiting the ODE (16.127) for $T_m(\mu)$, we obtain the following representation of the vertical flux:

$$\Psi^{\rm ex}_{\rm vf} = \tfrac{1}{2} B_{\rm v} R^2 \approx \frac{1}{\sqrt{D}}\, \tfrac{1}{2} B_{\rm v} R_0^2\, \frac{\sinh^2\mu}{D^{3/2}}$$

$$= -\frac{\sqrt{2}}{\pi\sqrt{D}} \sum_{m=0}^{\infty} (2-\delta_{m0})\, B_{\rm v} R_0^2\, T_m(\mu)\cos m\eta = \Psi^{\rm ex}(a_m = 0) \tag{16.142}$$

$$\Rightarrow\quad b_m = -\frac{\sqrt{2}}{\pi} B_{\rm v} R_0^2 \quad (\text{all } m)\,. \tag{16.143}$$

Hence, all coefficients b_m should be equal! However, from the relation (16.135) it follows that only b_0 and b_1 are needed to determine B_v. Substituting them, the required relation for the vertical field results:

$$ B_v \approx -\frac{\pi}{4\sqrt{2}R_0^2}(b_0 + 3b_1) = \frac{I_\varphi}{4\pi R_0}\left[\beta_p + \ln(8R_0/a) + \tfrac{1}{2}\ell_i - \tfrac{3}{2}\right], \qquad (16.144) $$

justifying the heuristic expression (16.62) derived in Section 16.1.4; QED.

This illustrates the *intermediate asymptotic* character of the ill-posed equilibrium problem. Since the exact expression (16.142) for the vertical field involves all coefficients b_m, but only two of them are determined from the present low-β tokamak expansion, the sum $\Psi^{\mathrm{ex}} = \Psi^{\mathrm{ex}}_{\mathrm{pl},\infty} + \Psi^{\mathrm{ex}}_{\mathrm{vf}}$ of the expressions (16.138) and (16.142) represents the solution "far away" from the plasma, *but not too far away* (in particular not at $\mu \ll 1$, see small text below) so that the functions $S_0(\mu)$, $T_0(\mu)$ and $T_1(\mu)$ suffice to approximate the solution. Once this is established, one can also extract the contribution of the vertical field from the solution (16.137) at the plasma boundary, as Shafranov does [408]:

$$ \Psi^{\mathrm{ex}}_{\mathrm{vf}} \equiv \Psi^{\mathrm{ex}}(a_m = 0) \approx -\frac{\pi}{2\sqrt{2}}\left[b_0 + \tfrac{1}{2}(b_0 + 3b_1)\frac{r}{R_0}\cos\theta\right] $$

$$ = \tfrac{1}{2}B_v R^2 \approx \tfrac{1}{2}B_v R_0^2\left(1 + 2\frac{r}{R_0}\cos\theta\right). \qquad (16.145) $$

This again yields the expression (16.144) for B_v.

Since the external solution (16.128) is exact, the method may be generalized to arbitrary aspect ratio. The coefficients a_m and b_m are then determined by the plasma current distribution; see Zakharov and Shafranov [494], Section 3.3.

▷ **Supplementary expressions** From Eq. (16.141), one may obtain exact relations of the functions $T_m(\mu)$ in terms of elliptic integrals, alternative to the hypergeometric series:

$$ T_0(\mu) = -\tfrac{1}{4}\int_0^\pi \sqrt{2D(\mu,\eta')}\,d\eta' = -\frac{1}{\lambda}E(\lambda^2), \qquad \lambda^2 \equiv [\cosh(\mu/2)]^{-2}, $$

$$ T_1(\mu) = \tfrac{3}{4}\int_0^\pi \sqrt{2D(\mu,\eta')}\cos\eta'\,d\eta' = \frac{(2-2\lambda^2)K(\lambda^2) - (2-\lambda^2)E(\lambda^2)}{\lambda^3}. \qquad (16.146) $$

These provide checks on the expansions (16.131), and (16.147) below.

The approximations of S_m and T_m for $\mu \ll 1$ may be derived from the functions P of Morse and Feshbach [341], p. 1329, and the functions Q of Bateman [25], p. 149:

$$ \left\{ \begin{array}{l} S_m(\mu) \approx \tfrac{1}{2}(m^2 - \tfrac{1}{4})\mu^2, \\[2ex] T_m(\mu) \approx -1 + \tfrac{1}{2}(m^2 - \tfrac{1}{4})\left[\mu^2\ln(8/\mu) + \left(\tfrac{1}{2} - \sum_{n=1}^m \frac{2}{2n-1}\right)\mu^2\right]. \end{array} \right. \qquad (16.147) $$

The logarithmic terms demonstrate the ill-posedness from another angle: to give finite results for the flux Ψ^{ex} and its derivatives in the limit $\mu \to 0$ (e.g. on the axis of symmetry), the constants b_m should be such that $\sum b_m T_m$ cancels, which is a hopeless task to accomplish when the solution is obtained by integrating from the inside to the outside. ◁

16.3 Exact equilibrium solutions

16.3.1 Poloidal flux scaling

Recall the analysis of the scaling properties of the MHD equations in Volume [1], Section 4.1.2. We found that in any MHD problem three trivial quantities appear that have no other function than to provide magnitude and dimensions for the occurring lengths, magnetic field strengths and time scales. Hence, they can, and should, be scaled out by dividing all occurring variables and parameters by appropriate powers of them, so that the resulting problem becomes dimensionless. Since equilibrium problems are time-independent, only two such trivial scaling quantities occur, viz. a length and a magnetic field strength. After these have been taken out, we can concentrate on the quantities of real physical interest, e.g. the field line geometry, represented by the parameter q_1, the amount of plasma confined, represented by the parameter β/ϵ, and the "fatness" of the torus, represented by the parameter ϵ. In Section 16.1.4, we have already demonstrated that a more effective parameterization is obtained by replacing the parameter pair β/ϵ, q_1 by the parameter pair $\epsilon\beta_\mathrm{p}$, q^*. We will now complete that analysis by systematically subjecting the Grad–Shafranov equation to this scaling, where also the degrees of freedom residing in the free flux function profiles will be accounted for.

This procedure involves the following steps.

(a) Construct *a minimum number of dimensionless quantities* of order unity by dividing lengths by the half-width a of the plasma, and magnetic fields by the strength B_0 of the *external* vacuum magnetic field at $R = R_0$, i.e. at the middle of the plasma, but on the outside (indicated by B_0 in Fig. 16.10(a)). This yields dimensionless poloidal coordinates (Fig. 16.10(b)):

$$x \equiv (R - R_0)/a \,, \qquad y \equiv Z/a \,, \tag{16.148}$$

the *inverse aspect ratio*:

$$\epsilon \equiv a/R_0 \quad (\ll 1 \text{ in asymptotic expansions}) \,, \tag{16.149}$$

the *dimensionless poloidal flux*:

$$\overline{\Psi} \equiv \Psi/(a^2 B_0) \,, \tag{16.150}$$

a separate parameter for the *inverse of the total poloidal flux through the plasma*, which is proportional to the modified safety factor q^*:

$$\alpha \equiv a^2 B_0/\Psi_1 \equiv 1/\overline{\Psi}_1 \ (\sim q^*) \,, \tag{16.151}$$

and a dimensionless "radial" *flux coordinate of unit range*:

$$\psi \equiv \Psi/\Psi_1 \equiv \alpha\overline{\Psi} \quad \Rightarrow \quad 0 \le \psi \le 1 \,. \tag{16.152}$$

From now on, we will leave this construction understood and drop the bars.

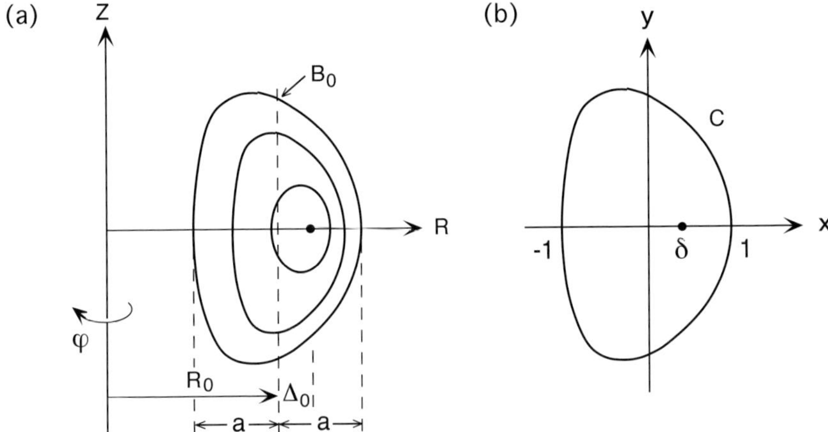

Fig. 16.10 Cross-sectional geometry of flux surfaces: (a) in R, Z, φ coordinates; (b) in the dimensionless poloidal x-y plane.

(b) Introduce *scaled dimensionless profiles* for the pressure and the "diamagnetism",

$$P(\psi) \equiv \frac{\alpha^2}{\epsilon B_0^2} p(\Psi) , \qquad Q(\psi) \equiv -\frac{\epsilon \alpha^2}{a^2 B_0^2} \tfrac{1}{2}\left[I^2(\Psi) - R_0^2 B_0^2\right] , \qquad (16.153)$$

so that the Grad–Shafranov equation transforms into

$$\psi_{xx} + \psi_{yy} - \frac{\epsilon}{1 + \epsilon x}\,\psi_x = \frac{1}{\epsilon}\left[Q' - (1 + \epsilon x)^2 P'\right] . \qquad (16.154)$$

To get a RHS of order unity, we introduce yet another profile that expresses the mentioned fact that tokamak equilibria at "high" β ($\sim \epsilon$) become θ-pinch like so that the pressure and diamagnetism profiles must approximately balance:

$$G(\psi) \equiv -\frac{1}{\epsilon}\left[Q(\psi) - P(\psi)\right] . \qquad (16.155)$$

The Grad–Shafranov equation then transforms into the desired form where the order of magnitude of each term is manifest:

$$\psi_{xx} + \psi_{yy} - \frac{\epsilon}{1 + \epsilon x}\,\psi_x = -G' - 2x(1 + \tfrac{1}{2}\epsilon x)P' . \qquad (16.156)$$

Notice that, even in the limit $\epsilon \to 0$, there is still poloidal asymmetry resulting in an outward shift of the flux surfaces due to the term $-2xP'$. This is the essence of the *high β tomakak ordering*, but with $\epsilon \neq 0$ the equation is exact.

(c) We are not content yet: the arbitrary profiles $G'(\psi)$ and $P'(\psi)$ represent an infinite amount of freedom which makes it hard to compare experimental with theoretical stability results obtained for different equilibria. We should distinguish the amplitudes of these functions (related to the global parameters $\epsilon\beta_{\rm p}$ and q^*) and their shapes.

Therefore, we introduce *unit profiles* $\Gamma(\psi)$ and $\Pi(\psi)$, or rather $\gamma(\psi)$ and $\pi(\psi)$, and amplitudes A and B:

$$ G' \equiv -A\Gamma(\psi) , \qquad \Gamma(\psi) = \Gamma_1 + (1 - \Gamma_1)\gamma(\psi) , \tag{16.157} $$

$$ \text{with boundary values} \quad \Gamma(0) = \gamma(0) = 1 , \quad \Gamma(1) = \Gamma_1 , \quad \gamma(1) = 0 , $$

$$ P' \equiv -\tfrac{1}{2}B\Pi(\psi) , \quad \Pi(\psi) = \Pi_1 + (1 - \Pi_1)\pi(\psi) , \tag{16.158} $$

$$ \text{with boundary values} \quad \Pi(0) = \pi(0) = 1 , \quad \Pi(1) = \Pi_1 , \quad \pi(1) = 0 , $$

where the parameters Γ_1 and Π_1 roughly represent a possible non-vanishing current density and pressure gradient at the plasma edge. When G' and P' are integrated with respect to ψ, two additional arbitrary integration constants appear that represent possible non-vanishing surface currents; they will be neglected here. Hence,

$$ G(\psi) = A \int_{\psi}^{1} \Gamma(\psi')\, d\psi' , \quad P(\psi) = \tfrac{1}{2}B \int_{\psi}^{1} \Pi(\psi')\, d\psi' , \quad Q(\psi) = P(\psi) - \epsilon G(\psi) . \tag{16.159} $$

The *scaled Grad–Shafranov equation* then becomes:

$$ \Delta^{*}\psi \equiv \psi_{xx} + \psi_{yy} - \frac{\epsilon}{1 + \epsilon x}\, \psi_x = A\Gamma(\psi) + Bx(1 + \tfrac{1}{2}\epsilon x)\Pi(\psi) , \tag{16.160} $$

where $\Gamma(\psi)$ and $\Pi(\psi)$ approximately represent the shapes of the toroidal current density and the pressure gradient profiles. This appears to be a final, definite, form of the Grad–Shafranov equation, with no arbitrariness in the normalizations.

(d) Finally, we transform the original boundary value problem (16.81)–(16.82) into a *non-standard boundary value problem* by considering not only the cross-sectional shape C, but also the dimensionless position $\delta \equiv \Delta_0/a$ of the magnetic axis (Fig. 16.10(b)) *as given*, so that the boundary conditions become

$$ \psi = 1 \qquad (\text{at the plasma boundary } C\!: \ r \equiv \sqrt{x^2 + y^2} = f(\theta)) , \tag{16.161} $$

$$ \psi = \psi_x = \psi_y = 0 \qquad (\text{at the magnetic axis: } x = \delta , \ y = 0) . \tag{16.162} $$

The boundary value problem (16.160)–(16.162) is then over-determined, so that the global parameters A and B become *eigenvalues* that should be determined together with the solution ψ. Effective methods exist to do this (see Section 16.3.3).

Now count! Since the Grad–Shafranov equation is a nonlinear equation, it is impossible to distinguish between cause and effect. This translates into arbitrariness of the choice of input and output variables of a numerical equilibrium code that solves this problem. It is essential, though, that the number of independent global and shape parameters is carefully counted in order to be able to compare results from different equilibrium solvers and to avoid spurious parameter scans. This is true in general, but more pertinent in MHD equilibrium because of the two-fold infinite freedom of the equilibrium profiles. Therefore, in Section 16.1.4, we have introduced a distinction between four kinds of parameter stressing the difference

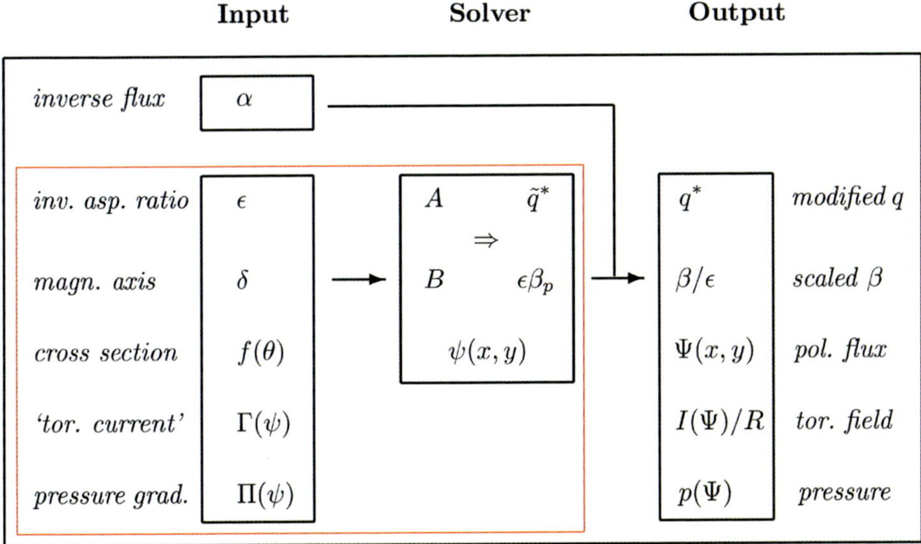

Fig. 16.11 Independent parameters and profiles in a numerical equilibrium solver. Inside the red box: core part of the solver, which is independent of the inverse flux parameter $\alpha \equiv 1/\bar{\Psi}_1$ or the modified safety factor q^*.

between precisely known external parameters and parameters that require knowledge of the plasma profiles that is only indirectly known from diagnostics.

The mentioned distinction between the four kinds of parameters has now been completed by the associated construction for the Grad–Shafranov equation:

(1) we have eliminated the *trivial scaling parameters* a and B_0;
(2) we have defined the *geometry parameters* ϵ and the shape $r = f(\theta)$ of the plasma cross-sectional boundary curve C;
(3) we have (temporarily) replaced the *global confinement parameters* $\epsilon\beta_p$ and q^* by the shift δ of the magnetic axis and the inverse poloidal flux α;
(4) we have defined the *distribution functions* $\Gamma(\psi)$ and $\Pi(\psi)$ for the toroidal plasma current and the pressure profile.

The peculiar construction of item (3) requires some explanation. Whereas $\epsilon\beta_p$ and q^* remain the global equilibrium parameters of interest, in a numerical solution procedure it is expedient to consider δ and α as the primary, input, parameters. Specifying δ then fixes one of the major geometric features of the solution, viz. the position of the magnetic axis, and specifying α fixes the other major geometric feature of the magnetic geometry, viz. the total poloidal flux. There is another, more fundamental, reason for the procedure put forward: the *poloidal flux scaling*, described by the parameter α, permits the solution of the core of the equilibrium problem, described by the scaled Grad–Shafranov equation (16.160), without actually specifying it! This is evident from the fact that the latter equation does not contain α. The eigenvalue parameters A and B of this equation will be completely determined, and hence the solution $\psi(x, y)$ as well, when all item (1), (2) and (4) parameters are given, but from the item (3) parameters only δ, which then uniquely determines the parameter $\epsilon\beta_{\rm p}$. However, the parameter q^* is not determined by the scaled equations, but only its relative magnitude with respect to α,

$$\widetilde{q}^* \equiv q^*/\alpha \,. \tag{16.163}$$

Hence, *the poloidal flux scaling,* described by the parameter α, *and the toroidal current scaling,* introduced in Section 16.1.4 and described by the parameter q^*, *are one and the same.* This is illustrated in Fig. 16.11 by the red box, which contains the α or q^* independent part of the equilibrium solver. In tokamak stability studies, constructing stability diagrams as schematically indicated in Fig. 16.8, usually q_1, or rather q^*, is varied while the rest of the equilibrium is kept fixed as much as possible. With the poloidal flux scaling, this computer time saving procedure is now uniquely defined.

If the core problem for the flux ψ and the eigenvalues A and B is solved, the explicit spatial dependence of the physical variables follows by substituting the value of α in the following expressions:

$$\psi = \psi(x, y) \quad \Rightarrow \quad (\alpha/\epsilon)\mathbf{B}_{\rm p} = (1 + \epsilon x)^{-1}\mathbf{e}_\varphi \times \nabla\psi \,,$$

$$(\alpha^2/\epsilon)p = P(\psi)\,, \qquad B_\varphi = (1 + \epsilon x)^{-1}\Big[1 - 2(\epsilon/\alpha^2)Q(\psi)\Big]^{1/2} \,,$$

$$\alpha\mathbf{j}_{\rm p} = Q'(\psi)\Big[1 - 2(\epsilon/\alpha^2)Q(\psi)\Big]^{-1/2}, \qquad (\alpha/\epsilon)j_\varphi = (1 + \epsilon x)^{-1}\Delta^*\psi \,. \tag{16.164}$$

The global parameters defined in Eqs. (16.66) [(16.163)] and (16.67),

$$\widetilde{q}^* = \frac{L^2}{2\pi}\Big[\iint_S (1 + \epsilon x)^{-1}\Delta^*\psi \, dx\, dy\Big]^{-1} \,, \tag{16.165}$$

$$\epsilon\beta_{\rm p} = 8\pi \iint_S P(\psi)\, dx\, dy \Big[\iint_S (1 + \epsilon x)^{-1}\Delta^*\psi \, dx\, dy\Big]^{-2} \,, \tag{16.166}$$

do not involve α, but just the eigenvalues A and B, according to Eqs. (16.159) and (16.160), so that they may be determined inside the red box of Fig. 16.11. At this level, the scaled counterpart of Eq. (16.69) may also be computed:

$$\widetilde{\beta}/\epsilon \equiv \alpha^2\beta/\epsilon = 2S^{-1}\iint_S P(\psi)\, dx\, dy = C\epsilon\beta_{\rm p}/\widetilde{q}^{*2} \,. \tag{16.167}$$

However, the scaled safety factor profile depends on α and, hence, does not belong to the core solver:

$$\widetilde{q}(\psi) \equiv q(\psi)/\alpha = \frac{1}{2\pi} \oint_\psi \frac{\sqrt{1 - 2(\epsilon/\alpha^2)Q(\psi)}}{(1 + \epsilon x)|\nabla\psi|}\, d\ell\,. \tag{16.168}$$

Notice that it requires accurate calculation over a flux surface: another reason not to exploit the boundary value q_1 as a global parameter (nor the arbitrary parameter q_{95}, that is sometimes used to indicate the value of q at the flux surface containing 95% of the flux in order to avoid the infinity when a separatrix occurs). To avoid misunderstanding: this does not imply that the safety factor profile $q(\psi)$ loses its central importance, in particular not because it is crucial in the construction of straight field line coordinates for stability analysis.

Summarizing, we have enumerated the freedom in the choice of MHD equilibrium parameters and profiles by prescribing three geometric quantities, viz. the inverse aspect ratio ϵ, the shift δ of the magnetic axis, the plasma cross-sectional shape C, and two arbitrary unit profiles $\Gamma(\psi)$ and $\Pi(\psi)$, corresponding to the toroidal current and the pressure gradient. Solving the Grad–Shafranov equation with these independent input parameters turns A and B into eigenvalues which should be computed together with the solution $\psi(x, y)$. These parameters are directly related to the scaled physical parameters $\widetilde{q}^*$ and $\epsilon\beta_p$. The unscaled parameters q^* and β/ϵ, and all other parameters and functions of interest, are obtained by means of a simple scaling in terms of the parameter $\alpha \equiv \Psi_1^{-1}$. As a bonus, in the *"high"-beta tokamak ordering*, where $\epsilon \ll 1$ while $\epsilon\beta_p \sim 1$ and $q^* \sim 1$ are kept finite, the scaling becomes trivial: all RHSs of Eqs. (16.164)–(16.168) become independent of both α and ϵ, whereas the LHSs scale as simple powers of those parameters.

16.3.2 Soloviev equilibrium

A useful special solution of the original Grad–Shafranov equation (16.81) was obtained by Soloviev [413] by assuming linear profiles for $I^2(\Psi)$ and $p(\Psi)$,

$$\tfrac{1}{2}I^2(\Psi) = \tfrac{1}{2}I_0^2 - E\Psi\,, \quad I_0 \equiv R_0 B_0 \quad \Rightarrow \quad \tfrac{1}{2}I^{2\prime} = -E\,,$$

$$p(\Psi) = p_0 - F\Psi\,, \quad p_0 \equiv p(0) \quad \Rightarrow \quad p' = -F\,, \tag{16.169}$$

so that the Grad–Shafranov equation becomes a linear, inhomogeneous, PDE:

$$R\frac{\partial}{\partial R}\left(\frac{1}{R}\frac{\partial\Psi}{\partial R}\right) + \frac{\partial^2\Psi}{\partial Z^2} = E + FR^2\,. \tag{16.170}$$

If one ignores the elliptic character of this equation, by replacing the BC (16.82) by the condition that Ψ should vanish at the magnetic axis, which is accepted to be located at some position determined by

$$\Psi = \Psi_R = \Psi_Z = 0 \quad \text{(at the magnetic axis } R = R_{\mathrm{m}},\ Z = 0),\qquad (16.171)$$

one easily checks (just substitute) that the Grad–Shafranov equation is solved by the following polynomial expression:

$$\Psi(R, Z) = (C - DR^2)^2 + \tfrac{1}{2}\big[E + (F - 8D^2)R^2\big]Z^2, \qquad R_{\mathrm{m}} = \sqrt{C/D}.\qquad (16.172)$$

Hence, one obtains a four-parameter family of solutions of the original equilibrium problem (16.81)–(16.82), where the position of the magnetic axis (at $R = R_{\mathrm{m}}$) is controlled by the parameters C and D, but the cross-section of the outer wall can only be specified a posteriori by cutting out some suitable part of the solution where the flux surfaces are still nested around the magnetic axis. Obviously, this is not what one needs in the control room of a tokamak experiment, but the Soloviev solution is, nevertheless, important to test equilibrium solvers and the accuracy of the associated construction of flux coordinates for stability analysis.

In order to develop intuition about the parameterization introduced in the previous section, let us now consider the scaled counterpart of the Soloviev construction. The flux functions $\Gamma(\psi)$ and $\Pi(\psi)$ corresponding to the linear profiles (16.169) become constants, $\Gamma_1 = \Pi_1 = 1$, and the scaled Grad–Shafranov equation (16.160) becomes

$$\psi_{xx} + \psi_{yy} - \frac{\epsilon}{1 + \epsilon x}\psi_x = A + Bx(1 + \tfrac{1}{2}\epsilon x).\qquad (16.173)$$

The scaled Soloviev solution now involves only three arbitrary parameters, viz. the inverse aspect ratio ϵ, the elongation $\sigma = b/a$ of the outer flux surface $\psi = 1$ at $x = 0$, and the triangularity τ of the outer flux surface. It may be written as

$$\psi(x, y) = \Big[x - \tfrac{1}{2}\epsilon(1 - x^2)\Big]^2 + (1 - \tfrac{1}{4}\epsilon^2)\Big[1 + \epsilon\tau x(2 + \epsilon x)\Big]\Big(\frac{y}{\sigma}\Big)^2,\qquad (16.174)$$

where the position of the magnetic axis at $x = \delta$, $y = 0$, found a posteriori from the "boundary condition" $\psi = \psi_x = \psi_y = 0$, is directly related to ϵ:

$$\delta = \frac{1}{\epsilon}\Big[\sqrt{1 + \epsilon^2} - 1\Big] \quad \Big(\approx \tfrac{1}{2}\epsilon,\ \text{if}\ \epsilon \ll 1\Big).\qquad (16.175)$$

The eigenvalue parameters A and B of the scaled Grad–Shafranov equation are found by substitution of Eq. (16.174) into Eq. (16.173):

$$A = 2\Big[1 + \frac{1 - \tfrac{1}{4}\epsilon^2}{\sigma^2}\Big], \qquad B = 4\epsilon\Big[1 + \frac{1 - \tfrac{1}{4}\epsilon^2}{\sigma^2}\tau\Big].\qquad (16.176)$$

They are independent of α, as expected, and they determine the scaled parameters $\tilde{q}^*$ and $\epsilon\beta_{\mathrm{p}}$ by the integrals (16.165) and (16.166). Also specifying the parameter

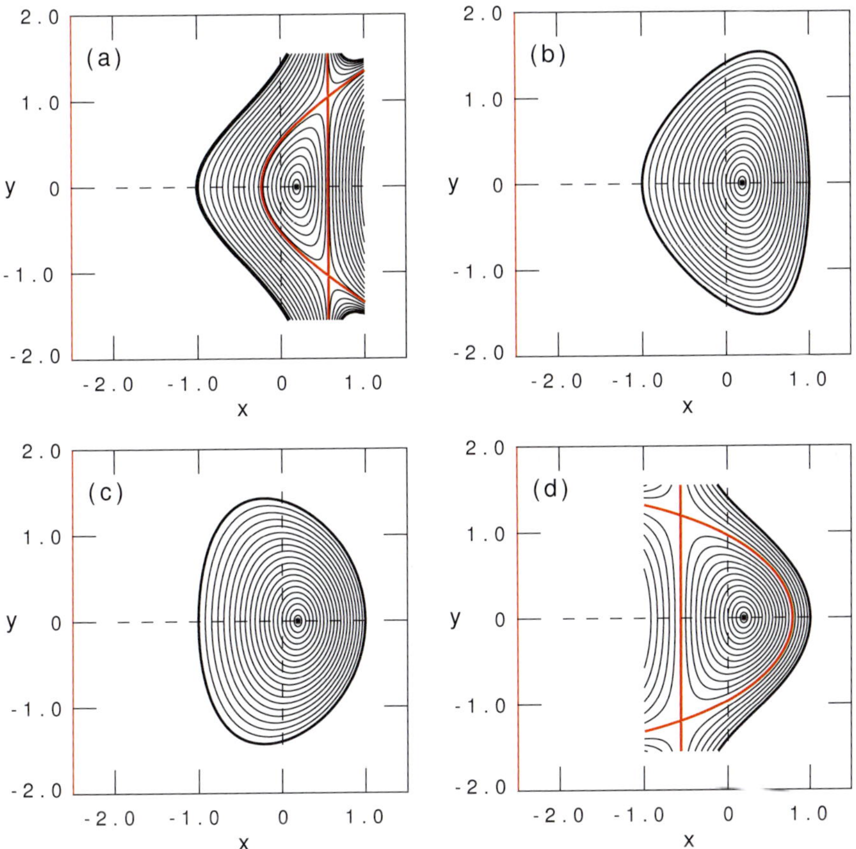

Fig. 16.12 Flux contours $0 \leq \sqrt{\psi} \leq 1$ for Soloviev equilibria, $\epsilon = 0.4$, $\sigma = 1.4$, (a) open configuration with separatrix on the right, $\tau = -2.0$; (b) closed configuration with negative triangularity, $\tau = -0.5$; (c) closed configuration with positive triangularity, $\tau = 1.0$; (d) open configuration with separatrix on the left, $\tau = 2.5$. Plasma boundary ($\psi = 1$) in thick black, axis of symmetry and separatrices in red.

α, the three parameters ϵ, σ and τ then finally completely determine the global parameters q^* and β/ϵ, and the safety factor profile according to Eq. (16.168).

According to our normalization, the plasma boundary curve C of the Soloviev equilibria is given by

$$\psi(x, y) = 1, \tag{16.177}$$

determined by the expression (16.174) in terms of the parameters ϵ, σ and τ. Resulting flux contour plots are shown in Fig. 16.12 for some representative parameter values. The qualitative geometry of the flux contours is determined by the extrema of the flux, $\psi_x = \psi_y = 0$, which are elliptic or hyperbolic points depending on the value of the discriminant,

$$D \equiv \psi_{xx}\psi_{yy} - 4\psi_{xy}^2 \begin{cases} > 0 & \textit{(elliptic point)} \\[2mm] < 0 & \textit{(hyperbolic point)} \end{cases} . \tag{16.178}$$

This yields the following characteristic features of the Soloviev equilibria:

(a) the central point on the symmetry axis, $x = -1/\epsilon$, $y = 0$, is elliptic for $\tau > 1$ and hyperbolic for $\tau < 1$;

(b) the magnetic axis at $x = \delta$, $y = 0$, given by Eq. (16.175), is elliptic if $\tau > -1/\epsilon^2$;

(c) the separatrices through the hyperbolic points

$$x_\mathrm{s} = -(1/\epsilon)[1 - \sqrt{1 - 1/\tau}], \quad y_\mathrm{s} = \pm[\sigma/(\epsilon\tau)]\sqrt{\tfrac{1}{2}(1 + \epsilon^2\tau)/(1 - \tfrac{1}{4}\epsilon^2)} \tag{16.179}$$

enter the plasma domain from the right ($x_\mathrm{s} < 1$) if $\tau < -1/[\epsilon(2 + \epsilon)]$ and from the left if $\tau > 1/[\epsilon(2 - \epsilon)]$. A system of closed flux contours requires the value of τ to be restricted within these limits. These features are manifest in Fig. 16.12.

These restrictions of the Soloviev equilibria should not be taken to be illustrative of the ultimate limit $\epsilon\beta_\mathrm{p,max} \sim 1$ of tokamaks at high-beta, defined in Eq. (16.71) of Section 16.1.4. This is evident from the orders of magnitude $\delta \sim \epsilon$ and $B \sim \epsilon$, which imply that Soloviev equilibria are essentially low-β equilibria with artificial limits $\epsilon\beta_\mathrm{p,max} \sim \epsilon$ due to the way in which they were constructed. If one chooses $\delta = \mathcal{O}(1)$ and solves Eq. (16.173) with the same boundary (16.177), one obtains valid high-beta equilibria, but they will not be of the Soloviev polynomial type.

To analyze the waves and instabilities of a toroidal equilibrium, we need to be able to construct the poloidal flux coordinates corresponding to the magnetic surfaces. One of the merits of the Soloviev equilibrium is that this can be done explicitly by means of a Pythagorean decomposition of ψ:

$$\left. \begin{array}{ll} f \equiv \sqrt{x - \tfrac{1}{2}\epsilon(1 - x^2)} & \psi(x, y) = f^2(x) + g^2(x, y) \\[4mm] g \equiv \sqrt{(1 - \tfrac{1}{4}\epsilon^2)[1 + \epsilon\tau x(2 + \epsilon x)]}\,\dfrac{y}{\sigma} \quad \Rightarrow \quad & \theta(x, y) \equiv \arctan(g/f) \end{array} \right\}, \tag{16.180}$$

with the inverse:

$$\begin{cases} x(\psi, \theta) = \dfrac{1}{\epsilon}\left[\sqrt{1 + 2\epsilon\sqrt{\psi}\cos\theta + \epsilon^2} - 1\right] \\[4mm] y(\psi, \theta) = \dfrac{\sigma\sqrt{\psi}\sin\theta}{\sqrt{(1 - \tfrac{1}{4}\epsilon^2)(1 + 2\epsilon\tau\sqrt{\psi}\cos\theta + \epsilon^2\tau)}} \end{cases} . \tag{16.181}$$

All variables can now be computed in these coordinates. In the limit $\epsilon \to 0$, they yield straight field line coordinates ψ, ϑ, with the inverse $x = \sqrt{\psi}\cos\vartheta$, $y = \sigma\sqrt{\psi}\sin\vartheta$, describing "straight low-beta tokamaks" with elliptical cross-section.

16.3.3 Numerical equilibria*

In this section, we will discuss two numerical methods for solving the Grad–Shafranov equation, where the plasma boundary shape is fixed and given a priori as a bounding flux contour. The emphasis is on computing the nested flux contours interior to this boundary to a high degree of accuracy, ensuring among other things a continuous representation of ψ and $\nabla\psi$. The essential form of the Grad–Shafranov equation can be written as

$$\Delta^*\psi = \widetilde{F}(\psi)\,,\tag{16.182}$$

where the nonlinear ψ dependence is found in the right hand side. It is then convenient to use Picard iteration to converge onto the solution such that, from a starting guess $\psi^{(0)}(x,y)$, the iterate solves the linearized problem

$$\Delta^*\psi^{(n+1)} = \widetilde{F}(\psi^{(n)})\,,\tag{16.183}$$

which is stopped when e.g. $\max|\psi^{(n+1)} - \psi^{(n)}| \le \epsilon_{\text{tol}}$. The max-norm runs over all grid points and tolerances ϵ_{tol} can be taken as low as, or smaller than, 10^{-8}, depending on machine precision.

Conformal mapping

One of the earlier numerical approaches to tokamak equilibrium exploited a conformal mapping to transform the circular plasma cross-section of a high-beta tokamak onto itself while shifting the image of the magnetic axis to the centre of the coordinates [142, 162]. The Moebius transformation $z \equiv x + iy \to w$, and its inverse, effecting this are given by

$$w(z) = \frac{z-\delta}{1-\delta z} \quad\Leftrightarrow\quad z(w) = \frac{w+\delta}{1+\delta z}\,.\tag{16.184}$$

It is then expedient to exploit polar coordinates in the mapped plane, $w \equiv s\exp(it)$, so that the two-dimensional Laplacian part of $\Delta^*\psi$ simply transforms through the explicitly known scale factor $h(s,t)$:

$$\psi_{xx} + \psi_{yy} = \frac{1}{h^2}\left(\frac{1}{s}\frac{\partial}{\partial s}s\frac{\partial\psi}{\partial s}\right)\,,\qquad h(s,t) \equiv |dz/dw|\,.\tag{16.185}$$

To fully exploit the power of analytic functions, it is essential to modify the Picard iterate of the scaled Grad–Shafranov equation (16.160) such that the two-dimensional Laplacian appears on the LHS,

$$\frac{1}{s}\frac{\partial}{\partial s}s\frac{\partial}{\partial s}\psi^{(n+1)} + \frac{1}{s^2}\frac{\partial^2}{\partial t^2}\psi^{(n+1)} = F^{(n)}(s,t)$$

$$\equiv h^2\left[A^{(n)}\Gamma(\psi^{(n)}) + B^{(n)}x(1 + \tfrac{1}{2}\epsilon x)\Pi(\psi^{(n)}) + \frac{\epsilon}{1+\epsilon x}\psi_x^{(n)}\right]\,,\tag{16.186}$$

whereas all the terms on the RHS are explicitly known on the (s, t) grid from the previous step, including the eigenvalues $A^{(n)}$ and $B^{(n)}$, as we will see. (A further simplification, which we will not exploit here, is to assume the high-beta tokamak approximation, so that the two terms with ϵ on the RHS vanish.)

Although the flux surfaces in the mapped plane are not circular (except for the outermost one), it is clear that mapping the magnetic axis onto the origin is the crucial step to facilitate Fourier analysis of the solutions with only few harmonics. This is very much like the Shafranov shifted circle representation, except that the present one is not restricted to small shifts. Exploiting a truncated Fourier series,

$$\psi^{(n+1)}(s, t) = \sum_{m=0}^{M} (1 - \tfrac{1}{2}\delta_{m0})\hat{\psi}_m^{(n+1)}(s) \cos mt, \tag{16.187}$$

and similarly for $F^{(n)}(s, t)$, the Fourier components of the Picard iterate become simple ODEs,

$$\frac{1}{s}\frac{d}{ds} s \frac{d}{ds} \hat{\psi}_m^{(n+1)} - \frac{m^2}{s^2} \hat{\psi}_m^{(n+1)} = \hat{F}_m^{(n)}, \tag{16.188}$$

whereas the two boundary conditions (16.161), prescribing the value $\psi = 1$ at the plasma boundary, and (16.162), prescribing $\psi = \psi_x = \psi_y = 0$ at the position $x = \delta$, $y = 0$ of the magnetic axis, become conditions on the harmonics:

$$\hat{\psi}_m^{(n+1)}(1) = 2\delta_{m0}, \qquad \hat{\psi}_m^{(n+1)}(0) = \hat{\psi}_m^{(n+1)\prime}(0) = 0. \tag{16.189}$$

The solutions of the ODEs (16.188) satisfying all these BCs, except the first one for $m = 0$ and the second one for $m = 1$, are simple integrals:

$$\hat{\psi}_0^{(n+1)}(s) = \int_0^s ds' \, s'^{-1} \int_0^{s'} ds'' s'' \hat{F}_0^{(n)}(s''),$$

$$\hat{\psi}_m^{(n+1)}(s) = s^m \int_1^s ds' \, s'^{-2m-1} \int_0^{s'} ds'' s''^{m+1} \hat{F}_m^{(n)}(s'') \quad (m \neq 0). \tag{16.190}$$

These expressions still contain the unknown values of A and B. They are determined by the BC on $m = 0$,

$$\hat{\psi}_0^{(n+1)}(1) = 2 \quad \Rightarrow \quad A, \tag{16.191}$$

and the BC on $m = 1$,

$$\hat{\psi}_1^{(n+1)\prime}(0) = -\tfrac{1}{2}\int_0^1 (1 - s^2)\hat{F}_1^{(n)}(s) \, ds = 0 \quad \Rightarrow \quad B. \tag{16.192}$$

Thus, the full solution $\psi^{(n+1)}(s, t)$ at the $(n + 1)$th step is found, requiring no additional iterations to determine the eigenvalues A and B.

This completes the solution of the Grad–Shafranov equation with conformal mapping. For reasonable profiles $\Gamma(\psi)$ and $\Pi(\psi)$, the convergence is fast and accurate with few harmonics. No further operations are needed than fast Fourier transforms in the angular variable and integrations over the radial variable. Important for stability analysis is that the first and second derivatives (magnetic field and currents) are obtained with the same accuracy as the flux function $\psi(s, t)$ itself.

For the generalization of the conformal mapping technique to arbitrary cross-sections, pre-mappings could be applied that are known from the design of wings for airplanes and biplanes. These map an arbitrary simply connected domain onto a circle, or of a doubly-connected domain onto an annulus, by means of the nonlinear integral equations of Theodorsen or Garrick, respectively. These integral equations were converted by Henrici [223] into effective numerical algorithms introducing the fast Hilbert transform (FHT), a powerful counterpart of the fast Fourier transform (FFT), for conjugate periodic functions. This results in beautiful analysis of the equilibrium and stability of high-beta tokamaks, in particular when the plasma current is confined to the surface so that both the plasma region and the vacuum are described by two-dimensional Laplace equations [163]. In that case, complex analysis provides the full solution in terms of a one-dimensional variational principle involving a double angular integral over the plasma boundary. For diffuse current distributions, the radial dependence of the perturbations could be effectively described by means of a new set of polynomials [167] that, together with FFTs, provide the general solution of incompressible fluid flow in arbitrary domains. Stability of tokamaks at high beta was extensively analyzed with the resulting program HBT [153, 179].

One complication of the conformal mapping of an arbitrary domain onto a circular disk is the fact that the distribution of angular points is not free. For example, mapping of an elliptical domain onto a circle results in a distribution which is sparse on the curved parts and crowded on the flat parts of the ellipse (just like the orthogonal Ψ, χ, φ coordinates described in Section 16.1.2), precisely opposite to what is needed in stability analysis. One could remedy this by angular grid accumulation, but this spoils most of the beauty of obtaining explicit results from complex analysis. This is one of the reasons to consider the more flexible technique of the next sub-section.

Two-dimensional finite elements

Later numerical approaches of solving the Grad–Shafranov equation still exploit the Picard iteration scheme (16.183), but they are more tuned to general solution strategies, exploiting two-dimensional finite elements and Galerkin methods with the essential elements of proven implementations, already encountered in Section 15.1.3, as found in [238, 30, 29, 320].

To specify the newer solution algorithms, we need to explain the choice of the grid on which the solution is computed, the way in which the solution is approximated numerically on this grid, and the manner in which the PDE (16.182) is turned into a discrete (linear algebraic) problem. A proven strategy which allows for a continuous $\nabla\psi$ is to use a third order finite element method (FEM). On a unit square $[-1, 1]^2$, we can define up to sixteen bicubic Hermite polynomials, four per corner $(x_0, y_0) = (\pm 1, \pm 1)$, namely

$$
\begin{aligned}
H_{00}(x, y) &= \tfrac{1}{16}(x + x_0)^2(xx_0 - 2)(y + y_0)^2(yy_0 - 2)\,, \\
H_{10}(x, y) &= -\tfrac{1}{16}x_0(x + x_0)^2(xx_0 - 1)(y + y_0)^2(yy_0 - 2)\,, \\
H_{01}(x, y) &= -\tfrac{1}{16}(x + x_0)^2(xx_0 - 2)y_0(y + y_0)^2(yy_0 - 1)\,, \\
H_{11}(x, y) &= \tfrac{1}{16}x_0(x + x_0)^2(xx_0 - 1)y_0(y + y_0)^2(yy_0 - 1)\,. \quad (16.193)
\end{aligned}
$$

This allows us to approximate any function $f(x, y)$ on $[-1, 1]^2$ by the expansion

$$
\begin{aligned}
f(x, y) = \sum_{x_0, y_0} H_{00}(x, y)f(x_0, y_0) + H_{10}(x, y)\frac{\partial f}{\partial x}(x_0, y_0) \\
+ H_{01}(x, y)\frac{\partial f}{\partial y}(x_0, y_0) + H_{11}(x, y)\frac{\partial^2 f}{\partial y \partial x}(x_0, y_0)\,. \quad (16.194)
\end{aligned}
$$

The function then has a prescribed functional dependence on the square as encoded in the bicubic expansion polynomials, and the sixteen expansion coefficients are the local corner values of the function f and its derivatives.

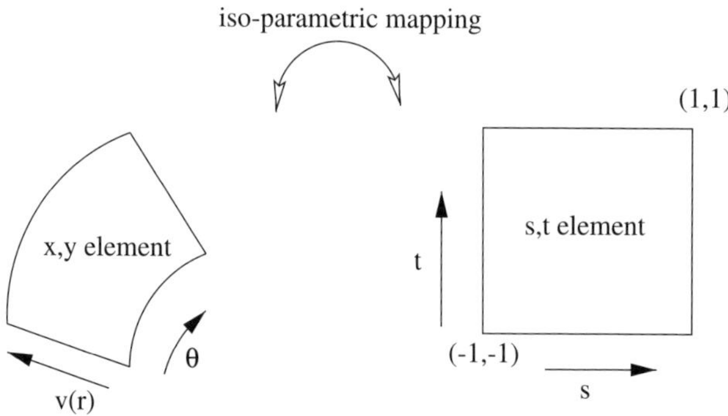

Fig. 16.13 Iso-parametric mapping to the local s, t coordinates for a curved quadrilateral on the square $[-1, 1]^2$.

For a given cross-sectional shape, the grid should allow alignment with the usually curved boundary. In the (x, y) plane, polar coordinates (r, θ) centered at $(x, y) = (0, 0)$ can be used to represent a given boundary curve in a (suitably truncated) Fourier series $r = f_{\rm b}(\theta) = \sum_m a_m \exp(im\theta)$. We can then use an arbitrary "radial" function $v(r)$ from 0 (center) to 1 (boundary) in global coordinates

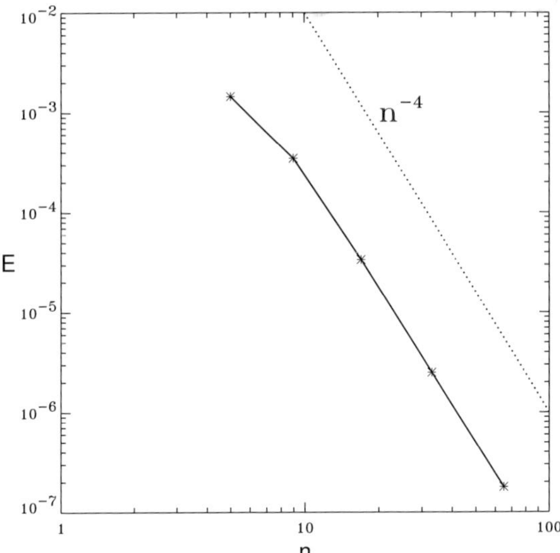

Fig. 16.14 Fourth order convergence behavior as observed in the max-norm of the error $E \equiv \max|\psi - \psi_{\mathrm{exact}}|$ in a computed Soloviev equilibrium for increasing grid resolution $n \equiv n_r = n_\theta$.

following the boundary shape where

$$x = v(r)f_{\mathrm{b}}(\theta)\cos\theta\,, \qquad y = v(r)f_{\mathrm{b}}(\theta)\sin\theta\,. \tag{16.195}$$

Our grid is then formed by taking a discrete number of n_r radial points r_i, and of n_θ angular points θ_j. The grid consists of curved quadrilateral elements, while we discussed the FEM representation of a function on a unit element $[-1,1]^2$. The numerical representation of the solution on the boundary-fitted grid is complete by making use of an iso-parametric mapping, where each curved quadrilateral is mapped onto $[-1,1]^2$ by changing from (x,y) to local (s,t) coordinates (see Fig. 16.13). In fact, this mapping consists of representing both the solution and the coordinates x,y in the same FEM representation, i.e.

$$\psi(x,y) = \sum H_{00}(s,t)\psi(s_0,t_0) + H_{10}(s,t)\frac{\partial\psi}{\partial s}(s_0,t_0)$$

$$+ H_{01}(s,t)\frac{\partial\psi}{\partial t}(s_0,t_0) + H_{11}(s,t)\frac{\partial^2\psi}{\partial s\partial t}(s_0,t_0)\,,$$

$$x(s,t) = \sum H_{00}(s,t)x(s_0,t_0) + H_{10}(s,t)\frac{\partial x}{\partial s}(s_0,t_0) + \cdots\,,$$

$$y(s,t) = \sum H_{00}(s,t)y(s_0,t_0) + H_{10}(s,t)\frac{\partial y}{\partial s}(s_0,t_0) + \cdots\,. \tag{16.196}$$

Identifying $v(r) \equiv s$ and $\theta \equiv t$, it is possible to compute the consistent FEM

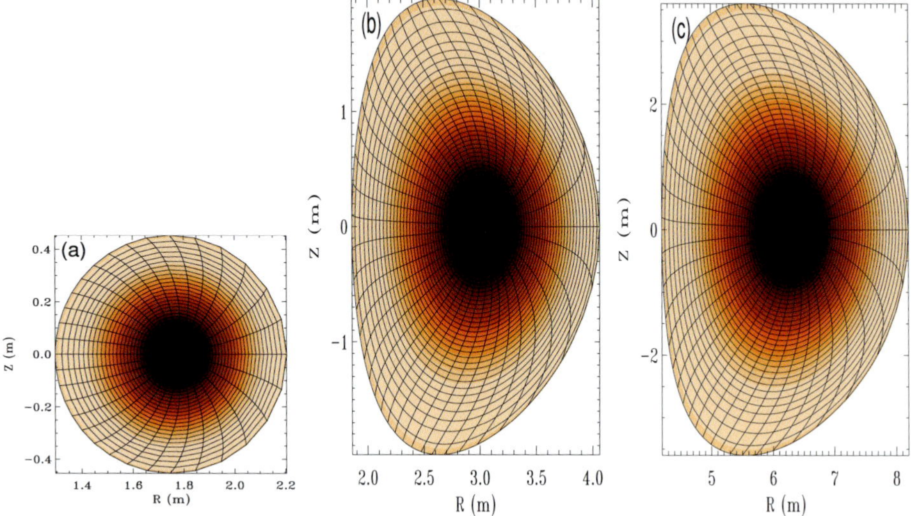

Fig. 16.15 The equilibrium flux surfaces of (a) TEXTOR, (b) JET and (c) ITER, normalized to the same width and calculated for the same normalized pressure and current distribution.

representation of the coordinates (as one can calculate x, $\partial x/\partial s$ etc. at all (s_0, t_0)). One can use this flexibility (after each Picard iterate or after Picard convergence) to take the local s coordinate to become some function of ψ, and align the grid with ψ flux contours.

Within each Picard step, one solves the discrete equivalent of Eq. (16.183). In FEM methodology, we formulate the problem in its weak form, where we select a space of test-functions χ and look for the solution ψ such that for all test functions we have

$$\int_V \chi \, \nabla \cdot (R^{-2} \nabla \psi) \, dV = \int_V \chi \, (R^{-2} \widetilde{F}) \, dV \, , \qquad (16.197)$$

where the integral is taken over the plasma volume. In the Galerkin method, the test functions are simply the finite elements (H_{00}, etc.) which are already used in the representation for ψ. This reduces the problem to a linear system $\mathbf{K}\mathbf{x} = \mathbf{b}$, where the vector $\mathbf{x}$ represents the unknown coefficients in the FEM representation, while the $\mathbf{K}$ matrix and $\mathbf{b}$ vector elements are integrals of known functions (involving the FEM and their derivatives). These integrals can e.g. be evaluated numerically by Gaussian quadrature. Figure 16.14 shows the expected quartic convergence reached when a Soloviev solution is determined numerically for increasing grid resolution $n \equiv n_r = n_\theta$ going from 5 up to 65. As the exact solution is known, the error can be quantified precisely and the fourth order convergence in ψ shows that very accurate solutions are already obtained for relatively coarse grids.

Figure 16.15 shows the flux surface distributions for three "generations" of tokamaks, calculated with the program HELENA [238]-FINESSE [29] described in this section. The three cases assumed the same flux profiles and correspond to low-beta configurations that differ in their geometric parameters: TEXTOR has a circular cross-section and $a = 0.45\,\mathrm{m}$, $R_0 = 1.75\,\mathrm{m}$; JET has a D-shaped cross-section (finite ellipticity and triangularity) with $a = 1.1\,\mathrm{m}$, $R_0 = 2.96\,\mathrm{m}$; while ITER will scale up to $a = 2\,\mathrm{m}$, $R_0 = 6.2\,\mathrm{m}$. Shown are computed pressure distributions and the nested flux surfaces, where the angular coordinate lines shown represent the straight field line coordinates. The cover of this book displays the corresponding 3D view for the ITER tokamak case: it combines density iso-surfaces with a pressure contour plot in a cross-sectional view, as well as selected field lines (magnetic axis in red, with progressively outwards a blue and green field line visualizing the varying safety factor). At left and right, an impression of the FEM grid is shown, together with magnetic field vectors (yellow).

16.4 Extensions

16.4.1 Toroidal rotation

The purely static, axi-symmetric, equilibrium can be generalized in several ways. Avoiding the significant complications due to poloidal flow (see Chapter 18), a first non-trivial extension of the Grad–Shafranov equilibrium includes the effect of toroidal rotation and the associated centrifugal force. The toroidal component of the stationary induction equation $\nabla \times (\mathbf{v} \times \mathbf{B}) = 0$ then dictates that each flux surface rotates at fixed angular velocity $\Omega(\Psi) = v_\varphi/R$.

The azimuthal component of the force balance equation still prescribes the current stream function I to be a flux function $I(\Psi) = RB_\varphi$. However, the poloidal part now requires two equations to be satisfied simultaneously. In the $\nabla\Psi$ direction, a Grad–Shafranov-like equation is obtained, namely

$$
R\frac{\partial}{\partial R}\left(\frac{1}{R}\frac{\partial\Psi}{\partial R}\right) + \frac{\partial^2\Psi}{\partial Z^2} = -I\frac{dI}{d\Psi} - \frac{\partial p}{\partial\Psi}R^2 = Rj_\varphi,
\tag{16.198}
$$

where the pressure is no longer a flux function. The dependence of the pressure $p = p(\Psi, R)$ is such that along the poloidal flux contours the force balance is ensured [454] by

$$
\left.\frac{\partial p}{\partial R}\right|_\Psi = \rho R\Omega^2(\Psi).
\tag{16.199}
$$

The latter equation can, e.g., be solved analytically under the additional assumption that the entropy is a flux function, $S \equiv p\rho^{-\gamma} = S(\Psi)$, a result which generalizes to all stationary MHD equilibria since $\mathbf{v} \cdot \nabla S = 0$ (with poloidal flows as well).

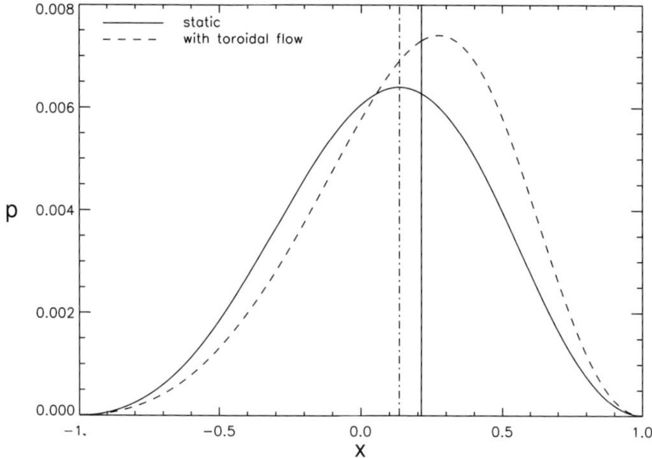

Fig. 16.16 The radial pressure profile for a toroidally rotating equilibrium compared to that of a static one. Note the outward shift of the magnetic axis (vertical dashed lines), and the separation of the pressure maximum from the magnetic axis when a large rotation is present.

The pressure can then be written as

$$
p(\Psi, R) = \left[R^2 \frac{\gamma - 1}{\gamma} \frac{\Omega^2}{2S^{1/\gamma}} + f \right]^{\gamma/(\gamma-1)} ,
\tag{16.200}
$$

so that for toroidally rotating equilibria four flux functions can be freely chosen: besides $I(\Psi)$ and $\Omega(\Psi)$ we get, e.g., $S(\Psi)$ and $f(\Psi)$.

For the purpose of equilibrium reconstruction for toroidally rotating tokamak plasmas where diagnostic information on density and temperature profiles is available, a convenient parameterization uses the corresponding static pressure $p_{\mathrm{st}}(\Psi)$ and density $\rho_{\mathrm{st}}(\Psi)$ profiles with the same entropy variation $S(\Psi)$, as follows:

$$
p(\Psi, R) = p_{\mathrm{st}}(\Psi) \left[(R^2 - R_0^2) \frac{\gamma - 1}{\gamma} \frac{\Omega^2 \rho_{\mathrm{st}}}{2p_{\mathrm{st}}} + 1 \right]^{\gamma/(\gamma-1)} .
\tag{16.201}
$$

This is useful as long as the toroidal rotation is low and its small influence on the equilibrium properties is to be quantified in comparison with a similar static reconstruction. The pressure can be shown to have its maximum shifted radially outward with respect to the magnetic axis, which in turn is shifted outward due to toroidal rotation.

As an example, consider the influence of a relatively large toroidal rotation on the pressure profile, as shown in Fig. 16.16 for a circular tokamak cross-section of inverse aspect ratio $\epsilon = 0.26$, and parameterized as follows:

$$I^2 = A\left[1 - 0.01\psi + 0.005\psi^2\right], \qquad \Omega = 0.075\left[1 - 0.999\psi\right],$$

$$p_{\mathrm{st}} = 0.0125\,A\left[1 - 2\psi + 1.001\psi^2\right], \qquad \rho_{\mathrm{st}} = A\left[1 - \psi + 0.6\psi^2 - 0.5\psi^4\right].$$

$$(16.202)$$

The last two profiles only enter in the computation when rotation is considered. The eigenvalue A determined numerically is 155 for the rotating equilibrium, versus 164 in the static case. The rotation would correspond to a maximal sonic Mach number $M_{\mathrm{s}} = v_\varphi/\sqrt{\gamma p/\rho} = 1.06$, which is rather high but illustrates well the influence of toroidal rotation.

16.4.2 Gravitating plasma equilibria*

In astrophysical contexts, the influence of external or self-gravity can become very important in the overall force balance. In accretion disks, for example, the pure Keplerian disk balances a central gravitational field with centrifugal forces. In the solar corona, quiescent prominences can be observed suspended above the solar limb, characterized by much denser and much cooler (typical factors 10–100) plasma than the coronal environment. There, the main force balance is again determined by Lorentz forces, now opposing gravity and pressure gradients. A recent analytical solution representative for quiescent prominences can be found in [318]. A similar analysis of the static force balance, now assuming translational invariance in the z-direction, finds for force balance in the $\nabla\Psi$ direction

$$\frac{\partial^2\Psi}{\partial x^2} + \frac{\partial^2\Psi}{\partial y^2} = -I\frac{dI}{d\Psi} - \frac{\partial p}{\partial\Psi} = j_z. \qquad (16.203)$$

Again, $I(\Psi) = B_z$ is a free flux function and $p = p(\Psi, y)$. Along a poloidal flux contour the equation

$$\left.\frac{\partial p}{\partial y}\right|_\Psi = -\rho g \qquad (16.204)$$

should be satisfied when considering a constant external gravitational acceleration $\mathbf{g} = -g\mathbf{e}_y$. Assuming the temperature to be a flux function $T(\Psi)$, three free flux functions can be chosen as the pressure is then $p(\Psi, y) = k(\Psi)\exp[-g\,y/T(\Psi)]$. Using a scaling with respect to half the horizontal diameter a, the outermost flux value Ψ_1 and a reference field strength B_0, a suitable scaling writes the governing Grad–Shafranov-like equation as

$$\frac{\partial^2\psi}{\partial x^2} + \frac{\partial^2\psi}{\partial y^2} = -\alpha^2\left[I\frac{dI}{d\psi} + \left(\frac{d\ell}{d\psi} + \frac{\ell\,y}{H^2}\frac{dH}{d\psi}\right)e^{-y/H}\right], \qquad (16.205)$$

where $\alpha = a\,B_0/\Psi_1$, and the pressure $p = \ell(\psi)\mathrm{e}^{-y/H}$ uses the scale height $H = T(\psi)/(ag)$. All quantities appearing are again dimensionless, e.g., $(x/a, (y - y_0)/a) \to (x, y)$ with y_0 a reference height in the prominence.

Numerical magneto-hydrostatic solutions describing the gravitationally stratified equilibrium of cool prominence plasma embedded in a near-potential coronal field are described by Petrie *et al.* [369]. The solutions are calculated using the FINESSE equilibrium solver. They describe the morphologies of the magnetic field distributions in and around prominences and the cool prominence plasma that these fields support. It reproduces the three-part structure encountered in observations: a cool dense prominence within a cavity/flux rope embedded in a hot corona.

16.4.3 Challenges

To be able to describe perturbations with minimum field line bending, it is necessary to represent the parallel gradient operator (16.29) as accurately as possible. An effective method is to replace the poloidal angle θ, which could be any angle as long as it increases by 2π after one revolution the short way around the torus, by a poloidal angle ϑ such that the field lines become straight lines in the φ–ϑ plane (Fig. 16.2). This construction has to be carried out on each flux surface labeled by ψ. To exploit the constructed flux variable ψ and the new angle ϑ as coordinates for the description of instabilities, inversion of the coordinates is needed,

$$
\left.\begin{array}{l} \psi = \psi(x_i, y_j) \\[2mm] \vartheta = \vartheta(x_i, y_j) \end{array}\right\} \quad \Rightarrow \quad \left\{\begin{array}{l} x = x(\psi_i, \vartheta_j) \\[2mm] y = y(\psi_i, \vartheta_j) \end{array}\right. , \tag{16.206}
$$

as exemplified by Eq. (16.181) for the Soloviev equilibrium. All quantities occurring in the stability analysis are then to be transformed to ψ, ϑ, φ coordinates, e.g. the normal field line curvature κ_n, the geodesic curvature κ_g, the toroidal current j_φ, etc. (see Section 17.1.2 of the next chapter). These quantities involve second derivatives with respect to ψ, so that the *equilibrium solutions need to be surprisingly accurate* for a reliable stability analysis.

In this chapter, we have presented a theoretical description of toroidal equilibrium, which involves the required accurate solution of the Grad–Shafranov equation and a careful choice of the parameters. The experimental counterpart is the production of *accurate equilibrium data* by means of all available diagnostics. To get agreement between these two is a long-term iterative process. The essential point to notice is that, at the present time, the situation is far from satisfactory: important parameters and profiles are only known to about 10%. Consequently, tokamak diagnostics need to be improved if we ever are to arrive at the agreement between theory and experiment that is standard in physics.

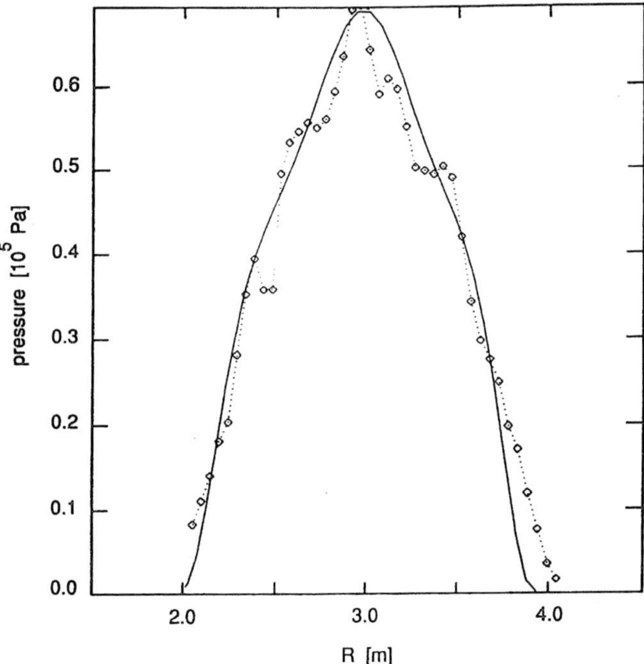

Fig. 16.17 Uncertainty: electron pressure profile obtained by LIDAR diagnostic at JET (dashed curve with diamonds) and reconstructed pressure profile by means of an equilibrium solver (drawn curve). (From Huysmans *et al.* [239].)

A systematic way of approaching this problem has been called *MHD spectroscopy* [180]; see Section 7.2.4 of Volume [1]. Here, the frequency spectrum of MHD waves is used to obtain equilibrium information (the inverse spectral problem). This spectrum is calculated by means of a spectral code with input of an equilibrium obtained by fitting experimental data to get an approximation of the free profiles. However, this involves considerable *uncertainty* as to how to handle deviations from nested flux surfaces, see e.g. the rather crude method of fitting diagnostic data with a symmetric flux function profile in Fig. 16.17.

We have extensively discussed the problem of the *freedom* in the choice of equilibrium profiles. This should be well distinguished from the just mentioned inaccuracy and uncertainty. In principle, the latter freedom can be eliminated by experimental data. More substantial are the problems of *three-dimensionality* of the equilibrium, associated with deviations from axi-symmetry due to field errors, saturated nonlinear instabilities, possibly leading to non axi-symmetric equilibrium states, stochastic field lines, etc. Another area of study is the effect of *stationary flow* ($v_0 \neq 0$) on equilibria, i.e. departure from static equilibrium in connection with the effects of neutral beams, pumped divertors, etc. (see Chapter 18). Here, three more free equilibrium profiles appear (the velocity and the density). The problem of infinite freedom remains with us!

16.5 Literature and exercises

Notes on literature

Axi-symmetric equilibrium

- The original papers on tokamak equilibrium by the "founding father" Shafranov [406, 407, 408] remain beautiful examples of down-to-earth theoretical analysis with inescapable experimental consequences. Over the years, his approach of general calculations interspersed with explicit applications has widely expanded, as presented in subsequent volumes of *Reviews of Plasma Physics* by himself [409], and with collaborators Soloviev [414], and Zakharov [494].

- Classical treatises on the structure of magnetic fields by Morozov & Soloviev [338], in *Reviews of Plasma Physics*, Volume 2, and on closed magnetic configurations by Soloviev [413], in Volume 6, contain a wealth of information on stability of magnetic fields, natural coordinates, equilibria with shaped cross-sections, etc.

Basic equilibrium concepts

- The older textbook *Ideal Magnetohydrodynamics* by Freidberg [140], Chapters 4–7, first presents the basic equilibrium considerations and then extends this to cylindrical, 2D (tokamaks), and 3D equilibria (stellarators). The new textbook on *Plasma Physics and Fusion Energy* [141], in particular Chapter 11, continues this practical presentation, always keeping eventual application to fusion energy in mind.

- Axi-symmetric equilibrium theory is presented in the textbooks *Plasma Physics for Nuclear Fusion*, Chapter 7, and *Plasma Physics and Controlled Fusion*, Chapter 6, by Miyamoto [334, 335]. The latter also discusses the different experimental approaches of tokamak, reversed field pinch, stellarator and inertial confinement to fusion.

- Chapter 3 of the "manual" on tokamaks by Wesson and collaborators [481] contains the essential concepts of flux functions, safety factor, beta, and many simple one-dimensional model calculations (à la Shafranov) on tokamak equilibrium problems.

- Summerschool proceedings [231, 122] present equilibrium theory together with many experimental contributions, demonstrating the lively research field of controlled thermonuclear fusion.

Exercises

[16.1] *Thin plasma slab*

In this exercise you will take a look at a thin plasma slab about the mid-plane. In this slab, it is reasonable to assume that all physical quantities depend only on the radius R. Show that the Grad–Shafranov equation reduces to

$$\frac{d}{dR}\left(p + \tfrac{1}{2}B^2\right) + \frac{B_\varphi^2}{R} = 0,$$

which is the force balance equation of cylindrical plasma as discussed in Chapter 12.

[16.2] *"Shafranov" shift*

In this exercise you are going to derive an analytical expression for the "Shafranov" shift. This is the difference between the center of each flux surface and the geometric axis. Assume that the inverse aspect ratio ϵ is small ($\ll 1$) and that the outer plasma boundary is circular to first order. In this case the magnetic flux function Ψ can be represented as $\Psi(\hat{r}, \hat{\theta}) = \Psi(\hat{r})$, where $(\hat{r}, \hat{\theta})$ are non-orthogonal polar coordinates defined with respect to the center of the flux surface. The connection with the cylindrical coordinates is given by

$$R = R_0 + r\cos(\hat{\theta}) + \Delta(\hat{r}), \qquad Z = \hat{r}\sin(\hat{\theta}),$$

where $\Delta(\hat{r})$ is the "Shafranov" shift. This shift has the boundary conditions $\Delta(0) = \Delta_0$ and $\Delta(1) = 0$, where Δ_0 is the shift of the magnetic axis with respect to $R = R_0$.

- Derive expressions for $R_{\hat{r}}$, $R_{\hat{\theta}}$, $Z_{\hat{r}}$ and $Z_{\hat{\theta}}$.

- Derive the co- and contravariant metric tensors in these new coordinates $(\hat{r}, \hat{\theta})$.
- Derive an expression for the Jacobian J.
- Derive an expression for the covariant magnetic field components.
- Show that

$$R^2 \nabla \cdot \left(\frac{1}{R^2}\nabla\Psi\right) \approx \frac{\partial^2\Psi}{\partial\hat{r}^2} + \frac{1}{\hat{r}}\frac{\partial\Psi}{\partial\hat{r}} - \cos\hat{\theta}\left[2\Delta'\frac{\partial^2\Psi}{\partial\hat{r}^2} - \frac{1}{\hat{r}}\left(\epsilon\hat{r} + \Delta' + \Delta''\hat{r}\right)\frac{\partial\Psi}{\partial\hat{r}}\right],$$

where all variables are dimensionless.

- Show that

$$-\frac{1}{2}\frac{dI^2}{d\Psi} \approx -\frac{1}{2R_0 B_{p0}}\frac{dI^2}{d\hat{r}}, \qquad -R^2\frac{dp}{d\Psi} \approx -\frac{R_0}{B_{p0}}\left(1 + 2\frac{\hat{r}}{R_0}\cos\hat{\theta}\right)\frac{dp}{d\hat{r}},$$

where B_{p0} is the poloidal magnetic field component.
- Show that the zeroth order reduces to

$$\frac{d}{d\hat{r}}\left(p + \tfrac{1}{2}B_{\varphi0}^2 + \tfrac{1}{2}B_{p0}^2\right) + \frac{B_{p0}^2}{\hat{r}} = 0.$$

Compare this with the equation found in Exercise[16.1] and discuss the differences.
- Show that the first order yields the equation for the "Shafranov" shift,

$$\frac{d\Delta}{d\hat{r}} = -\frac{1}{\hat{r}R_0 B_{p0}^2}\int_0^{\hat{r}}\left(\hat{r}B_{p0}^2 - 2\hat{r}^2\frac{dp}{d\hat{r}}\right)d\hat{r}.$$

- Derive the expression for the Shafranov shift defined by Shafranov himself.

[16.3] *"Shafranov" shift and Wesson profiles*

In this exercise, you will derive an analytical expression for the "Shafranov" shift using the profiles defined by Wesson [481] for the pressure and toroidal current density,

$$p = p_0(1 - \bar{r}^2), \qquad j_{\varphi0} = j_0(1 - \bar{r}^2)^\nu.$$

Here, $\bar{r} \equiv \hat{r}/a$, where a is the plasma radius. For this exercise we set $\nu = 1$, which means that both the pressure and the toroidal current density are parabolic functions.

- Derive the expression for the poloidal magnetic field B_p.
- Derive the expression for the toroidal magnetic field B_φ.

- From the toroidal magnetic field, derive the expression for the safety factor q.
- Find the expression for the poloidal beta, β_p.
- Finally, derive the expression for the derivative of the shift, Δ'.

[16.4] *Equilibria with toroidal flow and gravity*

In this exercise, you will extend the equilibrium with toroidal flow and gravity. The extension with flow is relevant in present tokamak experiments because the plasma rotates due to neutral beam injection. On the other hand, toroidal flow and gravity are essential ingredients of accretion disks. The additional inclusion of poloidal flow will be discussed in Chapter 18. The equilibrium has to satisfy Eqs. (12.24)–(12.27).

- Show that $\Omega = v_\varphi/R$ is a flux function, where v_φ is the toroidal velocity.
- Show that the projection of the momentum equation parallel to the poloidal magnetic field leads to the following two equations:

$$\left.\frac{\partial p}{\partial R}\right|_{\Psi=\mathrm{const}} = \rho\left(R\Omega^2 - \frac{\partial\Phi_{\mathrm{gr}}}{\partial R}\right), \qquad \left.\frac{\partial p}{\partial Z}\right|_{\Psi=\mathrm{const}} = -\rho\frac{\partial\Phi_{\mathrm{gr}}}{\partial Z}.$$

- Show that the projection perpendicular to the poloidal magnetic field lines results in the extended Grad–Shafranov equation (note the partial derivative on p!)

$$R^2\nabla\cdot\left(\frac{1}{R^2}\nabla\Psi\right) = -I\frac{dI}{d\Psi} - R^2\frac{\partial p}{\partial\Psi}.$$

- This extended Grad–Shafranov equation can be further specialized by assuming that the temperature, the density or the entropy is a flux function. Derive the equation for the pressure for all three cases.
- Derive the specialized extended Grad–Shafranov equation for all three choices.

[16.5] *Shafranov shift and toroidal flow*

In exercise [16.2] you have derived the "Shafranov" shift for static toroidal plasmas. Repeat this exercise, including toroidal flow now. Show that the three specialized extended Grad–Shafranov equations just derived result in the same equation for the "Shafranov" shift, i.e.

$$\frac{d\Delta}{d\hat{r}} = -\frac{1}{\hat{r}R_0 B_{\mathrm{p0}}^2}\int_0^{\hat{r}}\left[\hat{r}B_{\mathrm{p0}}^2 - \hat{r}^2\frac{d}{d\hat{r}}\left(2p + \rho v_{\varphi 0}^2\right)\right]d\hat{r}.$$

Comment on this result.

17

Linear dynamics of static toroidal plasmas

17.1 "Ad more geometrico"

17.1.1 Alfvén wave dynamics in toroidal geometry

It was shown in Chapters 6–11 of Volume [1] and 12–14 of this volume that spectral theory of MHD waves and instabilities essentially concerns the *dynamics of Alfvén waves* in the environment of magnetized plasmas. Since Alfvén waves travel along the magnetic field lines, and the field lines in turn are constrained to the nested magnetic surfaces in an axi-symmetric toroidal plasma, this implies that *the geometry of the magnetic field lines and of the constraining magnetic surfaces becomes the all-determining factor for MHD spectral theory of toroidal plasmas.* Recalling the "grand vision" of Section 12.1.1 on magnetized plasmas occurring everywhere in the Universe, it is appropriate at this point to call upon the great examples of general relativity, where light waves propagate along geodesics, and upon the dream of philosophers (Spinoza) to construct the theoretical understanding of the world "ad more geometrico" (in the geometrical manner). As we will see, even when fusion applications are the main concern, it pays off to exploit the ready-made concepts of geometry expanded by the great scientists of the past.

Recall that Alfvén wave dynamics is dominated by the gradient operator parallel to the magnetic field lines, $\mathbf{B} \cdot \nabla$. In toroidal geometry, this leads to such intricate dynamics that a very accurate description is needed of the geometry of the field lines and of the magnetic surfaces, i.e. of the equilibrium, if one wishes to study stability. Hence, we will start this chapter by describing the mapping to straight field line coordinates (Section 17.1.2), which is a nearly compelling representation for the (numerical) analysis of the stability of toroidal plasmas. We then list the different characteristics of the equilibrium, which turn out to be really geometrical properties, like the curvatures of the field lines and of the magnetic surfaces (Section 17.1.3). These play a central role in the waves and instabilities of toroidal plasmas described in the Sections 17.2–17.3. Broadly speaking, for fusion applica-

307

tions, the MHD waves of the system are desirable, as they provide information on the equilibrium distribution through MHD spectroscopy, and the MHD instabilities are to be avoided, as they lead to premature loss of the plasma.

17.1.2 Coordinates and mapping

To enable accurate description of Alfvén wave dynamics in toroidal geometry, the following steps should be taken in the analysis (see Fig. 17.1).

(a) *Solve the Grad–Shafranov equation* (16.81), or the scaled version (16.160) derived in Section 16.3.1. For the present purpose, we will exploit the unscaled poloidal flux Ψ (rather than the scaled flux $\psi \equiv \alpha \Psi$), and we will assume that the required solution

$$\Psi = \Psi(x, y) \tag{17.1}$$

is known in the relevant portion of the poloidal plane. In practice, this implies that a numerical equilibrium program has to be available, such as the finite element code described in Section 16.3.3.

(b) Rather than exploiting an arbitrary poloidal angle θ, or the angle χ belonging to the orthogonal flux coordinate triad, *construct the straight field line coordinates* Ψ, ϑ, φ, introduced in Section 16.1.2 (Fig. 16.2). Again, for the present purpose, we will assume that the solution of this problem, i.e. the distribution of the poloidal angle coordinate

$$\vartheta = \vartheta(x, y) \,, \tag{17.2}$$

is explicitly known on a grid in the x, y-plane. For these coordinates, all tangential information on the field lines has been lumped into the definition of the coordinates, while the normal behavior is described by the safety factor $q(\Psi)$. As shown below, for the harmonic dependence $\sim \exp \mathrm{i}(m\vartheta + n\varphi)$ of perturbations, the parallel gradient operator then becomes a simple multiplier,

$$F \equiv -\mathrm{i}\,\mathbf{B} \cdot \nabla \quad \Rightarrow \quad F \mathrm{e}^{\mathrm{i}(m\vartheta + n\varphi)} = \mathcal{J}^{-1}(m + nq)\,\mathrm{e}^{\mathrm{i}(m\vartheta + n\varphi)} \,, \tag{17.3}$$

so that it can be made to vanish exactly for rational field lines and surfaces (see Section 16.1.2). Consequently, straight field line coordinates are the optimal representation of the geometry of the field lines for the description of Alfvén wave dynamics. This establishes the Ψ, ϑ, φ coordinates as an important analytical tool, but it is not enough to exploit them for the explicit numerical construction of waves and instabilities. For that purpose one more step is needed.

(c) *Invert the poloidal coordinates* (17.1)–(17.2), as already announced in Eq. (16.206),

$$\left.\begin{array}{l} \Psi = \Psi(x, y) \\[2mm] \vartheta = \vartheta(x, y) \end{array}\right\} \;\Rightarrow\; \left\{\begin{array}{l} x = x(\Psi, \vartheta) \\[2mm] y = y(\Psi, \vartheta) \end{array}\right. , \tag{17.4}$$

so that we can represent all geometrical and physical quantities on a computational

$$
\boxed{
\begin{array}{l}
\textbf{EQUIL} \\[4pt]
\text{Grad–Shafranov} \quad \Rightarrow \quad
\left\{
\begin{array}{l}
\Psi = \Psi(x, y) \\[4pt]
\vartheta = \vartheta(x, y)
\end{array}
\right. \\[10pt]
\hphantom{xxxxx} + \quad \text{equil. variables} \quad V_0(x, y)
\end{array}
}
$$

$$\Downarrow$$

$$
\boxed{
\begin{array}{l}
\textbf{MAP} \\[4pt]
\text{Coord. inversion} \quad \Rightarrow \quad
\left\{
\begin{array}{l}
x = x(\Psi, \vartheta) \\[4pt]
y = y(\Psi, \vartheta)
\end{array}
\right. \\[10pt]
\hphantom{xxxxx} + \quad \text{equil. variables} \quad V_0(\Psi, \vartheta)
\end{array}
}
$$

$$\Downarrow$$

$$
\boxed{
\begin{array}{l}
\textbf{STAB} \\[4pt]
\text{Spectral equation} \quad \Rightarrow \quad \text{Eigenvalues} \ \{\omega_i\} \\[10pt]
\hphantom{xxxxx} + \quad \text{perturbations} \ V_1(\Psi, \vartheta)\mathrm{e}^{in\varphi}
\end{array}
}
$$

Fig. 17.1 The three parts of a numerical stability analysis of toroidal plasmas.

grid $\{\Psi_i, \vartheta_j\}$ in the poloidal plane. Once this has been established by construction (i.e., again having written a numerical program to do this), we can go ahead and study the waves and instabilities of the system.

17.1.3 Geometrical–physical characteristics

After the *straight-field line* (SFL) *coordinates* have been constructed, there really is no distinction anymore between physical and geometrical quantities. Moreover, the basic nonlinearity of the equilibrium implies that these quantities do not have the simple relationship of cause and effect to each other. Therefore, in somewhat arbitrary order, we now list the different geometrical–physical characteristics that will be needed in the analysis of MHD waves and instabilities in axi-symmetric toroidal systems.

(a) It is expedient to exploit the general tensor machinery for *curvilinear coordinates* (see Appendix A.3) by denoting the three coordinates as

$$
x^1 \equiv \Psi, \quad x^2 \equiv \vartheta, \quad x^3 \equiv \varphi, \tag{17.5}
$$

so that the two sets of basis vectors $\{\mathbf{a}^i\}$ and $\{\mathbf{a}_i\}$ determine the *contravariant and covariant elements of the metric tensor,*

$$\mathbf{a}^i \equiv \nabla x^i \,, \quad \mathbf{a}_i \equiv \frac{\partial \mathbf{r}}{\partial x^i} \quad \Rightarrow \quad g^{ij} = \mathbf{a}^i \cdot \mathbf{a}^j \,, \quad g_{ij} = \mathbf{a}_i \cdot \mathbf{a}_j \,, \tag{17.6}$$

where the latter are directly available in the SFL coordinates:

$$\begin{pmatrix} g_{11} & g_{12} & 0 \\ g_{12} & g_{22} & 0 \\ 0 & 0 & g_{33} \end{pmatrix} = \begin{pmatrix} x_\Psi^2 + y_\Psi^2 & x_\Psi x_\vartheta + y_\Psi y_\vartheta & 0 \\ x_\Psi x_\vartheta + y_\Psi y_\vartheta & x_\vartheta^2 + y_\vartheta^2 & 0 \\ 0 & 0 & R^2 \end{pmatrix} \,,$$

$$R = \epsilon^{-1}(1 + \epsilon x)\,. \tag{17.7}$$

The characteristic difference in length scales of the poloidal and toroidal coordinates is manifest from the expression for the *complete Jacobian* $\mathcal{J}$ in terms of the product of the distance R from the axis of symmetry and the Jacobian $\mathcal{D}$ of the poloidal plane,

$$\mathcal{J} \equiv (\nabla \Psi \times \nabla \vartheta \cdot \nabla \varphi)^{-1} = R \mathcal{D}\,,$$

$$\mathcal{D} \equiv \frac{\partial(x,y)}{\partial(\Psi,\vartheta)} = x_\Psi y_\vartheta - x_\vartheta y_\Psi = \sqrt{g_{11} g_{22} - g_{12}^2}\,, \tag{17.8}$$

from which the contravariant elements of the metric tensor may be obtained:

$$g^{11} = g_{22}/\mathcal{D}^2\,, \quad g^{12} = -g_{12}/\mathcal{D}^2\,, \quad g^{22} = g_{11}/\mathcal{D}^2\,, \quad g^{33} = 1/R^2\,. \tag{17.9}$$

We can now compute all geometric–physical quantities of interest.

(b) From Eq. (16.84) of Section 16.2.1, the *magnetic field* is expressed by

$$\mathbf{B} = \nabla \varphi \times \nabla \Psi + I \nabla \varphi = \mathbf{a}^3 \times \mathbf{a}^1 + I \mathbf{a}^3 = \mathcal{J}^{-1} \mathbf{a}_2 + I \mathbf{a}^3\,, \tag{17.10}$$

so that the physical components are given by

$$B_{\mathrm{p}} = |\nabla \Psi|/R = \sqrt{g^{11}}/R = \mathcal{J}^{-1}\sqrt{g_{22}}\,, \qquad B_\varphi = I/R\,, \tag{17.11}$$

whereas the contravariant and covariant components are related to them by

$$B^1 = 0\,, \quad B^2 = \mathcal{J}^{-1} = B_{\mathrm{p}}/\sqrt{g_{22}}\,, \quad B^3 = I/R^2 = B_\varphi/R\,,$$

$$B_1 = \mathcal{J}^{-1} g_{12} = (g_{12}/\sqrt{g_{22}}) B_{\mathrm{p}}\,, \quad B_2 = \mathcal{J}^{-1} g_{22} = \sqrt{g_{22}} B_{\mathrm{p}}\,, \quad B_3 = I = R B_\varphi\,. \tag{17.12}$$

The field line equation (16.14) of Section 16.1.2 then provides the *safety factor in SFL coordinates,*

$$q(\Psi) \equiv \frac{d\varphi}{d\vartheta}\Big|_{\mathrm{fl}} = \frac{B^3}{B^2} = \frac{\sqrt{g_{22}} B_\varphi}{R B_{\mathrm{p}}} = \frac{\mathcal{J} B_\varphi}{R} = \frac{\mathcal{J} I}{R^2}\,, \tag{17.13}$$

so that the expression for the parallel gradient operator becomes

$$F \equiv -\mathrm{i} \mathbf{B} \cdot \nabla = -\mathrm{i}(B^2 \partial_\vartheta + B^3 \partial_\varphi) = -\mathrm{i} \mathcal{J}^{-1}(\partial_\vartheta + q \partial_\varphi)\,. \tag{17.14}$$

This produces the desired effect on harmonic perturbations presented in Eq. (17.3).

(c) From Eqs. (16.84) and (16.81) of Section 16.2.1, the *current density* is expressed by

$$\mathbf{j} = -I'\nabla\varphi \times \nabla\Psi + Rj_\varphi\nabla\varphi, \tag{17.15}$$

from which the contravariant and covariant components may be constructed, analogous to the above derivation of the components of **B**. The toroidal component j_φ is related to the profiles $p(\Psi)$ and $I(\Psi)$ in the RHS of the original Grad–Shafranov equation (16.81). That equation is supposed to be solved at this stage, so that the *transformed Grad–Shafranov equation in SFL coordinates* becomes a first order differential relation between the different geometrical–physical quantities that is satisfied exactly (of course, to the extent that the equilibrium is solved exactly):

$$\Delta^*\Psi(\Psi,\vartheta) = \frac{R^2}{\mathcal{J}}\left(\partial_\psi - \partial_\vartheta\frac{g_{12}}{g_{22}}\right)\mathcal{J}B_{\mathrm{p}}^2 = -II' - R^2 p' = Rj_\varphi. \tag{17.16}$$

Recall that the functions $p = p(\Psi)$ and $I = I(\Psi)$ should be considered arbitrary.

(d) For the propagation of waves and instabilities, the geometry of the magnetic field lines and magnetic surfaces is important, in particular the *different curvatures* that appear. This involves projections onto three different triads of unit vectors, that need to be distinguished carefully. They refer to the field lines by themselves, the magnetic surfaces by themselves, and the field lines with respect to the magnetic surfaces, respectively:

(1) The traditional *Serret–Frenet triad for the field lines* $\{\mathbf{b}, \boldsymbol{\nu}, \boldsymbol{\beta}\}$, consisting of the tangent, the normal and the binormal, is defined by

$$\mathbf{b} \equiv \mathbf{B}/|\mathbf{B}|, \qquad \boldsymbol{\nu} = \mathbf{b}\cdot\nabla\mathbf{b}/|\mathbf{b}\cdot\nabla\mathbf{b}|, \qquad \boldsymbol{\beta} \equiv \mathbf{b}\times\boldsymbol{\nu}. \tag{17.17}$$

Only the first of the three Serret–Frenet formulas for the tangential derivatives is needed. This defines the *curvature vector $\boldsymbol{\kappa}$ of the field line* as the tangential derivative of the tangent,

$$\boldsymbol{\kappa} \equiv \kappa\boldsymbol{\nu} \equiv \frac{d\mathbf{b}}{ds} \equiv \mathbf{b}\cdot\nabla\mathbf{b}, \tag{17.18}$$

which, by definition, is pointing in the direction of the normal to the field line.

(2) The *magnetic surface symmetry triad* $\{\mathbf{n}, \mathbf{t}, \mathbf{e}_\varphi\}$, consisting of the triad of the normal to the magnetic surface, the tangent to the magnetic surface in the poloidal plane and the unit vector in the toroidal direction, is defined by

$$\mathbf{n} \equiv \frac{\nabla\Psi}{|\nabla\Psi|} = \frac{\mathbf{a}^1}{RB_{\mathrm{p}}}, \qquad \mathbf{t} \equiv \mathbf{e}_\varphi\times\mathbf{n} = \frac{\mathbf{a}_2}{\mathcal{J}B_{\mathrm{p}}}, \qquad \mathbf{e}_\varphi \equiv \frac{\nabla\varphi}{|\nabla\varphi|} = R\mathbf{a}^3. \tag{17.19}$$

This determines the principal curvatures, viz. the *poloidal and toroidal curvature of the magnetic surface:*

$$\kappa_{\mathrm{p}} \equiv \mathbf{t}\cdot(\nabla\mathbf{n})\cdot\mathbf{t} \quad = -\mathbf{t}\cdot(\nabla\mathbf{t})\cdot\mathbf{n} \quad = RD^\dagger B_{\mathrm{p}}, \tag{17.20}$$

$$\kappa_\varphi \equiv \mathbf{e}_\varphi\cdot(\nabla\mathbf{n})\cdot\mathbf{e}_\varphi = -\mathbf{e}_\varphi\cdot(\nabla\mathbf{e}_\varphi)\cdot\mathbf{n} = B_{\mathrm{p}}DR, \tag{17.21}$$

where D and $D^\dagger$ indicate the normal derivative and its adjoint:

$$D \equiv \frac{\mathbf{n}}{RB_{\mathrm{p}}} \cdot \nabla = \partial_\Psi - \frac{g_{12}}{g_{22}}\partial_\vartheta \,, \qquad D^\dagger \equiv \nabla \cdot \frac{\mathbf{n}}{RB_{\mathrm{p}}} = \frac{1}{\mathcal{J}}\left(\partial_\Psi - \partial_\vartheta \frac{g_{12}}{g_{22}}\right)\mathcal{J}\,.$$

$$(17.22)$$

The equalities for κ_{p} and κ_φ are proved below, under Eqs. (17.29)–(17.32).

(3) The *mixed field line/magnetic surface triad* $\{\mathbf{b}, \mathbf{n}, \boldsymbol{\pi}\}$, consisting of the tangent to the field line, the normal to the magnetic surface, and the tangent to the magnetic surface normal to the field line, is defined by

$$\mathbf{b} \equiv \frac{\mathbf{B}}{|\mathbf{B}|} = \frac{1}{\mathcal{J}B}(\mathbf{a}_2 + q\mathbf{a}_3)\,, \qquad \mathbf{n} \equiv \frac{\nabla\Psi}{|\nabla\Psi|} = \frac{\mathbf{a}^1}{RB_{\mathrm{p}}}\,,$$

$$\boldsymbol{\pi} \equiv \mathbf{b} \times \mathbf{n} = \frac{1}{\mathcal{J}B}\left(\frac{B_\varphi}{B_{\mathrm{p}}}\mathbf{a}_2 - q\frac{B_{\mathrm{p}}}{B_\varphi}\mathbf{a}_3\right)\,. \qquad (17.23)$$

We have encountered this projection frequently in our spectral studies, and we will continue to exploit it (in the cyclically permuted order $\mathbf{n}$, $\boldsymbol{\pi} \equiv \mathbf{e}_\perp$, $\mathbf{b} \equiv \mathbf{e}_\|$) to distinguish the different components of perturbations. It determines the two essential projections of the field line curvature vector with respect to the magnetic surface, viz. the *normal and the geodesic curvature of the magnetic field line*:

$$\kappa_n \equiv \boldsymbol{\kappa} \cdot \mathbf{n} \equiv \mathbf{b} \cdot (\nabla\mathbf{b}) \cdot \mathbf{n} = -(B_{\mathrm{p}}^2/B^2)\kappa_{\mathrm{p}} - (B_\varphi^2/B^2)\kappa_\varphi\,, \qquad (17.24)$$

$$\kappa_{\mathrm{g}} \equiv \boldsymbol{\kappa} \cdot \boldsymbol{\pi} \equiv \mathbf{b} \cdot (\nabla\mathbf{b}) \cdot \boldsymbol{\pi} = \frac{B_\varphi}{\mathcal{J}B_{\mathrm{p}}B^2}\frac{\partial B}{\partial\vartheta}\,. \qquad (17.25)$$

These equalities are proved below, under Eqs. (17.36)–(17.37).

It should be noted that the magnetic surface curvatures κ_{p} and κ_φ and the field line curvatures κ_n and κ_{g}, though exploiting the same stem symbol κ, refer to entirely different geometrical objects. What is worse, to stick to the conventions of the literature, the sign of the magnetic surface curvatures κ_{p} and κ_φ is chosen positive when the surface is concave with respect to the plasma, whereas the sign of the normal field line curvature κ_n is chosen negative in that case (see Fig. 17.2). The reader be aware!

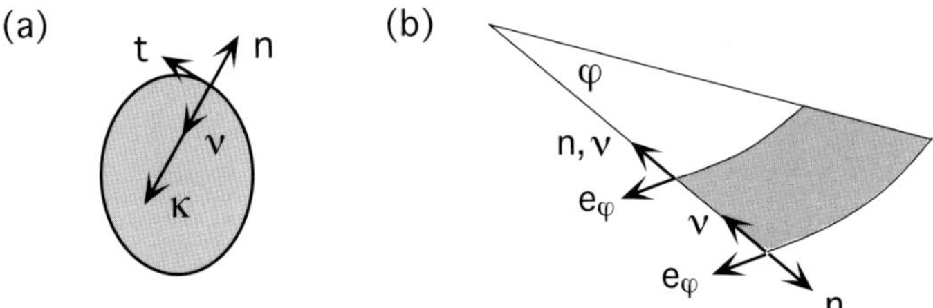

Fig. 17.2 Relation of the magnetic surface curvatures κ_{p} and κ_φ to the normal field line curvature κ_n: (a) for the poloidal field component, $\kappa_{\mathrm{p}} = -\kappa_n > 0$; (b) for the toroidal field component, $\kappa_\varphi = -\kappa_n < 0$ (inside) and $\kappa_\varphi = -\kappa_n > 0$ (outside).

▷ **Christoffel symbols and curvatures** To obtain the expressions for the curvatures of the magnetic surfaces and of the field lines, we need to calculate the derivatives of the basis vectors, i.e. the Christoffel symbols (see Appendix A.3):

$$\Gamma_{ij}^{k} \equiv \mathbf{a}^{k} \cdot \partial_i \mathbf{a}_j = \tfrac{1}{2} g^{kl}\left(\partial_i g_{lj} + \partial_j g_{il} - \partial_l g_{ij}\right). \tag{17.26}$$

There are ten non-vanishing Christoffel symbols, viz. Γ_{11}^{1}, Γ_{12}^{1}, Γ_{22}^{1}, Γ_{33}^{1}, Γ_{11}^{2}, Γ_{12}^{2}, Γ_{22}^{2}, Γ_{33}^{2}, Γ_{13}^{3} and Γ_{23}^{3}, which may be computed from the coordinate definitions (17.4), e.g.

$$\Gamma_{11}^{1} = (1/\mathcal{D})(y_\vartheta x_{\Psi\Psi} - x_\vartheta y_{\Psi\Psi}), \qquad \Gamma_{33}^{1} = -(R/\mathcal{D})y_\vartheta, \quad \text{etc.} \tag{17.27}$$

Consequently, all necessary information about the coordinates $\{x(\Psi,\vartheta), y(\Psi,\vartheta), \varphi\}$, i.e. the first and second derivatives of x and y, is contained in the four metric elements and the six Christoffel symbols not containing the index 3:

$$x\,,\ x_\Psi\,,\ x_\vartheta\,,\ y_\Psi\,,\ y_\vartheta \qquad \Rightarrow \qquad g_{11}\,,\ g_{12}\,,\ g_{22}\,,\ g_{33}\,,$$

$$x_{\Psi\Psi}\,, x_{\Psi\vartheta}\,, x_{\vartheta\vartheta}\,, y_{\Psi\Psi}\,, y_{\Psi\vartheta}\,, y_{\vartheta\vartheta} \ \Rightarrow\ \Gamma_{11}^{1}\,,\ \Gamma_{12}^{1}\,,\ \Gamma_{22}^{1}\,,\ \Gamma_{11}^{2}\,,\ \Gamma_{12}^{2}\,,\ \Gamma_{22}^{2}\,. \tag{17.28}$$

Thus, the expressions for the poloidal and toroidal magnetic surface curvatures reduce to

$$\kappa_{\mathrm{p}} \equiv \mathbf{t} \cdot (\nabla \mathbf{n}) \cdot \mathbf{t} = \frac{1}{\mathcal{J}^2 R B_{\mathrm{p}}^3}\frac{\partial \mathbf{a}^1}{\partial \vartheta} \cdot \mathbf{a}_2 = -\frac{1}{\mathcal{J}^2 R B_{\mathrm{p}}^3}\mathbf{a}^1 \cdot \frac{\partial \mathbf{a}_2}{\partial \vartheta} \equiv -\frac{1}{\mathcal{J}^2 R B_{\mathrm{p}}^3}\Gamma_{22}^{1}, \tag{17.29}$$

$$\kappa_{\varphi} \equiv \mathbf{e}_{\varphi} \cdot (\nabla \mathbf{n}) \cdot \mathbf{e}_{\varphi} = \frac{1}{R^3 B_{\mathrm{p}}}\frac{\partial \mathbf{a}^1}{\partial \varphi} \cdot \mathbf{a}_3 = -\frac{1}{R^3 B_{\mathrm{p}}}\mathbf{a}^1 \cdot \frac{\partial \mathbf{a}_3}{\partial \varphi} \equiv -\frac{1}{R^3 B_{\mathrm{p}}}\Gamma_{33}^{1}. \tag{17.30}$$

The explicit expressions for the latter two Christoffel symbols read:

$$\Gamma_{22}^{1} = \tfrac{1}{2}g^{1\ell}(\partial_2 g_{\ell 2} + \partial_2 g_{2\ell} - \partial_\ell g_{22}) = R^2 B_{\mathrm{p}}^2\left(\partial_2 g_{12} - \tfrac{1}{2}\partial_1 g_{22} - \tfrac{1}{2}\frac{g_{12}}{g_{22}}\partial_1 g_{22}\right)$$

$$= -\mathcal{J}R^2 B_{\mathrm{p}}^3\left(\partial_\Psi - \partial_\vartheta \frac{g_{12}}{g_{22}}\right)\mathcal{J}B_{\mathrm{p}} \quad \Rightarrow \quad \kappa_{\mathrm{p}} = RD^{\dagger}B_{\mathrm{p}}\,, \tag{17.31}$$

$$\Gamma_{33}^{1} = \tfrac{1}{2}g^{1\ell}(\partial_3 g_{\ell 3} + \partial_3 g_{3\ell} - \partial_\ell g_{33}) = -\tfrac{1}{2}R^3 B_{\mathrm{p}}^2\left(\partial_1 R^2 - \frac{g_{12}}{g_{22}}\partial_2 R^2\right)$$

$$= -R^3 B_{\mathrm{p}}^2\left(\partial_\Psi - \frac{g_{12}}{g_{22}}\partial_\vartheta\right)R \quad \Rightarrow \quad \kappa_{\varphi} = B_{\mathrm{p}}DR\,, \tag{17.32}$$

which proves the assertions (17.20) and (17.21).

The curvature of the magnetic field lines may be written in the alternative form

$$\boldsymbol{\kappa} \equiv \mathbf{b} \cdot \nabla \mathbf{b} \overset{(\text{A.10})}{=} -\mathbf{b} \times (\nabla \times \mathbf{b}) \qquad \Rightarrow \qquad \boldsymbol{\kappa} \cdot \mathbf{b} = 0\,, \tag{17.33}$$

so that we need to express the curls of the tangent vector,

$$(\nabla \times \mathbf{b})^1 = \mathcal{J}^{-1}\partial_2 b_3\,, \quad (\nabla \times \mathbf{b})^2 = \mathcal{J}^{-1}\partial_1 b_3\,, \quad (\nabla \times \mathbf{b})^3 = \mathcal{J}^{-1}(\partial_1 b_2 - \partial_2 b_1)\,, \tag{17.34}$$

in terms of the co- or contravariant components of the unit vectors (17.23):

$$n_1 = 1/(RB_{\mathrm{p}})\,, \qquad n_2 = n_3 = 0\,,$$

$$b^1 = 0\,, \qquad b^2 = 1/(\mathcal{J}B)\,, \qquad b^3 = q/(\mathcal{J}B)\,,$$

$$\pi^1 = 0\,, \qquad \pi^2 = B_\varphi/(\mathcal{J}B_{\mathrm{p}}B)\,, \qquad \pi^3 = -B_{\mathrm{p}}/(RB)\,. \tag{17.35}$$

Substitution of these equalities leads to the two desired expressions for the normal and the geodesic curvature of the field lines:

$$\kappa_n = \boldsymbol{\kappa} \cdot \mathbf{n} = -(\mathbf{b} \times (\nabla \times \mathbf{b})) \cdot \mathbf{n} = (\mathbf{b} \times \mathbf{n}) \cdot (\nabla \times \mathbf{b}) = \boldsymbol{\pi} \cdot (\nabla \times \mathbf{b})$$

$$= \mathcal{J}^{-1}[\pi_1 \partial_2 b_3 - \pi_2 \partial_1 b_3 + \pi_3(\partial_1 b_2 - \partial_2 b_1)]$$

$$= -(B_{\mathrm{p}}^2/B^2)\kappa_{\mathrm{p}} - (B_{\varphi}^2/B^2)\kappa_{\varphi}\,, \tag{17.36}$$

$$\kappa_{\mathrm{g}} = \boldsymbol{\kappa} \cdot \boldsymbol{\pi} = -(\mathbf{b} \times (\nabla \times \mathbf{b})) \cdot \boldsymbol{\pi} = -(\boldsymbol{\pi} \times \mathbf{b}) \cdot (\nabla \times \mathbf{b}) = -\mathbf{n} \cdot (\nabla \times \mathbf{b})$$

$$= -n_1(\nabla \times \mathbf{b})^1 = -\frac{1}{\mathcal{J} R B_{\mathrm{p}}} \frac{\partial}{\partial \vartheta}\left(\frac{qR^2}{\mathcal{J}B}\right) = \frac{B_{\varphi}}{\mathcal{J}B_{\mathrm{p}}B^2}\frac{\partial B}{\partial \vartheta}\,, \tag{17.37}$$

where we used the fact that the functions $I(\Psi)$ and $q(\Psi)$ are independent of ϑ in the last step. This proves the assertions (17.24) and (17.25). ◁

As compared to cylindrical plasmas (Chapter 9 [1]), two additional curvatures have now entered the picture, viz. the geodesic curvature κ_{g} and the toroidal curvature κ_{φ}. They both play an important role in spectral theory of toroidal plasmas. The geodesic curvature of the field lines indicates that field lines are no longer geodesics in toroidal geometry,[1] due to the fact that, in general, the magnitude of the magnetic field is not constant in the magnetic surfaces. Hence, as first noticed in [161], one may expect that Alfvén waves will have to decide whether to follow the shortest path (the geodesic) or the field line. As we will see in Section 17.2.3, they "make the best of it" doing both through coupling to the slow magneto-sonic modes. This is related to the *geodesic acoustic modes* (GAMs) [485], which have become a subject of intense research recently [245] (see Section 17.2.3).

The distinguishing feature of the toroidal curvature is its change in sign on the inside ("high field side") of the torus compared to that on the outside. As we have stressed many times, interchange modes driven by pressure gradients in toroidal plasmas are only possible if the overwhelming stabilizing field line bending energy contribution of the Alfvén waves is minimized by modes for which the parallel gradient operator vanishes. This requires localization to rational magnetic surfaces. Interchange modes in toroidal geometry, satisfying this requirement, will become unstable if a counterpart to Suydam's cylindrical stability criterion is violated, viz. Mercier's toroidal stability criterion (Section 17.2.5). However, a much more subtle localization of the modes, called ballooning, was theoretically proposed [91] (and, of course, utilized by the plasma) that couples a pressure gradient to the negative toroidal curvature on the inside. This leads to much more severe stability limitations of pressure driven toroidal modes since they include *ballooning instabilities*, which are absent in a cylinder (Section 17.2.5).

---

[1] To avoid possible misunderstanding: the fact that the field lines are straight in the SFL representation is just a convenience that has nothing to do with geodesics, which are of course independent of the coordinates. This is manifest from the occurrence of the coordinate independent length element $d\ell = \mathcal{J}\,d\vartheta$ in the expressions.

17.2 Analysis of waves and instabilities in toroidal geometry

17.2.1 Spectral wave equation

We will now present the main stages of the analysis of wave propagation and stability of toroidal plasmas, in close analogy with the analysis of cylindrical plasmas (Chapter 9 of Volume [1]). Since toroidal equilibria are two-dimensional, we will obtain partial differential equations, instead of the ordinary differential equations, like the Hain–Lüst equation, that were obtained for one-dimensional equilibria.

As always, we start from the linearized equation of motion

$$ \mathbf{F}(\boldsymbol{\xi}) = \rho \frac{\partial^2 \boldsymbol{\xi}}{\partial t^2} \,, \tag{17.38} $$

with the static MHD force operator defined in Eqs. (6.23)–(6.25) [1], and assume *Fourier-normal mode solutions* of the form

$$ \boldsymbol{\xi}(\mathbf{r}, t) = \hat{\boldsymbol{\xi}}(\Psi, \vartheta) e^{i(n\varphi - \omega t)} \,, \tag{17.39} $$

where $\hat{\boldsymbol{\xi}}$ is the amplitude of the normal modes and n is the toroidal mode number. As in Chapters 7 and 9 [1], we exploit the *field line/magnetic surface projection*, expressed by the field line triad $\{\mathbf{n}, \boldsymbol{\pi}, \mathbf{b}\}$ defined in Eq. (17.23). In this projection, the gradient operators become

$$ D \equiv \frac{1}{RB_{\rm p}} \mathbf{n} \cdot \nabla \qquad = \partial_\Psi - (g_{12}/g_{22})\, \partial_\vartheta \,, $$

$$ G \equiv -iRB_{\rm p} B \boldsymbol{\pi} \cdot \nabla = -\frac{iRB_\varphi}{\mathcal{J}} \left(\partial_\vartheta - q(B_{\rm p}^2/B_\varphi^2)\partial_\varphi \right) , $$

$$ F \equiv -iB\mathbf{b} \cdot \nabla \qquad = -\frac{i}{\mathcal{J}} (\partial_\vartheta + q\partial_\varphi) \,. \tag{17.40} $$

(Watch out when converting these expressions for vectors $\boldsymbol{\xi}$: the partial derivatives ∂_ϑ and ∂_φ act not only on the vector components but also on the basis vectors and the unit vectors!) The projections of the displacement vector will be denoted as

$$ X \equiv RB_{\rm p} \boldsymbol{\xi} \cdot \mathbf{n} = \xi^1 \,, $$

$$ Y \equiv \frac{iB}{RB_{\rm p}} \boldsymbol{\xi} \cdot \boldsymbol{\pi} = \frac{iB_\varphi}{\mathcal{J} RB_{\rm p}^2} \left[\xi_2 - q(B_{\rm p}^2/B_\varphi^2)\xi_3 \right] , $$

$$ Z \equiv \frac{i}{B} \boldsymbol{\xi} \cdot \mathbf{b} \qquad = \frac{i}{\mathcal{J} B^2} (\xi_2 + q\xi_3) \,, \tag{17.41} $$

where the factors i in the components Y and Z are introduced to obtain real expressions in the final Fourier analyzed form of the spectral equation, whereas the factors $RB_{\rm p}$ and B, found by trial and error, just simplify that equation.

At this stage in our exposition, we will not give the complete derivations anymore, but just indicate the steps and give the final result. (The student wishing to enter this field is best advised anyway to spend some time deriving these equations by her- or himself.) The important steps are to first Fourier analyze the spectral equation (17.38) with respect to φ, i.e. to replace ∂_φ by in, then to project the equation onto the orthogonal triad $\{\mathbf{n}, \boldsymbol{\pi}, \mathbf{b}\}$, and finally to eliminate all second order derivatives of equilibrium quantities by exploiting the properties listed in Section 17.1.3. For example, the pressure terms entering $\mathbf{F}$ and the magnetic field perturbation $\mathbf{Q}$ become

$$\boldsymbol{\xi} \cdot \nabla p = p'X, \qquad \nabla \cdot \boldsymbol{\xi} = D^\dagger X + GB^{-2}Y + FZ,$$

$$Q_n = \frac{\mathrm{i}}{RB_\mathrm{p}}X, \qquad Q_\perp = \frac{RB_\mathrm{p}}{B}\left\{\left[D^\dagger\left(\frac{B_\varphi}{R}\right)\right]X + FY\right\},$$

$$Q_\parallel = -BDX - \frac{RB_\varphi}{B}\left[D^\dagger\left(\frac{B_\varphi}{R}\right)\right]X - \frac{1}{B}GY. \tag{17.42}$$

After some tedious, but straightforward, algebra one then finds the following formulation of *the spectral problem*:

$$\begin{pmatrix} \mathcal{A}_{11} & \mathcal{A}_{12} & \mathcal{A}_{13} \\ \mathcal{A}_{21} & \mathcal{A}_{22} & \mathcal{A}_{23} \\ \mathcal{A}_{31} & \mathcal{A}_{32} & \mathcal{A}_{33} \end{pmatrix} \begin{pmatrix} X \\ Y \\ Z \end{pmatrix} = -\rho\omega^2 \begin{pmatrix} \mathcal{B}_{11}\,X \\ \mathcal{B}_{22}\,Y \\ \mathcal{B}_{33}\,Z \end{pmatrix}, \tag{17.43}$$

This equation was first derived by Goedbloed [161] exploiting orthogonal Ψ, χ, φ coordinates. (Note that the expression (9) of that paper for the geodesic curvature should be multiplied by B_p/B to get the correct expression (17.25).)

▷ **Matrix elements** The explicit expressions for the matrix elements read:

$$\mathcal{A}_{11} \equiv D(\gamma p + B^2)D^\dagger - F\frac{1}{R^2 B_\mathrm{p}^2}F - E,$$

$$\mathcal{A}_{21} \equiv -\frac{\gamma p + B^2}{B^2}GD^\dagger - 2\left(\frac{\mathrm{i}}{\mathcal{J}}\partial_\vartheta\frac{B_\varphi \kappa_\varphi}{B_\mathrm{p}} + n\frac{B_\mathrm{p}\kappa_\mathrm{p}}{R}\right),$$

$$\mathcal{A}_{31} \equiv -F\gamma p D^\dagger,$$

$$\mathcal{A}_{12} \equiv DG\frac{\gamma p + B^2}{B^2} - 2\left(\frac{B_\varphi \kappa_\varphi}{B_\mathrm{p}}\frac{\mathrm{i}}{\mathcal{J}}\partial_\vartheta + \frac{B_\mathrm{p}\kappa_\mathrm{p}}{R}n\right),$$

$$\mathcal{A}_{22} \equiv -\frac{1}{B^2}G\gamma p G\frac{1}{B^2} - G\frac{1}{B^2}G - F\frac{R^2 B_\mathrm{p}^2}{B^2}F,$$

$$\mathcal{A}_{32} \equiv -F\gamma p G\frac{1}{B^2},$$

$$\mathcal{A}_{13} \equiv D\gamma p F, \qquad\qquad \mathcal{B}_{11} \equiv \frac{1}{R^2 B_\mathrm{p}^2},$$

$$\mathcal{A}_{23} \equiv -\frac{1}{B^2}G\gamma p F, \qquad\qquad \mathcal{B}_{22} \equiv \frac{R^2 B_\mathrm{p}^2}{B^2},$$

$$\mathcal{A}_{33} \equiv -F\gamma p F, \qquad\qquad \mathcal{B}_{33} \equiv B^2. \tag{17.44}$$

where

$$E \equiv 2\left[D\left(\frac{B_{\rm p}}{R}\kappa_{\rm p}\right) + \frac{B_\varphi}{R}{}^\dagger D\left(\frac{B_\varphi}{B_{\rm p}}\kappa_\varphi\right) \right],\tag{17.45}$$

with $D^\dagger$ and (for lack of a better symbol) ${}^\dagger D$ defined by

$$D^\dagger \equiv \frac{1}{\mathcal{J}}D\mathcal{J} = \frac{1}{\mathcal{J}}\left(\partial_\Psi - \partial_\vartheta\frac{g_{12}}{g_{22}}\right)\mathcal{J},\qquad {}^\dagger D \equiv \mathcal{J}\left(\partial_\Psi + \partial_\vartheta\frac{g_{12}}{g_{22}}\right)\frac{1}{\mathcal{J}}.\tag{17.46}$$

The square brackets in the definition of E indicate that the action of the differential operator is to be restricted to the terms inside those brackets. $\triangleleft$

The two-dimensional spectral equation (17.43) may be considered as the generalization of the cylindrical MHD wave equation (9.28) of Volume [1]. The latter equation is recovered by exploiting the leading order relationship (16.96) between Ψ and $B_{\rm p}$ of the Shafranov shifted circle approximation. This yields the following translation recipe to the cylindrical wave equation:

$$\Psi \to R_0\int B_\theta\, dr\,,\quad \vartheta \to \theta\,,\quad \varphi \to z/R_0\,,\quad \mathcal{J} \to r/B_\theta\,,$$

$$R \to R_0\,,\quad n \to kR_0\,,\quad \kappa_{\rm p} \to 1/r\,,\quad \kappa_\varphi \to 0\,,\quad g_{12} \to 0\,,$$

$$D \to \frac{1}{R_0 B_\theta}\frac{d}{dr}\,,\quad D^\dagger \to \frac{1}{R_0 r}\frac{d}{dr}\frac{r}{B_\theta}\,,\quad G \to R_0 B_\theta G_{\rm cyl}\,,\quad F \to F_{\rm cyl}\,,$$

$$X \to R_0 B_\theta \xi_{\rm cyl}\,,\quad Y \to (B/R_0 B_\theta)\eta_{\rm cyl}\,,\quad X \to (1/B)\zeta_{\rm cyl}\,.\tag{17.47}$$

However, the next step, to the construction of a scalar wave equation for the normal component of $\boldsymbol{\xi}$, i.e. the analog of the Hain–Lüst equation, is not taken now: the elimination of the tangential components is no longer algebraic but involves PDEs. If one wishes to continue along this line, one usually exploits an ordering in a small parameter, like the low-β tokamak ordering. This involves the counterpart of the equilibrium expansion described in Section 16.2.2, i.e. expanding the vector wave equation (17.43) to the first non-trivial order (considering the cylindrical solution as the trivial leading order; after all, one only goes to toroidal corrections if the cylindrical basis is fully understood). This generally leads to a system of coupled ODEs describing the Fourier harmonics of the vector $\boldsymbol{\xi}$ in the angle ϑ. An example along this line may be found in Section 18.3.3 of the next chapter.

The equation of motion may be used as a starting point for further analysis. For numerical work, the corresponding quadratic forms of the next section are to be preferred. In the following Sections 17.2.3 and 17.2.5, we will then use the representation (17.43) to derive the analytical "core structure" for toroidal systems, consisting of the ODEs describing the continuous spectra and ballooning stability, and the explicit Mercier criterion.

17.2.2 Spectral variational principle

Recall from Section 6.1.1 of Volume [1] that there are two equivalent approaches to spectral theory, one based on differential equations and the other one on quadratic forms (as in the Schrödinger and Heisenberg pictures of quantum mechanics). The quadratic forms approach for MHD was formulated in Eq. (6.90)[1] as a variational principle for the eigenvalues and eigenfunctions in terms of the Rayleigh quotient,

$$\delta\Lambda = 0, \qquad \Lambda[\boldsymbol{\xi}] \equiv \frac{W[\boldsymbol{\xi}]}{K[\boldsymbol{\xi}]}, \tag{17.48}$$

where W is the potential energy and K is the norm of the perturbations. (The notation for the norm is changed here to K since I is used for another purpose.) The eigenvalues ω^2 are the stationary values of Λ. Exploiting the same techniques as in the derivation of the spectral wave equation (17.43), the diligent student will be able to derive the following explicit expressions for the quadratric forms of the potential energy and of the norm of the perturbations in the field line projection:

$$
\begin{aligned}
W = \pi \iint \Bigg[&\frac{1}{R^2 B_{\mathrm{p}}^2}|FX|^2 + B^2\Big|D^\dagger X + \frac{2\kappa_n}{RB_{\mathrm{p}}}X + \frac{1}{B^2}GY\Big|^2 \\
&+ \frac{R^2 B_{\mathrm{p}}^2}{B^2}\Big|FY + \frac{2B_\varphi(\kappa_{\mathrm{p}} - \kappa_\varphi)}{R^2 B_{\mathrm{p}}}X\Big|^2 + \gamma p\Big|D^\dagger X + G\frac{1}{B^2}Y + FZ\Big|^2 \\
&+ \frac{2}{RB_{\mathrm{p}}}\Big\{p'\kappa_{\mathrm{p}} + \frac{II'}{R^2}(\kappa_{\mathrm{p}} - \kappa_\varphi)\Big\}|X|^2 \Bigg] \mathcal{J}\,d\Psi\,d\vartheta ,
\end{aligned}
\tag{17.49}
$$

$$
K = \pi \iint \rho\Big[\frac{1}{R^2 B_{\mathrm{p}}^2}|X|^2 + \frac{R^2 B_{\mathrm{p}}^2}{B^2}|Y|^2 + B^2|Z|^2 \Big]\mathcal{J}\,d\Psi\,d\vartheta .
\tag{17.50}
$$

As discussed in Chapter 6[1], these expressions are completely equivalent to the spectral wave equation when used with the variational principle (17.48). The cylindrical expression (9.102)[1] for W may be obtained from the toroidal expression (17.49) by again exploiting the translation recipe (17.47) (and an integration by parts to combine the second and third term).

Since W has been the starting point for a substantial number of investigations in ideal MHD stability of tokamaks, it is of some interest to consider different alternative expressions for the term in curly brackets, which potentially gives rise to instabilities:

$$
\begin{aligned}
U &\equiv \frac{2}{RB_{\mathrm{p}}}\Big\{p'\kappa_{\mathrm{p}} + \frac{II'}{R^2}(\kappa_{\mathrm{p}} - \kappa_\varphi)\Big\} = -\frac{2}{R^2 B_{\mathrm{p}}}\Big\{j_\varphi(\kappa_{\mathrm{p}} - \kappa_\varphi) - Rp'\kappa_\varphi\Big\} \\
&= -\frac{2}{R^2 B_{\mathrm{p}}}\Big\{j_\parallel \frac{B_\varphi}{B}(\kappa_{\mathrm{p}} - \kappa_\varphi) + Rp'\kappa_n\Big\} .
\end{aligned}
\tag{17.51}
$$

For the reduction, the first order differential form (17.16) of the Grad–Shafranov equation has been exploited. Clearly, all the ingredients for instability are there, e.g. the parallel current $j_\parallel$, driving *kink modes*, and the pressure gradient–curvature term $p'\kappa_n$ ($\approx -p'\kappa_\varphi$), driving *ballooning modes*, but it is not so clear in what form they should be presented. There is also a connection with the curvature terms of the equation of motion (17.43), in particular the expression E, but it is not instructive enough to reproduce it here. One may wonder what all these expressions are good for. Actually, the sole purpose of this paragraph is to exhibit the ambiguity of certain ingrained terminology in tokamak literature. The terms "kink" and "ballooning" do not have a unique meaning. There are many ways of transforming the terms using the equilibrium conditions and which presentation is the most meaningful depends entirely on the approximations made to study a particular case.

The expression for the energy has been used extensively in analytic studies of MHD stability of internal modes, e.g. the *internal kink mode*, involving a singular perturbation at a rational $q = 1$ surface, which requires delicate balancing of terms; see e.g. [72], [105]. This has its counterpart in extreme conditions on numerical accuracy of the equilibrium, its inversion and the spectral representation needed to describe these modes; see e.g. [272].

17.2.3 Alfvén and slow continuum modes

Recall that the continuous spectra in cylindrical plasmas are obtained by considering perturbations that are *localized to a particular magnetic surface*. This construction can be generalized to toroidal geometry, as first shown by Goedbloed [161] and, independently, by Pao [359]. Since the equation of motion (17.43) is written in the field line projection, which already incorporates one of the essential properties of the singular Alfvén and slow continuum modes, the construction is relatively straightforward. The first step is to consider the limit $D \to \infty$ so as to obtain modes localized to a single magnetic surface. To leading order, the first component of Eq. (17.43) then becomes a derivative with respect to Ψ, which can be integrated once to give

$$D^\dagger X \approx -\frac{1}{\gamma p + B^2}\, G \frac{\gamma p + B^2}{B^2}\, Y - \frac{\gamma p}{\gamma p + B^2}\, FZ\,. \qquad (17.52)$$

Substitution of this expression into the second and third component of Eq. (17.43), and the use of commutation relations like

$$GB^{-1} - B^{-1}G = \mathrm{i}RB_\mathrm{p}\kappa_\mathrm{g}\,, \qquad (17.53)$$

leads to a system of two ordinary differential equations for Y and Z where the normal derivatives of the equilibrium (i.e. the curvatures κ_p and κ_φ of the magnetic surfaces) no longer appear. As a result, a system of equations is obtained that is *intrinsic to every magnetic surface:* a two-dimensional creature living on such a surface will not notice the three-dimensional embedding. Consequently, we have effectively obtained modes that are localized about a single magnetic surface, i.e. of the form

$$\boldsymbol{\xi}(\Psi, \vartheta, \varphi) \approx -\mathrm{i}\delta(\Psi - \Psi_0)\left[\eta(\vartheta)\boldsymbol{\pi} + \zeta(\vartheta)\mathbf{b}\right]\mathrm{e}^{\mathrm{i}n\varphi}, \qquad (17.54)$$

where the variables

$$\eta \equiv \mathrm{i}\boldsymbol{\xi}\cdot\boldsymbol{\pi} \equiv (RB_\mathrm{p}/B)Y \quad \text{and} \quad \zeta \equiv \mathrm{i}\boldsymbol{\xi}\cdot\mathbf{b} \equiv BZ$$

are more convenient than Y and Z for the present purpose. These variables describe the *Alfvén and slow magnetic surface resonances*, satisfying a system of two coupled ODEs [161]:

$$\begin{pmatrix} \alpha_{11} & \alpha_{12} \\ \alpha_{21} & \alpha_{22} \end{pmatrix}\begin{pmatrix} \eta \\ \zeta \end{pmatrix} = \rho\omega^2 \begin{pmatrix} \eta \\ \zeta \end{pmatrix}, \qquad (17.55)$$

where

$$\alpha_{11} \equiv \frac{B}{RB_\mathrm{p}}F\frac{R^2B_\mathrm{p}^2}{B^2}F\frac{B}{RB_\mathrm{p}} + \frac{4\gamma pB^2}{\gamma p + B^2}\kappa_\mathrm{g}^2, \quad \alpha_{12} \equiv -\frac{2\mathrm{i}\gamma pB^2}{\gamma p + B^2}\kappa_\mathrm{g}F\frac{1}{B},$$

$$\alpha_{21} \equiv \frac{\mathrm{i}}{B}F\frac{2\gamma pB^2}{\gamma p + B^2}\kappa_\mathrm{g}, \qquad\qquad \alpha_{22} \equiv \frac{1}{B}F\frac{\gamma pB^2}{\gamma p + B^2}F\frac{1}{B}.$$

$$(17.56)$$

The appropriate boundary conditions to impose on these equations are poloidal periodicity of the variables η and ζ and of their first derivatives.

Notice that the only curvature that has survived in this representation is the geodesic curvature κ_g of the field lines inside a particular magnetic surface. The two-dimensional creature introduced above will be able to draw the correct geometrical and, hence, physical conclusions about the dynamics of the waves on the magnetic surface, in particular their anisotropy with respect to the field lines. From the representation (17.55)–(17.56) it is clear that the MHD continuum modes approximately behave like such two-dimensional creatures. It is also clear, from the occurrence of the off-diagonal matrix elements $\sim \gamma p\kappa_\mathrm{g}$ in the ODEs (17.55), that the Alfvén waves are no longer polarized purely perpendicular to the field lines (as they are in the one-dimensional plane slab and cylindrical equilibria), but that they get a parallel component by coupling to the slow magneto-acoustic modes. Vice versa, the polarization of the slow waves will no longer be purely parallel to the field lines due to geodesic coupling to the Alfvén waves. Consequently, in toroidal configurations, the Alfvén and slow magneto-acoustic modes can no longer be distinguished on the basis of their polarization.

Since the determination of the continuous spectra has now been turned into the construction of the discrete, doubly infinite, set of eigenvalues $\{\omega^2_{A/S,i}(\Psi_0)\}$ on each magnetic surface, we may also obtain those eigenvalues from a variational principle in terms of the two-vector $\mathbf{v} \equiv (\eta, \zeta)^{\mathrm{T}}$:

$$\delta\hat{\Lambda} = 0\,,$$

$$\hat{\Lambda}[\mathbf{v}] \equiv \frac{\displaystyle\int \left\{ \left| \frac{RB_{\mathrm{p}}}{B} F \frac{B}{RB_{\mathrm{p}}} \eta \right|^2 + \frac{\gamma p}{\gamma p + B^2} \left| BF \frac{1}{B} \zeta + 2\mathrm{i}\kappa_{\mathrm{g}} B\eta \right|^2 \right\} \mathcal{J}\, d\vartheta}{\displaystyle\int \rho(\eta^2 + \zeta^2)\mathcal{J}\, d\vartheta}\,. \tag{17.57}$$

This expression shows that the continuous spectra of static toroidal plasmas are exclusively stable, $\omega^2_{A/S,i} \geq 0$: geodesic curvature does not create instability. An additional contribution from the normal curvature is needed to create instability, but this involves another kind of localization (see Section 17.2.5). On the other hand, in plasmas with background flow, the continua may become unstable due to a generalization of the geodesic curvature term (see Section 18.3).

▷ **Geodesic acoustic modes** A special example of the Alfvén–slow wave coupling due to geodesic curvature is the geodesic acoustic mode (GAM), first described by Winsor, Johnson and Dawson [485]. This is an electrostatic wave in low-beta plasmas ($\gamma p \ll B^2$), where the perpendicular displacement $[B/(RB_{\mathrm{p}})]\eta \approx$ const in the magnetic surfaces so that the magnetic (Alfvén wave) perturbations, indicated by the first term of the variational principle (17.57), are negligible. Note that this does not imply that η is negligible: the GAM is a slow wave having both perpendicular and parallel components, constrained by the electrostatic condition. Therefore, it is expedient to convert η and ζ to the density perturbations $\widetilde{\rho}$ that are relevant for these modes:

$$\widetilde{\rho} = -\nabla \cdot (\rho\boldsymbol{\xi}) = -\rho(D^\dagger X + GB^{-2}Y + FZ) - (RB_{\mathrm{p}})^{-2}(\nabla\rho \cdot \nabla\Psi)X$$

$$\approx -\rho \frac{B^2}{\gamma p + B^2}(2\mathrm{i}\kappa_{\mathrm{g}}\eta + FB^{-1}\zeta) \approx -\rho(2\mathrm{i}\kappa_{\mathrm{g}}\eta + FB^{-1}\zeta)\,, \tag{17.58}$$

where the localization assumption (17.52) and the commutation relation (17.53) were used for the reductions to the last line. The ODEs (17.55) then transform into

$$4\gamma p\kappa^2_{\mathrm{g}}\eta - 2\mathrm{i}\gamma p\kappa_{\mathrm{g}} FB^{-1}\zeta = 2\mathrm{i}\gamma p\kappa_{\mathrm{g}}(\widetilde{\rho}/\rho) = \rho\omega^2\eta\,,$$

$$2\mathrm{i}\gamma pB^{-1}F\kappa_{\mathrm{g}}\eta + B^{-1}F\gamma pFB^{-1}\zeta = -\gamma pB^{-1}F(\widetilde{\rho}/\rho) = \rho\omega^2\zeta\,. \tag{17.59}$$

Inserting these expressions into the variational principle (17.57) yields the expression for the GAM frequencies derived by Winsor *et al.* [485], except for a different normalization factor:

$$\omega^2 = \frac{\gamma p}{\rho} \frac{\displaystyle\int \left[|B^{-1}F\widetilde{\rho}|^2 + |2\kappa_{\mathrm{g}}\widetilde{\rho}|^2 \right] \mathcal{J}\, d\vartheta}{\displaystyle\int |\widetilde{\rho}|^2 \mathcal{J}\, d\vartheta}\,. \tag{17.60}$$

As a companion to MHD spectroscopy, Itoh *et al.* proposed *GAM spectroscopy* [245], exploiting these modes to determine the velocity profiles of the different ion species. ◁

In the equivalent approach by Pao [359], the continua in toroidal geometry are obtained by starting from the six first order ODEs for the primitive variables $\mathbf{v}_1$, p_1 and $\mathbf{B}_1$. In producing a set of equations for the normal component of $\mathbf{v}_1$ and the total pressure perturbation (a generalization of the one-dimensional Appert–Gruber–Vaclavik [10] representation in terms of ξ and Π), a four by four matrix operator acting on the tangential components of $\mathbf{v}_1$ and $\mathbf{B}_1$ has to be inverted. The continuous spectra are obtained for those frequencies where this inverse does not exist; see also Kieras and Tataronis [275].

17.2.4 Poloidal mode coupling

In the cylindrical limit, all equilibrium quantities are constant on a magnetic surface, so that the tangential dependence of the perturbations reduces to just the Fourier amplitude dependence $\exp(im\vartheta)$ with the familiar result of uncoupled Alfvén and slow continua:

$$\omega_{\mathrm{A}} = \pm \frac{(n + m/q)B_{\varphi 0}}{R_0\sqrt{\rho}}, \qquad \eta_{\mathrm{A}} \neq 0, \quad \zeta_{\mathrm{A}} = 0 \quad (Alfvén),$$

$$\omega_{\mathrm{S}} = \pm \sqrt{\frac{\gamma p}{\gamma p + B^2}}\,\frac{(n + m/q)B_{\varphi 0}}{R_0\sqrt{\rho}}, \qquad \eta_{\mathrm{S}} = 0, \quad \zeta_{\mathrm{S}} \neq 0 \quad (slow). \quad (17.61)$$

In toroidal geometry, as we have noted above, the Alfvén and slow continua are coupled through the combination of finite compressibility and finite beta (γp terms) and geodesic curvature (κ_{g}). In low-beta plasmas, this coupling is weak. However, an entirely different, usually much stronger, coupling arises from the fact that all continua are degenerate at the cylindrical cross-over points $r = r_{\mathrm{cross}}$, where q_{cross} has a rational value determined by

$$n + m/q = -n - m'/q \quad \Rightarrow \quad q_{\mathrm{cross}} = -\frac{m + m'}{2n}. \quad (17.62)$$

At those values, two Alfvén wave branches (and two slow wave branches, at much lower frequency) cross, creating the possibility of lifting of the degeneracy by mode coupling due to poloidal variation of the equilibrium. In particular, the poloidal modulation of the equilibrium with the large radius R creates coupling between modes separated by $\Delta m \equiv |m - m'| = 1$, the ellipticity of the plasma cross-section causes $\Delta m = 2$, the triangularity causes $\Delta m = 3$, etc. This coupling produces *gaps in the continua* (see Fig. 17.3), very analogous to the band structure of electrons in a crystal lattice [121].

An illuminating example of poloidal mode coupling is shown in Fig. 17.4, which presents the results of an analytic calculation of the spectrum of free-boundary kink modes (a particular kind of Alfvén waves) of a skin-current low-beta tokamak,

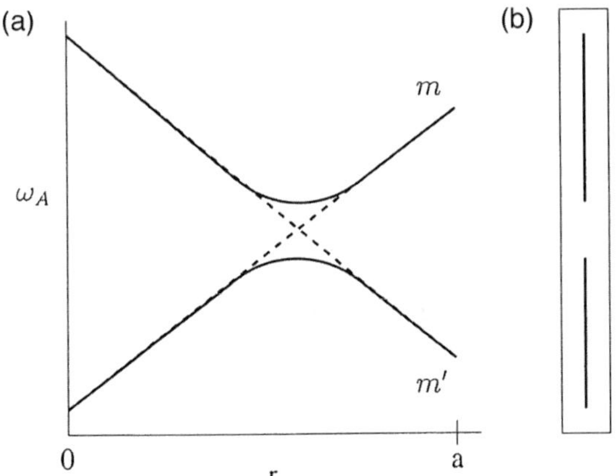

Fig. 17.3 (a) Poloidal mode coupling of two cylindrical modes, labeled m and m', creates (b) a "gap" in the continuous spectrum of a toroidal plasma.

where the toroidal current flows exclusively on the plasma–vacuum surface. The top part of the figure shows the spectrum for the "zero-order" cylinder, where the different modes (labeled by the poloidal mode number m) are uncoupled:

$$\bar{\omega}_m^2 = (nq^*)^2 - |m| + (nq^* + m)^2 . \tag{17.63}$$

The bottom part shows the results of standard first-order perturbation theory of the coupling between those modes due to the toroidicity or, what amounts to the same in the low-beta tokamak ordering, due to beta (causing $\Delta m = 1$ splitting) and due to the ellipticity of the plasma cross-section (causing $\Delta m = 2$ splitting). These couplings are due to a surface energy integral (unstable when $S > 0$) with an integrand of the form

$$S(\theta) \equiv [b_{\rm p}^2(\theta)\kappa_{\rm p}(\theta) + \epsilon\beta_{\rm p}\kappa_\varphi(\theta)]/h(\theta)$$

$$\approx \tfrac{1}{2}(3 - b^2/a^2) + 3\epsilon\beta_{\rm p}\cos\theta - (b^2/a^2 - 1)\cos 2\theta , \tag{17.64}$$

where the three terms correspond to *kink, ballooning and ellipticity*, respectively. Due to these splittings, there is a complete reordering of the structure of the spectrum, with important effects on the $|m| = 1$ kink mode (recall that the Kruskal–Shafranov limit is at $q^* = 1$) and elliptical splitting for the axi-symmetric modes (at $q^* = 0$). Mode coupling spectra for diffuse high-beta tokamak equilibria, exploiting the techniques of conformal mapping and polynomials described in Section 16.3.3, and exhibiting even more striking similarity with the electron band structures, may be found in [280].

Since the continuous spectra really come about from the diffuse inhomogeneity of the equilibrium, let us now consider the full spectrum of a diffuse low-beta toka-

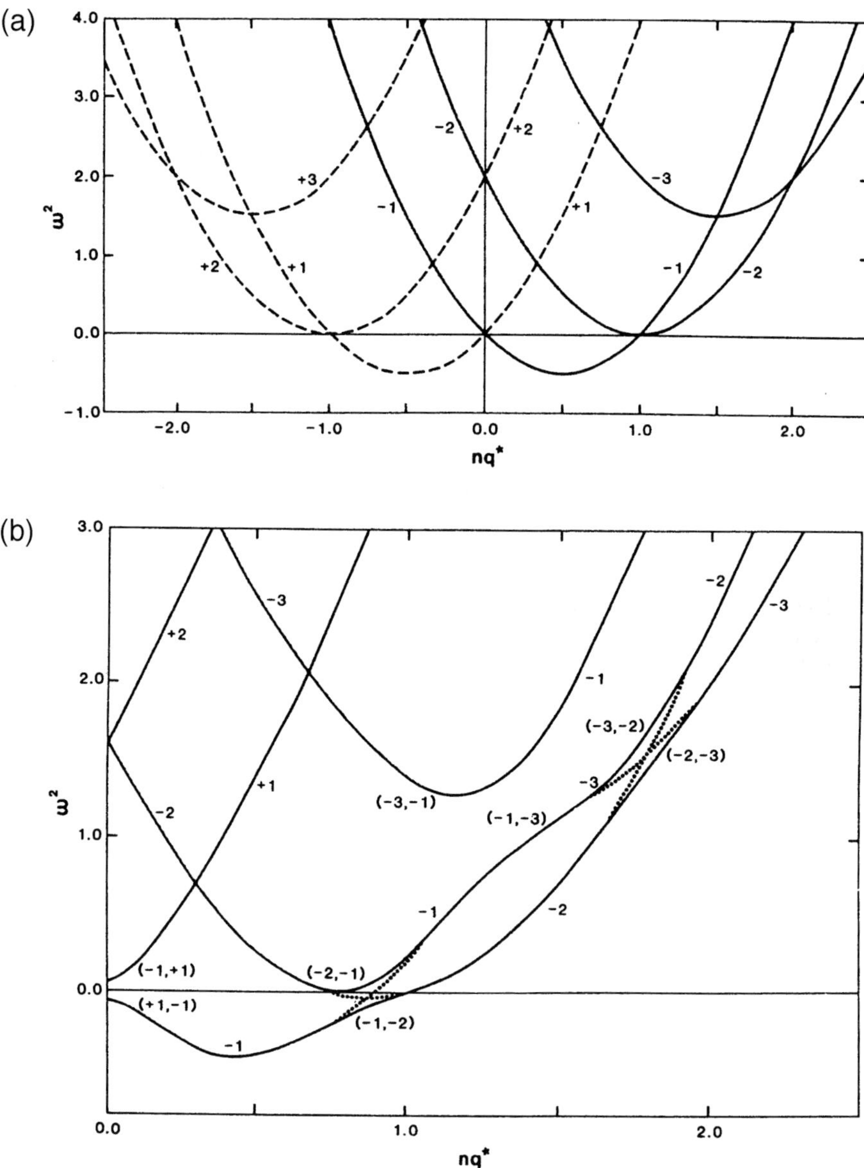

Fig. 17.4 Mode coupling of free-boundary kink modes in a skin-current model of a low-beta tokamak: (a) spectrum of uncoupled modes for circular cylinder, $\epsilon\beta_{\mathrm{p}} = 0$; (b) spectrum of coupled modes for low-beta tokamak with elliptical cross-section, $\epsilon\beta_{\mathrm{p}} = 0.1$, $b/a = 1.2$. (From D'Ippolito and Goedbloed [121].)

mak. Here, "full" is meant is the sense of collecting all values of the continua over the full range $0 \le \psi \le 1$ (in plots, usually ψ is replaced by the quasi-radial variable $s \equiv \sqrt{\psi}$). In the calculation of the continuum gaps, important global modes,

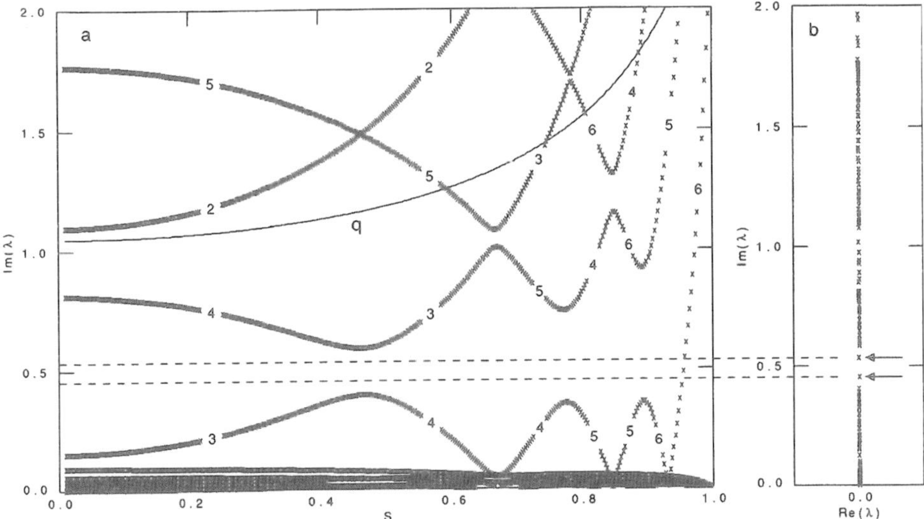

Fig. 17.5 Continuous spectrum and gap modes in a model tokamak equilibrium. (a) The continua, with frequency parameter $\lambda \equiv -i\omega$, are plotted as a function of $s \equiv \sqrt{\psi}$; the labels refer to the poloidal mode number m; slow continua crowd at the bottom of the diagram. (b) Discrete representation of the complete spectrum with two "gap modes" (indicated by the arrows). (From Poedts *et al.* [379].)

named *toroidal Alfvén eigenmodes* (TAEs), were found by Cheng and Chance [87]. These modes are not continuum modes, but discrete global modes with frequencies located inside the gaps, again very much like the discrete modes that occur in the forbidden bands of solid state physics. The TAEs should not be confused with the global Alfvén eigenmodes (GAEs), which are part of possible cluster sequences that occur at the extrema of the continua, as extensively discussed in Chapters 7 and 9 of Volume [1]. Whereas the GAEs already occur in 1D cylindrical equilibria, the TAEs require poloidal mode coupling to produce the continuum gaps. The different kinds of TAE caused by cross-sectional shaping, finite beta, and toroidal rotation are called ellipticity induced Alfvén eigenmodes (EAEs) [38], beta induced Alfvén eigenmodes (BAEs) [448] and toroidal flow induced Alfvén eigenmodes (TFAEs) [453, 454]. Their importance resides in the fact that they may be driven unstable by fusion-produced alpha particles [149, 86] and, thus, pose a threat to alpha particle confinement in future fusion reactors. Their calculation requires large-scale computations of the low mode number gaps ($n, m \sim 1$) and of the global TAE modes. An example is given in Fig. 17.5, taken from work by Poedts *et al.* [379] (exploiting a simple technique to compute the continua with a finite element spectral code [381]), which shows both the gaps in the continua and the discrete "gap modes" located in those gaps.

17.2.5 Alfvén and slow ballooning modes

Alfvén and slow continuum modes are constrained to "live" on the magnetic surfaces, where they necessarily remain stable, as is manifest from the expression for the potential energy in the numerator of the Rayleigh quotient (17.57). However, this does not exhaust the ways in which MHD waves can follow the magnetic field lines. In particular, notice that vanishing of the parallel gradient operator expression, $F \sim m + nq \approx 0$, leads to completely degenerate continua in the origin $\omega^2 = 0$. In the study of MHD stability, this is always the sign that higher order contributions should be considered. The obvious constraint to be dropped is the restriction of the displacement vector $\boldsymbol{\xi}$ to be tangential to the magnetic surfaces. This leads to a different kind of localized modes, called ballooning modes, which may be considered as *Alfvén and slow field line resonances.*

The ballooning mode analysis in tokamaks centers about the description of *instabilities at finite beta, driven by a combination of the pressure gradient p' and the curvature of the field lines* which produces a negative potential energy on the outside of the torus. Hence the name "ballooning". To properly describe this requires a subtle localization about single magnetic field lines, developed by Connor, Hastie and Taylor [90, 91], and others (see, e.g. Coppi *et al.* [93, 94]), where the main difficulty to be resolved is the basic incompatibility of field line localization with poloidal and toroidal periodicity. This is resolved by the so-called ballooning transformation to an extended domain $-\infty < \vartheta < \infty$ of the poloidal angle or, equivalently, by the consideration of a covering space; see Dewar and Glasser [118]. In this section, we will follow the analysis of the latter paper.

To describe modes that are localized to field lines, we exploit a representation of the magnetic field in terms of the two Clebsch potentials α and Ψ,

$$\mathbf{B} = \nabla\alpha \times \nabla\Psi \,, \qquad \alpha \equiv \varphi - q\vartheta \,, \tag{17.65}$$

where α has been chosen such that it varies linearly perpendicular to the straight field lines in the ϑ–φ plane, so that α and Ψ may be considered as field line labels (see footnote on Section 7.3.2 [1]). Substitution of α in the expression for $\mathbf{B}$ reproduces the basic magnetic field representation (17.10) for axi-symmetric equilibria:

$$\mathbf{B} = \nabla(\varphi - q\vartheta) \times \nabla\Psi = \nabla\varphi \times \nabla\Psi + I(\Psi)\nabla\varphi \,. \tag{17.66}$$

With this representation, the angles φ and ϑ can be continued indefinitely: the *covering space* [118] corresponding to a magnetic surface. The ballooning equations will turn out to be a system of two second order differential equations in terms of the extended angular variable ϑ, for each field line indicated by Ψ_0 and α_0.

Field line localization is effected by the assumption of large mode numbers, in particular the toroidal wave number, $n \gg 1$. Hence, we need to expand our equations in powers of the small parameter $n^{-1} \ll 1$. Such expansions are usually more conveniently carried out starting from the quadratic forms than from the differential equations, since they typically require expanding one order higher for the latter than is needed with the quadratic forms. Hence, we will work out the consequences of field line localization from the two quadratic forms (17.49) and (17.50) for W and K.

We now consider WKB solutions of the plasma displacement of the form

$$X(\Psi, \vartheta, \varphi) = \tilde{X}(\Psi, \vartheta)\, e^{inS(\Psi, \vartheta, \varphi)}, \qquad (17.67)$$

and similarly for the components Y and Z, with a split in a slowly varying amplitude $\tilde{X}$ and a rapidly varying phase nS, where S is known as the *eikonal* [184]. An obvious choice for the eikonal is

$$S(\Psi, \alpha) = \alpha + q(\Psi)\vartheta_0 \equiv \varphi - q(\Psi)(\vartheta - \vartheta_0) = S(\Psi, \vartheta, \varphi), \qquad (17.68)$$

where ϑ_0 is an arbitrary constant. Associated with this eikonal is a local wave vector $\mathbf{k} = n\nabla S$, which is normal to the field lines since $\mathbf{b} \cdot \nabla S = 0$ and, hence, normal to the eikonal wave fronts, that we will project onto the field line/magnetic surface triad $\{\mathbf{n}, \boldsymbol{\pi}, \mathbf{b}\}$. The possible confusion of the distinction between the normal vector $\mathbf{n}$ and the wave number n we eliminate right away by renormalizing $\mathbf{k}$ onto a wave vector $\tilde{\mathbf{k}}$ of unit order of magnitude:

$$\mathbf{k} = n\nabla S \quad \Rightarrow \quad \tilde{\mathbf{k}} \equiv n^{-1}\mathbf{k} = \nabla S = \tilde{k}_n \mathbf{n} + \tilde{k}_\pi \boldsymbol{\pi}, \qquad (17.69)$$

$$\tilde{k}_n = \mathbf{n} \cdot \nabla S = \mathbf{n} \cdot [-q'(\vartheta - \vartheta_0)\nabla\Psi - q\nabla\vartheta]$$

$$= -RB_{\mathrm{p}}[q'(\vartheta - \vartheta_0) - (g_{12}/g_{22})q],$$

$$\tilde{k}_\pi = \boldsymbol{\pi} \cdot \nabla S = \boldsymbol{\pi} \cdot [-q\nabla\vartheta + \nabla\varphi] = -B/(RB_{\mathrm{p}}). \qquad (17.70)$$

These wave numbers are local wave numbers in the WKB sense, i.e. they depend on the coordinates, $\tilde{k}_n = \tilde{k}_n(\Psi, \vartheta)$ and $\tilde{k}_\pi = \tilde{k}_\pi(\Psi, \vartheta)$, and, hence, they become capricious quantities twisting with the field lines.

▷ **Meaning of the constant** ϑ_0 One should well distinguish between the two coordinate dependencies $S(\Psi, \alpha)$ and $S(\Psi, \vartheta, \varphi)$ indicated in the equalities (17.68). From the former, we obtain the following, equally meaningful, expressions for the components of $\tilde{\mathbf{k}}$:

$$\tilde{\mathbf{k}} = \frac{\partial S}{\partial \Psi}\nabla\Psi + \frac{\partial S}{\partial \alpha}\nabla\alpha = q'\vartheta_0\nabla\Psi + \nabla\alpha = \vartheta_0\nabla q + \nabla\alpha \quad \Rightarrow \quad \tilde{k}_{\hat{q}} = \vartheta_0, \quad \tilde{k}_{\hat{\alpha}} = 1,$$
$$(17.71)$$

where $\tilde{k}_{\hat{q}}$ and $\tilde{k}_{\hat{\alpha}}$ are the covariant components of $\tilde{\mathbf{k}}$ with respect to the coordinates q and α. Here, the coordinate Ψ has been replaced by the equivalent coordinate $q(\Psi)$. Hence, ϑ_0 is the covariant component of $\tilde{\mathbf{k}}$ with respect to q, considered as a radial coordinate.   ◁

For the analysis of ballooning modes it is expedient to exploit a projection based

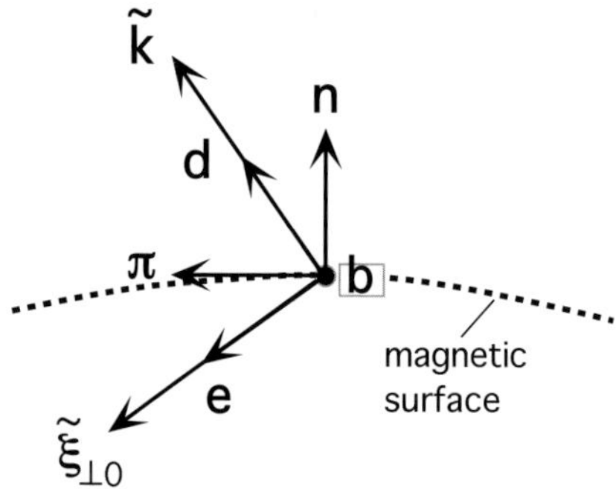

Fig. 17.6 Ballooning triad $\{\mathbf{b}, \mathbf{d}, \mathbf{e}\}$, based on the local wave vector $\widetilde{\mathbf{k}} \equiv n^{-1}\mathbf{k}$, with dominant perpendicular component $\widetilde{\boldsymbol{\xi}}_{\perp 0}$ of the ballooning perturbation.

on a new orthogonal triad $\mathbf{d}, \mathbf{e}, \mathbf{b}$, where $\mathbf{d} \equiv \mathbf{k}/|\mathbf{k}|$ and $\mathbf{e} \equiv \mathbf{b} \times \mathbf{d}$, with corresponding components $\widetilde{\xi}_d$, $\widetilde{\xi}_e$, $\widetilde{\xi}_b$ of the vector $\widetilde{\boldsymbol{\xi}}$. This is illustrated in Fig. 17.6, where the result of the analysis below ($\widetilde{\xi}_d \approx 0$) is already indicated.

We now work out the effects of the operators D, G, F, defined in Eqs. (17.40), acting on the perturbations X, Y, Z, defined in Eqs. (17.41):

$$
\begin{aligned}
DX &= (RB_{\mathrm{p}})^{-1}\mathbf{n} \cdot \nabla(\widetilde{X}\mathrm{e}^{inS}) = \left[\left(D + in\widetilde{k}_n/(RB_{\mathrm{p}})\right)\widetilde{X}\right]\mathrm{e}^{inS} \\
&\equiv [\widetilde{D}\widetilde{X}]\mathrm{e}^{inS} \quad \Rightarrow \quad \widetilde{D} \equiv D + in\widetilde{k}_n/(RB_{\mathrm{p}})\,,
\end{aligned}
\tag{17.72}
$$

$$
\begin{aligned}
GX &= -\mathrm{i}RB_{\mathrm{p}}B\boldsymbol{\pi} \cdot \nabla(\widetilde{X}\mathrm{e}^{inS}) = -\mathrm{i}RB_{\mathrm{p}}B\left[\boldsymbol{\pi} \cdot \nabla\widetilde{X} + in\widetilde{k}_\pi\widetilde{X}\right]\mathrm{e}^{inS} \\
&\equiv [\widetilde{G}\widetilde{X}]\mathrm{e}^{inS} \quad \Rightarrow \quad \widetilde{G} \equiv -\mathrm{i}RB_\varphi\mathcal{J}^{-1}\partial_\vartheta - nB^2\,,
\end{aligned}
\tag{17.73}
$$

$$
\begin{aligned}
FX &= -\mathrm{i}B\mathbf{b} \cdot \nabla(\widetilde{X}\mathrm{e}^{inS}) = -\mathrm{i}B\left[(\mathcal{J}B)^{-1}\partial_\vartheta\widetilde{X}\right]\mathrm{e}^{inS} \\
&\equiv [\widetilde{F}\widetilde{X}]\mathrm{e}^{inS} \quad \Rightarrow \quad \widetilde{F} \equiv -\mathrm{i}\mathcal{J}^{-1}\partial_\vartheta\,,
\end{aligned}
\tag{17.74}
$$

and, of course, similarly for Y and Z. The new operators $\widetilde{D}$, $\widetilde{G}$ and $\widetilde{F}$ act on the slowly varying amplitude functions $\widetilde{X}$, $\widetilde{Y}$ and $\widetilde{Z}$ only, which is indicated by enclosing them by square brackets. Expanding in the small parameter n^{-1}, it is obvious that the leading order expressions will be obtained from the terms proportional to n in the operators $\widetilde{D}$ and $\widetilde{G}$.

We now substitute these expressions into the quadratic forms (17.49) and (17.50) for W and K, and reduce them order by order according to the following scheme:

$$
\widetilde{X} = \widetilde{X}_0 + n^{-1}\widetilde{X}_1 + \cdots\,, \quad \text{and similarly for } \widetilde{Y} \text{ and } \widetilde{Z},
$$

$$
\Rightarrow \quad W = n^2 W_0 + W_2 + \cdots\,, \qquad K = K_0 + \cdots\,.
\tag{17.75}
$$

This yields the following lowest order expressions:

$$W_0 = \pi \iint (\gamma p + B^2) \left| i[\tilde{k}_n/(RB_{\rm p})]\tilde{X}_0 - \tilde{Y}_0 \right|^2 \mathcal{J} \, d\Psi \, d\vartheta \,, \tag{17.76}$$

$$K_0 = \pi \iint \rho \left[\frac{1}{R^2 B_{\rm p}^2} |\tilde{X}_0|^2 + \frac{R^2 B_{\rm p}^2}{B^2} |\tilde{Y}_0|^2 + B^2 |\tilde{Z}_0|^2 \right] \mathcal{J} \, d\Psi \, d\vartheta \,, \tag{17.77}$$

with the obvious minimizing relation

$$i\mathbf{k} \cdot \tilde{\boldsymbol{\xi}}_0 \equiv i(\tilde{k}_n \tilde{\xi}_{n0} + \tilde{k}_\pi \tilde{\xi}_{\pi 0}) = i[\tilde{k}_n/(RB_{\rm p})]\tilde{X}_0 - \tilde{Y}_0 = 0$$

$$\Rightarrow \quad W_0 = 0 \,. \tag{17.78}$$

Hence, to leading order, the ballooning perturbation $\tilde{\xi}_{d0} = 0$ (see Fig. 17.6). The next order, W_2, contains the zero order variables $\tilde{X}_0$, $\tilde{Y}_0$ and $\tilde{Z}_0$, as well as the first order variables $\tilde{X}_1$ and $\tilde{Y}_1$, in the combination $\tilde{\xi}_{d1}$. The latter variable is eliminated by minimizing W_2 with respect to it. This involves first completing the squares of the second and fourth term (involving $\tilde{D}^\dagger \tilde{X}_0$) of the expression (17.49) for W, then combining them into a new square, and finally minimizing the resulting expression for W_2 with respect to $\tilde{\xi}_{d1}$, which now remains finite:

$$i\mathbf{k} \cdot \tilde{\boldsymbol{\xi}}_1 \equiv i[\tilde{k}_n/(RB_{\rm p})]\tilde{X}_1 - \tilde{Y}_1 = -\frac{B^2}{\gamma p + B^2}\left(\tilde{D}^\dagger \tilde{X}_0 + \frac{2\kappa_n}{RB_{\rm p}}\tilde{X}_0 + \frac{I}{B^2}\tilde{F}\tilde{Y}_0 \right)$$

$$- \frac{\gamma p}{\gamma p + B^2}\left(\tilde{D}^\dagger \tilde{X}_0 + I\tilde{F}\frac{1}{B^2}\tilde{Y}_0 + \tilde{F}\tilde{Z}_0 \right). \tag{17.79}$$

Substitution of this expression back into W_2 gives the final result:

$$W_2 = \pi \iint \left\{ \frac{1}{R^2 B_{\rm p}^2} |\tilde{F}\tilde{X}_0|^2 + \frac{R^2 B_{\rm p}^2}{B^2} \left| \tilde{F}\tilde{Y}_0 + \frac{2B_\varphi(\kappa_{\rm p} - \kappa_\varphi)}{R^2 B_{\rm p}}\tilde{X}_0 \right|^2 \right.$$

$$+ \frac{\gamma p B^2}{\gamma p + B^2} \left| \tilde{F}\tilde{Z}_0 - \frac{2\kappa_n}{RB_{\rm p}}\tilde{X}_0 + 2i\frac{RB_{\rm p}}{B}\kappa_{\rm g}\tilde{Y}_0 \right|^2 + U|\tilde{X}_0|^2 \bigg\} \mathcal{J} \, d\Psi \, d\vartheta$$

$$= \pi \iint \left\{ \frac{1}{R^2 B_{\rm p}^2}(1 + \tilde{k}_n^2/\tilde{k}_\pi^2)|\tilde{F}\tilde{X}_0|^2 - \frac{2}{RB_{\rm p}}[\kappa_n - (\tilde{k}_n/\tilde{k}_\pi)\kappa_{\rm g}]p'|\tilde{X}_0|^2 \right.$$

$$+ \frac{\gamma p B^2}{\gamma p + B^2} \left| \tilde{F}\tilde{Z}_0 - \frac{2}{RB_{\rm p}}[\kappa_n - (\tilde{k}_n/\tilde{k}_\pi)\kappa_{\rm g}]\tilde{X}_0 \right|^2 \bigg\} \mathcal{J} \, d\Psi \, d\vartheta \,, \tag{17.80}$$

where U is defined in Eq. (17.51), and the last expression follows by substituting $\tilde{Y}_0$ from (17.78), integration by parts and use of the relations of Section 17.1.3.

In conclusion, the variational principle (17.48) has been reduced to

$$\delta\Lambda_2 = 0 \,, \qquad \Lambda_2[\tilde{\boldsymbol{\xi}}_0] \equiv \frac{W_2[\tilde{\boldsymbol{\xi}}_0]}{K_0[\tilde{\boldsymbol{\xi}}_0]} \,, \tag{17.81}$$

with W_2 and K_0, given by the expressions (17.80) and (17.77), converted into expressions in terms of the leading order displacement vector $\widetilde{\boldsymbol{\xi}}_0$ of the ballooning perturbations. Since this vector is directed perpendicular to $\mathbf{k}$, it is appropriate to convert $(\widetilde{u}, \widetilde{v}, \widetilde{\zeta})$ to a two-component representation in terms of $\widetilde{v}$ and $\widetilde{\zeta}$:

$$\widetilde{\boldsymbol{\xi}}_0(\Psi, \vartheta) = -\mathrm{i}\left[\widetilde{v}(\Psi, \vartheta)\mathbf{e} + \widetilde{\zeta}(\Psi, \vartheta)\mathbf{b}\right],$$

$$\widetilde{u} \equiv \mathrm{i}\widetilde{\boldsymbol{\xi}}_0 \cdot \mathbf{d} = (1/\widetilde{k})[\mathrm{i}(\widetilde{k}_n/(RB_\mathrm{p}))\widetilde{X}_0 - \widetilde{Y}_0] = 0\,,$$

$$\widetilde{v} \equiv \mathrm{i}\widetilde{\boldsymbol{\xi}}_0 \cdot \mathbf{e} = (1/\widetilde{k})[(B/(R^2 B_\mathrm{p}^2))\widetilde{X}_0 - \mathrm{i}(RB_\mathrm{p}/B)\widetilde{Y}_0] = \widetilde{k}\widetilde{X}_0\,,$$

$$\widetilde{\zeta} \equiv \mathrm{i}\widetilde{\boldsymbol{\xi}}_0 \cdot \mathbf{b} = B\widetilde{Z}_0\,. \tag{17.82}$$

This yields the most symmetric representation of the Euler equations associated with the variational principle (17.81) in terms of two coupled second order ordinary differential equations in the extended variable ϑ, with parametric dependence of the coefficients on Ψ, as derived by Dewar and Glasser [118] (see also [308]):

$$\begin{pmatrix} \widetilde{\alpha}_{11} & \widetilde{\alpha}_{12} \\ \widetilde{\alpha}_{21} & \widetilde{\alpha}_{22} \end{pmatrix} \begin{pmatrix} \widetilde{v} \\ \widetilde{\zeta} \end{pmatrix} = \rho\omega^2 \begin{pmatrix} \widetilde{v} \\ \widetilde{\zeta} \end{pmatrix}, \tag{17.83}$$

where

$$\widetilde{\alpha}_{11} \equiv \frac{B}{\widetilde{k}}\widetilde{F}\frac{\widetilde{k}^2}{B^2}\widetilde{F}\frac{B}{\widetilde{k}} + \frac{4\gamma p B^2}{\gamma p + B^2}\kappa_e^2 - 2\frac{B}{\widetilde{k}}\kappa_e p'\,, \quad \widetilde{\alpha}_{12} \equiv -\frac{2\mathrm{i}\gamma p B^2}{\gamma p + B^2}\kappa_e \widetilde{F}\frac{1}{B}\,,$$

$$\widetilde{\alpha}_{21} \equiv \frac{\mathrm{i}}{B}\widetilde{F}\frac{2\gamma p B^2}{\gamma p + B^2}\kappa_e\,, \qquad\qquad \widetilde{\alpha}_{22} \equiv \frac{1}{B}\widetilde{F}\frac{\gamma p B^2}{\gamma p + B^2}\widetilde{F}\frac{1}{B}\,.$$

$$\tag{17.84}$$

Here, $\kappa_e \equiv \mathbf{e} \cdot \boldsymbol{\kappa} = -(\widetilde{k}_\pi/\widetilde{k})\kappa_n + (\widetilde{k}_n/\widetilde{k})\kappa_\mathrm{g}$ is the component of the field line curvature perpendicular to $\mathbf{k}$ and the operator

$$\widetilde{F} \equiv -\mathrm{i}\mathbf{B} \cdot \widetilde{\nabla} = -(\mathrm{i}/\mathcal{J})\,\partial_\vartheta \tag{17.85}$$

is just the part of the parallel gradient operator which acts on the slowly varying amplitude functions. Since the functions $\widetilde{v}$ and $\widetilde{\zeta}$ are defined on an infinite domain for periodic configurations, the appropriate boundary conditions at $\vartheta \to \pm\infty$ are that they should decay rapidly enough to remain square integrable.

As compared to the magnetic-surface-localized Alfvén and slow continua of Eq. (17.56), due to the additional freedom of field line localization, there is now a new, potentially negative term, proportional to $-\kappa_e p'$, in the upper left diagonal element of the matrix (17.83). This is the driving term of high-n ballooning modes, corresponding to the low-n ballooning term $\epsilon\beta_\mathrm{p}\kappa_\varphi$ encountered in the coupling term (17.64) of the high-beta tokamak model of the previous section.

Finally, the ballooning equation of Connor, Hastie and Taylor [90, 91] is obtained from the two ODEs (17.83) by neglecting the compressibility terms ($\gamma p \ll B^2$, appropriate for low-beta) so that $\widetilde{\zeta} = 0$ (or $\widetilde{Z}_0 = 0$). This gives:

$$ -\frac{B^2}{\mathcal{J}\widetilde{k}^2}\frac{\partial}{\partial\vartheta}\left(\frac{\widetilde{k}^2}{\mathcal{J}B^2}\frac{\partial\widetilde{X}_0}{\partial\vartheta}\right) - \frac{2B}{\widetilde{k}}\kappa_e p'\widetilde{X}_0 = \rho\omega^2\widetilde{X}_0 . \tag{17.86} $$

This appears to yield a very simple prescription for the study of the stability of tokamaks with respect to high-n ballooning modes: just investigate the Sturmian properties of the eigenfunctions $\widetilde{X}_0$ on the infinite domain $-\infty < \vartheta < \infty$ and determine the lowest eigenvalues $\omega^2(\Psi, \vartheta_0)$. If they are positive, the configuration is locally stable, if there is a negative one the configuration is unstable.

One aspect of the ballooning theory presented so far is entirely unclear. How does this analysis reconcile poloidal periodicity with field line localization? Following Dewar and Glasser [118] again (see also [117], [220], [367] for the associated small-scale radial periodicity), this question is answered by observing that the eigenvalue problem (17.83), or (17.86), is infinitely degenerate with respect to poloidal periodicity. In particular, notice that the Rayleigh quotient Λ_2 is invariant under a change of the poloidal angle ϑ over a period 2π, provided that the value of the parameter ϑ_0 is replaced by $\vartheta_0 + 2\pi$ as well. Exploiting the coordinates Ψ and α to indicate the field lines, and considering the variable X only (dropping the subscript 0), this implies the following:

$$ \omega^2(\Psi, \alpha - 2\pi q, \vartheta_0 + 2\pi) = \omega^2(\Psi, \alpha, \vartheta_0) \qquad \text{(eigenvalues)}, $$

$$ X(\vartheta + 2\pi; \Psi, \alpha - 2\pi q, \vartheta_0 + 2\pi) = X(\vartheta; \Psi, \alpha, \vartheta_0) \quad \text{(eigenfunctions)} . \tag{17.87} $$

Hence, instead of the single non-periodic mode (17.67), the superposition

$$ X(\Psi, \vartheta, \varphi) = \sum_{\ell=-\infty}^{\infty} \widetilde{X}(\Psi, \vartheta + 2\pi\ell)\,e^{in[\varphi - q(\vartheta + 2\pi\ell - \vartheta_0)]} \tag{17.88} $$

is a periodic solution of the ballooning equations, with the eigenvalue ω^2, provided that the amplitude $\widetilde{X}$ falls off rapidly enough at infinity to leave the quadratic forms finite. This behavior is dictated by the indicial equation, which turns out to be associated with the Mercier criterion at marginal stability (see below).

▷ **Analogies between continuum and ballooning equations** The analogy between the coupled ODEs (17.55), describing the Alfvén and slow continua, and (17.83), describing the Alfvén and slow ballooning modes, is striking but misleading. Formally, one may obtain Eq. (17.55) from Eq. (17.83), as noted by Lifschitz [308], by taking the limit $\vartheta_0 \to \infty$ so that $\widetilde{k} \approx \widetilde{k}_n \approx q'\vartheta_0 \cdot RB_\mathrm{p} \to \infty$, $\kappa_e \to \kappa_g$, $\widetilde{v} \to \widetilde{\eta}$ and $\kappa_e p'/\widetilde{k} \to 0$. However, we then get an equation in terms of $\widetilde{F}$ acting on $\widetilde{\eta}$, rather than F acting on η, whereas the relationship between η and $\widetilde{\eta}$, viz.

$$ \eta(\vartheta) = e^{inq\vartheta_0} \cdot \widetilde{\eta}(\vartheta)\,e^{-inq\vartheta} , \tag{17.89} $$

becomes a tricky one in view of the wildly oscillating factor involving ϑ_0. In order to permit the consideration of $n \sim 1$ modes one would have to re-interpret a posteriori the ballooning approximation, which assumed $n \gg 1$. Physically, this is related to the absence in Eq. (17.55) of the most important term in the ballooning equation, viz. the one that drives instabilities: $\kappa_e p'$. In order to derive it, it was necessary to relax on the constraint of displacements lying in the magnetic surfaces (necessary for Eq. (17.55)), on the one hand, but to restrict the displacements to be localized around the field lines (implicit in the Ansatz (17.67)), on the other. The latter constraint is necessary to minimize field line bending, which is quite appropriate in stability studies but completely spurious for the study of Alfvén wave heating, where the global exciting wave just enforces the field line bendings that correspond to their scale lengths. For example, in the case of Alfvén wave heating of the solar corona, it is a safe assumption that these scale lengths are arbitrary, in the sense of being dictated by the photospheric convection patterns and not by the field line structure of the coronal loops. We conclude that each of the formalisms associated with the two types of equation has its own domain of application, where the typical ballooning term $\kappa_e p'$ is crucial for stability studies but a minor correction in heating scenarios and, vice versa, the low mode number structure of the continua (with gaps in the case of tokamaks) is crucial for heating but strictly lost in the ballooning formalism. ◁

Mercier criterion The Mercier criterion was originally derived by Mercier [331] as a generalization of Suydam's criterion. With the introduction of the ballooning formalism, it became clear that it can also be interpreted as a condition on the solutions of the ballooning equation for $\vartheta \to \infty$. This leads again to a cluster point analysis, but now in the stretched angular variable ϑ, rather than the radial flux variable Ψ. The derivation is put in small print.

▷ **Derivation of the Mercier criterion** The Mercier criterion may be obtained from the marginal ($\omega^2 = 0$) version of the ballooning equation (17.86),

$$\frac{1}{\mathcal{J}} \frac{d}{d\vartheta} \left[\frac{1}{\mathcal{J} R^2 B_{\mathrm{p}}^2} \left(1 + \frac{\widetilde{k}_n^2}{\widetilde{k}_\pi^2} \right) \frac{d\widetilde{X}_0}{d\vartheta} \right] + \frac{2}{R B_{\mathrm{p}}} \left(\kappa_n - \frac{\widetilde{k}_n}{\widetilde{k}_\pi} \kappa_{\mathrm{g}} \right) p' \widetilde{X}_0 = 0, \qquad (17.90)$$

where the equilibrium functions are periodic, but $\widetilde{k}_n / \widetilde{k}_\pi$ has a secular dependence on ϑ:

$$\widetilde{k}_n / \widetilde{k}_\pi = (R^2 B_{\mathrm{p}}^2 / B^2) z, \qquad z \equiv q'(\vartheta - \vartheta_0) - (g_{12}/g_{22}) q. \qquad (17.91)$$

Splitting periodic and secular dependencies, the ballooning equation is written as

$$\frac{d}{d\vartheta} \left[(az^2 + b) \frac{d\widetilde{X}_0}{d\vartheta} \right] + (\dot{c} z + d) \widetilde{X}_0 = 0, \qquad (17.92)$$

$$a \equiv \frac{R^2 B_{\mathrm{p}}^2}{\mathcal{J} B^2}, \qquad b \equiv \frac{1}{\mathcal{J} R^2 B_{\mathrm{p}}^2}, \qquad c \equiv \frac{I p'}{B^2}, \qquad d \equiv \frac{2 \mathcal{J} \kappa_n p'}{R B_{\mathrm{p}}},$$

where the dot indicates differentiation with respect to ϑ. We try solutions of the form

$$\widetilde{X}_0 = z^\nu \left[f_0(\vartheta) + \frac{f_1(\vartheta)}{z} + \frac{f_2(\vartheta)}{z^2} + \cdots \right], \qquad (17.93)$$

where the f_is are periodic functions of ϑ. The task is to determine the index ν (similar to the derivation of Suydam's criterion in Section 9.4.1 [1], but now on the extended angular domain). This is done by substituting (17.93) into (17.92) and balancing the powers of z:

$$z^{\nu+2}: \quad \frac{d}{d\vartheta}\left(a\frac{df_0}{d\vartheta}\right) = 0 \quad \Rightarrow \quad f_0 = C_1\,, \tag{17.94}$$

$$z^{\nu+1}: \quad \frac{d}{d\vartheta}\left(a\frac{df_1}{d\vartheta}\right) = -\left[\nu(a\dot{z})^{\cdot}+\dot{c}\right]f_0$$

$$\Rightarrow \quad f_1 = -C_1\left[\nu z + \int (c/a)\,d\vartheta\right] + C_2\int (1/a)\,d\vartheta + C_3\,,$$

$$- C_1\left[2\pi\nu q' + \oint (c/a)\,d\vartheta\right] + C_2\oint (1/a)\,d\vartheta = 0\,, \tag{17.95}$$

$$z^{\nu}: \quad \frac{d}{d\vartheta}\left(a\frac{df_2}{d\vartheta}\right) = -2\nu a\dot{z}\frac{df_1}{d\vartheta} - \left[(\nu-1)(a\dot{z})^{\cdot}+\dot{c}\right]f_1 - \left[\nu(\nu+1)a\dot{z}^2 + d\right]f_0\,,$$

$$- C_1\oint \left[c\dot{z} - (c^2/a) - d\right]d\vartheta + C_2\left[(\nu+1)2\pi q' - \oint (c/a)\,d\vartheta\right] = 0\,. \tag{17.96}$$

Here, periodicity of f_0 is manifest, periodicity of f_1 requires the relationship (17.95) between the constants C_1 and C_2, whereas periodicity of f_2 requires the relationship (17.96). Compatibility of the latter two conditions yields the *indicial equation:*

$$\nu(\nu+1) + D = 0\,, \quad D \equiv \frac{1}{4\pi^2 q'^2}\left\{\oint (c/a)\,d\vartheta\left[2\pi q' - \oint (c/a)\,d\vartheta\right]\right.$$

$$\left. - \oint (1/a)\,d\vartheta\oint \left[c\dot{z} - (c^2/a) - d\right]d\vartheta\right\}\,. \tag{17.97}$$

This implies infinitely oscillatory solutions (local instability with respect to interchanges) when the indices are complex ($D < 1/4$) and stability when the indices are real. ◁

By exploiting the Grad–Shafranov relation (17.16) and the other relations of Section 17.1.3, many equivalent forms of the Mercier criterion $D > 1/4$ can be derived. The following one, due to Pao [358], is probably the most appealing one:

$$\tfrac{1}{2}\pi^2 q'^2 + p'\left[\oint \frac{B^2}{R^2 B_{\mathrm{p}}^2}\mathcal{J}\,d\vartheta\oint \frac{\kappa_{\mathrm{p}}}{RB_{\mathrm{p}}}\mathcal{J}\,d\vartheta\right.$$

$$\left. + \oint \frac{1}{R^2 B_{\mathrm{p}}^2}\mathcal{J}\,d\vartheta\oint \frac{(-\kappa_{\mathrm{p}} + \kappa_{\varphi})B_{\varphi}^2}{RB_{\mathrm{p}}}\mathcal{J}\,d\vartheta\right] > 0\,. \tag{17.98}$$

In this form, it clearly exhibits the competing contributions of the magnetic shear (first term), which is stabilizing, and the pressure gradient, which drives interchange instabilities if the term in square brackets is positive. In the cylindrical limit, $\kappa_{\varphi} = 0$ and $B^2 = \mathrm{const}$, that term is positive definite and Suydam's criterion is recovered. In the toroidal case, the term may change sign through the contribution of the toroidal curvature. In general, numerical analysis of specific equilibria is needed to determine stability with respect to interchange modes.

17.3 Computation of waves and instabilities in tokamaks

We now turn to numerical theory to provide us with the procedures that produce the explicit answers needed in fusion research. Once large-scale computing is embraced, some of the analytical methods lose their attraction and new possibilities present themselves. Thus, we will only discuss ideal MHD in short, since it is contained in resistive MHD, as far as the numerics is concerned, and its numerical implementation appears to be too restrictive at the present time.

17.3.1 Ideal MHD versus resistive MHD in computations

There is a long history on the subject of ideal MHD stability of tokamaks. There appears to be no point trying to summarize this in a few lines. Excellent texts on this subject exist [140], [481]. The reader is advised to consult those. The present chapter is more concerned with the structures of the different theoretical methods than with specific results. We will merely use the results to illustrate the methods.

Troyon beta limit There is also a long history on the development and subsequent use of toroidal ideal MHD stability programs, the most well known of them being ERATO [198, 197] and PEST [193]. Again, we will not attempt to summarize the results reached. There is one famous result, though, that illustrates one of the points of this section. This is the Troyon beta limit [447] for stability of tokamaks with respect to ideal MHD kink modes. It summarizes the results of a great number of numerical optimization studies with respect to high-β stability of tokamaks. The result is a quite simple scaling law of the maximum value of β with respect to kink mode stability as a function of the toroidal current I_φ, the vacuum magnetic field B_0 and the plasma radius a (with the units indicated in brackets):

$$\beta\,(\%) < g_{\rm T} \frac{I_\varphi\,({\rm MA})}{a\,({\rm m})\,B_0\,({\rm T})}, \tag{17.99}$$

where $g_{\rm T}$ is a factor obtained from the numerical studies, originally put at $g_{\rm T} \approx 3$. How is it that all the complicated Alfvén wave dynamics in the curved environment of toroidal confinement systems just leads to such a simple answer? Is it a genuine scaling law or just a very stimulating way of getting experimentalists to improve tokamak performance? (Like Moore's law in computer chips: not a law of nature, but demonstration of the possibility of obtaining desirable scaling by intense technological efforts.) Figure 17.7 appears to point in the latter direction, with the swarm of experimental points continually moving up, in JET reaching $\beta \approx 6\%$ in 1990, but soon afterwards (1995) put into the shade by the record values of $\beta \approx 12.6\%$ obtained in the DIII-D tokamak [432], with some experimental points far above the curve. Not surprisingly, the slope of the curve also turned out

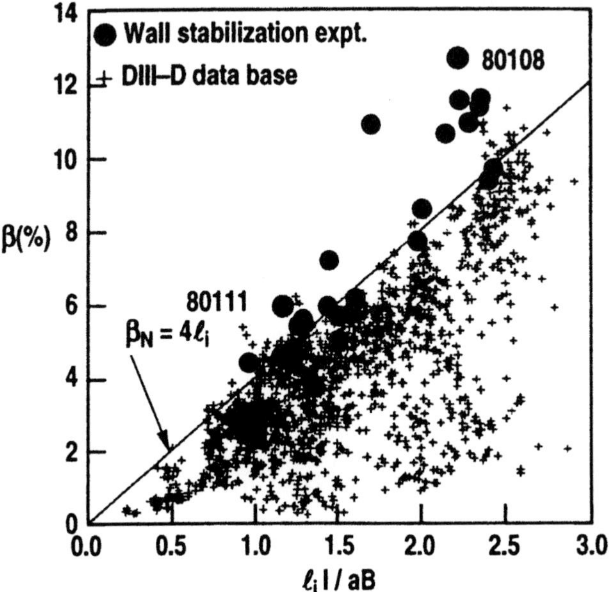

Fig. 17.7 Toroidal beta as a function of normalized current in DIII-D discharges. The drawn line is a modified form of the Troyon limit. (From Taylor *et al.* [432]).

to depend on details of the current distribution, parameterized by the internal inductance ℓ_i. Thus, ideal MHD computations have served not only as a theoretical framework for the interpretation of tokamak stability results, but also as a stimulus for experimental progress. (In this respect, the comparison with Moore's law should be taken as positive evaluation of the Troyon scaling law.)

▷ **High-beta tokamak scaling** The high-beta scaling of Section 16.1.4 actually yields a quadratic dependence on the toroidal current, following from Eqs. (16.69) and (16.66):

$$\beta = \frac{0.04\pi a^2 \beta_{\rm p}}{S}\left(\frac{10^{-6}I_\varphi}{aB_0}\right)^2 \qquad \text{(all in mks units)}. \tag{17.100}$$

For limiting values $\beta_{\rm p} \sim \epsilon^{-1}$, roughly independent of the toroidal current, this expression appears not to conflict with the general tendency of the points shown in Fig. 17.7. ◁

Turning to resistive MHD calculations Progress in computer equipment and computational methods not only served to narrow down the gap between theory of highly simplified analytical models and numerical analysis of more realistic models of the magnetic confinement geometry of tokamaks, but it also revealed a false dilemma between ideal (conservative) and resistive (dissipative) MHD. Historically, the vast majority of tokamak stability investigations has been done by means of the energy principle of ideal MHD [35, 36, 203], and its spectral generalizations, extensively discussed in this volume and the preceding one. The reason is obvious: that theory is the simplest and also the most appealing one with respect to mathematical and physical properties (the power of spectral theory combined with the beauty of conservation laws). Also, with the velocity representation (or the displacement vector $\boldsymbol{\xi}$), reduction to the smallest number of unknowns was effected, a very desirable feature in the early days of computing with very restricted central memory sizes. However, it falsely created the impression that computing for the more extended MHD models, including resistive MHD, would necessarily be less accurate and also much slower than for ideal MHD. In order to appreciate this misunderstanding, we first need to introduce the newer numerical methods used in resistive MHD calculations. At the end of this section, we will then return to an explanation of why and how the superiority of the conservative model over the dissipative model turned out to be a false dilemma.

Moving to dissipative MHD, we immediately realize that the basic variable $\boldsymbol{\xi}$ is no longer available since its definition is strictly based on flux conservation. Therefore, let us return to the basic equations of resistive MHD, Eqs. (4.122)–(4.125) of Volume [1], already exploited in Section 14.2, but now replacing the pressure by the temperature using the relation $p = (\gamma - 1)C_v \rho T$:

$$\frac{\partial \rho}{\partial t} = - \nabla \cdot (\rho \mathbf{v}) \,, \tag{17.101}$$

$$\rho \frac{d\mathbf{v}}{dt} = - (\gamma - 1)C_v \nabla(\rho T) + \mu_0^{-1}(\nabla \times \mathbf{B}) \times \mathbf{B} \,, \tag{17.102}$$

$$C_v \rho \frac{dT}{dt} = - (\gamma - 1)C_v \rho T \, \nabla \cdot \mathbf{v} + \mu_0^{-2}\eta(\nabla \times \mathbf{B})^2 \,, \tag{17.103}$$

$$\frac{\partial \mathbf{B}}{\partial t} = \nabla \times (\mathbf{v} \times \mathbf{B}) - \mu_o^{-1}\nabla \times (\eta \nabla \times \mathbf{B}) \,, \tag{17.104}$$

$$\nabla \cdot \mathbf{B} = 0 \,. \tag{17.105}$$

We will eliminate the factors C_v and μ_0 by defining $\bar{T} \equiv (\gamma - 1)C_v T$, $\bar{\eta} \equiv \eta/\mu_0$ and $\bar{\mathbf{B}} \equiv \mathbf{B}/\sqrt{\mu_0}$, but we will drop the bars immediately. The basic equations (17.101)–(17.104) govern the temporal evolution of the density ρ, the velocity $\mathbf{v}$, the temperature T and the magnetic field $\mathbf{B}$, whereas Eq. (17.105) is a constraint on $\mathbf{B}$ that should be satisfied with the same accuracy as the other equations. (See Tóth [440], and Section 19.3.1, for a discussion of the numerical aspects.)

Let us now linearize these equations according to the scheme

$$f(\mathbf{r}, t) \approx f_0(\psi, \vartheta) + f_1(\psi, \vartheta)e^{i(n\varphi - \omega t)} \,, \tag{17.106}$$

where the f_0s describe the equilibrium quantities, the f_1s are the amplitudes of the normal mode solutions and n is the toroidal mode number. Recall that for ideal MHD, dropping the terms with η and writing $\mathbf{v}_1 \equiv \partial\boldsymbol{\xi}/\partial t$, three of the four equations can be integrated directly and the spectral problem

$$\mathbf{F}(\boldsymbol{\xi}) = -\rho\omega^2\boldsymbol{\xi} \tag{17.107}$$

is obtained, where $\mathbf{F}$ is the force operator, which is self-adjoint, and the eigenvalues ω^2 are real. For resistive MHD, this scheme does not work and we are forced to consider the eight components ρ_1, $\mathbf{v}_1$, T_1, $\mathbf{B}_1$ describing the perturbed state. In general, the eigenvalues ω will then have both real and imaginary parts. Instead of ω, it is customary to exploit the complex eigenvalue parameter $\lambda \equiv -i\omega$, the real part of which measures the exponential growth rate of instabilities.

We will simply remove the initial condition $\nabla \cdot \mathbf{B}_1 = 0$ from the eigenvalue problem by exploiting the vector potential $\mathbf{A}_1$ as a variable, $\mathbf{B}_1 = \nabla \times \mathbf{A}_1$, and choosing the gauge condition $\Phi_1 = 0$ on the scalar potential. This elimination of the initial condition will introduce spurious eigenvalues $\lambda = 0$, which have to be removed later on. However, this is a relatively easy numerical task.

The equations of the resulting eigenvalue problem, already presented as Eqs. (15.63)–(15.66) of Section 15.2.1, are repeated here for convenience:

$$\lambda\rho_1 = -\nabla \cdot (\rho\mathbf{v}_1)\,, \tag{17.108}$$

$$\lambda\rho\mathbf{v}_1 = -\nabla\left(\rho T_1 + (p/\rho)\,\rho_1\right) + (\nabla \times \mathbf{B}) \times (\nabla \times \mathbf{A}_1)$$
$$-\mathbf{B} \times (\nabla \times \nabla \times \mathbf{A}_1)\,, \tag{17.109}$$

$$\lambda\rho T_1 = -\rho\mathbf{v}_1 \cdot \nabla(p/\rho) - p\nabla \cdot \mathbf{v}_1$$
$$+2\eta\,(\gamma - 1)\,(\nabla \times \mathbf{B}) \cdot (\nabla \times \nabla \times \mathbf{A}_1)\,, \tag{17.110}$$

$$\lambda\mathbf{A}_1 = -\mathbf{B} \times \mathbf{v}_1 - \eta\,\nabla \times \nabla \times \mathbf{A}_1\,. \tag{17.111}$$

Here, explicit numerical knowledge of the equilibrium, characterized by the functions $\rho(\Psi, \vartheta)$, $p(\Psi, \vartheta)$ and $\mathbf{B}(\Psi, \vartheta)$, is presupposed. Hence, the equilibrium part of the calculation (Section 16.3.3), the inversion of the coordinates (Section 17.1.2) and the equilibrium properties (Section 17.1.3) remain in effect, we just discuss a different implementation of the box labeled "STAB" in Fig. 17.1. Introducing the basic state eight-vector

$$\mathbf{u} \equiv (\,\rho_1, \mathbf{v}_1, T_1, \mathbf{A}_1\,)^{\mathrm{T}}\,, \tag{17.112}$$

the system of equations (17.108)–(17.111) may be written in matrix form as

$$\mathbf{L} \cdot \mathbf{u} = \lambda\mathbf{R} \cdot \mathbf{u}\,, \tag{17.113}$$

exploiting the same operators $\mathbf{L}$ and $\mathbf{R}$ as in Eq. (15.67). (Note again the interchange of LHS and RHS!) This is the basic eigenvalue problem of resistive MHD, to be compared with Eq. (17.107) of ideal MHD.

Before we discuss the numerical implementation of this analysis, we observe some simple facts.

(a) The resistive eigenvalue problem (17.108)–(17.111) in terms of state vectors $\mathbf{u}$ with eight rather than three components, implies *a substantial increase of necessary memory size:* a distinct disadvantage of resistive MHD compared to ideal MHD. This is one of the reasons that dissipative spectral computations have become feasible only recently, with the advent of cheap memory in large-scale computing.

(b) Since no special tricks have been used in the derivation of the resistive equations, *the extension with other dissipation mechanisms like viscosity, heat conduction, etc. is easy:* a distinct advantage of the resistive MHD computations.

(c) The eigenvalue λ appears linearly in the resistive eigenvalue problem (17.113), whereas the ideal MHD problem (17.107) has a quadratic eigenvalue ω^2. For a given accuracy of the computed equilibrium and of the eigenvalue solvers exploited, the number of accurate digits obtained in λ with resistive calculations is of the same order as that in ω^2 with ideal calculations. Hence, for a given time scale of interest, it appears that *the resistive calculations are trivially more accurate than the ideal ones!* (A sobering thought after twenty years of intensive numerical research in ideal MHD.)

(d) Likewise, the spatial derivatives of the equilibrium quantities appearing in the dissipative system are typically first order, versus second order in the ideal case: a distinct disadvantage for ideal MHD computations. Again, in this respect as well, *the resistive calculations are trivially more accurate than the ideal ones!*

To sum it up: we appear to lose a powerful tool, but we get a much more powerful numerical tool in return.

Let us now see how it works in practice; we here summarize some of the numerical implementations of these ideas by Kerner [268, 269, 270]. The eigenvalue problem (17.108)–(17.111) is solved by the Galerkin method. *A weak form* of Eq. (17.113) is constructed by multiplying with an arbitrary test function $\mathbf{v}$ and integrating over the domain of interest, i.e. the plasma interior (restricting the analysis to internal modes for the time being):

$$\int \mathbf{v}^{\mathrm{T}} \cdot \mathbf{L} \cdot \mathbf{u} \, dV = \lambda \int \mathbf{v}^{\mathrm{T}} \cdot \mathbf{R} \cdot \mathbf{u} \, dV . \qquad (17.114)$$

$\mathbf{v}$ is a solution of Eq. (17.113) in the weak sense if Eq. (17.114) is satisfied for every test function $\mathbf{v}$ of the appropriate space, i.e. the space of functions that satisfy the pertinent boundary conditions.

The set of test functions $\mathbf{v}$ in the weak form (17.114) is chosen to be the same as used in the discretization of the physical variables $\mathbf{u}$. For both of them, we exploit a finite element representation, extensively discussed in Section 15.1.3, with pairs

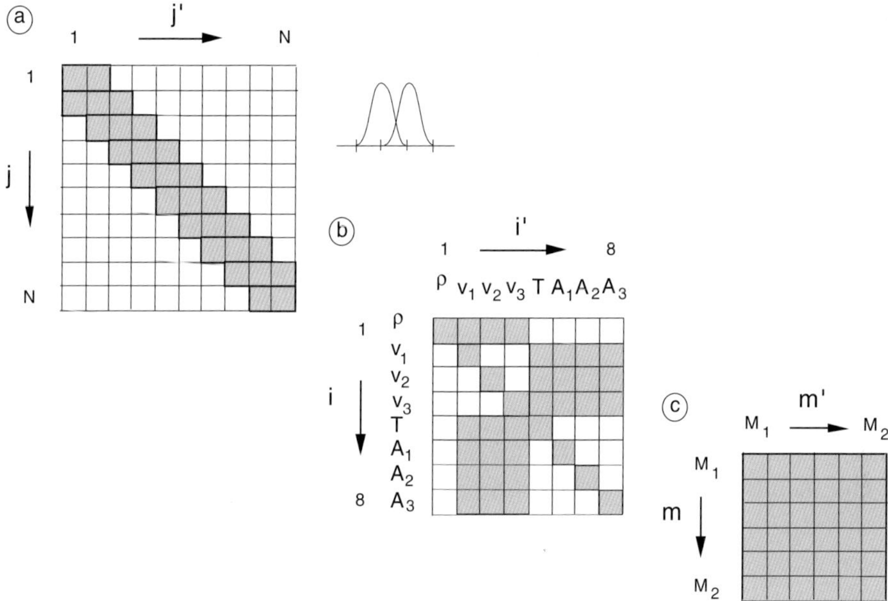

Fig. 17.8 Block structure of the CASTOR eigenvalue matrices: (a) tridiagonal structure due to finite element radial discretization; the finite elements can be accumulated at positions where the modes tend to localize (e.g. at the locations of the ideal MHD singularities); (b) block structure due to cross-products of the physical variables; (c) block structure due to poloidal mode coupling. The blocks of (b) and (c) are essentially full, sparseness is solely due to the finite element discretization in the radial direction. This discretization reflects the difference of wave propagation inside and across magnetic surfaces. On each grid point two kinds of finite element are exploited so that there are $(2 \times 8 \times (M_2 - M_1 + 1))^2 \times 3 \times N$ non-zero complex matrix elements, where N is the number of radial points, and M_1 and M_2 are the lower and upper limit of the range of the poloidal mode number m.

of quadratic and cubic finite elements on each grid point labeled j. For simplicity, we first consider the 1D cylindrical case with radial variations only. Hence, if the ith component of $\mathbf{u}$ is approximated by

$$u^i \simeq \tilde{u}^i = \sum_{j=1}^{2N} x_j^i \, h_j^i(r) \qquad (i = 1, \ldots, 8), \tag{17.115}$$

where N is the number of pairs of finite elements employed (which is the same as the number of radial grid points), the following set of equations is obtained for the coefficients x_j^i:

$$\sum_{i'} \sum_{j'} \left(\int h_j^i \, L^{ii'} h_{j'}^{i'} \, dV \right) \cdot x_{j'}^{i'} = \lambda \sum_{i'} \sum_{j'} \left(\int h_j^i \, R^{ii'} h_{j'}^{i'} \, dV \right) \cdot x_{j'}^{i'}. \tag{17.116}$$

These equations lead to the non-symmetric eigenvalue problem

$$\mathbf{A} \cdot \mathbf{x} = \lambda \, \mathbf{B} \cdot \mathbf{x}, \qquad\qquad (17.117)$$

where $\mathbf{A}$ and $\mathbf{B}$ are large *non-Hermitian matrices* and $\lambda \equiv -i\omega$ is the complex eigenvalue. The ideal MHD spectral problem is contained as a special case ($\eta = 0$), with Hermitian matrices and λ purely real or purely imaginary.

In the toroidal case, except for the radial finite elements, one also needs a discretization of the poloidal variation. Because of poloidal periodicity, it is most effective to exploit fast Fourier transforms for this purpose, so that the volume integrations of Eq. (17.116) also contain angular integrations over products of Fourier harmonics involving the poloidal variation of the equilibrium quantities. The resulting block-tridiagonal structure of the matrix $\mathbf{A}$ is illustrated in Fig. 17.8 for the particular implementation used in the pair of codes HELENA-CASTOR [238, 272] for the calculation of static axi-symmetric equilibria in tokamaks and their ideal or resistive spectra of waves and instabilities. For the computation of external kink modes, extension with an external vacuum region surrounded by a conducting wall, or a boundary of a prescribed shape, has been constructed using the same numerical discretization methods [239]. Those methods have also been used in the succeeding pair of codes FINESSE-PHOENIX [29, 52] for toroidally and poloidally rotating plasmas, that are exploited in the next chapter.

In conclusion, numerical computation starting from the resistive spectral set of equations (17.108)–(17.111) is much more flexible, and even more accurate, than starting from the corresponding ideal MHD spectral equation (17.107). There is also no need to develop separate codes for ideal MHD since it is contained by means of the simple switch $\eta = 0$. This does not imply that ideal MHD stability is now superceded. Actually, quite the opposite: as we will see in the following sections, all ideal MHD instabilities remain present, and even much more pronounced, since accurately computable.

17.3.2 Edge localized modes

Edge-localized modes (ELMs) typically occur in the *H-mode confinement regime* in tokamaks. This operating regime, discovered in ASDEX [471], is characterized by a factor of two improved energy confinement as compared to the L-mode regime, due to the suppression of plasma edge turbulence and large edge gradients of the temperature and the density (and also in the shear flow and the associated radial electric field) [195]. The H-mode may be degraded by ELMs of large magnitude, destroying the desired transport barrier. On the other hand, ELMs of small magnitude may have the effect of providing a control mechanism of the impurity content, so that quasi-stationary state H-mode operation is facilitated.

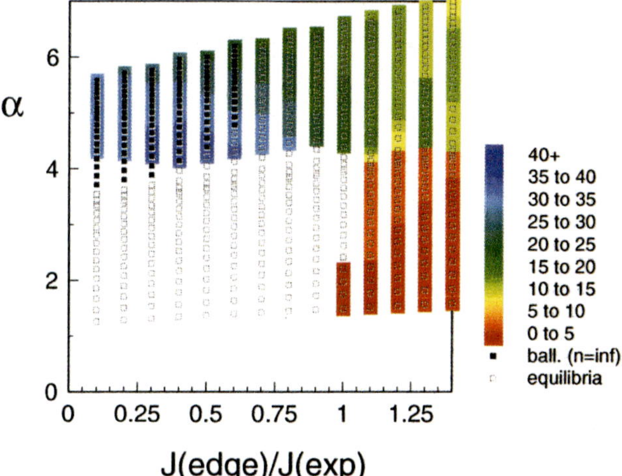

Fig. 17.9 MHD stability limits for toroidal mode number $n = 1$–40 as a function of the edge pressure gradient α and the edge current density, relative to the experimental value, for a low triangularity JET H-mode discharge. Black squares indicate $n = \infty$ ballooning instability, colored squares indicate peeling–ballooning instability with the most unstable n indicated by the color. (From Huysmans [236].)

The traditional, over-simplified, approach to MHD stability of tokamaks is to separately consider the global external kink modes, with toroidal mode number $n = 1$, and the highly localized ballooning modes, with $n = \infty$. However, to properly describe ELMs in the H-mode regime, the full variety of combined modes should be considered. In particular, the large gradients of the pressure and of the current density at the plasma edge give rise to mixed *peeling–ballooning modes with a wide range of intermediate values of* n; see Huysmans [236]. The peeling mode proper is an external kink mode, with extreme localization at the edge, that is driven by a finite current density, or its derivatives, at the edge of the plasma; see Frieman *et al.* [146]. In the intermediate n range, these modes couple to pressure gradient driven ballooning modes, so that the distinction between the two becomes meaningless; hence, the terminology "peeling–ballooning".

The stability diagram for these modes is shown in Fig. 17.9. Whereas $n = \infty$ ballooning modes become unstable if the pressure gradient parameter $\alpha > 3.7$ (black squares) and become stable again when the edge current density becomes larger than 65% of the experimental value, the intermediate n peeling–ballooning modes clearly are unstable for parameters that are experimentally relevant for the onset of ELMs (colored squares). It should be noted, though, that, whereas the maximum pressure gradient observed before the occurrence of ELMs is a rather well known quantity, the edge current density is really dependent on details of the equilibrium reconstruction that are much less known. Furthermore, the shape of the plasma boundary, in particular its triangularity, may significantly enhance stability with respect to these modes.

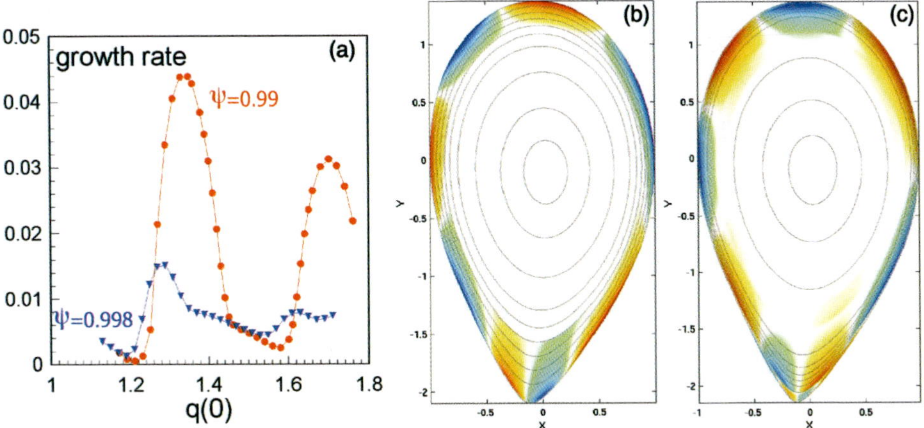

Fig. 17.10 (a) Growth rate of the $n = 1$ peeling–tearing instability for two shapes of the plasma boundary: $\psi_b = 0.99$ (modest triangularity, red dots) and $\psi_b = 0.998$ (high triangularity, blue triangles), resistivity $\eta = 2 \times 10^{-8}$; (b) ideal peeling mode for $\psi_b = 0.99$, $q_0 = 1.45$, $q_b = 4.06$; (c) resistive peeling–tearing mode for $\psi_b = 0.998$, $q_0 = 1.54$, $q_b = 4.4$. (From Huysmans [235].)

Obviously, this does not exhaust the dynamics of different MHD modes in the plasma edge. For example, *low n external kink modes* have been observed during an initially ELM free period. Because these modes are driven by the edge current density, or its gradient, they should be classified as peeling modes as well. Since their instability hinges on the proximity of a rational surface in the vacuum, the plasma boundary shape strongly influences the growth rate of these instabilities; see Fig. 17.10(a). For example, the presence of the X point of a divertor close to the plasma boundary stabilizes the ideal $n = 1$ peeling modes (mode structure shown in Fig. 17.10(b)), whereas it has little influence on resistive $n = 1$ peeling–tearing modes (mode structure shown in Fig. 17.10(c)). Hence, *peeling–tearing instabilities* may be considered as a possible mechanism for the more benign, small magnitude, ELMs.

The parameterization of Fig. 17.10 requires some explanation. In stability codes exploiting straight field line coordinates, like CASTOR, the separatrix region itself cannot be studied because the Jacobian diverges. Instead, the geometry close to an X point is mimicked by considering boundaries parameterized by the flux ψ_b as a fraction of the flux $\psi = 1$ at the X point. Also, because q_b is a rather poor measure for kink modes (q^* would have been better; see Section 16.1.4), the total current

has been parameterized by the safety factor on axis, q_0. One then finds instability
of ideal peeling modes in the range $1.2 < q_0 < 1.45$ (red peak in Fig. 17.10(a)) and
of resistive peeling modes in the range $1.45 < q_0 < 1.54$ (blue curve, indicating
even increased growth rate for higher triangularity). As illustrated in Fig. 17.10(b)
and (c), these two types of mode are similar in the bulk of the plasma, but exhibit a
distinct parity difference in the immediate neighborhood of the X point.

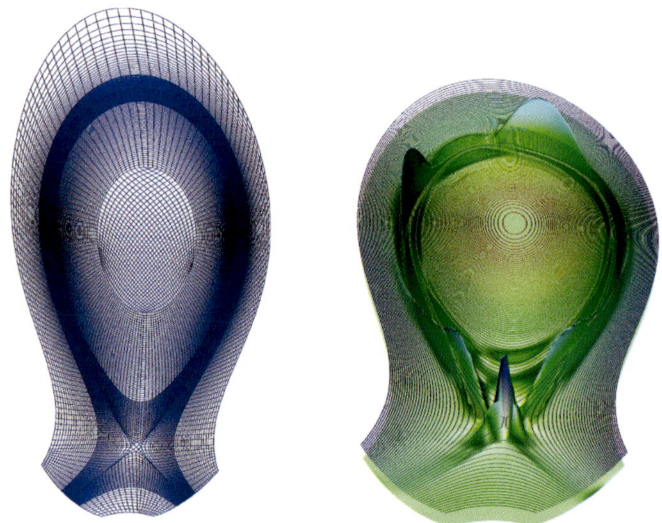

Fig. 17.11 Separatrix geometry and a peeling mode current perturbation computed
with the nonlinear evolution code JOREK. (From Huysmans & Czarny [237].)

The eventual effect of ELMs on plasma confinement has to be studied for mode
amplitudes that have grown into the nonlinear regime. This requires a nonlinear
evolution code, possibly with input of initial data from an equilibrium-spectral
code such as HELENA-CASTOR. A popular model to reduce the complexity of the
nonlinear calculation is Strauss' *reduced MHD model* [425], further developed by
many authors, see e.g. [246], [69]. A direct extension of the above analysis into the
nonlinear regime has been initiated by the development and application of the nonlinear evolution code JOREK by Huysmans and Czarny [237], based on such a reduced resistive MHD model. The equations solved are evolution equations for the
poloidal flux, the poloidal vorticity, the density and the temperature, with a number
of transport coefficients, such as resistivity, viscosity and heat conductivity. Significant is the fact that most of the discretization methods, like two-dimensional finite
elements for the poloidal plane, Fourier transformation for the toroidal direction,
grid refinement techniques, etc. are the same as, or developments of, the methods
used in the spectral code. The time evolution is treated fully implicitly (see Sections 15.4 and 19.4). Most important is the incorporation of the X point geometry

characteristic for divertor tokamaks, like JET and ITER, since ELMs usually consist of an external kink mode component of the peeling mode kind, as discussed above. Consequently, these modes are very sensitively dependent on the separatrix geometry. Fig. 17.11 shows the flux aligned finite element grid used in JOREK and the localized current perturbation of a peeling mode that is ideally stable, but resistively still unstable, in agreement with the linear results [235]. An example of a peeling–ballooning mode computed with this code is shown in Fig. 17.15.

17.3.3 Internal modes

Tearing modes Plasmas with a fixed boundary display *internal modes,* like the tearing modes and resistive interchanges discussed in Sections 14.2.2 and 14.2.3. These modes can be stabilized by the combined effect of good average curvature of the field lines and finite pressure of the plasma [158]. In toroidal geometry, as shown by Glasser *et al.* in [158], the dependence of the growth rate on the resistivity is much more complicated than suggested by the simple tearing scaling (14.84) or the interchange scaling (14.90) of Section 14.2. In particular, the eigenvalues of the resistive modes become complex and exhibit an intriguing pattern of coalescence and splitting in the complex λ plane, as shown in Fig. 17.12. For $\eta = 10^{-6}$, the frequency of one stable mode (imaginary $\lambda \Rightarrow$ real ω) decreases as η decreases and the frequency of a second stable mode increases as η decreases. At $\eta = 2.3 \times 10^{-7}$, the two modes coalesce and split into two overstable modes (complex conjugate in terms of ω) which become purely exponential for $\eta = 4 \times 10^{-8}$. Finally, if the resistivity is still further decreased, the two modes approach the origin with the resistive interchange power $\eta^{1/3}$ of Eq. (14.90).

The influence of the pressure (parameterized by β_{p}) on the growth rate of the internal $m = 2, n = 1$ tearing mode is shown in Fig. 17.13. As is evident from this figure, the pressure has a stabilizing influence on the internal tearing modes, so that an increase of β_{p} results in a decrease of the instability window in the plot of the growth rate versus the safety factor at the edge. In contrast, the pressure has almost no effect on the external, free boundary, tearing modes, which means that those modes are essentially incompressible. At zero pressure the instability window of the external tearing modes is much larger than for the internal modes, and at $\beta_{\mathrm{p}} = 0.26$ the growth rates are four times larger than for the internal, fixed boundary, modes. Thus, the stabilizing influence on the tearing mode, which can be very effective for modes in the plasma center, is not effective for tearing modes at the plasma edge [239].

Infernal modes The long term prospect of steady state operation in tokamaks, facilitated by bootstrap currents (see Wesson [481], Section 4.9), has led to intense

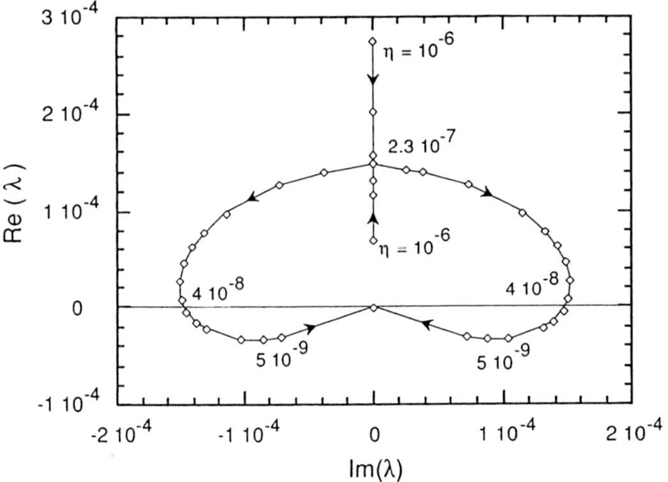

Fig. 17.12 Locus of the eigenvalue λ ($\equiv -i\omega$) of the internal, $m = 2, n = 1$, resistive tearing mode in the complex plane as a function of resistivity; $q_1 = 2.65$, $\beta_{\rm p} = 0.26$. (From Huysmans *et al.* [239].)

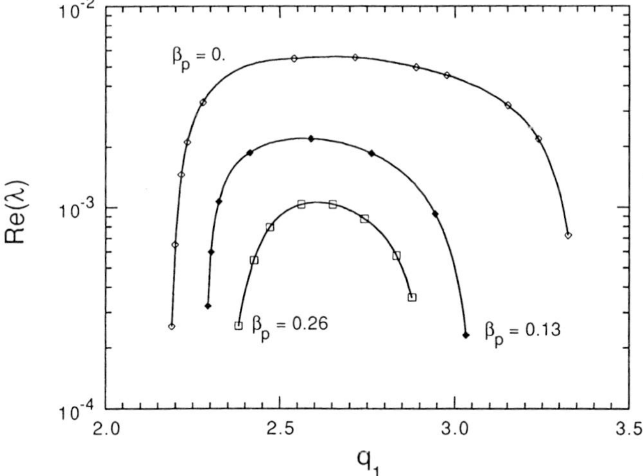

Fig. 17.13 Growth rate versus total current for an internal, $m = 2, n = 1$, resistive tearing mode for three values of the poloidal beta, for fixed resistivity $\eta = 10^{-6}$. (From Huysmans *et al.* [239].)

research of the advanced Tokamak (AT) scenario (see e.g. Freidberg [141], Section 13.7.4). This scenario requires unusual (hollow) current and q-profiles, i.e. negative shear in the plasma center, and, thus, reopened many questions of MHD stability of tokamaks with respect to these new profiles. We discuss one example.

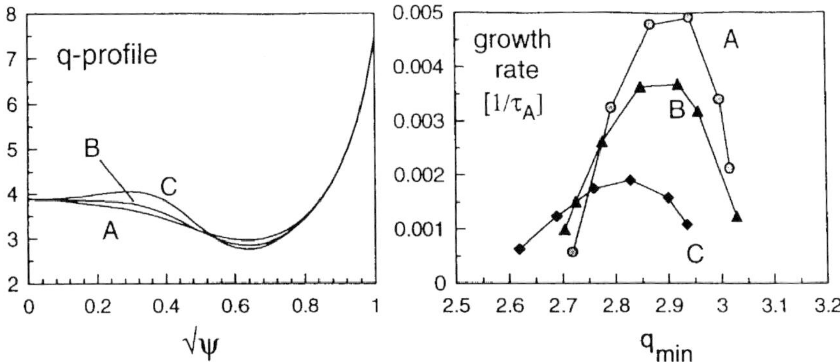

Fig. 17.14 Safety factor profiles and corresponding growth rates of the infernal mode. (From Holties *et al.* [228, 230].)

The standard ballooning theory, discussed in Section 17.2.5, leads to the one-dimensional problem of solving an ODE for the ballooning perturbation, which may be investigated for each magnetic surface separately. This approach is justified for toroidal mode numbers $n \gg 1$, and may be extended to lower values of n, provided that the magnetic shear is large enough (the standard tokamak scenario). In that case, growth rates of ballooning instabilities decrease with decreasing n, so that establishing high-n ballooning stability is sufficient. However, for the low shear region that is unavoidable in the AT scenarios, stability with respect to intermediate values of n is completely unrelated to the standard ballooning conditions. For those low-shear q-profiles, even when the ballooning stability criterion is satisfied, a new class of unstable pressure-driven modes, called *infernal modes* (a play on words), was found by Manickam *et al.* [325]. For those modes, the growth rate is a wildly oscillating function of n with instability in low-n regions sensitively dependent on the value of q.

The infernal modes were investigated for advanced tokamak regimes in JET obtained with pellet injection [228, 230]. The improved confinement turned out to be transient and ended in a collapse due to an MHD instability. The *infernal* mode was considered a likely candidate for this MHD instability since it is driven by the large pressure gradient in the region of negative shear. Increasing the shear around q_{min} reduces the growth rates of the infernal mode but enlarges the instability window as the infernal mode becomes more localized and gives a less deteriorating effect on the plasma (Fig. 17.14). Broadening the pressure profile, i.e. moving the pressure gradient away from q_{min}, has a stabilizing effect. It is possible to completely stabilize the infernal mode by making the pressure gradient equal to zero in a large enough region around q_{min}. One might hope that the effect of the instability is precisely this switching off (therefore called "self healing").

Ballooning modes A complete nonlinear evolution study of medium-n ideal ballooning instabilities was performed by Huysmans and Czarny [237] by means of the JOREK code. The initial, linearly unstable, mode structure is shown in Fig. 17.15. The initial equilibrium has a large pressure gradient (the edge pedestal) inside the separatrix. However, as time proceeds in the computation, the modes more and more stretch out into the "vacuum", so that the distinction between internal and external modes becomes meaningless. In effect, the medium-n ballooning modes develop into "blobs" or density filaments, very similar to the experimentally observed ELMs.

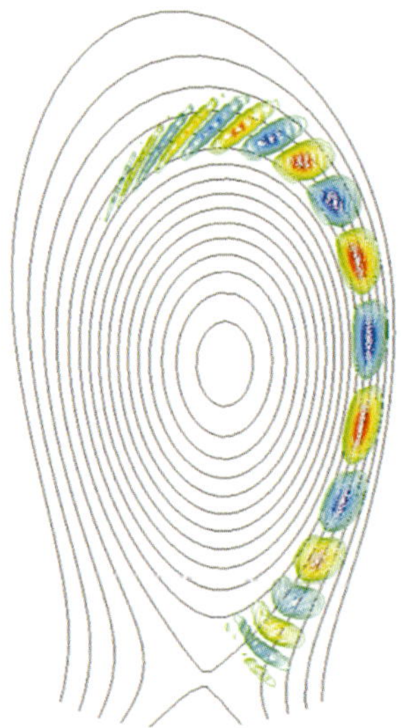

Fig. 17.15 Initially unstable linear mode structure of an $n = 6$ ballooning mode. (From Huysmans & Czarny [237].)

17.3.4 Toroidal Alfvén eigenmodes and MHD spectroscopy

Toroidal Alfvén eigenmodes We have encountered the TAE modes in connection with the poloidal mode coupling of the Alfvén and slow continua in toroidal systems (Section 17.2.4, Fig. 17.3). Due to this coupling, gaps appear in the continua in which the TAEs may be found (see Fig. 17.5). They are naturally excited by energetic particles (neutral beam injection, fusion α-particles) and, in turn, these destabilized TAEs may cause severe losses of α-particles in future ignited plasmas. Hence, a lot of research is devoted to possible damping mechanisms [449].

  TAE modes not only pose a threat to plasma confinement for future fusion machines, but they may also be used for a positive purpose, viz. to diagnose the plasma

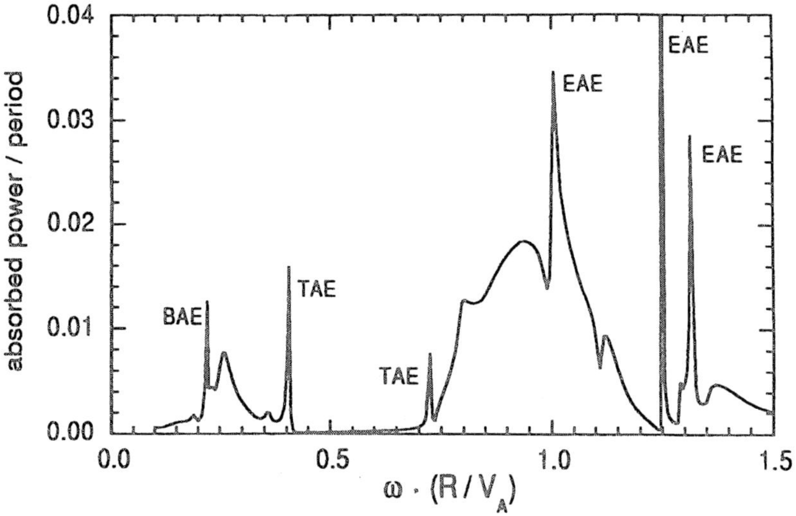

Fig. 17.16 Power absorbed by the plasma as a function of the normalized antenna frequency ($n = 1$, $\eta = 10^{-7}$, $V_{\mathrm{A}}/R = 332\,\mathrm{kHz}$). (From Huysmans *et al.* [240].)

by means of measurements of the magnetic signals they produce. This activity, proposed by Goedbloed *et al.* [165, 180], has been called *MHD spectroscopy*. Recall from our discussion of the example of *helioseismology* in Section 7.2.4 of Volume [1] how agreement between observed Doppler shifts of spectral lines, due to solar oscillations [89], could be brought into agreement with the calculated ones for a standard solar model to within 0.1%! This impressive agreement may serve as an example for what is possible in a purely classical, fluid dynamical, kind of spectroscopy and, hence, in the MHD spectroscopy of plasmas.

External excitation of TAEs At JET, TAEs could be artificially excited by means of an external antenna. To that end, the saddle coils for disruption control were adapted to permit scanning of the driving frequency in the Alfvén frequency range of 30–500 kHz. The frequencies of the TAE modes then showed up as resonances of the power absorbed by the plasma, as measured by the antenna impedance.

The theoretical counterpart of this activity was undertaken by means of a modification of CASTOR. Replacing the spectral problem (17.117) by a representation for external driving,

$$(\mathbf{A} + \mathrm{i}\omega_{\mathrm{d}}\mathbf{B}) \cdot \mathbf{x} = \mathbf{f}\,, \tag{17.118}$$

where ω_{d} is the driving frequency and $\mathbf{f}$ is the driving term, the response of the continuous spectra and other modes could be calculated. The response of the plasma due to an $n = 1$ magnetic field perturbation which would be induced by the saddle coils is shown in Fig. 17.16. On top of the broad continuum shoulders several sharp peaks are found labeled with TAE (toroidicity induced: $\Delta m = 1$ couplings dominate in the eigenfunctions), EAE (ellipticity induced: $\Delta m = 2$ couplings) and BAE (β induced, actually compressibility induced: γp). As may be deduced from the corresponding eigenfunctions, these peaks in the antenna impedance are indicative of the background plasma equilibrium profiles [241, 271]. Observing such peaks and correlating them with the background profiles would be an example of *active MHD spectroscopy*. We will see below how a more effective, passive, way of MHD spectroscopy was advanced later at JET.

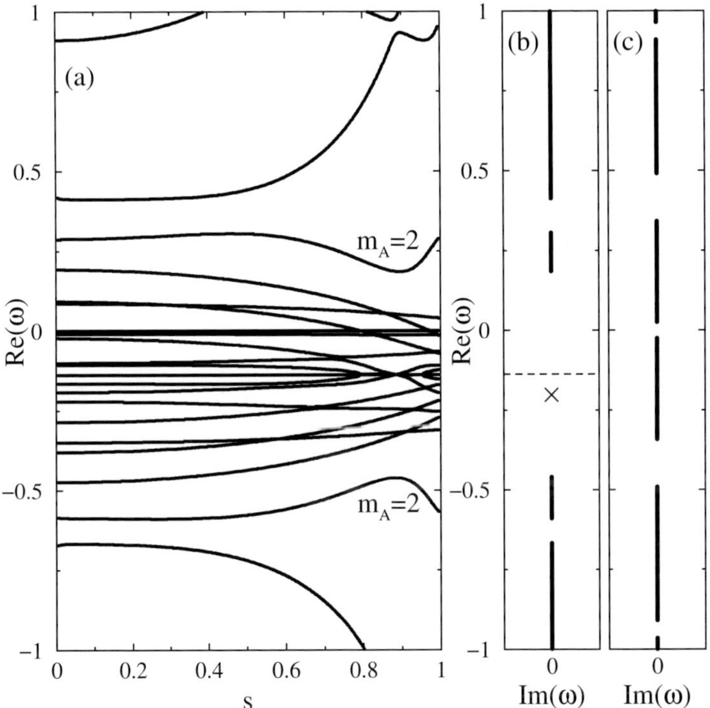

Fig. 17.17 Toroidal flow induced Alfvén eigenmode (TFAE): (a) Alfvén and slow continua plotted against $s = \sqrt{\psi}$; (b) large gap in the Alfvén spectrum with TFAE (indicated by the cross); (c) for the corresponding static equilibrium, the gap is much narrower and no TFAE occurs. (From van der Holst *et al.* [453].)

Toroidal flow induced Alfvén eigenmodes One particular kind of TAE mode, which might become important in MHD spectroscopy because it occurs at very low frequencies (in the slow magneto-sonic range), is the *toroidal flow induced Alfvén eigenmode* (TFAE), found by van der Holst *et al.* [453, 454]. It is illustrated in Fig. 17.17. The mode occurs in the $\Delta m = 0$ gap caused by the coupling of the Alfvén and slow continua, similar to the geodesic coupling described in Section 17.2.3. However, this gap is not caused by geodesic coupling but by centrifugal and Coriolis forces, and it is significantly wider than the static $\Delta m = 0$ gap. Because of the flow, the continua are Doppler shifted and the symmetry with respect to the Doppler shifted frequency $n\Omega$ is broken through the Coriolis effect. In Fig. 17.17(b), the Alfvén part of the spectrum is highlighted by omitting the slow continua. Inside this gap, the TFAE is found with a frequency $\mathrm{Re}(\omega) = -0.202$. For comparison, the gaps for the corresponding static case are shown in Fig. 17.17(c) (omitting two symmetric BAEs in the narrow $\Delta m = 0$ gap). It is evident that toroidal flow changes the low-frequency part of the continuous spectra completely. The gap is determined by a three mode interaction involving a central Alfvén mode and two sideband slow modes.

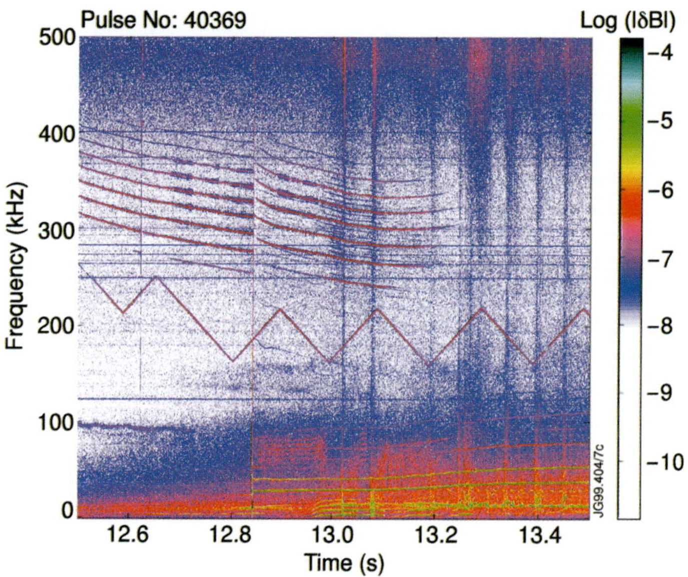

Fig. 17.18 MHD spectroscopy: magnetic perturbations measured at the vessel wall for determination of the toroidal mode numbers of TAEs in JET discharge #40369. (From Sharapov *et al.* [411].)

MHD spectroscopy in rotating plasmas In principle, measurement of the frequencies of TAEs yields information about the safety factor profile $q(\Psi)$ since the TAE frequency is determined by plasma parameters in a narrow range of width associated with the magnetic cross-over surface (Fig. 17.3), as shown in many studies, see e.g. Holties *et al.* [229]. Neutral beam injection in tokamaks causes the plasma to rotate in the toroidal direction and, thus, the TAE frequencies are Doppler shifted

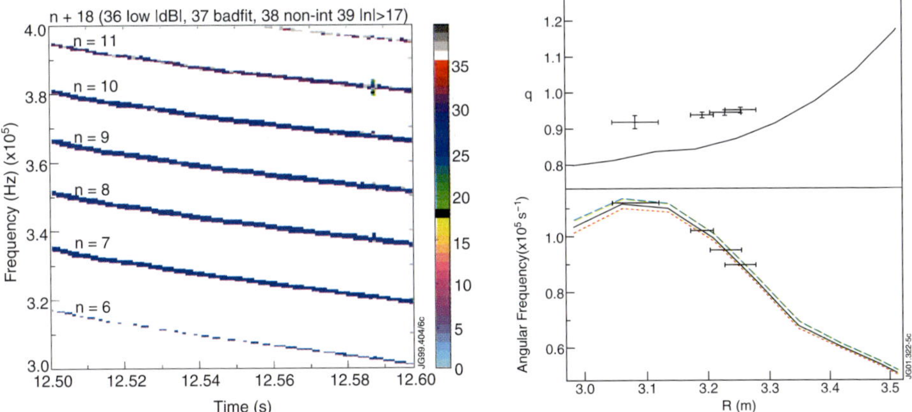

Fig. 17.19 MHD spectroscopy: TAEs observed with magnetic pick-up coils (left) and resulting safety factor and rotation profiles (right) for JET discharge #40369. (From Sharapov *et al.* [411].)

with a factor that depends on the toroidal mode number n times the rotation frequency. Information about the radial profile of the rotation frequency may be obtained from independent (charge-exchange) measurements. From these combined data, the q profile may be obtained, as shown by Sharapov *et al.* [411]. Figure 17.18 shows the measured magnetic field perturbations at the vessel wall in JET for different values of n. The TAEs occur in the frequency range 300–400 kHz, whereas low frequency MHD instabilities occur in the range 0–80 kHz. The profiles of the safety factor and of the rotation profile that were obtained by inversion of these data are shown in Fig. 17.19.

One may also use the unstable Alfvén waves in the lower frequency range, excited by energetic ions, for the purpose of determining the q profile [411]. In tokamaks with a non-monotonic profile of the safety factor, these occur in the form of *Alfvén wave cascades* where the frequency changes upward or downward when the minimum value of the safety factor decreases in time during the discharge. This recently discovered class of modes offers an additional way of identifying the plasma parameters from the dependence on the Alfvén wave spectrum, another example of MHD spectroscopy [411].

At the end of Chapter 16, we noted the lack of accurate information on the profile of the safety factor q, a crucial parameter for the operation of future fusion reactors. This deficiency is presently being addressed by a rapidly growing body of research devoted to the mentioned Alfvén wave cascades, also called *chirping modes*, see e.g. Edlund *et al.* [129], which provide information on the precise location of the $q = 1$ surface in reversed shear profiles (relevant for advanced tokamak scenarios). Clearly, MHD spectroscopy is well underway in tokamaks.

17.4 Literature and exercises

Notes on literature

Basic papers on MHD stability

- 'Zur Stabilität eines Plasmas' by Hain, Lüst & Schlüter [203] and 'An energy principle for hydromagnetic stability problems' by Bernstein, Frieman, Kruskal & Kulsrud [35] are the first (and very unevenly cited) papers on the subject.
- 'Hydromagnetic stability of a plasma' in Volume 2 of *Reviews of Plasma Physics* by Kadomtsev [252] is one of the first overviews of the subject.
- 'Kink instabilities in a high-β tokamak' by Freidberg & Haas [143].
- 'Study of the MHD spectrum of an elliptic plasma column' by Chance, Greene, Grimm & Johnson [82].

Textbook chapters on MHD stability of toroidal plasmas

- *Ideal Magnetohydrodynamics* (Chapter 10) by Freidberg [140], and *Plasma Physics and Fusion Energy* (Chapter 12 and Section 13.7) by Freidberg [141].
- *Theory of Toroidally Confined Plasmas* (Chapter 4) by White [483].
- *Magnetohydrodynamics and Spectral Theory* (Chapter 9) by Lifschitz [308].
- *Plasma Confinement* (Chapter 7) by Hazeltine and Meiss [221].
- *Tokamaks* (Chapters 6 and 7) by Wesson [481].

Ballooning modes

- 'Ballooning effects and plasma stability in tokamaks' in Volume 11 of *Reviews of Plasma Physics* by Pogutse & Yurchenko [382] is a very broad overview of ballooning effects on tokamak stability with many references up to 1980.
- Clear expositions of the ballooning representation may be found in *Ideal Magnetohydrodynamics* (Sections 10.5.3–10.5.5) by Freidberg [140] and *Plasma Confinement* (Sections 7.10–7.12) by Hazeltine and Meiss [221].

Exercises

[17.1] *Mapping for stability**

According to Section 17.1.2, stability analysis preferably should be done in straight field line coordinates. In this exercise, we will investigate how one can invert the coordinates.

- Derive the co- and contravariant metric tensor by making use of the transformation

$$g_{ij} = \frac{\partial \tilde{x}^\mu}{\partial x^i} \frac{\partial \tilde{x}^\nu}{\partial x^j} \tilde{g}_{\mu\nu} \,,$$

where $\tilde{x}^\mu$ and $\tilde{g}_{\mu\nu}$ are the coordinates and the metric tensor of the old coordinate system (R, Z, φ), while x^i are the coordinates of the new system $(\Psi, \vartheta, \varphi)$.
- Express the covariant elements of the magnetic field in the new coordinate system in terms of the old one, by means of

$$B_i = \frac{\partial \tilde{x}^\mu}{\partial x^i} \tilde{B}_\mu \,,$$

and also the other way around.

- Derive expressions for R_Ψ, R_ϑ, Z_Ψ, Z_ϑ, and for the co- and contravariant elements of the metric tensor.

- Derive an expression for the Jacobian $\mathcal{J}$. Is it defined everywhere?

- Now we have derived the basic geometric elements for stability analysis in straight field line coordinates. As an extra, also derive the expressions for the co- and contravariant magnetic field components in terms of the metric elements.

- For toroidal plasmas, the safety factor $q(\Psi)$ is defined as

$$q(\Psi) = \frac{\mathbf{B} \cdot \nabla\varphi}{\mathbf{B} \cdot \nabla\vartheta} \, .$$

Show how that this can be written in terms of the equilibrium quantities derived.

[17.2] *The magnetic axis*

In this chapter, the powerful spectral equation (17.43) has been derived using straight field line coordinates. Special care is required at the magnetic axis. Here, you will see why.

- What kind of properties does the magnetic field have on the magnetic axis?
- What consequences does this have for the metric elements g_{ij} and the Jacobian $\mathcal{J}$?
- What does this all mean for the spectral equation (17.43)?
- How would you solve this problem?

[17.3] *Projections of the displacement field*

In this exercise, you will derive expressions for the projections X, Y, Z of the displacement vector in straight field line coordinates. This is the first step in the derivation of the spectral equation (17.43).

- Derive the expressions for the co- and contravariant components of the vector $\mathbf{n}$. Show that the toroidal contravariant component is equal to zero.

- Do the same for the vectors $\boldsymbol{\pi}$ and $\mathbf{b}$. Show that one contravariant component of each vector vanishes.

- Convert the unit vector pair $\{\boldsymbol{\pi}, \mathbf{b}\}$ into $\{\mathbf{n}, \mathbf{e}_\varphi\}$. Show that the projections X, Y, Z can be written in terms of the physical components $\xi_n, \xi_p, \xi_\varphi$ as

$$X = RB_\mathrm{p}\xi_n , \quad Y = \frac{i}{RB_\mathrm{p}}(B_\varphi\xi_p - B_p\xi_\varphi) , \quad Z = \frac{i}{B^2}(B_p\xi_p + B_\varphi\xi_\varphi) .$$

Also write the inverse relations.

[17.4] *Thin plasma slab*

Similar to Exercise[16.1], you are going to consider a thin plasma slab about the mid-plane. You will show that the spectral equation (17.43) reduces to the spectral equation (9.28)[1]. It is reasonable to assume that all physical quantities only depend on the radius R in this approximation. Furthermore, $R \approx R_0 + r$, $Z \approx a\theta$, where a is small.

- Derive the expression for R_Ψ, R_θ, Z_Ψ and Z_θ.
- Derive the expressions for the elements of the metric tensor g_{ij}.
- Show that the Jacobian reduces to $\mathcal{J} = a/B_Z$.
- Show that the poloidal and toroidal curvature are zero and $1/R$, respectively. Explain why the poloidal curvature is zero.

- Reduce the projections X, Y, Z of the displacement field for this case.
- Show that the matrix elements $\mathcal{A}_{ij}$ and $\mathcal{B}_{ij}$ reduce to the ones of Eq. (9.28) of Volume [1], up to a factor. Make use of $k = m/a$.

[17.5] *Small inverse aspect ratio*

In the previous exercise, you have investigated a thin plasma slab about the mid-plane. Now, you do the same for a plasma with small inverse aspect ratio, $\epsilon \ll 1$. As stated in Exercise [16.2], assume that the outer plasma boundary is circular up to first order. Under these two conditions, the magnetic flux function Ψ can be represented as $\Psi(\hat{r}, \hat{\theta}) = \Psi(\hat{r})$. Here, $(\hat{r}, \hat{\theta})$ are non-orthogonal polar coordinates defined with respect to the center of the flux surfaces. The relation with the cylindrical coordinates is the following:

$$ R = R_0 + \hat{r} \cos \hat{\theta}, \quad Z = \hat{r} \sin \hat{\theta} . $$

Furthermore, the relation between the angle $\hat{\theta}$ and the straight field line angle ϑ is $\hat{\theta} = \vartheta$.

- Derive the expressions for R_Ψ, R_ϑ, Z_Ψ, Z_ϑ and the elements of the metric tensor $\hat{g}_{ij}$.
- Show that the Jacobian reduces to $J = \hat{r}/B_p$.
- Show that the poloidal and toroidal curvature are $1/\hat{r}$ and zero, respectively. Explain.
- Derive the expressions for the projections X, Y, Z in these coordinates.
- Derive the matrix elements $\mathcal{A}_{ij}$ and $\mathcal{B}_{ij}$ in these coordinates. When it is not really necessary do NOT replace R with the expression above.
- Show that these matrix elements are the same as the ones of Eq. (9.28) [1] using that $R_0 \gg \hat{r}$ and defining $k \equiv n/R_0$.

[17.6] *Alfvén and slow continuum modes*

In Chapter 12, a new method was developed to determine the eigenvalues of the Frieman–Rotenberg equation, which was applied in Chapter 13 to a 1D equilibrium. Here, you will apply the same ideas to the Alfvén and slow continuum equations.

- Show that

$$ \oint \mathcal{J} v^*(Fw) \, d\vartheta = \oint \mathcal{J}(Fv)^* w \, d\vartheta - \mathrm{i}\, [v^* w]_0^{2\pi} , $$

$$ \oint \mathcal{J} v^*(F\alpha F w) \, d\vartheta = \oint \mathcal{J}(Fv)^* \alpha(Fw) \, d\vartheta - \mathrm{i}\, [v^* \alpha F w]_0^{2\pi} , $$

where v and w are complex functions depending on the straight field line angle ϑ, and $\alpha = \alpha(\Psi_0, \vartheta)$ is a real function. The latter function only depends on the equilibrium quantities. For what kind of function does the term in square brackets vanish?

- Now we define an inner product as follows:

$$ \langle \mathbf{v}^*, \mathbf{w} \rangle \equiv \oint \mathbf{v}^* \cdot \mathbf{w} \mathcal{J} d\vartheta . $$

Substitute the vector $(\eta, \zeta)^{\mathrm{T}}$ for $\mathbf{v}$ and $\mathbf{w}$, and show that the spectral equation for the MHD continua can be written as

$$ \omega^2 = \beta(\gamma_1 - \mathrm{i}\gamma_2) . $$

Derive expressions for β, γ_1 and γ_2. What are their properties?

- What do you know of the eigenfrequency ω?

18

Linear dynamics of stationary toroidal plasmas*

18.1 Transonic toroidal plasmas

Waves and instabilities of transonically rotating axi-symmetric plasmas is a highly complex problem that is of interest for the two unrelated fields of laboratory plasma confinement, aimed at eventual thermonuclear energy production, and the dynamics of a vast number of astrophysical plasmas rotating about compact objects, broadly indicated as accretion disks. The complexity comes from the *transonic transitions of the poloidal flow* which causes the character of the rotating equilibrium states to change dramatically, from elliptic to hyperbolic or vice versa, when the poloidal velocity surpasses certain critical speeds. Associated with these transitions the different types of magnetohydrodynamic shocks may appear (see Chapter 20). Obviously, at such transitions the possible waves and instabilities of the system also change dramatically. We here describe these changes for the two mentioned classes of physical systems, starting from the point of view that the continuous spectrum of ideal MHD presents the best organizing principle for the structure of the complete spectrum of waves and instabilities since it is the most robust part of it. It provides the simplest approach to local waves and instabilities of the system and, possibly, to the onset of MHD turbulence.

The equilibrium problem of translation symmetric and axi-symmetric plasmas was formulated by Goedbloed and Lifschitz [181] in terms of three generic functions (see Section 18.2) that permit analysis of the different singularities and the resolution of the concomitant discontinuities that occur in transonic MHD flows. In the present chapter, we exploit that equilibrium formulation and summarize whatever expressions are needed to perform a spectral analysis of the waves and instabilities of these plasmas.

A cartoon of the physical system is sketched in Fig. 18.1, which shows the compact central object generating the gravitational acceleration $\mathbf{g} = -\nabla\Phi_{\mathrm{gr}}$ (absent for the tokamak case), and a toroidal plasma with poloidal (indicated by the subscript p) and toroidal (subscript φ) velocity and magnetic field components. For the purpose of the present study, treating rotating laboratory and astrophysical plasmas on an equal footing, two important restrictions have to be made.

(a) We neglect the dynamics external to the toroidal plasma and assume the outer plasma boundary to be fixed in space. For tokamaks, this condition is evidently justified by the presence of a conducting wall and current-carrying coils fixed to the laboratory which absorbs the mechanical forces. For accretion disks, this may be justified by considering the toroidal plasma to be embedded in a gas which toroidally rotates at approximately the Kepler velocity. This assumption amounts to the neglect of the accretion flow itself, i.e. we assume that the external accretion process proceeds on a much slower time scale than the internal plasma rotations. In this manner, we concentrate our study on the *internal* plasma dynamics as influenced by the toroidal and poloidal rotations and the gravitational field.

(b) We assume the PDEs describing the background flow to be *elliptic*. To our knowledge, this assumption has always been made, explicitly or implicitly, in wave and stability studies of axi-symmetric systems. In fact, this appears to be basic to the classical paradigm of splitting the study of the dynamics of a system in a time-independent equilibrium and the linear perturbations of it [168]. Certainly, all numerical programs for studying linear stability of tokamaks start from an equilibrium with nested flux surfaces, where prescription of the shape of the outer boundary is sufficient to determine the solutions. In this chapter, using the numerical programs FINESSE [29] and PHOENIX [173, 52] which exploit algorithmic techniques explained in Chapters 15–17, we will stay with this paradigm. Entering the hyperbolic flow regime appears to require a full nonlinear evolution study, without making the split in equilibrium and perturbations. Such studies, exploiting the Versatile Advection Code (VAC) [439], are also carried out at present [257, 80]. However, here we push the spectral approach as far as possible by studying the linear dynamics in the elliptic flow regions while approaching the boundaries of transition to hyperbolic flow. We will see that this method produces convincing evidence of qualitatively new linear dynamics caused by the transonic transitions.

In current astrophysical terminology, configurations like that of Fig. 18.1 are called "thick accretion disks" or "accretion tori" [361, 139]. However, whereas the cited authors are concerned with accretion disks that are thick because of pressure effects, so that the thermal speed approaches the toroidal rotation velocity ($v_{\text{th}} \sim v_\varphi$), we are here concerned with *magnetically dominated accretion tori*, where the magnetic pressure causes the accretion disk to be thick ($v_{\text{th}} \sim \sqrt{\beta}\,v_{\text{A}} \ll v_\varphi \ll v_{\text{A}}$). As opposed to the extreme high-β regime where the magneto-rotational instability operates [18, 19] (see discussion for 1D disk equilibria in Secton 13.4), in this regime, where magnetic fields dominate ($\beta \ll 1$), anomalous dissipation mechanisms due to local MHD instabilities have not been investigated. Such dissipation processes are needed to set up turbulence that would cause the co-moving condition of plasma and magnetic field to be broken so that jets could emerge from

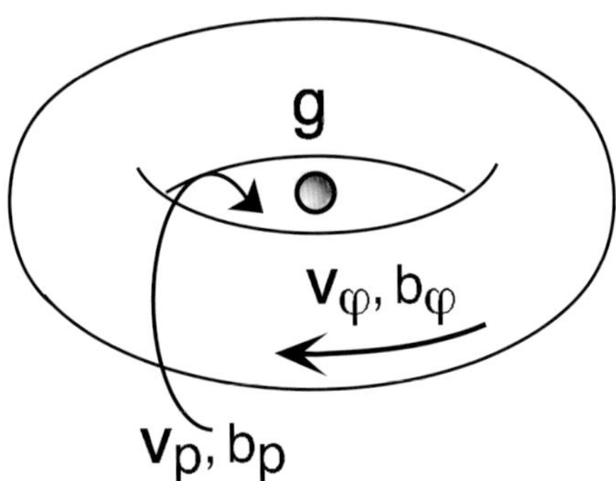

Fig. 18.1 Magnetized plasma rotating about a compact object (forming a thick accretion disk) or inside a tokamak (where $g = 0$). Lower case b indicates the Alfvén speed in the direction of the magnetic field.

the disk (discussed in more detail in Sections 20.4 and 21.3). We analyze the high-β regime from a fundamental point of view, i.e. we formulate the equations for tokamaks and astrophysical objects on the same theoretical footing and investigate their similarities and differences with respect to local modes of the continuous spectrum.

Whereas transonic poloidal flow substantially upsets "intuition" obtained from static tokamak equilibrium, one extremely important feature remains intact: the poloidal velocity and magnetic field are parallel so that *flow surfaces and magnetic surfaces coincide!* This feature alone justifies the study of modes of the continuous spectrum since these modes are preferentially localized on these surfaces.

On first reading, the reader may want to jump to the concluding Section 18.4.3.

18.2 Axi-symmetric equilibrium of transonic stationary states⋆

18.2.1 General equations and toroidal rescalings⋆

(a) General equations We start from the MHD equations for stationary equilibrium, Eqs. (12.24)–(12.27), and exploit a right-handed system of cylinder coordinates R, Z, φ, where R is the distance to the symmetry axis (see Appendix A.2.4). For stationary axi-symmetric equilibrium flows, the poloidal components of the magnetic field and of the velocity may be expressed in terms of the poloidal magnetic flux function Ψ and the poloidal velocity stream function χ :

$$\mathbf{B} = \frac{1}{R}\,\mathbf{e}_\varphi \times \nabla\Psi + B_\varphi \mathbf{e}_\varphi\,, \tag{18.1}$$

$$\mathbf{v} = \frac{1}{\rho R}\,\mathbf{e}_\varphi \times \nabla\chi + v_\varphi \mathbf{e}_\varphi\,. \tag{18.2}$$

The six physical variables ρ, p, B_{p}, B_φ, v_{p} and v_φ, which are 2D functions of R and Z, are then determined by the unknown flux function $\Psi(R, Z)$ and five arbitrary, but considered to be known, 1D functions of Ψ alone [486, 210]:

$$\chi'(\Psi) \equiv \rho v_{\mathrm{p}}/B_{\mathrm{p}}\,, \tag{18.3}$$

$$H(\Psi) \equiv \tfrac{1}{2}v_{\mathrm{p}}^2 \frac{B^2}{B_{\mathrm{p}}^2} + \frac{\gamma}{\gamma-1}\frac{p}{\rho} - \tfrac{1}{2}R^2\Omega^2 + \Phi_{\mathrm{gr}}\,, \quad \Phi_{\mathrm{gr}} = -\frac{GM_*}{\sqrt{R^2+Z^2}}\,, \tag{18.4}$$

$$S(\Psi) \equiv \rho^{-\gamma}p\,, \tag{18.5}$$

$$K(\Psi) \equiv R[v_\varphi - (1/\chi')B_\varphi]\,, \tag{18.6}$$

$$\Omega(\Psi) \equiv \frac{1}{R}[v_\varphi - (\chi'/\rho)\,B_\varphi]\,. \tag{18.7}$$

Here, $\chi' \equiv d\chi/d\Psi$ is the derivative of the poloidal velocity stream function, H is the Bernoulli function, S is the entropy, K is the *poloidal vorticity–current density stream function* (see below), $\Omega \equiv -\Phi'_{\mathrm{el}}$ is minus the derivative of the electric potential and M_* is the mass of the central object. For the final determination of the physical variables, GM_* is to be considered as an additional (constant) flux function. As we will see, judicious exploitation of the intricate relationship between the 2D physical variables and the 1D flux functions is the key to a proper analysis of equilibrium and stability of axi-symmetric transonic flows.

▷ **Poloidal vorticity–current density stream function** We propose to abandon the misnomer "specific angular momentum" that has been used to indicate the flux function K by quite a number of authors [475, 49, 386, 317, 224, 452, 55], including ourselves [256]. The point is that the angular momentum density vector $\mathbf{L} \equiv \mathbf{r} \times \rho\mathbf{v}$ (or the specific angular momentum when divided by ρ) has no magnetic contributions. Such contributions enter the general expression for the angular momentum of relativistic matter and field, but they are negligible in non-relativistic MHD; see Jackson [247], p. 288 (2nd edition, p. 264). Also, in two-fluid plasma theory [169], the canonical angular momenta of the separate electron and ion fluids obtain a contribution of the toroidal component of the vector potential, i.e. of the poloidal magnetic flux. However, in one-fluid MHD, these contributions add up to the angular momentum ($Rv_\varphi \neq K$) where the magnetic contributions cancel out because of quasi charge-neutrality. The magnetic field does enter in a relevant manner in the completely different physical quantity of the angular momentum *flux* tensor $\mathbf{M} \equiv -\mathbf{T} \times \mathbf{r}$, where $\mathbf{T} \equiv \rho\mathbf{vv} + (p + \tfrac{1}{2}B^2)\mathbf{I} - \mathbf{BB}$ is the stress tensor. In equilibrium, omitting the gravitational term, the torque $\mathbf{N} \equiv \mathbf{r} \times \mathbf{F} = \partial\mathbf{L}/\partial t = -\nabla \cdot \mathbf{M}$ vanishes, where the vertical component yields

$$N_Z = -\frac{1}{R}\frac{\partial}{\partial R}(RM_{RZ}) - \frac{\partial M_{ZZ}}{\partial Z}$$

$$= -\frac{1}{R}\frac{\partial}{\partial R}\left(\frac{\partial\chi}{\partial Z}K\right) + \frac{1}{R}\frac{\partial}{\partial Z}\left(\frac{\partial\chi}{\partial R}K\right) = \mathbf{v}\cdot\nabla K = \chi'\mathbf{B}\cdot\nabla K = 0, \tag{18.8}$$

showing that K must be a flux function and that there is a relation with two components of the angular momentum flux tensor. However, this is just the toroidal force balance, $F_\varphi = -(\nabla \cdot \mathbf{T})_\varphi = -(1/R)\mathbf{v} \cdot \nabla K = 0$, which also exhibits a relation with two components of a tensor, viz. the stress tensor. All this does not reveal the essential physical connection with vorticity and current density, which K expresses. The poloidal components of the vorticity $\mathbf{w} \equiv \nabla \times \mathbf{v}$ and the current density $\mathbf{j} \equiv \nabla \times \mathbf{B}$ may be derived from stream functions V and I,

$$\mathbf{w}_\mathrm{p} = -\frac{1}{R}\mathbf{e}_\varphi \times \nabla V \,, \quad V \equiv Rv_\varphi \left[= \frac{(\chi'^2/\rho)K - R^2\Omega}{\chi'^2/\rho - 1} \right],$$

$$\mathbf{j}_\mathrm{p} = -\frac{1}{R}\mathbf{e}_\varphi \times \nabla I \,, \quad I \equiv RB_\varphi \left[= \frac{\chi'(K - R^2\Omega)}{\chi'^2/\rho - 1} \right], \tag{18.9}$$

which, separately, are *not* flux functions, but the combination

$$K \equiv V - I/\chi' \quad \left(\Rightarrow \ K\nabla\chi = V\nabla\chi - I\nabla\Psi \right) \tag{18.10}$$

is a flux function. This function exhibits the proper limits to both hydrodynamics (HD), where $I = 0$, and static MHD, where $V = 0$ and angular momentum is not a relevant concept. Hence, the new name: *poloidal vorticity–current density stream function.*　　$\triangleleft$

The transonic poloidal flow field is best characterized by the value of the square of the poloidal Alfvén Mach number:

$$M^2 \equiv \frac{\rho v_\mathrm{p}^2}{B_\mathrm{p}^2} = \frac{\chi'^2}{\rho} \,, \tag{18.11}$$

which is another unknown function, with different dependence on the poloidal coordinates R and Z than Ψ has. Exploiting this quantity, the relationship between the plasma velocity and the vectorial Alfvén velocity $\mathbf{b} \equiv \mathbf{B}/\sqrt{\rho}$ may be written as

$$\mathbf{v} = M\mathbf{b} + R\Omega\mathbf{e}_\varphi \,, \tag{18.12}$$

so that $\mathbf{v}_\mathrm{p} = M\mathbf{b}_\mathrm{p}$, but the toroidal velocity consists of a Mach number component Mb_φ and a supplementary component $R\Omega$ (Fig. 18.2). Hence, the physical interpretation of the quantity $R\Omega$ is quite different here from both HD and MHD in the absence of poloidal rotation (where it represents the complete toroidal rotation velocity): $R\Omega$ is the toroidal velocity of the rotating frame in which the plasma velocity becomes parallel to the Alfvén speed.

We now scale the flux functions H, S, K, Ω and GM_* by means of χ' to obtain five *scaled flux functions* $\Lambda_i(\Psi)$:

$$\Lambda_1 \equiv \chi'^2 H \,, \quad \Lambda_2 \equiv \frac{\gamma}{\gamma - 1}\left(\chi'^2\right)^\gamma S \,,$$

$$\Lambda_3 \equiv \frac{1}{\sqrt{2}}\chi' K \,, \quad \Lambda_4 \equiv \frac{1}{\sqrt{2}}\chi'\Omega \,, \quad \Lambda_5 \equiv GM_*\chi'^2 \,, \tag{18.13}$$

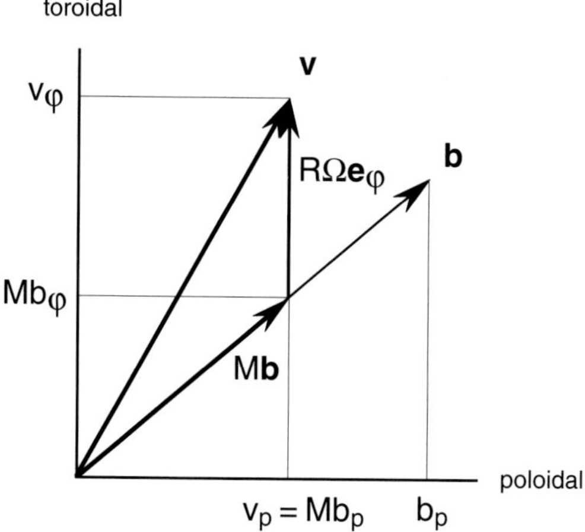

Fig. 18.2 Representation of the plasma velocity relative to the vectorial Alfvén speed $\mathbf{b} \equiv \mathbf{B}/\sqrt{\rho}$. The poloidal velocity is measured in terms of the poloidal Alfvén Mach number: $v_{\mathrm{p}} = Mb_{\mathrm{p}}$, but the toroidal velocity has an additional contribution: $v_\varphi = Mb_\varphi + R\Omega$.

where only the first four are needed to describe the equilibrium of laboratory plasmas. Prescribing the (completely arbitrary) functions $\Lambda_i(\Psi)$, the only remaining constraints on the physical variables come from the two poloidal components of the momentum equation. They may be cast in the form of a variational principle,

$$\delta \int \left[\frac{1}{2R^2}(1 - M^2)\,|\nabla\Psi|^2 - \frac{\Pi_1}{M^2} + \frac{\Pi_2}{\gamma M^{2\gamma}} + \frac{\Pi_3}{M^2 - 1} \right] dV = 0\,, \qquad (18.14)$$

giving as Euler–Lagrange equations a nonlinear PDE for the poloidal flux $\Psi(R, Z)$:

$$\nabla \cdot \left(\frac{1 - M^2}{R^2} \nabla\Psi \right) + \frac{1}{M^2}\frac{\partial\Pi_1}{\partial\Psi} - \frac{1}{\gamma M^{2\gamma}}\frac{\partial\Pi_2}{\partial\Psi} - \frac{1}{M^2 - 1}\frac{\partial\Pi_3}{\partial\Psi} = 0\,, \qquad (18.15)$$

and an algebraic Bernoulli equation for the squared poloidal Alfvén Mach number $M^2(R, Z)$:

$$\frac{1}{2R^2}|\nabla\Psi|^2 - \frac{\Pi_1}{M^4} + \frac{\Pi_2}{M^{2(\gamma+1)}} + \frac{\Pi_3}{(M^2 - 1)^2} = 0\,. \qquad (18.16)$$

Here, three generic functions Π_j appear that depend on the five scaled flux functions and on the coordinates as follows:

$$\Pi_1(\Psi; R, Z) \equiv \Lambda_1 + R^2\Lambda_4^2 + \frac{\Lambda_5}{\sqrt{R^2 + Z^2}} = M^2 \left(\tfrac{1}{2}M^2 B^2 + \frac{\gamma}{\gamma - 1}p \right),$$

$$\Pi_2(\Psi) \qquad \equiv \Lambda_2 \qquad\qquad\qquad = \frac{\gamma}{\gamma - 1}M^{2\gamma}p\,,$$

$$\Pi_3(\Psi; R) \quad \equiv (R^{-1}\Lambda_3 - R\Lambda_4)^2 \qquad = \tfrac{1}{2}(M^2 - 1)^2\, B_\varphi^2 \,. \qquad (18.17)$$

The utmost right expressions show the relationship with the physical variables. The equations (18.15) and (18.16) determine the spatial dependence of $B_{\rm p}(R, Z)$, and of the five remaining physical variables through the five flux functions. Hence, solution of these two equations constitutes the core of the equilibrium problem. The rest is trivial algebra, except for the choice of meaningful magnitudes for the free functions. This is our next topic.

(b) Toroidal rescalings We introduce a reference point $R = R_0$, $Z = 0$ somewhere in the plasma, and exploit the plasma half-width a and the magnetic field B_0 at the reference point as units of length and magnetic field strength. Dimensionless coordinates with respect to the reference point are then given by

$$x \equiv (R - R_0)/a\,, \quad y \equiv Z/a\,, \qquad (18.18)$$

$$\Rightarrow \ R = R_0(1 + \epsilon x)\,, \quad Z = ay\,, \qquad \epsilon \equiv a/R_0\,, \qquad (18.19)$$

where ϵ may be used for small inverse aspect ratio expansions ($\epsilon \ll 1$). For the time being, we will not make this assumption. However, it is clear that the dependence on this geometric factor has a big influence on what the magnitudes of the five flux functions physically represent. Since there is hardly any observational information on those functions, it is expedient to separate their influence as much as possible from that of the geometric factors that occur. This is done similar to the poloidal flux scaling of Section 16.3.1.

(1) All quantities are made dimensionless by putting $a = 1$ and $B_0 = 1$:

$$a\nabla \ \to \ \nabla \equiv \mathbf{e}_x\partial_x + \mathbf{e}_y\partial_y\,, \quad (a^2 B_0)^{-1}\psi \ \to \ \psi\,, \quad B_0^{-2}\Pi_j \ \to \ \Pi_j \quad (j = 1, 2, 3)\,, \qquad (18.20)$$

i.e. we indicate dimensionless variables by the same symbol as the original dimensional variables. This trivial operation is well to be distinguished from the non-trivial rescalings with ϵ and the total dimensionless flux Ψ_1, to be discussed now. Here (in contrast to Chapter 16), we fix the reference point to be the (unknown) location $x = \delta$, $y = 0$ of the magnetic axis, where δ is the shift relative to the geometric center of the plasma, and assume the plasma shape to be up–down symmetric. Since $\Psi = \Psi_0 = 0$ at this reference point, we obtain a dimensionless flux variable of unit range by normalizing with respect to the flux Ψ_1 at the plasma boundary:

$$\psi \equiv \Psi/\Psi_1\,. \qquad (18.21)$$

Next, we rescale the generic functions and the flux functions with respect to ϵ and Ψ_1:

$$\bar{\Pi}_j \equiv (\epsilon\Psi_1)^{-2}\Pi_j \ (j = 1, 2, 3)\,, \qquad \bar{\Lambda}_{1,2} \equiv (\epsilon\Psi_1)^{-2}\Lambda_{1,2}\,,$$

$$\bar{\Lambda}_3 \equiv \Psi_1^{-1}\Lambda_3\,, \quad \bar{\Lambda}_4 \equiv (\epsilon^2\Psi_1)^{-1}\Lambda_4\,, \quad \bar{\Lambda}_5 \equiv (\epsilon\Psi_1^2)^{-1}\Lambda_5\,. \qquad (18.22)$$

In effect, Eqs. (18.15) and (18.16) transform to the following dimensionless representation of the core equations for $\psi(x, y)$ and $M^2(x, y)$:

$$\nabla \cdot \left[\frac{1 - M^2}{(1 + \epsilon x)^2} \nabla \psi \right] + \frac{1}{M^2} \frac{\partial \bar{\Pi}_1}{\partial \psi} - \frac{1}{\gamma M^{2\gamma}} \frac{\partial \bar{\Pi}_2}{\partial \psi} - \frac{1}{M^2 - 1} \frac{\partial \bar{\Pi}_3}{\partial \psi} = 0, \quad (18.23)$$

$$\frac{1}{2(1 + \epsilon x)^2} |\nabla \psi|^2 - \frac{\bar{\Pi}_1}{M^4} + \frac{\bar{\Pi}_2}{M^{2(\gamma+1)}} + \frac{\bar{\Pi}_3}{(M^2 - 1)^2} = 0, \quad (18.24)$$

where the relationship of the rescaled generic functions and flux functions will be given below, after a further rescaling. In this manner, the arbitrary shapes of the five rescaled flux functions $\bar{\Lambda}_i(\psi)$ become well-defined, the parameter ϵ only describes the (usually small) effects of toroidicity, and the value of Ψ_1 *appears* to have vanished from the equilibrium core equations. (This looks analogous to the similar scaling with respect to Ψ_1 developed for static equilibria [164], as explained in Section 16.3.1. There, by means of that scaling, the value q_0 ($\sim \Psi_1^{-1}$) of the safety factor on axis could be fixed after solving the rescaled Grad–Shafranov equation for ψ. However, in the present case, a particular choice of the rescaled flux functions indirectly also fixes the value of Ψ_1 through the Bernoulli equation since the magnetic field on the magnetic axis is normalized to unity, $B_\varphi(x = \delta, y = 0) \equiv B_0 = 1$. Hence, counter to appearances, the rescaled equilibrium equations are not independent of Ψ_1 now. See the paragraph below in small print for how this may be used for an optimal parameterization of the flux functions.)

(2) The problem (18.23)–(18.24) still suffers from cancellations of large factors associated with the gravitational potential and the toroidal rotation. This may be cured by extracting the separate horizontal (x) and vertical (y) dependencies of those factors to facilitate maximal orderings for *thin accretion disks* (both effects appearing in leading order), *thick accretion disks* (vertical gravity negligible), and *tokamaks* (horizontal gravity absent as well). This is effected by redefining the first flux function,

$$\bar{\Lambda}_1^* \left(\equiv (\epsilon \Psi_1)^{-2} \Lambda_1^* \equiv (\epsilon \Psi_1)^{-2} \chi'^2 H^* \right) \equiv \bar{\Lambda}_1 + \bar{\Lambda}_4^2 + \bar{\Lambda}_5, \quad (18.25)$$

i.e. by introducing a *modified Bernoulli function* H^* that incorporates the large leading order rotational and gravitational flux function contributions, and that may be assumed to be positive definite:

$$H^* \equiv H + \tfrac{1}{2} R_0^2 \Omega^2 + R_0^{-1} G M_*$$

$$= \tfrac{1}{2} v_{\rm p}^2 \frac{B^2}{B_{\rm p}^2} + \frac{\gamma}{\gamma - 1} \frac{p}{\rho} + \tfrac{1}{2}(R_0^2 - R^2)\Omega^2 + \left(\frac{1}{R_0} - \frac{1}{\sqrt{R^2 + Z^2}} \right) G M_*.$$
$$(18.26)$$

The rescaled generic functions then become

$$\bar{\Pi}_1(\psi; x, y) \equiv \bar{\Lambda}_1^* + \epsilon e \left(2\bar{\Lambda}_4^2 - \bar{\Lambda}_5 \right) + \epsilon^2 f \bar{\Lambda}_5,$$

$$\bar{\Pi}_2(\psi) \quad\equiv \bar{\Lambda}_2,$$

$$\bar{\Pi}_3(\psi; x) \quad\equiv \left[\bar{\Lambda}_3 - \bar{\Lambda}_4 - \epsilon \left(g\bar{\Lambda}_3 + h\bar{\Lambda}_4 \right) \right]^2, \quad (18.27)$$

where four geometry functions appear,

$$e(x) \equiv x(1 + \tfrac{1}{2}\epsilon x) \approx x\,,$$

$$f(x,y) \equiv \epsilon^{-2}\left\{\left[(1+\epsilon x)^2 + \epsilon^2 y^2\right]^{-1/2} - 1 + \epsilon x + \tfrac{1}{2}\epsilon^2 x^2\right\} \approx \tfrac{1}{2}(3x^2 - y^2)\,,$$

$$g(x) \equiv x(1+\epsilon x)^{-1} \approx x\,,$$

$$h(x) \equiv x = x\,, \tag{18.28}$$

that have order unit magnitude as shown by the approximations (valid for $\epsilon \ll 1$) on the right. In the limit of vanishing toroidicity ($\epsilon \to 0$), the three generic functions become flux functions and the previous translation symmetric equations of Goedbloed and Lifschitz [181] are recovered. This completes the separation of flux functions and geometry dependencies as far as possible without solving the differential equations.

We will use the rescaled flux functions $\bar{\Lambda}_i$ in the rest of the analysis, where we exploit ϵ as a small parameter to obtain analytical results but we keep ϵ arbitrary in the numerical analysis. Since $e \sim f \sim 1$, assuming $\bar{\Lambda}_i \sim 1$ implies that horizontal gravity and toroidal rotation effects (factor e) will enter to first order, but vertical gravity effects (factor f) only to second order. To get horizontal gravity and toroidal rotation to appear in leading order (for tokamaks and thick accretion disks), we may keep $\bar{\Lambda}_1^* \sim \bar{\Lambda}_2 \sim 1$ but we need to push the orders of magnitude of the other flux functions:

$$\bar{\Lambda}_3 \approx \bar{\Lambda}_4 \sim \epsilon^{-1/2}\,, \quad \bar{\Lambda}_5 \sim \epsilon^{-1}\,, \quad \text{such that} \quad \bar{\Lambda}_3 - \bar{\Lambda}_4 \sim 1\,. \tag{18.29}$$

To get vertical gravity to appear in the leading order as well (for thin accretion disks), we need to push the orders of magnitude of those flux functions even further:

$$\bar{\Lambda}_3 \approx \bar{\Lambda}_4 \sim \epsilon^{-1}\,, \quad 2\bar{\Lambda}_4^2 \approx \bar{\Lambda}_5 \sim \epsilon^{-2} \quad \Rightarrow \quad \bar{\Lambda}_3 - \bar{\Lambda}_4 \sim 1\,, \quad 2\bar{\Lambda}_4^2 - \bar{\Lambda}_5 \sim \epsilon^{-1}\,. \tag{18.30}$$

In this manner, we guarantee that explicit results from $\epsilon \ll 1$ analysis may be generalized immediately to "exact" results by means of the numerical codes.

Note that the physical meaning of the different geometrical terms is not straightforward because of the flux function constraints on the variables. In particular, the horizontal force balance, which requires approximate Keplerian toroidal rotation in HD, is determined by the flux function combination $2\bar{\Lambda}_4^2 - \bar{\Lambda}_5$ in MHD:

$$(\epsilon \Psi_1 / \chi')^2\, (2\bar{\Lambda}_4^2 - \bar{\Lambda}_5) \equiv R_0^2 \Omega^2 - R_0^{-1} GM_*$$

$$\to v_\varphi^2 - R_0^{-1} GM_* \quad \text{when either } M = 0 \text{ or } B_\varphi = 0\,. \tag{18.31}$$

Only in the absence of poloidal flow or toroidal magnetic field, one obtains the standard Keplerian expression on the second line of Eq. (18.31). In general, horizontal force balance is determined by the expression on the first line of Eq. (18.31), i.e.

instead of v_φ only the toroidal deviation $R\Omega$ from the Alfvén speed (see Fig. 18.2) appears to count.

The transonic equilibrium problem (18.23), (18.24), (18.27), (18.28) is solved by first choosing the functional dependencies of the five rescaled flux functions $\bar{\Lambda}_i(\psi)$ (based on whatever theoretical or observational evidence is available) and then iteratively solving the nonlinear PDE (18.23) for the flux $\psi(x, y)$ in an elliptic flow domain, subject to the boundary conditions

$$\psi = 0 \ \text{ on the magnetic axis}, \qquad \psi = 1 \ \text{ on the plasma boundary}, \qquad (18.32)$$

while inserting the algebraic solutions $M^2(\psi(x, y); x, y)$ of the Bernoulli equation (18.24) for each step of the iteration. This internal equilibrium problem is strictly determined by the shape of the boundary curve, which, in turn, should be determined by the external dynamics. The latter part we take for granted at present and we simply prescribe it to be a circle or an ellipse (to study the influence of external flattening of the disk). Thus, the external flattening dominates the internal one described by the geometry function f and the flux function $\bar{\Lambda}_5$.

The reader will have noticed that the elliptic problem (18.23), (18.32) is overdetermined. The implication for the parameterization of the flux functions $\bar{\Lambda}_i(\psi)$ is discussed below.

▷ **Parameterization of the flux functions** $\bar{\Lambda}_i(\psi)$ The price for the convenience of dealing with a poloidal magnetic flux ψ of unit range is overdetermination of the boundary value problem (18.23), (18.32) by two conditions on ψ. One of the amplitudes of the $\bar{\Lambda}_i'$s is no longer a free parameter but becomes an *eigenvalue of the flux equation*, to be determined together with the solution ψ. Two other complications in the coupled flux-Bernoulli equations appear to be the absence of control over the value of M^2 to remain in a certain chosen flow domain and the indirect dependence on Ψ_1 mentioned above. These apparent complications may be turned into an advantage by building them into the parameterization of the flux functions $\bar{\Lambda}_i(\psi)$, as follows.

For the parameter to be used as an eigenvalue we take the amplitude of $-\bar{\Lambda}_1^{*\prime}$ and call it C. To isolate it from the rest, we introduce shape functions $\sigma_i(\psi)$ of unit amplitude on axis (or some other convenient normalization, e.g. at the plasma boundary), but arbitrary otherwise. These functions enter with amplitudes A_i (where $A_1 = 1$) in the functions $\bar{\Lambda}_i$ in the flux equation. In the Bernoulli equation, the integrals of the functions $\sigma_i(\psi)$,

$$\lambda_i(\psi) \equiv \int_0^\psi \sigma_i(\psi)\, d\psi, \qquad \sigma_i(0) \equiv 1, \qquad (18.33)$$

then appear with integration constants B_i and a common multiplication factor C:

$$\bar{\Lambda}_1^* = C\left[B_1 - \lambda_1(\psi)\right],$$

$$\bar{\Lambda}_{2,5} = C\left[B_{2,5} - A_{2,5}\,\lambda_{2,5}(\psi)\right],$$

$$\bar{\Lambda}_{3,4} = \sqrt{C}\left[B_{3,4} - A_{3,4}\,\lambda_{3,4}(\psi)\right]. \qquad (18.34)$$

In this manner, we obtain control over the different orders of M^2 appearing in the flux equation and the Bernoulli equation. On the magnetic axis, the constants A_i determine the orders of magnitude of the flux functions derivatives $\bar\Lambda'_i$, and the constants B_i determine the orders of magnitude of the flux functions $\bar\Lambda_i$ themselves.

Since $B_{\mathrm{p}} \sim |\nabla\psi| = 0$ on the magnetic axis, the Bernoulli equation (18.24) gives the following relationship between the constants B_i and $M_0^2 \equiv M^2(\delta, 0)$, not involving the unknown eigenvalue C:

$$\frac{B_1 + \epsilon e_0(2B_4^2 - B_5) + \epsilon^2 f_0 B_5}{M_0^4} - \frac{B_2}{M_0^{2(\gamma+1)}} - \frac{[B_3 - B_4 - \epsilon(g_0 B_3 + h_0 B_4)]^2}{(M_0^2 - 1)^2} = 0\,,$$

$$(18.35)$$

where e_0, etc., are the values of e, etc., on the magnetic axis. Rather than solving for M_0^2, this relationship may be used to *prescribe* M_0^2 (giving control over the flow regime) by eliminating one of the constants B_i (e.g. B_1). On the other hand, since the toroidal magnetic field on the magnetic axis is normalized to unity,

$$B_0 \equiv B_\varphi(x = \delta, y = 0) = \frac{\sqrt{2}\epsilon\Psi_1}{|1 - M_0^2|}\sqrt{C}\left|\frac{B_3}{1 + \epsilon\delta} - (1 + \epsilon\delta)B_4\right| = 1\,, \qquad (18.36)$$

we obtain *control over the value of the safety factor on axis*, $q_0 = (\Psi_1)^{-1}(r/\psi')_0 \sim \sqrt{C}CB_3$, by means of the parameter B_3 (or B_4, or both). This requires knowledge of the equilibrium solutions ψ and C so that prescribing q_0 can only be done by iterating on the value of B_3 in the numerical solution procedure of the equilibrium equations. Similarly, one may *prescribe the position δ of the magnetic axis* by iterating on one of the constants A_i (e.g. the amplitude A_2 associated with the pressure gradient). In this manner, we have obtained a complete classification of the free parameters with a clear relationship to the original physical flux functions (18.3)–(18.7) through the definitions (18.13) and (18.22).

Finally, the sixth flux function $\bar\Lambda_0(\psi)$, associated with χ'^2 by the definition (18.109), may be parameterized as

$$\bar\Lambda_0 = 1 - A_0\lambda_0(\psi)\,, \qquad \lambda_0(0) = 0\,. \qquad (18.37)$$

For accretion disks, this parameterization represents no further freedom of choice since it has to be consistent with that of $\bar\Lambda_5(\psi)$, so that $A_0 \equiv A_5/B_5$ and $\lambda_0(\psi) \equiv \lambda_5(\psi)$.   ◁

18.2.2 Elliptic and hyperbolic flow regimes*

As stated in Section 18.1, we will have to restrict the analysis to flow regimes described by *elliptic partial differential equations*. This is a complication since the condition for ellipticity or hyperbolicity is determined by the solutions ψ and M^2 of Eqs. (18.23) and (18.24) themselves, according to the following expression involving the Alfvén Mach numbers [181]:

$$\Delta_{\mathrm{char}} = (M^2 - 1)\frac{\bar\Pi_1 - \frac{1}{2}[\gamma + 1 - (\gamma - 1)M^2]M^{-2(\gamma-1)}\bar\Pi_2}{\bar\Pi_1 - \frac{1}{2}(\gamma + 1)M^{-2(\gamma-1)}\bar\Pi_2 - M^6(M^2 - 1)^{-3}\bar\Pi_3}$$

$$= \frac{\gamma p + B^2}{B_{\mathrm{p}}^2}\frac{(M^2 - 1)^2(M^2 - \widetilde{M}_{\mathrm{c}}^2)}{(M^2 - \widetilde{M}_{\mathrm{s}}^2)(M^2 - \widetilde{M}_{\mathrm{f}}^2)}\left\{\begin{array}{l} < 0 : \text{elliptic} \\ = 0 : \text{parabolic}\,, \\ > 0 : \text{hyperbolic} \end{array}\right. \qquad (18.38)$$

$$\widetilde{M}_c^2 \equiv \frac{\gamma p}{\gamma p + B^2}, \qquad \widetilde{M}_{f,s}^2 \equiv \frac{\gamma p + B^2}{2B_p^2}\left[1 \pm \sqrt{1 - \frac{4\gamma p B_p^2}{(\gamma p + B^2)^2}}\,\right]. \qquad (18.39)$$

The quantity Δ_{char} appears as a square root in the directional derivatives of the characteristics, giving two real characteristics in the hyperbolic flow regimes and no characteristics at all in the elliptic flow regimes. Hence, the transitions from ellipticity to hyperbolicity [495, 210, 4] occur at positions on the magnetic/flow surfaces where either the numerator or the denominator of Δ_{char} vanishes. The tildes on the quantities indicate that the explicit expressions for the transition values of M^2 are only obtained after solving an implicit relation, e.g.

$$M^2 - \widetilde{M}_c^2(M^2, \psi; x, y) = 0 \quad \Rightarrow \quad M^2 = M_c^2(\psi; x, y), \qquad (18.40)$$

from the numerator expression on the first line of Eq. (18.38), and, similarly, $M^2 = M_{f,s}^2(\psi; x, y)$ are obtained from the denominator expression. The numerical aspects of this problem are discussed in [29] and implemented in FINESSE.

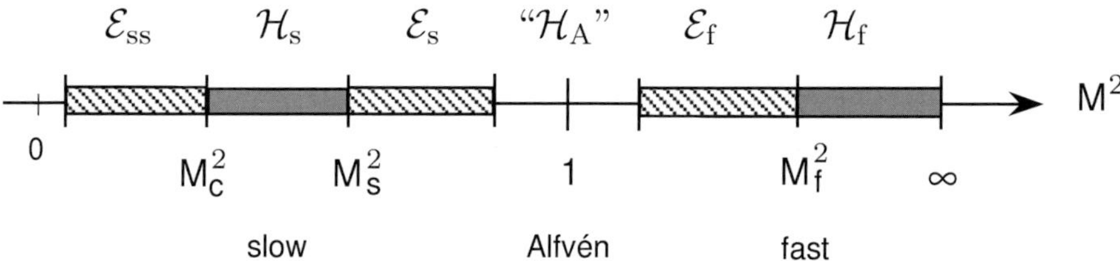

Fig. 18.3 Elliptic (hatched) and hyperbolic (dark) flow regimes corresponding to the value of the poloidal Alfvén Mach number M. Flow should be in the 1st elliptic ($\mathcal{E}_{\mathrm{ss}}$: sub-slow), the 2nd elliptic ($\mathcal{E}_{\mathrm{s}}$: slow), or the 3rd elliptic ($\mathcal{E}_{\mathrm{f}}$: fast) flow regime for the spectral results of this chapter to be valid.

As follows from the definitions (18.39), the transition values for M^2 are well-ordered in magnitude giving five different flow regimes (Fig. 18.3). We just need to specify the flow to be either in the first (sub-slow) elliptic flow regime ($\mathcal{E}_{\mathrm{ss}}$, where $M^2 < M_c^2$), or the second (slow) elliptic flow regime ($\mathcal{E}_{\mathrm{s}}$, where $M_s^2 < M^2 < 1$), or the third (fast) elliptic flow regime ($\mathcal{E}_{\mathrm{f}}$, where $1 < M^2 < M_f^2$). Since the effects of transition through a hyperbolic flow regime should be first noticeable in the regime $\mathcal{E}_{\mathrm{s}}$, we will pay special attention to equilibria in that regime.

18.2.3 *Expansion of the equilibrium in small toroidicity*[*]

(a) Small inverse aspect ratio expansion We assume the inverse aspect ratio to be small, $\epsilon \ll 1$, and impose the outer plasma boundary to be circular, and try to find solutions of the equilibrium equations (18.23) and (18.24) in the form of the *Shafranov shifted circle approximation*, where the poloidal magnetic flux/flow

surfaces are assumed to have circular cross-sections that are shifted by an amount $\Delta(\hat{r})$ with respect to the center of the plasma (see Section 16.2.2, Fig. 16.9). To first order, these solutions may then be represented as

$$\psi(\hat{r}, \hat{\theta}) \approx \psi_0(\hat{r}), \tag{18.41}$$

$$M^2(\hat{r}, \hat{\theta}) \approx M_0^2(\hat{r}) + \mu(\hat{r}) \cos\theta, \tag{18.42}$$

where $\hat{r}$, $\hat{\theta}$ are non-orthogonal polar coordinates defined with respect to the centers of the flux surfaces. Note that $\psi_1(\hat{r}) \equiv 0$ since the perturbation of the flux surfaces is absorbed in the definition of the coordinates, i.e. it is represented by the shift $\Delta(\hat{r})$. The latter quantity satisfies the boundary conditions $\Delta(0) = \delta$ and $\Delta(1) = 0$, where δ is the shift of the magnetic axis. The perturbation of the squared poloidal Alfvén Mach number is denoted by $\mu \equiv M_1^2(\hat{r})$. Due to the shifted circle approximation (18.41), the flux functions will now be determined by the radial dependence $\psi_0(\hat{r})$:

$$\bar{\Lambda}_i(\psi) \approx \bar{\Lambda}_i(\psi_0(\hat{r})) \equiv \bar{\Lambda}_{i,0}(\hat{r}). \tag{18.43}$$

Since we have uniquely settled the notation, we will drop the subscript 0 on ψ_0 and M_0^2 from now on. (This is not only done for convenience, but also because we will need the subscripts 0 and 1 later on to indicate equilibrium values on the magnetic axis and at the boundary.) The expansion procedure will then consist of a zeroth order part determining the radial dependencies of $\psi(\hat{r})$ and $M^2(\hat{r})$ and a first order part determining the shift $\Delta(\hat{r}) \sim \epsilon$ (and, hence, δ) and the Alfvén Mach number perturbation $\mu(\hat{r}) \sim \epsilon$.

As in Section 16.2.2, the connection with the Cartesian coordinates is given by

$$x = x(\hat{r}, \hat{\theta}) = \hat{r}\cos\theta + \Delta(\hat{r}),$$

$$y = y(\hat{r}, \hat{\theta}) = \hat{r}\sin\theta. \tag{18.44}$$

To convert the equations to the $\hat{r}$, $\hat{\theta}$, φ coordinates, we insert these into Eq. (18.19):

$$R = R(\hat{r}, \hat{\theta}) = \epsilon^{-1} + \hat{r}\cos\theta + \Delta,$$

$$Z = Z(\hat{r}, \hat{\theta}) = \hat{r}\sin\theta, \tag{18.45}$$

from which the metric tensor, poloidal field and Jacobian follow, to first order:

$$\hat{g}^{ij} \approx \begin{pmatrix} 1 - 2\Delta'\cos\theta & (\Delta'/\hat{r})\sin\hat{\theta} & 0 \\ (\Delta'/\hat{r})\sin\theta & 1/\hat{r}^2 & 0 \\ 0 & 0 & \epsilon^2(1 - 2\epsilon\hat{r}\cos\theta) \end{pmatrix}, \tag{18.46}$$

$$B_{\mathrm{p}} \approx \hat{r}\Psi_1\psi'/\hat{J}, \qquad \hat{J} \equiv (\nabla\hat{r} \times \nabla\hat{\theta} \cdot \nabla\varphi)^{-1} \approx \epsilon^{-1}\hat{r}[1 + (\epsilon\hat{r} - \Delta')\cos\hat{\theta}]. \tag{18.47}$$

The hats distinguish these metric coefficients from the ones for the straight-field-line coordinates, that we will eventually exploit in Section 18.3.

(b) Zeroth and first order solutions Choosing $\bar{\Lambda}_1^* \sim \bar{\Lambda}_2 \sim \cdots \sim \bar{\Lambda}_5 \sim 1$, the expansions of the generic functions, accurate to first order, are given by

$$\bar{\Pi}_1(\hat{r}, \hat{\theta}) \approx \bar{\Lambda}_{1,0}^* + \epsilon x(2\bar{\Lambda}_{4,0}^2 - \bar{\Lambda}_{5,0}),$$

$$\bar{\Pi}_2(\hat{r}) \quad \approx \bar{\Lambda}_{2,0},$$

$$\bar{\Pi}_3(\hat{r}, \hat{\theta}) \approx (\bar{\Lambda}_{3,0} - \bar{\Lambda}_{4,0})^2 - 2\epsilon x(\bar{\Lambda}_{3,0}^2 - \bar{\Lambda}_{4,0}^2), \tag{18.48}$$

where three flux function combinations appear in zeroth order:

$$(\epsilon\Psi_1)^2\bar{\Lambda}_{1,0}^* = \tfrac{1}{2}M^4B^2 + \frac{\gamma}{\gamma-1}M^2p,$$

$$(\epsilon\Psi_1)^2\bar{\Lambda}_{2,0} = \frac{\gamma}{\gamma-1}M^{2\gamma}p,$$

$$(\epsilon\Psi_1)^2(\bar{\Lambda}_{3,0} - \bar{\Lambda}_{4,0})^2 = \tfrac{1}{2}(M^2-1)^2B_\varphi^2, \tag{18.49}$$

and two in first order:

$$(\epsilon\Psi_1)^2(2\bar{\Lambda}_{4,0}^2 - \bar{\Lambda}_{5,0}) = M^2[(\sqrt{\rho}v_\varphi - MB_\varphi)^2 - \rho R_0^{-1}GM_*],$$

$$(\epsilon\Psi_1)^2(\bar{\Lambda}_{3,0}^2 - \bar{\Lambda}_{4,0}^2) = -\tfrac{1}{2}(M^2-1)B_\varphi[(1+M^2)B_\varphi - 2M\sqrt{\rho}v_\varphi]. \tag{18.50}$$

We have again omitted the subscript 0 on the original physical variables.

As compared to the original 2D problem, where the flux surfaces $\psi(\hat{r}, \hat{\theta}) = $ const and the Bernoulli surfaces $M^2(\hat{r}, \hat{\theta}) = $ const intersect, in the present 1D problem $\psi(\hat{r})$ and $M^2(\hat{r})$ trivially label the same surfaces. This implies that there is a much closer relationship now between the free flux functions and the physical variables (both being functions of $\hat{r}$ alone). In the zeroth order, the choice of the three functions (18.49) determines three of the four physical variables p, $B_{\rm p}$, B_φ and M^2 (leaving room for only one extra condition: see below). In the first order, the choice of the two functions (18.50) determines the two remaining physical variables ρ and v_φ (except for the contribution of GM_*).

Inserting the metric coefficients (18.46)–(18.47) and the flux function expressions (18.48)–(18.50) into the equilibrium equations (18.23) and (18.24), we obtain two simple equations for ψ and M^2 in the zeroth order and two more complicated ones for Δ and μ in the first order. The *zeroth order flux equation* is a second order ODE for $\psi(\hat{r})$:

$$\frac{\psi'}{\hat{r}}[(1-M^2)\hat{r}\psi']' + \frac{\bar{\Lambda}_{1,0}^{*\prime}}{M^2} - \frac{\bar{\Lambda}_{2,0}'}{\gamma M^{2\gamma}} - \frac{(\bar{\Lambda}_{3,0} - \bar{\Lambda}_{4,0})^{2\prime}}{M^2-1} = 0, \tag{18.51}$$

where the primes now denote differentiation with respect to $\hat{r}$. Since $B_{\mathrm{p}} \approx (\epsilon\Psi_1)\psi'$ to leading order, this yields the cylindrical equilibrium relation (cf. Eq. (12.30)):

$$\frac{d}{d\hat{r}}(p + \tfrac{1}{2}B^2) - (M^2 - 1)\frac{B_{\mathrm{p}}^2}{\hat{r}} = 0. \tag{18.52}$$

However, the zeroth order Bernoulli equation reduces to an identity so that there is no longer an equilibrium restriction on the choice of $M^2(\hat{r})$. Hence, Eq. (18.52) is the only condition now on the six physical equilibrium functions (of $\hat{r}$).

After some lengthy, but straightforward, reductions the *first order magnetic flux and Bernoulli equations* yield two coupled differential equations for Δ and μ:

$$[(1 - M^2)\hat{r}B_{\mathrm{p}}^2\Delta']' + (\hat{r}B_{\mathrm{p}}^2\mu)' = -\epsilon\hat{r}E, \tag{18.53}$$

$$M^2(M^2 - 1)B_{\mathrm{p}}^2\Delta' + \frac{B_{\mathrm{p}}^2}{M^2}(M^2 - M_{\mathrm{s}}^2)(M^2 - M_{\mathrm{f}}^2)\mu = -\epsilon\hat{r}F, \tag{18.54}$$

where the two expressions on the RHS (with respect to their dimension recall that $\epsilon \equiv R_0^{-1}$) are defined by

$$E \equiv -(M^2 - 1)B_{\mathrm{p}}^2 - \hat{r}[2p + \rho(v^2 - R_0^{-1}GM_*)]', \tag{18.55}$$

$$F = (M^2 - 1)\rho(v^2 - R_0^{-1}GM_*) + 2MB_\varphi(\sqrt{\rho}v_\varphi - MB_\varphi). \tag{18.56}$$

The appearance in Eq. (18.54) of the transition values of M^2 defined in Eqs. (18.39) is no accident, as will be explained below. (Because $M^2 = M^2(\hat{r})$ to leading order, according to Eq. (18.42), there is no need to solve an implicit equation in this case so that the tildes have been dropped.) Rather than integrating Eqs. (18.53) and (18.54) directly, it is more instructive to first decouple them to a separate second order ODE for $\Delta(\hat{r})$ and a first order ODE for $\mu(\hat{r})$:

$$\frac{d}{d\hat{r}}\left[\frac{(\gamma p + B^2)(M^2 - 1)(M^2 - M_{\mathrm{c}}^2)}{(M^2 - M_{\mathrm{s}}^2)(M^2 - M_{\mathrm{f}}^2)}\,\hat{r}\frac{d\Delta}{d\hat{r}}\right]$$
$$= \epsilon\left\{\hat{r}E + \left[\frac{M^2\hat{r}^2 F}{(M^2 - M_{\mathrm{s}}^2)(M^2 - M_{\mathrm{f}}^2)}\right]'\right\}, \tag{18.57}$$

$$\frac{d}{d\hat{r}}\left[\frac{(\gamma p + B^2)(M^2 - M_{\mathrm{c}}^2)}{M^4}\,\hat{r}\mu\right] = -\epsilon\left\{\hat{r}E + \left(\frac{\hat{r}^2 F}{M^2}\right)'\right\}. \tag{18.58}$$

Integration yields

$$\Delta' = -\epsilon\hat{r}\frac{(M^2 - M_{\mathrm{s}}^2)(M^2 - M_{\mathrm{f}}^2)I + M^2 F}{(\gamma p + B^2)(M^2 - 1)(M^2 - M_{\mathrm{c}}^2)}, \tag{18.59}$$

$$\frac{\mu}{M^2} = -\frac{\rho_1}{\rho} = -\epsilon\hat{r}\frac{M^2 I + F}{(\gamma p + B^2)(M^2 - M_{\mathrm{c}}^2)}, \tag{18.60}$$

where

$$I \equiv \frac{1}{\hat{r}^2} \int_0^{\hat{r}} E\hat{r}\, d\hat{r} \,. \tag{18.61}$$

In the spectral analysis of the following sections, only Δ' and μ appear, so that no further integration is needed. Of course, for the limiting case of static tokamak equilibria, the well-known expression for Δ' is recovered [481].

At this point, a truly amazing confluence of apparently completely unrelated topics may be noted. The coefficient in front of the highest derivative of the differential equation (18.57) for Δ is of exactly the same form as the analogous coefficient appearing in the spectral Hain–Lüst equation for cylindrical plasmas [204, 160], where the eigenvalue ω^2 of the latter is now replaced by the squared poloidal Alfvén Mach number, M^2. As we have seen in Section 18.2.2, the transonic transitions from elliptic to hyperbolic flows occur at the values $M^2 = 1$ (recall that this 1 just corresponds to the normalization with respect to the poloidal Alfvén speed), $M^2 = M_c^2$, $M^2 = M_s^2$, and $M^2 = M_f^2$, defined in Eq. (18.39). On the other hand, in the Hain–Lüst equation, the spectrum of MHD waves is concentrated about continua [450, 187] at the Alfvén and slow (or cusp) frequencies $\omega^2 = \omega_A^2$ and $\omega^2 = \omega_S^2$, whereas the special values $\omega^2 = \omega_s^2$ and $\omega^2 = \omega_f^2$ do not correspond to continua [10] but to monotonicity transitions in the spectrum [183, 166]. Apparently, there is a deep correspondence between the linear waves and the non-linear stationary states. This is another example of the recent insight that equilibrium and perturbations are not really separate issues in transonic magnetohydrodynamics [168]. We will see more of this in the following sections.

Obviously, for the shifted circle approximation ($\Delta' \ll 1$) to be valid, the singularities $M^2 = 1$ and $M^2 = M_c^2$ must be avoided. However, as in the analogous spectral case, the denominator zeros of Eq. (18.57) just constitute apparent singularities of the differential equation so that approaching $M^2 \downarrow M_s^2$ or $M^2 \uparrow M_f^2$ from within the elliptic flow regimes poses no fundamental problem. We will use this fact in the following sections where we will investigate the effect of the transonic transition from sub-slow to slow flow on changes of the continuous spectra in the second elliptic flow regime.

(c) Approximate solutions in the second elliptic flow regime We will simplify the solutions obtained above by means of approximations that are valid in the second elliptic flow regime by extending the usual low-β tokamak approximation ($B_\varphi \sim 1$, $B_p \sim \epsilon$, $\beta \equiv p/(2B^2) \sim \epsilon^2$) with poloidal rotation effects. So far, we have made no assumption on the magnitude of M^2. Now, we will push the value of M^2 in the region $M_s^2 \le M^2 < 1$ as much as possible within the low-β approximation. Since $M_s^2 \approx M_c^2(1 + \mathcal{O}(\epsilon^4)) \sim \epsilon^2$ and $M_f^2 \sim \epsilon^{-2}$, the validity of the expressions

(18.59) and (18.60) for Δ' and μ/M^2 demands that M^2 is larger than $\mathcal{O}(\epsilon^2)$ but smaller than $\mathcal{O}(\epsilon)$:

$$M^2 \sim \epsilon^{2-\nu} \quad (0 \le \nu < 1). \tag{18.62}$$

Hence, we may choose ν such that the poloidal Alfvén Mach number squared, M^2, represents flows much faster than the slow but much slower than the Alfvén speed:

$$\tfrac{1}{2}\gamma\beta \approx M_c^2 \ll M^2 \ll 1. \tag{18.63}$$

As we will see, this ordering is quite effective to study a particular type of violent instability caused by the coupling of slow and Alfvén modes and driven by poloidal flows in the second elliptic flow domain. Let us call it the *trans-slow poloidal flow ordering* (the adjective "super-slow" being technically more precise, but intuitively contradictory).

From the equilibrium equations (18.23) and (18.24) and the definitions (18.27) of the generic functions, it is obvious that we must then choose

$$\bar{\Lambda}_1^{*\prime} \sim M^2(\bar{\Lambda}_3 - \bar{\Lambda}_4)^{2\prime}, \qquad \bar{\Lambda}_1^* \sim M^4(\bar{\Lambda}_3 - \bar{\Lambda}_4)^2, \tag{18.64}$$

in order to balance the magnetic terms. This dictates how the orders of magnitude of the different parameters appearing in the flux functions $\bar{\Lambda}_i$ (see paragraph in small print at the end of Section 18.2.1) have to be chosen to be consistent with the trans-slow ordering.

For our present purpose, we will consider a particular class of equilibria with negligible pressure ($p \approx 0$) and negligible non-parallel velocity ($\Omega \approx 0$). These additional assumptions are not necessary for the trans-slow ordering, but they simplify the equilibrium expressions considerably:

$$\Delta' \approx \frac{\epsilon}{\hat{r}B_{\mathrm{p}}^2} \int_0^{\hat{r}} \hat{r}^2 \left[M^2 B^2 (1 - \Gamma)\right]' d\hat{r}, \tag{18.65}$$

$$\frac{\mu}{M^2} = -\frac{\rho_1}{\rho} \approx \epsilon\hat{r}(1 - \Gamma). \tag{18.66}$$

Here, the flux function $\Gamma(\psi(\hat{r}))$ measures the relative strength of the gravitational interaction:

$$\Gamma(\psi) \equiv \frac{\bar{\Lambda}_5(\psi)}{2\bar{\Lambda}_1^*(\psi)} \approx \frac{\rho G M_*}{R_0 M^2 B^2}, \tag{18.67}$$

where we have restored the dimensional factors in the last expression.

Since we have assumed approximately parallel flow ($\Omega \approx 0$), M^2 now represents both the poloidal and the toroidal Alfvén Mach number, $M^2 \equiv \rho v_{\mathrm{p}}^2/B_{\mathrm{p}}^2 \approx \rho v_\varphi^2/B_\varphi^2$, so that $\Gamma \approx G M_*/(R v_\varphi^2)$ becomes a measure for the deviation from Keplerian flow (where $\Gamma = 1$). However, there is no need to stick to Keplerian flow in the presence of poloidal rotation, so that we obtain a class of equilibria that

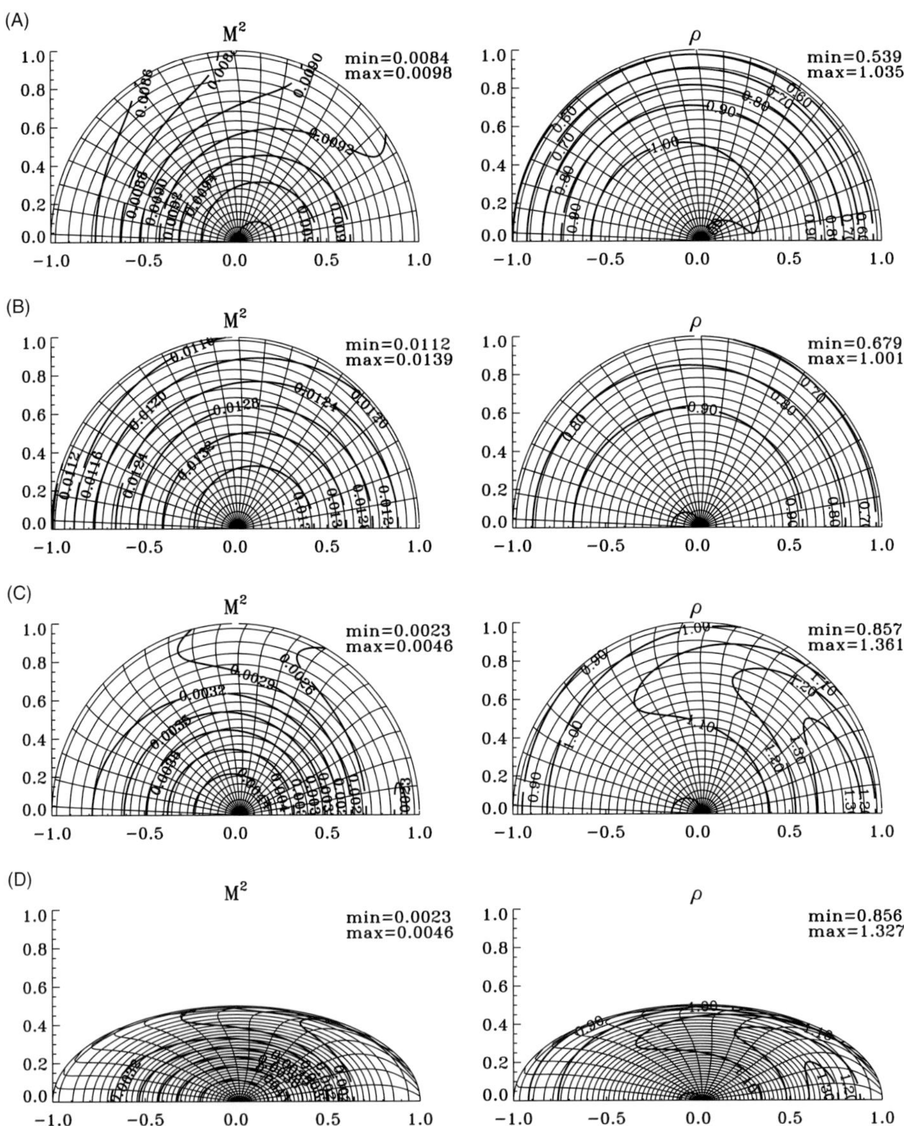

Fig. 18.4 Contours of the squared poloidal Alfvén Mach number M^2 and of the density ρ: (A) tokamak ($\Gamma_0 = 0$); (B) thick accretion disk, small central mass ($\Gamma_0 = 0.25$); (C) thick accretion disk, large central mass ($\Gamma_0 = 2$); (D) flat accretion disk, large central mass ($\Gamma_0 = 2$).

may be changed continuously from tokamak ($\Gamma = 0$) to accretion disk ($\Gamma \neq 0$). From Eqs. (18.65) and (18.66) it is clear that weak gravitational "tokamak-like" equilibria ($\Gamma < 1$) have a density that is larger on the inside than on the outside of the torus ($\rho_1 < 0$), as is well known for poloidal flows in the second elliptic

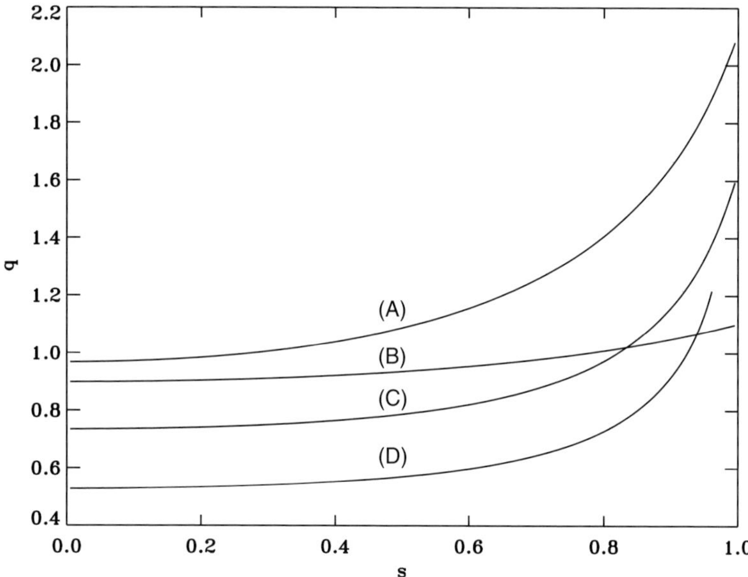

Fig. 18.5 Safety factor q versus $s \equiv \sqrt{\psi}$ for the equilibria of Fig. 18.4.

regime [496]. For strong gravitational interaction ($\Gamma > 1$), the opposite holds: the density is larger on the outside, i.e. in the region of smaller gravity in order to balance the stronger gravitational pull of the inner region. Note that the trans-slow ordering (18.63) permits the study of the effects of very massive central objects, as long as the expressions (18.65) and (18.66) for Δ' and μ remain small, e.g. by considering $\Gamma \sim \epsilon^{-1/2} \gg 1$.

The mentioned features are confirmed by the solutions shown in Figs. 18.4 and 18.5, obtained using the transonic equilibrium program FINESSE [29]. In the code, no small ϵ approximations were made, but the parameters of the flux functions were chosen to agree with the trans-slow ordering. (Explicit parameters may be found in [173], where the functions $\bar{\Lambda}_3$ and $\bar{\Lambda}_4$ should be corrected to have opposite sign.) As a result, there is perfect agreement between the numerical results and the present analysis. The solutions shown are representative examples of transonic equilibria:

(A) *Tokamak* ($\Gamma_0 = 0$);

(B) *Thick accretion disk with a small central mass* ($\Gamma_0 = 0.25$);

(C) *Thick accretion disk with a massive central object* ($\Gamma_0 = 2$);

(D) *Flat accretion disk with a massive central object* ($b/a = 0.5, \Gamma_0 = 2$).

The stability of these equilibria will be discussed in Sections 18.3 and 18.4, after we have developed the theory of continuous spectra for transonic equilibria.

18.3 Equations for the continuous spectrum*

18.3.1 Reduction for straight-field-line coordinates*

The perturbations of the stationary equilibria of Section 18.2 will be described by
the force-operator formalism in terms of the displacement $\boldsymbol{\xi}$ developed by Frieman and Rotenberg [147], as discussed in Chapters 12 and 13. Recall from Section 12.2.2 that, for normal modes $\sim e^{-i\omega t}$, the spectral equation takes the form

$$\mathbf{F}(\boldsymbol{\xi}) + \nabla \cdot (\rho \boldsymbol{\xi} \mathbf{v} \cdot \nabla \mathbf{v} - \rho \mathbf{v} \mathbf{v} \cdot \nabla \boldsymbol{\xi}) + 2i\rho\omega \mathbf{v} \cdot \nabla \boldsymbol{\xi} + \rho\omega^2 \boldsymbol{\xi} = 0 , \qquad (18.68)$$

where

$$\mathbf{F}(\boldsymbol{\xi}) \equiv \nabla(\gamma p \nabla \cdot \boldsymbol{\xi} + \boldsymbol{\xi} \cdot \nabla p) - \mathbf{B} \times (\nabla \times \mathbf{Q})$$
$$+ (\nabla \times \mathbf{B}) \times \mathbf{Q} + (\nabla \Phi_{\mathrm{gr}}) \nabla \cdot (\rho \boldsymbol{\xi}) , \qquad (18.69)$$

with $\mathbf{Q} \equiv \nabla \times (\boldsymbol{\xi} \times \mathbf{B})$, the usual force-operator for static equilibria [35]. Note that
the eigenvalue problem (18.68) is *quadratic* (involving both ω and ω^2), in contrast
to the static spectral problem, which is linear in the eigenvalue ω^2.

The Frieman–Rotenberg equation (18.68) will be expressed in straight-field-line
coordinates Ψ, ϑ, φ, with Jacobian $\mathcal{J} \equiv (\nabla \Psi \times \nabla \vartheta \cdot \nabla \varphi)^{-1}$, where the poloidal
angle ϑ is constructed such that the magnetic field lines are represented by straight
lines in the ϑ–φ plane so that the inverse rotational transform of the field lines (the
"safety factor") is constant on the magnetic/flow surfaces:

$$q(\Psi) \equiv \frac{d\varphi}{d\vartheta}\bigg|_{\text{field line}} = \frac{\mathcal{J} B_\varphi}{R} = \frac{\mathcal{J}}{R^2} \frac{\sqrt{\rho} M (K - R^2 \Omega)}{M^2 - 1} . \qquad (18.70)$$

As compared to static plasmas or plasmas with toroidal flow only, where $q =
(\mathcal{J}/R^2)I$ with the constant factor $I \equiv RB_\varphi$, there is now a complicated quotient
which is not constant on the magnetic/flow surfaces. In the straight-field-line coordinates, the flow lines are not represented by straight lines in general (except when
$\Omega = 0$, see Fig. 18.2) so that their inverse rotational transform is not constant on
the magnetic/flow surfaces:

$$u(\Psi, \vartheta) \equiv \frac{d\varphi}{d\vartheta}\bigg|_{\text{flow line}} = \frac{\mathcal{J} \sqrt{\rho} v_\varphi}{MR} = \frac{K - M^{-2} R^2 \Omega}{K - R^2 \Omega} q(\psi) . \qquad (18.71)$$

Obviously, one could also construct a poloidal angle such that the flow lines become straight. In that case, the roles of q and u would be interchanged. However,
the field lines turn out to play the more fundamental role in transonic MHD, as will
become clear from the spectral analysis in the next section.

We now project the spectral equation (18.68) onto the three preferred physical
directions expressed by the magnetic surface/field line triad

$$\mathbf{n} \equiv (RB_{\mathrm{p}})^{-1} \nabla \Psi , \quad \boldsymbol{\pi} \equiv \mathbf{b} \times \mathbf{n} , \quad \mathbf{b} \equiv \mathbf{B}/B . \qquad (18.72)$$

The displacement vector $\boldsymbol{\xi}$ will be represented by the three components

$$ X \equiv RB_{\mathrm{p}}\boldsymbol{\xi}\cdot\mathbf{n}\,, \quad Y \equiv \mathrm{i}B(RB_{\mathrm{p}})^{-1}\boldsymbol{\xi}\cdot\boldsymbol{\pi}\,, \quad Z \equiv \mathrm{i}B^{-1}\boldsymbol{\xi}\cdot\mathbf{b}\,, \qquad (18.73) $$

and the gradient operator, for toroidal harmonics $\sim \mathrm{e}^{-\mathrm{i}n\varphi}$, will be represented by the three scalar operators

$$ D \equiv (RB_{\mathrm{p}})^{-1}\mathbf{n}\cdot\nabla = \frac{\partial}{\partial\psi} - \frac{g_{12}}{g_{22}}\frac{\partial}{\partial\vartheta}\,, $$

$$ G \equiv -\mathrm{i}RB_{\mathrm{p}}B\boldsymbol{\pi}\cdot\nabla = -\mathrm{i}\frac{RB_\varphi}{\mathcal{J}}\frac{\partial}{\partial\vartheta} - nB_{\mathrm{p}}^2\,, $$

$$ F \equiv -\mathrm{i}B\mathbf{b}\cdot\nabla \quad = \frac{1}{\mathcal{J}}\left(-\mathrm{i}\frac{\partial}{\partial\vartheta} + nq\right)\,, \qquad (18.74) $$

where g_{ij} is the covariant metric tensor of the Ψ, ϑ, φ coordinate system. Recall that the main reason for exploiting straight-field-line coordinates is the fact that the operator F has the simple representation $\mathcal{J}^{-1}(m + nq)$ for single poloidal harmonics $\mathrm{e}^{\mathrm{i}m\vartheta}$ so that resonant modes are well represented.

Associated with the magnetic field and flow lines, two parallel gradient operators enter the formalism, viz. the *parallel field operator* $F \equiv -\mathrm{i}\mathbf{B}\cdot\nabla$, just defined, and the *parallel flow operator*

$$ \mathcal{E} \equiv -\mathrm{i}\mathbf{v}\cdot\nabla = -\frac{\mathrm{i}M}{\mathcal{J}\sqrt{\rho}}\frac{\partial}{\partial\vartheta} + \frac{nv_\varphi}{R} = \frac{M}{\mathcal{J}\sqrt{\rho}}\left(-\mathrm{i}\frac{\partial}{\partial\vartheta} + nu\right) = (M/\sqrt{\rho})F + n\Omega\,. $$

$$ (18.75) $$

Here, the very last expression represents a bit of luck since $n\Omega$ is constant on magnetic surfaces. Whereas the parallel gradient operator F enters as usual when converting the expression for the static force operator part $\mathbf{F}$, the qualitatively new features of rotating plasmas originate from the inertial term $\mathbf{v}\cdot\nabla\mathbf{v}$ and the Doppler shift operator $\omega + \mathrm{i}\mathbf{v}\cdot\nabla$ (second and third term of the first line of Eq. (18.68)). The latter expression involves the parallel flow operator $\mathcal{E}$, but it may be transformed in terms of F by means of Eq. (18.75):

$$ \omega + \mathrm{i}\mathbf{v}\cdot\nabla\mathbf{v} = \omega - \mathcal{E} = \tilde{\omega} - (M/\sqrt{\rho})F\,, \qquad (18.76) $$

where we have introduced a frequency $\tilde{\omega}$ that represents the Doppler shifted frequency for a frame rotating with the angular frequency Ω:

$$ \tilde{\omega} \equiv \omega - n\Omega\,. \qquad (18.77) $$

Note that this frequency is not the one corresponding to the full toroidal rotation velocity v_φ, but only the part that represents the deviation from the Alfvén speed (see Fig. 18.2). Consequently, the optimal representation of the parallel field operator F for straight-field-line coordinates is also fully exploited for the parallel flow operator. Moreover, for the present calculation of continuous spectra, where modes are localized on a single magnetic surface, by using the Doppler shifted frequency $\widetilde{\omega}$ we may eliminate most of the terms with Ω so that the characteristic dependence on the poloidal Alfvén Mach number M becomes manifest.

By means of the projection (18.72), the eigenvalue problem (18.68) becomes a 3×3 matrix equation, involving the gradient operators D, G and F, acting on the three-vector $\mathbf{X} \equiv (X, Y, Z)^{\mathrm{T}}$:

$$\mathbf{A} \cdot \mathbf{X} = \mathbf{B} \cdot \mathbf{X}, \tag{18.78}$$

where the matrix $\mathbf{B}$ contains all terms involving $\omega - \mathcal{E}$. To get the equations for the continuous spectrum, we exploit the same technique as used in Section 17.2.3 for the similar problem in the static toroidal case [161] (where the alternative method of exploiting the primitive variables [359] leads to the same result). We consider modes that are localized around a particular magnetic surface $\Psi = \Psi_0$ by taking the limit $\partial / \partial \Psi \to \infty$. For finite $\widetilde{\omega}$, the normal component of the spectral equation (18.78) then reduces to a derivative with respect to Ψ that may be integrated to give the following leading order relation, similar to Eq. (17.52) of the static case:

$$D^\dagger X \approx -\frac{1}{B^2} G Y + \mathrm{i} \frac{\gamma p}{\gamma p + B^2} \frac{1}{\rho} \left[\partial \left(\frac{\rho R B \varphi}{B^2} \right) \right] Y - \frac{\gamma p}{\gamma p + B^2} \frac{1}{\rho} F \rho Z. \tag{18.79}$$

Here, we exploit a short-hand notation for the invariant tangential derivative:

$$[\partial f] \equiv [\mathbf{B}_{\mathrm{p}} \cdot \nabla f] = \frac{1}{\mathcal{J}} \frac{\partial f}{\partial \vartheta}, \tag{18.80}$$

where the square brackets are used to indicate that the derivative only acts on the quantity inside. Eq. (18.79) implies that $\partial X / \partial \Psi$, Y and Z are of the same order, so that X itself is small compared to Y and Z. Hence, these modes are dominantly polarized tangential to the flux surfaces.

Substituting Eq. (18.79) into the tangential components of the spectral equation (18.78) then leads to a 2×2 matrix eigenvalue problem in terms of the two-vector $\mathbf{V} \equiv (Y, Z)^{\mathrm{T}}$ that only involves the parallel operators F and $\mathcal{E}$, i.e. it no longer contains derivatives of $\mathbf{V}$ with respect to Ψ. Due to this property, the reduced 2×2 problem is non-singular. Therefore, it is advantageous to separate the improper Ψ-dependence from the proper tangential dependence:

$$\begin{pmatrix} Y(\Psi, \vartheta) \\ Z(\Psi, \vartheta) \end{pmatrix} \mathrm{e}^{\mathrm{i}n\varphi} \approx \delta(\Psi - \Psi_0) \begin{pmatrix} \hat{Y}(\vartheta) \\ \hat{Z}(\vartheta) \end{pmatrix} \mathrm{e}^{\mathrm{i}n\varphi}. \tag{18.81}$$

In this manner, a proper eigenvalue problem is obtained *for each magnetic/flow surface*. The collection of discrete eigenvalues obtained from this reduced one-dimensional problem in terms of $\hat{Y}(\vartheta)$, $\hat{Z}(\vartheta)$ will map out the continuous spectra of the original two-dimensional problem in terms of $Y(\psi, \vartheta)$, $Z(\psi, \vartheta)$.

We will omit the details of the formidable amount of manipulations needed to reduce the expressions to the compact form of the continuum equations that will be presented in the following subsection. The reductions basically consist of commuting the parallel gradient operators with equilibrium quantities while consistently exploiting the invariance properties of the magnetic/flow surfaces based on the five basic equilibrium flux functions (18.3)–(18.7). Some of the crucial geometrical properties needed are presented in the paragraph in small print below. Here, the basic handicap in the derivation has been the bias of the previously derived expressions for the case of toroidal flow in the absence of poloidal flow [454]. Only when it was realized that the case of poloidal flow is fundamentally different (since it has entropy conservation built into the equilibrium equations so that the previous possibility of entropy instabilities is absent now), the necessary freedom of expression was obtained to produce the beautiful expressions (18.88)–(18.90) where poloidal flow completely changes the picture obtained previously for static, toroidally rotating, and gravitating equilibria [453, 454].

▷ **Intrinsic properties of the magnetic/flow surfaces** A two-dimensional creature (cf. Section 17.2.3) living on a magnetic/flow surface would be able to test the validity of:

(a) relationships between the physical variables due to the flux functions (18.3)–(18.7):

$$\partial M^2 = -\left(M^2/\rho\right)\partial\rho\,,$$

$$\partial\left(\tfrac{1}{2}M^2B^2/\rho - \tfrac{1}{2}\Omega^2 R^2 + \Phi_{\mathrm{gr}}\right) = -(1/\rho)\,\partial p\,,$$

$$\partial p = \left(\gamma p/\rho\right)\partial\rho\,,$$

$$(M^2 - 1)\,\partial(RB_\varphi) = -RB_\varphi\,\partial M^2 - \sqrt{\rho}M\Omega\,\partial R^2\,,$$

$$\partial\Omega = 0\,; \tag{18.82}$$

(b) the presence of a normal vorticity and a normal current density (i.e. non-divergence-free vorticity and current density in the two dimensions of the surface):

$$w_{\mathrm{n}} = \frac{1}{RB_{\mathrm{p}}}\,\partial(Rv_\varphi)\,; \qquad j_{\mathrm{n}} = \frac{1}{RB_{\mathrm{p}}}\,\partial(RB_\varphi) = \sqrt{\rho}Mw_{\mathrm{n}}\,; \tag{18.83}$$

(c) a relationship between the coupling factor of the matrix $\hat{\mathbf{A}}$ and the Coriolis coupling factor of the matrix $\hat{\mathbf{B}}$ occurring in the eigenvalue problem (18.88)–(18.90) below:

$$\frac{M^2B^2}{\rho}\,\partial\left(\frac{\rho RB_\varphi}{B^2}\right) = M\alpha - \partial(rB_\varphi)\,; \tag{18.84}$$

(d) the geodesic curvature of the magnetic field lines:

$$\kappa_{\mathrm{g}} \equiv \mathbf{b}\cdot(\nabla\mathbf{b})\cdot\boldsymbol{\pi} = -\frac{1}{RB_{\mathrm{p}}}\,\partial\left(\frac{RB_\varphi}{B}\right)\,; \tag{18.85}$$

(e) and, finally, the most surprising intrinsic, Gaussian, curvature of the magnetic/flow
surface:

$$K \equiv \kappa_{\mathrm{p}}\kappa_\varphi = -\frac{1}{RB_{\mathrm{p}}}\,\partial\Big(\frac{1}{B_{\mathrm{p}}}\,\partial R\Big)\,, \tag{18.86}$$

where the two principal curvatures of the surface,

$$\kappa_{\mathrm{p}} \equiv R\,D^\dagger B_{\mathrm{p}} = \frac{R}{J}\Big(\frac{\partial}{\partial\Psi} - \frac{\partial}{\partial\vartheta}\frac{g_{12}}{g_{22}}\Big)\mathcal{J}B_{\mathrm{p}}\,, \quad \kappa_\varphi \equiv B_{\mathrm{p}}DR = B_{\mathrm{p}}\Big(\frac{\partial}{\partial\Psi} - \frac{g_{12}}{g_{22}}\frac{\partial}{\partial\vartheta}\Big)R\,, \tag{18.87}$$

would be beyond his understanding, but he would highly appreciate the last part of
Eq. (18.86). And so did we, when we tried to derive the equations for the transonic
continua and got stuck with products of normal derivatives, that should not be there in
tangential equations, until we realized the beauty of this equation.

With these geometric counterparts of Section 17.1.3, we are now prepared to analyze the
continuous spectra of stationary transonic equilibra. $\qquad\triangleleft$

18.3.2 Continua of poloidally and toroidally rotating plasmas*

By means of the magnetic surface localization technique described in the previous sub-section, we arrive at the following formulation. The continuous spectra of
axi-symmetric plasmas, transonically rotating in both the toroidal and the poloidal
direction and gravitating about a massive object, are obtained by solving the following set of ordinary differential equations on each magnetic/flow surface:

$$\hat{\mathbf{A}}\cdot\hat{\mathbf{V}} = \hat{\mathbf{B}}\cdot\hat{\mathbf{V}}\,, \qquad \hat{\mathbf{V}} \equiv \begin{pmatrix} \hat{Y} \\ \hat{Z} \end{pmatrix}\,, \tag{18.88}$$

where the matrix operator $\hat{\mathbf{A}}$ represents the magnetic contributions,

$$\hat{\mathbf{A}} \equiv \begin{pmatrix} F\dfrac{R^2 B_{\mathrm{p}}^2}{B^2}F - (M^2 - M_{\mathrm{c}}^2)B^2 U^2 & -\mathrm{i}(M^2 - M_{\mathrm{c}}^2)\dfrac{B^2}{\rho}UF\rho \\[2ex] \mathrm{i}\rho F(M^2 - M_{\mathrm{c}}^2)\dfrac{B^2}{\rho}U & FM_{\mathrm{c}}^2 B^2 F + V \end{pmatrix}\,,$$

$$U \equiv \frac{1}{\rho}\Big[\partial\Big(\frac{\rho R B_\varphi}{B^2}\Big)\Big]\,, \qquad V \equiv \rho\Big[\partial\Big((M^2 - M_{\mathrm{c}}^2)\frac{B^2}{\rho^2}\partial\rho\Big)\Big]\,, \tag{18.89}$$

and the matrix operator $\hat{\mathbf{B}}$ represents the hydrodynamic ones,

$$\hat{\mathbf{B}} \equiv \begin{pmatrix} (\sqrt{\rho}\,\tilde{\omega} - FM)\dfrac{R^2 B_{\mathrm{p}}^2}{B^2}(\sqrt{\rho}\,\tilde{\omega} - MF) & -\mathrm{i}\alpha\sqrt{\rho}\,\tilde{\omega} \\[2ex] \mathrm{i}\alpha\sqrt{\rho}\,\tilde{\omega} & (\sqrt{\rho}\,\tilde{\omega} - FM)B^2(\sqrt{\rho}\,\tilde{\omega} - MF) \end{pmatrix}\,. \tag{18.90}$$

The off-diagonal elements of the latter matrix contain the *Coriolis coupling factor*

$$\alpha \equiv 2\sqrt{\rho}\Big[Rv_{\rm p}\partial(\arctan{(B_\varphi/B_{\rm p})}) + v_\varphi\partial R \Big]$$

$$= M\frac{B^2}{RB_\varphi}\Big[\partial\Big(\frac{R^2 B_\varphi^2}{B^2}\Big)\Big] + \sqrt{\rho}\,\Omega[\partial R^2]\,, \tag{18.91}$$

where we have transformed the expression involving $v_{\rm p}$ and v_φ to an equivalent form in terms of M and Ω (as we have consistently done in all of the analysis; see above paragraph in small print for further intrinsic properties of the magnetic/flow surfaces). For economy of representation, the parallel field operator F, rather than the parallel flow operator $\mathcal{E}$, has been exploited in the matrix $\hat{\mathbf{B}}$ so that the Doppler shifted frequency $\widetilde{\omega} \equiv \omega - n\Omega$ appears instead of ω itself. Transforming the diagonal elements of $\hat{\mathbf{B}}$ back to the original form,

$$(\sqrt{\rho}\,\widetilde{\omega} - FM)\cdots(\sqrt{\rho}\,\widetilde{\omega} - MF) = \rho(\omega - \mathcal{E})\cdots(\omega - \mathcal{E})\,, \tag{18.92}$$

clearly demonstrates that these contributions are, in fact, the hydrodynamic ones: the eigenvalue parameter ω is Doppler shifted by an amount determined by the operator $\mathcal{E}$. Recall the extensive analysis of Chapters 12 and 13 on this topic.

In the matrix $\hat{\mathbf{A}}$, the critical "cusp" value of the square of the poloidal Alfvén Mach number,

$$M_{\rm c}^2 \equiv \frac{\gamma p}{\gamma p + B^2} \tag{18.93}$$

(omitting the tilde for simplicity), has been introduced to highlight the main issue of this chapter. When $M^2 > M_{\rm c}^2$, there is a big negative definite contribution in the upper diagonal element, i.e. precisely in the location where the driving terms of all usual MHD instabilities (including ballooning modes) are to be found. Considering field-aligned modes, giving small contributions of the F operator terms, appears to be sufficient to bring out the instability mechanism: no subtle ballooning-type localizations are necessary. It is not quite that simple, though, because the Coriolis coupling factor α in the matrix $\hat{\mathbf{B}}$ not only strongly couples the two possible polarizations $\hat{Y}$ and $\hat{Z}$ of the modes (as do the off-diagonal elements of the matrix $\hat{\mathbf{A}}$), but also makes the system *non-Hermitian*. We will investigate the peculiar consequences of this fact in the explicit analytical and numerical examples of Sections 18.3.3 and 18.4.

Where does the gravitational potential reside in this formulation? It has been eliminated by making use of the invariance of the Bernoulli function, $\partial H = 0$ (see paragraph in small print in Section 18.3.1), so that it is mainly hidden in the resultant density variations over the magnetic/flow surfaces. However, those cause very significant differences in the stability properties of tokamaks and accretion disks, as we will see.

The previously derived expressions for the continuous spectra of static toroidal equilibria [161] (Section 17.2.3) are recovered in the limit of vanishing poloidal and toroidal velocities ($M \to 0$, $\Omega \to 0$). In that case, the functions ρ and RB_φ become constant on the magnetic surfaces and the remaining poloidal variation ∂B^2 becomes proportional to the geodesic curvature (17.37), which becomes the sole cause of coupling between the two polarizations $\hat{Y}$ and $\hat{Z}$. In the presence of poloidal flow, the expression $\partial(\rho RB_\varphi/B^2)$ no longer represents the geodesic curvature (see paragraph in small print in Section 18.3.1) and the density variations become the more important effect. The other relevant limits of static toroidal equilibria with gravity [377], toroidally rotating θ-pinches [222] and tokamaks [454] are not fully recovered from the present formulation since those cases admit equilibria with $\partial S \neq 0$, which may drive a kind of Brunt–Väisälää instability of the continuum modes. As in the convection zone of the Sun, one might expect these instabilities to switch off when they have generated a constant entropy distribution on the magnetic/flow surfaces. Such instabilities are absent from the present analysis since the poloidal flow itself imposes the constraint $\partial S = 0$. The mentioned cases are recovered for equilibria that obey this additional constraint.

The present analysis is actually closer to the one by Hameiri and Hammer [213] for incompressible plasmas. They showed that a poloidally rotating straight θ-pinch with a non-circular cross-section has an unstable continuous spectrum for super-Alfvénic speeds, $M^2 > 1$, which is caused by parametric instability due to coupling of the continuum modes with the plasma rotation. The generalization to toroidal plasmas, and to general cylindrical plasmas with a poloidal magnetic field component [351], may be obtained from Eqs. (18.88)–(18.90) by taking the limit $\gamma \to \infty$, so that $M_c^2 \to 1$. In the incompressible limit, the hyperbolic flow regimes disappear and the elliptic flow regimes coalesce, with $M^2 = 1$ in the middle. Hence, our present instability threshold $M^2 = M_c^2$ moves to $M^2 = 1$ and the poloidal variation of the poloidal magnetic field, $\partial B_{\mathrm{p}} \neq 0$ due to the non-circular cross-section, becomes the driving force of the instability.

(a) Cylindrical limits of the continua With respect to the meaning of the dominant diagonal terms in the spectral equations (18.88)–(18.90), recall the asymptotic spectral structure of one-dimensional configurations in the limit of large normal "wavenumbers" (rapid variations in the direction of inhomogeneity). For static plasmas, the structure of the spectrum then proves to center about the slow continua ω_{S}^2, the Alfvén continua ω_{A}^2, and the fast continua $\omega_{\mathrm{F}}^2 \equiv \infty$, where the three MHD waves become purely polarized in the three orthogonal directions associated with the magnetic surfaces and the field lines. Considering the limit of an infinitely slender torus ($\epsilon \to 0$), or a periodic cylinder with longitudinal coordinate $\varphi = z/R_0$, the tangential dependence of the modes becomes $\mathrm{e}^{\mathrm{i}(m\vartheta + n\varphi)}$, where m

and n do not couple, so that

$$\omega_S \equiv M_c\,\omega_A\,, \qquad \omega_A \equiv (\sqrt{\rho})^{-1}F\,, \quad F \equiv \mathcal{J}^{-1}(m+nq) = \epsilon(m/q+n)\,, \tag{18.94}$$

where F is the algebraic multiplication factor resulting from the parallel field operator F. For stationary plasmas, the left–right symmetry with respect to the origin $\omega = 0$ is lifted by the Doppler shift Ω_0, and the continuum frequencies become (see Chapter 13):

$$\Omega_S^\pm \equiv \Omega_0 \pm \omega_S\,, \qquad \Omega_A^\pm \equiv \Omega_0 \pm \omega_A\,, \qquad \Omega_F^\pm \equiv \pm\infty\,. \tag{18.95}$$

In plane slab geometry, $\Omega_0 \equiv \mathbf{k}\cdot\mathbf{v}$, where $\mathbf{k}$ is the horizontal wavenumber. In the periodic cylinder, according to Eq. (18.75), Ω_0 is the result of the parallel flow operator $\mathcal{E}$:

$$\Omega_0 \equiv E \equiv M(\mathcal{J}\sqrt{\rho})^{-1}(m+nu) = n\Omega + M(\sqrt{\rho})^{-1}F = n\Omega + M\omega_A\,. \tag{18.96}$$

For fixed m and n, the spectrum becomes schematically as indicated in Fig. 13.3, with essential asymmetry between the continuum modes propagating in the forward $(+)$ and in the backward $(-)$ directions with respect to the background flow. We stress that Fig. 13.3 is essentially restricted to small inhomogeneity in 1D systems (as obtained in a thin slab) so that the different parts of the spectrum do not overlap and the continua remain purely real. Under these restrictions, monotonicity of the discrete spectra on the real axis outside the continua and separator frequencies was proved by a modification of the proof for static 1D systems [183]; see Section 13.1.3.

In the limit of infinitely large aspect ratio ($\epsilon \to 0$), all poloidal variations disappear from the matrices $\hat{\mathbf{A}}$ and $\hat{\mathbf{B}}$. This directly yields the two continua associated with singular perturbations tangential to the magnetic/flow surfaces:

(1) the *Alfvén continua* $A_m^\pm$, with polarizations $\hat{Y} \neq 0$, $\hat{Z} = 0$ and frequencies

$$(\sqrt{\rho}\,\tilde\omega - MF)^2 - F^2 = 0 \quad\Rightarrow\quad \omega = \Omega_A^\pm \equiv \Omega_0 \pm \omega_A = n\Omega + (M \pm 1)\,\omega_A\,; \tag{18.97}$$

(2) the *slow continua* $S_m^\pm$, with polarizations $\hat{Y} = 0$, $\hat{Z} \neq 0$ and frequencies

$$(\sqrt{\rho}\,\tilde\omega - MF)^2 - M_c^2 F^2 = 0 \quad\Rightarrow\quad \omega = \Omega_S^\pm \equiv \Omega_0 \pm \omega_S = n\Omega + (M \pm M_c)\,\omega_A\,. \tag{18.98}$$

Note that, in the limit of vanishing magnetic field ($\omega_A \to 0$), the matrix $\hat{\mathbf{A}}$ disappears and the matrix $\hat{\mathbf{B}}$ yields the usual hydrodynamic flow continua [77] at Ω_0, which may be considered as degeneracies of the Alfvén and slow continua in that limit. In other words: contrary to statements in the literature [416, 222, 40], *there are no separate, additional, flow continua in MHD, but they are absorbed by the Alfvén and slow continua when a magnetic field is present.* As already discussed in Section 13.1.3, this is consistent with the fact that the Alfvén and slow polarizations represent all possible degrees of freedom tangential to the magnetic/flow surfaces. Accordingly, in MHD, the frequencies Ω_0 are not continua but just act as separator frequencies, similar to the frequency ranges $\Omega_{s0}^{\pm}$ and $\Omega_{f0}^{\pm}$ indicated by the grey boxes with labels below the axis in Fig. 13.3. (Recall that the elimination of the latter two frequency ranges as possible continua also has been the subject of prolonged controversy; see Goedbloed [166] for a final account on this.)

One more fundamental issue (also discussed in Section 13.1.3) remains to be addressed. Describing plasma perturbations by means of the primitive Eulerian variables ρ_1, $\mathbf{v}_1$, p_1, $\mathbf{B}_1$, instead of in terms of the Lagrangian variable $\boldsymbol{\xi}$, an additional class of continua is obtained, viz. continua with entropy perturbations $S_1 \sim p_1 - (\gamma p_0/\rho_0)\rho_1 \neq 0$. This class of continua is excluded in the Frieman–Rotenberg formalism because they are not representable in terms of $\boldsymbol{\xi}$ and they just represent initial entropy perturbations that are passively carried by the plasma flow, without influencing the rest of the dynamics ($\mathbf{v}_1 = 0$, $\mathbf{B}_1 = 0$). Hence, these continua do not represent important physics. Nevertheless, they are extremely useful for bookkeeping purposes, as we will see below when we compare the results of our Lagrangian analysis with the numerical results from the PHOENIX code, which exploits an Eulerian representation. Hence, we also list this third class of continua:

(3) the *Eulerian entropy continua* E_m, with $\delta S \neq 0$, $\hat{Y} = \hat{Z} = 0$, and frequencies

$$\omega = \Omega_{\mathrm{E}} \equiv \Omega_0 = n\Omega + M\omega_{\mathrm{A}}\,. \tag{18.99}$$

Although their frequencies coincide with the mentioned frequencies of the hydrodynamic flow continua, they should not be confused with the latter. The flow continua are transformed into the Alfvén and slow continua when $B \neq 0$, but the Eulerian entropy continua are unaffected by the magnetic field. (Actually, the latter were even absent in the original analysis of the HD flow continua by Case [77] because he assumed incompressibility.)

(b) Poloidal mode couplings We now have the necessary labels $\mathrm{A}_m^{\pm}$, $\mathrm{S}_m^{\pm}$ and E_m at our disposal to discuss the different mode couplings that occur in stationary toroidal plasmas. These couplings are caused by the poloidal variation of the quantities inside the square brackets in Eqs. (18.89) and (18.91), which will be explicitly calculated in Sections 18.3.3 and 18.4. Anticipating those calculations, we just present one result of a full numerical calculation with PHOENIX (or, rather, CSPHOENIX, see Section 18.4) of the continuous spectra of a tokamak equilibrium in the second elliptic flow regime, viz. equilibrium (A) of Fig. 18.4. Fig. 18.6 shows the real part of the normalized continuum frequencies $\bar{\omega} \equiv \epsilon^{-1}\omega$, together with the labels of the modes wherever this is possible (i.e. away from mode crossings). Every point in the plot (the appearance of curves is just the result of many runs of the code with

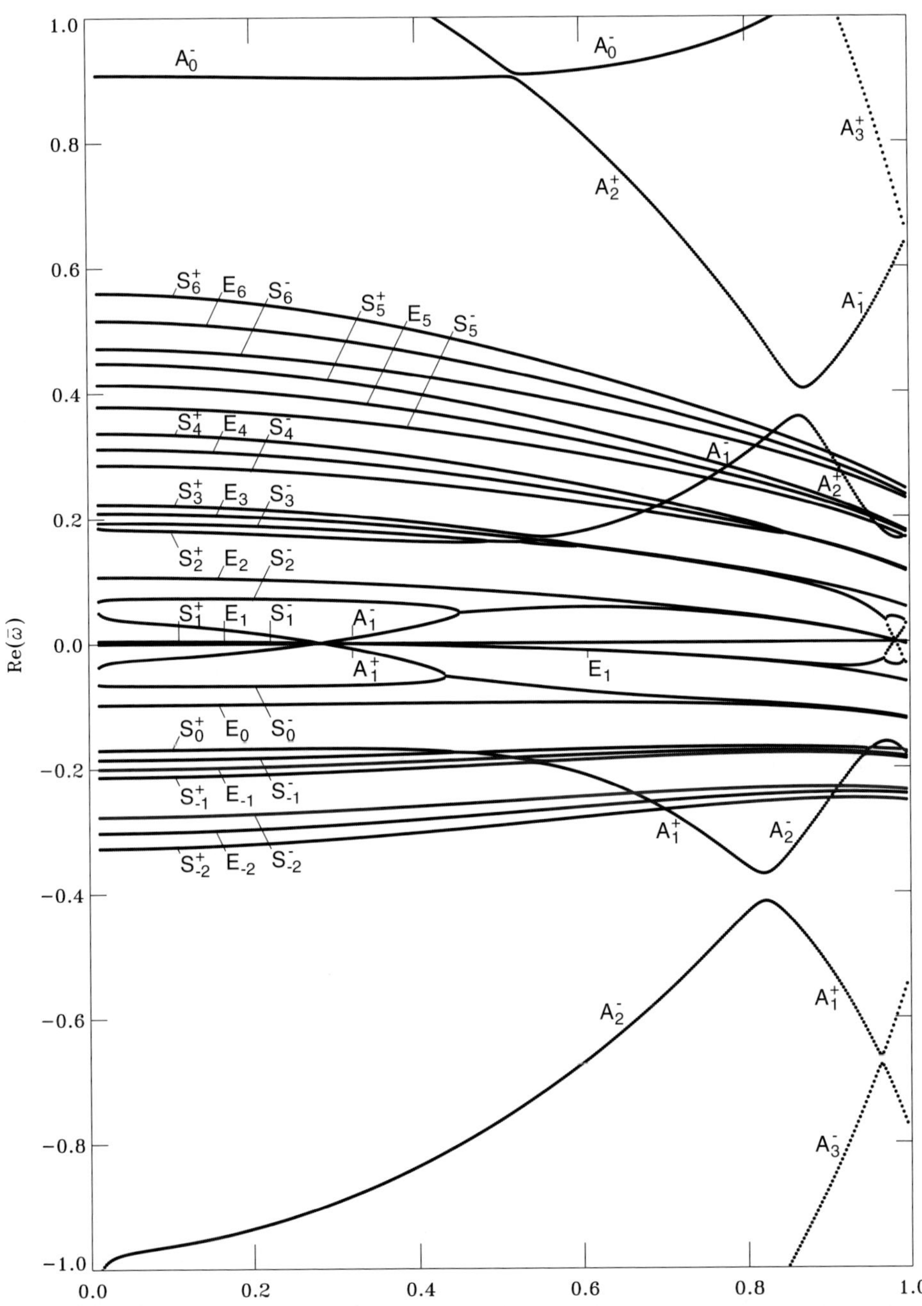

Fig. 18.6 Poloidal mode labeling and coupling of the real parts of the continuum frequencies $\bar{\omega} \equiv \epsilon^{-1}\omega$ as a function of the radial flux coordinate $s \equiv \sqrt{\psi}$ for equilibrium (A) of Figs. 18.4 and 18.5; toroidal mode number $n = -1$, poloidal mode numbers $-2 \leq m \leq 6$.

small increments of s) corresponds to one constituent frequency of the continuous spectrum corresponding to an improper mode that becomes singular at the indicated position s. Note that the curves for the entropy continua E_m do not interact with the other ones, as expected, but they do provide a very convenient indication of the radial variation of Ω_0 (mainly due to $\rho(s)$ and $q(s)$), which is the central frequency for both the Alfvén and the slow modes, according to Eqs. (18.97) and (18.98). For all the numerical examples of the equilibria A–D (corresponding to Figs. 18.4 and 18.5), parameters have been chosen such as to agree with the trans-slow ordering (18.63), whereas the contribution Ω of the non-parallel toroidal flow has been chosen to be negligible so that

$$\omega_S/\omega_A \equiv M_c \ (\approx 0.008) \ \ll \ \Omega_0/\omega_A \approx M \ (\approx 0.1) \ \ll \ 1 \,. \tag{18.100}$$

As a result, the spacing $\Delta\Omega_E \sim M$ of the entropy curves E_m is just intermediate between the narrow spacing $\Delta\Omega_S \sim M_c$ of the slow curves $S_m^\pm$ and the very wide spacing $\Delta\Omega_A \sim 1$ of the Alfvén curves $A_m^\pm$: triplets S_m^+, E_m, S_m^- are easily distinguished, but the complete quintuplets A_m^-, S_m^+, E_m, S_m^-, A_m^+ are hardly visible on the scale of Fig. 18.6.

Of course, the main point of Fig. 18.6 is the illustration of how poloidal mode coupling completely upsets the near cylindrical radial variation of the triplets S_m^+, E_m, S_m^- by the crossing of the lower m Alfvén branches at the rational surfaces $q = 1$ (at $s \approx 0.3$) and $q = 2$ (at $s \approx 0.98$), where ω_A vanishes for the $m = 1$ and $m = 2$ harmonics, respectively. In a wide neighborhood of the $q = 1$ rational surface, the slow modes $S_2^\pm$ and $S_0^\pm$ are pushed far apart by the $A_1^\pm$ harmonics and, similarly, strong mode interaction around the $q = 2$ surface pushes the $S_3^\pm$ and $S_1^\pm$ modes apart by the $A_2^\pm$ modes. More spectacular yet is the merging of the S_2^- and A_1^- branches (at $s = 0.46$) and of the A_1^+ and S_0^- branches (at $s = 0.43$), where the frequency $\bar\omega$ becomes complex (the reader may want to glance at Fig. 18.10) and the instabilities originate which are the main subject of this chapter. *Through the six mode interaction $S_{m+1}^\pm$, $A_m^\pm$, $S_{m-1}^\pm$ at the rational surfaces, the continuous spectra of transonically rotating plasmas become unstable.* (Fortunately, for the analysis, the slow modes $S_m^\pm$, which are also degenerate at the rational surfaces, $\omega_S = 0$, appear not to interact with the six modes.) We will investigate these six-mode interactions analytically in Section 18.3.3 and numerically in Section 18.4.

Another interesting feature of the continuous spectra shown in Fig. 18.6 is the avoided crossings of the A_m^+ and A_{m-1}^- Alfvén modes and of the A_m^- and A_{m-1}^+ Alfvén modes at

$$q = -\frac{1}{n}\left[m - \tfrac{1}{2}(1 \mp M)\right], \tag{18.101}$$

resp., i.e., for $m = 1$: $q = 1.55$ (at $s = 0.86$) and $q = 1.45$ (at $s = 0.82$). The resulting $|\Delta m| = 1$ gaps in the continuous Alfvén spectra are well-known for static tokamak equilibria. Equation (18.101) shows that poloidal rotation removes the forward/backward degeneracy of the gap frequencies with important consequences for the discrete gap modes that may be found there.

18.3.3 Analysis of trans-slow continua for small toroidicity*

(a) Fourier representation (exact) We exploit the equilibria of Section 18.2.3 to study the transonic continuous spectra by means of the formalism of Section 18.3.2. To that end, we write the matrix eigenvalue problem (18.88) as

$$\begin{pmatrix} \mathcal{K} & \mathcal{M} \\ \mathcal{N} & \mathcal{L} \end{pmatrix} \begin{pmatrix} \hat{Y} \\ \hat{Z} \end{pmatrix} = 0\,, \tag{18.102}$$

where the elements

$$\mathcal{K} \equiv \mathcal{J}^{-1}[(\tilde{\omega} - \mathcal{J}F f_2)\, f_1\, (\tilde{\omega} - f_2\mathcal{J}F) - \mathcal{J}F f_3\, \mathcal{J}F + f_4]\,,$$

$$\mathcal{M} \equiv \mathcal{J}^{-1}[-\mathrm{i} f_5\,\tilde{\omega} + \mathrm{i} f_6\, \mathcal{J}F + f_7]\,,$$

$$\mathcal{N} \equiv \mathcal{J}^{-1}[\mathrm{i} f_5\,\tilde{\omega} - \mathrm{i}\mathcal{J}F f_6 + f_7]\,,$$

$$\mathcal{L} \equiv \mathcal{J}^{-1}[(\tilde{\omega} - \mathcal{J}F f_2)\, f_8\, (\tilde{\omega} - f_2\mathcal{J}F) - \mathcal{J}F f_9\, \mathcal{J}F - f_{10}] \tag{18.103}$$

need evaluation of the following functions of ϑ:

$$f_1 \equiv \frac{\mathcal{J}\rho R^2 B_\mathrm{p}^2}{B^2}\,, \quad f_2 \equiv \frac{M}{\mathcal{J}\sqrt{\rho}}\,, \quad f_3 \equiv \frac{R^2 B_\mathrm{p}^2}{\mathcal{J}B^2}\,, \quad f_4 \equiv \mathcal{J}(M^2 - M_\mathrm{c}^2)B^2 U^2,$$

$$f_5 \equiv \mathcal{J}\alpha\sqrt{\rho}\,, \quad f_6 \equiv (M^2 - M_\mathrm{c}^2)B^2 U\,, \quad f_7 \equiv \mathcal{J}(M^2 - M_\mathrm{c}^2)\frac{B^2}{\rho}\, U\,[\partial\rho],$$

$$f_8 \equiv \mathcal{J}\rho B^2\,, \quad f_9 \equiv \frac{M_\mathrm{c}^2 B^2}{\mathcal{J}}\,, \quad f_{10} \equiv \mathcal{J}V\,. \tag{18.104}$$

Fourier analysis of Eq. (18.102) yields an infinite matrix eigenvalue problem,

$$\sum_{m'} \begin{pmatrix} K_{mm'} & M_{mm'} \\ N_{mm'} & L_{mm'} \end{pmatrix} \begin{pmatrix} Y_{m'} \\ Z_{m'} \end{pmatrix} = 0\,, \tag{18.105}$$

where $Y_{m'}$ and $Z_{m'}$ are the Fourier components of $\hat{Y}(\vartheta)$ and $\hat{Z}(\vartheta)$, and the matrix elements

$$K_{mm'} \equiv \frac{1}{2\pi} \oint d\vartheta\, \mathrm{e}^{-\mathrm{i}m\vartheta}\, \mathcal{J}\mathcal{K}\mathrm{e}^{\mathrm{i}m'\vartheta} \equiv \frac{1}{2\pi} \oint K(m, m'; \vartheta)\mathrm{e}^{-\mathrm{i}(m-m')\vartheta}\, d\vartheta\,, \quad \text{etc.},$$

$$\tag{18.106}$$

involve integrands with the operators $\mathcal{K}$, etc., that are converted into algebraic expressions $K(m, m'; \vartheta)$, etc., by means of integration by parts:

$$K \equiv f_1[\widetilde{\omega} - f_2(m + nq)][\widetilde{\omega} - f_2(m' + nq)] - f_3(m + nq)(m' + nq) + f_4 ,$$

$$M \equiv -\mathrm{i}f_5\,\widetilde{\omega} + \mathrm{i}f_6(m' + nq) + f_7 ,$$

$$N \equiv \mathrm{i}f_5\,\widetilde{\omega} - \mathrm{i}f_6(m + nq) + f_7 , \tag{18.107}$$

$$L \equiv f_8[\widetilde{\omega} - f_2(m + nq)][\widetilde{\omega} - f_2(m' + nq)] - f_9(m + nq)(m' + nq) - f_{10} .$$

Obviously, M should not be confused here with the Alfvén Mach number.

Before we can continue with the evaluation of the matrix elements for small toroidicity, we need to consider the way in which the density enters the problem. Recall that ρ did not enter the equilibrium core equations (18.23) and (18.24). However, the spectral equation (18.88) manifestly depends on ρ, both through the weight factor $\sqrt{\rho}$ in front of the eigenvalue parameter $\widetilde{\omega}$ and through the angular derivative $\partial\rho$, that is crucial for the instabilities analyzed here. At this point, we introduce a third dimensional unit (after length a and magnetic field strength B_0), viz. the density ρ_0 on the magnetic axis. Similar to the transformations (18.20), we will exploit ρ_0 to make all density-dependent quantities dimensionless by putting $\rho_0 = 1$ without indicating this by further subscripts:

$$\rho/\rho_0 \to \rho , \qquad (a\sqrt{\rho}/B_0)\,\omega \to \omega . \tag{18.108}$$

The radial and angular dependencies of the dimensionless density are determined by M^2 and χ'^2, according to Eq. (18.11). Whereas M^2 should be a solution of the Bernoulli equation (18.24), $\chi'^2(\psi)$ is an arbitrary function, except that its amplitude is given by the value of M^2 on the magnetic axis: $\chi'^2(0) \equiv M_0^2$. Hence, it is expedient to introduce a sixth flux function, with unit amplitude:

$$\bar{\Lambda}_0(\psi) \equiv M_0^{-2}\chi'^2(\psi) \left[\equiv \bar{\Lambda}_5(\psi)/\bar{\Lambda}_5(0) \right] , \qquad \bar{\Lambda}_0(0) = 1 , \tag{18.109}$$

where we have indicated in brackets that this flux function is not really an additional free function, for accretion disks, since it has to have the same shape as $\bar{\Lambda}_5(\psi)$ according to Eq. (18.13). The density is then determined by

$$\rho = \bar{\Lambda}_0(\psi)\,(M^2/M_0^2)^{-1} , \tag{18.110}$$

so that the radial variation is due to both $\bar{\Lambda}_0$ and M^2, but the more important angular variation is solely due to M^2.

Since we are concerned with continuum modes in the Alfvénic range of frequencies, according to Eq. (18.94) we need to rescale ω with ϵ to get $\mathcal{O}(1)$ expressions:

$$\bar{\omega} \equiv \epsilon^{-1}\omega \left[\equiv \frac{R_0\sqrt{\rho}}{B_0}\,\omega \right] . \tag{18.111}$$

We have not used the smallness of ϵ yet, so that the expressions so far are exact.

(b) Expansion in small toroidity We now apply the trans-slow poloidal flow ordering (18.63) to the equilibrium expressions f_i defined in Eqs. (18.104). This requires conversion of the metric derived in Section 18.2.3 to the straight-field-line coordinates Ψ, ϑ, φ:

$$\vartheta \approx \hat\theta - (\epsilon\hat{r} - \Delta' - \sigma)\sin\theta. \tag{18.112}$$

Here, σ is the angular variation of RB_φ,

$$RB_\varphi \sim 1 + \sigma\cos\vartheta, \qquad \sigma \approx -2\epsilon\hat{r}\,\frac{\bar\Lambda_{4,0}}{\bar\Lambda_{3,0} - \bar\Lambda_{4,0}}, \tag{18.113}$$

which we may neglect in the present case ($\bar\Lambda_4 \ll \bar\Lambda_3$). The metric tensor and the Jacobian then become

$$\hat{g}^{11} \equiv \psi'^2(1 - 2\Delta'\cos\theta), \qquad \hat{g}^{12} \equiv -\psi'^2[\epsilon - \hat{r}^{-1}(\hat{r}\Delta')' - \sigma']\sin\vartheta,$$

$$\hat{g}^{22} \equiv \hat{r}^{-2}[1 - 2(\epsilon\hat{r} - \Delta' - \sigma)\cos\vartheta], \qquad \hat{g}^{33} \equiv \epsilon^2(1 - 2\epsilon\hat{r}\cos\theta),$$

$$J \equiv (\nabla\psi \times \nabla\vartheta \cdot \nabla\varphi)^{-1} \approx (\epsilon\Psi')^{-1}\hat{r}[1 + (2\epsilon\hat{r} - \sigma)\cos\vartheta]. \tag{18.114}$$

In these coordinates, we may remove all factors ψ' in favor of the safety factor q:

$$q \equiv JB_\varphi/R \approx \hat{r}/\Psi', \qquad B_{\rm p} \approx \epsilon r q^{-1}[1 - (\epsilon\hat{r} + \Delta')\cos\vartheta]. \tag{18.115}$$

Finally, the most crucial quantity is the perturbed density,

$$\rho(\hat{r})[1 - (\mu/M^2)\cos\vartheta] = \rho(\hat{r})[1 - \epsilon(1 - \Gamma)\hat{r}\cos\vartheta], \tag{18.116}$$

where now, in contrast to the previous subsection, $\rho(\hat{r})$ only indicates the zeroth order radial density profile. The gravitational interaction coefficient Γ, defined in Eq. (18.67), determines the first order angular variation of the density.

We evaluate the factors f_i of Eqs. (18.104) to first order in terms of Δ' and Γ:

$$f_1 \approx (\epsilon q)^{-1}\rho\hat{r}^2\,(1 + \epsilon a_1\hat{r}\cos\vartheta), \qquad a_1 \equiv 3 + \Gamma - 2(\epsilon r)^{-1}\Delta',$$

$$f_2 \approx \epsilon M(q\sqrt{\rho})^{-1}\,(1 + \epsilon a_2\hat{r}\cos\vartheta), \qquad a_2 \equiv -1 - \Gamma,$$

$$f_3 \approx \epsilon\hat{r}^2 q^{-3}\,(1 + \epsilon a_3\hat{r}\cos\vartheta), \qquad a_3 \equiv -2(\epsilon\hat{r})^{-1}\Delta',$$

$$f_4 \approx \epsilon M^2 q^{-1}\,a_4^2\hat{r}^2\sin^2\vartheta, \qquad a_4 \equiv -1 - \Gamma,$$

$$f_5 \approx -(\sqrt{\rho}M + \epsilon^{-1}\rho\Omega)\,a_5\hat{r}\sin\vartheta, \qquad a_5 \equiv 2,$$

$$f_6 \approx \epsilon M^2 q^{-1}\,a_6\hat{r}\sin\vartheta, \qquad a_6 \equiv -1 - \Gamma,$$

$$f_7 \approx \epsilon^2 M^2 q^{-1}\,a_6 a_7\hat{r}^2\sin^2\vartheta, \qquad a_7 \equiv 1 - \Gamma,$$

$$ f_8 \approx \epsilon^{-1} \rho q \left(1 + \epsilon a_8 \hat{r} \cos \vartheta\right), \qquad\qquad a_8 \equiv -1 + \Gamma, $$

$$ f_9 \approx \epsilon q^{-1} M_{\rm c}^2, $$

$$ f_{10} \approx \epsilon^2 M^2 q^{-1} a_{10} \hat{r} \cos \vartheta, \qquad\qquad a_{10} \equiv 1 - \Gamma. \tag{18.117} $$

Scaling the matrix elements and the eigenvector as

$$ \bar{K} \equiv (\epsilon \hat{r}^2)^{-1} q\, K, \quad \bar{M} \equiv (\epsilon \hat{r})^{-1} M, \quad \bar{N} \equiv (\epsilon \hat{r})^{-1} N, \quad \bar{L} \equiv (\epsilon q)^{-1} L, $$

$$ \bar{Y} \equiv \hat{r} q^{-1} Y, \quad \bar{Z} \equiv Z, \tag{18.118} $$

the transformed matrix eigenvalue problem (18.105) will contain the following matrix elements (accurate to first order):

$$ \bar{K}_{mm'} \approx \left\{ [\sqrt{\rho}\,\bar{\omega} - M(m/q + n)]^2 - (m/q + n)^2 \right\} \delta_{mm'} $$
$$ + \tfrac{1}{2} M^2 a_4^2 (\delta_{mm'} - \delta_{m,m'-2} - \delta_{m,m'+2}), $$

$$ \bar{M}_{mm'} \approx -\tfrac{1}{2} \left\{ M a_5 \sqrt{\rho}\,\bar{\omega} + M^2 a_6(m'/q + n) \right\} (\delta_{m,m'-1} - \delta_{m,m'+1}), $$

$$ \bar{N}_{mm'} \approx \tfrac{1}{2} \left\{ M a_5 \sqrt{\rho}\,\bar{\omega} + M^2 a_6(m/q + n) \right\} (\delta_{m,m'-1} - \delta_{m,m'+1}), $$

$$ \bar{L}_{mm'} \approx \left\{ [\sqrt{\rho}\,\bar{\omega} - M(m/q + n)]^2 - M_{\rm c}^2(m/q + n)^2 \right\} \delta_{mm'}. \tag{18.119} $$

Since Ω is negligible in that order for the equilibria we are considering, $\tilde{\omega} \approx \omega$, so that we do not need to distinguish the two in the normalized eigenvalue $\bar{\omega}$.

(c) "Dispersion equation" for trans-slow Alfvén continuum modes The matrix elements (18.119) give rise to two separate coupling schemes, with either the Alfvén waves or the slow waves as the central perturbation. Since the most important instabilities result from the first scheme, we will concentrate on that. The Alfvén wave coupling scheme

$$ \begin{pmatrix} \bar{L}_{m-1,m-1} & \bar{N}_{m-1,m} & 0 \\ \bar{M}_{m,m-1} & \bar{K}_{m,m} & \bar{M}_{m,m+1} \\ 0 & \bar{N}_{m+1,m} & \bar{L}_{m+1,m+1} \end{pmatrix} \begin{pmatrix} \bar{Z}_{m-1} \\ \bar{Y}_m \\ \bar{Z}_{m+1} \end{pmatrix} = 0, \tag{18.120} $$

precisely provides us with the means to analyze the interaction of the six modes $A_m^{\pm}$, $S_{m-1}^{\pm}$ and $S_{m+1}^{\pm}$ that we encountered in Fig. 18.6 of Section 18.3.2. In the neighborhood of the Alfvén resonance $m/q + n \sim \epsilon$ this gives the following "dispersion equation" (quotation marks to remind us of the contradiction in terms):

$$\begin{vmatrix} (\sqrt{\rho}\,\bar{\omega} + M/q)^2 & C & 0 \\ C & \rho\,\bar{\omega}^2 - (m/q+n)^2 + \tfrac{1}{2}M^2\,a_4^2 & C \\ 0 & C & (\sqrt{\rho}\,\bar{\omega} - M/q)^2 \end{vmatrix} = 0,$$

$$C \equiv \tfrac{1}{2}Ma_5\sqrt{\rho}\,\bar{\omega} - \tfrac{1}{2}(M^2/q)\,a_6\,. \tag{18.121}$$

Inserting the explicit expressions for a_4, a_5, a_6 derived in Eq. (18.117), and defining a rescaled eigenvalue $\hat{\omega} \equiv (\sqrt{\rho q}/M)\,\bar{\omega}$, and a measure $Q \equiv (n/M)(m/n+q)$ for the "distance" to the rational value of q, the "dispersion equation" becomes

$$\begin{vmatrix} (\hat{\omega}+1)^2 & q[\hat{\omega} + \tfrac{1}{2}(1+\Gamma)] & 0 \\ q[\hat{\omega} + \tfrac{1}{2}(1+\Gamma)] & \hat{\omega}^2 - Q^2 + \tfrac{1}{2}q^2(1+\Gamma)^2 & -q[\hat{\omega} - \tfrac{1}{2}(1+\Gamma)] \\ 0 & -q[\hat{\omega} - \tfrac{1}{2}(1+\Gamma)] & (\hat{\omega}-1)^2 \end{vmatrix} = 0,$$

$$\tag{18.122}$$

which turns out to be a cubic equation in $\hat{\omega}^2$,

$$\hat{\omega}^6 - [2+Q^2+\tfrac{1}{2}q^2(1-\Gamma)(3+\Gamma)]\hat{\omega}^4 + [1+2Q^2+\tfrac{1}{2}q^2(1-\Gamma)(1+3\Gamma)]\hat{\omega}^2 - Q^2 = 0,$$

$$\tag{18.123}$$

that may be solved by means of the Cardano expressions.

At a rational surface, $Q^2 = 0$ and we get two marginal modes $\hat{\omega}^2 = 0$ and four solutions

$$\hat{\omega}^2 = 1 + \tfrac{1}{4}q^2(1-\Gamma)(3+\Gamma) \pm q|1-\Gamma|\sqrt{1 + \tfrac{1}{16}q^2(3+\Gamma)^2}\,. \tag{18.124}$$

These correspond to four stable waves ($\hat{\omega}^2 > 0$) when $\Gamma < 1$, but two solutions may become exponentially unstable ($\hat{\omega}^2 < 0$) when $\Gamma > 1$. Clearly, $\Gamma = 1$ marks a qualitative change in the continuous spectrum, corresponding to the qualitative change in the equilibria given by Eq. (18.66) and illustrated in Fig. 18.4.

The solutions $\bar{\omega} \equiv M(\sqrt{\rho q})^{-1}\hat{\omega}$ of Eq. (18.123) are plotted as a function of q in Figs. 18.7 and 18.8 for four representative values of Γ, roughly corresponding to the equilibria (A) and (C) of Fig. 18.4, for which the "exact" numerical continuous spectra will be discussed in Section 18.4 (Figs. 18.9–18.12). Starting with $\Gamma = 0$ (tokamak, i.e. no gravitating central object, Fig. 18.7(a)), the six-mode interaction $S_2^\pm$, $A_1^\pm$, $S_0^\pm$ describes the real part of the "exact" spectrum shown in Fig. 18.6 quite well around the $q = 1$ resonance (where $\omega_A = 0$ for $n = -1$, $m = 1$). Away from the resonance (increasing Q^2), the curves $\mathrm{Re}\,(\bar{\omega})$ of the pair S_2^-, A_1^- and of the pair A_1^+, S_0^- coalesce (at the same value of q: the asymmetry present in the "exact" solution of Fig. 18.6 is due to higher order contributions that are

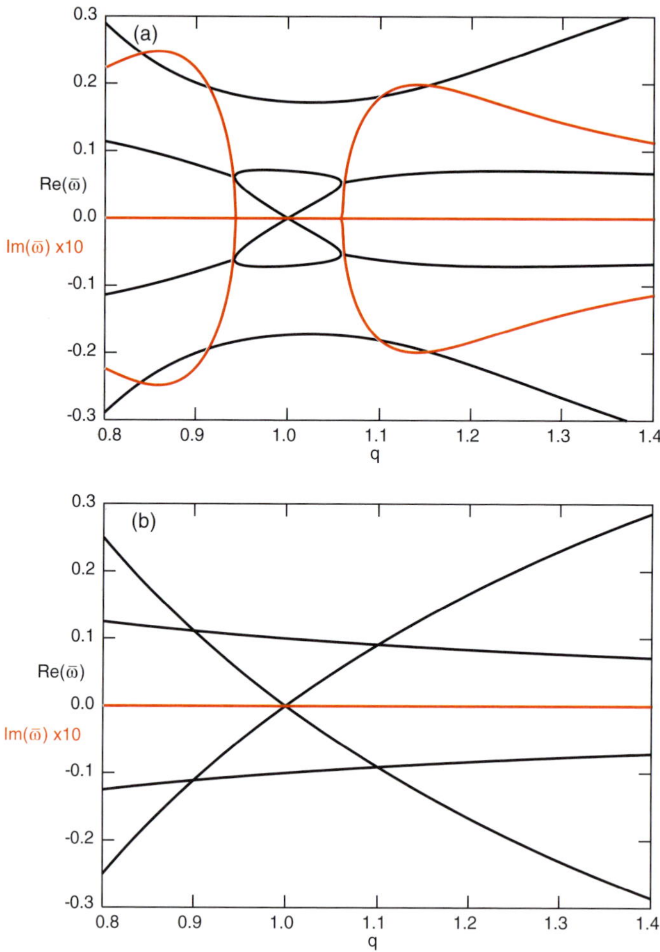

Fig. 18.7 Real and imaginary part of the roots of the "dispersion equation" (18.123) for the trans-slow Alfvénic continuous spectrum of (a) tokamak ($\Gamma = 0$) and (b) Keplerian accretion disk ($\Gamma = 1$) for $n = -1$, $m = 1$, $M = 0.1$, $\rho = 1$.

neglected here) and an imaginary contribution $\mathrm{Im}\,(\bar{\omega})$ emerges (indicated in red): *the trans-slow continuous spectrum becomes overstable*. The growth rate of the instability is roughly $M\times$ the real part of the Alfvén frequency: quite sizeable! The spectrum for an accretion disk with moderate central mass (rougly corresponding to equilibrium (B) with $\Gamma = 0.25$, not shown) is similar to that of the tokamak, but has smaller imaginary parts (corresponding to the reduction of the density variation expressed by Eq. (18.66)). This tendency continues until $\Gamma = 1$, which is completely stable (Fig. 18.7(b)): Eq. (18.123) degenerates into two solutions $\bar{\omega} = \pm(\sqrt{\rho})^{-1}(m/q + n)$ corresponding to the Alfvén modes $\mathrm{A}_1^{\pm}$ and four so-

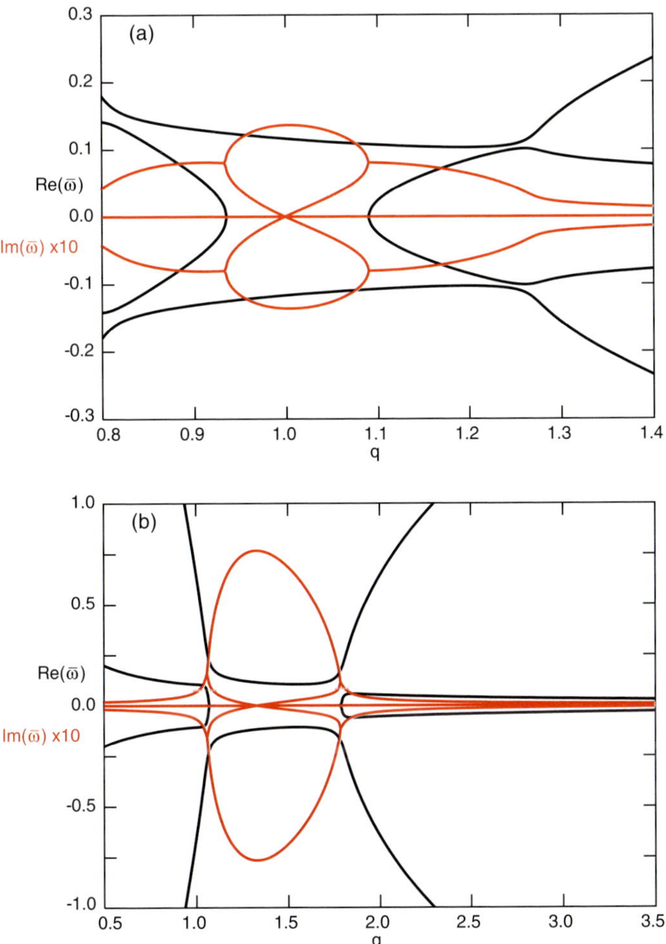

Fig. 18.8 Real and imaginary part of the roots of the "dispersion equation" (18.123) for the trans-slow Alfvénic continuous spectrum of accretion disks with massive central object: (a) $\Gamma = 2$ for $n = -1$, $m = 1$ and (b) $\Gamma = 10$ for $n = -3$, $m = 4$; $M = 0.1$, $\rho = 1$.

lutions $\bar\omega = \pm(\sqrt{\rho}q)^{-1}M$ corresponding to the degenerate slow modes $\mathrm{S}_2^\pm$, $\mathrm{S}_0^\pm$ (since the contribution of M_c has been neglected). Beyond $\Gamma = 1$, an entirely different spectral structure emerges. For $\Gamma = 2$ (roughly corresponding to equilibrium (C), Fig. 18.8(a)), far away from the resonance, two real and four complex roots (i.e. four overstable modes) are found. In the neighborhood of the resonance $q = 1$, the complex roots coalesce two by two to produce two purely imaginary roots, i.e. instabilities with maximum growth rate at the resonance. These growth rates are enormous, of the same order of magnitude as the Alfvén frequency!

Note that the solution $\hat\omega^2 = \hat\omega^2(Q, q, \Gamma)$ of the "dispersion equation" (18.123) involves the equilibrium parameters q, M, Γ, but that Q also brings in the mode numbers n and m in a physically significant manner. This is shown by taking the limits

$$|n| \to \infty, \quad \frac{m}{n} + q \to 0, \quad \text{for fixed} \quad Q \equiv \frac{n}{M}\left(\frac{m}{n} + q\right), \tag{18.125}$$

which implies that the instabilities of the plots shown, obtained for certain values of the parameter Q, will be obtained for any sextuple $\mathrm{A}_m^\pm$, $\mathrm{S}_{m-1}^\pm$, $\mathrm{S}_{m+1}^\pm$: *all toroidal mode numbers n are unstable close to or at the rational surfaces $q = -m/n$ for trans-slow ($M^2 > M_c^2$) Alfvénic continuum modes.* Their growth rates are just determined by the values of the equilibrium parameters ρ (the density), q (the inverse rotational transform of the magnetic field lines), M (the poloidal Alfvén Mach number of the flow) and Γ (the strength of the gravitational interaction) at that particular location.

To demonstrate this, we have plotted the roots of the "dispersion equation" for mode numbers corresponding to a rational surface, at $q = 4/3$, for $\Gamma = 10$ (Fig. 18.8(b)). In that case, the very massive central object drives the growth of the instability to values far in excess of the Alfvén frequency. We obtain the following estimate of the asymptotic growth rate for $\Gamma \gg 1$ from Eq. (18.123):

$$\bar\omega \approx \pm \mathrm{i}\, M\Gamma/\sqrt{2\rho}\,. \tag{18.126}$$

In this limit, the modes become dominantly Alfvénic ($|Y_m| \gg |Z_{m-1}|, |Z_{m+1}|$) and the growth rates are so large that nonlinear effects soon should take over. Hence, these modes might produce the levels of MHD turbulence in thick magnetically dominated accretion disks with trans-slow poloidal velocities that are needed to break the co-moving constraint of the poloidal flow and magnetic field in order to produce accretion and jets.

18.4 Trans-slow continua in tokamaks and accretion disks*

In this section, we turn to a full numerical study of the trans-slow Alfvénic continuous spectrum for the sequence of equilibria (A)–(D) of Figs. 18.4 and 18.5, where the gravitational strength parameter $\Gamma_0 \equiv \Gamma(0)$ is increased from 0 (tokamak) to 2 (thick accretion disk with massive central object). The effect of flattening of the cross-section of the disk was also studied, but it is not shown here (see [173]).

We exploit the numerical programs FINESSE [29] and PHOENIX [173, 52], developed in the context of MHD spectroscopy for laboratory and astrophysical plasmas [180], relying on the techniques discussed in Chapters 15–17. The first code computes the equilibrium of an axi-symmetric plasma with transonic rotations in both the toroidal and poloidal direction in the presence of the gravitational field of a central object. The discretization techniques for the flux equation are the same as exploited in the HELENA code [238] for static equilibria. However, the additional implementation of the Bernoulli equation required a careful monitoring of the parameters of the new flux functions $\bar{\Lambda}_i(\psi)$ to remain in a particular elliptic flow regime. The second code, PHOENIX, computes the complete ideal or resistive MHD spectrum of this equilibrium, exploiting an Eulerian representation so that the Eulerian entropy continua are computed as well, as we have seen in Section 18.3.2. This spectral code is a further development of the CASTOR code [272] for static plasmas, with the addition of the rotational and gravitational contributions and implementation of the powerful Jacobi–Davidson algorithm [412] (see Section 15.3.4) for scanning the spectrum of non-Hermitian operators in the complex ω-plane. Here, the resistivity was switched off. For the computation of the continuous spectrum, the auxiliary program CSPHOENIX was exploited. Through a special limiting procedure [381] (also exploited to construct Fig. 17.5), this program uses the same discretized matrix elements as PHOENIX, but evaluated on a single magnetic/flow surface only. In this manner, we obtain the continuous spectrum of transonic equilibria, characterized by the flux functions $\bar{\Lambda}_i(\psi)$, in the form of plots of $\mathrm{Re}(\bar{\omega})$ and $\mathrm{Im}(\bar{\omega})$ as a function of the radial flux coordinate $s \equiv \sqrt{\psi}$.

These plots are shown in two ways, viz. (1) as separate plots of $\mathrm{Re}(\bar{\omega})$ and $\mathrm{Im}(\bar{\omega})$ versus s or q (Figs. 18.9 and 18.11); (2) as plots in the complex $\bar{\omega}$ plane, with s as parameter, in color coding (Figs. 18.10 and 18.12). We have restricted the analysis to one toroidal harmonic ($n = -1$), with a band of nine coupling poloidal harmonics ($-2 \leq m \leq 6$), and a flux range that usually contains only one rational surface $q = 1$ (except for the tokamak case of Fig. 18.9, which also contains the $q = 2$ rational surface). The numerical results clearly exhibit the central six-mode interaction analyzed in Section 18.3.3, but without neglect of the higher order terms in M^2 and M_c^2. Hence, asymmetries arise in the continuous spectra and many more unstable roots become visible. When comparing the numerical values of $\bar{\omega}$ shown in Fig. 18.9 with the analytical solutions of the "dispersion equation" (18.123) shown in Fig. 18.6, one should realize that, in the latter plots, only q varies with the radial position whereas the equilibrium quantities ρ, M^2 and Γ have been kept constant. Of course, in the numerical solutions of the present section, the full 2D dependencies of the equilibrium quantities have been used. When this is taken into account, the quantitative agreement between the two approaches is excellent.

*18.4.1 Tokamaks and magnetically dominated accretion disks**

The continuous spectra for the tokamak equilibrium (A), where $\Gamma_0 = 0$, are shown in Fig. 18.9. Figure 18.9(a) represents the central interaction region of the larger plot already shown in Fig. 18.6 with the mode labels. The imaginary parts of the

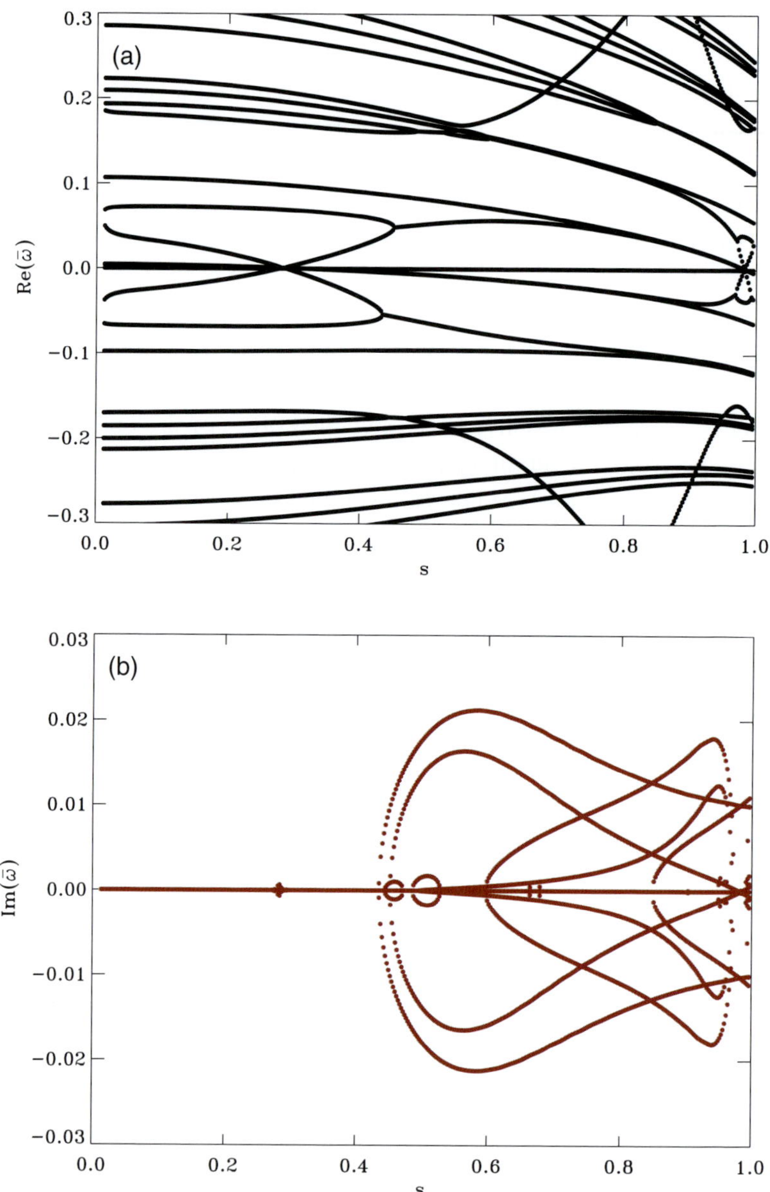

Fig. 18.9 (a) Real and (b) imaginary parts of the frequencies $\bar{\omega}$ of the trans-slow continuous spectrum as a function of the radial flux coordinate $s \equiv \sqrt{\psi}$ for a tokamak with circular cross-section, equilibrium (A); $n = -1$.

continuum frequencies in Fig. 18.9(b) show the sudden onset of instability, with complex values of $\bar{\omega}$ (i.e. overstability), at $s \approx 0.43$–0.45 that we have already interpreted as due to the six-mode interaction $S_2^{\pm}$, $A_1^{\pm}$, $S_0^{\pm}$. However, some addi-

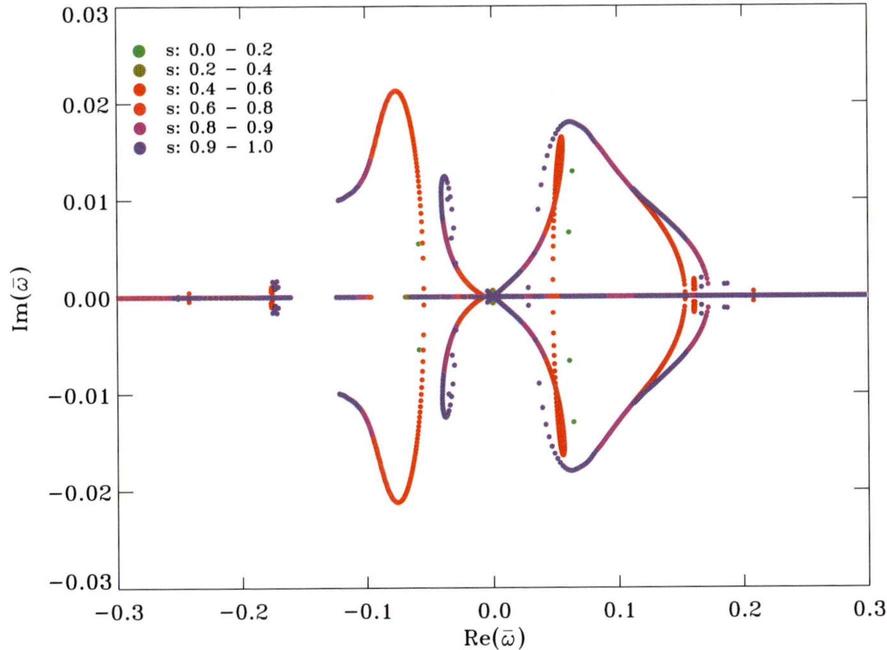

Fig. 18.10 Complex spectrum corresponding to Fig. 18.9, equilibrium (A).

tional structure is to be seen in this figure. Going from $s = 0.28$ ($q = 1$) to the right in Fig. 18.9(b), we subsequently encounter the following features.

(a) The degenerate six-mode interaction analyzed in Section 18.3.3 splits apart in the three-mode interaction A_1^+, $S_0^\pm$, which becomes unstable at $s = 0.43$ and remains unstable in the rest of the interval, and the three-mode interaction A_1^+, $S_2^\pm$, which becomes unstable at $s = 0.45$ and becomes stable again at $s = 0.98$ ($q = 2$). Due to this lifting of the degeneracy, the first triple is more unstable than the second one;

(b) Just to the right of the onset of these instabilities, $\mathrm{Re}(\bar{\omega})$ of the slow constituent S_0^+ of the first triple coalesces with that of S_{-1}^+ to produce a secondary instability with $\mathrm{Im}(\bar{\omega}) \sim 10^{-3}$ (small circle starting at $s = 0.44$). Similarly, a secondary instability is produced by the S_2^+ constituent of the second triple, which gets entangled in a higher order slow mode interaction with S_3^- (bigger circle starting at $s = 0.48$);

(c) Moving further to the right, we encounter the two triples of the six modes associated with the $q = 2$ rational surface, viz. the triple A_2^+, $S_3^\pm$, which becomes unstable at $s = 0.60$, has its maximum growth rate $\mathrm{Im}(\bar{\omega}) \approx 0.018$ at $s = 0.94$, and becomes stable again just left of $s = 0.98$ ($q = 2$), and the triple A_2^+, $S_1^\pm$ with the smaller growth rate, which becomes unstable earlier and also becomes stable again to the left of $q = 2$;

(d) Finally, the $\mathrm{Re}(\bar\omega)$'s of the slow modes $\mathrm{S}_4^\pm$ coalesce at $s = 0.85$ and the first unstable triple A_3^+, $\mathrm{S}_4^\pm$ of the $q = 3$ rational surface is encountered. Even though that rational surface falls far outside the domain, the effects of the associated six-mode interaction $\mathrm{S}_4^\pm$, $\mathrm{A}_3^\pm$, $\mathrm{S}_2^\pm$ are already clearly noticed.

In Fig. 18.10, the composite continuous spectrum is shown in the complex $\bar\omega$-plane with the radial position s in color coding. The plot clearly exhibits the cause of the lifting of the degeneracy of the six-mode interaction, viz. asymmetry with respect to left and right propagation of the continuum modes. Here, left and right are to be interpreted as follows. For $m > 0$ and $n < 0$, lines of constant phase $\Phi \equiv m\vartheta + n\varphi - \omega t$ in the φ-ϑ plane (representing the magnetic/flow surface) move to the left when $\mathrm{Re}(\bar\omega) < 0$ and to the right when $\mathrm{Re}(\bar\omega) > 0$. Hence, the triples A_1^+, $\mathrm{S}_0^\pm$ (*L*) and A_2^-, $\mathrm{S}_1^\pm$ (*L*) correspond to continuum modes that rotate leftward, and the triples A_2^+, $\mathrm{S}_3^\pm$ (*R*) and A_1^-, $\mathrm{S}_2^\pm$ (*R*) correspond to continuum modes that rotate rightward.

Whereas the frequencies of the continuous spectrum can be calculated one-by-one for each magnetic/flow surface separately, without the radial dependencies of $\mathrm{Re}(\bar\omega)$ and $\mathrm{Im}(\bar\omega)$ shown in the plots, it would have been extremely difficult to make sense of all these interactions. In the remaining plots, for the equilibrium (C), we use $q(\psi)$ instead of s as the radial flux parameter. The connection between the two representations is given by the functions $q(s)$ shown in Fig. 18.5.

18.4.2 Gravity dominated accretion disks*

We have seen in Section 18.3.3 that the structure of the continuous spectra changes dramatically at $\Gamma = 1$. In order to avoid misunderstanding about this transition, note that (1) the function $\Gamma(\psi)$ will not be constant in realistic equilibria, so that a condition $\Gamma = 1$ can, in general, be satisfied only at one particular magnetic/flow surface, leaving the other surfaces unstable to either $\Gamma < 1$ or $\Gamma > 1$ types of instabilities; (2) the condition $\Gamma = 1$, though associated with Keplerian rotation, does not represent a privileged equilibrium state in the case of transonic poloidal flows. It just implies that the poloidal density variation changes from tokamak-like (with density maxima on the inside of the torus) to gravity-dominated (with density maxima on the outside), as indicated by Eq. (18.66) and illustrated in Fig. 18.4. Rather than unduly complicating the discussion by presenting the instabilities of the continuous spectrum for equilibria having a local $\Gamma = 1$ condition, we move on to gravity-dominated equilibria with $\Gamma > 1$ everywhere. We skip presentation of the spectra of equilibria (B) and (D) that may be found in Ref. [173].

The continuous spectra of the gravity-dominated accretion disk equilibrium (C), where $\Gamma_0 = 2$, are shown in Fig. 18.11. Comparison with the solutions of the dispersion equation of Section 18.3.3 shows that the spectrum does, in fact, exhibit the six-mode instability structure characteristic of $\Gamma > 1$ equilibria, with much increased growth rates: $\mathrm{Re}(\bar{\omega}) \sim \mathrm{Im}(\bar{\omega})$ now. Again, the left–right symmetry of the rotating modes is broken in the numerical equilibrium, as clearly demonstrated by the asymmetric [with respect to $\mathrm{Re}(\bar{\omega}) = 0$] spectrum of Fig. 18.12. New for this class of equilibria is that, moving to the rational surface $q = 1$, the real part of the frequencies of the leftward and rightward rotating continuum modes decreases until there is *mode locking* at $q = 1$ when $\bar{\omega}$ becomes imaginary: a purely exponentially growing instability is found with a huge growth rate, in agreement with the estimate given in Eq. (18.126). At this point, the calculation might be repeated for $n = -2$, $n = -3$, etc., but the analysis of Section 18.3.3 already indicates that all of these modes will be unstable with approximately the same growth rate and localized at the respective rational surfaces $q = -m/n$.

Finally, the continuous spectra of the gravity-dominated accretion disk equilibrium (D), with $\Gamma_0 = 2$ and elliptical cross-section ($b/a = 0.5$), were also computed (not shown here, see [173]). The mode-locked instabilities found in equilibrium (C) are now completely swamped by other instabilities that grow an order of magnitude faster. The instabilities remain to be analyzed in detail, but it is already clear that flattening of the disk makes the trans-slow Alfvén continuum modes even more unstable. From these results it appears likely that the instabilities in flat accretion disks involve localization at the most curved sections of the magnetic/flow surfaces, i.e. on the inside and on the outside of the torus. Together with the finding of Section 18.3.3 that all higher toroidal harmonics n with their associated poloidal harmonics m are unstable, as expressed by Eqs. (18.125) and (18.126), this shows that the instabilities of the trans-slow Alfvén continua do not require that the magnetic/flow surfaces are closed, but just that they exist. In general accretion flows on open magnetic/flow surfaces, the same instabilities will operate as the ones found here for closed surfaces, in particular on the highly curved inside where jet formation is assumed to take place.

18.4.3 A new class of transonic instabilities

In summary, we have studied "transonic" axi-symmetric MHD equilibria and stability of toroidal plasmas without (tokamak) and with (accretion disk) the gravitational field of a massive central object (Fig. 18.1). "Transonic" here refers to the poloidal flow speed, measured by the value M^2 of the poloidal Alfvén Mach number squared defined in Eq. (18.11), when it surpasses the characteristic slow magnetosonic values M_c^2 and M_s^2, the Alfvén value 1, or the fast magnetosonic value M_f^2. For the validity of the analysis, the hyperbolic flow regimes $M_c^2 < M^2 < M_s^2$ and $M^2 > M_f^2$ of Fig. 18.3 should be excluded. This leaves three elliptic flow

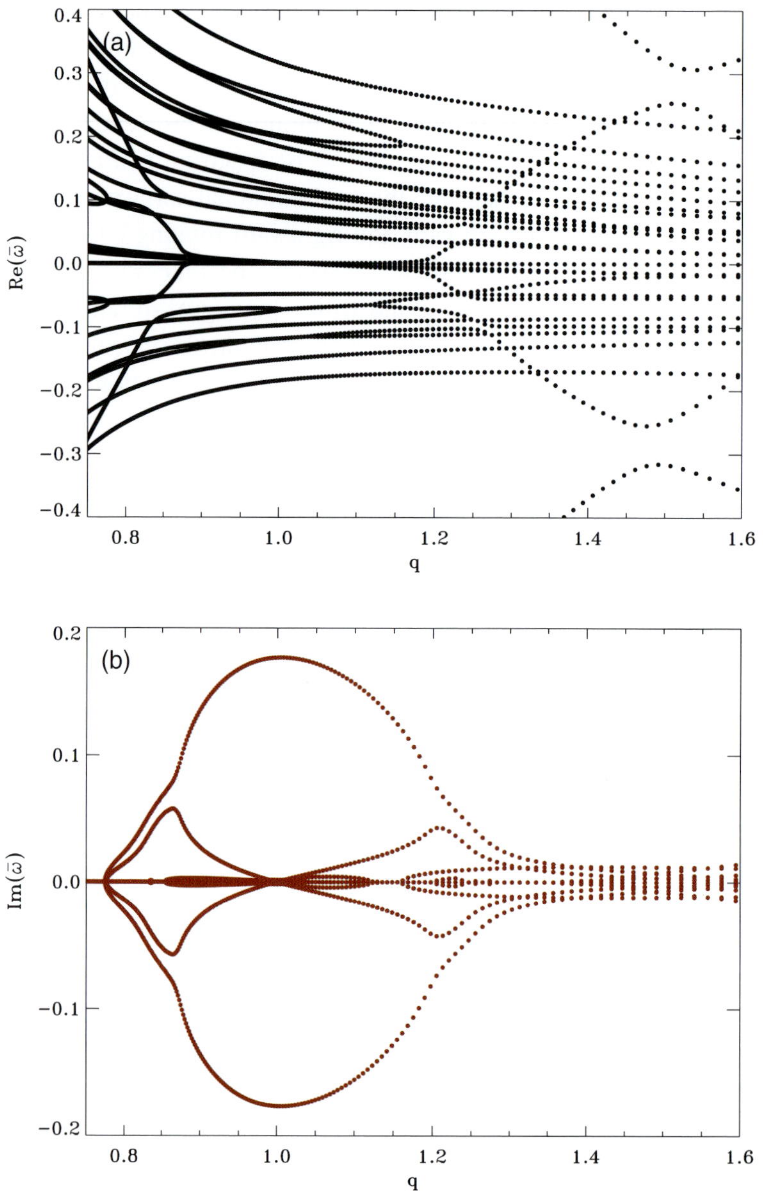

Fig. 18.11 (a) Real and (b) imaginary parts of the frequencies $\bar{\omega}$ of the trans-slow continuous spectrum as a function of $q(\psi)$ for an accretion disk with circular cross-section, $\Gamma_0 = 2$, equilibrium (C); $n = -1$.

regimes, called slow ($0 < M^2 < M_c^2$), trans-slow ($M_s^2 < M^2 < 1$), and fast ($1 < M^2 < M_f^2$). We have studied the trans-slow regime in detail since it is the first "transonic" one encountered when increasing the poloidal flow speed.

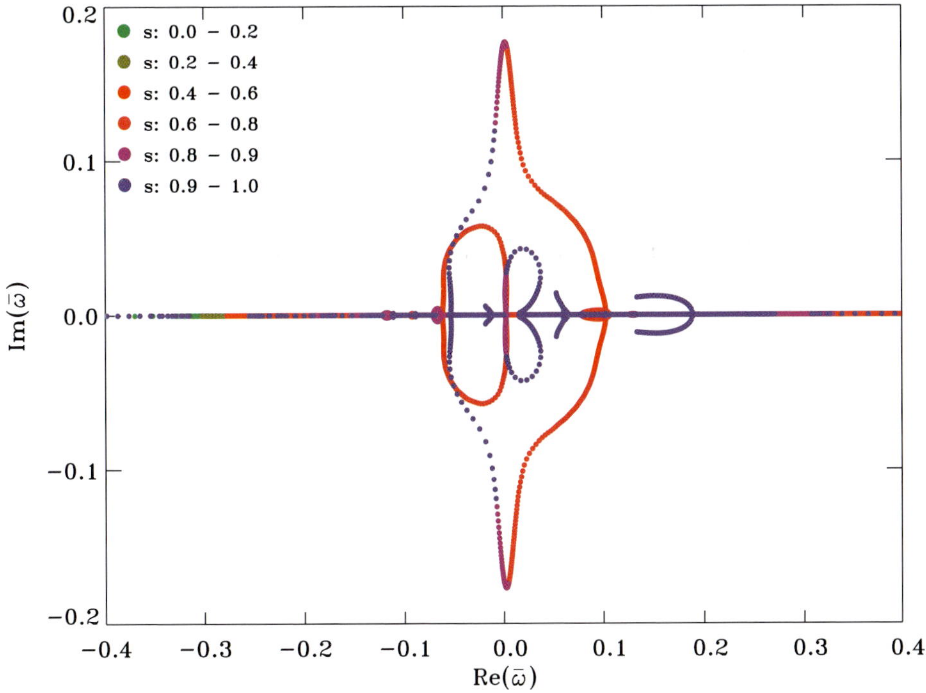

Fig. 18.12 Complex spectrum corresponding to Fig. 18.11, equilibrium (C).

The equilibria are characterized by five flux functions, where we have eliminated misunderstanding in the literature about the nature of the poloidal vorticity–current density stream function $K(\psi)$, defined in Eqs. (18.6) and (18.10), previously associated with the specific angular momentum. These functions are rescaled in Eqs. (18.13) and (18.22) to facilitate the analysis of the core equations (18.23) and (18.24) for ψ and M^2 when the centrifugal and gravitational forces approximately balance. The analysis is done both by means of an extension of the low-β tokamak ordering, called the trans-slow ordering defined in Eq. (18.63), and by numerical means, exploiting the FINESSE code [29]. The former yields the second order ODE (18.57) for the Shafranov shift $\Delta(\hat{r})$ and the first order ODE (18.58) for the toroidal perturbation μ of M^2. The structure of the ODE for Δ (in terms of M^2) turns out to be completely analogous to the structure of the Hain–Lüst equation [204] (in terms of the eigenfrequency parameter ω^2), in agreement with recent insight on transonic MHD [168]. The trans-slow equilibria computed with FINESSE (Fig. 18.4) are characterized by the relative strength $\Gamma(\psi)$ of the gravitational interaction, defined in Eq. (18.67), where $\Gamma \ll 1$ refers to tokamak-like equilibria and $\Gamma \gg 1$ to accretion disks with a very massive central object.

Next, we derived the equations (18.88)–(18.90) governing the radial asymptotics of the slow and Alfvén continuous spectra associated with perturbations of a single magnetic/flow surface. They clearly exhibit a major change of the coupling between Alfvén and slow modes when the critical value $M^2 = M_c^2$ is surpassed. Again, the analysis of the continuous spectra is performed both by means of an expansion in small inverse aspect ratio of the equations (18.88)–(18.90) for the Lagrangian displacement vector and by means of the full spectral code PHOENIX [173, 52] exploiting the primitive Eulerian variables. This enabled labeling of all spectral curves obtained (Fig. 18.6) with perfect agreement between the Lagrangian analysis and the Eulerian numerics for the coupling between the Alfvén and slow continuum modes, defined in Eqs. (18.97) and (18.98), whereas the PHOENIX code adds the Eulerian entropy continuum modes, defined in Eq. (18.99), that do not couple to the other modes.

The small inverse aspect ratio expansion yields the dispersion equation (18.122) for the coupling of six modes, labeled $A_m^\pm$, $S_{m-1}^\pm$, and $S_{m+1}^\pm$, which yield unstable Alfvénic continuum modes in the neighborhood of rational surfaces when $M^2 > M_c^2$. This is demonstrated in Figs. 18.7 and 18.8 for different values of Γ. For accretion disks with a very massive central object ($\Gamma \gg 1$), the instability growth rates become huge (larger than the Alfvén frequency) when the toroidal and poloidal mode number n and m are large. These results are confirmed by the PHOENIX code where no assumption on the aspect ratio needs to be made. Figures 18.9–18.12 show the complex frequencies of the trans-slow Alfvén continuum modes for the equilibria depicted in Fig. 18.4. Increasing the gravitational parameter Γ from 0 (tokamak) to 2 (accretion disk) and changing the cross-sectional shape from circular to elliptical ($b/a = 0.5$), the growth rate increases from fractions of the Alfvén frequency to exceeding it significantly.

In conclusion, one of the important issues in accreting flows about massive central objects is to explain how there can be accretion of magnetized plasma, and even ejection of jets, in the first place. If one takes the achievements of tokamak research seriously, plasma flow and magnetic field should remain confined near the toroidal magnetic/flow surfaces. Some violent dissipative mechanism appears to be needed to break this constraint. Obviously, for tokamaks, such a mechanism is not desired and, if it were found to operate under fusion conditions, one would certainly need to find ways to eliminate it. Fortunately, two important features of stationary flows in toroidal plasmas are absent in tokamaks but dominant in accretion disks, viz. the existence of large, "transonic", poloidal flows and the presence of a strong gravitational field. Since the MHD description of plasmas is independent of the size of the magnetic configuration, it appeared to be attractive to find a class of equilibria common to tokamaks and thick accretion disks, but distinctly different with respect to their stability properties. We have found such a class, as described in this chapter.

The instabilities found are due to the continuous spectrum, which is stable for the sub-slow poloidal flow speeds usually encountered in tokamaks, but which becomes unstable when the poloidal flow surpasses the critical threshold associated with the slow magneto-sonic cusp speed. The instability becomes explosive when the gravitational field is increased to the values encountered in accretion disks about massive back holes. Thus, these instabilities turn out to have all the requisite requirements mentioned.

For magnetically dominated thick accretion disks, the trans-slow poloidal flow ordering turns out to be quite effective to describe the instabilities. It assumes low β and sub-Alfvénic poloidal flow,

$$\beta \ll M^2 \equiv \frac{\rho v_{\rm p}^2}{B_{\rm p}^2} \left(\sim \frac{\rho v_\varphi^2}{B_\varphi^2} \sim \frac{\rho G M_*}{R B_\varphi^2} \right) \ll 1 \qquad (18.127)$$

(where the estimates in brackets are for flow parallel to the magnetic field, $\Omega = 0$, and for Keplerian toroidal rotation, $\Gamma = 1$, resp.). Under these conditions, the continuous spectrum becomes unstable over most of the plasma volume. The instability is switched on whenever the poloidal Alfvén Mach number exceeds the critical value $M_{\rm c}^2 \approx \frac{1}{2}\gamma\beta$. Hence, it appears to be unavoidable in accretion flows onto a compact object. At this point, it may be objected that we have not treated an actual accretion flow, but just a poloidal rotation. However, the discussion of the "dispersion equation" in Section 18.3.3 shows that the instabilities operate for *all* mode number pairs n, m, including the higher ones. The latter represent localization on the magnetic/flow surfaces independent on whether or not these surfaces are closed or open, or whether the flow is directed towards or away from the gravitational center. Whereas the use of state-of-the-art numerical tools for the computation of equilibrium and perturbations required us to stay away from the hyperbolic flow regimes, it appears not too farfetched now to assume that transition of the critical value $M^2 = M_{\rm c}^2$ in accretion flows will turn on the trans-slow Alfvén continuum instabilities precisely at the location of that transition. That would be the location where the flow becomes detached from the magnetic field due to a high level of MHD turbulence. Obviously, this conjecture needs further investigation by means of nonlinear numerical codes, along the lines of recent efforts by various authors (see [218, 242, 81] and references quoted there).

We conclude that we have found a very wide class of local instabilities of the magnetic/flow surfaces of transonic axi-symmetric plasmas. These instabilities would exclude operation of tokamaks in the trans-slow poloidal flow regimes (recall the caveat of Section 18.1 though), except possibly for a thin outer layer in divertor-operated fusion machines. For magnetically dominated accretion tori, these instabilities may provide a route to MHD turbulence and associated anomalous dissipation, breaking the co-moving condition of plasma and magnetic field, completely independent of the magneto-rotational instability.

18.5 Literature and exercises[*]

Notes on literature

Equilibrium of rotating axi-symmetric plasmas

- Zehrfeld & Green, 'Stationary toroidal equilibria at finite beta' [495].
- Morozov & Soloviev, 'Steady-state plasma flow in a magnetic field', in *Reviews of Plasma Physics, Vol. 8* [339].
- Hameiri, 'The equilibrium and stability of rotating plasmas' [210].

Transonic MHD equilibria and instabilities

- This chapter is based on the paper 'Unstable continuous spectra of transonic axi-symmetric plasmas' by Goedbloed *et al.* [173], where more examples and the detailed numerical data of the equilibria can be found.

Transonic two-fluid equilibria

- A generalization to the equations for transonic two-fluid equilibria was initiated by McClements & Thyagaraja [328]. The equations were cast into the form of a general variational principle with a Lagrangian for seven fields (electron and ion densities and stream functions, magnetic flux, electric potential and gravitational potential) and the "Bernoulli nightmare" for these equations was pointed out by Goedbloed [169]; see also the comment [435] and the response [170].

Exercises

[18.1] *Thin plasma slab: equilibrium*

In Exercises [16.1] and [17.4] you have exploited the thin plasma slab approximation for the equilibrium as well as for the perturbations. Here, you will apply this approximation to the generalized Grad–Shafranov equation (18.15). This equation describes the axi-symmetric equilibrium of a tokamak plasma or accretion disk.

- Explain why you cannot use the utmost right expressions of Eq. (18.17).
- Show that the generalized Grad–Shafranov equation (18.15) in the thin slab approximation reduces to

$$\frac{d}{dR}\left(p + \tfrac{1}{2}B^2\right) = \rho\left(\frac{v_\varphi^2}{R} - \frac{GM_*}{R^2}\right).$$

What does this equation describe?

[18.2] *Thin plasma slab: continuous spectrum*

In the previous exercise, you have looked at the equilibrium in the thin plasma slab approximation. Now, you do the same for equations (18.88) describing the continuous spectra for axi-symmetric plasmas with both toroidal and poloidal flow. Show that these equations decouple in this approximation and that the eigenvalues are the Doppler shifted Alfvén and slow frequencies.

[18.3] *Connection with a static tokamak plasma: equilibrium*

In this exercise, you will show that the generalized Grad–Shafranov equation (18.15) reduces to the ordinary Grad–Shafranov equation (16.81) in the limit of no gravity and no toroidal and poloidal flow. As you will see, these limits have to be taken with special care.

- Show that the limit of no poloidal flow is the same as $M^2 \to 0$.
- Show that

$$\lim_{M^2 \to 0} \frac{\Pi_1}{M^2} = \lim_{M^2 \to 0} \frac{\Pi_2}{M^{2\gamma}}.$$

- Now take a look at the Bernoulli equation (18.16). Show that this equation reduces to a trivial equation in the no poloidal flow limit.
- Before further reducing the generalized Grad–Shafranov equation (18.15), take a closer look at Π_1 and Π_3. Show that Π_1 reveals that the temperature becomes a flux function in the no flow and no gravity limit.
- Again, using the no flow and no gravity limit, show that Π_3 can be written as

$$\Pi_3 = \tfrac{1}{2} \chi'^2 (K/R)^2.$$

- Notice that the limits of no gravity and no toroidol flow can be taken without any problem. However, the no poloidal flow limit is a very special one. In this limit, the poloidal Alfvén Mach number also vanishes, which upsets an assumption of the generalized Grad–Shafranov equation (18.15). Show that

$$\lim_{M^2 \to 0} \frac{1}{\chi'^2} \frac{d\chi'^2}{d\Psi} = 0.$$

- Now, show that the generalized Grad–Shafranov equation (18.15) reduces to the ordinary Grad–Shafranov equation (16.81). What extra assumption do you have to use?

[18.4] *Connection with a static tokamak plasma: continuous spectrum*

In this exercise, you will find the connection between the equation (18.88) for the continua in rotating plasmas and the corresponding equation (17.55) for the static case. Reduce the first equation to the latter one by setting the flow to zero. What extra assumption do you have to use to get this result? (Hint: What are the flux functions for a static axi-symmetric equilibrium?)

[18.5] *Dispersion equation*

Reproduce the plots shown in Figs. 18.7 and 18.8. This can be done using IDL, Matlab or a similar program. For the computation of the roots, you can use the Cardano expressions or Laguerre's method. Make the program such that you can easily use both of them.

- Make a function which computes the coefficients of the dispersion equation (18.123) for given poloidal mode number m, toroidal mode number n, poloidal Alfvén Mach number M, gravitational interaction Γ and safety factor q.
- Make a function which computes the roots using the Cardano expressions. The input of this function should be the coefficients of a third order polynomial.

- Do the same for Laguerre's method. If you use IDL or Matlab, this method is part of their standard library, so you can skip this task.
- Make a function which finds the roots of the dispersion equation (18.123). The input for this should be the poloidal mode number m, the toroidal mode number n, the poloidal Alfvén Mach number M, the density ρ, the gravitational interaction Γ, the minimum and maximum value of the safety factor q and the method to be used. The function should return the roots found.
- Reproduce the plots shown in Fig. 18.7 and 18.8.
- Explain why, for example, the first plot of Fig. 18.8 does not look the same as the plots shown in Fig. 18.11.

Part V

Nonlinear dynamics

19

Computational nonlinear MHD

In Chapter 15, we introduced basic concepts for solving linear MHD problems computationally. In principle, the various options discussed there for discretizing a set of partial differential equations in space and time can all be adopted for simulating nonlinear MHD phenomena. Many suitable combinations of spatial (finite difference, finite element, spectral) and temporal (explicit or implicit) discretizations have successfully been applied in problem-specific contexts. However, the conservation law nature of the ideal nonlinear MHD equations poses additional challenges when discontinuous, shock-dominated evolutions are computed. In this chapter, we pay particular attention to certain variants of "shock-capturing" schemes, which have proven to be of general use for nonlinear hyperbolic equations. We explain fundamental strategies and some of the difficulties encountered upon application of these schemes to the MHD system. This is then again illustrated with examples of their use, including simulations determining the nonlinear evolution of MHD instabilities, as well as advanced computations of astrophysically relevant MHD processes, up to modern solar system space weather models. A brief discussion of alternative algorithmic approaches is included as well, complemented with representative applications.

For simulations involving a hierarchy of temporal scales, one must again handle the time scale problem by some (semi-)implicit time integration method. We limit their discussion to some exemplary treatments in computational nonlinear MHD, focused on resistive mode developments and wave dynamics in externally driven systems. We conclude with an impression of state-of-the-art global simulations of laboratory tokamak plasmas, where even extended MHD models are beginning to be applied.

19.1 General considerations for nonlinear conservation laws

As discussed extensively in Volume [1], Chapter 4, the ideal MHD equations represent a set of *nonlinear* conservation laws. In the nonlinear dynamics, shocks can form spontaneously out of continuous initial data by wave steepening, and contact as well as tangential discontinuities may be encountered. All these discontinuities constitute moving internal interfaces across which the Rankine–Hugoniot conditions must hold, requiring an appropriate numerical representation. In what follows, we introduce various general concepts for conservative systems, which will be useful to design and validate proper numerical treatments.

19.1.1 Conservative versus primitive variable formulations

In general, a system of conservation laws can be expressed in conservation form:

$$\frac{\partial \mathbf{U}}{\partial t} + \nabla \cdot \mathcal{F}(\mathbf{U}) = 0. \tag{19.1}$$

This form exploits the set of conservative variables $\mathbf{U}$, and specifies that any local Eulerian change in a conservative variable is due to spatially diverging (or converging) fluxes $\mathcal{F}(\mathbf{U})$. In Section 4.3[1], we derived the conservation form (4.71)–(4.74) for the ideal MHD equations, and denoted the conservative variables as $\mathbf{U} \equiv (\rho, \boldsymbol{\pi}, \mathcal{H}, \mathbf{B})^{\mathrm{T}}$. In turn, these conservative variables and their fluxes were specific functions of the density ρ, velocity $\mathbf{v}$, pressure p and magnetic field $\mathbf{B}$, or of any other set involving $\mathbf{v}$, $\mathbf{B}$ and two thermodynamic quantities (such as the internal energy e, temperature T, specific entropy s, or the related $S \equiv p\rho^{-\gamma} = f(s)$, $\ln(\rho)$, etc.). Writing the ideal MHD equations in terms of any latter set of so-called primitive variables is mathematically and physically equivalent, but the equations then deviate from the conservation form (19.1). It was pointed out in Section 4.5 [1] that only from the conservation form, a simple substitution recipe could be followed to obtain jump conditions, relating changes in the dynamical variables across moving interfaces, i.e. across shock fronts and co-moving plasma–plasma interfaces. When solving the equations numerically, one must obey the conservation laws also at the discrete level, or at least be consistent with their conservation properties within discretization errors.

Quasi-linear forms, flux Jacobians and characteristic speeds Simplifying the discussion to 1D, a set of conservation laws writes as

$$\frac{\partial \mathbf{U}}{\partial t} + \frac{\partial \mathbf{F}(\mathbf{U})}{\partial x} = 0, \tag{19.2}$$

when expressed in terms of the conservative variables $\mathbf{U}(x, t)$. Many equivalent formulations may exist in terms of other, primitive variables $\mathbf{V}(x, t)$, which will have a quasi-linear form

$$\frac{\partial \mathbf{V}}{\partial t} + \mathbf{W}\frac{\partial \mathbf{V}}{\partial x} = 0 \tag{19.3}$$

involving a square matrix $\mathbf{W}(\mathbf{V})$. If the number of (conservative or primitive) variables is indicated by n, $\mathbf{W}$ will be an $n \times n$ matrix. The change in variables from $\mathbf{U}$ to $\mathbf{V}$ is quantified by a transformation matrix $\mathbf{U_V}$ found from $d\mathbf{U} = \mathbf{U_V}d\mathbf{V}$. This is also an $n \times n$ square matrix, and we will assume that this transformation is well-defined and thus invertible, $d\mathbf{V} = \mathbf{U_V}^{-1}d\mathbf{U}$. The conservation law can similarly be written in quasi-linear form by exploiting the flux Jacobian matrix $\mathbf{F_U}$, such that

$$\frac{\partial \mathbf{U}}{\partial t} + \mathbf{F_U}\frac{\partial \mathbf{U}}{\partial x} = 0\,. \tag{19.4}$$

For any equivalent set of primitive variables $\mathbf{V}$ with their coefficient matrix $\mathbf{W}(\mathbf{V})$, we then necessarily find

$$\mathbf{F_U} = \mathbf{U_V}\mathbf{W}(\mathbf{V})\mathbf{U_V}^{-1}\,. \tag{19.5}$$

This similarity relation between matrices $\mathbf{F_U}$ and $\mathbf{W}(\mathbf{V})$ guarantees that they have identical eigenvalues. These can thus be computed from either $|\mathbf{F_U} - \lambda_p\mathbf{I}| = 0$ or $|\mathbf{W} - \lambda_p\mathbf{I}| = 0$. We deduce from dimensional analysis of Eq. (19.4) that these eigenvalues indicate velocities, which are called the characteristic speeds. For hyperbolic equations, like the ideal MHD equations, these n eigenvalues λ_p for $p = 1, \ldots, n$ are real and introduce corresponding sets of right $\mathbf{r}^p$ as well as left eigenvectors $\mathbf{l}^p$ from

$$\mathbf{F_U}\mathbf{r}^p = \lambda_p\mathbf{r}^p\,,$$
$$\mathbf{l}^p\mathbf{F_U} = \mathbf{l}^p\lambda_p\,. \tag{19.6}$$

As always, eigenvectors are determined up to a multiplicative factor only, so that we are free to use any appropriate scaling for these eigenvectors. We can write the set of right eigenvectors $\mathbf{r}^p$ as columns of a matrix $\mathbf{R}$, and the first set of the relations (19.6) then writes as

$$\mathbf{F_U}\mathbf{R} = \mathbf{R}\mathbf{\Lambda}\,, \tag{19.7}$$

where we introduced the diagonal matrix $\mathbf{\Lambda} \equiv \mathrm{diag}(\lambda_1, \ldots, \lambda_n)$ containing the eigenvalues. If all these eigenvalues are always distinct, the system of equations is said to be strictly hyperbolic. We then usually assume the ordering $\lambda_1 < \lambda_2 < \ldots < \lambda_n$. When identical eigenvalues occur, their degeneracy can be accounted for, e.g. by reducing the matrix rank and regrouping variables with the same characteristic velocities together. For the ideal MHD equations, we will have to face such additional complications, since the occurring eigenvalues may coincide and the system is not strictly hyperbolic. In the strictly hyperbolic case, we are sure to have linearly independent right eigenvectors in $\mathbf{R}$. This means that we can compute the left eigenvectors $\mathbf{l}^p$ by inverting the matrix $\mathbf{R}$, and identifying the rows of $\mathbf{R}^{-1}$ with the left eigenvectors, since $\mathbf{R}^{-1}\mathbf{F_U} = \mathbf{\Lambda}\mathbf{R}^{-1}$. This makes left and right eigenvectors orthonormal, i.e. $\mathbf{l}^q \cdot \mathbf{r}^p = \delta_{qp}$. With each primitive variable formulation, different (albeit related) sets of right and left eigenvectors are found. Denoting $\hat{\mathbf{R}}$ as the matrix with the right eigenvectors $\hat{\mathbf{r}}^p$ for $\mathbf{W}(\mathbf{V})$ in its columns, we find from (19.5) that $\hat{\mathbf{R}} = \mathbf{U_V}^{-1}\mathbf{R}$. Its left eigenvectors form the rows of $\hat{\mathbf{R}}^{-1} = \mathbf{R}^{-1}\mathbf{U_V}$.

Finally, we may try to find a set of variables $\tilde{\mathbf{R}}$ for which the governing equation (19.2) rewrites to

$$\frac{\partial \tilde{\mathbf{R}}}{\partial t} + \mathbf{\Lambda}\frac{\partial \tilde{\mathbf{R}}}{\partial x} = 0. \tag{19.8}$$

Such so-called Riemann invariants would remain constant on curves $dx = \lambda_p dt$ in the (x,t) plane. These curves are the characteristics of the hyperbolic partial differential equations at hand, in the same sense as discussed for the ideal MHD case in Section 5.4[1]. When dealing with a system of $n = 2$ unknowns, one can always find both Riemann invariants, one for each characteristic $p = 1, 2$, and write them collectively as in Eq. (19.8). They can be used to solve the initial value problem by drawing the pair of characteristics (with slopes locally determined by λ_p) from each point on the x-axis at $t = 0$ into the (x,t) plane, and use the fact that they "propagate" a constant value of their respective Riemann invariant. Using this method of characteristics, the value of $\mathbf{U} = (U_1(x,t), U_2(x,t))^{\mathrm{T}}$ at any later time and location can in principle be determined by back tracing the local values of the Riemann invariants along their corresponding characteristics to $t = 0$. This is schematically illustrated in Fig. 19.1. The pair of Riemann invariants $\tilde{\mathbf{R}} = (\tilde{R}_1(U_1, U_2), \tilde{R}_2(U_1, U_2))^{\mathrm{T}}$ provides a kind of mapping from (U_1, U_2) space to the (x,t) plane, in which the characteristic curves are drawn. An extended region in (x,t) space in which one of the two Riemann invariants remains constant is then called a simple wave region, and through this mapping we then have $U_2(U_1)$ in this region.

When there are $n > 2$ conserved quantities, one may not always be able to find a full set of Riemann invariants obeying Eq. (19.8). A full set of Riemann invariants should obey the proportionalities

$$\frac{\partial \tilde{\mathbf{R}}}{\partial t} \sim \mathbf{R}^{-1}\frac{\partial \mathbf{U}}{\partial t} = \hat{\mathbf{R}}^{-1}\frac{\partial \mathbf{V}}{\partial t},$$
$$\frac{\partial \tilde{\mathbf{R}}}{\partial x} \sim \mathbf{R}^{-1}\frac{\partial \mathbf{U}}{\partial x} = \hat{\mathbf{R}}^{-1}\frac{\partial \mathbf{V}}{\partial x}. \tag{19.9}$$

The related variables found by writing $\mathbf{R}^{-1}\mathbf{U}$ or equivalently $\hat{\mathbf{R}}^{-1}\mathbf{V}$ are termed

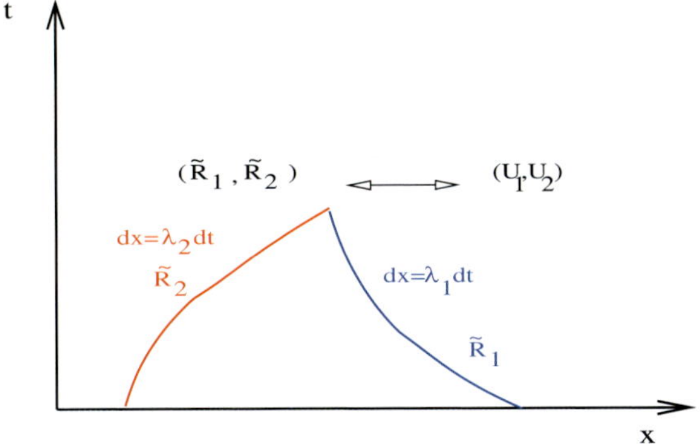

Fig. 19.1 The method of characteristics, schematically illustrated for the case of two variables $(U_1(x,t), U_2(x,t))$.

characteristic variables. In order to generalize the simple wave concept mentioned above for the case of two unknowns to a system of $n > 2$ variables $\mathbf{U} = (U_1(x,t), \ldots, U_n(x,t))^{\mathrm{T}}$, we can again ask whether an extended region of (x,t) space can correspond to an essentially one-parameter variation where $U_i(U_1)$ for all $i = 1, \ldots, n$. One can appreciate that there will be n such simple wave constructions possible, one for each eigenvalue λ_p. In fact, in the (x,t) plane, a pth simple wave region will have straight lines for the pth characteristic curve $dx = \lambda_p dt$ (of varying slope $\lambda_p(U_1)$), along which the solution $\mathbf{U}$ is constant (since $U_i(U_1)$). For the simple wave associated with the pth characteristic field, one arrives at a set of $n - 1$ *generalized* Riemann invariants, defined from

$$\frac{dU_i}{r_i^p} = \frac{dU_j}{r_j^p}, \quad \text{or equivalently from} \quad \frac{dV_i}{\hat{r}_i^p} = \frac{dV_j}{\hat{r}_j^p}, \quad \text{for } i \neq j \in \{1, \ldots, n\}.$$

$$(19.10)$$

When integrated, these $n - 1$ relations yield $n - 1$ functions $J^p(\mathbf{U})$ which obey $\nabla_{\mathbf{U}} J^p \cdot \mathbf{r}^p = 0$, or equivalently $\nabla_{\mathbf{V}} J^p \cdot \hat{\mathbf{r}}^p = 0$. Simple wave regions can be proven to be essential ingredients in nonlinear wave problems. In particular, a region in (x,t) bordering a constant state $\mathbf{U} = \mathbf{U}_0$ region will be a simple wave region (or another constant state, which can be regarded as a trivial simple wave).

In what follows, we will encounter specific examples for the various matrices, their eigenvalue and eigenvector pairs, and the (generalized) Riemann invariants. They usually play a prominent role in the design of shock-capturing numerical algorithms for integrating nonlinear hyperbolic systems. They are also useful for thoroughly checking numerically obtained solutions.

Iso-thermal MHD in 1.5D and Riemann invariants The simplest MHD problem where we can directly illustrate the above concepts is posed by the 1.5D iso-thermal MHD equations. The assumption of iso-thermal conditions can be justified in various astrophysical contexts, when dealing with the dynamics of a gas whose cooling time due to radiative processes is very short with respect to all other dynamical time scales. These conditions can be met in dilute environments which are constantly fed energy from a central source (e.g. irradiation from a star). To keep the gas at a constant temperature, one must thus relax energy conservation. Instead of explicitly handling source and sink terms in an equation governing the energy evolution, the iso-thermal magnetohydro system avoids the solution of this equation altogether, and assumes that the net effect is to maintain a constant temperature. The $\nabla \cdot \mathbf{B} = 0$ condition in a 1D configuration with quantities varying along x turns B_x into a constant parameter. Therefore, the first non-trivial MHD counterpart of a 1D iso-thermal gas problem needs to account for a non-vanishing component $B_y(x)$ (and hence also v_y) in a translationally invariant y-direction. This is typically referred to as a 1.5D problem.

The governing equations write in the conservation form (19.2) as

$$
\begin{pmatrix} \rho \\ m_x \\ m_y \\ B_y \end{pmatrix}_t + \begin{pmatrix} m_x \\ m_x^2/\rho - B_x^2 + c_{\mathrm{i}}^2\rho + \frac{1}{2}(B_x^2 + B_y^2) \\ m_x m_y/\rho - B_x B_y \\ B_y m_x/\rho - B_x m_y/\rho \end{pmatrix}_x = 0\,. \tag{19.11}
$$

This introduces the four conserved variables consisting of density, both momentum components and the translational magnetic field component. In the x-momentum equation, the gas pressure is written as $p = c_{\mathrm{i}}^2\rho$, with c_{i}^2 denoting the squared iso-thermal sound speed, a constant proportional to the fixed temperature. The flux Jacobian matrix is then found to be

$$
\mathbf{F_U} = \begin{pmatrix} 0 & 1 & 0 & 0 \\ c_{\mathrm{i}}^2 - m_x^2/\rho^2 & 2m_x/\rho & 0 & B_y \\ -m_x m_y/\rho^2 & m_y/\rho & m_x/\rho & -B_x \\ (-m_x B_y + B_x m_y)/\rho^2 & B_y/\rho & -B_x/\rho & m_x/\rho \end{pmatrix}\,. \tag{19.12}
$$

A primitive variable formulation (19.3) for the same system exploits the velocity components and is

$$
\begin{pmatrix} \rho \\ v_x \\ v_y \\ B_y \end{pmatrix}_t + \begin{pmatrix} v_x & \rho & 0 & 0 \\ c_{\mathrm{i}}^2/\rho & v_x & 0 & B_y/\rho \\ 0 & 0 & v_x & -B_x/\rho \\ 0 & B_y & -B_x & v_x \end{pmatrix} \begin{pmatrix} \rho \\ v_x \\ v_y \\ B_y \end{pmatrix}_x = 0\,. \tag{19.13}
$$

The eigenvalues from both the flux Jacobian $\mathbf{F_U}$ and the matrix $\mathbf{W}(\mathbf{V})$ in this equation are found to be

$$\lambda_1 = v_x - c_f, \quad \lambda_2 = v_x - c_s, \quad \lambda_3 = v_x + c_s, \quad \lambda_4 = v_x + c_f, \tag{19.14}$$

where the familiar slow and fast magneto-acoustic speeds are computed from

$$c_{f,s}^2 = \tfrac{1}{2}\left[c_i^2 + (B_x^2 + B_y^2)/\rho\right] \pm \tfrac{1}{2}\sqrt{\left[c_i^2 + (B_x^2 + B_y^2)/\rho\right]^2 - 4c_i^2 B_x^2/\rho}. \tag{19.15}$$

As already mentioned, some of the four eigenvalues coincide in certain limits, such that the MHD system is not strictly hyperbolic.

▷ **Eigenvectors** The right eigenvectors for the Jacobian matrix are the columns of

$$\mathbf{R} = \begin{pmatrix} 1 & 1 & 1 & 1 \\ v_x - c_f & v_x - c_s & v_x + c_s & v_x + c_f \\ v_y + \dfrac{c_f B_x B_y}{\rho c_f^2 - B_x^2} & v_y + \dfrac{c_s B_x B_y}{\rho c_s^2 - B_x^2} & v_y - \dfrac{c_s B_x B_y}{\rho c_s^2 - B_x^2} & v_y - \dfrac{c_f B_x B_y}{\rho c_f^2 - B_x^2} \\ \dfrac{c_f^2 B_y}{\rho c_f^2 - B_x^2} & \dfrac{c_s^2 B_y}{\rho c_s^2 - B_x^2} & \dfrac{c_s^2 B_y}{\rho c_s^2 - B_x^2} & \dfrac{c_f^2 B_y}{\rho c_f^2 - B_x^2} \end{pmatrix}. \tag{19.16}$$

Note that we are free to adopt another scaling for these eigenvectors. This will be needed for proper handling of the indeterminacies where $\rho c_{s,f}^2 = B_x^2$. The left eigenvectors of $\mathbf{F_U}$ can be found in the rows of $\mathbf{R}^{-1}$. This matrix is given by

$$\begin{pmatrix} -\dfrac{v_y c_f}{2c_i^2}\dfrac{B_y}{B_x}\dfrac{c_s^2}{c_f^2 - c_s^2} + \dfrac{v_x c_f + c_i^2}{2c_i^2}\alpha_f^2 & -\alpha_f^2\dfrac{c_f}{2c_i^2} & \dfrac{B_y}{B_x}\dfrac{2c_s^2 c_f}{(c_f^2 - c_s^2)c_i^2} & \dfrac{B_y}{2(c_f^2 - c_s^2)} \\ \dfrac{v_y c_s}{2c_i^2}\dfrac{B_y}{B_x}\dfrac{c_f^2}{c_f^2 - c_s^2} + \dfrac{v_x c_s + c_i^2}{2c_i^2}\alpha_s^2 & -\alpha_s^2\dfrac{c_s}{2c_i^2} & -\dfrac{B_y}{B_x}\dfrac{2c_f^2 c_s}{(c_f^2 - c_s^2)c_i^2} & \dfrac{-B_y}{2(c_f^2 - c_s^2)} \\ -\dfrac{v_y c_s}{2c_i^2}\dfrac{B_y}{B_x}\dfrac{c_f^2}{c_f^2 - c_s^2} - \dfrac{v_x c_s - c_i^2}{2c_i^2}\alpha_s^2 & \alpha_s^2\dfrac{c_s}{2c_i^2} & \dfrac{B_y}{B_x}\dfrac{2c_f^2 c_s}{(c_f^2 - c_s^2)c_i^2} & \dfrac{-B_y}{2(c_f^2 - c_s^2)} \\ \dfrac{v_y c_f}{2c_i^2}\dfrac{B_y}{B_x}\dfrac{c_s^2}{c_f^2 - c_s^2} - \dfrac{v_x c_f - c_i^2}{2c_i^2}\alpha_f^2 & \alpha_f^2\dfrac{c_f}{2c_i^2} & -\dfrac{B_y}{B_x}\dfrac{2c_s^2 c_f}{(c_f^2 - c_s^2)c_i^2} & \dfrac{B_y}{2(c_f^2 - c_s^2)} \end{pmatrix}. \tag{19.17}$$

In these expressions, we introduced the positive parameters

$$\alpha_f^2 = \frac{c_i^2 - c_s^2}{c_f^2 - c_s^2}, \qquad \alpha_s^2 = \frac{c_f^2 - c_i^2}{c_f^2 - c_s^2}. \tag{19.18}$$

For the primitive variable formulation, the right eigenvectors are found to be

$$\hat{\mathbf{R}} = \begin{pmatrix} 1 & 1 & 1 & 1 \\ -c_f/\rho & -c_s/\rho & c_s/\rho & c_f/\rho \\ \dfrac{c_f}{\rho}\dfrac{B_x B_y}{\rho c_f^2 - B_x^2} & \dfrac{c_s}{\rho}\dfrac{B_x B_y}{\rho c_s^2 - B_x^2} & -\dfrac{c_s}{\rho}\dfrac{B_x B_y}{\rho c_s^2 - B_x^2} & -\dfrac{c_f}{\rho}\dfrac{B_x B_y}{\rho c_f^2 - B_x^2} \\ \dfrac{c_f^2 B_y}{\rho c_f^2 - B_x^2} & \dfrac{c_s^2 B_y}{\rho c_s^2 - B_x^2} & \dfrac{c_s^2 B_y}{\rho c_s^2 - B_x^2} & \dfrac{c_f^2 B_y}{\rho c_f^2 - B_x^2} \end{pmatrix}. \tag{19.19}$$

Finally, the left eigenvectors for $\mathbf{W}(\mathbf{V})$ are the rows of

$$
\hat{\mathbf{R}}^{-1} =
\begin{pmatrix}
\frac{1}{2}\alpha_f^2 & -\frac{1}{2}\alpha_f^2\frac{\rho c_f}{c_i^2} & \frac{1}{2}\dfrac{B_x B_y}{c_f(c_f^2 - c_s^2)} & \dfrac{B_y}{2(c_f^2 - c_s^2)} \\[2ex]
\frac{1}{2}\alpha_s^2 & -\frac{1}{2}\alpha_s^2\frac{\rho c_s}{c_i^2} & -\frac{1}{2}\dfrac{B_x B_y}{c_s(c_f^2 - c_s^2)} & \dfrac{-B_y}{2(c_f^2 - c_s^2)} \\[2ex]
\frac{1}{2}\alpha_s^2 & \frac{1}{2}\alpha_s^2\frac{\rho c_s}{c_i^2} & \frac{1}{2}\dfrac{B_x B_y}{c_s(c_f^2 - c_s^2)} & \dfrac{-B_y}{2(c_f^2 - c_s^2)} \\[2ex]
\frac{1}{2}\alpha_f^2 & \frac{1}{2}\alpha_f^2\frac{\rho c_f}{c_i^2} & -\frac{1}{2}\dfrac{B_x B_y}{c_f(c_f^2 - c_s^2)} & \dfrac{B_y}{2(c_f^2 - c_s^2)}
\end{pmatrix}.
\tag{19.20}
$$

Using the columns in $\hat{\mathbf{R}}$ of Eq. (19.19), we can derive generalized Riemann invariants. ◁

Following the definition from Eq. (19.10), we get for the first wave family associated with $\lambda_1 = v_x - c_f$ the system

$$
\begin{aligned}
c_f\,d\rho + \rho\,dv_x &= 0\,, \\
B_x(c_f^2 - c_i^2)\,dv_x + B_y c_f^2\,dv_y &= 0\,, \\
\rho c_f\,dv_y - B_x\,dB_y &= 0\,.
\end{aligned}
\tag{19.21}
$$

Changing to dimensionless variables $r = \rho c_i^2/B_x^2$ and $q = c_f^2/c_i^2$, we find

$$
\begin{aligned}
dv_x &= -\,c_i\frac{\sqrt{q}}{r}\,dr\,, \\[1ex]
dv_y &= c_i\frac{\sqrt{(q-1)(qr-1)}}{r(qr-1)}\,\mathrm{sgn}(B_x B_y)\,dr\,, \\[1ex]
dq &= \frac{q^2(q-1)}{q^2 r - 1}\,dr\,.
\end{aligned}
\tag{19.22}
$$

The function $J = r/(q-1) + \int dq\, q^{-2}(q-1)^{-2}$ can then be verified to obey $dJ = \partial J/\partial r\, dr + \partial J/\partial q\, dq = 0$. This essentially integrates the third equation in the set (19.22). Since this generalized Riemann invariant does not depend on the velocity components, it will be referred to as the magneto-acoustic Riemann invariant. The two other generalized Riemann invariants for simple waves associated with λ_1 are found from integrating the first two equations in (19.22). Note the requirement $q > 1$ and $qr > 1$, restricting (r, q) state space, so that better parameterizations may exist [345, 346]. Similar expressions can be found for the three generalized Riemann invariants associated with each simple wave family for $\lambda_{2,3,4}$. For the iso-thermal system considered, we can explicitly evaluate the integral appearing in the fast and slow magneto-acoustic (generalized) Riemann invariants, to get the analytic forms

$$
J_{s,f} = \frac{-c_{s,f}^2 + \rho c_i^4/B_x^2}{c_{s,f}^2 - c_i^2} + \frac{c_{s,f}^2 - c_i^2}{c_{s,f}^2} - 2\ln\left|\frac{c_{s,f}^2 - c_i^2}{c_{s,f}^2}\right|.
\tag{19.23}
$$

These can be used to verify numerical solutions to Riemann problems, as will be explained and demonstrated further on.

Linear hyperbolic PDE system It is instructive to briefly reconsider the linear system case, where the flux is given by $\mathbf{F} = \mathbf{A}\mathbf{U}$ and the Jacobian is a constant $n \times n$ matrix $\mathbf{A} = \mathbf{F}_{\mathbf{U}}$. Again assuming strict hyperbolicity, such that $\mathbf{A}$ has real and distinct eigenvalues, we then find immediately from Eq. (19.9) that Riemann invariants and characteristic variables coincide, $\tilde{\mathbf{R}} = \mathbf{R}^{-1}\mathbf{U}$, since matrix $\mathbf{R}^{-1}$ is then also constant. In terms of these characteristic variables, the system of equations becomes a set of n decoupled advection equations

$$\frac{\partial \tilde{R}_p}{\partial t} + \lambda_p \frac{\partial \tilde{R}_p}{\partial x} = 0, \quad p = 1, \ldots, n. \tag{19.24}$$

Each of these equations has a trivial, exact solution in terms of its initial condition,

$$\tilde{R}_p(x, t) = \tilde{R}_p(x - \lambda_p t, 0). \tag{19.25}$$

The exact solution to the linear hyperbolic PDE system in terms of the conservative variables is then found from $\mathbf{U} = \mathbf{R}\tilde{\mathbf{R}}$, or

$$\mathbf{U}(x, t) = \sum_{p=1}^{n} \tilde{R}_p(x - \lambda_p t, 0)\mathbf{r}^p. \tag{19.26}$$

For every point x, the solution at any time t thus depends only on the value of the initial data at n discrete points. In particular, these points are at the intersection of the $t = 0$ axis with all the p-characteristics through (x, t). The p-characteristics are in this linear case all straight lines $x = x_0 + \lambda_p t$. Since we know the exact solution when the initial condition is translated to the characteristic variables, the linear hyperbolic system can be used to validate numerical approaches for handling, in particular, discontinuous data.

19.1.2 Scalar conservation law and the Riemann problem

The consequences of nonlinearity in a system of conservation laws can already be illustrated in the case of a single scalar conservation law. A general nonlinear scalar conservation law for the quantity $u(x, t)$ is written as

$$u_t + (f(u))_x = 0, \tag{19.27}$$

where the flux function $f(u)$ depends nonlinearly on u. The inviscid Burgers equation is obtained for $f(u) = u^2/2$. Assuming differentiability, Burgers' equation can be written as

$$u_t + u\,u_x = 0, \tag{19.28}$$

so that we can speak of a local "advection" speed u, which is also the characteristic speed quantified by the Jacobian, i.e. derivative, $f_u = u \equiv f'(u)$. Identifying u as a density, the local density determines the local advection speed, with denser regions traveling faster than more rarified ones. Note further that the spatial integral $\int_{x_1}^{x_2} u\,dx$ determining the total "mass" over the interval $[x_1, x_2]$ will be the same for all times if the fluxes $f(u)$ at the edges x_1 and x_2 vanish at all times. This is the conservation property of Eq. (19.27). From these considerations, it is now easily understood how wave steepening and shock formation can occur. Envision a triangular pulse as initial data, given by

$$
u(x,0) = \begin{cases} u_0 & (x \leq -x_0), \\ u_0 + h_0(x_0 + x)/x_0 & (-x_0 < x \leq 0), \\ u_0 + h_0(x_0 - x)/x_0 & (0 < x \leq x_0), \\ u_0 & (x > x_0). \end{cases} \tag{19.29}
$$

Since the characteristic speed is given by the local value of u, the tip of the triangle experiences the fastest rightward advection (we assume $u_0 > 0$ and $h_0 > 0$). In conserving the total area underneath the triangle, the front edge steepens. At the space-time point when the tip of the triangle has caught up with the rightmost point of the front edge, a discontinuity appears in the solution. This happens at the time when their characteristics meet, namely at time $t_s = x_0/h_0$. From this time of shock formation onwards, conservation now demands the discrete equivalent of the conservation law to hold across the discontinuity, this means in terms of the adjacent left u_ℓ and right u_r values that

$$
f(u_\ell) - f(u_r) = s\,(u_\ell - u_r). \tag{19.30}
$$

Here, the discontinuity is assumed to travel with shock speed s, and for the inviscid Burgers case, we find that $s = (u_\ell + u_r)/2$. Precisely at time t_s, we then have $s(t_s) = u_0 + h_0/2$. From this time on, the base of the triangle will widen due to the speed difference between the left edge traveling with u_0 and the shocked right edge traveling at speed $s(t)$. In accord with conservation, the height of the triangle must therefore decrease in time. The full solution for times $t > t_s = x_0/h_0$ thus works out to be

$$
u(x,t) = \begin{cases} u_0 & (x \leq -x_0 + u_0 t), \\ u_0 + 2h_0 \dfrac{x + x_0 - u_0 t}{\left(\sqrt{h_0 t} + \sqrt{x_0}\right)^2} & (-x_0 + u_0 t < x \leq u_0 t + \sqrt{x_0 h_0 t}), \\ u_0 & (x > u_0 t + \sqrt{x_0 h_0 t}). \end{cases} \tag{19.31}
$$

The shock speed for the right edge is then $s(t) = u_0 + h_0\sqrt{x_0}/\left(\sqrt{x_0} + \sqrt{h_0 t}\right)$. The evolution is schematically indicated in Fig. 19.2.

Fig. 19.2 Evolution of an initially triangular pulse in the inviscid Burgers equation.

The relation (19.30) is the jump condition or Rankine–Hugoniot relation for the simple scalar conservation law. Note that it is symmetric in its two arguments representing a left and right state u_ℓ, u_r. Obviously, though, for Burgers' equation we appreciate immediately that a different solution must emerge when we interchange the two states in the particular discontinuous initial condition where

$$u = u_\ell \quad \text{for} \quad x \le 0,$$
$$u = u_r \quad \text{for} \quad x > 0. \tag{19.32}$$

This represents the Riemann problem where two constant states in contact are left to evolve. When $u_\ell > u_r$, we can expect a pure shock solution, with shock speed found from (19.30). Conversely, when $u_\ell < u_r$, we expect the right state to "run away" from the left one, turning the initial discontinuity into a continuously increasing profile. Therefore, the jump relation (19.30) is not sufficient to discriminate between allowed (weak) discontinuous solutions. A physically admissible shock must obey another relation as well, known as the *Lax entropy condition*. This Lax entropy condition states that a shock with speed s satisfying the Rankine–Hugoniot relations must additionally obey

$$f'(u_\ell) > s > f'(u_r). \tag{19.33}$$

We identified $f'(u)$ from the scalar conservation law as the characteristic speed, so this expresses that the shock speed must lie in between the characteristic speeds of the two adjacent states. This condition is asymmetric in left versus right state. For Burgers' equation, we find $u_\ell > \frac{1}{2}(u_\ell + u_r) > u_r$, as expected. The space-time characteristics for Burgers' equation on either side of this admissible shock have constant slopes $u_\ell > u_r$, so that the characteristics "go into the shock" [302].

▷ **Generalization to nonlinear systems** We can readily generalize the Rankine–Hugoniot relations for shocks to nonlinear systems. From the conservation form (19.2), we obtain the equivalent of Eq. (19.30), expressing conservation across a moving discontinuity,

$$\mathbf{F}(\mathbf{U}) - \mathbf{F}(\mathbf{U}^*) = s(\mathbf{U} - \mathbf{U}^*). \tag{19.34}$$

For a fixed state $\mathbf{U}^*$, relation (19.34) defines its Hugoniot locus, consisting of all states $\mathbf{U}$ which can be connected to $\mathbf{U}^*$ via a discontinuous shock moving at (scalar) speed s. For a system of n equations, we will then find n one-parameter families. When writing $\mathbf{U}(\zeta, \mathbf{U}^*)$ and $s(\zeta, \mathbf{U}^*)$, where ζ is chosen such that $\mathbf{U}(0, \mathbf{U}^*) = \mathbf{U}^*$, we can look in particular at weak shocks. Then $\mathbf{F}(\mathbf{U}(\zeta, \mathbf{U}^*)) \approx \mathbf{F}(\mathbf{U}(0, \mathbf{U}^*)) + \mathbf{F_U}(\mathbf{U}(0, \mathbf{U}^*))(\mathbf{U}(\zeta, \mathbf{U}^*) - \mathbf{U}^*)$. We then differentiate relation (19.34) to ζ, writing $d\mathbf{U}/d\zeta = \mathbf{U}'$, and evaluate it in $\zeta = 0$:

$$\mathbf{F_U}(\mathbf{U}(0, \mathbf{U}^*))\mathbf{U}'(0, \mathbf{U}^*) = s(0, \mathbf{U}^*)\mathbf{U}'(0, \mathbf{U}^*) \,. \tag{19.35}$$

Since this expresses that $\mathbf{U}'(0, \mathbf{U}^*)$ must be an eigenvector of the flux Jacobian evaluated in $\mathbf{U}^*$, with eigenvalue $s(0, \mathbf{U}^*)$, the n parameter families for shocks can be associated with the n eigenvalues of $\mathbf{F_U}$. For weak shocks, we find $s_p = \lambda_p(\mathbf{U}^*)$ so that shock speeds are given by the characteristic speeds. However, this correspondence for weak shocks does not hold for strong shocks, where the nonlinear character of Eq. (19.34) can cause significant deviations. Furthermore, we again will need selection criteria to distinguish which part of the Hugoniot locus represents physically admissible shocks. One such criterion is the generalization of the Lax entropy condition. This states that a jump in the pth wave family obeying Rankine–Hugoniot which travels with speed s is allowed when

$$\lambda_p(\mathbf{U}_\ell) > s > \lambda_p(\mathbf{U}_r) \,. \tag{19.36}$$

The p-characteristics then enter the p-shock from both sides. From every point on the shock, one can then travel along characteristics backward in time. This indicates how information reaches the shock from the past, not from the future, as required by "causality". This time-irreversibility argument leads to the nomenclature of "entropy" condition for (19.36). However, additional arguments may be needed to select physically admissible shocks, since for a system of n nonlinear equations, a discontinuity traveling at speed s can have more than one of the n characteristic families on each side converge into the shock. This situation represents so-called overcompressive shocks.   ◁

Rarefaction waves Returning to Burgers' equation, we can also analyze the case where $u_\ell < u_r$, and a continuously increasing profile is expected. In fact, there are infinitely many weak solutions in this case, including a pure shock one with shock speed $s = (u_\ell + u_r)/2$. But now the characteristics go out of the shock, which makes this solution unstable to perturbations. A small change in initial data can yield a completely different solution in this "entropy-violating" shock [302]. The physically correct continuous solution to the Riemann problem will obey an x/t self-similarity in the (x, t) plane. Writing $u(x, t) = u(x/t) \equiv u(\xi)$, we find that the conservation law (19.27) translates into

$$f'(u)\frac{du}{d\xi} = \xi\frac{du}{d\xi} \,, \tag{19.37}$$

or we must have $\xi = f'(u)$. For Burgers' equation, this means that the solution to the Riemann problem, for which $u_\ell < u_r$ with the discontinuity initially at $x = 0$, is given by

$$u(x, t) = u(x/t) = \begin{cases} u_\ell & (x < u_\ell t) \\ x/t & (u_\ell t < x < u_r t) \\ u_r & (x > u_r t) \end{cases} \,. \tag{19.38}$$

This is a so-called rarefaction wave, as u decreases (in terms of a density: the medium gets rarefied) when the signal passes. Hence, the solution to the Riemann problem for the inviscid Burgers' equation is either a shock wave (case $u_\ell > u_r$), or a rarefaction wave (case $u_\ell < u_r$).

▷ **Generalization to nonlinear systems** The self-similar rarefaction wave solution for the nonlinear system starts with the quasi-linear form (19.4), as it analyzes the possibility for continuously varying solutions where $\mathbf{U}(x,t) = \mathbf{U}(x/t) \equiv \mathbf{U}(\xi)$. Note that this is the simple wave construction mentioned earlier, with the dependence on x/t sometimes referred to as a centered simple wave. We deduce that, for such self-similar solutions, $\partial \mathbf{U}/\partial t = -x\mathbf{U}'/t^2$ and $\partial \mathbf{U}/\partial x = \mathbf{U}'/t$, leading to

$$\mathbf{F_U}\mathbf{U}' = \xi\mathbf{U}'. \tag{19.39}$$

This expression means that ξ must be an eigenvalue λ_p of the flux Jacobian $\mathbf{F_U}$, and that $\mathbf{U}'$ (with the prime indicating the derivative with respect to ξ) must be proportional to the corresponding right eigenvector $\mathbf{r}^p$. Note that we already found this relation for the *local* Hugoniot locus (weak shock case), but here it is true along the entire parameterized $\xi = x/t$ range. Curves in state space $\mathbf{U}$ along which the tangent always coincides with an eigenvector are termed integral curves. Hence, from Eq. (19.39), we can write

$$\xi = \lambda_p(\mathbf{U}(\xi)) \quad \text{and} \quad \mathbf{U}' = \alpha(\xi)\mathbf{r}^p(\mathbf{U}(\xi)). \tag{19.40}$$

Differentiating the first expression with respect to ξ we find

$$1 = (\nabla_\mathbf{U}\lambda_p) \cdot \mathbf{U}'. \tag{19.41}$$

Using the correspondence between $\mathbf{U}'$ and $\mathbf{r}^p$, the proportionality constant is found from

$$\alpha(\xi) = \frac{1}{(\nabla_\mathbf{U}\lambda_p) \cdot \mathbf{r}^p}. \tag{19.42}$$

Obviously, the construction fails for wave families where the denominator in Eq. (19.42), the so-called structure coefficient $s_p = (\nabla_\mathbf{U}\lambda_p) \cdot \mathbf{r}^p$, vanishes identically. In such a case, the pth wave field is termed linearly degenerate. When it is always strictly positive (or strictly negative), the field is genuinely nonlinear. In MHD, we will see that slow and fast magneto-sonic wave families are neither linearly degenerate nor genuinely nonlinear. This is in contrast with the Euler system for gas dynamics, where all fields are either linearly degenerate or genuinely nonlinear, which is the characterizing property of a convex system of conservation laws. Note that we can also compute the structure coefficients from the primitive eigenvectors through $s_p = (\nabla_\mathbf{V}\lambda_p) \cdot \hat{\mathbf{r}}^p$. These coefficients relate to the tendency of a wave family to steepen or spread, and when $s_p = 0$ identically, the pth wave mode propagates by means of finite discontinuities. Finally, when a rarefaction wave solution connects a left constant state $\mathbf{U}_\ell$ with a right constant state $\mathbf{U}_r$, we need $\xi = \lambda_p$ to increase monotonically from its value $\xi_\ell = \lambda_p(\mathbf{U}_\ell) < \xi = x/t < \xi_r = \lambda_p(\mathbf{U}_r)$, which also acts as a selection criterion for admissible rarefaction waves. Note the reversal in the order as compared to the Lax entropy condition (19.36). Through a p-rarefaction, all $n-1$ generalized Riemann invariants for the pth wave family remain constant. ◁

Compound waves: a scalar example The flux $f(u)$ appearing in the nonlinear conservation law (19.27) is said to be convex when the second derivative $f''(u)$ has the same sign everywhere, expressing that the first derivative $f'(u)$ varies monotonically. Burgers' equation with flux function $f(u) = u^2/2$ has this property, and the Riemann problem gave rise to either a rarefaction solution or a shock. When we take as nonlinear flux function $f(u) \equiv u^3$, the flux is non-convex, as $f''(u) = 6u$ changes sign at $u = 0$. The characteristic speed is now locally $3u^2$. In the solution of the Riemann problem for a scalar equation with a non-convex flux function, a new possible outcome arises. The familiar cases return as long as we stay on the monotonic side by e.g. restricting $u > 0$. Then, we conclude by analogy with Burgers' equation that for $f(u) \equiv u^3$ the rarefaction wave will occur when $0 < u_\ell < u_r$ with solution

$$u(x,t) = u(x/t) = \begin{cases} u_\ell & (x < 3\,u_\ell^2 t) \\ \sqrt{x/(3t)} & (3\,u_\ell^2 t < x < 3\,u_r^2 t) \\ u_r & (x > 3\,u_r^2 t) \end{cases} . \qquad (19.43)$$

In contrast, when $0 < u_r < u_\ell$, the discontinuity is maintained and travels at the shock speed $s = u_\ell^2 + u_\ell\,u_r + u_r^2$.

When we allow for negative states u, we can find a "compound" solution as follows. Taking $u_\ell > 0 > u_r$, we can seek an intermediate state $u_m < 0$ which connects to u_ℓ discontinuously, hence at shock speed $s = u_\ell^2 + u_\ell\,u_m + u_m^2$, while at the same time u_m is connecting to the right state u_r by a rarefaction solution. This compound solution then has

$$u(x,t) = u(x/t) = \begin{cases} u_\ell & (x < s\,t = 3\,u_m^2 t) \\ -\sqrt{x/(3t)} & (3\,u_m^2 t < x < 3\,u_r^2 t) \\ u_r & (x > 3\,u_r^2 t) \end{cases} . \qquad (19.44)$$

For the case considered here, $u_m = -u_\ell/2$, and thus this compound solution will only emerge when $u_\ell > 0 > -u_\ell/2 > u_r$. We will see that the MHD system allows for such compound solutions in both fast and slow wave families.

19.1.3 *Numerical discretizations for a scalar conservation law*

Turning to numerical treatments for conservation laws, we would like the employed discretization to allow for a proper treatment of discontinuous solutions as well, and only give rise to physically admissible shocks. To stress the importance of using a discretization which is able to handle discontinuities properly, suppose we were to apply the following first order scheme to the inviscid Burgers equation, directly discretizing its quasi-linear form (19.28) as

$$u_i^{n+1} - u_i^n + \frac{\Delta t}{\Delta x} u_i^n \left(u_i^n - u_{i-1}^n \right) = 0, \qquad (19.45)$$

where we assumed $u \geq 0$. The particular initial data $[1, 1, 1, 0, 0, 0]$ is seen [302] to remain a discrete solution that does not change from time t^n to $t^{n+1} = t^n + \Delta t$.

However, we know from the discussion above that the correct solution should have the discontinuous jump traveling rightward, at the shock speed $s = 0.5$. The reason why this discretization fails for discontinuous solutions is that it cannot be written in a conservative form. A conservative scheme for a set of conservation laws with conserved variables $\mathbf{U}$ is of the form

$$\mathbf{U}_i^{n+1} = \mathbf{U}_i^n - \frac{\Delta t}{\Delta x}\left[\mathbf{F}_{i+1/2} - \mathbf{F}_{i-1/2}\right], \qquad (19.46)$$

where the numerical fluxes $\mathbf{F}_{i+1/2}$ are interpreted as time-average fluxes over cell edges $x_i + \Delta x/2 \equiv x_{i+\frac{1}{2}}$. Note that the non-conservative scheme (19.45) would do fine for smooth solutions to the inviscid Burgers equation, but because it is not in conservation form, the discrete equivalent of the Rankine–Hugoniot relation (19.30) is not guaranteed. In fact, a rigorous mathematical analysis by Hou and Le Floch [232] demonstrates that using any finite difference scheme in non-conservative form to numerically solve a scalar conservation law will lead to significant errors which grow in time. A demonstration on how bad a non-conservative MHD code may perform on discontinuity dominated test problems can be found in [131]. The Lax–Wendroff theorem ensures that a convergent conservative scheme does converge to a weak solution of the conservation law (see, e.g. [302] and references therein).

Lax–Friedrichs, Lax–Wendroff and MacCormack schemes In the case of a single, scalar conservation law as expressed by (19.27), we now revisit and readily generalize to the nonlinear case some of the conditionally stable explicit discretizations encountered in Chapter 15, which are conservative. The first order Lax–Friedrichs scheme (15.110) is written as

$$u_i^{n+1} = \tfrac{1}{2}\left(u_{i+1}^n + u_{i-1}^n\right) - \tfrac{1}{2}\frac{\Delta t}{\Delta x}\left(f_{i+1}^n - f_{i-1}^n\right). \qquad (19.47)$$

This scheme is stable when the CFL condition

$$\left|\frac{\Delta t}{\Delta x}f'(u_i)\right| \le 1, \qquad (19.48)$$

is satisfied at all grid points. This Lax–Friedrichs scheme is conservative, since we can write it in the generic form (19.46) when we identify the numerical flux as

$$F_{i+1/2}^{\mathrm{LF}} = \tfrac{1}{2}\left[f_{i+1} + f_i - \frac{\Delta x}{\Delta t}(u_{i+1} - u_i)\right]. \qquad (19.49)$$

To generalize the second order Lax–Wendroff scheme (15.117) to a nonlinear equation solver, the most popular formulation is in the form of a predictor–corrector type method as proposed by Richtmyer. The Richtmyer two–step version

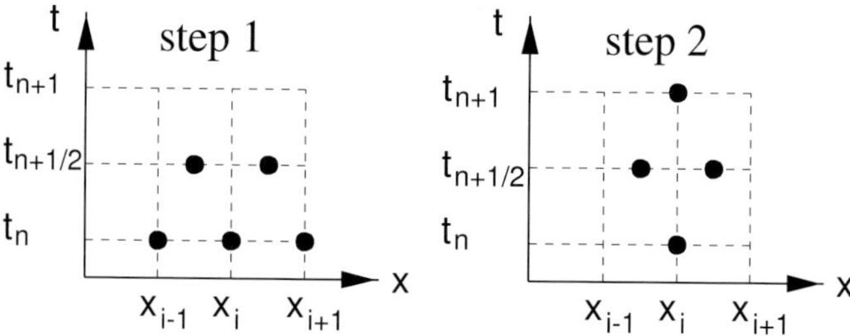

Fig. 19.3 Stencils of the predictor (left) and corrector (right) steps in the two-step Lax–Wendroff method.

of the Lax–Wendroff method reads:

$$\text{``predictor''}: \quad u_{i+\frac{1}{2}}^{n+\frac{1}{2}} = \tfrac{1}{2}(u_{i+1}^n + u_i^n) - \tfrac{1}{2}\frac{\Delta t}{\Delta x}(f_{i+1}^n - f_i^n),$$

$$\text{``corrector''}: \quad u_i^{n+1} = u_i^n - \frac{\Delta t}{\Delta x}\left(f_{i+\frac{1}{2}}^{n+\frac{1}{2}} - f_{i-\frac{1}{2}}^{n+\frac{1}{2}}\right). \tag{19.50}$$

Hence, in the predictor step the *intermediate* values at half time step and at half mesh points, $u_{i+1/2}^{n+1/2}$, are calculated by the Lax–Friedrichs scheme and these intermediate values are used in the corrector step. Clearly, the scheme is conservative as it uses the fluxes $f_{i\pm\frac{1}{2}}^{n+\frac{1}{2}}$ and determines the values u_i^{n+1} in the next time step with the leapfrog scheme (15.115). The stencils of the two steps are given in Fig. 19.3.

It is this two-step version of the Lax–Wendroff scheme that has been used extensively in MHD simulations. As a concrete example, Ofman and Davila [355] used it to address the nonlinear evolution of both standing and traveling Alfvén waves in 3D slab models of driven coronal "loops". The loop equilibria were approximated as channels with a density depletion in uniformly magnetized zero-β slabs. Their nonlinear resistive MHD simulations demonstrated that nonlinear effects play a crucial role in the resonant absorption of the wave energy (discussed in Chapter 11 [1] for the linear regime). In the vicinity of the resonant dissipation layers, highly sheared flows were formed, and Kelvin–Helmholtz instabilities developed. We will present a closely related 3D nonlinear study for cylindrical loop models further on in this chapter, in Section 19.4.2 (see Fig. 19.30).

Another multi-level method is the two-step MacCormack method:

$$\text{``predictor''}: \quad u_i^* = u_i^n - \frac{\Delta t}{\Delta x}(f_{i+1}^n - f_i^n),$$

$$\text{``corrector''}: \quad u_i^{n+1} = \tfrac{1}{2}(u_i^n + u_i^*) - \tfrac{1}{2}\frac{\Delta t}{\Delta x}(f_i^* - f_{i-1}^*). \tag{19.51}$$

Note that the step size in the predictor step is now Δt and that the spatial derivative

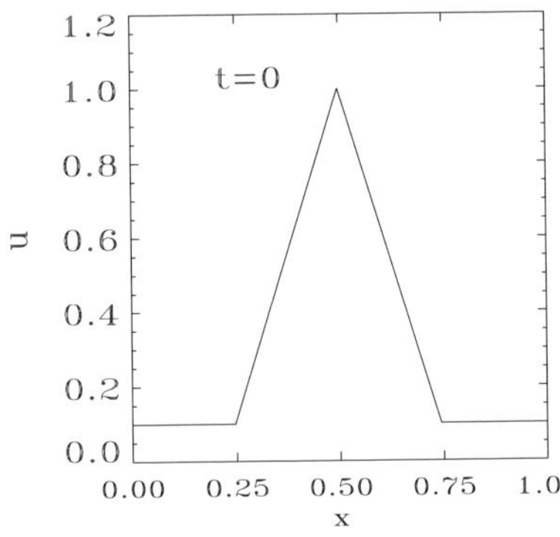

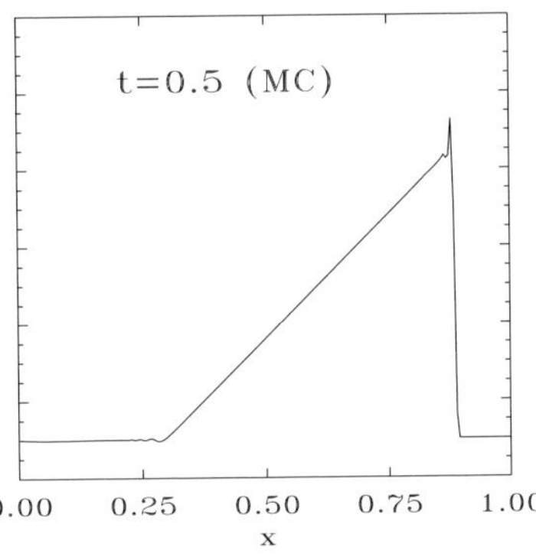

Fig. 19.4 Initial triangular pulse (left) and shocked numerical solution of Burgers' equation as obtained with the second order MacCormack method.

switches from left- to right-sided in the partial steps. One typically would permute this order from one full time step to the next, to better maintain the symmetry. The intermediate value u^* is used to determine the fluxes f_i^* and f_{i-1}^*, which "correct" it and replace it by the final value u_i^{n+1}. The MacCormack method is equivalent to the two-step Lax–Wendroff scheme for linear problems but somewhat better for nonlinear problems. This scheme is also second order accurate. An example numerical solution by the MacCormack method for solving Burgers' equation is shown in Fig. 19.4, where the initial condition is the triangular pulse discussed earlier. For the particular pulse chosen, shock formation started at $t_s = 0.277778$. At the time $t = 0.5$ shown, the numerical solution has propagated in full agreement with the analytic outcome. Note that a small oscillatory tail exists at the trailing edge of the pulse, as well as immediately behind the shock transition.

▷ **Exercise** Show that for a linear advection equation with $f(u) = vu$, the substitution of the first step in the second step of the two-step Lax–Wendroff scheme yields the original Lax–Wendroff scheme (15.117). (Be careful about the step sizes!) For the general nonlinear scalar conservation law, verify that the MacCormack method is conservative and give the expression for its numerical flux $F_{i+\frac{1}{2}}$. ◁

Both the Lax–Wendroff (19.50) and MacCormack (19.51) schemes are better approximations to the PDE (19.27) than the Lax–Friedrichs scheme (19.47), since the former two schemes are second order accurate. Second order accuracy in both space and time translates in local truncation errors given by $\mathcal{O}(C^3)$ or $\mathcal{O}(C\delta^2)$ in the dimensionless parameters $\delta = |(\Delta x/u)\,\partial u/\partial x|$ and the Courant number $C = f'(u)\,\Delta t/\Delta x$. These schemes yield satisfactory results as long as the solution $u(x)$ and the coefficient function $f'(u(x))$ are smooth. However, in the neighborhood of shocks and steep gradients $\delta \sim \mathcal{O}(1)$ and in regions where the characteristic speed $f'(u)$ is large, we can have $C \sim \mathcal{O}(1)$. As a result, in such re-

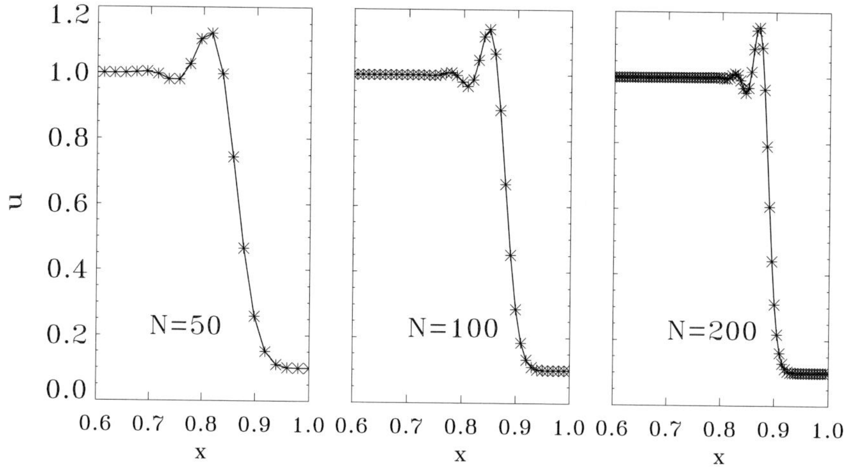

Fig. 19.5 Gibbs phenomenon, as illustrated for the linear advection of a single discontinuity, as obtained with the MacCormack method for different resolutions. The initial discontinuity was at $x = 0.5$, and the solution corresponds to time $t = 0.4$. The exact solution should jump at $x = 0.9$.

gions the truncation errors can become as large as the solution itself, yielding "numerical pollution" in the form of short-wavelength oscillations. These can already be demonstrated for a linear advection problem where $f(u) = vu$ with constant v, where Lax–Wendroff and MacCormack schemes coincide, by numerically solving its Riemann problem (19.32). This is illustrated in Fig. 19.5, where we show the effect in combination with increasing the number of grid points. The result of solving a Riemann problem with $u_\ell = 1$ and $u_r = 0.1$, initially at $x = 0.5$ with $v = 1$, is shown at time $t = 0.4$. The initial discontinuity is captured numerically by a steep negative gradient region, immediately trailed by an oscillatory part. This oscillation does not disappear when more grid points are used, but covers roughly the same number of grid points throughout. This so-called *Gibbs phenomenon* is an indication of the *dispersive* properties of this scheme, where numerical wave signals of different wavelengths travel at different speeds. Note that the Riemann problem has an initially monotone (although discontinuous) variation with x, i.e. there is no internal extremum of $u(x, t = 0)$. The Lax–Wendroff and MacCormack methods can introduce internal extrema (as those seen in Fig. 19.4) from initially monotone data, and are therefore "non-monotonicity preserving". In fact, for a scalar nonlinear conservation law, the exact solution should maintain monotonicity if the initial data is monotone (even including discontinuities), so it would be nice to guarantee "monotonicity preserving" schemes. We return to this issue when discussing total variation diminishing schemes, but first explain how it is done in a flux corrected transport approach.

Flux corrected transport The flux corrected transport or FCT method was one of the first high-resolution methods [62]. Flux corrected transport rather refers to a *"technique"* which is applicable to different numerical schemes. When applied to the finite difference schemes discussed in Chapter 15, the schemes are modified to predictor/corrector-type. In the "predictor" step an artificial damping or *diffusion* term is added so that the computed solution ensures both conservation and positivity (an initial positive density profile remains positive under pure advection). In the "corrector" step a certain amount of damping that may have been "excessive" is removed again in an *anti-diffusion* stage. This anti-diffusion is necessarily nonlinear, and designed to preserve monotone profiles: no new local extrema in the solution are created, and existing extrema do not get accentuated. The overall method is stable under the usual CFL-type condition.

As a concrete example, an FCT variant of the Lax–Wendroff scheme (15.117) for the linear advection equation (15.101), $u_t = -v\,u_x$ with constant v, gives

$$u_i^* = u_i^n - \frac{\Delta t}{2\,\Delta x}\,v\,(u_{i+1}^n - u_{i-1}^n) + \left[\frac{1}{2}\frac{(\Delta t)^2}{(\Delta x)^2}\,v^2 + \epsilon_1\right](u_{i+1}^n - 2u_i^n + u_{i-1}^n) \quad (19.52)$$

for the predictor step. Notice the additional second order damping term with coefficient ϵ_1 in this step. To undo this added diffusion, one would like to subtract the diffusion in a corrector step of the form

$$u_i^{n+1} = u_i^* - \epsilon_2\left(u_{i+1}^* - 2u_i^* + u_{i-1}^*\right). \quad (19.53)$$

To be consistent with the time-invariant solution when $v = 0$, one must have $\epsilon_1(v = 0) = \epsilon_2(v = 0)$. The total artificial diffusion added to the scheme, i.e. after both the diffusion and the anti-diffusion steps, equals $\epsilon_1 - \epsilon_2$ and thus, in general, *does not vanish*. Boris and Book [62, 63] analyzed amplitude and phase errors for different choices of these coefficients for the linear advection equation extensively. While originally taking $\epsilon_1 = \epsilon_2 = 1/8$, an improved choice with $\epsilon_1 = \epsilon_2 = (1 - C^2)/6$ where $C = v\Delta t/\Delta x$ reduces phase errors to fourth order accuracy. The key observation of the FCT method is that in practice a nonlinear effectively variable numerical diffusion ϵ_2 is required in Eq. (19.53) in order not to introduce new extrema in the anti-diffusion stage. The artificial diffusion term is then written in the form $\epsilon_2(u^*; i + 1/2)(u_{i+1}^* - u_i^*) - \epsilon_2(u^*; i - 1/2)(u_i^* - u_{i-1}^*)$, where the notation indicates that ϵ_2 depends on some finite number of values of u^*. Writing $\Delta u_{i+1/2}^* = u_{i+1}^* - u_i^*$, the FCT corrector step (19.53) gets modified to

$$u_i^{n+1} = u_i^* - F_{i+1/2}^A + F_{i-1/2}^A,$$

$$F_{i+1/2}^A = \mathrm{sgn}\,\Delta u_{i+1/2}^* \max\left\{0, \min\left(\Delta u_{i-1/2}^*\,\mathrm{sgn}\,\Delta u_{i+1/2}^*,\right.\right.$$
$$\left.\left. |\epsilon_2\,\Delta u_{i+1/2}^*|, \Delta u_{i+3/2}^*\,\mathrm{sgn}\,\Delta u_{i+1/2}^*\right)\right\}. \quad (19.54)$$

The locally effective anti-diffusion coefficient is then determined by means of a nonlinear "flux limiter" or filter and, hence, the FCT method is a special case of the so-called "flux-limiter methods". Note that the stencil of the method widens as a result.

▷ **Exercise** Rewrite the first, predictor, stage of the FCT method given by Eq. (19.52) as a conservative update of the solution involving a transport and diffusion stage expressed as

$$u_i^* = u_i^n - F_{i+1/2}^{\mathrm{T}} + F_{i-1/2}^{\mathrm{T}} + F_{i+1/2}^{\mathrm{D}} - F_{i-1/2}^{\mathrm{D}}. \tag{19.55}$$

The transport flux will be $F_{i+1/2}^{\mathrm{T}} = v u_{i+1/2}^n \Delta t / \Delta x$ with $u_{i+1/2}^n = \frac{1}{2}(u_i^n + u_{i+1}^n)$, and the diffusion flux reads $F_{i+1/2}^{\mathrm{D}} = (\epsilon_1 + C^2/2)\Delta u_{i+1/2}^n$. Implement and test this FCT scheme, as well as the standard Lax–Wendroff method for numerically solving the linear advection equation. Consider both smooth and discontinuous initial profiles. ◁

The FCT method is readily generalized from the linear case discussed above to the case of a system of nonlinear conservation laws as given by Eq. (19.2) for the 1D case. The flux $\mathbf{F}(\mathbf{U})$ for each conserved quantity is then split into a transport $\mathbf{U}v$ part and a "source term" contribution, where the transport velocity $v(\mathbf{U})$ is the fluid velocity for Euler and MHD equations. One combines a predictor–corrector approach for reaching second order temporal accuracy with the transport, diffusion and anti-diffusion stages in each partial step as

$$\mathbf{U}_i^{\mathrm{T}} = \mathbf{U}_i^n - \frac{\Delta t^{\mathrm{p,c}}}{\Delta x}\left(\mathbf{U}_{i+1/2}^n v_{i+1/2}^{\mathrm{p,c}} - \mathbf{U}_{i-1/2}^n v_{i-1/2}^{\mathrm{p,c}}\right) + \Delta t^{\mathrm{p,c}} \mathbf{S}_i^{\mathrm{p,c}},$$

$$\mathbf{U}_i^{\mathrm{D}} = \mathbf{U}_i^{\mathrm{T}} + \nu_{i+1/2}^{\mathrm{p,c}}\Delta\mathbf{U}_{i+1/2}^n - \nu_{i-1/2}^{\mathrm{p,c}}\Delta\mathbf{U}_{i-1/2}^n,$$

$$\mathbf{U}_i^{\mathrm{A}} = \mathbf{U}_i^{\mathrm{D}} - F_{i+1/2}^{\mathrm{A;p,c}} + F_{i-1/2}^{\mathrm{A;p,c}}. \tag{19.56}$$

The diffusion coefficient is taken in analogy with the above best choice for ϵ_1 as

$$\nu_{i+1/2}^{\mathrm{p,c}} = \frac{1}{6} + \frac{1}{3}\left(\frac{\Delta t^{\mathrm{p,c}}}{\Delta x} v_{i+1/2}^{\mathrm{p,c}}\right)^2. \tag{19.57}$$

The anti-diffusion involves the nonlinear filter operation of Eq. (19.54), where it uses the diffused solution differences $\Delta\mathbf{U}^{\mathrm{D}}$ from that partial step, except in the term with ϵ_2 where it uses the transported values $\epsilon_2\Delta\mathbf{U}^{\mathrm{T}}$. The coefficient ϵ_2 gets replaced by the anti-diffusion coefficient

$$\nu_{i+1/2}^{\mathrm{p,c}} = \frac{1}{6} - \frac{1}{6}\left(\frac{\Delta t^{\mathrm{p,c}}}{\Delta x} v_{i+1/2}^{\mathrm{p,c}}\right)^2. \tag{19.58}$$

In the predictor step $\Delta t^{\mathrm{p}} = \Delta t/2$ is used, while the corrector uses the full time step $\Delta t^{\mathrm{c}} = \Delta t$. It should be noted that various variants of the FCT method can be conceived, and the more advanced make use of a multi-dimensional variant of the nonlinear limiter employed in the anti-diffusion stage.

Examples of the use of the FCT method in an MHD context are ample in the literature. Noteworthy early examples are: astrophysical jet simulations by Kössl, Müller and Hillebrandt [286], assuming axial symmetry, but allowing for helical equipartition magnetic fields, and a paper by DeVore [115] discussing FCT techniques for compressible MHD. In the solar physics context, we mention applications to magneto-convection and magnetic flux tubes by Steiner *et al.* [420, 419] (discussed and illustrated further on in this chapter, see Figure 19.23), and Alfvén wave heating of coronal loops by Poedts and Boynton [374]. The method is still in use for MHD simulations to date, although it is a somewhat outdated scheme.

Total variation diminishing schemes Another way to ensure a monotonicity preserving scheme for a scalar nonlinear conservation law, is to require a more stringent property, namely that the scheme is total variation diminishing, or TVD. The total variation of a function $u(x)$ on its domain, e.g. $[0, 1]$, is defined as

$$TV(u) \equiv \int_0^1 \left| \frac{du}{dx} \right| \, dx \,. \tag{19.59}$$

In case of discontinuous profiles $u(x)$, the derivative is to be interpreted in the sense of distribution functions. In analogy with this definition, the total variation of the numerical approximation of u is then

$$TV(u^n) = \sum_{i=0}^{N-1} \left| u_{i+1}^n - u_i^n \right| \,. \tag{19.60}$$

A scheme is said to be *"total variation diminishing"* (TVD), or actually has a non-increasing total variation in time, if and only if for every discrete time level n

$$TV(u^{n+1}) \leq TV(u^n) \,. \tag{19.61}$$

Again, it can be shown that the true solution of a scalar conservation law has this TVD property, i.e. $TV(u(x, t_2)) \leq TV(u(x, t_1))$, for all $t_2 > t_1$. When a scheme is TVD as defined by (19.61), it is rather obvious that it will also be monotonicity preserving: if a local extremum in the discrete solution appeared from an initially monotone sequence u_i with $i = 0, \ldots N$, it would naturally raise the total variation too. Hence, we may build in the TVD property (19.61) into the scheme, and we are then sure that no spurious oscillations will be introduced when numerically solving the Riemann problem. A result due to Harten [216] is thereby very useful. It states that any scheme which can be written in the general form

$$u_i^{n+1} = u_i^n + A_{i+1/2} \underbrace{\left(u_{i+1}^n - u_i^n \right)}_{\Delta u_{i+1/2}^n} - B_{i-1/2} \underbrace{\left(u_i^n - u_{i-1}^n \right)}_{\Delta u_{i-1/2}^n} \tag{19.62}$$

is TVD when the scheme dependent coefficients $A_{i+1/2}$ and $B_{i-1/2}$ obey at all grid indices i

$$A_{i+1/2} \geq 0,$$

$$B_{i-1/2} \geq 0, \tag{19.63}$$

$$0 \leq A_{i+1/2} + B_{i+1/2} \leq 1.$$

As a particular example, the first order Lax–Friedrichs scheme (19.47) has this property, as we may formally manipulate the discrete formula to

$$u_i^{n+1} = u_i^n + \tfrac{1}{2}\left(1 - \frac{\Delta t}{\Delta x}\frac{f_{i+1}^n - f_i^n}{\Delta u_{i+\frac{1}{2}}}\right)\Delta u_{i+\frac{1}{2}} - \tfrac{1}{2}\left(1 + \frac{\Delta t}{\Delta x}\frac{f_i^n - f_{i-1}^n}{\Delta u_{i-\frac{1}{2}}}\right)\Delta u_{i-\frac{1}{2}}. \tag{19.64}$$

The requirements (19.63) translate into the CFL condition

$$\left|\frac{\Delta t}{\Delta x}\frac{f_{i+1}^n - f_i^n}{u_{i+1}^n - u_i^n}\right| \leq 1, \tag{19.65}$$

as stated in (19.48).

We can improve the rather diffusive nature of the first order Lax–Friedrichs scheme by changing the numerical flux (19.49) to the following:

$$F_{i+1/2}^{\mathrm{LLF}} = \tfrac{1}{2}\left\{f_{i+1} + f_i - |\alpha_{i+\frac{1}{2}}|\,[u_{i+1} - u_i]\right\}, \tag{19.66}$$

in which the coefficient $\alpha_{i+\frac{1}{2}}$ is found from

$$\alpha_{i+1/2} \equiv \begin{cases} \Delta x/\Delta t & \text{if } u_{i+1}^n - u_i^n = 0, \\[2mm] \dfrac{f_{i+1}^n - f_i^n}{u_{i+1}^n - u_i^n} & \text{if } u_{i+1}^n - u_i^n \neq 0. \end{cases} \tag{19.67}$$

This definition ensures that the derivatives are taken in the "upwind" direction, meaning that we effectively switch between one-sided left or right derivative evaluation depending on the sign of $\alpha_{i+\frac{1}{2}}$. Indeed, we get

$$u_i^{n+1} = u_i^n - \frac{\Delta t}{\Delta x}\begin{cases} (f_{i+1}^n - f_i^n) & \text{for } \alpha_{i+1/2}, \alpha_{i-1/2} < 0, \\[2mm] (f_i^n - f_{i-1}^n) & \text{for } \alpha_{i+1/2}, \alpha_{i-1/2} > 0. \end{cases} \tag{19.68}$$

In the linear advection problem, the sign of the constant advection speed v determines which of these discretizations represents the upwind scheme. The full scheme with numerical flux (19.66) can be rewritten in different forms, e.g.

$$u_i^{n+1} = u_i^n - \tfrac{1}{2}\frac{\Delta t}{\Delta x}\Big[\alpha_{i+1/2}\Delta u_{i+1/2}^n - |\alpha_{i+1/2}|\Delta u_{i+1/2}^n$$

$$+ \alpha_{i-1/2}\Delta u_{i-1/2}^n + |\alpha_{i-1/2}|\Delta u_{i-1/2}^n\Big]. \tag{19.69}$$

This is of the general form (19.62) with

$$\begin{cases} A_{i+1/2} = \frac{1}{2}\frac{\Delta t}{\Delta x}\left(|\alpha_{i+1/2}| - \alpha_{i+1/2}\right) \geq 0, \\[2mm] B_{i-1/2} = \frac{1}{2}\frac{\Delta t}{\Delta x}\left(|\alpha_{i-1/2}| + \alpha_{i-1/2}\right) \geq 0. \end{cases} \qquad (19.70)$$

Hence, two of the three sufficient conditions (19.63) for the scheme to have the TVD property are satisfied. The third condition reads

$$0 \leq |\alpha_{i+1/2}|\frac{\Delta t}{\Delta x} \leq 1, \qquad (19.71)$$

which is a CFL-type condition on the time step for a given spatial resolution. When it is satisfied, the upwind scheme has the TVD property.

So far, we only gave examples of first order accurate TVD schemes, which lead to oscillation-free solutions. In what follows, we will present second order TVD schemes. However, the scheme then needs a truly nonlinear dependence on the discrete data, even for the case of a linear advection equation. This is a direct consequence of the Godunov theorem, which states that a linear monotonicity preserving scheme is at most first order accurate. Second order TVD schemes can be developed by introducing nonlinear flux limiters, which we already encountered for the flux corrected transport method. Such schemes are second order only in "smooth" regions and first order at extrema and shocks. Still, in comparison with fully first order schemes, the shock smearing will be reduced. An example second order TVD type solution for the same Burgers problem shown in Fig. 19.4 is illustrated in Fig. 19.6. Clearly, erroneous oscillations are now totally absent, making the solution visually superior. The same second order TVDLF method (discussed fully below) is used in Fig. 19.7, where we demonstrate the three possible outcomes of the Riemann problem for the non-convex equation $u_t + (u^3)_x = 0$ discussed earlier. The numerical solutions agree in all cases with the known analytic solutions.

Higher order TVD schemes are thus a type of "hybrid" scheme, combining a higher order scheme in smooth regions with a first order scheme in regions where high gradients occur. In what follows, we will discuss such *hybrid* schemes, also called "high-resolution methods", that circumvent the Godunov theorem. TVD methods, as well as the flux corrected transport method, are widely used in MHD computations. Other modern schemes somewhat relax the requirement of monotone schemes, and achieve higher than first order at extrema. Popular categories are known as (weighted) essentially non-oscillatory, or (W)ENO, schemes and discontinuous Galerkin methods. All these schemes are frequently exploited in combination with so-called finite volume treatments, the ideas of which are briefly discussed next.

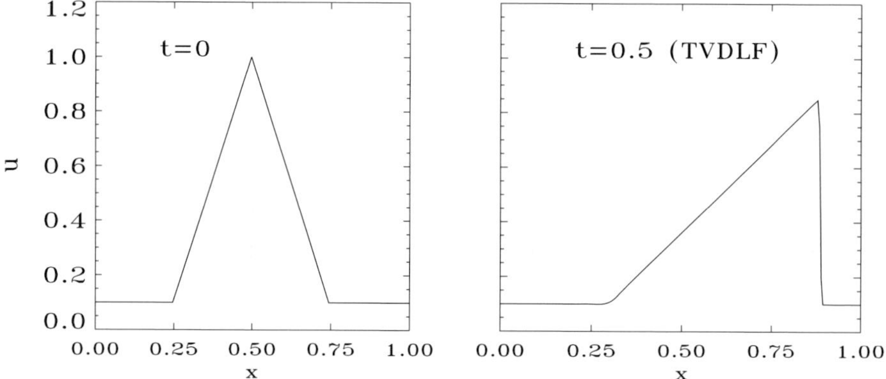

Fig. 19.6 Initial triangular pulse (left) and shocked numerical solution of Burgers' equation as obtained with the second order TVDLF method.

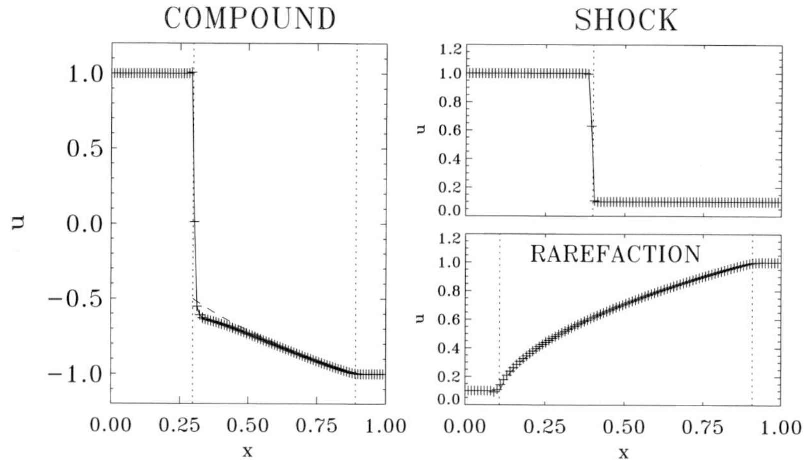

Fig. 19.7 Riemann problems as solved with the TVDLF scheme for a non-convex scalar equation. The discontinuity was placed at $x = 0.1$ at $t = 0$, and the frames shown are at time $t = 0.27$. The vertical dotted lines in all frames correspond to the analytic locations of shocks or rarefaction edges. In the left panel, the dashed line indicates the exact solution for the compound wave structure.

19.1.4 Finite volume treatments

In many "high resolution shock capturing" schemes, finite volume methods are used for the spatial discretization. This discretization technique refers directly to the integral form of the conservation laws. For a single scalar conservation law as given by Eq. (19.27), the frequently used differential form actually follows from the integral form *only when the solution is smooth or regular enough* (as it must e.g. be continuously differentiable). This integral form states that a conserved scalar quantity $u(x,t)$ changes in the space-time volume $[x_1, x_2] \times [t_1, t_2]$ in accord with

$$\int_{x_1}^{x_2} u(x,t_2)\, dx = \int_{x_1}^{x_2} u(x,t_1)\, dx + \int_{t_1}^{t_2} f(x_1,t)\, dt - \int_{t_1}^{t_2} f(x_2,t)\, dt. \quad (19.72)$$

The instantaneous volume average changes only through the temporally varying fluxes across the domain boundaries. This is also true in general, when we deal with a multi-dimensional set of conservation laws as in Eq. (19.1),

$$\frac{\partial \mathbf{U}}{\partial t} + \nabla \cdot \mathcal{F} = 0, \quad (19.73)$$

where $\mathbf{U}$ is the state vector of n conserved variables and $\mathcal{F}$ is the flux. For the MHD system, where $n = 8$, recall from the analysis of Section 4.3.2 [1] the explicit expressions for $\mathbf{U} = (\rho, \boldsymbol{\pi}, \mathcal{H}, \mathbf{B})^{\mathrm{T}}$ and $\mathcal{F} = (\boldsymbol{\pi}, \mathbf{T}, \tilde{\mathbf{U}}, \mathbf{Y})^{\mathrm{T}}$, where

$$\boldsymbol{\pi} \equiv \rho \mathbf{v} \qquad\qquad\qquad\qquad \textit{(momentum density)}, \quad (19.74)$$

$$\mathbf{T} \equiv \rho \mathbf{v}\mathbf{v} + (p + \tfrac{1}{2}B^2)\,\mathbf{I} - \mathbf{B}\mathbf{B} \qquad \textit{(stress tensor)}, \quad (19.75)$$

$$\mathcal{H} \equiv \tfrac{1}{2}\rho v^2 + \frac{p}{\gamma - 1} + \tfrac{1}{2}B^2 \qquad \textit{(total energy density)}, \quad (19.76)$$

$$\tilde{\mathbf{U}} \equiv \left(\tfrac{1}{2}\rho v^2 + \frac{\gamma}{\gamma - 1}p\right)\mathbf{v} + B^2\mathbf{v} - \mathbf{v} \cdot \mathbf{B}\,\mathbf{B} \quad \textit{(energy flow)}, \quad (19.77)$$

$$\mathbf{Y} \equiv \mathbf{v}\mathbf{B} - \mathbf{B}\mathbf{v} \qquad\qquad\qquad \textit{(no name)}. \quad (19.78)$$

Note that in 3D, the term $\nabla \cdot \mathcal{F}$ in Eq. (19.73) can be expressed in terms of fluxes along three Cartesian coordinate axes, by writing

$$\nabla \cdot \mathcal{F} = \frac{\partial \mathbf{F}_x}{\partial x} + \frac{\partial \mathbf{F}_y}{\partial y} + \frac{\partial \mathbf{F}_z}{\partial z}. \quad (19.79)$$

When we discretize space in control volumes V_i with bounding surfaces ∂V_i, with outward unit normal $\mathbf{n} = (n_x, n_y, n_z)$, we find from Gauss' theorem that

$$\frac{d}{dt}\int_{V_i} \mathbf{U}(\mathbf{x},t)\, dx = -\int_{\partial V_i} \mathcal{F} \cdot \mathbf{n}\, dS = -\int_{\partial V_i} (\mathbf{F}_x n_x + \mathbf{F}_y n_y + \mathbf{F}_z n_z)\, dS. $$
$$(19.80)$$

This formula can easily be generalized to account for source and sink terms as well, if they exist for certain components of $\mathbf{U}$. We now introduce the matrix $\mathbf{T}(\mathbf{n})$ which rotates all occurring vector quantities to a local orthogonal coordinate system formed by $\mathbf{n}$, $\mathbf{t}$, $\mathbf{s}$ ($\equiv \mathbf{n} \times \mathbf{t}$), where the latter are tangential unit vectors within the bounding surface ∂V_i. For the MHD system as above, we will have $\mathbf{T}$ given by

$$\begin{pmatrix}
1 & 0 & 0 & 0 & 0 & 0 & 0 & 0 \\
0 & \sin\theta\cos\phi & \sin\theta\sin\phi & \cos\theta & 0 & 0 & 0 & 0 \\
0 & \cos\theta\cos\phi & \cos\theta\sin\phi & -\sin\theta & 0 & 0 & 0 & 0 \\
0 & -\sin\phi & \cos\phi & 0 & 0 & 0 & 0 & 0 \\
0 & 0 & 0 & 0 & 1 & 0 & 0 & 0 \\
0 & 0 & 0 & 0 & 0 & \sin\theta\cos\phi & \sin\theta\sin\phi & \cos\theta \\
0 & 0 & 0 & 0 & 0 & \cos\theta\cos\phi & \cos\theta\sin\phi & -\sin\theta \\
0 & 0 & 0 & 0 & 0 & -\sin\phi & \cos\phi & 0
\end{pmatrix}.$$

$$\tag{19.81}$$

In the latter expression, we introduced the spherical coordinate angles (θ,ϕ) fully determining the direction of the normal $\mathbf{n}$ and the tangent vectors. One can then verify, or understand from the fact that the MHD equations must be unchanged under rotation, that

$$\mathbf{F}_x n_x + \mathbf{F}_y n_y + \mathbf{F}_z n_z = \mathbf{T}^{-1}(\mathbf{n})\mathbf{F}_x\left(\mathbf{T}(\mathbf{n})\mathbf{U}\right). \tag{19.82}$$

Ultimately, this means that we obtain an essentially 1D problem in the direction normal to the control volume boundary, since Eq. (19.80) now becomes

$$\frac{d\int_{V_i}\mathbf{U}(\mathbf{x},t)\,d\mathbf{x}}{dt} = -\int_{\partial V_i}\mathbf{T}^{-1}(\mathbf{n})\mathbf{F}_x\left(\mathbf{T}(\mathbf{n})\mathbf{U}\right)\,dS. \tag{19.83}$$

We then usually employ control volumes with multiple flat surface segments, turning the integral over the boundary into a discrete sum over its sides. This means that, in practice, the only information of the grid needed in the process consists of the volumes V_i of the grid cells and the geometry of the cells such as the number of bounding surface segments, their surface area and their normal directions.

As a simple example, consider a 2D problem for scalar quantity u with a two-dimensional flux vector $\mathcal{F}(u)$, discretized on a triangular mesh as in Fig. 19.8. Clearly, in the two-dimensional case the "volumes" become surfaces S_i. If we write u_i^{n+1} for the volume average value $S_i^{-1}\int_{S_i} u\,d\mathbf{x}$ of u over cell i at time $(n+1)\Delta t$, we get using a forward Euler temporal discretization

$$u_i^{n+1} = u_i^n - \frac{\Delta t}{S_i}\sum_{m=1}^{3}\mathcal{F}(u_m^n)\cdot\mathbf{l}_m. \tag{19.84}$$

The sum in the right hand side is now an approximation of the flux, using values at time level $n\Delta t$ only, while $\mathbf{l}_m$ denotes a vector that is normal to the cell edge with length equal to the length of the cell edge. Writing $\mathcal{F}(u_m^n)$ involves evaluating the flux using values for u on the cell edge. From Fig 19.8, an accurate choice for this average would be $\frac{1}{2}(u_i^n + u_{i+1}^n)$. However, *this choice leads to numerical*

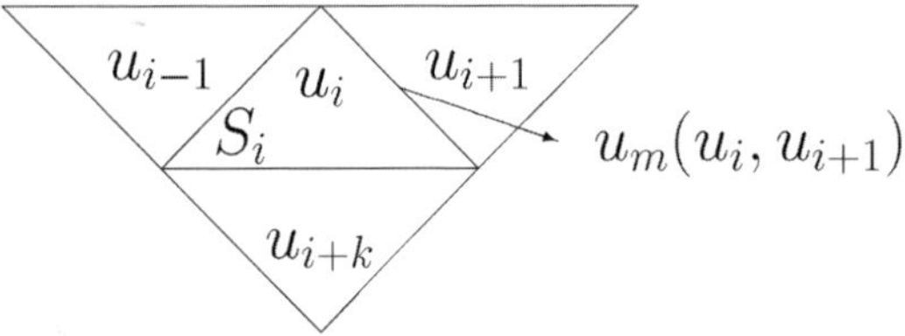

Fig. 19.8 Fraction of a triangular mesh. S_i denotes the surface of the ith cell and u_m the average or mean value of u on the interface between two cells.

instabilities for the very same reason as the choice of the centered space derivative combined with a forward time discretization did in the forward in time centered in space scheme discussed in Chapter 15: it does not take into account from where the physical information comes. In other words, this "natural" choice gives the same "weight" to the information coming from the "left" and the "right". We will therefore need to generalize the idea of upwinding to systems of nonlinear equations. Generally speaking, though, in finite volume methods, the temporal evolution of the dependent variables in some discrete points (called "nodes") of the control volumes or grid cells is determined by applying the conservation laws *on each cell*, since the dependent variables are the conserved quantities themselves. The nodes can be at the cell centers but could also lie on the cell interfaces or on cell corners or vertices. In fact, there are many possibilities and the decoupling of the cells and the nodes allows more freedom in an FVM than typically encountered in an FDM or even an FEM. One may say that, by the decoupling of cells and nodes, the FVM combines an advantage of the FEM, viz. geometric flexibility, with an advantage of the FDM, viz. the flexibility in the definition of discrete values of the dependent variables (i.e. the free choice of the nodes). At the same time, the FVM does not as easily generalize to higher order than FDM or FEM.

19.2 Upwind-like finite volume treatments for 1D MHD

The FVM turned the solution of the conservation laws into, in essence, 1D flux updates across the segmented boundaries of control volumes. We will therefore now pay further attention to the 1D case in particular. Recapitulating, in a 1D configuration, a shock capturing finite volume scheme thus interprets $\mathbf{U}_i$ as the average value of the solution $\mathbf{U}(x, t)$ in the interval $[x_{i-1/2}, x_{i+1/2}]$:

$$\mathbf{U}_i(t) \equiv \frac{1}{\Delta x_i} \int_{x_{i-1/2}}^{x_{i+1/2}} \mathbf{U}(x, t) \, dx \,, \tag{19.85}$$

where the domain is now subdivided in a set of equal-sized "cells" covering the whole domain, as shown in Fig. 19.9. The update of these volume averages be-

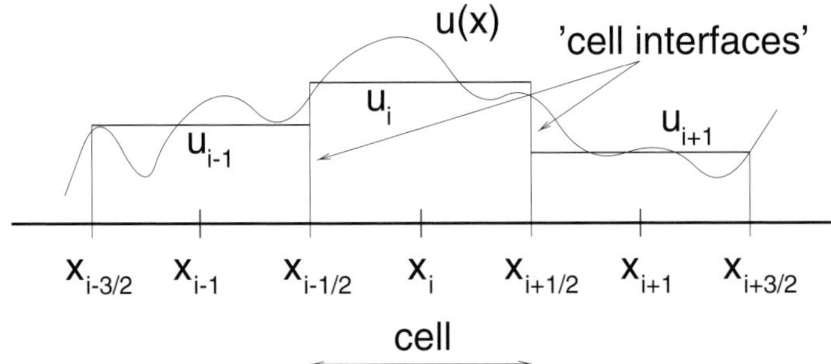

Fig. 19.9 Schematic representation of the finite volume method in 1D.

comes simply

$$\frac{d\mathbf{U}_i}{dt} + \frac{1}{\Delta x_i}\left(\mathbf{F}_{i+1/2} - \mathbf{F}_{i-1/2}\right) = 0\,. \tag{19.86}$$

The advantage of the volume average interpretation is that the discretized equation itself can be seen as an integral law, rather than a differential law, and that its weak solutions will obey conservation by construction. In 1D MHD where we can allow for translational invariance in ignorable y and z directions, the normal B_x component remains constant, so that we have at most seven components for $\mathbf{U}$.

19.2.1 The Godunov method

A method originally proposed by Godunov was to consider the cell averaged values $\mathbf{U}_i^n$ for time $t = t_n$ as piecewise constant data throughout the cells, and concentrate on the discontinuous Riemann problems that then arise at the cell interfaces. The solution of each local Riemann problem is in essence self-similar in x/t, so one can properly restrict the time step such that no wave interaction occurs in one cell of size Δx within Δt^{n+1}. This is achieved by imposing $\Delta t^{n+1} < \Delta x/(2\max|\lambda_p^n|)$ on the discrete (but variable) time step, with the maximum taken over all eigenvalues of the flux Jacobian $\mathbf{F_U}$. For 1D MHD, this is set by the fast eigenvalues and is $|v_x|\pm c_{\mathrm{f}}$. The Godunov method consists of using the exact nonlinear solution of the Riemann problems at the cell interfaces in the numerical flux. Denoting the exact Riemann problem solution for state $\mathbf{U}_i^n$ and $\mathbf{U}_{i+1}^n$ as $\hat{\mathbf{U}}((x-x_{i+1/2})/t, \mathbf{U}_i^n, \mathbf{U}_{i+1}^n)$, the Godunov scheme then uses as numerical flux

$$\mathbf{F}_{i+1/2}\left(\mathbf{U}_i, \mathbf{U}_{i+1}\right) = \mathbf{F}(\hat{\mathbf{U}}\left(0, \mathbf{U}_i^n, \mathbf{U}_{i+1}^n\right))\,. \tag{19.87}$$

Due to its piecewise constant representation of cell values, this original Godunov scheme is at best first order accurate. In any case, it requires an exact solution to the nonlinear Riemann problem. For 1D MHD, where seven wave families are involved, this is a far from trivial exercise involving adequate procedures to handle the nonlinearities, whereby one must face existence and uniqueness issues as well. We take in what follows a more pragmatic approach, discussing various upwind-like methods which avoid the full solution of the nonlinear Riemann problem.

The linearized MHD system and the Riemann problem A full solution to the Riemann problem can directly be given for any linear hyperbolic system, as was explained before. Specifying how this works for 1D MHD, we can, e.g., look at the primitive variable formulation $\mathbf{V}_t + \mathbf{W}\mathbf{V}_x = 0$, where

$$
\begin{pmatrix} \rho \\ v_x \\ v_y \\ v_z \\ p \\ B_y \\ B_z \end{pmatrix}_t + \begin{pmatrix} v_x & \rho & 0 & 0 & 0 & 0 & 0 \\ 0 & v_x & 0 & 0 & 1/\rho & B_y/\rho & B_z/\rho \\ 0 & 0 & v_x & 0 & 0 & -B_x/\rho & 0 \\ 0 & 0 & 0 & v_x & 0 & 0 & -B_x/\rho \\ 0 & \rho c^2 & 0 & 0 & v_x & 0 & 0 \\ 0 & B_y & -B_x & 0 & 0 & v_x & 0 \\ 0 & B_z & 0 & -B_x & 0 & 0 & v_x \end{pmatrix} \begin{pmatrix} \rho \\ v_x \\ v_y \\ v_z \\ p \\ B_y \\ B_z \end{pmatrix}_x = 0.
$$

$$(19.88)$$

To turn this into a linear hyperbolic PDE, we will evaluate $\mathbf{W}$ in a fixed state $\mathbf{V}^*$, and assume that the eigenvalues for the matrix $\mathbf{W}(\mathbf{V}^*)$ are distinct. They are the familiar ordered set

$$
v_x^* - c_f^*, \quad v_x^* - b_x^*, \quad v_x^* - c_s^*, \quad v_x^*, \quad v_x^* + c_s^*, \quad v_x^* + b_x^*, \quad v_x^* + c_f^*, \tag{19.89}
$$

where the Alfvén speed $b_x^* = |B_x^*|/\sqrt{\rho^*}$, and the slow and fast magneto-acoustic speeds are as in Eq. (19.15), with the iso-thermal sound speed squared c_i^2 replaced by the sound speed squared c^{*2}. We wish to solve the Riemann problem for this linear system exactly, with given initial constant left and right states $\mathbf{V}_\ell$ and $\mathbf{V}_r$. We will need the left $\hat{\mathbf{l}}^p$ and right $\hat{\mathbf{r}}^p$ eigenvectors of matrix $\mathbf{W}$ in the process, and as explained before, these are collected in the matrices $\hat{\mathbf{R}}^{-1}$ and $\hat{\mathbf{R}}$, respectively. The latter $\hat{\mathbf{R}}$, with columns $\hat{\mathbf{r}}^p$, is given by

$$
\begin{pmatrix}
\alpha_f\rho & 0 & \alpha_s\rho & 1 & \alpha_s\rho & 0 & \alpha_f\rho \\
-\alpha_f c_f & 0 & -\alpha_s c_s & 0 & \alpha_s c_s & 0 & \alpha_f c_f \\
\alpha_s c_s\beta_y\beta_x & -\beta_z & -\alpha_f c_f\beta_y\beta_x & 0 & \alpha_f c_f\beta_y\beta_x & \beta_z & -\alpha_s c_s\beta_y\beta_x \\
\alpha_s c_s\beta_z\beta_x & \beta_y & -\alpha_f c_f\beta_z\beta_x & 0 & \alpha_f c_f\beta_z\beta_x & -\beta_y & -\alpha_s c_s\beta_z\beta_x \\
\alpha_f\rho c^2 & 0 & \alpha_s\rho c^2 & 0 & \alpha_s\rho c^2 & 0 & \alpha_f\rho c^2 \\
\alpha_s c\beta_y\sqrt{\rho} & -\beta_z\sqrt{\rho}\beta_x & -\alpha_f c\beta_y\sqrt{\rho} & 0 & -\alpha_f c\beta_y\sqrt{\rho} & -\beta_z\sqrt{\rho}\beta_x & \alpha_s c\beta_y\sqrt{\rho} \\
\alpha_s c\beta_z\sqrt{\rho} & \beta_y\sqrt{\rho}\beta_x & -\alpha_f c\beta_z\sqrt{\rho} & 0 & -\alpha_f c\beta_z\sqrt{\rho} & \beta_y\sqrt{\rho}\beta_x & \alpha_s c\beta_z\sqrt{\rho}
\end{pmatrix},
$$

$$(19.90)$$

while its inverse $\hat{\mathbf{R}}^{-1}$, with $\hat{\mathbf{l}}^p$ in the rows, is

$$
\begin{pmatrix}
0 & -\dfrac{c_{\mathrm{f}}\alpha_{\mathrm{f}}}{2c^2} & \dfrac{\alpha_{\mathrm{s}}c_{\mathrm{s}}}{2c^2}\beta_y\beta_x & \dfrac{\alpha_{\mathrm{s}}c_{\mathrm{s}}}{2c^2}\beta_z\beta_x & \dfrac{\alpha_{\mathrm{f}}}{2c^2\rho} & \dfrac{\alpha_{\mathrm{s}}}{2c\sqrt{\rho}}\beta_y & \dfrac{\alpha_{\mathrm{s}}}{2c\sqrt{\rho}}\beta_z \\[2ex]
0 & 0 & -\tfrac{1}{2}\beta_z & \tfrac{1}{2}\beta_y & 0 & -\beta_z\dfrac{\beta_x}{2\sqrt{\rho}} & \beta_y\dfrac{\beta_x}{2\sqrt{\rho}} \\[2ex]
0 & -\dfrac{c_{\mathrm{s}}\alpha_{\mathrm{s}}}{2c^2} & -\dfrac{\alpha_{\mathrm{f}}c_{\mathrm{f}}}{2c^2}\beta_y\beta_x & -\dfrac{\alpha_{\mathrm{f}}c_{\mathrm{f}}}{2c^2}\beta_z\beta_x & \dfrac{\alpha_{\mathrm{s}}}{2c^2\rho} & -\dfrac{\alpha_{\mathrm{f}}}{2c\sqrt{\rho}}\beta_y & -\dfrac{\alpha_{\mathrm{f}}}{2c\sqrt{\rho}}\beta_z \\[2ex]
1 & 0 & 0 & 0 & -\dfrac{1}{c^2} & 0 & 0 \\[2ex]
0 & \dfrac{c_{\mathrm{s}}\alpha_{\mathrm{s}}}{2c^2} & \dfrac{\alpha_{\mathrm{f}}c_{\mathrm{f}}}{2c^2}\beta_y\beta_x & \dfrac{\alpha_{\mathrm{f}}c_{\mathrm{f}}}{2c^2}\beta_z\beta_x & \dfrac{\alpha_{\mathrm{s}}}{2c^2\rho} & -\dfrac{\alpha_{\mathrm{f}}}{2c\sqrt{\rho}}\beta_y & -\dfrac{\alpha_{\mathrm{f}}}{2c\sqrt{\rho}}\beta_z \\[2ex]
0 & 0 & \tfrac{1}{2}\beta_z & -\tfrac{1}{2}\beta_y & 0 & -\beta_z\dfrac{\beta_x}{2\sqrt{\rho}} & \beta_y\dfrac{\beta_x}{2\sqrt{\rho}} \\[2ex]
0 & \dfrac{c_{\mathrm{f}}\alpha_{\mathrm{f}}}{2c^2} & -\dfrac{\alpha_{\mathrm{s}}c_{\mathrm{s}}}{2c^2}\beta_y\beta_x & -\dfrac{\alpha_{\mathrm{s}}c_{\mathrm{s}}}{2c^2}\beta_z\beta_x & \dfrac{\alpha_{\mathrm{f}}}{2c^2\rho} & \dfrac{\alpha_{\mathrm{s}}}{2c\sqrt{\rho}}\beta_y & \dfrac{\alpha_{\mathrm{s}}}{2c\sqrt{\rho}}\beta_z
\end{pmatrix}.
$$

$$(19.91)$$

In these expressions, we introduced dimensionless parameters

$$
\alpha_{\mathrm{s}}^2 = \frac{c_{\mathrm{f}}^2 - c^2}{c_{\mathrm{f}}^2 - c_{\mathrm{s}}^2}, \quad
\alpha_{\mathrm{f}}^2 = \frac{c^2 - c_{\mathrm{s}}^2}{c_{\mathrm{f}}^2 - c_{\mathrm{s}}^2}, \quad
\beta_x = \frac{B_x}{|B_x|}, \quad
\beta_y = \frac{B_y}{B_\perp}, \quad
\beta_z = \frac{B_z}{B_\perp},
$$

$$(19.92)$$

where $B_\perp = \sqrt{B_y^2 + B_z^2}$. In the form given in (19.90), the right eigenvectors have the same dimension as the primitive variables $\mathbf{V}$. Also, the fast eigenvectors can be artificially decomposed in an acoustic ($\sim \alpha_{\mathrm{f}}$) and a magnetic ($\sim \alpha_{\mathrm{s}}$) contribution, and a similar argument is true for the slow eigenvectors. To solve the linear Riemann problem posed above, we write the constant left and right states $\mathbf{V}_{l,r}$ as linear combinations of the right eigenvectors of $\mathbf{W}(\mathbf{V}^*)$. The orthonormal set $\hat{\mathbf{r}}^{*p}$ and $\hat{\mathbf{l}}^{*p}$ allows us to write

$$
\mathbf{V}_\ell = \sum_{p=1}^{7}(\hat{\mathbf{l}}^{*p} \cdot \mathbf{V}_\ell)\hat{\mathbf{r}}^{*p} = \sum_{p=1}^{7}\beta_p\hat{\mathbf{r}}^{*p}, \quad
\mathbf{V}_r = \sum_{p=1}^{7}(\hat{\mathbf{l}}^{*p} \cdot \mathbf{V}_r)\hat{\mathbf{r}}^{*p} = \sum_{p=1}^{7}\gamma_p\hat{\mathbf{r}}^{*p}.
$$

$$(19.93)$$

The solution of the Riemann problem can then use the general expression given in Eq. (19.26), which due to our initial data becomes

$$
\mathbf{V}(x,t) = \sum_{x/t<\lambda_p^*}(\hat{\mathbf{l}}^{*p} \cdot \mathbf{V}_\ell)\hat{\mathbf{r}}^{*p} + \sum_{x/t>\lambda_p^*}(\hat{\mathbf{l}}^{*p} \cdot \mathbf{V}_r)\hat{\mathbf{r}}^{*p}. \qquad (19.94)
$$

In the (x,t) plane, we can graphically illustrate this solution as shown in Fig. 19.10. At $t = 0$, the two constant states $\mathbf{V}_\ell$ and $\mathbf{V}_r$ are denoted by their seven coefficients from the expansion (19.93), $(\beta_1, \beta_2, \ldots, \beta_7)$ and $(\gamma_1, \gamma_2, \ldots, \gamma_7)$, respectively. Assuming distinct and ordered eigenvalues $\lambda_1^* < \lambda_2^* < \cdots < \lambda_7^*$, the seven characteristic curves $x = \lambda_p(\mathbf{V}^*)t$ through the origin (the position $x = 0$ of the discontinuity at $t = 0$) divide the half plane $t > 0$ in eight regions. In each of those,

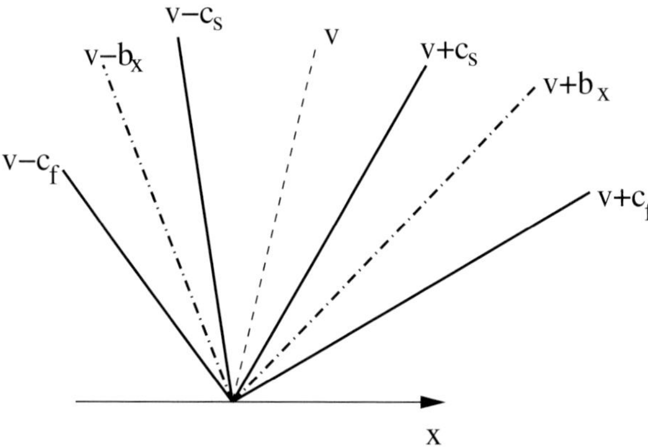

Fig. 19.10 Schematic (x,t) diagram of the solution of the linear MHD Riemann problem in 1D: backward and forward traveling fast, Alfvén and slow discontinuities are separated by a contact discontinuity traveling at speed v. Out of a single discontinuity separating two constant states, six new constant states may emerge.

the solution of the linear Riemann problem is a constant state. Indeed, according to the formula (19.94), the coefficients used in writing $\mathbf{V}(x,t)$ as a linear combination of the (constant) right eigenvectors $\hat{\mathbf{r}}^p$ can be found by drawing the seven p-characteristics backwards in time from the point (x,t). Their intersection with the initial data determines which of the two coefficients β_p or γ_p prevails. Hence, out of one discontinuity separating two constant MHD states for a system of seven linear equations, up to a maximum of eight constant states (six on top of the given left and right state) separated by seven discontinuities emerge. The discontinuities travel at the characteristic speeds given by the eigenvalues of the coefficient matrix, while the jumps across the seven discontinuities are proportional to the right eigenvectors of the matrix. This yields the generic structure shown in Fig. 19.10.

▷ **Exercise** Note that the scaling of the eigenvectors given here for the full MHD equations is different from the one used in the expressions (19.19)–(19.20) for the 1.5D iso-thermal MHD system. It is a matter of algebra to show how a similarly scaled system can be used there, and this is left as an exercise. This rescaling will be needed for a convenient numerical handling of the degeneracies in the eigenvalue expressions. Historically, eigenvector expressions scaled as in Eq. (19.19) were given in [250], but in this form the expressions contain indefinite limits when strict hyperbolicity is lost. Better behaved scaled versions were introduced by [70], and the slightly different scaling used here follows [393]. ◁

Degeneracies of the MHD characteristic speeds The seven wave speeds (19.89) (dropping the $*$-superscript) are degenerate if one of the following situations arises.

- When both tangential field components vanish, i.e. $B_y = 0 = B_z$, and $b_x^2 \neq c^2$, either of both fast ($b_x > c$) or both slow ($b_x < c$) characteristic speeds coincide with the Alfvén signals at $v_x \pm b_x$. This is termed a *double umbilic point.*

- When tangential field components vanish as well as $b_x^2 = c^2$, the slow, Alfvén and fast wave speeds coincide, making $v_x \pm b_x$ *triple umbilic points.*

- When $B_x = 0$, Alfvén and slow pairs collapse to a *quintuple umbilic point* v_x.

Therefore, the MHD equations are not strictly hyperbolic and this complicates the strategy to design a true nonlinear MHD Riemann solver. Similar degeneracies are already present in the 1.5D iso-thermal MHD system discussed earlier, where only four wave speeds are at play. For this simplified system, the outcome of the true nonlinear Riemann problem can be analyzed to potentially lead to 289 distinct (mathematically allowed) outcomes, in which the trivial case of equal left and right states is included too [345, 346]. In fact, both fast and slow wave families can arise as shocks, rarefaction waves or compound waves. Therefore, $256 = 4^4$ cases allow four "waves" with each wave either absent or present in one of its three manifestations. The other 33 possibilities account for overcompressive shock situations, with (1) both left-going waves merged and 16 possibilities arising out of the two right-going waves, (2) both right-going waves merged, yielding 16 possibilities for the left pair of waves, and (3) both left and right going wave pairs merged into overcompressive shocks. Since the full MHD system additionally includes the entropy and Alfvén wave families, allowing for contact and rotational discontinuities, it is practically impossible to appropriately handle that many distinct possibilities. As a result, the nonlinear Riemann solvers available in the literature necessarily handle the MHD Riemann problem in a more tractable simplified manner. One way to do so is to allow only discontinuous shock transitions in all wave families. This was done by Dai and Woodward [103] who, in an all-shock nonlinear Riemann solver, exploited the nonlinear Rankine–Hugoniot relations at the up to seven transitions, conveniently written in a Lagrangian mass coordinate. A more elaborate nonlinear 1D MHD Riemann solver was presented in [399], where, in addition to shocks, the solver allowed for either slow or fast rarefaction waves. To do so, the differential equations governing the various Riemann invariants were used through the rarefactions. Still, these nonlinear Riemann solvers which only use regular (shock or rarefaction) waves are fairly computationally involved, and a tendency emerged to avoid using them as the basic building block for a shock-capturing method. Instead, approximate Riemann solvers gained in popularity. These will be discussed later on.

▷ **Eigenvectors of the flux Jacobian** Further on in the discussion, we will also need the eigenvectors of the flux Jacobian. The right eigenvectors can be found directly from the relation $\mathbf{R} = \mathbf{U_V}\hat{\mathbf{R}}$, giving column vectors $\mathbf{r}^p$. The gives the following expressions for the fast eigenvectors $\mathbf{r}_{\mathrm{f}}^{1,7}$, the slow eigenvectors $\mathbf{r}_{\mathrm{s}}^{3,5}$, the Alfvén eigenvectors $\mathbf{r}_{\mathrm{a}}^{2,6}$ and the entropy eigenvectors $\mathbf{r}_e^4$, respectively:

$$
\begin{pmatrix}
\alpha_f \rho \\
\alpha_f \rho (v_x \mp c_f) \\
\alpha_f \rho v_y \pm \alpha_s \rho c_s \beta_y \beta_x \\
\alpha_f \rho v_z \pm \alpha_s \rho c_s \beta_z \beta_x \\
g_f^{\mp} \\
\alpha_s c \beta_y \sqrt{\rho} \\
\alpha_s c \beta_z \sqrt{\rho}
\end{pmatrix},
\begin{pmatrix}
\alpha_s \rho \\
\alpha_s \rho (v_x \mp c_s) \\
\alpha_s \rho v_y \mp \alpha_f \rho c_f \beta_y \beta_x \\
\alpha_s \rho v_z \mp \alpha_f \rho c_f \beta_z \beta_x \\
g_s^{\mp} \\
-\alpha_f c \beta_y \sqrt{\rho} \\
-\alpha_f c \beta_z \sqrt{\rho}
\end{pmatrix},
\begin{pmatrix}
0 \\
0 \\
\mp \rho \beta_z \\
\pm \rho \beta_y \\
g_a^{\mp} \\
-\beta_z \sqrt{\rho} \beta_x \\
\beta_y \sqrt{\rho} \beta_x
\end{pmatrix},
\begin{pmatrix}
1 \\
v_x \\
v_y \\
v_z \\
\frac{1}{2} v^2 \\
0 \\
0
\end{pmatrix},
$$

$$
\tag{19.95}
$$

where we introduced the symbols

$$
\begin{aligned}
g_f^{\mp} &= \alpha_f \rho \left(\tfrac{1}{2} v^2 + \tfrac{1}{\gamma-1} c^2 \mp c_f v_x \right) \pm \alpha_s \rho c_s \beta_x \left(v_y \beta_y + v_z \beta_z \right) + \alpha_s c \sqrt{\rho} B_\perp \,, \\
g_a^{\mp} &= \mp \rho (v_y \beta_z - v_z \beta_y) \,, \\
g_s^{\mp} &= \alpha_s \rho \left(\tfrac{1}{2} v^2 + \tfrac{1}{\gamma-1} c^2 \mp c_s v_x \right) \mp \alpha_f \rho c_f \beta_x \left(v_y \beta_y + v_z \beta_z \right) - \alpha_f c \sqrt{\rho} B_\perp \,.
\end{aligned}
$$

$$
\tag{19.96}
$$

One can then verify that the structure coefficients

$$
\left(\frac{\partial \lambda_p}{\partial \rho}, \frac{\partial \lambda_p}{\partial m_x}, \frac{\partial \lambda_p}{\partial m_y}, \frac{\partial \lambda_p}{\partial m_z}, \frac{\partial \lambda_p}{\partial \mathcal{H}}, \frac{\partial \lambda_p}{\partial B_y}, \frac{\partial \lambda_p}{\partial B_z} \right) \cdot \mathbf{r}^p = 0 \,,
\tag{19.97}
$$

for the entropy $p = 4$ and the Alfvén wave $p = 2, 6$ families. They are thus linearly degenerate and do not allow for rarefaction wave constructions using Eq. (19.42). They do allow for discontinuous shock-type solutions, and the generalized Riemann invariants for these characteristic fields then directly demonstrate

$$
\text{entropy:} \quad \lambda_4 = \frac{m_x}{\rho} \,, \quad dv_x = dv_y = dv_z = dp = dB_y = dB_z = 0 \,,
$$

$$
\text{Alfvén:} \quad \lambda_{2,6} = \frac{m_x}{\rho} \mp \frac{\beta_x B_x}{\sqrt{\rho}} \,, \quad d\rho = dv_x = dp = dB_\perp = 0 \,, \quad dv_{y,z} = \pm \frac{\beta_x}{\sqrt{\rho}} dB_{y,z} \,.
$$

$$
\tag{19.98}
$$

The entropy family thus leads to contact discontinuities, where only a jump in density (or entropy, temperature or total energy $\mathcal{H}$) occurs. For all wave families other than the entropy one, one true Riemann invariant can directly be identified with the entropy $S = p\rho^{-\gamma}$, since we know that in ideal MHD this variable obeys

$$
\frac{\partial S}{\partial t} + \mathbf{v} \cdot \nabla S = 0 \,.
\tag{19.99}
$$

From Eq. (19.98), we recognize that the Alfvén families yield rotational discontinuities, where the total field magnitude is conserved and the tangential field $\mathbf{B}_\perp$ gets rotated. For the fast and slow wave families, it is easier to compute the structure coefficient from the equivalent $(\nabla_V \lambda_p) \cdot \hat{\mathbf{r}}^p$ from the primitive right eigenvector set. To compute partial derivatives with respect to density ρ, pressure p and both tangential magnetic field components, one can use the identity

$$
c_{s,f}^4 - (c^2 + b_x^2 + b_\perp^2) c_{s,f}^2 + c^2 b_x^2 = 0 \,,
\tag{19.100}
$$

which one differentiates to obtain expressions for e.g. $\partial c_{s,f}/\partial\rho$. This expression uses $b_\perp = B_\perp/\sqrt{\rho}$. The end result for the structure coefficients yields

$$s_1 = -\alpha_f c_f\left(\tfrac{1}{2}\gamma\alpha_f^2 + \alpha_s^2 + \tfrac{1}{2}\right), \quad s_3 = -\alpha_s c_s\left(\tfrac{1}{2}\gamma\alpha_s^2 + \alpha_f^2 + \tfrac{1}{2}\right),$$

$$s_5 = \alpha_s c_s\left(\tfrac{1}{2}\gamma\alpha_s^2 + \alpha_f^2 + \tfrac{1}{2}\right), \qquad s_7 = \alpha_f c_f\left(\tfrac{1}{2}\gamma\alpha_f^2 + \alpha_s^2 + \tfrac{1}{2}\right). \quad (19.101)$$

These were also presented in [393], and they reduce to the gas dynamic expression when either a fast or a slow wave becomes a purely acoustic wave (e.g. when $\alpha_f = 1, c_f = c$ and $\alpha_s = 0$). The non-convex nature of these waves is now hidden in the parameters α_f^2 and α_s^2, which always obey $\alpha_f^2 + \alpha_s^2 = 1$. With these expressions, we can write out the slow and fast rarefaction wave (centered simple wave) integral curves, as given by Eq. (19.40). There, the integral curves are formulated as curves in state space $\mathbf{U}$ everywhere tangent to the eigenvectors $\mathbf{r}^p$. Similarly, we can use the primitive state space $\mathbf{V}$ and write integral curves from $\hat{\mathbf{r}}^p$. Using a dimensionless parameter θ along the integral curve instead of the dimensional $\xi = x/t$, we find for the fast integral curve for the left going wave ($p = 1$) the governing differential equations

$$\rho' = -\rho, \qquad p' = -\gamma p, \qquad v_x' = c_f,$$

$$\mathbf{v}_t' = \frac{c_f}{1 - c_f^2/b_x^2}\frac{\mathbf{B}_t}{B_x}, \qquad \mathbf{B}_t' = \frac{c_f^2/b_x^2}{1 - c_f^2/b_x^2}\mathbf{B}_t. \quad (19.102)$$

The $'$ stands for derivation with respect to θ, and we denote the transverse vector components with subscript t. Expressions for all six generalized Riemann invariants associated with either fast and slow wave families can be obtained in a way similar to the one outlined for the 1.5D iso-thermal MHD system [250]. Apart from the entropy, not many can be given in easily verified closed form (not involving integral factor) expressions, with the exception of fast and slow magneto-acoustic Riemann invariants for an ideal gas with the fixed value of $\gamma = 5/3$ (monatomic gas). Since

$$\frac{dv_y}{dv_z} = \frac{dB_y}{dB_z} = \frac{B_y}{B_z}, \quad (19.103)$$

the generalized Riemann invariants show that, through fast or slow simple waves, the transverse field and flow vary in magnitude but do not rotate, as opposed to Alfvén waves. $\triangleleft$

19.2.2 A robust shock-capturing method: TVDLF

Before presenting numerical schemes which, to varying degree, exploit the detailed knowledge of the characteristic fields in sets of conservation laws, we can easily generalize the TVD upwind method presented in Eq. (19.66) from a scalar conservation law to any system of conservation laws. We know that we need to impose a conservative scheme, of the form given in Eq. (19.46). We recognize that the coefficient $\alpha_{1+\frac{1}{2}}$ in the numerical flux function given by Eq. (19.66) is a (numerical) proxy for the characteristic speed $f'(u)$. A straightforward generalization to the case of a nonlinear system then adopts the numerical flux

$$\mathbf{F}_{i+\frac{1}{2}} = \tfrac{1}{2}\left[\mathbf{F}(\mathbf{U}^{\mathrm{L}}_{i+\frac{1}{2}}) + \mathbf{F}(\mathbf{U}^{\mathrm{R}}_{i+\frac{1}{2}}) - |c^{\max}_{i+1/2}|\left(\mathbf{U}^{\mathrm{R}}_{i+\frac{1}{2}} - \mathbf{U}^{\mathrm{L}}_{i+\frac{1}{2}}\right)\right]. \qquad (19.104)$$

In this expression, $\mathbf{U}^{\mathrm{L}}$ and $\mathbf{U}^{\mathrm{R}}$ denote the state on the cell interface $i + \frac{1}{2}$, to the left or right, respectively. The scalar coefficient $c^{\max}$ then denotes the maximal physical propagation speed for the system under consideration. Obviously, for 1D MHD we get $c^{\max} = |v_x| + c_{\mathrm{f}}$, the maximal speed set for fast magneto-acoustic disturbances. This scheme has the advantage that we may even exploit it for a zero-beta plasma.

Minor variations of formula (19.104) can be found easily, and depend on how one computes $c^{\max}$ at the cell interface. One can use the maximal physical propagation speed of the arithmetic average $(\mathbf{U}^{\mathrm{L}} + \mathbf{U}^{\mathrm{R}})/2$ (which could also average the primitive variables), or the maximum of each state separately, i.e.

$$\max\left(|c^{\max}(\mathbf{U}^{\mathrm{L}})|, |c^{\max}(\mathbf{U}^{\mathrm{R}})|\right). \qquad (19.105)$$

The TVDLF method is then also referred to as local Lax–Friedrichs method, since it uses a local value for the maximal propagation speed. In fact, a first order variant was exploited as early as in 1961 by Rusanov [398], to numerically solve several shock-dominated 2D gas dynamic problems. A first order scheme indeed results when we take the states $\mathbf{U}^{\mathrm{L}} = \mathbf{U}_i$ and $\mathbf{U}^{\mathrm{R}} = \mathbf{U}_{i+1}$, combined with the single step temporal advance written in Eq. (19.46). A temporally second order variant is easily obtained by using a two-step predictor–corrector approach, and use specifically the time-centered (i.e. at $n + \frac{1}{2}$) fluxes (19.104) in the corrector step. For full second order accuracy, one also needs to raise the spatial order. To that effect, a linear reconstruction from cell-center to cell edge quantities $\mathbf{U}^{\mathrm{L,R}}$ needs to be performed as well, which ideally remains in agreement with the TVD concept. For MHD, this method was introduced and compared in [444] on a variety of hydro- and magnetohydrodynamic problems with other approaches, such as FCT and various characteristic based schemes.

Linear reconstruction and slope limiters In the finite volume discretization, the evaluation of fluxes at the cell edges implies that some extrapolation within the cell i is needed from the volume averaged value $\mathbf{U}_i$. Simply using a constant extrapolation where $\mathbf{U}(x \in [x_{i-1/2}, x_{i+1/2}]) = \mathbf{U}_i$ is of course consistent with $\mathbf{U}_i$ as the cell average, but leads to first order accuracy where averaged edge fluxes $\mathbf{F}_{i+1/2} = (\mathbf{F}(\mathbf{U}_i) + \mathbf{F}(\mathbf{U}_{i+1}))/2$ are used. Better, and still consistent with $\mathbf{U}_i$ as a volume average, is to use a linear extrapolation within the cell with slope σ_i, hence

$$\mathbf{U}(x \in [x_{i-1/2}, x_{i+1/2}]) = \mathbf{U}_i + \sigma_i(x - x_i)/\Delta x_i. \qquad (19.106)$$

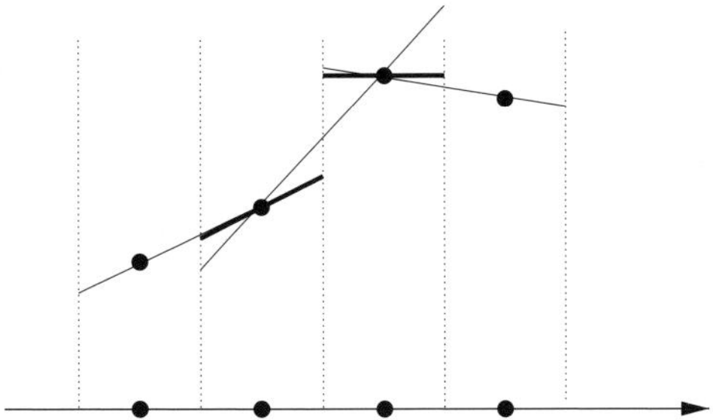

Fig. 19.11 Slope limited linear reconstruction process.

The slope is then the difference

$$\boldsymbol{\sigma}_i = \mathbf{U}_{i+1/2} - \mathbf{U}_{i-1/2} \equiv \Delta\mathbf{U}_i. \tag{19.107}$$

This linear reconstruction is exploited to get a *left* and *right* edge centered state,

$$\mathbf{U}^{\mathrm{L}}_{i+1/2} = \mathbf{U}_i + \Delta\mathbf{U}_i/2 \quad \text{and} \quad \mathbf{U}^{\mathrm{R}}_{i+1/2} = \mathbf{U}_{i+1} - \tfrac{1}{2}\Delta\mathbf{U}_{i+1}. \tag{19.108}$$

The flux at the cell edge then takes the average from $\mathbf{F}_{i+1/2} = (\mathbf{F}(\mathbf{U}^{\mathrm{L}}_{i+1/2}) + \mathbf{F}(\mathbf{U}^{\mathrm{R}}_{i+1/2}))/2$. This process of linear reconstruction raises the spatial order of accuracy to second order. In practice, one must limit the slopes used in the linear reconstruction in order to avoid the introduction of spurious oscillations. This means that we rather write, denoting this slope limiting process by an overbar:

$$\mathbf{U}^{\mathrm{L}}_{i+\frac{1}{2}} = \mathbf{U}_i^{n+1/2} + \tfrac{1}{2}\Delta\bar{\mathbf{U}}_i^{n+1/2},$$

$$\mathbf{U}^{\mathrm{R}}_{i+\frac{1}{2}} = \mathbf{U}_{i+1}^{n+1/2} - \tfrac{1}{2}\Delta\bar{\mathbf{U}}_{i+1}^{n+1/2}. \tag{19.109}$$

On comparing the slopes as obtained from using neighboring left or right cell values, one typically needs to take the least steep slope of the two, and to fall back on constant extrapolation within a cell when these slopes conflict in sign. Schematically, this is shown in Fig. 19.11. Different slope limiters exist which ensure the total variation diminishing (TVD) concept, thereby avoiding the creation of spurious oscillations in the numerical solution to the Riemann problem. However, these statements are strictly speaking only true for a single scalar nonlinear equation. A particularly robust (albeit diffusive) slope limiter is the "minmod" limiter, where

$$\Delta\bar{\mathbf{U}}_i = \mathrm{sgn}(\mathbf{U}_i - \mathbf{U}_{i-1})\max\big[0,$$
$$\min\big\{|\mathbf{U}_i - \mathbf{U}_{i-1}|, (\mathbf{U}_{i+1} - \mathbf{U}_i)\,\mathrm{sgn}(\mathbf{U}_i - \mathbf{U}_{i-1})\big\}\big]. \tag{19.110}$$

By suitably generalizing the minmod idea where left and right-sided slopes are compared, the monotonized central-difference (MC) limiter compares three slopes, and is in practice a better choice. These and various other slope limiter flavors can be found in [303]. The stencil of the resulting second order TVDLF scheme is five cells wide, through the slope limited linear reconstruction.[1] In the second order variant, slight improvements (in terms of robustness) may occur when performing the linear reconstruction on the primitive variables $\mathbf{V}$ instead of the conservative variables $\mathbf{U}$.

Example Riemann problems for 1.5D iso-thermal MHD Using the second order TVDLF method, we can now generate numerical solutions for 1D MHD problems. In particular, we solve the 1.5D iso-thermal MHD equations, and generate numerically the solution for the following initial conditions. In the examples, we actually use an excessive amount of grid points (order 1000 to 50000), with an emphasis on the fully converged end state.

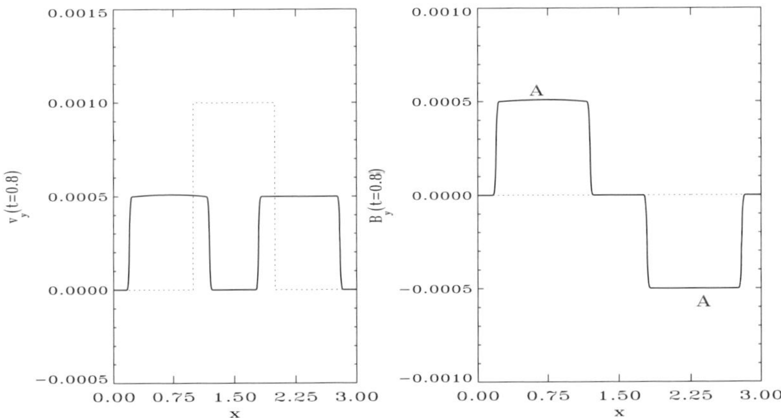

Fig. 19.12 An Alfvén pulse simulated in 1.5D zero-beta MHD.

A first test considers linear shear Alfvén waves, and takes $B_x = 1$, in a zero-beta plasma, i.e. $p = 0$. This zero-beta limit will pose difficulties for characteristic based schemes, since the slow magneto-sonic speed vanishes and the matrices collecting the eigenvectors become singular. The density $\rho = 1$, and we set up a pulse in transverse velocity $v_y = 0.001$ at $x \in [1, 2]$, on a total domain extending from $x \in [0, 3]$. The pulse will split into two equal sized Alfvén signals, shown at $t = 0.8$ in Fig. 19.12. The figure shows both the initial (dotted) and final numerical solution, for the transverse velocity and magnetic field. The propagation speed

---

[1] Note that this wider stencil becomes a severe complication when computing on multi-dimensional unstructured grids. There, the use of compact schemes employing only nearest neighbor information while retaining higher order have led to the residual distribution schemes, also successfully adopted to MHD.

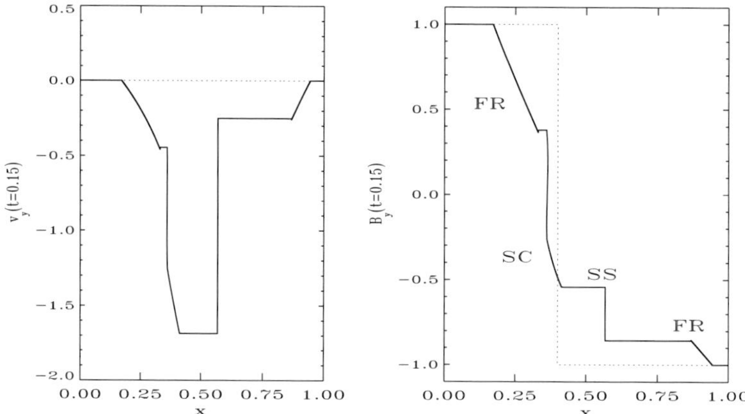

Fig. 19.13 A Riemann problem in 1.5D iso-thermal MHD, giving rise to a solution containing a left going fast rarefaction, a slow compound wave, a slow shock and a right going fast rarefaction.

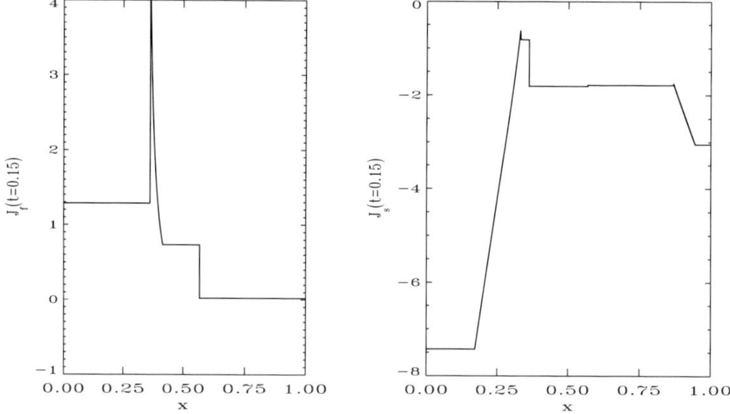

Fig. 19.14 The fast and slow magneto-acoustic Riemann invariants as computed from the numerical solutions.

and polarization are in exact accord with the expected outcome, where both pulses travel at the Alfvén speed along the x-directed field. Note that we actually set up a double Riemann problem.

The next two tests consider Riemann problems, chosen to demonstrate all flavors of the nonlinear wave patterns one may encounter for this system. On a unit domain, a discontinuity is placed at $x = 0.4$ separating $\mathbf{V}_\ell = (1, 0, 0, 1)$ from $\mathbf{V}_r = (0.125, 0, 0, -1)$. The constant parameters are set to $B_x = 0.75$ and $c_i^2 = 1$. This test is analogous to that introduced by Brio and Wu [70] for the 1.5D MHD

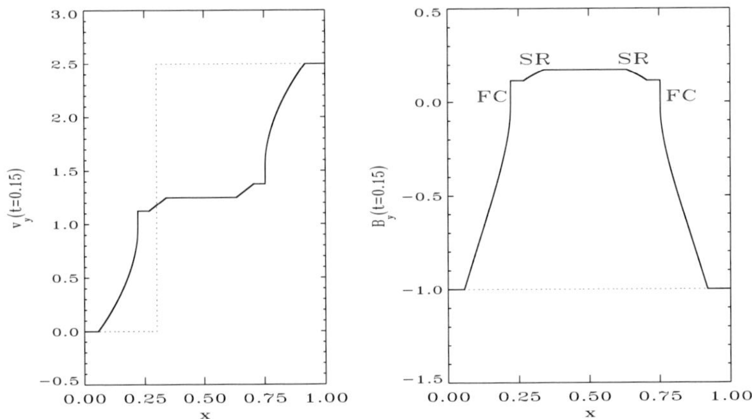

Fig. 19.15 A Riemann problem in 1.5D iso-thermal MHD, giving rise to a solution containing a left going fast compound wave, two slow rarefactions, and a right going fast compound wave.

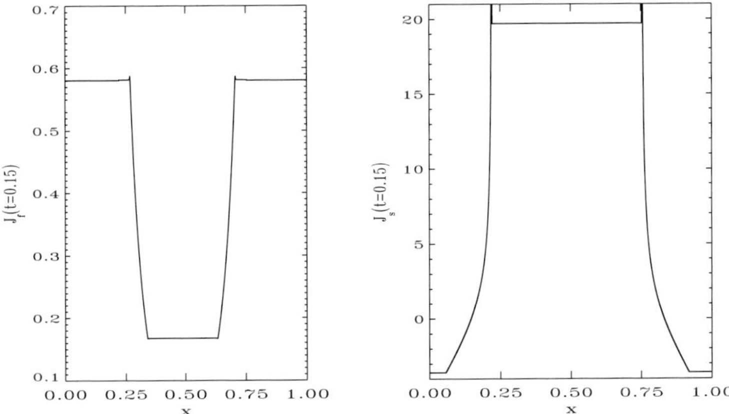

Fig. 19.16 The fast and slow magneto-acoustic Riemann invariants as computed from the numerical solutions.

system. It leads to a left going fast rarefaction, a slow compound wave (shock with rarefaction wave attached to it), a right going slow shock and fast rarefaction. These can be seen in Fig. 19.13 at $t = 0.15$, again showing the transverse vector components. Fig. 19.14 plots the fast and slow magneto-acoustic Riemann invariants from Eq. (19.23). Through the rarefactions, we indeed find constant values as evaluated from the numerical solutions. They change discontinuously at shocks, while the fast invariant stays constant through fast rarefactions, and the slow invariant is constant through the rarefaction part of the slow compound wave.

A final Riemann problem, analogous to a test described in [345, 346], leads to fast compound waves. The discontinuity, initially at $x = 0.3$, brings the states $\mathbf{V}_\ell = (1, 0, 0, -1)$ and $\mathbf{V}_r = (1, 2.5, 2.5, -1)$ into contact. The parameters fix $B_x = 1$ and $c_i^2 = 1$, and we generate a solution up to $t = 0.15$. We find a fast compound wave, with a rarefaction part running ahead of the fast shock attached to it, two slow rarefaction waves, and a similar fast compound wave at right. Fig. 19.15 plots the solutions obtained together with initial conditions, while the evaluation of the magneto-acoustic Riemann invariants is shown in Fig. 19.16. One may note slight overshoots in the solutions, despite the excessive resolution used. However, the main point demonstrated here is the feasibility of using a simple TVDLF scheme for generating numerical solutions in excellent agreement with theory.

19.2.3 Approximate Riemann solver type schemes

When discussing the Godunov method, we noted that the basic ingredient is the (exact) solution of the nonlinear Riemann problem encountered at cell edges. It was soon realized that one might just as well solve these local Riemann problems in an approximate fashion. *Approximate Riemann solver* based schemes use a linearization of the nonlinear problem, and capitalize on the fact that the exact analytic solution for a linear hyperbolic system is known. In hydrodynamics, a method initiated by Roe has proven to be very powerful, and a useful review on characteristic based schemes for the Euler equations is given in [392]. We present its extension to MHD, which has been initiated by the seminal work of Brio and Wu [70].

Roe-type approximate Riemann solver As a general procedure to numerically solve the system $\mathbf{U}_t + (\mathbf{F}(\mathbf{U}))_x = 0$, one considers again the local Riemann problem obtained from a left and right cell interface value $\mathbf{U}_\ell$ and $\mathbf{U}_r$. Instead of using the exact nonlinear solution, Roe suggested solving a linear Riemann problem instead, namely

$$\mathbf{U}_t + (\mathbf{G}(\mathbf{U}))_x = 0, \tag{19.111}$$

where $\mathbf{G}(\mathbf{U}) = \mathbf{F}(\mathbf{U}_r) + \mathbf{A}(\mathbf{U} - \mathbf{U}_r)$ includes a constant matrix $\mathbf{A} = \mathbf{A}(\mathbf{U}_\ell, \mathbf{U}_r)$. This matrix must satisfy the conditions

- $\mathbf{F}(\mathbf{U}_\ell) - \mathbf{F}(\mathbf{U}_r) = \mathbf{A}(\mathbf{U}_\ell, \mathbf{U}_r)(\mathbf{U}_\ell - \mathbf{U}_r)$,
- $\mathbf{A}(\mathbf{U}_\ell, \mathbf{U}_r) \rightarrow \mathbf{F}_\mathbf{U}(\mathbf{U}_r)$ as $\mathbf{U}_\ell \rightarrow \mathbf{U}_r$,
- $\mathbf{A}(\mathbf{U}_\ell, \mathbf{U}_r)$ has only real eigenvalues,
- $\mathbf{A}(\mathbf{U}_\ell, \mathbf{U}_r)$ has a complete system of eigenvectors.

These conditions ensure that the exact Riemann problem solution is obtained when the initial states obey the Rankine–Hugoniot relations, i.e. are separated by a single shock transition or contact discontinuity. Indeed, when $(\mathbf{U}_\ell, \mathbf{U}_r)$ are such that $\mathbf{F}(\mathbf{U}_\ell) - \mathbf{F}(\mathbf{U}_r) = s\,(\mathbf{U}_\ell - \mathbf{U}_r)$, the first condition ensures that $\mathbf{U}_\ell - \mathbf{U}_r$ is an eigenvector of $\mathbf{A}$ with eigenvalue equal to the shock speed s. The solution of the linear Riemann problem merely maintains this jump at the physically correct shock speed. The second condition ensures consistency with the original nonlinear equations. The last two conditions guarantee solvability of the linear Riemann problem.

Provided such a Roe matrix $\mathbf{A}$ can be found, the Roe scheme uses the numerical Roe flux expression at the interface between cells $i, i+1$ given by

$$\mathbf{F}_{i+1/2}\left(\mathbf{U}_i, \mathbf{U}_{i+1}\right) = \mathbf{F}(\mathbf{U}_i) + \mathbf{A}_{i+1/2}\left(\hat{\mathbf{U}} - \mathbf{U}_i\right). \tag{19.112}$$

In the above expression (19.112), $\hat{\mathbf{U}} = \hat{\mathbf{U}}(0, \mathbf{U}_i, \mathbf{U}_{i+1})$ indicates the exact solution of the linear Riemann problem and the matrix depends on $\mathbf{A}_{i+1/2}(\mathbf{U}_i, \mathbf{U}_{i+1})$. Let us write the eigenvalues λ_p and right eigenvectors $\mathbf{r}^p$ of the local Roe matrix $\mathbf{A}_{i+1/2}(\mathbf{U}_i, \mathbf{U}_{i+1})$, hence $\mathbf{A}_{i+1/2}\mathbf{r}^p = \lambda_p \mathbf{r}^p$. More explicit expressions for the Roe flux can then be obtained as follows. As encountered in the discussion of the linear hyperbolic system, one decomposes both states in terms of the eigenvectors $\mathbf{r}^p$, writing $\mathbf{U}_i = \sum \beta_p \mathbf{r}^p$ and $\mathbf{U}_{i+1} = \sum \gamma_p \mathbf{r}^p$. The same decomposition of the difference $\mathbf{U}_{i+1} - \mathbf{U}_i$ then introduces coefficients α_p from

$$\mathbf{U}_{i+1} - \mathbf{U}_i = \sum \left(\gamma_p - \beta_p\right)\mathbf{r}^p \equiv \sum \alpha_p \mathbf{r}^p. \tag{19.113}$$

As found from the analytic solution for the Riemann problem for the linear hyperbolic system, the solution along the (x, t) ray $(x - x_{i+\frac{1}{2}})/t = 0$ is thus

$$\hat{\mathbf{U}} = \sum_{0 < \lambda_p} \beta_p \mathbf{r}^p + \sum_{0 > \lambda_p} \gamma_p \mathbf{r}^p. \tag{19.114}$$

This can be written as

$$\hat{\mathbf{U}} = \tfrac{1}{2}(\mathbf{U}_i + \mathbf{U}_{i+1}) + \tfrac{1}{2}\left[\sum_{\lambda_p < 0} - \sum_{\lambda_p > 0}\right]\alpha_p \mathbf{r}^p. \tag{19.115}$$

The Roe flux (19.112) then becomes, as a result of the first Roe condition imposed on the matrix $\mathbf{A}$,

$$\mathbf{F}_{i+1/2} = \tfrac{1}{2}\left(\mathbf{F}(\mathbf{U}_i) + \mathbf{F}(\mathbf{U}_{i+1})\right) - \tfrac{1}{2}\sum \mid \lambda_p \mid \alpha_p \mathbf{r}^p. \tag{19.116}$$

The solver is then completely determined once the matrix $\mathbf{A}_{i+1/2}$ that satisfies the Roe conditions is constructed. For the Euler system of compressible gas dynamics, the matrix satisfying all conditions is the flux Jacobian of the system $\mathbf{F}_{\mathbf{U}}(\bar{\mathbf{U}})$, evaluated in a special, so-called Roe-average, state computed from the edge values $\bar{\mathbf{U}}(\mathbf{U}_\ell, \mathbf{U}_r)$. When generalized to the MHD system, Brio and Wu [70] identified a similar Roe-average for the special case of $\gamma = 2$ only. A general procedure to construct Roe matrices for systems of conservation laws, including the gas dynamic and ideal MHD systems, was presented in [76]. There, an MHD Roe solver without hypothesis on the value of γ is presented. For the MHD case, Brio and Wu already noted that one may use the flux Jacobian, merely evaluated in a state obtained from a simple averaging (e.g. the arithmetic average of both edge states). This then no longer captures stationary discontinuities as exact steady numerical solutions, but instead resolves these jumps with few grid points. In practice, one thus uses expression (19.116) for computing the flux, which evaluates the right eigenvectors of the flux Jacobian in the chosen average state $\mathbf{r}^p(\bar{\mathbf{U}})$, evaluates the eigenvalues for this state $\lambda_p(\bar{\mathbf{U}})$, and computes the wave strength coefficients $\alpha_p = \mathbf{l}^p \cdot (\mathbf{U}_{i+1} - \mathbf{U}_i)$, using the left eigenvectors of the Jacobian $\mathbf{l}^p(\bar{\mathbf{U}})$. The latter can also be computed from the jumps in (any choice of) the associated primitive variables $\mathbf{V}$, where we write $\alpha_p = \hat{\mathbf{l}}^p \cdot (\mathbf{V}_{i+1} - \mathbf{V}_i)$. Earlier, we gave the needed expressions for the right eigenvectors $\mathbf{r}^p$ in Eq. (19.95)–(19.96), for the eigenvalues in Eq. (19.89), and for the left eigenvectors (of the primitive formulation) in Eq. (19.91), so that all ingredients to the above flux formula (19.116) are known. This Roe-type approximate Riemann solver is similarly extended to second order by using a limited linear reconstruction from cell center to cell edge values, and replacing $\mathbf{U}_{i+1}$ with $\mathbf{U}^{\mathrm{R}}_{i+1/2}$ and $\mathbf{U}_i$ with $\mathbf{U}^{\mathrm{L}}_{i+1/2}$. The limiting can actually be done on the characteristic wave contributions separately, too, and again exists in many flavors.

▷ **Roe average, entropy fixes and further details**　The Roe average state, as presented for $\gamma = 2$ by Brio and Wu [70], has been used for the general case as well, and reduces to the gas dynamical Roe average for vanishing magnetic field. For left and right states $\mathbf{U}^{\mathrm{L}}$ and $\mathbf{U}^{\mathrm{R}}$, all average variables are computed from

$$\bar{\rho} = \sqrt{\rho^{\mathrm{L}}\rho^{\mathrm{R}}}, \qquad\qquad \bar{\mathbf{v}} = \frac{\sqrt{\rho^{\mathrm{L}}}\mathbf{v}^{\mathrm{L}} + \sqrt{\rho^{\mathrm{R}}}\mathbf{v}^{\mathrm{R}}}{\sqrt{\rho^{\mathrm{L}}} + \sqrt{\rho^{\mathrm{R}}}},$$

$$\bar{\mathbf{B}} = \frac{\sqrt{\rho^{\mathrm{R}}}\mathbf{B}^{\mathrm{L}} + \sqrt{\rho^{\mathrm{L}}}\mathbf{B}^{\mathrm{R}}}{\sqrt{\rho^{\mathrm{L}}} + \sqrt{\rho^{\mathrm{R}}}}, \qquad \bar{h} = \frac{\sqrt{\rho^{\mathrm{L}}}h^{\mathrm{L}} + \sqrt{\rho^{\mathrm{R}}}h^{\mathrm{R}}}{\sqrt{\rho^{\mathrm{L}}} + \sqrt{\rho^{\mathrm{R}}}}. \qquad (19.117)$$

The latter quantity is the specific enthalpy $h = (\mathcal{H} + p + B^2)/\rho$, and can be used to obtain the average squared sound speed $\bar{c}^2 = (\gamma - 1)(\bar{h} - \bar{v}^2/2 - \bar{B}^2/\bar{\rho})$. Roe and Balsara [393] analyzed the sensitivity of the approximate Riemann solver to the precise choice of averaging near the triple umbilic point, and concluded that any reasonable averaging would perform well. Therefore, less computationally elaborate (e.g. arithmetic) averagings have found widespread use. As was quickly noted upon application of the Roe solver to the Euler gas dynamic system, the solver may fail in cases where a transonic rarefaction is part of the solution, in a sense that it would rather detect an entropy violating weak shock solution. For this purpose, entropy fixes are used, which merely replace the near vanishing eigenvalue by some small value. For the wave speeds appearing in (19.116) for the MHD system, in particular for the fast and slow families $p = 1, 3, 5, 7$, we thus write

$$\lambda_p \to \tfrac{1}{2}\left(\lambda_p^2/\epsilon_{\mathrm{e}} + \epsilon_{\mathrm{e}}\right), \qquad (19.118)$$

if $|\lambda_p| < \epsilon_e$. This is thus only effective when transonic solutions $|v_x| \approx c_{s,f}$ occur, and introduces a small parameter ϵ_e. In the numerical evaluation of the various terms in the eigenvector expressions, we still encounter difficulties in certain degenerate cases. In particular, the eigenvectors contain the ratios $\beta_y = B_y/B_\perp$ and β_z, which become ill-defined when the transverse field vanishes completely. This is remedied by replacing these factors with

$$\beta_y = \frac{B_y + \epsilon}{\sqrt{B_y^2 + B_z^2 + 2\epsilon^2}}, \qquad \beta_z = \frac{B_z + \epsilon}{\sqrt{B_y^2 + B_z^2 + 2\epsilon^2}}. \tag{19.119}$$

With ϵ a very small number above machine precision, we then in effect replace the indeterminacies by $1/\sqrt{2}$ constants. Similarly, when $c_s^2 = c_f^2$, one uses $\alpha_f = \alpha_s = 1/\sqrt{2}$ which still leaves $\alpha_f^2 + \alpha_s^2 = 1$. $\quad\triangleleft$

HLL and HLLC discretizations The approximate Riemann solver just discussed uses a linearization of the Riemann problem encountered at cell interfaces, which is at the heart of all Godunov-type schemes. The Roe-type solver clearly incorporates much knowledge of the underlying physical problem, and in effect uses the proper characteristic decomposition to "weight" the information propagating in different directions. It approximates the time-averaged flux appropriate for that state of the Riemann fan sketched schematically in Fig. 19.10 which is found along the vertical t-axis. This "upwind" concept is a key issue for generating numerically stable conservative schemes. Characteristic based schemes are by far the most accurate numerical methods for handling discontinuity dominated problems, and various variants have successfully improved on the work initiated in MHD by Brio and Wu [70]. Zachary and Collela [493] constructed an MHD Godunov method in which the approximation to the upwind flux was build up in the primitive variable space, organizing the states in the Riemann fan about some mean entropy wave speed. A TVD Roe-type upwind scheme was presented by Ryu and Jones [399], where its results were benchmarked against the predictions of a nonlinear Riemann solver. A large amount of 1D MHD Riemann problem tests are documented there.

The methods just mentioned all incorporate the full 7-wave structure which is a natural outcome of the MHD Riemann problem. Contrasting the flux expression (19.116) to the TVDLF method with flux given by (19.104), we may reinterpret the latter as one where the 7-wave structure is essentially replaced by an approximation where only the maximal (in absolute value) wave speed enters, replacing all $|\lambda_p|$ by $|c^{\max}|$. This effectively retains a crude form of upwinding, but introduces more numerical diffusion. The latter can in fact be advantageous to deal with some of the robustness issues that plague the approximate Riemann solvers in particular. For the Euler system, a catalog of failings on shock-dominated problems has been presented by Quirk [389]. Some of these flaws can be cured by appropriate fixes, but a fairly common strategy is to add numerical dissipation

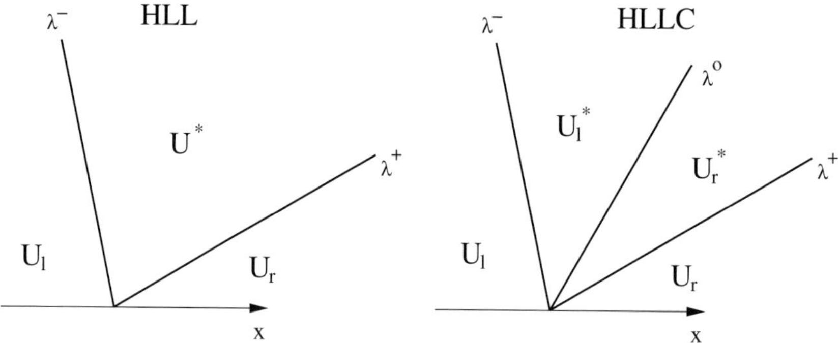

Fig. 19.17 The approximate representation of the Riemann fan employed in the HLL (left) and HLLC (right) methodology.

in some form or other. A trend away from using the full 7-wave structure towards the other extreme of using only the fastest wave (TVDLF) is noticeable in the extensive literature on the subject. Versions of solvers originated in gas dynamics have been adapted to MHD, which in effect replace the Riemann fan by a 2-wave or 3-wave structure. Appropriate recipes for computing meaningful intermediate states then ensure desirable properties like positivity (ensuring always positive pressures and densities), the ability to capture isolated discontinuities, etc. A 2-wave approximation is used in the HLL solver (originally due to Harten, Lax and van Leer [217]), and this involves an intermediate state $\mathbf{U}^*$ which generalizes the Rankine–Hugoniot relation (19.34) to ensure

$$\mathbf{F}(\mathbf{U}_\ell) - \mathbf{F}(\mathbf{U}_r) = \lambda^-(\mathbf{U}_\ell - \mathbf{U}^*) + \lambda^+(\mathbf{U}^* - \mathbf{U}_r), \tag{19.120}$$

where λ^- and λ^+ are both fastest wave speeds still retained in the Riemann fan approximation. This method then switches the flux evaluation between $\mathbf{F}(\mathbf{U}_\ell)$ when $\lambda^- \geq 0$, $\mathbf{F}(\mathbf{U}_r)$ when $\lambda^+ \leq 0$, or a weighted flux expression

$$\mathbf{F}^* = \frac{\lambda^+\mathbf{F}(\mathbf{U}_\ell) - \lambda^-\mathbf{F}(\mathbf{U}_r) + \lambda^-\lambda^+(\mathbf{U}_r - \mathbf{U}_\ell)}{\lambda^+ - \lambda^-} \tag{19.121}$$

for the case where the intermediate state should apply. Note that $\mathbf{F}^* \neq \mathbf{F}(\mathbf{U}^*)$. In practical application, this HLL method has proven to work very similarly to the TVDLF method, demonstrating minor improvements over the latter. For contact discontinuities, both methods introduce fairly large diffusion, but at the same time the schemes are very robust and computationally simple.

In an attempt to remedy the smearing of contact discontinuities, the HLLC approach "restores the contact surface in the HLL-solver" [436], hence the added C for "contact". Its adaptation to MHD has been undertaken by various authors, e.g. by [311] and [305]. One can readily generalize the relation (19.120) to introduce a third central wave λ^0 separating a left $\mathbf{U}_\ell^*$ and right $\mathbf{U}_r^*$ middle state, and choose a well-motivated construction to uniquely compute these two additional states. Four corresponding flux expressions are in use, depending on the orientation of the three waves in the Riemann fan with respect to the time axis in the (x, t) plane. Schematically then, the Riemann fan for HLL and HLLC type methods is shown in Fig. 19.17.

Other variants of the HLL family of solvers exist, and they all share a relative straightforward application to any system of conservation laws (with more than two unknowns) without requiring full decompositions in the many $(n > 2)$ wave families. In particular, HLL(C) has been adapted to relativistic MHD computations as well, a subject which will return in Chapter 21. A final noteworthy solver for ideal MHD is the HLLD (where the D stands for discontinuity) variant by Miyoshi and Kusano [336]. Depending on whether B_x vanishes or not, HLLD switches between the use of two intermediate states (like HLLC, to which it reduces for vanishing magnetic field) versus four intermediate states. This solver then exactly resolves isolated discontinuities and shocks, and tests demonstrate that it is more robust than using a full approximate Riemann solver.

19.2.4 Simulating 1D MHD Riemann problems

A vast number of 1D MHD Riemann problem tests can be found throughout the literature, with the paper by Ryu and Jones [399] collecting a good number of distinct possibilities. One distinguishes the discontinuous transitions according to their type, as will be discussed in Chapter 20. We can have the already mentioned *contact and rotational discontinuities*, but also *tangential discontinuities* as B_x tends to zero, which can be regarded as degeneracies where a contact, two slow, and two rotational discontinuities coincide. *Fast and slow shocks*, as well as *rarefaction waves*, can be of special type, called *switch-off* (or *switch-on*) when they reduce (increase) the tangential magnetic field to (from) zero, and a fast shock is of pure *magneto-sonic* kind when B_x vanishes. Slow and fast signals can become of *compound* nature, where they become of *"intermediate"* type, i.e. cross the unit Alfvén Mach number and include a $180°$ degree rotation of the transverse field. Whether the latter are physically realizable transitions has received a lot of attention since they were first discovered by the Brio and Wu numerical solutions. In the coplanar 1.5D MHD case considered there, one can hardly avoid them in numerical solutions, such as those generated for the iso-thermal case above. In principle, they may then be replaced by a π-rotational wave, combined with a slow or fast shock, so that they represent cases where the Riemann problem has a non-unique solution. A clear discussion of the subject is found in Torrilhon [437], where it was shown that any non-planar Riemann problem will have a unique solution containing only regular waves if the initial jump in transverse velocity vanishes. Conversely, there exist also non-planar Riemann problems, which can have both a regular and a non-regular solution, similar to some coplanar problems. A thorough discussion, where on the basis of evolutionary arguments (i.e. structural stability against small perturbations) the admissibility of intermediate shocks is disputed, is found in [132].

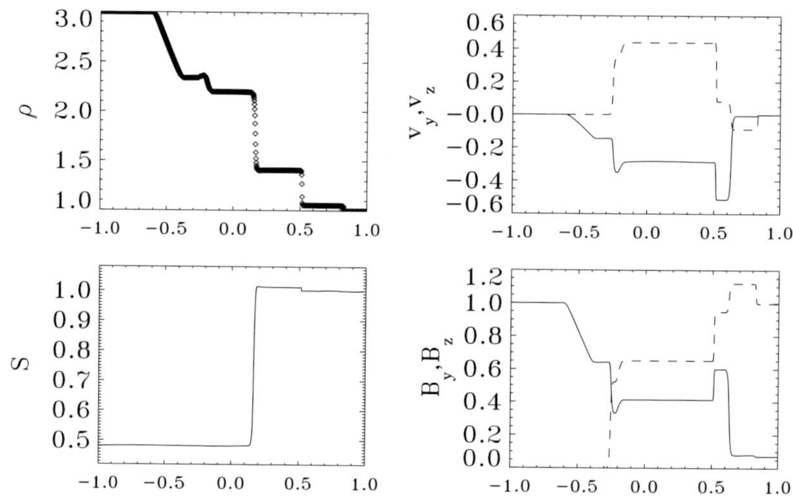

Fig. 19.18 A Riemann problem in 1D MHD, numerically solved with a TVD scheme using a Roe-type approximate Riemann solver. We encounter from left to right: a fast rarefaction, a rotational discontinuity, a slow rarefaction, a contact discontinuity, a slow shock, a rotational discontinuity and a fast shock.

We give now two examples of Riemann problems, using an approximate Riemann solver based scheme to generate the solution. The first shown in Fig. 19.18 takes initial conditions also presented in Torrilhon [437], namely a left state $\mathbf{V}_\ell = (3, 0, 0, 0, 3, 1, 0)$ adjacent to a right state $\mathbf{V}_r = (1, 0, 0, 0, 1, \cos(1.5), \sin(1.5))$ where $\gamma = 5/3$ and $B_x = 3/2$. Our numerical solution is shown in the same format as Fig. 2 from [437], except for the bottom left panel where we show entropy $S = p\rho^{-\gamma}$ instead of pressure. This latter shows its invariance through fast and slow rarefactions, and Alfvén discontinuities. Further, we plotted the solution in the state space spanned by the transverse magnetic field components. The latter can be compared with Fig. 3 in [437]. One can detect from left to right in Fig. 19.18: a fast rarefaction, a rotational discontinuity, a slow rarefaction, a contact discontinuity, a slow shock, a rotational discontinuity and a fast shock. The representation of the numerical solution in the state space (Fig. 19.19) shows the expected outcome: only Alfvén signals rotate the transverse field, while fast and slow shocks and rarefactions alter the magnitude of the transverse field. Obviously, the solution is not perfect, in the sense that several grid points are needed to represent the discontinuities, and one may detect overshoots in density and a diffused contact discontinuity. Nevertheless, a very satisfactory agreement with the exact solution is obtained, and this is generally true for all modern upwind-type algorithms.

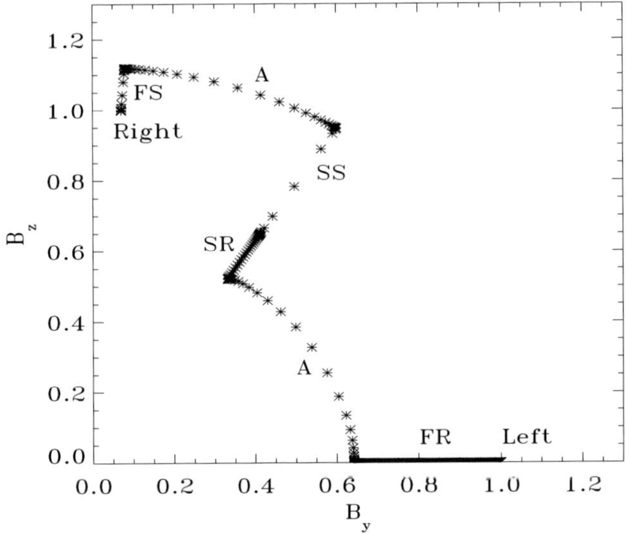

Fig. 19.19 The same numerical solution from Fig. 19.18, plotted in the (B_y, B_z) state plane. Fast and slow signals change the magnitude of the transverse field, while Alfvén signals rotate the field. In reality, the parts corresponding to shocks and rotational discontinuities should contain no points, the arcs and line segments seen relate to the numerical representation.

A second and final Riemann problem shown is taken from Torrilhon [438], where a non-coplanar Riemann problem which has a unique, regular, solution was solved numerically using a variety of modern upwind algorithms. The initial conditions take $\mathbf{V}_\ell = (1, 0, 0, 0, 1, 1, 0)$ and $\mathbf{V}_r = (0.2, 0, 0, 0, 0.2, \cos(3), \sin(3))$ with $B_x = 1$ and $\gamma = 5/3$. This is an almost coplanar configuration (exact coplanar when 3 is replaced by π) without jump in transverse velocity, where the unique exact solution has a fast rarefaction, a rotational discontinuity, a slow shock, a contact discontinuity, a slow shock, a rotational discontinuity and a fast rarefaction. It was demonstrated convincingly how all modern methods pseudo-converge to a solution containing a compound wave structure for up to several 1000 grid points. Fig. 19.20 shows how the numerical solution significantly errs at $x \approx -0.25$. Only after heavy refinement, the pseudo-convergence changes to true convergence to the exact regular solution of the Riemann problem. This is a serious shortcoming

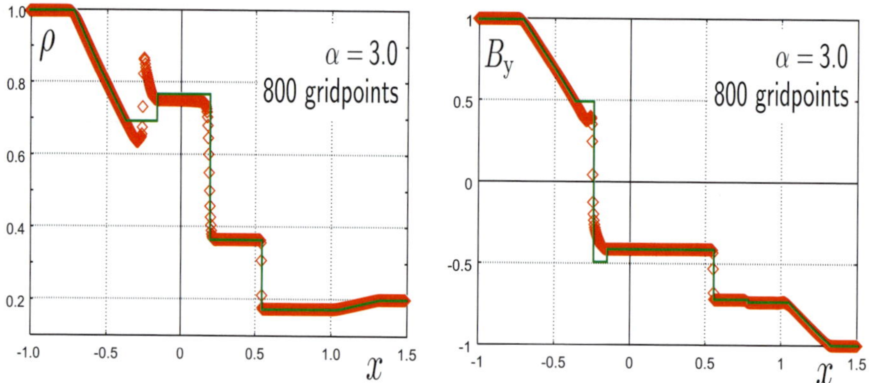

Fig. 19.20 A Riemann problem in 1D MHD, with a unique regular solution (solid line), and a typical "high" resolution numerical result already exploiting 800 grid points: a wrong compound structure plagues the scheme to pseudo-converge. (From Torrilhon [438].)

of all modern shock-capturing schemes, which can only be remedied thus far by combining these methods with some form of grid-adaptivity.

19.3 Multi-dimensional MHD computations

The description of numerical methods has thus far concentrated on 1D problems. Conceptually, multi-D simulations can be reduced to a succession of 1D problems in the various coordinate directions. As already explained earlier, even finite volume methods on general (unstructured) grids can also be reduced to successive 1D problems only, in directions normal to the grid cell faces. A common strategy is to use a Strang type dimensional splitting, which for a 2D case in a Cartesian (x, y) setting writes as

$$\mathbf{U}^{n+1} = L^x_{\Delta t/2} L^y_{\Delta t} L^x_{\Delta t/2} \mathbf{U}^n \,, \tag{19.122}$$

where the operator $L^x_{\Delta t/2}$ indicates an update with a half time step $\Delta t/2$ taking account of only the fluxes in the x-direction. Alternatively, we can alternate coordinate directions using the so-called Godunov splitting

$$\mathbf{U}^{n+2} = L^x_{\Delta t} L^y_{\Delta t} L^y_{\Delta t} L^x_{\Delta t} \mathbf{U}^n \,. \tag{19.123}$$

These dimensional splitting strategies retain the second order accuracy of the individual dimensional sweeps.

The relative ease with which multi-dimensional codes are developed from 1D solvers has led to a rapid surge in multi-dimensional MHD modeling, once shock-capturing algorithms for the MHD system emerged in earnest. Dai and Woodward [104] used their own approximate, nonlinear all-shock Riemann solver, described in [103], as a basic ingredient in a multi-dimensional ideal MHD solver. They used Strang splitting and exploited an essentially third order scheme, using parabolic instead of linear reconstructions (the piecewise parabolic method, or PPM). In a variety of shock-dominated 2D tests, they demonstrated robustness of their scheme, despite the fact that no additional measures for controlling numerical monopole errors were employed. A multi-dimensional TVD scheme, also using dimensional splitting, was described in Ryu *et al.* [400]. A serious comparison between different shock-capturing schemes (FCT, TVD, TVDLF) for 1D and 2D hydro- and MHD problems can be found in [444]. Both these latter works in essence use second order accurate, shock-capturing schemes.

It should be emphasized that the assumption of directional splitting strategies may not always be justified: characteristic based schemes then implicitly assume that waves travel orthogonal to grid cell interfaces, which is typically not the case. Multi-dimensional upwind strategies may need to be incorporated. Other modern efforts, such as [315], introduce higher order interpolations with a fully consistent strategy for Maxwell's law $\nabla \cdot \mathbf{B} = 0$ incorporated in the scheme.

19.3.1 $\nabla \cdot \mathbf{B} = 0$ *condition for shock-capturing schemes*

For multi-D MHD, one also needs to handle the non-trivial $\nabla \cdot \mathbf{B} = 0$ constraint. Even if satisfied exactly at $t = 0$, one can still numerically generate $\nabla \cdot \mathbf{B} \neq 0$ due to the nonlinearities of the various shock-capturing methods. Besides the fact that this is clearly undesirable physically, it became clear that when no corrective action is taken at all, the accumulated errors may cause fatal numerical instabilities. However, exact solenoidal fields may not be needed in numerical simulations, as one always faces discretization and machine precision errors.

In practice, it turns out to be very difficult, although possible [441], to insist on (1) a conservative form which is needed for correctly handling shocks, (2) ensuring solenoidal $\mathbf{B}$ in some discrete sense, *and* (3) having the discretized Lorentz force orthogonal to the magnetic field in the cell centers. The conservative form of the MHD equations uses the divergence of the Maxwell stress tensor, which is equal to the Lorentz force provided $\nabla \cdot \mathbf{B} = 0$ since

$$\nabla \cdot \left(\tfrac{1}{2} B^2 \mathbf{I} - \mathbf{BB} \right) = -(\nabla \times \mathbf{B}) \times \mathbf{B} - \mathbf{B} (\nabla \cdot \mathbf{B}) . \tag{19.124}$$

In the context of shock-capturing schemes discussed thus far, many strategies to cope with the solenoidal constraint have been developed. A thorough comparison of seven different strategies on a series of nine 2D MHD tests can be found in [440], where the same high resolution base scheme (a TVD method) was used in combination with varying strategies for numerical monopoles. In what follows, we briefly discuss some of the most frequently used methods to handle the $\nabla \cdot \mathbf{B}$ problem in state-of-the-art MHD codes.

Vector potential One way of handling $\nabla \cdot \mathbf{B}$ is by replacing the magnetic field variable by a corresponding vector potential defined from

$$\mathbf{B} = \nabla \times \mathbf{A}. \tag{19.125}$$

Analytically, this guarantees a solenoidal magnetic field, but when evaluating $\mathbf{B}$ in some discrete manner from $\mathbf{A}$, the operators should still be chosen to satisfy $\nabla \cdot (\nabla \times) = 0$. Using $\mathbf{A}$ unavoidably increases the order of the occurring spatial derivatives to evaluate the Lorentz force. The need to express boundary conditions on $\mathbf{A}$ instead of $\mathbf{B}$ when using a ghost cell prescription for boundaries is also a non-trivial complication. This can be avoided by directly prescribing fluxes at the boundary instead. From the discussion of the characteristic based solvers, it is clear that they "conflict" with using the vector potential as a basic variable. In many solvers, though, the vector potential can successfully be exploited.

Projection method for $\nabla \cdot \mathbf{B}$ Originally mentioned by Brackbill and Barnes [68],[2] one can combine any multi-dimensional method with a projection scheme strategy, which controls the numerical value of $\nabla \cdot \mathbf{B}$ in a particular discretization to a given accuracy. The basic idea [68] is to correct the $\mathbf{B}^*$ computed by a scheme with $\nabla \cdot \mathbf{B}^* \neq 0$, by projecting it on the sub-space of zero divergence solutions. Hence, we modify $\mathbf{B}^*$ by subtracting the gradient of a scalar field ϕ, to be computed from the Poisson equation

$$\nabla^2 \phi = \nabla \cdot \mathbf{B}^*. \tag{19.126}$$

By construction, this yields a solenoidal $\mathbf{B} = \mathbf{B}^* - \nabla \phi$ which is then used in the next time step. This process can be repeated after each or only after a certain number of time steps. It is important to note that the accuracy up to which the Poisson problem (19.126) is solved need not be machine precision (one can use iterative methods as those discussed in Chapter 15), and that this approach keeps the order of accuracy of the base scheme, while not violating its conservation properties [440]. Note that the projection does not change the current density $\mathbf{j} = \nabla \times \mathbf{B}$. This "elliptic divergence cleaning" indirectly affects the local balance of thermal to kinetic contributions to the total energy. It is possible to rather keep thermal energy unmodified, at the cost of given up exact conservation of total energy.

    The projection scheme in essence uses numerical concepts from incompressible fluid flow (where $\nabla \cdot \mathbf{v} = 0$), and on uniform Cartesian grids was shown [440] to

---

[2] In [68], this option of using a projection scheme is only mentioned, while the authors then go on to advocate the use of a non-conservative formulation of the momentum equation.

make the smallest possible correction to remove the divergence created by the base scheme: it corresponds to minimizing $|\mathbf{B} - \mathbf{B}^*|$ under the solenoidal constraint. For cases where the initial field is not known analytically, it may even be useful to eliminate finite numerical divergence errors of the discrete initial $\mathbf{B}$.

Powell's source terms and divergence wave Another popular means of handling monopole errors was introduced by Powell [383]. In essence, the method uses additional source terms proportionate to $\sim \nabla \cdot \mathbf{B}$ in the original set of ideal MHD equations in conservation form (19.73). These terms are

$$-\nabla \cdot \mathbf{B} \begin{pmatrix} 0 \\ \mathbf{B} \\ \mathbf{v} \cdot \mathbf{B} \\ \mathbf{v} \end{pmatrix}. \tag{19.127}$$

The source term for the momentum equation, $-\nabla \cdot \mathbf{BB}$, directly follows from the relation (19.124). One of the reasons to add these terms is that the resulting modified set of PDEs restores Galilean invariance, as originally already noted by Godunov [159]. In Section 5.2[1], we also noted that the unmodified system has eight eigenvalues, one of which is zero while all others lie symmetrically about the fluid velocity v. It is the spurious zero eigenvalue which conflicts with the $\nabla \cdot \mathbf{B} = 0$ constraint, as it carries a jump in the normal component of the field. When adding the source terms (19.127), the zero eigenvalue gets replaced by an additional v. In a characteristic based scheme, it involves an additional "eight wave" contribution. The effect of this operation can be clearly illustrated on the (ideal) induction equation. The original form of this equation can be written as

$$\frac{\partial \mathbf{B}}{\partial t} = \nabla \times (\mathbf{v} \times \mathbf{B}), \quad \text{so that} \quad \frac{\partial \nabla \cdot \mathbf{B}}{\partial t} = 0 \tag{19.128}$$

and $\nabla \cdot \mathbf{B}$ errors remain undamped in time. With the source term included, however, we have

$$\frac{\partial \mathbf{B}}{\partial t} - \nabla \times (\mathbf{v} \times \mathbf{B}) + (\nabla \cdot \mathbf{B})\mathbf{v} = 0, \tag{19.129}$$

so that numerical errors in $\nabla \cdot \mathbf{B}$ become compensated, when we note the equivalence of $\nabla \times (\mathbf{v} \times \mathbf{B}) = \mathbf{B} \cdot \nabla \mathbf{v} + \mathbf{v} \nabla \cdot \mathbf{B} - \mathbf{v} \cdot \nabla \mathbf{B} - \mathbf{B} \nabla \cdot \mathbf{v}$. We can further deduce from the divergence of (19.129) and mass conservation that

$$\frac{\partial}{\partial t}\left(\frac{\nabla \cdot \mathbf{B}}{\rho}\right) + \mathbf{v} \cdot \nabla \left(\frac{\nabla \cdot \mathbf{B}}{\rho}\right) = 0. \tag{19.130}$$

The latter equation is an advection equation for $\nabla \cdot \mathbf{B}/\rho$ and it ensures that $\nabla \cdot \mathbf{B}$ errors are advected away with the flow (however, it fails to do anything in stagnation points). The use of these source terms was exploited in combination with different base schemes (FCT, TVD, TVDLF) in [444], with a clearly stabilizing influence overall.

However, just like any method which does not obey the strict conservation form of the equations, it can potentially introduce incorrect jumps across discontinuities, and an example was given in [440]. One can, of course, easily restore conservation of energy and momentum, by taking only the source term for the induction equation along, and additional arguments for doing so were given by [248] and [109]. Any of these source term strategies carry over easily to grid-adaptive computations, such as demonstrated in Powell *et al.* [384].

Constrained transport Another commonly used strategy is to insist on *machine precision accuracy* in maintaining $\nabla \cdot \mathbf{B}$ *in one particular discretization*. The original concept for this so-called constrained transport is due to Evans and Hawley [130], and has since been used in many MHD codes. One needs to ensure an initial magnetic field representation that fully complies with zero divergence in the chosen discretization, and make sure that boundary conditions remain compatible with this constraint. Originally, typical constrained transport approaches used a staggering with thermodynamic quantities at cell center, magnetic fields at cell face centers (at the middle of the edges in 2D), and electric field vectors (i.e. $\mathbf{E} = -\mathbf{v} \times \mathbf{B}$) at cell vertices (corners in 2D). By updating the magnetic field using a discrete evaluation of Stokes' law, the initial divergence of the magnetic field is maintained. In Fig. 19.21(a), this staggering is shown for a 2D Cartesian grid, where $\Omega = E_z$. The update for $\mathbf{B} = (B^x, B^y)$ from the induction equation writes as

$$
\frac{B^{x,n+1}_{j+\frac{1}{2},k} - B^{x,n}_{j+\frac{1}{2},k}}{\Delta t} = -\frac{\Omega_{j+\frac{1}{2},k+\frac{1}{2}} - \Omega_{j+\frac{1}{2},k-\frac{1}{2}}}{\Delta y},
$$

$$
\frac{B^{y,n+1}_{j,k+\frac{1}{2}} - B^{y,n}_{j,k+\frac{1}{2}}}{\Delta t} = +\frac{\Omega_{j+\frac{1}{2},k+\frac{1}{2}} - \Omega_{j-\frac{1}{2},k+\frac{1}{2}}}{\Delta x}. \tag{19.131}
$$

One can verify that, with this update, the particular discretized divergence found from

$$
(\nabla \cdot \mathbf{B})_{j,k} = \frac{B^x_{j+\frac{1}{2},k} - B^x_{j-\frac{1}{2},k}}{\Delta x} + \frac{B^y_{j,k+\frac{1}{2}} - B^y_{j,k-\frac{1}{2}}}{\Delta y} \tag{19.132}
$$

is maintained at its original value, i.e. $\nabla \cdot \mathbf{B}^{n+1} = \nabla \cdot \mathbf{B}^n$. Note that truncation errors still persist in any other discretization (certainly at shock locations or at other discontinuities). Tóth [440] realized that this constrained transport concept can be recast fully in a finite volume sense, i.e. in terms of (average) fluxes across grid cell edges for *cell-centered magnetic field* evaluations involving spatio-temporal averaging of the original staggered variables. This could be reversed, so that one no

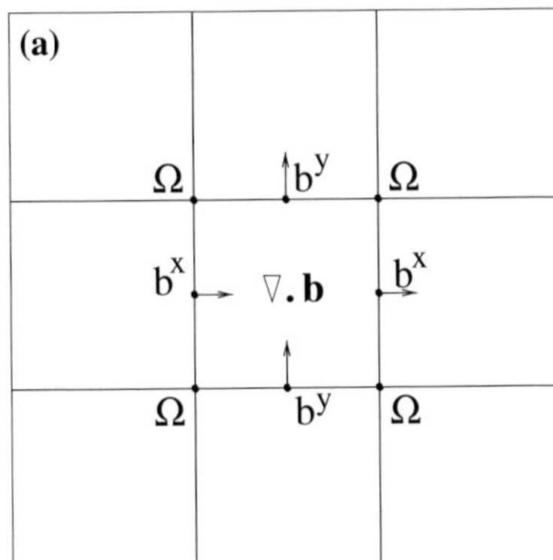

Fig. 19.21 (a) The staggering typically employed in constrained transport codes. (b) An alternative, fully cell-centered oriented scheme was introduced by Tóth: the field interpolated central difference scheme. (From Tóth [440].)

longer needs to stagger the variables in actual implementations. Once this realization was made, a fair variety of variants on the CT idea could be introduced. One of these in essence "zooms out", to employ the cell-centered variables at neighboring grid points, as shown in Fig. 19.21(b). This eliminates the need for spatial interpolations needed in CT schemes (as magnetic fields and velocities need to be evaluated at corners and cell centers), and the corresponding field interpolated central difference scheme [440] only uses temporal interpolations.

Parabolic and hyperbolic cleaning A last strategy mentioned here is a parabolic and hyperbolic variant of the initial elliptic strategy to use the numerical truncation errors in $\nabla \cdot \mathbf{B}$ as source terms. One can add diffusion type source terms to both the energy equation and the induction equations, or to the induction equation alone, which diffuse $\nabla \cdot \mathbf{B}$ at a maximal rate. This maximal rate is one that does not further constrain the discrete time step beyond the usual CFL condition. For a purely parabolic equation with a diffusion coefficient η_{D} as in

$$\frac{\partial \mathbf{B}}{\partial t} = \eta_{\mathrm{D}} \nabla^2 \mathbf{B} \,, \tag{19.133}$$

one gets an additional constraint for $\Delta t < \Delta x^2 / \eta_{\mathrm{D}}$. Taking $\eta_{\mathrm{D}} \sim \Delta x^2$ then suggests the use of parabolic source terms for (19.73) given by

$$\begin{pmatrix} 0 \\ 0 \\ \mathbf{B} \cdot \nabla C_{\mathrm{d}} \Delta x^2 \left(\nabla \cdot \mathbf{B} \right) \\ \nabla C_{\mathrm{d}} \Delta x^2 \left(\nabla \cdot \mathbf{B} \right) \end{pmatrix} \,. \tag{19.134}$$

The coefficient C_d is a numerical factor of order unity (depending on the scheme to treat source additions and to evaluate the gradients). This strategy can be easily adapted to curvilinear hierarchically nested grids, as shown in [456]. Obviously, like any source term strategy, it can potentially cause violations of the discrete Rankine–Hugoniot relations across shock fronts, so one best uses only the term for the induction equation. In grid-adaptive computations, this strategy compared favorably with all other source term approaches on a variety of shock dominated MHD problems [261].

While this parabolic treatment tries to diffuse numerical monopole errors, and the elliptic approach with Powell's source terms passively advects $\nabla \cdot \mathbf{B}/\rho$, a new *hyperbolic cleaning* variant was proposed by Dedner *et al.* [107]. The idea is to combine the two ideas, namely to transport $\nabla \cdot \mathbf{B}$ at a maximal admissible speed and damp the error simultaneously. This comes at the price of one additional scalar quantity, whose spatio-temporal evolution is coupled with $\nabla \cdot \mathbf{B}$. Various variants can be found in [107].

▷ **Splitting strategy for strong potential background fields** In a seminal paper by Tanaka [431], one of the first finite volume TVD schemes for 3D MHD simulations was presented and applied to model the Earth's magnetosphere in the supersonic solar wind. Figure 19.22 shows a 3D impression of the field lines obtained in the steady state solution, with the sunward direction at right. A key feature of the method was to account

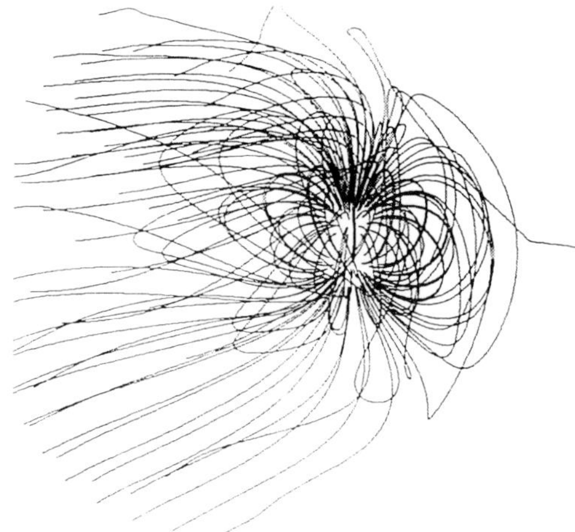

Fig. 19.22 A 3D model of the earth's magnetosphere embedded in supersonic solar wind flow. The Sun direction is towards the right. (From Tanaka [431].)

for the fact that the strong dipole field of the Earth will cause serious difficulties when the total magnetic field is taken as dependent variable, as both high- and low-β regions need to be computed accurately in the shock-dominated interaction. An elegant solution to this problem is to split the magnetic field into a static potential background field (e.g. only the dipole) and solve for the deviation from this field. Hence, we write

$$\mathbf{B} = \mathbf{B}_0 + \mathbf{B}_1 \,, \quad \text{where} \quad \frac{\partial \mathbf{B}_0}{\partial t} = 0, \ \ \nabla \cdot \mathbf{B}_0 = 0, \ \ \nabla \times \mathbf{B}_0 = \mathbf{0} \,, \tag{19.135}$$

and take $\mathbf{B}_1$ as dependent variable. Also, the equation for total energy conservation needs replacing by an equation governing the partial energy,

$$\mathcal{H}_1 = \tfrac{1}{2}\rho v^2 + \frac{p}{\gamma - 1} + \tfrac{1}{2}B_1^2 \,. \tag{19.136}$$

In combination with Powell's strategy for handling monopole errors, the method replaces the basic conservation law (19.73), now written for $\mathbf{U} = (\rho, \rho\mathbf{v}, \mathcal{H}_1, \mathbf{B}_1)^{\mathrm{T}}$, with

$$\frac{\partial \mathbf{U}}{\partial t} + \nabla \cdot \mathcal{F} + \nabla \cdot \mathcal{G} = \mathbf{S} \,, \tag{19.137}$$

where the fluxes $\mathcal{F}$ have *exactly the same form as for the full system under replacing* $\mathcal{H} \leftrightarrow \mathcal{H}_1$ *and* $\mathbf{B} \leftrightarrow \mathbf{B}_1$, but with additional fluxes and source terms given by

$$\mathcal{G} = \begin{pmatrix} 0 \\ \mathbf{B}_1 \cdot \mathbf{B}_0 \mathbf{I} - (\mathbf{B}_0\mathbf{B}_1 + \mathbf{B}_1\mathbf{B}_0) \\ \mathbf{B}_1 \cdot \mathbf{B}_0 \mathbf{v} - \mathbf{B}_1 \cdot \mathbf{v}\mathbf{B}_0 \\ \mathbf{v}\mathbf{B}_0 - \mathbf{B}_0\mathbf{v} \end{pmatrix}, \quad \mathbf{S} = \begin{pmatrix} 0 \\ -(\mathbf{B}_0 + \mathbf{B}_1)\nabla \cdot \mathbf{B}_1 \\ -\mathbf{v} \cdot \mathbf{B}_1 \nabla \cdot \mathbf{B}_1 \\ -\mathbf{v}\nabla \cdot \mathbf{B}_1 \end{pmatrix}. \tag{19.138}$$

Note that the method involves just a few extra flux and source terms to be implemented. With e.g. a basic TVDLF solver, this is easily incorporated. The CFL limit still needs to account for the full magnetic field influence. In the original paper, the magnetic monopole treatment was actually done by a projection scheme, and the solver was a full characteristic based scheme. The linearized Riemann solver was suitably modified for the changed choice of variables, and it was implemented for unstructured grids. The corresponding set of left and right eigenvectors for the modified system is given in Powell *et al.* [384]. ◁

19.3.2 Example nonlinear MHD scenarios

The aforementioned shock-capturing algorithms for nonlinear MHD have been successfully applied in many studies of magnetized plasmas. In this section, we briefly discuss three exemplary applications. Nowadays, many mostly freely distributed codes exist, with more advanced ones offering a fair variety of choice to the user with respect to the employed spatial and/or temporal discretizations, as well as the option to select additional physics modules that go beyond ideal MHD. With an early example set by the Versatile Advection Code (VAC) [439, 444, 440, 441, 443, 257, 264, 261], the incomplete and continuously growing code list includes FCTMHD3D [116], BATS-R-US [186, 384], PLUTO [333], ATHENA [154], FLASH [73], CO5BOLD [476], NIRVANA [497], Clawpack [303], RAMSES [148], ASTRO-BEAR [100], Racoon [125], etc. Many of these codes even allow for automated grid adaptivity, where finer grids are created and eliminated dynamically during the computation, in direct correspondence with the evolving flow features.

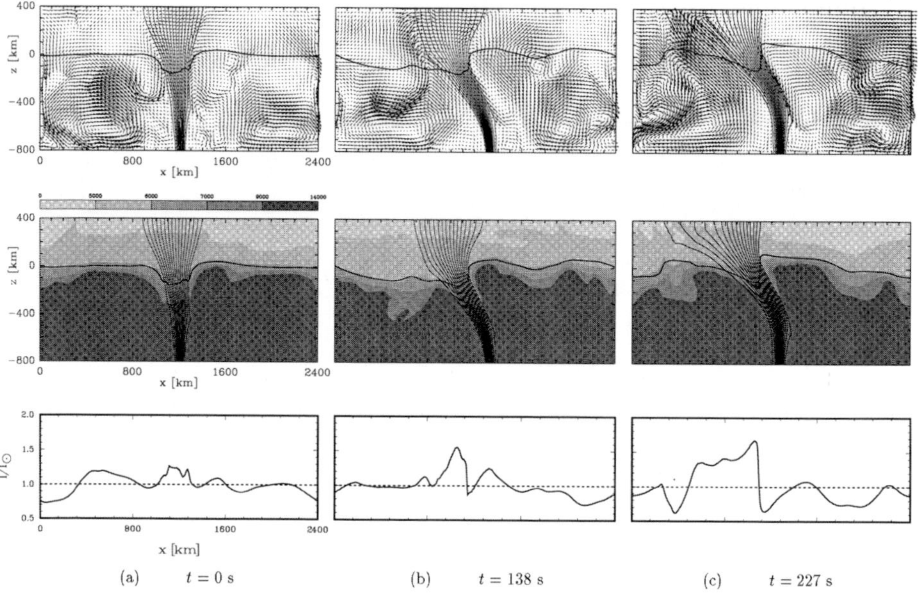

Fig. 19.23 Snapshots of a translation symmetric magnetic flux sheet in a simulation of solar magneto-convection: magnetic field lines with velocity arrows (top), with temperature contours (middle), and simulated radiation intensity (bottom). (From Steiner *et al.* [420].)

This adaptive mesh refinement (AMR) approach [34] provides an efficient means to perform very high-resolution simulations at affordable computational cost.

Solar magneto-convection and flux sheet dynamics As an example where the FCT method was applied in MHD context, Figure 19.23 shows snapshots of an early simulation [420] of solar magneto-convection interacting with a "magnetic element", i.e. a thin flux sheet of a few hundred kilometers extent at optical depth unity. FCT was used in a formulation exploiting density, momentum and entropy density ρS as conserved variables. The induction equation used a constrained transport type approach (with staggering of the variables) to ensure divergence free magnetic fields at machine precision. This planar simulation already incorporated the effects of radiative energy transfer (in a grey approximation), as well as conductive, viscous and resistive effects. Since the resolution (of order 240×120) was recognized to be too coarse to treat the latter effects properly, a sub-grid-scale prescription related viscous effects with the mesh width, and visco-resistive effects were inactivated altogether in regions with field strengths exceeding 50 G. These and similar simulations underlined the need for higher resolution observations (spa-

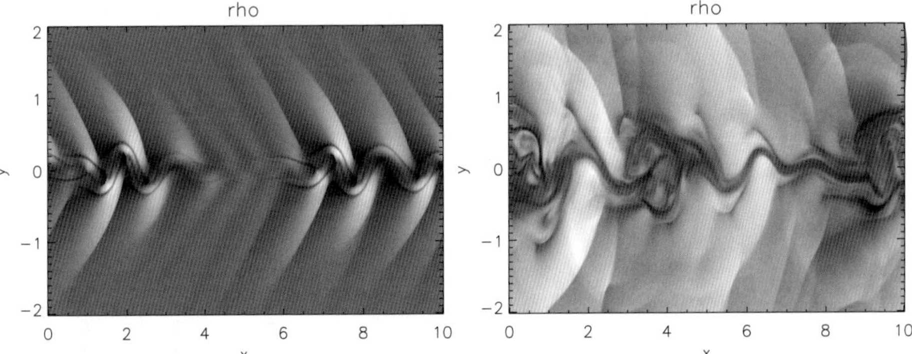

Fig. 19.24 Evolution of the density for a narrow supersonic jet of plasma beta $\beta = 6.54$. After few sound crossing times of the jet with radius $R_{\rm j} = 0.125$, a sudden sinuous deformation occurs, with shock fronts extending far away from the jet boundary. (From Baty and Keppens [27].)

tially and temporally) than those available at the time, since they predicted the excitation of upward traveling shock waves both within and exterior to the magnetic element, which contribute to the heating of chromospheric and coronal layers.

Magnetized, supersonic jet evolutions A second example, taken from Baty and Keppens [27], considers the nonlinear evolution of a magnetized, supersonic, planar jet in a pure ideal MHD approximation. The jet configuration is idealized by a "radially" varying velocity

$$\mathbf{v}(y) = \tfrac{1}{2} V \left[1 - \tanh \left(\frac{|y|}{\epsilon} - \frac{R_{\rm j}^2}{\epsilon |y|} \right) \right] \mathbf{e}_x \, . \tag{19.139}$$

This planar sheared flow profile is embedded in a uniform plasma, with the magnetic field $\mathbf{B} = B_0 \mathbf{e}_x$ aligned with the flow direction. This highly simplified model of a magnetized jet is fully specified by three dimensionless parameters, namely (1) the ratio $F = R_{\rm j}/\epsilon$ which quantifies the jet radius as compared to the thickness of the shear flow layer bounding the jet, (2) the sonic Mach number $M = V/c_{\rm s}$ where $c_{\rm s}$ denotes the (constant) initial sound speed, and (3) the initial Alfvén Mach number $M_{\rm A} = V/v_{\rm A}$. In [27], a parametric study of this jet model was presented, where first the linear instabilities were quantified in terms of their linear growth rate and eigenfunction characteristics, augmented with nonlinear simulations of extended jet segments. Using a linear MHD code of the type discussed in Chapter 15, a narrow supersonic and super-Alfvénic jet with $F = 1.25$, $M_{\rm s} = 3$ and $M_{\rm A} = 7$ was found unstable to a dominant kink (sinuous) deformation over a wide range of horizontal wavelengths. The eigenfunction of this sinuous instability has an asymmetric density perturbation $\rho_1(y) = -\rho_1(-y)$, and is characterized by significant density variation extending far beyond the jet radius. Figure 19.24 shows two consecutive snapshots of the corresponding nonlinear simulation, taken after several sound crossing times of the narrow jet diameter $2R_j = 0.35$. The density grey scale clearly shows the strong sudden sinuous deformation of the jet segment, whose horizontal extent is taken to span several wavelengths of the most unstable linear mode. The "wings" of the density perturbations extending into the jet surroundings are seen to steepen into shocks, which are of fast magneto-sonic type. The jet evolution shows a shock-dominated disruption of the original jet, where jet kinetic energy gets converted to thermal energy. This nonlinear simulation used a Roe-type approximate Riemann solver in a second order accurate TVD scheme with a 1600×1600 resolution. In their parametric survey of such planar jet configurations, the authors used both fixed high-resolution grids and AMR strategies. In the latter, parabolic cleaning for the magnetic monopoles was enforced. Moreover, the AMR grid hierarchy used different shock-capturing discretizations per grid level: TVDLF on lower levels was combined with a TVD scheme using an approximate Riemann solver on the highest grid levels.

Space weather applications A final example represents a modern space weather application, where 3D MHD simulations of coronal mass ejections (CMEs) play an essential role. Space weather is concerned with the effects of such CMEs on the Earth's magnetosphere: besides giving rise to the beauty of the northern lights, CME-driven perturbations and the currents they induce can be quite devastating to electrical power plants, communication satellites, modern navigation systems, pipelines, etc. Continuous monitoring of the solar corona using space and ground based solar telescopes has dramatically increased our awareness and understanding of these events. To a fair degree of accuracy, instantaneous solar coronal data gives us telltale signatures for pre-eruptive events. Ultimately, one may envision a space weather forecasting scenario where both past and instantaneous pre-eruption conditions from observations are used to initialize a 3D computational model of the solar corona and the inner heliosphere. In order to have relevance for forecasting, the simulation must yield a prediction of the eventual local heliospheric conditions at Earth orbit, prior to the actual arrival of the CME itself. The latter translates to being able to compute 3D MHD dynamics on an enormous range of length scales (resolving the coronal field structure in the corona, covering a distance of 1 AU, and having sufficiently detailed resolution near the Earth's orbit. This is currently heavily pursued using massively parallel, grid-adaptive MHD codes.

A pioneering example is shown in Fig. 19.25, obtained from Manchester *et al.* [324]. The authors used the BATS-R-US code, acronym for block adaptive tree solar wind Roe upwind scheme, which features an octree block based AMR method in a parallel (MPI-based) implementation. The code solves the ideal MHD

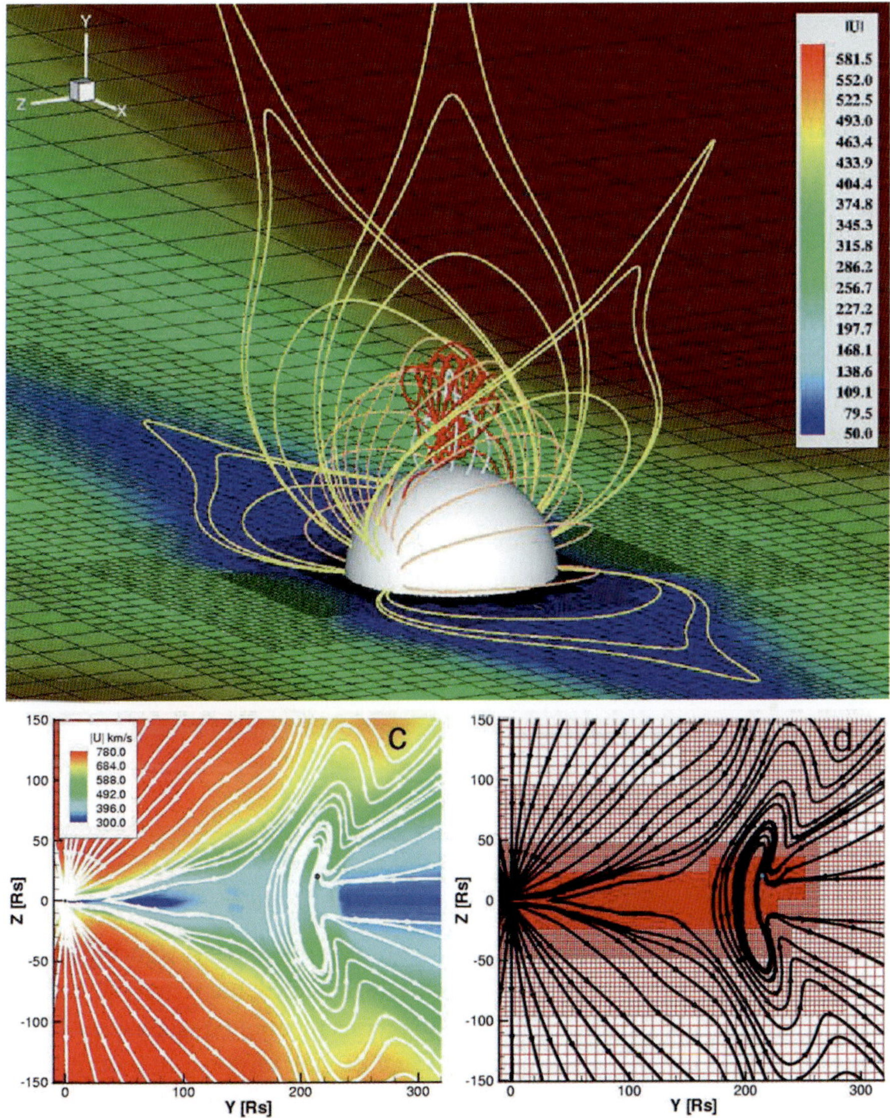

Fig. 19.25 Top: close-up view of an initial coronal magnetic field configuration. Orange and yellow lines correspond to the steady-state equatorial streamer belt. The inserted flux rope is shown with red and light blue field lines. Bottom: global view of the simulation when the CME arrives at Earth (little blue dot at right). Shown are the velocity magnitude, field lines and the grid in a planar cross-section. (From Manchester *et al.* [324].)

equations, augmented with the inclusion of the solar graviofofoftational field and a parameterized volumetric heating term in the energy equation. The latter is designed to give a good agreement with the latitudinal variation of the solar wind measured by the Ulysses spacecraft, for representative solar minimum conditions. The code uses the Powell source term strategies for handling magnetic monopole errors and employs an approximate Riemann solver of Roe-type. The local $t = 0$ field configuration is shown in the top panel of Fig. 19.25: the simulation starts by inserting an analytic model for a helically magnetized flux rope (red field lines) into steady-state solar wind conditions. The background wind is for a predominantly dipolar solar magnetic field, and this yields an equatorial streamer belt character-ized by a helmet streamer configuration (shown in the yellow field lines), together with open field lines starting from both polar regions. The fast wind then originates from these coronal holes. This initial configuration is not in force balance and the flux rope expands and gives rise to a CME with initial speeds exceeding 1000 km/s. Figure 19.25 shows the global view of the end result, at the time when the CME arrives at Earth's orbit. It also shows the grid in a cross-section through the 3D domain, highlighting the need for grid refinement when these simulations are used to obtain conditions at or even within the Earth's magnetosphere.

Naturally, the model can still be improved with more realistic initial conditions, and should be interfaced with codes that better represent the Earth's magneto-sphere, all the way down to ionospheric layers. This kind of activity is nowadays pursued in very sophisticated frameworks such as the Space Weather Modeling Framework [445], where different codes are run concurrently to model the dynam-ics from Sun to Earth. At the core of several framework components, one finds MHD solvers employing shock-capturing algorithms.

19.3.3 Alternative numerical methods

We emphasized in our discussion the use of conservative shock-capturing schemes and finite volume treatments for nonlinear MHD scenarios, but it must be clear that virtually all other (combinations of) discretizations mentioned in Chapter 15 have found widespread use in the astrophysical and plasma physical literature. In the following, we point out various alternative approaches, along with brief summaries of typical example applications.

MHD solvers using primitive variable formulations While we stressed the need for conservative formulations, there are many instances where primitive variable formulations are adequate. In particular, one of the earliest codes designed to tackle nonlinear MHD problems, the ZEUS code [423], is directly solving the MHD equations in a primitive variable formulation. In essence, ZEUS solves the gas dy-namic part of the MHD equations using upwinding for the advection terms only (the terms of the form $\nabla \cdot \rho \mathbf{v}$), and then augments this with finite difference treat-ments to add Lorentz force and gas pressure "source" contributions. To handle

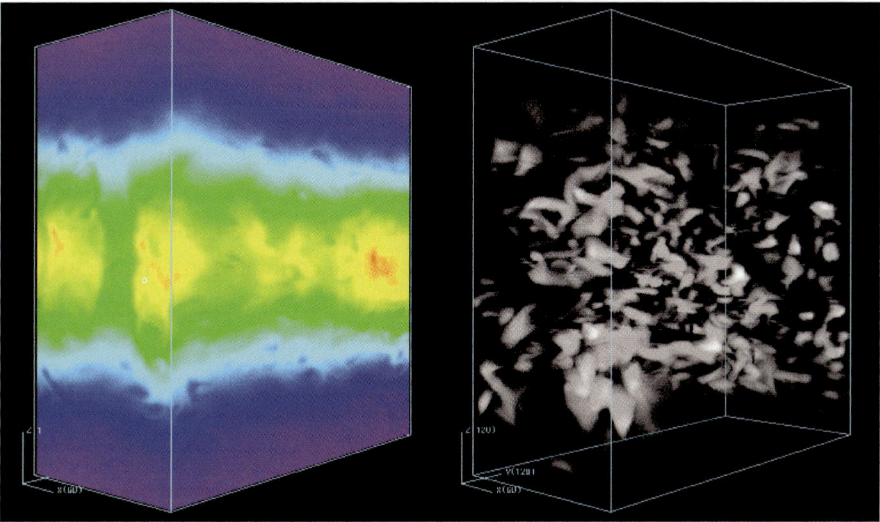

Fig. 19.26 3D rendering of the turbulent local dynamics in a magneto-rotationally unstable gravitationally stratified and magnetized accretion disk. After about 20 full orbital revolution periods, the left panel shows the density structure, while the right iso-surfaces characterize the magnetic pressure. (From Stone *et al.* [422].)

shocks, artificial viscosity terms are then needed which automatically take effect in compression zones only, both for stability and for physical reasons (restoring the correct shock propagation speeds since the conservative form is not obeyed discretely). These involve usually problem-tailored parameter settings, and it was demonstrated in [131] that erroneous shocks can be easily produced, a statement which holds generally for all non-conservative treatments. In ZEUS, the induction equation is treated separately, in its Stokes' type form, using a constrained transport approach. The precise form of the latter approach pioneered in ZEUS is referred to as MOC-CT (method of characteristics–constrained transport) [423], which specifically takes the Alfvén wave space-time characteristics into account to evaluate the electromotive force (i.e. the cell edge located electric field $-\mathbf{v} \times \mathbf{B}$). This allowed for many nonlinear MHD computations, and one of the first test suites, still used for numerical MHD simulations, was presented in [421].

An early, impressive application of the MOC-CT scheme concentrated on the nonlinear evolution of the magneto-rotational instability (MRI) encountered in Section 13.4. In the radially sheared near Keplerian flows in accretion disks, the MRI provides a linear MHD route to instability, which requires the presence of a small seed magnetic field. Only by numerical simulations could one assess the nonlinear evolution of the MRI, and demonstrate that it would lead to sustained MHD turbulence, accompanied by enhanced outward angular momentum transport. The latter could explain the observationally inferred high accretion rates, which translate into "anomalously high" effective viscosities needed in simplified Navier–Stokes accretion disk models (the so-called α-disk prescriptions). Realizing that it is practically impossible to perform global simulations while assessing turbulent dynamics, Hawley *et al.* [219] introduced the local "shearing box" model in which the 3D dynamics is simulated in a patch located at a global disk radius R_0, and the equation of motion is then formulated in a co-rotating frame where Coriolis, centrifugal and gravitational forces are accounted for. The local nature justifies an expansion of the effective potential due to centrifugal and gravitational effects, and the radial velocity shear becomes a linear profile. The "shearing box" is thought to be surrounded by similar local slabs which, due to the velocity profile, slide past each other in the "toroidal" direction in direct relation to the linear velocity variation. This means that a kind of shifted periodic boundary condition can be adopted in the radial direction (artificially connecting front and back planes of the shearing box), while other boundaries can be taken as standard periodic pairs. In [219], the vertical component of gravity was neglected, and it was found that a saturated, turbulent state indeed forms, with increased outward angular momentum. The angular momentum flux in the saturated state was dominated by magnetic stresses. The code used an MOC-CT scheme, and was rather similar to the ZEUS implementation. A follow-up study with ZEUS [422] extended this work to include vertical gravitational stratification. Figure 19.26 shows a snapshot of the nonlinear end state from that study, as seen in density (left) and magnetic pressure iso-surfaces (right). Anisotropic MHD turbulence has developed, as a result of the MRI. The simulation shown in Fig. 19.26 started with a weak vertical magnetic seed field, and shows the local shearing box dynamics after about twenty orbital revolution times.

Spectral and pseudo-spectral MHD simulations Spectral methods are extensively used in MHD turbulence studies, specifically for 3D incompressible simulations. An example of a compressible MHD code based on Fourier collocation methods is the CRUNCH3D code, which addressed MHD turbulence and magnetic reconnection in Cartesian triple periodic domains (demonstrating the distinct possibility for magnetic flux tube tunneling, see [101]). In cylindrical flux-tube like configurations, Fourier representations along and about the loop can be used in conjunction with finite differences for the radial direction. This mixed discretization was employed in a parallel *pseudo-spectral* code, which was used in [265] to study Kelvin–Helmholtz instabilities in 3D magnetized jet flows. Figure 19.27 shows a visualization of the disruption of such jets, for different initial perturbations selecting linearly unstable modes of prescribed axial k_z and poloidal m mode numbers. The linear growth rates were found to be in exact correspondence with linear anal-

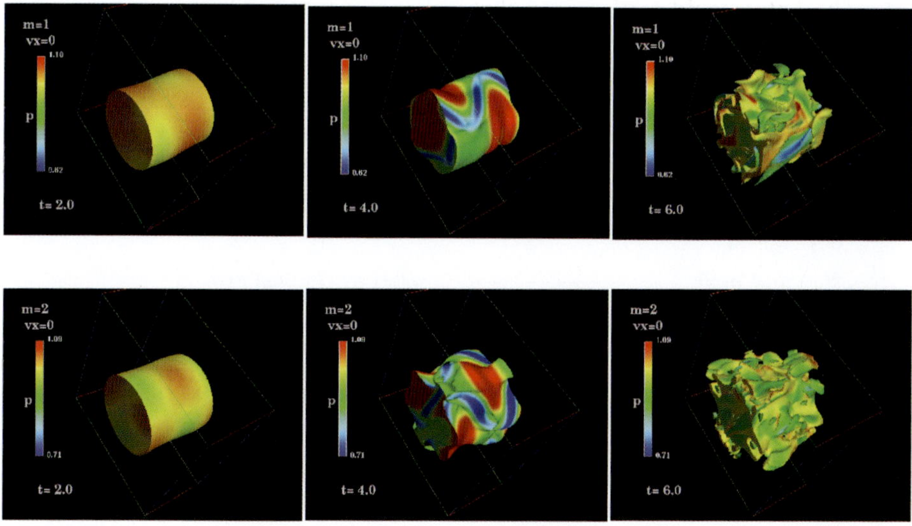

Fig. 19.27 As studied in [265], 3D jets deform and disrupt due to Kelvin–Helmholtz instabilities originating at the jet surface. Top: the jet iso-surface is colored by thermal pressure, at consecutive times during the evolution under a given initial $m = \pm 1$ perturbation. Bottom: the same for an $m = \pm 2$ initial perturbation.

ysis, while the knowledge on the evolution of all individual Fourier mode pairs (k_z, m) provided by the pseudo-spectral code allowed us to identify and distinguish the most dominant nonlinear interactions.

▷ **Pseudo-spectral method** For numerically solving nonlinear PDEs, the spectral method employing Fourier representations becomes very expensive in terms of CPU time. This is a consequence of the convolutions required to evaluate the nonlinear terms. Suppose, e.g., that two functions $f(x)$ and $g(x)$ are both approximated by means of Fourier harmonics. Their product $f(x)g(x)$ then requires the convolution of the two Fourier sums,

$$p(x) \equiv f(x)g(x) \approx \hat{f}(x)\hat{g}(x)$$

$$= \sum_{k=-N}^{N} f_k e^{ikx} \times \sum_{k'=-N}^{N} g_{k'} e^{ik'x} = \sum_{k=-N}^{N} \sum_{k'=-N+k}^{N} f_{k'} g_{k-k'} e^{ikx}, \quad (19.140)$$

which involves $\sim N^2$ calculations. Higher order terms become even more CPU time consuming, so that the spectral method is often no longer practical in nonlinear problems.

For nonlinear problems, the pseudo-spectral method often provides an attractive combination of accuracy and computational speed. This method makes handy use of the fact that the Fourier transform of a convolution product is equal to $\sqrt{2\pi}$ times the product of the Fourier transforms. In the pseudo-spectral method, the terms in the equations are computed in real space but the unknowns are updated in Fourier space (Fig. 19.28). This involves FFTs from Fourier space to real space, and back, which is much faster for the determination of the nonlinear terms. For example, for quadratic terms, the CPU time for convolution of the two truncated Fourier series is proportional to N^2 in the spectral method

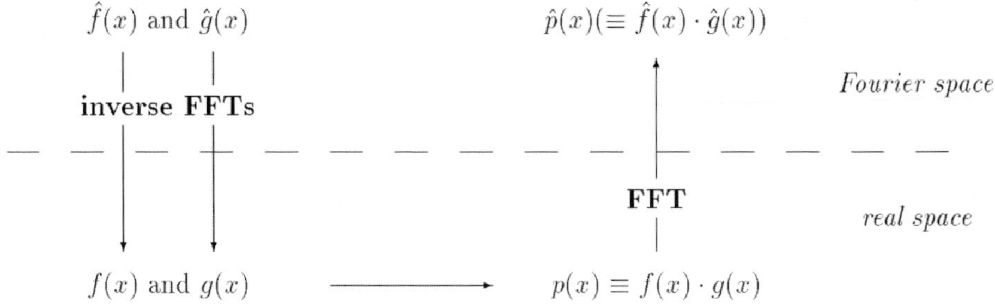

Fig. 19.28 Schematic representation of the determination of a quadratic term in the pseudo-spectral method. All quantities are updated in Fourier space but the products are calculated in real space.

and to $N \log N$ for the pseudo-spectral method. Hence, for large N, the pseudo-spectral method is much more efficient than the spectral method. ◁

Lagrangian versus Eulerian approaches In essentially all methods mentioned thus far, we implicitly assumed a static grid on which we wish to follow the temporal variations of the basic MHD variables. Hence, a Eulerian viewpoint was adopted all along. In Volume [1], Chapter 4, both Eulerian and Lagrangian formalisms were discussed, and it is also possible to directly solve the Lagrangian equations and thus obtain the evolving MHD dynamics in the co-moving frame. In practice, an initial grid on which one expresses the initial conditions then moves and deforms, and volume changes directly yield density changes from mass conservation. However, numerical errors may then eventually lead to grid entangling in more than 1D. An effective way to avoid this problem, known as arbitrary Lagrangian–Eulerian, or ALE, is to use a hybrid approach where one or more Lagrangian steps are taken, consecutively remapped to a less deformed grid, and then the next cycle of Lagrangian updates is started. An early, related approach by Brackbill [67] used particle-in-cell type methods for MHD simulations. A modern example of a 3D Lagrangian-Eulerian remap code, Lare3D [12], applies this to nonlinear shock-dominated MHD. In Lare3D, a single Lagrangian step is followed by a remap to a fixed Eulerian grid, and the latter remapping is fully conservative. The Lagrangian update is done at second order accuracy and the use of artificial viscosity (and resistivity) to cope with shocks. Lare3D employs a staggered prescription with magnetic field components on cell faces, scalars in cell volumes, and velocities at cell vertices. The $\nabla \cdot \mathbf{B} = 0$ constraint is then maintained using constrained transport, applied to the advection, as well as to the remap stage. Although the implementation does not guarantee perfect conservation of total energy, it was shown by Arber *et al.* [12] that it can give results as accurate as approximate Riemann solver based methods for a suite of shock dominated problems. Another advantage is that it maintains pressure positivity, so that one can handle very low values of beta.

Solar physics applications with higher order centered differences As a final example of an alternative to the Riemann-solver-type shock-capturing schemes discussed in this chapter, we now turn to a modern MHD computation of solar photospheric magneto-convection taken from Vögler *et al.* [469]. The employed code MURaM (acronym for Max-Planck-Institute for Aeronomy/University of Chicago Radiation Magnetohydrodynamics) is tailored to solve the non-ideal 3D MHD equations in a Cartesian box, and in essence employs straightforward fourth order centered difference formulas, in combination with an explicit fourth order Runge–Kutta time stepping. These, and similar higher than second order CD approaches, have found widespread use in many solar and astrophysics applications. This is presumably because the current emphasis is more on coping with more realistic (i.e. beyond ideal MHD) physics than on handling very strong shocks accurately. MURaM includes detailed non-local radiative transfer effects, and even accounts fully for partial ionization effects. The latter introduces the need for interpolations through pre-computed tabulated equations of state, to determine pressure and temperature from density and internal energy values, and in MURaM this allows for the eleven most abundant elements in the solar photosphere. This is in sharp contrast to the simple ideal equation of state assumed in most of this and the preceding volume, and would clearly complicate the use of approximate Riemann solvers. The central fourth order differencing in MURaM operates on the usual set of conservative variables, and is in fact equivalent to a conservative finite volume scheme. However, the price to pay for the CD approach is the need to incorporate artificial diffusion terms, even for the mass conservation equation. These terms are not totally undesirable, since one needs them to dissipate energy at the unresolved sub-grid scales anyway. These artificial diffusion coefficients then have a shock-resolving contribution that is effective in converging flow regions alone, and a contribution for so-called "hyper-diffusion", which merely serves to detect and dissipate small-scale fluctuations (in essence depending on numerical approximations of third to first order derivative ratios). For modest Mach number flows, such as typical in the solar photosphere regions, this approach has proven its worth many times.

With the aim of studying the upper solar convection zone as well as the directly observable photospheric dynamics, present state-of-the-art simulations spend most computational effort on handling the radiative transfer as accurately as possible. As an example simulation result from [469], which in many ways generalizes the

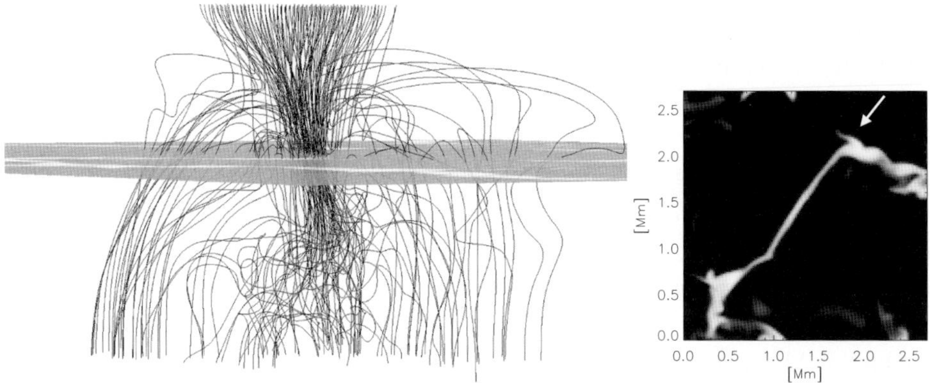

Fig. 19.29 3D view of the magnetic field in a kG magnetic sheet, in a radiative MHD simulation spanning a vertical scale of 800 km down into optical depth unity (grey iso-surface) up to 600 km above the solar photosphere. The view angle is along the arrow as indicated in the photospheric field intensity plot at right. (From Vögler *et al.* [469].)

2D results shown in Figure 19.23 from Steiner *et al.* [420] discussed earlier, Figure 19.29 shows a 3D view of a spontaneously formed thin magnetic sheet of kilo Gauss field strength. This represents a face-on view (in the direction as indicated on the right panel of field intensity at the indicated photospheric "plane") of the sub- to upper photospheric field configuration. As a general tendency, magnetic field gets expelled from up-welling adiabatically expanding convection flows and concentrates along the narrow downdraft lanes bordering the cellular convection patterns. These high resolution radiative MHD simulations (employing $100 \times 288 \times 288$ resolution) extend from 800 km down into optical depth unity (i.e. the photosphere) to 600 km above, and cover an area of 6000^2 km. The simulations actually followed an initially uniform 200 Gauss magnetic field, and gave unprecedented insights into evolving solar plage dynamics over many hours. The snapshot shown in Figure 19.29 is representative for the quasi-two-dimensional flux sheets occasionally found in the simulated time interval, and shows that they are in fact rather shallow phenomena, as the magnetic configuration is clearly disrupted in the deeper convective layers. This is fully consistent with the vertical variation of the plasma beta parameter in the strong field regions, found to vary from values above 100 at the bottom, to about 0.1 near the top boundary. Besides the sheet-like structures accumulating at the convection cell edges, the simulations also witnessed the formation of a more coherent "micropore" of kilo Gauss field strength at the vertices where several down flow lanes merge. This micropore appeared dark in intensity since temporarily the convective transport got almost completely suppressed. This structure only survived for about two convective turnover times, and showed remarkable internal structure on shorter time scales. This represents a current frontier of MHD modeling in which predictions for future generation solar telescopes can be made, as it gives insights into details below current spatial and temporal resolution limits.

19.4 Implicit approaches for extended MHD simulations

All examples and methods discussed thus far in this chapter use explicit time integration strategies, generalizing the basic explicit schemes mentioned in Section 15.4.1. In Section 15.4.2, we pointed out that explicit schemes may face insurmountable difficulties, due to the disparate time scales inherent in the MHD wave dynamics, and especially when it is necessary to study resistive phenomena occurring on the longest diffusion time scale. To this end, the use of implicit time integration methods was discussed briefly in Section 15.4.3, and linear MHD computations employing implicit time integration strategies were covered in Section 15.4.4. Also, for the full nonlinear MHD system, the use of implicit (or semi-implicit) time integration methods is needed to alleviate the stringent CFL limit on the time step. Examples of the typical use of implicit methods in MHD context for both steady and unsteady problems can be found in [443], and an extended description of the numerical algorithms is given in [266]. A key observation in the use of fully implicit schemes, such as those discussed in Section 15.4.4 to nonlinear MHD problems, is that (1) the nonlinearity of the system will call for some suitable (Newton) iteration strategy translating the problem into repeated solves of linear systems which involve the flux Jacobian; and (2) the solution of such huge algebraic linear systems. In 1D problems, the latter is feasible with the use of direct methods (of the type presented in Section 15.3.1) since we then deal with typically block-tridiagonal systems. However, in multi-dimensional problems, we face the need to solve many block penta-diagonal (in 2D) or block hepta-diagonal (3D) systems, which necessitate the use of iterative linear solvers and adequate preconditioners to accelerate their convergence. The computation of the flux Jacobian can be avoided altogether in an approach where the Newton iteration is performed in combination with iterative Krylov sub-space methods using a matrix-free approach (in the iterative scheme, only matrix–vector products are required). This is an area of active research, with a modern example found in Tóth *et al.* [442]. In what follows, we will restrict ourselves to pointing out the basic idea of alternating direction implicit strategies, reducing the problem to consecutive solves of "1D" problems. Furthermore, we review in more detail the widely used semi-implicit methods, as employed in state-of-the-art simulations of laboratory tokamak plasmas.

19.4.1 Alternating direction implicit strategies

Consider the extension of the hyperbolic model problem (15.101) to two spatial dimensions:

$$\frac{\partial u}{\partial t} = -v\left(\frac{\partial u}{\partial x} + \frac{\partial u}{\partial y}\right). \tag{19.141}$$

Application of the implicit "BTCS Euler scheme" (backward in time, centered in space, see Eq. (15.125)) yields

$$\frac{u_{i,j}^{n+1} - u_{i,j}^{n}}{\Delta t} = -\frac{1}{2}v\left(\frac{u_{i+1,j}^{n+1} - u_{i-1,j}^{n+1}}{\Delta x} + \frac{u_{i,j+1}^{n+1} - u_{i,j-1}^{n+1}}{\Delta y}\right), \tag{19.142}$$

from which it follows that

$$v\Delta y\Delta t\, u_{i+1,j}^{n+1} - v\Delta y\Delta t\, u_{i-1,j}^{n+1} + 2\Delta x\Delta y\, u_{i,j}^{n+1}$$

$$+ v\Delta x\Delta t\, u_{i,j+1}^{n+1} - v\Delta x\Delta t\, u_{i,j-1}^{n+1} = 2\Delta x\Delta y\, u_{i,j}^{n}. \tag{19.143}$$

Since the indices i and j run over all grid points, one obtains an algebraic system with a penta-diagonal coefficient matrix. The solution procedure for such penta-diagonal systems of equations is CPU time consuming. This problem can be overcome by using a "splitting method", e.g. *the alternating direction implicit* (ADI) method. This method is of order $\mathcal{O}((\Delta t)^2, (\Delta x)^2, (\Delta y)^2)$ and unconditionally stable. The ADI algorithm produces two sets of tridiagonal systems, and subsequent solution of both is much faster than solution of the above penta-diagonal system. In the ADI formulation the above finite difference equations become:

$$\frac{u_{i,j}^{n+\frac{1}{2}} - u_{i,j}^{n}}{\Delta t/2} = -\frac{1}{2}v\left(\frac{u_{i+1,j}^{n+\frac{1}{2}} - u_{i-1,j}^{n+\frac{1}{2}}}{\Delta x} + \frac{u_{i,j+1}^{n} - u_{i,j-1}^{n}}{\Delta y}\right), \tag{19.144}$$

$$\frac{u_{i,j}^{n+1} - u_{i,j}^{n+1/2}}{\Delta t/2} = -\frac{1}{2}v\left(\frac{u_{i+1,j}^{n+\frac{1}{2}} - u_{i-1,j}^{n+\frac{1}{2}}}{\Delta x} + \frac{u_{i,j+1}^{n+1} - u_{i,j-1}^{n+1}}{\Delta y}\right). \tag{19.145}$$

These yield tridiagonal systems for $u_{i+1,j}^{n+\frac{1}{2}}$, $u_{i,j}^{n+\frac{1}{2}}$, $u_{i-1,j}^{n+\frac{1}{2}}$ and $u_{i,j+1}^{n+1}$, $u_{i,j}^{n+1}$, $u_{i,j-1}^{n+1}$, respectively. The formulation of Eq. (19.144), referred to as the "x-sweep", is implicit in the x-direction and explicit in the y-direction and provides the necessary data to solve the tridiagonal system (19.145). The latter system is implicit in the y-direction and explicit in the x-direction and is called the "y-sweep".

The ADI method has been applied successfully to compute nonlinear ideal and resistive internal kink instabilities in 2D (Schnack and Killeen [404]), and even 3D (Finan and Killeen [135]) laboratory plasmas. However, since in MHD one has to deal with eight equations simultaneously, the ADI method still involves large block tridiagonal matrices. For wave driven problems, in which the time step is limited by the wave period, the gain in the size of the time step (due to the implicit scheme) is too small to make up for the huge amount of additional work. As a result, the ADI method is not often used then, and for such problems, the *semi-implicit scheme* yields a valuable alternative. This scheme was introduced in MHD by Harned and Kerner [214] and is discussed in the next subsection.

19.4.2 Semi-implicit methods

As already mentioned in Section 15.4.3, the term "semi-implicit" is used for several alternative schemes which are neither explicit nor fully implicit. Different options can be, e.g., to treat some dependent variables explicitly and other ones implicitly, or to update only specific terms in the equations implicitly. In the following discussion, semi-implicit refers to a means to relax the CFL condition somewhat by treating only the fastest sub-spectrum of waves implicitly. This has been applied in nonlinear MHD with a lot of success. In hot elongated plasmas, the fast magneto-sonic waves have extremely short time scales and, hence, cause a severe limit on the time step for explicit schemes due to the CFL condition for numerical stability. When these fast magneto-sonic waves are handled in an implicit way, the time step is limited by the CFL condition on the Alfvén waves, which is much less restrictive because of the time scale discrepancy between fast and Alfvén waves.

The implicit treatment of only the sub-spectrum of fast magneto-sonic waves can be achieved in a simple and elegant way. The procedure involves three steps: (1) identification of the mode that produces numerical instability when the time step is raised; (2) subtraction of a simple approximation of the term producing this mode from both sides of the equation; (3) evaluation of the new approximate term on the RHS explicitly and on the LHS implicitly. Below we will demonstrate this procedure on the simple linear advection model problem already considered in Chapter 15 and then apply it to MHD.

Demonstration of the procedure and physical interpretation Consider again the model problem equation

$$\frac{\partial u}{\partial t} = -v \frac{\partial u}{\partial x}.$$ (19.146)

This simple linear equation describes only one type of mode, but nevertheless we can demonstrate the idea of the semi-implicit method on it. The von Neuman stability analysis of the explicit Lax–Friedrichs scheme from Eq. (15.110), namely

$$u_i^{n+1} = \tfrac{1}{2}(u_{i+1}^n + u_{i-1}^n) - \tfrac{1}{2}v \frac{\Delta t}{\Delta x}(u_{i+1}^n - u_{i-1}^n),$$ (19.147)

yielded the CFL stability condition. This gave a restriction of the time step for a given spatial discretization, viz. $|v\Delta t/\Delta x| \leq 1$. In an implicit scheme, the spatial derivative is evaluated at the next time level:

$$u_i^{n+1} = u_i^n - v\Delta t \frac{\partial u}{\partial x}_i^{n+1}, \qquad (19.148)$$

which yields an unconditionally stable numerical scheme. For this simple model problem (which easily generalizes to nonlinear cases, with v a function of u or of (x, y, z), or both), the three steps of the semi-implicit procedure become very simple. The first step can be skipped since there is only one wave mode. The term that generates this mode is also obvious, since there is only one RHS term in Eq. (19.146). A semi-implicit method for this model problem is

$$u_i^{n+1} + \tfrac{1}{2}v_0 \frac{\Delta t}{\Delta x} (u_{i+1}^{n+1} - u_{i-1}^{n+1})$$

$$= \tfrac{1}{2}(u_{i+1}^n + u_{i-1}^n) - \tfrac{1}{2}v \frac{\Delta t}{\Delta x} (u_{i+1}^n - u_{i-1}^n) + \tfrac{1}{2}v_0 \frac{\Delta t}{\Delta x} (u_{i+1}^n - u_{i-1}^n),$$

$$(19.149)$$

where the last term on the LHS and on the RHS is additional to the original Lax–Friedrichs scheme (19.147). The coefficient v_0 is a constant which is to be chosen from numerical stability considerations. By consequence, the LHS operator to be inverted is linear and the obtained tridiagonal system is easy to solve.

▷ **Exercise** Show that the semi-implicit scheme (19.149) is unconditionally stable for $v_0 = \tfrac{1}{2}v$ when v is constant. Apply the von Neumann stability analysis of Chapter 15. ◁

Notice that the terms added to the LHS and to the RHS in Eq. (19.149) are identical, except for the fact that the LHS term is evaluated at time level $n + 1$ while the RHS term is evaluated at time level n. This comes down to adding some numerical dissipation which slows down the high-k (short wavelength) modes. As a matter of fact, the dispersion relation of the original model equation is $\omega/k = v$, so that there is no dispersion for a constant v. The semi-implicit equation (19.149), however, is a discretization of

$$\frac{\partial u}{\partial t} = -v \frac{\partial u}{\partial x} - v_0 \Delta t \frac{\partial}{\partial x} \frac{\partial u}{\partial t}, \qquad (19.150)$$

with the dispersion relation $\omega = kv - iv_0\omega k\Delta t$, or $\omega/k = v/(1 + iv_0 k\Delta t)$. Hence,

$$\left| \frac{\omega}{k} \right| = \frac{v}{\sqrt{1 + (kv_0\Delta t)^2}}, \qquad (19.151)$$

so that the high-k modes are indeed slowed down. In the MHD application discussed below, the semi-implicit term is chosen such that compressible modes propagating perpendicular to the magnetic field are slowed down in a similar way.

Application to nonlinear MHD The semi-implicit method is most powerful when used in conjunction with a spectral discretization, although it can also be used in finite difference schemes. In MHD applications for tokamak and solar coronal loop simulations, it has been successfully used in 3D, often applying a finite difference method in the direction normal to the initial magnetic flux surfaces and a (pseudo-)spectral method in the two spatial directions spanning these surfaces. Below, we describe the three-step procedure mentioned, in the way it was originally introduced to MHD by Harned and Kerner [214].

The first step, identification of the trouble mode, is obvious: the fast magneto-sonic mode is the one that causes the severe time step limitation in explicit schemes. This wave mode propagates in the direction normal to the flux surfaces and this is precisely the direction in which the mesh size has to be very small in order to resolve the singular layers that develop in e.g. kink instabilities or resonantly excited continuum modes. The simple approximation of the term that generates these modes follows from Eq. (5.51) of Section 5.2.3, where we do not assume normal modes, i.e. the linearized ideal MHD equations in terms of the amplitude $\hat{\mathbf{v}}$ for plane waves:

$$\left(\left[-\frac{\partial^2}{\partial t^2} - (\mathbf{k} \cdot \mathbf{b})^2\right]\mathbf{I} - (b^2 + c^2)\mathbf{kk} + \mathbf{k} \cdot \mathbf{b}(\mathbf{kb} + \mathbf{bk})\right) \cdot \hat{\mathbf{v}} = 0. \quad (19.152)$$

Retaining only the "fast compressional modes" ($\mathbf{k} \perp \mathbf{b}$), one obtains a representative equation for the fast evolution:

$$\frac{\partial^2 \hat{\mathbf{v}}}{\partial t^2} = -(b^2 + c^2)\mathbf{kk} \cdot \hat{\mathbf{v}}. \quad (19.153)$$

For a general (inhomogeneous) equilibrium, remembering that the ∇ operator was replaced by $i\mathbf{k}$, this equation becomes

$$\frac{\partial^2 \hat{\mathbf{v}}}{\partial t^2} = \left(\frac{\gamma p_0 + B_0^2}{\rho_0}\right) \nabla \nabla \cdot \hat{\mathbf{v}}. \quad (19.154)$$

The semi-implicit method then approximates this simply by replacing the coefficient by a constant A_0^2, introducing the linear operator

$$S(\hat{\mathbf{v}}_\perp) \equiv A_0^2 \nabla \nabla \cdot \hat{\mathbf{v}}_\perp, \quad (19.155)$$

where only the perpendicular component of $\mathbf{v}$ is included since the severest time step limitations occur when $k_\parallel \ll k_\perp$. In that case, the parallel component of the fast magneto-sonic wave is negligible. The constant A_0^2 is to be determined from stability considerations.

▷ **Semi-implicit predictor–corrector scheme for MHD** Harned and Kerner [214] used a predictor–corrector scheme to evolve the MHD equations. They applied the semi-implicit term in the corrector step only, which leads to the following scheme:

– predictor step:

$$\mathbf{v}^* = \mathbf{v}^n + (\alpha \Delta t / \rho^n)\, \mathbf{F}(\mathbf{v}^n, \mathbf{B}^n, p^n)\,,$$

$$\mathbf{B}^* = \mathbf{B}^n + \alpha \Delta t\, \nabla \times (\mathbf{v}^n \times \mathbf{B}^n)\,,$$

$$\rho^* = \rho^n - \alpha \Delta t\, \nabla \cdot (\rho^n \mathbf{v}^n)\,,$$

$$p^* = p^n - \alpha \Delta t\, (\mathbf{v}^n \cdot \nabla p^n + \gamma p^n \nabla \cdot \mathbf{v}^n)\,;$$

– corrector step:

$$v_z^{n+1} = v_z^n + (\Delta t / \rho^*)\, F_z(\mathbf{v}^*, \mathbf{B}^*, p^*)\,,$$

$$(1 - S)\, \mathbf{v}_\perp^{n+1} = \mathbf{v}_\perp^n + (\Delta t / \rho^*)\, \mathbf{F}_\perp(\mathbf{v}^*, \mathbf{B}^*, p^*) - S(\mathbf{v}_\perp^n)\,,$$

$$\mathbf{B}^{**} = \mathbf{B}^n + \tfrac{1}{2}\Delta t\, \nabla \times \left[(\mathbf{v}^{n+1} + \mathbf{v}^n) \times \mathbf{B}^* \right]\,,$$

$$\rho^{n+1} = \rho^n - \tfrac{1}{2}\Delta t\, \nabla \cdot \left[\rho^* (\mathbf{v}^{n+1} + \mathbf{v}^n) \right]\,,$$

$$p^{n+1} = p^n - \tfrac{1}{2}\Delta t\, \left[(\mathbf{v}^{n+1} + \mathbf{v}^n) \cdot \nabla p^* + \gamma p^* \nabla \cdot (\mathbf{v}^{n+1} + \mathbf{v}^n) \right]\,;$$

– diffusion step:

$$(1 - \eta_0 \Delta t \nabla^2)\, \mathbf{B}^{n+1} = \mathbf{B}^{**} + \Delta t\, \nabla \times (\eta \nabla \times \mathbf{B}^{**}) - \eta_0 \Delta t\, \nabla^2 \mathbf{B}^{**}\,.$$

In the above scheme, $\mathbf{F}$ denotes the force operator in the momentum equation. Notice that the semi-implicit term has been added to the corrector step in the momentum equation, *implicitly* in the LHS and *explicitly* in the RHS. The z-component of the momentum equation is treated separately because the semi-implicit term only involves the r- and θ-component of the velocity. The magnetic field is always advanced after the velocity field and requires no special treatment. The parameter α determines the time step size. For $\alpha = 0.5$ the ideal MHD part of the scheme (apart from the fast magneto-sonic waves) is second order accurate in time. In practice, one chooses α equal to 0.51 or 0.52 in order to add a small amount of numerical damping to make the scheme more stable. Also notice the Crank–Nicolson type treatment (see Section 15.4.3) of the velocity in the corrector step: all information on $\mathbf{v}$ is used as soon as it becomes available. ◁

The stability of the semi-implicit algorithm was analyzed in [214], where it was found that the following constraints must be satisfied throughout the plasma for linear stability:

$$A_0^2 > \frac{1}{16\rho}(B_z^2 + B_\theta^2 + \gamma p)(1 + 2\alpha)^2\,, \tag{19.156}$$

$$\frac{1}{\rho}\left(\frac{m}{r} B_\theta + \frac{2\pi n}{L} B_z \right)^2 (\Delta t)^2 < \frac{16}{(1 + 2\alpha)^2}\,. \tag{19.157}$$

The constraint (19.156) determines the constant A_0 and follows from the requirement of unconditional stability with respect to the fast magneto-sonic modes. The second constraint is a CFL-like condition on the time step following from the explicit treatment of the shear Alfvén term $\omega_A^2 = (\mathbf{k} \cdot \mathbf{v}_A)^2$ on the LHS.

The semi-implicit scheme was applied [214] in combination with a finite difference discretization for the direction normal to the flux surfaces and a spectral discretization in the other two spatial directions. The constant A_0 in the semi-implicit term then avoids the CPU time consuming convolutions that would complicate a fully implicit time advance. In fact, the only "overhead" the semi-implicit scheme produces involves the solution of N_F tri-diagonal systems in every time step, with N_F the number of Fourier modes. However, there are very efficient solvers for such simple systems. Kerner [269] derived the following expression for the maximal time step of the semi-implicit scheme, Δt_{SI}, compared to the CFL limited time step of an explicit scheme, Δt_{CFL}. From numerical simulations in a cylindrical plasma of finite length $2\pi R_0$ and radius a (i.e. a "straight tokamak"),

$$\frac{\Delta t_{SI}}{\Delta t_{CFL}} = 15 \, \frac{R_0}{a} \, \frac{N}{100} \, . \tag{19.158}$$

This means that the ratio is proportional to the aspect ratio of the cylinder and to the number of radial grid points used in the discretization. Indeed, the explicit time step size depends on this number because the fast magneto-sonic wave propagates in the radial direction, but the semi-implicit scheme treats these waves implicitly so that the gain is larger for higher radial resolutions. A high resolution run with $N = 1000$ yields a 1500 times larger time step for the semi-implicit scheme for an aspect ratio of 10.

The semi-implicit method is widely used in MHD and below we briefly discuss some successful MHD applications. Harned and Schnack [215] adjusted the semi-implicit term in order to treat both the fast magneto-sonic waves and the Alfvén waves implicitly. Their scheme is more complicated but allows even larger time steps as the shear Alfvén time step constraint is eliminated as well. Lerbinger and Luciani [301] went even further and used the full linearized MHD force operator in the semi-implicit term. As a result, the time step in their scheme is limited only by nonlinear physical phenomena and they show results on nonlinear tearing modes obtained with time steps as large as 10 Alfvén crossing times in the linear phase and 2 Alfvén transit times in the nonlinear saturated phase. Keppens *et al.* [262, 263] applied the semi-implicit term in both the predictor and corrector steps of the scheme and combined it with a pseudo-spectral discretization in a data-parallel implementation running on different platforms.

Example application: resonant Alfvén wave heating In Chapter 11 [1], we discussed the phenomenon of resonant absorption in the context of linear MHD. We remarked that the assumption of small amplitude perturbations may not be valid in the resonance layer itself when the model is applied to some very hot plasmas of interest, e.g. in the solar corona and in tokamaks. The high temperatures in

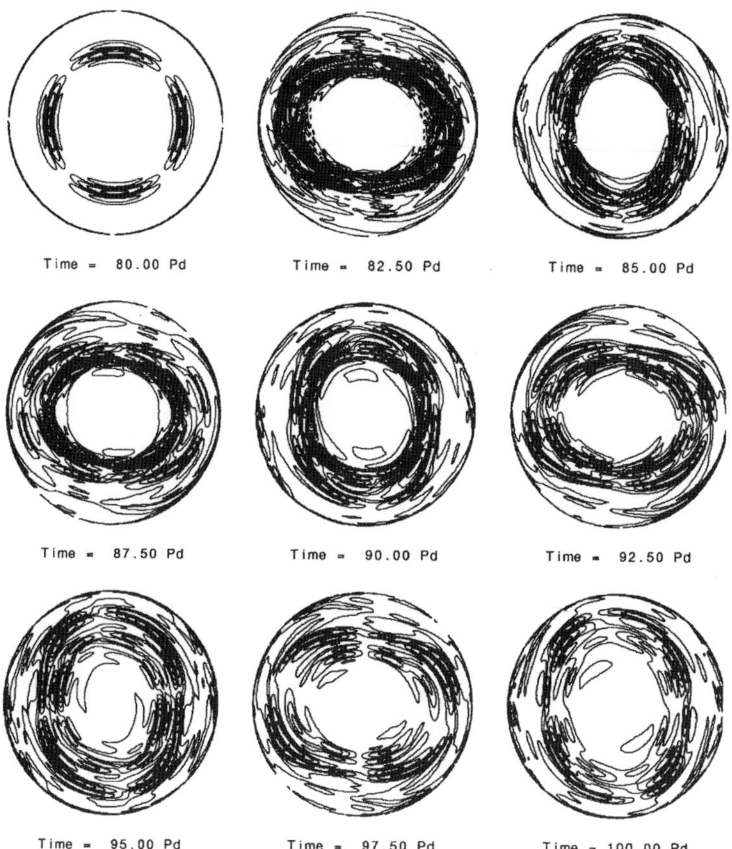

Fig. 19.30 Contour plots of snapshots of the current density for a nonlinear continuation of the linear steady state of a "straight line-tied coronal loop"; times given in units of the period $P_{\rm d}$ of the driving frequency. (From Poedts *et al.* [380].)

these plasmas in combination with relatively long length scales give rise to extremely high magnetic Reynolds numbers. As a result, the plasma response to the external periodic driver is close to the ideal plasma behavior, which is *singular*. In other words, the large amplitudes in the resonant layer(s) probably induce nonlinear effects. Poedts and Goedbloed [375] showed that the dynamics in the resonant layer is very nonlinear indeed. The semi-implicit nonlinear continuation of a linear stationary state shown in Fig. 19.30 gives a substantially different behavior from the linear MHD picture. The periodic shear flow in the dissipative layer turned out to be Kelvin–Helmholtz unstable, as predicted by Heyvaerts and Priest [225], and found previously in more simplified slab loop simulations in Ofman and Davila [355]. This instability deforms the resonant layers, but for the parameter domain studied, the instability is not violent enough to destroy these layers in half a driving period (the stationary flow is periodic). Therefore, the heating mechanism seems to survive in the nonlinear picture although the efficiency, time scales and heat deposition profiles are very different from the linear MHD results. This is because the background magnetic field and plasma density change in response to the heat deposition so that the resonance positions vary in time in the nonlinear picture. Clearly, the driven system does not evolve to a stationary state in nonlinear MHD.

19.4.3 Simulating ideal and resistive instability developments

The semi-implicit method is often used in MHD to determine the linear and non-linear growth rates and to study the nonlinear behavior and saturation of ideal and resistive instabilities. The method has been applied to the study of so-called "relaxation" instabilities, which occur periodically and each time transfer energy and plasma particles from one part of the plasma to another or even out of the plasma. There are two known relaxation instabilities: "sawtooth" instabilities, which affect only the plasma center and "edge localized modes" (ELMs) which only occur at the plasma surface. The latter have been discussed in Chapter 17 in the context of linear MHD. Sawtooth instabilities have been simulated [48, 71] using a semi-implicit scheme very similar to the original one of Harned and Kerner [214].

An exemplary challenging calculation involves the nonlinear saturation of the ideal internal kink mode, predicted analytically by Rosenbluth *et al.* [396]. The $m = 1$ internal kink mode develops nonlinearly into a helical equilibrium state with a singular current sheet on the $q = 1$ surface. This saturated nonlinear ideal equilibrium state turns out to be unstable when finite resistivity is taken into account. The singular current sheet, in which the magnetic field changes sign, is unstable in a resistive plasma and the magnetic field lines reconnect in this layer. Kadomtsev derived the time scale for this reconnection process and showed that it is proportional to $\eta^{-1/2}$. Numerical simulations [269] confirmed this property of Kadomtsev's model for sawtooth instabilities by semi-implicitly evolving the nonlinear ideal equilibrium in resistive MHD (Fig. 19.31).

The semi-implicit method is also used to study the nonlinear growth and saturation of ideal and resistive kink instabilities of line-tied coronal loops [26, 312]. Such loops become kink unstable when the twist exceeds a critical value depending on the radial profile and the loop length. The nonlinear evolution of such $m = 1$ kink modes is similar to their equivalents in tokamaks but different due to the stabilizing anchoring of the foot points of the loops in the dense photosphere. Here too current sheets develop in which magnetic reconnection occurs and releases part of the magnetic energy. This released energy may be dissipated and contribute to

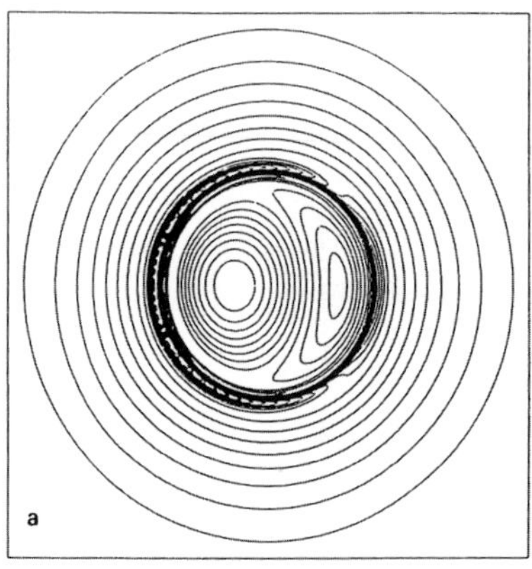 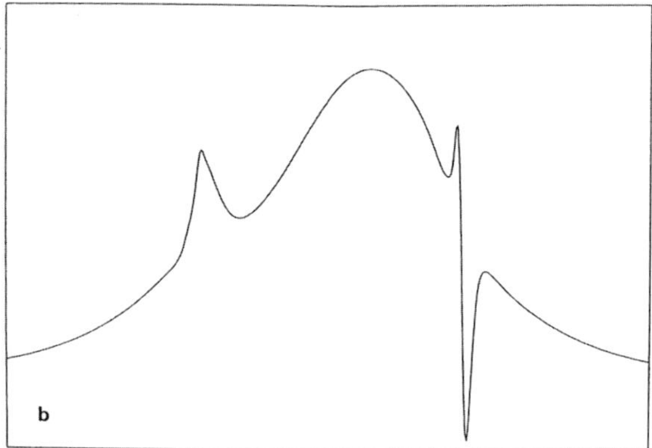

Fig. 19.31 (a) Contour plot of the magnetic flux and (b) current density as a function of radius through the symmetry plane for a nonlinearly saturated state of the $m = 1$ internal kink mode in a "straight tokamak". There is a current sheet at the rational surface $q = 1$. (From Kerner [269].)

the observed heating of the loops, providing an alternative for the wave heating mechanism discussed in the previous subsection.

19.4.4 Global simulations for tokamak plasmas

As a final example where nonlinear MHD simulations, using semi-implicit algorithms, are heavily used, we discuss a state-of-the-art study from laboratory fusion context. A US-based consortium, the NIMROD team (non-ideal MHD with rotation, open discussion), has progressed to simulating long-term plasma dynamics in global tokamak geometry. In NIMROD [415], the 3D toroidal geometry is discretized using a 2D FEM representation in the poloidal cross-section (as was found optimal for handling also the equilibrium problem for axi-symmetric tokamak plasmas, see Chapter 16), with a Fourier handling of the toroidal direction. The Galerkin method is used, whereby nonlinear terms which depend on φ are handled pseudo-spectrally in the periodic toroidal direction, using FFTs. The major challenge is to account for the wide range in time scales of interest (as already discussed earlier in this chapter), as well as for the extreme and anisotropic properties of high temperature, magnetized plasmas. The latter implies that one wishes to compute plasma evolution over resistive time scales while facing magnetic Reynolds numbers of order 10^9 or higher, in conjunction with thermal conduction processes qualified by ratios $\kappa_\parallel / \kappa_\perp \approx 10^9$. In order to handle some even more affordable numerical values, the temporal discretization must use implicit

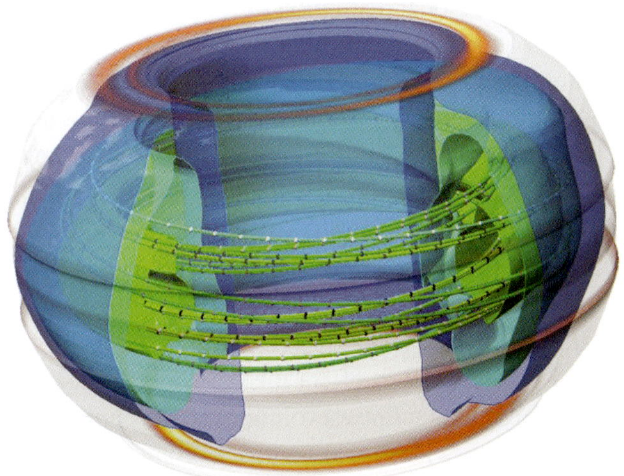

Fig. 19.32 Visualizing temperature iso-surfaces, various magnetic field lines colored by the local temperature, and the heat load on the wall (orange color scale), this 3D MHD simulation models a disruption which occurred in shot # 87009 of the DIII-D tokamak. (An animation is available at `nimrodteam.org`.)

strategies. A modification akin to the semi-implicit strategies described earlier is made to the momentum equation, involving the full self-adjoint linear ideal MHD force operator about a steady state. In fact, the nonlinear equations are also written in a manner which accounts for this strong background equilibrium field, reminiscent of the $\mathbf{B}_0 + \mathbf{B}_1$ strategy introduced by Tanaka [431] for shock-capturing methods. The $\nabla \cdot \mathbf{B} = 0$ constraint is handled by introducing a diffusive source term to the induction equation, recognized as the parabolic cleaning strategy mentioned in Section 19.3.1. The semi-implicit operator is modified in a manner which nearly guarantees no numerical dissipation for linear stable modes, while unstable modes will grow at their physically correct growth rates, if the employed time step is suitably restricted. In addition to this, physical dissipation terms (resistivity, anisotropic thermal conduction, viscosity) are handled fully implicitly.

The main strength of the NIMROD strategy is its demonstrated ability to solve for long term non-ideal evolution in global tokamak plasmas. The order of the FEM used in the poloidal discretization must be higher than 2, but provided such second or higher order elements are used, several stringent tests for tearing modes, and extreme anisotropic heat conduction conditions were demonstrated successfully in [415]. Figure 19.32 shows a snapshot of a simulation where a disruption (shot # 87009) in the DIII-D tokamak was simulated from beginning to end. The initial MHD equilibrium was found unstable to a helical (2/1 in poloidal/toroidal mode numbers) deformation with a prominent magnetic island structure. The instability grows and leads to a global disruption of the tokamak plasma. Figure 19.32 shows an impression of the plasma at the time of maximal heat load to the tokamak walls. The field lines are colored by local temperature and iso-surfaces of temperature show significant deformation from their initial nested tori configuration. The heat flux on the wall is also indicated: the reddish features at top and bottom indicate where most heat is deposited at this instant. In an animation of the disruption, one can follow the sudden loss of equilibrium, and see how eventually a single field line connects top and bottom areas of maximal heat load on the tokamak walls. The plasma as a whole cools rapidly: one obviously wants to avoid this kind of global disruption in actual fusion plasmas. Global tokamak modeling will thereby serve a crucial role: it offers the only way to quantify macroscopic plasma properties throughout nonlinear evolutions in the entire tokamak geometry. Combined with detailed knowledge and feedback from MHD equilibrium reconstructions and linear MHD spectral codes, one may hope to devise strategies for controlling self-burning fusion plasmas. Eventually, fluid models of tokamak plasmas can be expected to be replaced by more elaborate fluid-like treatments which gradually incorporate more and more kinetic effects. This requires detailed knowledge of the temporal and spatial scales inherent in the more elaborate closures, such that suitable numerical strategies can be tailored to the challenge. An up-to-date review of these issues for a hierarchy of ever more sophisticated fluid models can be found in [403], where various "extended MHD" models are discussed.

19.5 Literature and exercises

Notes on literature

Numerical methods

– Many excellent books cover computational hydrodynamics, e.g. Wesseling, *Principles of Computational Fluid Dynamics* [479]. Riemann solver based methods for gas dynamics are thoroughly discussed in Toro, *Riemann Solvers and Numerical Methods for Fluid Dynamics* [436]. Their application to MHD was pioneered in the paper by Brio and Wu, 'An upwind differencing scheme for the equations of ideal magnetohydrodynamics' [70].

– The lecture notes for the 1997 Saas Fee Advanced Course on *Computational Methods for Astrophysical Fluid Flow* [304] include material on finite volume methods for conservation laws, with a section on MHD, by LeVeque. The monograph by Murawski, *Analytical and Numerical Methods for Wave Propagation in Fluid Media* [343], also discusses aspects of modern shock-capturing schemes for the MHD system.

– Introductions to the very broad topics of computational plasma physics and computational astrophysics, with sections on numerical MHD, are the books by Tajima, *Computational Plasma Physics* [430] and Bodenheimer *et al.*, *Numerical Methods in Astrophysics* [54].

Exercises

[19.1] *Conservative versus primitive variable formulations for iso-thermal hydro*

Consider the 1D iso-thermal hydrodynamic equations given by

$$\frac{\partial \rho}{\partial t} + \frac{\partial (\rho v)}{\partial x} = 0,$$

$$\frac{\partial m}{\partial t} + \frac{\partial (mv + p)}{\partial x} = 0.$$

These express conservation of mass in terms of the density ρ, and momentum $m = \rho v$ conservation with velocity v. For an iso-thermal gas of uniform temperature, the gas pressure is directly proportional to the density, i.e. $p = c_i^2 \rho$, where c_i denotes the constant iso-thermal sound speed. We can alternatively write the governing equations in the form (19.3) in terms of primitive variables $\mathbf{V} = (\rho, v)^{\mathrm{T}}$.

– Compute the flux Jacobian $\mathbf{F_U}$ and verify that

$$\mathbf{F_U} = \begin{pmatrix} 0 & 1 \\ c_i^2 - m^2/\rho^2 & 2m/\rho \end{pmatrix}, \quad \mathbf{W}(\mathbf{V}) = \begin{pmatrix} v & \rho \\ c_i^2/\rho & v \end{pmatrix}, \quad \begin{cases} \lambda_1 = v - c_i \\ \lambda_2 = v + c_i \end{cases},$$

$$\mathbf{R} = \begin{pmatrix} 1 & 1 \\ v - c_i & v + c_i \end{pmatrix}, \quad \mathbf{R}^{-1} = \begin{pmatrix} \frac{1}{2}\dfrac{v + c_i}{c_i} & -\dfrac{1}{2c_i} \\ -\frac{1}{2}\dfrac{v - c_i}{c_i} & \dfrac{1}{2c_i} \end{pmatrix}, \quad \hat{\mathbf{R}} = \begin{pmatrix} 1 & 1 \\ -\dfrac{c_i}{\rho} & \dfrac{c_i}{\rho} \end{pmatrix},$$

$$\hat{\mathbf{R}}^{-1} = \begin{pmatrix} \frac{1}{2} & -\frac{1}{2}\rho/c_i \\ \frac{1}{2} & \frac{1}{2}\rho/c_i \end{pmatrix}, \quad \tilde{\mathbf{R}} = \begin{pmatrix} v - c_i \ln \rho \\ v + c_i \ln \rho \end{pmatrix}.$$

– Note that the equations for the Riemann invariants (generally found from Eq.(19.8)) can be manipulated to the following characteristic equations, as also found from the generalized Riemann invariants construction (19.10),

$$dv \pm (c_i/\rho)d\rho = 0, \quad \text{along curves } dx/dt = v \mp c_i.$$

In these equations, $dv = (\partial v/\partial t)dt + (\partial v/\partial x)dx$ denotes the total velocity change.

[19.2] *Linear hyperbolic systems*

Consider the 2×2 linear system

$$\begin{cases} u_t + 0.5u_x + w_x = 0, \\ w_t - 1.25u_x + 3.5w_x = 0. \end{cases}$$

Show that this system is strictly hyperbolic. Compute its eigenvalues and the left and right eigenvectors. Calculate the characteristic variables and verify that their evolution is governed by constant coefficient linear advection equations. Compute the analytic solution for the particular initial data $(u, w) = (4, 1)$ for $x < 0$ and $(u, w) = (2, 100)$ for $x > 0$.

[19.3] *Shallow water magnetohydrodynamics*

An interesting set of nonlinear conservation laws, known as the "shallow water" magnetohydrodynamic equations, is given by

$$\frac{\partial}{\partial t}\begin{pmatrix} h \\ h\mathbf{v} \\ h\mathbf{B} \end{pmatrix} + \nabla \cdot \begin{pmatrix} h\mathbf{v} \\ h\mathbf{v}\mathbf{v} - h\mathbf{B}\mathbf{B} + \frac{1}{2}gh^2\mathbf{I} \\ h\mathbf{v}\mathbf{B} - h\mathbf{B}\mathbf{v} \end{pmatrix} = 0.$$

These equations govern the two-dimensional motion of a narrow layer of local height $h(x,y)$ of conducting fluid, with purely planar $\mathbf{v} = (v_x, v_y)$ and $\mathbf{B} = (B_x, B_y)$. The magnetic field divergence constraint complements this with $\nabla \cdot (h\mathbf{B}) = 0$. The parameter g represents a constant gravitational acceleration, perpendicular to the (x,y) plane.

- Compute the flux Jacobian of this five component system. Derive an equivalent quasi-linear form in terms of primitive variables $\mathbf{V} = (h, v_x, v_y, B_x, B_y)^{\mathrm{T}}$.

- Compute all the characteristic speeds, and the left and right eigenvectors for both primitive and conservative variable formulations.

- Try to derive generalized Riemann invariants, and compute the structure coefficients for the five wave modes. Discuss the Rankine–Hugoniot relations and the equations governing (centered) simple waves.

You can verify your results and find physical explanations for the occurring wave modes in this system in [111].

[19.4] *FCT for Burgers' equation*

Implement and test the FCT method for numerically simulating the solution to the inviscid Burgers equation (19.28) for a given initial triangular pulse (19.29). Quantify your result against the correct analytic solution given by Eq. (19.31) for different resolutions.

[19.5] *A Roe scheme for iso-thermal MHD*

Construct a Roe-type Riemann solver based scheme for the 1.5D iso-thermal MHD system. This can then be compared to the algorithm presented in [277], where a TVD scheme of Roe-type was exploited for multi-dimensional iso-thermal MHD. Expressions for suitably scaled eigenvectors are given there as well, but note the difference between our coefficients α_{f}^2, α_{s}^2 and theirs.

20

Transonic MHD flows and shocks

20.1 Transonic MHD flows

20.1.1 Flow in laboratory and astrophysical plasmas

We started the study of the effects of background flow on waves and instabilities of laboratory and astrophysical plasmas in Chapter 12. We also considered the modifications of the equilibrium caused by the flow. These modifications were rather trivial for plane shear flows, but considerable for rotating plasmas due to centrifugal forces. However, except for the forebodings of Chapter 18, the most substantial effects have not been faced yet. The adjective "substantial" on background flows obviously should refer to some standard on what is a sizeable velocity. For transonic gas dynamics, it is clear that the appropriate standard velocity is the sound speed. For the macroscopic description of plasmas, which incorporates the dynamics of ordinary gases, the three MHD speeds (slow, Alfvén and fast) collectively take over the role of the sound speed. This implies that trans"sonic" MHD flows will be characterized by different flow regimes depending on the speed of the background flow relative to those three MHD speeds. In addition, the relative direction of the background velocity, $\mathbf{v}_0$, with respect to the direction of the background magnetic field, $\mathbf{B}_0$, introduces an anisotropy in plasma dynamics that is not present in ordinary gas dynamics.

Because of the important implications indicated above, the theory of MHD flows is being extensively investigated at present, both in laboratory and in astrophysical contexts. In the linear analysis, the usual approach of a split in background equilibrium and perturbations by waves and instabilities leads to substantial difficulties since these two topics become intermingled to some extent for transonic flows [169, 171], as discussed in Section 18.2.2. This was found out when a dual set of numerical codes FINESSE–PHOENIX [29, 173, 52] was developed, where the first code computes transonic equilibria and the second one the waves and instabilities of such equilibria. For the computation of transonic equilibria, the usual

487

assumption of ellipticity breaks down, with concomitant complexity of bookkeeping of solutions, whereas the appearance of new overstable modes due to transonic transitions (see Section 18.4.3) indicates that an enormous field of research opens up when the interaction of transonic flows and waves is explored. This may imply that, eventually, nonlinear analysis of the complete dynamics (without making the split in equilibrium and perturbations) is the more appropriate approach. This will necessarily involve *large-scale computing* with the tools of computational nonlinear MHD, expounded in Chapter 19. However, at present, the deep relationship between linear instabilities in transonic MHD flows and the resulting nonlinear 3D dynamics has hardly been investigated, let alone properly understood. Hence, let us now turn to the most basic part of magnetohydrodynamics, leaving the techniques of linearization and of a split in equilibrium and stability, and of solving differential equations, and just consider what effects nonlinearity might have on a single point in the flow. To our surprise, we will find out that, again, a rich structure is there, with some common features with linear dynamics that have not been investigated, even today. MHD never disappoints!

20.1.2 Characteristics in space and time

Recall the discussion of the characteristics of ideal MHD in Volume [1], Section 5.4.3, Eqs. (5.116)–(5.119). That discussion mostly centered on waves and instabilities of static equilibria, as illustrated by the Friedrichs diagrams of Fig 5.13 for the propagation of plane waves and point disturbances in plasmas without background flow. The generalization to *plasmas with background flow* (stationary equilibria) of Chapters 12 and 13, leads to *seven characteristic speeds* of the flow:

$$u_{\rm E} = v_{\rm n} \equiv \mathbf{n} \cdot \mathbf{v}\,,$$

$$u_{\rm s}^{\pm} = v_{\rm n} \pm v_{\rm sn}\,, \qquad v_{\rm sn} \equiv \left[\frac{1}{2\rho}\left(\gamma p + B^2 - \sqrt{(\gamma p + B^2)^2 - 4\gamma p B_{\rm n}^2}\right)\right]^{1/2}\,,$$

$$u_{\rm A}^{\pm} = v_{\rm n} \pm v_{\rm An}\,, \qquad v_{\rm An} \equiv B_{\rm n}/\sqrt{\rho}\,, \qquad B_{\rm n} \equiv \mathbf{n} \cdot \mathbf{B}\,,$$

$$u_{\rm f}^{\pm} = v_{\rm n} \pm v_{\rm fn}\,, \qquad v_{\rm fn} \equiv \left[\frac{1}{2\rho}\left(\gamma p + B^2 + \sqrt{(\gamma p + B^2)^2 - 4\gamma p B_{\rm n}^2}\right)\right]^{1/2}\,.$$

$$(20.1)$$

By varying the direction of the normal $\mathbf{n}$, this yields the space-time manifolds along which perturbations propagate. For one spatial dimension, they were shown in Fig. 5.14 [1], which we here repeat for convenience (Fig. 20.1).

Permitting two spatial dimensions, the temporal snapshots of the three MHD perturbations become the well-known figures of the Friedrichs group diagram. These figures may exhibit an interesting new feature, depending on the magnitude of the

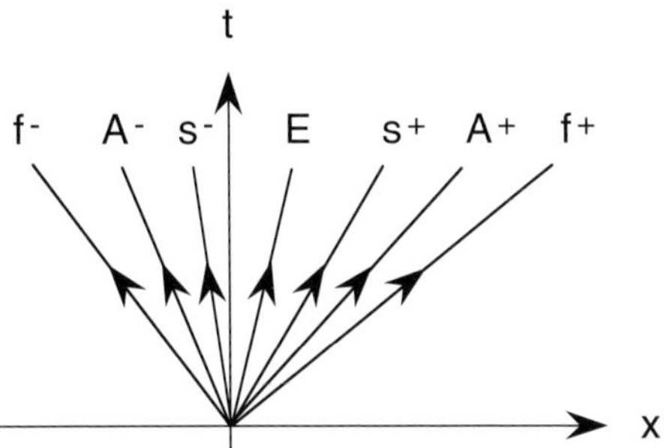

Fig. 20.1 Space-time characteristics of the three MHD waves ($s^\pm$, $A^\pm$, $f^\pm$), traveling in forward and backward directions, and the entropy disturbances (E), which are just carried with the plasma flow.

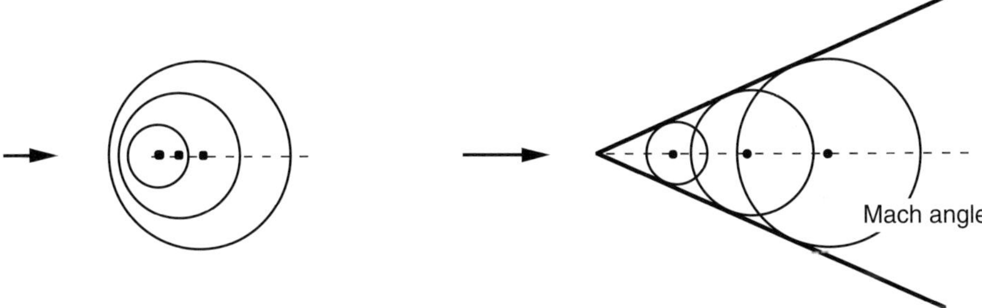

Fig. 20.2 Sound in (a) subsonic and (b) supersonic gas flow about a point source.

background flow. This is illustrated in Fig. 20.2 for the case of sound waves in ordinary fluids: when the flow velocity becomes supersonic, the spatial part of the characteristics forms envelopes where information accumulates and discontinuous solutions, *or shocks*, are formed. Whereas in subsonic flows the solutions propagate everywhere in space, in supersonic flows these discontinuities separate space in regions where the solutions propagate (*hyperbolic* regions) and regions where they do not propagate (*elliptic* regions).

One of the deep problems in transonic flows is that the transitions from ellipticity to hyperbolicity occur at locations that are not known a priori. Those locations are part of the solution of the problem. In MHD, we will use the term "transonic" as well to indicate flows that surpass one of the three characteristic speeds v_A, v_s, v_f defined in Eqs. (20.1). The theory of MHD shocks, associated with those transonic transitions, will be discussed in Sections 20.2 and 20.3. We will present the subject from a new angle, time reversal duality (Section 20.3.2), and we will show how this connects with the classification of MHD shocks in terms of converging characteristics, with a central role for the intermediate shocks; see Wu [489].

Another fundamental problem associated with the transonic transitions is the construction of the stationary equilibrium state itself. So far, the tacit assumption for the basic equilibrium state, providing the space in which the waves and instabilities "live" (see Section 17.1), has been that the governing nonlinear partial differential equation (for axi-symmetric equilibria, the Grad–Shafranov equation discussed in Chapter 16) is *elliptic*. The numerical techniques exploited need this property. In fact, all of the standard methods in use in MHD spectral analysis are based on the assumption that the equilibria are described by elliptic equations and the perturbations by hyperbolic ones. However, when the poloidal flow velocity increases beyond certain critical values, the stationary equilibrium equations become hyperbolic and both the classical paradigm of a split in equilibrium and perturbations and the numerical techniques based on it break down. As a result, the standard equilibrium solvers, as used in tokamak computations, diverge and we need to rethink the problem completely. It should be noted that we have evaded this problem in Chapter 18 by restricting the analysis to the elliptic flow regimes. As we have seen, in contrast to ordinary fluid dynamics, in MHD such "transonic" flow regimes exist.

The fundamental reason of the bankruptcy of the classical paradigm of equilibrium and perturbations is associated with the Lagrangian time derivative $D/Dt \equiv \partial/\partial t + \mathbf{v} \cdot \nabla$ in the MHD equations. Whereas, the Eulerian time derivative $\partial/\partial t$ produces the eigenfrequencies ω of the waves, the spatial derivative $\mathbf{v} \cdot \nabla$ produces not only the Doppler shifts of the perturbations but also the possibility of spatial discontinuities of the equilibria. However, since the two pieces of the Lagrangian time derivative really belong together, *the waves and the stationary equilibria, with transitions from ellipticity to hyperbolicity, are not separate issues.* This is the fundamental reason for the difficulties mentioned in the previous section in the study of stability of transonic MHD flows.

20.2 Shock conditions

One of the most striking physical phenomena in transonic flows is the occurrence of shocks, where virtually all physical variables are discontinuous. We have encountered discontinuities at plasma–plasma and plasma–vacuum interfaces for static and subsonic stationary flows in Chapter 4 [1] when discussing the different model problems of laboratory and astrophysical plasmas. In contrast, in the theory of transonic flows, one of the essential problems is the occurrence of *internal surfaces of discontinuity* where the same jump conditions provide boundary conditions that are to be imposed at the positions of the shock fronts, which are a priori unknown, to be determined together with the solution. In the present chapter, we just develop the subject of shock conditions so far that they could be implemented in one of the general Riemann solvers discussed in Chapter 19. Although straightforward in principle, this involves a surprising amount of algebra.

The jump conditions for the density, the velocity, the pressure and the magnetic field were derived from the MHD conservation equations in Volume [1], Section 4.5, Eqs. (4.151)–(4.156). The basic "trick" consisted of integrating the evolution equations across a thin layer, of thickness δ, in which nonlinear and dissipative effects smooth the discontinuities to large, but finite, variations in the normal direction, and then taking the limit $\delta \to 0$. This turns the partial differential equation into algebraic relations between the variables on the two sides of the shock. In a sense, this procedure translates the basic nonlinearities of the MHD equations into algebraic relations at just a single point. The jumps are denoted by the notation

$$[\![f]\!] \equiv f_1 - f_2 \,, \tag{20.2}$$

where f_1 indicates the value of a physical variable in the undisturbed, *upstream,* part of the fluid in front of the shock, and f_2 indicates the value in the shocked, *downstream,* part of the fluid behind the shock.

For convenience, we repeat this general form of the *jump conditions in the shock frame,* in which the shock is stationary and the fluid velocities $\mathbf{v}'$ are evaluated with respect to that frame. We only change the order of the equations to facilitate the reductions to the final shock relations of Section 20.3:

$$[\![\rho v'_n]\!] = 0 \qquad\qquad (mass)\,, \tag{20.3}$$

$$[\![B_n]\!] = 0 \qquad\qquad (normal\ flux)\,, \tag{20.4}$$

$$\rho v'_n [\![\mathbf{v}'_t]\!] = B_n [\![\mathbf{B}_t]\!] \qquad\qquad (tangential\ momentum)\,, \tag{20.5}$$

$$\rho v'_n [\![\mathbf{B}_t/\rho]\!] = B_n [\![\mathbf{v}'_t]\!] \qquad\qquad (tangential\ magnetic\ flux)\,, \tag{20.6}$$

$$[\![\rho v'^{2}_n + p + \tfrac{1}{2}B^2_t]\!] = 0 \qquad\qquad (normal\ momentum)\,, \tag{20.7}$$

$$\rho v'_n [\![\tfrac{1}{2}(v'^{2}_n + v'^{2}_t) + e + p/\rho + B^2_t/\rho]\!] = B_n [\![\mathbf{v}'_t \cdot \mathbf{B}_t]\!] \quad (energy)\,, \tag{20.8}$$

where $e = (\gamma - 1)^{-1} p/\rho$ for the ideal plasmas that we are concerned about. Recall that the momentum and magnetic flux equations have been projected in the directions normal (subscript n) and tangential (subscript t) to the shock front. As stated in [1], Eq. (4.150), to properly discuss shocks, we need to restrict the possible jumps by demanding that the entropy increases across the shock front:

$$[\![S]\!] \equiv [\![\rho^{-\gamma}p]\!] \le 0 \qquad\qquad (entropy)\,. \tag{20.9}$$

This additional condition models the effects of a thin dissipative layer of thickness δ, where the large gradients of the physical variables become jumps in the limit $\delta \to 0$ and entropy production is the only dissipative effect that remains.

20.2.1 Special case: gas dynamic shocks

Before analyzing the different MHD shocks on the basis of these equations, it is instructive to discuss the ordinary gas dynamic shocks, which are contained as a special case ($\mathbf{B} = 0$). For $\rho v'_\mathrm{n} \neq 0$, Eqs. (20.3)–(20.8) reduce to:

$$[\![\rho v'_\mathrm{n}]\!] = 0 \,, \qquad [\![\mathbf{v}'_\mathrm{t}]\!] = 0 \,, \tag{20.10}$$

$$[\![\rho {v'_\mathrm{n}}^2 + p]\!] = 0 \,, \tag{20.11}$$

$$[\![\tfrac{1}{2}{v'_\mathrm{n}}^2 + e + p/\rho]\!] = 0 \,, \qquad e = \frac{p}{(\gamma - 1)\rho} \,. \tag{20.12}$$

Since $[\![\mathbf{v}'_\mathrm{t}]\!] = 0$ now, we may transform to a coordinate system moving with the tangential flow so that $\mathbf{v}'_\mathrm{t} = 0$. The velocities in the laboratory and shock frames then have the relative magnitudes as illustrated in Fig. 20.3.

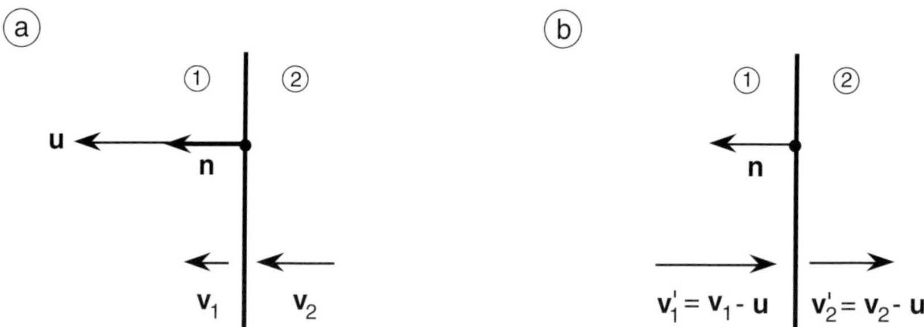

Fig. 20.3 Gas dynamic shock: (a) laboratory frame; (b) shock frame.

Dropping the primes for simplicity of the notation, the shock conditions become

$$\rho_1 v_1 = \rho_2 v_2 \,, \tag{20.13}$$

$$\rho_1 v_1^2 + p_1 = \rho_2 v_2^2 + p_2 \,, \tag{20.14}$$

$$\tfrac{1}{2}v_1^2 + e_1 + p_1/\rho_1 = \tfrac{1}{2}v_2^2 + e_2 + p_2/\rho_2 \,, \qquad e_{1,2} = \frac{p_{1,2}}{(\gamma - 1)\rho_{1,2}} \,. \tag{20.15}$$

These equations determine the values of the three downstream parameters ρ_2, v_2, p_2 in terms of the three upstream parameters ρ_1, v_1, p_1. They are easily reduced to more compact expressions involving only one parameter: *gas dynamic shocks are essentially a one-parameter family.* We will show this by two methods, one following the traditional exposition, as in Landau and Lifschitz, Fluid Mechanics [294], Chapter IX, and another one that anticipates the sequence of steps involved in the analogous, but more complex, reduction for shocks in magnetohydrodynamics.

First, consider the shock conditions from the thermodynamic point of view,[1] not necessarily restricted to ideal gases. Defining the *mass flow through the shock*, $m \equiv \rho v_n$, *the specific volume,* $V \equiv \rho^{-1}$, *internal energy,* e, and *enthalpy (or heat function),* $w \equiv e + pV$, the jump conditions (20.10)–(20.12) become:

$$[\![m]\!] = 0 \,, \tag{20.16}$$

$$m^2 = -[\![p]\!]/[\![V]\!] \,, \tag{20.17}$$

$$[\![\tfrac{1}{2}m^2V^2 + w]\!] = 0 \,, \quad \text{or} \quad e_1 - e_2 + \tfrac{1}{2}(p_1 + p_2)(V_1 - V_2) = 0 \,. \tag{20.18}$$

Since, obviously, $m^2 > 0$, the jump condition (20.17) permits solutions with $p_2 > p_1$ and $V_2 < V_1$ as well as with $p_2 < p_1$ and $V_2 > V_1$. The latter ones are eliminated by the entropy condition (20.9), as will be elaborated below. The second expression (20.18) is called the Hugoniot or shock "adiabatic", not to be

---

[1] Recall the following *definitions and relationships of the thermodynamical variables.* There are four basic thermodynamic parameters, viz. p, V $(\equiv \rho^{-1})$, s and T, of which only two are independent. Consequently, the relationships $s = s(p, V)$ and $T = T(p, V)$ permit the representation of adiabatic (constant s) and isothermal (constant T) processes in a p–V diagram. Characteristic functions like the specific internal energy $e = e(T, V)$ and the specific enthalpy $w = w(T, p)$ permit the consideration of thermodynamic processes at constant volume and constant pressure, with the associated specific heat coefficients,

$$C_V \equiv \partial e/\partial T|_V \,, \qquad C_p \equiv \partial w/\partial T|_p \,. \tag{i}$$

In the formulation of the first law of thermodynamics, the characteristic functions are considered as functions of the entropy, $e = e(s, V)$, $w = w(s, p)$, with

$$de = Tds - pdV \,, \qquad \text{so that} \quad T = \partial e/\partial s|_V \,, \quad p = -\partial e/\partial V|_s \,, \tag{ii}$$

$$dw = Tds + Vdp \,, \qquad \text{so that} \quad T = \partial w/\partial s|_p \,, \quad V = \partial w/\partial p|_s \,. \tag{iii}$$

Partial differentiation of the latter expressions yields the relations of Maxwell, $\partial T/\partial V|_s = \partial p/\partial s|_V$, etc.

For *the special case of ideal gases,* the coefficients C_p and C_V and, hence, their ratio $\gamma \equiv C_p/C_V$, are constant throughout the medium. The equation of state

$$pV = \mathcal{R}T \,, \qquad \mathcal{R} = C_p - C_V = (\gamma - 1)C_V \,, \tag{iv}$$

involving the gas constant $\mathcal{R}$ $(\approx (1 + Z)k/m_i$ for plasmas $)$, then implies that the internal energy and the enthalpy are simply proportional to the temperature:

$$e \equiv \frac{1}{\gamma - 1}pV = C_V T \,, \qquad w \equiv e + pV = \frac{\gamma}{\gamma - 1}pV = C_p T \,, \tag{v}$$

in agreement with (i). Exploiting these relations, integration of (ii) yields the expression for the specific entropy for ideal gases:

$$s = C_V \ln S + \text{const} \,, \qquad S \equiv pV^\gamma \,, \tag{vi}$$

where, by the choice of the initial conditions [294], the value of s (or S) is constant throughout the medium (of course, except for the jumps at the shock fronts).

confused with the ordinary Poisson adiabatics that relate p and V for genuine adiabatic processes, with constant entropy.

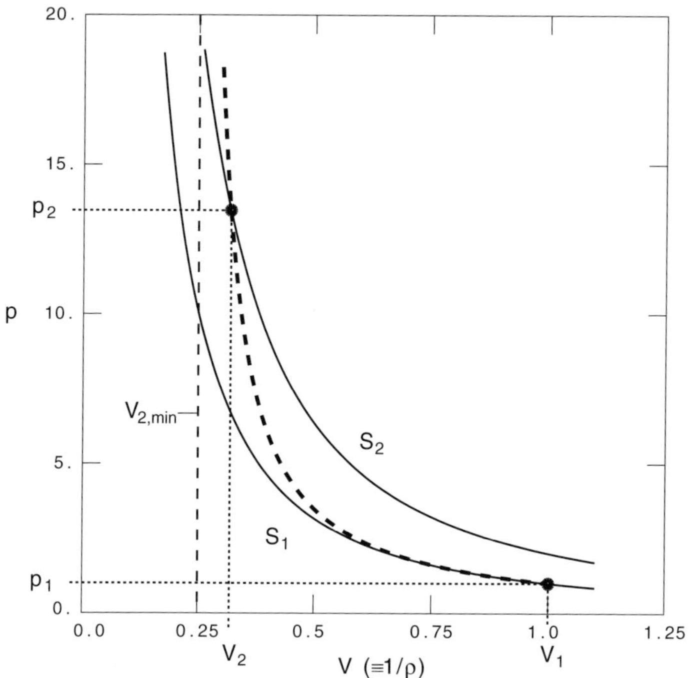

Fig. 20.4 Hugoniot adiabatic (dashed) connecting the upstream and downstream states on the Poisson adiabatics (S_1 and S_2). Maximum compression is indicated by the vertical dashed asymptote labeled $V_{2,\mathrm{min}}$.

As shown in Fig. 20.4 for the case of ideal gases, the Hugoniot adiabatic can also be represented by a curve in the p–V diagram, viz. the curve which connects the representative upstream point p_1, V_1 to all possible downstream points p_2, V_2 for the different values of the parameter m. According to Eq. (20.17), the slope of the chord connecting these points is just $-m^2$. The explicit expression of the Hugoniot adiabatic for ideal gases is obtained from the expression (20.18) by substituting the equation of state $e = pV/(\gamma - 1)$:

$$\frac{p_2}{p_1} = \frac{\gamma + 1 - (\gamma - 1)\, V_2/V_1}{(\gamma + 1)\, V_2/V_1 - \gamma + 1}. \tag{20.19}$$

This relation is exhibited in Fig. 20.4 (dashed line) together with the two Poisson adiabatics (labeled S_1 and S_2) for the upstream and downstream states. Notice the distinct difference between the Hugoniot and the Poisson adiabatic: the Hugoniot adiabatic "nestles" against the upstream Poisson adiabatic S_1 at p_1, V_1 (where the values of the two functions, and their first and second derivatives are all equal) but it intersects the downstream Poisson adiabatic S_2 at p_2, V_2. This illustrates an important property of shocks: due to the nonlinearity, they are not additive. If a point on a Hugoniot adiabatic is taken as an upstream state for a second shock, the corresponding Hugoniot adiabatic will be different from the first one and the new shock will overtake the old one.

Introducing the *shock strength*, $\sigma \equiv 1 - V_2/V_1$ $(\equiv \rho_1/\rho_2)$, which theoretically ranges from 0 for weak or infinitesimal shocks $(V_2 = V_1)$ to 1 for infinite compression $(V_2 = 0)$, it is noted from Eq. (20.19) that there is actually a limit to the amount of compression, viz. $(V_2/V_1)_{\min} = (\gamma - 1)/(\gamma + 1)$ $(= 1/4$ for $\gamma = 5/3)$ where $p_2 \to \infty$. This is indicated by the vertical asymptote to the Hugoniot adiabatic in Fig. 20.4 which corresponds to maximum shock strength, $\sigma \to \sigma_{\max} \equiv 1 - (V_2/V_1)_{\min} \equiv 2/(\gamma + 1)$ $(= 3/4$ for $\gamma = 5/3)$.

In our second approach, we stress the central role of sound waves in the formation of gas dynamic shocks. This is most conveniently expressed by means of the squared upstream and downstream *Mach numbers*,

$$M_1^2 \equiv \frac{v_1^2}{v_{s,1}^2} \equiv \frac{\rho_1 v_1^2}{\gamma p_1}, \qquad M_2^2 \equiv \frac{v_2^2}{v_{s,2}^2} \equiv \frac{\rho_2 v_2^2}{\gamma p_2}. \tag{20.20}$$

The mass and momentum conservation relations (20.13) and (20.14) then yield the following ratios of the primitive variables on the two sides of the shock:

$$\frac{\rho_2}{\rho_1} = \frac{v_1}{v_2} = \frac{M_1^2(\gamma M_2^2 + 1)}{M_2^2(\gamma M_1^2 + 1)} \quad \left[= \frac{(\gamma + 1)M_1^2}{(\gamma - 1)M_1^2 + 2} \right], \tag{20.21}$$

$$\frac{p_2}{p_1} = \frac{\gamma M_1^2 + 1}{\gamma M_2^2 + 1} \quad \left[= \frac{2\gamma M_1^2 - \gamma + 1}{\gamma + 1} \right]. \tag{20.22}$$

In addition to the trivial solution without jumps $(M_2^2 = M_1^2)$, the energy conservation relation (20.15) yields the *distilled energy jump condition*, relating the downstream to the upstream Mach number:

$$M_2^2 = \frac{(\gamma - 1)M_1^2 + 2}{2\gamma M_1^2 - \gamma + 1}. \tag{20.23}$$

Substituting this condition, Eqs. (20.21) and (20.22) reduce to the expressions in square brackets.

We now exploit the principle of *scale-independence* of the MHD equations, introduced in Section 4.1.2 [1], which also applies to the equations of hydrodynamics (HD). This principle implies that in any ideal MHD, or HD, problem three parameters can be found that should not be considered as free since they just fix the scales of length, time and mass, which can be eliminated by defining appropriate dimensionless variables. Here, since the problem is essentially time-independent, only

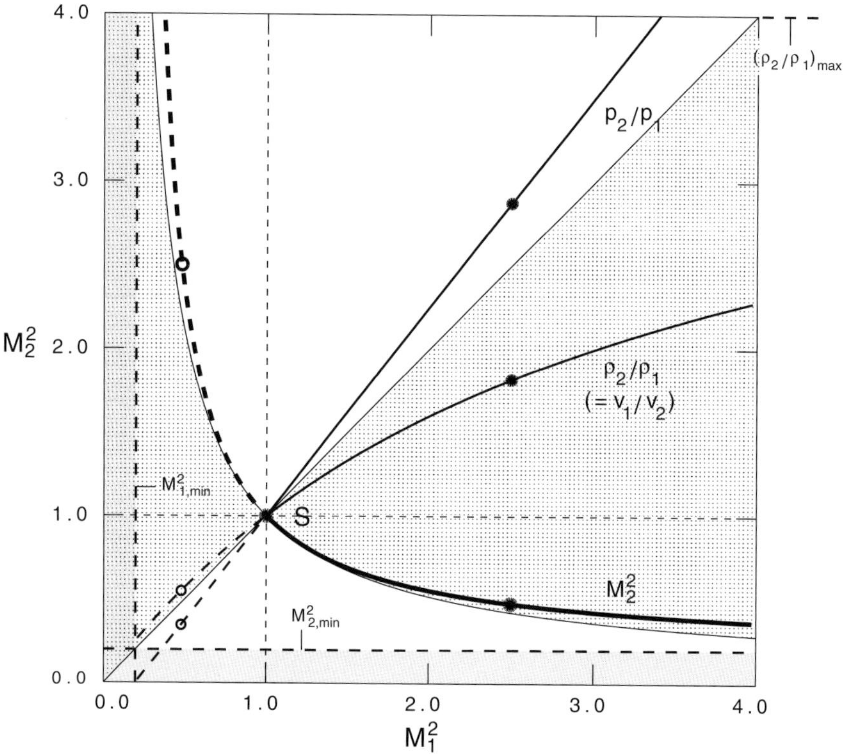

Fig. 20.5 Duality between entropy-forbidden jumps (dashed curve) and entropy-permitted shocks (solid curve) in hydrodynamics. An arbitrary point on the permitted part of the jump curve (filled circle) and its forbidden dual (open circle) are indicated together with corresponding points on the curves for p_2/p_1 and ρ_2/ρ_1. The point S indicates sound waves (HD shocks of infinitesimal strength).

two such scale parameters occur, viz. ρ_1 and p_1. In addition, two parameters have been eliminated by moving to a frame with vanishing tangential velocity. Hence, of the five arbitrary downstream parameters ρ_1, $\mathbf{v}_1$, p_1, only one should be counted as free. This is expressed by the upstream Mach number M_1. The downstream variables are then expressed by the dimensionless parameters ρ_2/ρ_1 and p_2/p_1, determined by the conditions (20.22), and the Mach number M_2 (the dimensionless velocity), determined by the distilled jump condition (20.23). This proves that M_1 is the only controlling parameter for gas dynamic shocks.

The distilled jump condition (20.23) is shown in Fig. 20.5 as a curve in the M_1^2–M_2^2 plane. It appears to be defined for all values $M_{min}^2 \leq M_1^2 < \infty$ and $M_{min}^2 \leq M_2^2 < \infty$, where $M_{min}^2 \equiv \frac{1}{2}(1 - 1/\gamma)$ $(= 1/5$ for $\gamma = 5/3)$. Here, $M_1^2 = M_{min}^2$ marks the boundary of an unphysical domain (indicated in grey) where $p_2 < 0$, and $M_2^2 = M_{min}^2$ marks the boundary of an unphysical domain (also in grey) beyond maximum compression, $(\rho_2/\rho_1)_{\max} \equiv (\gamma + 1)/(\gamma - 1)$ ($= 4$ for $\gamma = 5/3$). However, we still have to implement the entropy condition (20.9), similarly to above, leading to the *distilled entropy condition:*

$$\frac{S_2}{S_1} \equiv \frac{p_2}{p_1}\left(\frac{\rho_1}{\rho_2}\right)^\gamma = \left(\frac{M_2^2}{M_1^2}\right)^\gamma \left(\frac{\gamma M_1^2 + 1}{\gamma M_2^2 + 1}\right)^{\gamma+1} \geq 1 \tag{20.24}$$

$$\left[\Rightarrow \frac{2\gamma M_1^2 - \gamma + 1}{\gamma + 1}\left(\frac{(\gamma - 1)M_1^2 + 2}{(\gamma + 1)M_1^2}\right)^\gamma \geq 1\right].$$

This condition is only satisfied in the dotted areas of the M_1^2–M_2^2 plane of Fig. 20.5. The curve $M_2^2 = M_2^2(M_1^2)$ enters one of these for $M_1^2 \geq 1$, where $M_2^2 \leq 1$, $v_1/v_2 = \rho_2/\rho_1 \geq 1$, and $p_2/p_1 \geq 1$. Hence, entropy-permitted shocks (heavy solid curve) are only obtained for *supersonic upstream flow, corresponding to subsonic downstream flow.* Vice versa, entropy-forbidden jumps (dashed part of the jump curve) are obtained for $M_1^2 < 1$, where $M_2^2 > 1$. Note that this part of the distilled jump curve lies just outside the other dotted area of positive entropy jump, but it touches it at the central sonic point $M_1^2 = M_2^2 = 1$ indicated by S in Fig. 20.5. This point corresponds to sound waves, i.e. shocks of infinitesimal strength ($\sigma \to 0$). It marks the transition from forbidden jumps to permitted shocks. The symmetry of the jump curve with respect to the sonic point has important consequences, as we will show now.

The procedure of replacing entropy conservation of ideal HD or MHD by the dissipative concept of entropy increase severely restricts the permitted dynamics. (For example, the entropy waves introduced in [1], Section 5.2.1, that would occur at $M_1^2 = M_2^2 = 0$, are eliminated since weak entropy shocks would be a contradiction in terms.) However, since entropy increase may be considered as *the arrow of time,* the discarded entropy-forbidden jumps may be turned into physically acceptable solutions by just reversing this arrow, i.e. by reversing the direction of the flow. This does not affect the jump conditions since they do not involve the Mach numbers themselves, only their squares, but the roles of upstream and downstream states are interchanged and what previously was a forbidden jump becomes a permitted shock for the reversed flow, and vice versa. Exploiting the symmetry of the jump curve by defining parameters that measure the "distance" to the sonic point S,

$$\Delta_1 \equiv M_1^2 - 1, \qquad \Delta_2 \equiv M_2^2 - 1, \tag{20.25}$$

the distilled jump condition (20.23) is transformed into the more concise form

$$\Delta_2 = -\frac{\Delta_1}{1 - \Delta_1/\Delta_{\min}}, \qquad \Delta_{\min} \equiv -\tfrac{1}{2}(1 + 1/\gamma), \tag{20.26}$$

where $\Delta_{\min}$ ($= -4/5$ for $\gamma = 5/3$) corresponds to $M_{\min}^2$ defined above. The principle of *time reversal duality between entropy-forbidden jumps and entropy-permitted shocks* is now expressed by the correspondence

$$\Delta_2 = \Delta_2^{\pm}(\Delta_1) \qquad \Longleftrightarrow \qquad \Delta_1 = \Delta_1^{\mp}(\Delta_2)\,, \qquad (20.27)$$

with associated parameter ranges

$$\left\{ \begin{array}{ll} \Delta_2^+ : & 0 \le \Delta_1 < \infty \\ \Delta_2^- : \Delta_{\min} < \Delta_1 \le 0 \end{array} \right. \qquad \Longleftrightarrow \qquad \left\{ \begin{array}{ll} \Delta_1^- : \Delta_{\min} < \Delta_2 \le 0 \\ \Delta_1^+ : & 0 \le \Delta_2 < \infty \end{array} \right. \,.$$

Except that reflected shocks frequently occur in nature, this duality also may be exploited in problems without reflection to express shock conditions in terms of downstream instead of upstream controlling parameters. In HD this involves the simple operation of inverting the distilled jump relation (20.23), but in MHD this extremely useful operation is no longer simple at all, as we will see.

A final, parenthetical, remark: the subject of gas dynamic shocks is a beautiful one, too bad it is stained by the activities of war. Just note the dates of the original contributions to the field as cited in, e.g., Refs. [97], [200], [294] and [405].

20.2.2 *MHD discontinuities without mass flow*

We now discuss MHD discontinuities. Additional possibilities arise, both in HD and in MHD, when there is no mass flow across the discontinuity. In MHD, this gives rise to the contact and tangential discontinuities already introduced in Section 4.5.2[1]. The gas dynamic counterpart was skipped over in Section 20.2.1 since tangential velocities were neglected there, in agreement with the jump conditions (20.10). However, in the *absence of mass flow across the surface of discontinuity,* the original conditions (20.3)–(20.9) also admit the solutions

$$m \equiv \rho v_{\mathrm{n}} = 0 \quad \Rightarrow \quad [\![p]\!] = 0\,, \quad \text{but} \quad [\![\rho]\!] \neq 0\,, \quad [\![\mathbf{v}_{\mathrm{t}}]\!] \neq 0\,. \qquad (20.28)$$

The fluid density and the tangential velocity then display jumps of arbitrary magnitude, as is usual at the *interface between two fluids.*

Introducing a magnetic field, it is clear from the tangential jump conditions (20.5) and (20.6), both involving $\mathbf{v}_{\mathrm{t}}'$ and $\mathbf{B}_{\mathrm{t}}$, that a transformation to $\mathbf{v}_{\mathrm{t}}' = 0$, as illustrated in Fig. 20.3, is generally not possible. The tangential velocity is an essential constituent of MHD discontinuities, both for the present discontinuities without mass flow and for the ones with mass flow discussed in the next section. Recall that the jump conditions (20.3)–(20.9) refer to the shock frame so that the condition for absence of mass flow should be written with a prime: $v_{\mathrm{n}1,2}' \equiv v_{\mathrm{n}1,2} - u = 0$ (Fig. 20.3(b)), whereas $v_{\mathrm{n}1} = v_{\mathrm{n}2} = u$ in the laboratory frame (Fig. 20.3(a)). This implies that these discontinuities are just carried with the fluid flow. As in Section 20.2.1, we will consistently exploit the shock frame but drop the primes.

Two very different kinds of discontinuity without mass flow are obtained depending on whether $\mathbf{B}$ has a component normal to the discontinuity or not. In the former case ($B_n \neq 0$), discontinuity of the tangential velocity is not possible and only the density may jump, so that *contact discontinuities* (paragraph *(a)* below) are more restrictive in plasmas than in ordinary fluids. In the latter case ($B_n = 0$), both ρ and $\mathbf{v}_t$ may jump, as in the gas dynamic case, but continuity of p is replaced by continuity of $p + \frac{1}{2}B^2$ so that *tangential discontinuities* (paragraph *(b)* below) display a much wider variety in plasmas than in ordinary fluids.

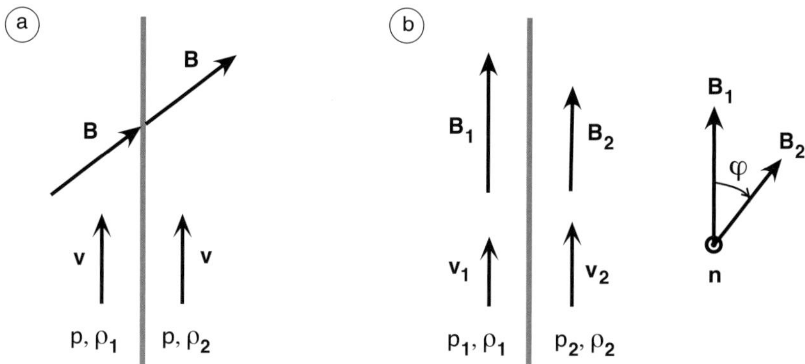

Fig. 20.6 (a) Contact discontinuities; (b) tangential discontinuities.

We now discuss the two discontinuities without mass flow separately.

(a) Contact discontinuities If the mass flow vanishes ($\rho v_n = 0$) and the normal magnetic field does not vanish ($B_n \neq 0$) at the discontinuity, then

$$\llbracket \mathbf{v}_t \rrbracket = 0\,, \qquad \llbracket p \rrbracket = 0\,, \qquad \llbracket \mathbf{B}_t \rrbracket = 0\,, \qquad \llbracket B_n \rrbracket = 0\,, \qquad (20.29)$$

$$\llbracket \rho \rrbracket \neq 0\,. \qquad (20.30)$$

In contrast to the gas dynamic contact discontinuity of Eq. (20.28), jumps in the tangential flow are now inhibited by the normal magnetic field. Hence, all quantities, except the density, are continuous across the discontinuity (Fig. 20.6(a)).

Since the density jumps, the entropy also jumps, $\llbracket S \rrbracket \neq 0$. Such a jump is just carried with the fluid without interaction with the other quantities. For small amplitudes, when linear theory applies, the perturbations just transform into the *entropy waves* discussed in Section 5.2.2, Eqs. (5.45) [1]. We encountered these jump conditions in Section 4.5 and 4.6 [1] as the boundary conditions for plasmas with a jump in the density (astrophysical plasma models IV–VI).

(b) Tangential discontinuities If both the mass flow and the normal magnetic field vanish at the discontinuity ($\rho v_n = 0$ and $B_n = 0$), then

$$[\![p + \tfrac{1}{2} B_t^2]\!] = 0 \,, \tag{20.31}$$

$$[\![\rho]\!] \neq 0 \,, \qquad [\![\mathbf{v_t}]\!] \neq 0 \,, \qquad [\![p]\!] \neq 0 \,, \qquad [\![\mathbf{B_t}]\!] \neq 0 \,. \tag{20.32}$$

Hence, nearly arbitrary jumps of ρ, $\mathbf{v_t}$, p, $\mathbf{B_t}$ are possible and the only restriction is that the total pressure should be constant across the discontinuity (Fig. 20.6(b)). Such discontinuities are of considerable interest in plasma physics since they provide a powerful tool for modeling the dynamics of plasma–plasma interfaces separating plasmas with different properties. We have come across this type of discontinuity in the context of plasma confinement when discussing the boundary conditions for the laboratory plasma models II and II* in Section 4.6 [1].

The jumps in the tangential velocity and magnetic field are due to singularities of the vorticity $\boldsymbol{\omega} \equiv \nabla \times \mathbf{v}$ and of the current density $\mathbf{j} = \nabla \times \mathbf{B}$, i.e. vortex and current sheets of strength

$$\boldsymbol{\omega}^* = \mathbf{n} \times [\![\mathbf{v_t}]\!] \neq 0 \,, \qquad \mathbf{j}^* = \mathbf{n} \times [\![\mathbf{B_t}]\!] \neq 0 \,. \tag{20.33}$$

They produce changes not only of the magnitudes, but also of the directions of $\mathbf{v_t}$ and $\mathbf{B_t}$, as indicated in Fig. 20.6(b) by the angle φ for $\mathbf{B_t}$.

The jumps in the pressure and the density here admit solutions both with and without entropy jump. This indicates that these discontinuities represent some kind of a degeneracy. In particular, for infinitesimal amplitudes, they transform into a superposition of *zero-frequency Alfvén and magneto-sonic waves*. This degeneracy is lifted by contact discontinuities as well as by the discontinuities of the next section.

20.2.3 MHD discontinuities with mass flow

The generalization of the gas dynamic shocks of Section 20.2.1 to plasmas with a magnetic field requires *mass flow across the surface of discontinuity*. It leads to the two major classes of MHD discontinuities that operate in transonic plasmas, viz. rotational (or Alfvén) discontinuities and magneto-acoustic shocks. Their analysis involves quite different descriptions, but we will keep it on the same footing as long as possible.

In this case, the full potential of the jump conditions (20.3)–(20.9) is realized, so that it is expedient to recast them into dimensionless form. This can be effected by means of the normal mass flow ρv_n and the normal magnetic field B_n, since these variables are constant across the discontinuity according to the first two conditions. Before we get rid of the dimensions, we first reformulate the remaining jump conditions (20.5)–(20.9):

$$\rho v_{\mathrm{n}} [\![\mathbf{v}_{\mathrm{t}}]\!] = B_{\mathrm{n}} [\![\mathbf{B}_{\mathrm{t}}]\!] \,, \tag{20.34}$$

$$B_{\mathrm{n}}^2 [\![\mathbf{B}_{\mathrm{t}}]\!] = \rho^2 v_{\mathrm{n}}^2 [\![\mathbf{B}_{\mathrm{t}}/\rho]\!] \,, \tag{20.35}$$

$$[\![p + \tfrac{1}{2} B_{\mathrm{t}}^2]\!] + \rho^2 v_{\mathrm{n}}^2 [\![1/\rho]\!] = 0 \,, \tag{20.36}$$

$$[\![e + p/\rho]\!] + \tfrac{1}{2}(\rho v_{\mathrm{n}}/B_{\mathrm{n}})^2 [\![B^2/\rho^2]\!] = 0 \,, \tag{20.37}$$

$$[\![\rho^{-\gamma} p]\!] \le 0 \,. \tag{20.38}$$

Here, the jump condition (20.35) is obtained by combining the tangential conditions (20.5) and (20.6), whereas the derivation of the energy condition (20.37) from Eq. (20.8) involves a number of steps that are given in small print below.

▷ **Reduction of the energy condition** The energy conservation relation (20.8) is usually transformed (see Landau and Lifschitz, *Electrodynamics of Continuous Media* [295], Chapter VIII) by generalizing the shock adiabatic (20.18) with magnetic contributions,

$$[\![e]\!] + \left\{ \tfrac{1}{2}(p_1 + p_2) + \tfrac{1}{4}(B_{\mathrm{t}1} - B_{\mathrm{t}2})^2 \right\} [\![1/\rho]\!] = 0 \,. \tag{20.39}$$

However, for the purpose of reducing the jump conditions to a minimum number of free parameters, it is more expedient to transform it into

$$\left[\!\!\left[e + \frac{p}{\rho} + \tfrac{1}{2} v_{\mathrm{n}}^2 + \tfrac{1}{2} \left| \mathbf{v}_{\mathrm{t}} - \frac{B_{\mathrm{n}}}{\rho v_{\mathrm{n}}} \mathbf{B}_{\mathrm{t}} \right|^2 + \frac{B_{\mathrm{t}}^2}{\rho} \left(1 - \frac{B_{\mathrm{n}}^2}{2\rho v_{\mathrm{n}}^2} \right) \right]\!\!\right] = 0 \,,$$

where the fourth term vanishes due to Eq. (20.5), and the third and fifth term combine to

$$\left[\!\!\left[\tfrac{1}{2} v_{\mathrm{n}}^2 + \frac{B_{\mathrm{t}}^2}{\rho} \left(1 - \frac{B_{\mathrm{n}}^2}{2\rho v_{\mathrm{n}}^2} \right) \right]\!\!\right] - \left[\!\!\left[\frac{v_{\mathrm{n}}^2 B^2}{2 B_{\mathrm{n}}^2} \right]\!\!\right] - \frac{B_{\mathrm{n}}^2}{2\rho^2 v_{\mathrm{n}}^2} \left[\!\!\left[(\rho v_{\mathrm{n}}^2/B_{\mathrm{n}}^2 - 1)^2 B_{\mathrm{t}}^2 \right]\!\!\right] = \left[\!\!\left[\frac{v_{\mathrm{n}}^2 B^2}{2 B_{\mathrm{n}}^2} \right]\!\!\right]$$

due to Eq. (20.35). This yields the energy conservation condition (20.37). ◁

We now exploit ρv_{n} and B_{n} to create dimensionless variables that characterize the states on the two sides of the shock. To that end, we first introduce the square of the *normal Alfvén Mach number*,

$$M_{\mathrm{An}}^2 \equiv \frac{v_{\mathrm{n}}^2}{v_{\mathrm{An}}^2} \equiv \frac{\rho v_{\mathrm{n}}^2}{B_{\mathrm{n}}^2} \left(= \frac{\rho v_{\mathrm{n}}}{B_{\mathrm{n}}^2} \cdot v_{\mathrm{n}} = \frac{\rho^2 v_{\mathrm{n}}^2}{B_{\mathrm{n}}^2} \cdot \frac{1}{\rho} \right) \,, \tag{20.40}$$

which is the MHD counterpart of the ordinary Mach number of hydrodynamics. Since the latter does not occur in MHD, there is no confusion if we simplify the notation by suppressing the subscripts and indicate the square of the normal Alfvén Mach number also by M^2. According to the equalities (20.40) in brackets, this quantity may be considered as the dimensionless normal speed (apart from the sign, see below), or as the dimensionless inverse density (or specific volume). Hence, the ratio of the normal speeds is proportional, and the ratio of the densities is inversely proportional, to the ratio of the squares of the normal Alfvén Mach number across the discontinuity,

$$\frac{v_{\mathrm{n}2}}{v_{\mathrm{n}1}} = \frac{\rho_1}{\rho_2} = \frac{M_2^2}{M_1^2}\,. \tag{20.41}$$

The jump conditions (20.34)–(20.38) simplify considerably by introducing the following dimensionless variables:

$$\bar{v}_{\mathrm{n}i} \equiv \frac{\rho|v_{\mathrm{n}}|}{B_{\mathrm{n}}^2}\,v_{\mathrm{n}i} = -M_i^2\,, \qquad \bar{B}_{\mathrm{n}} \equiv \frac{B_n}{|B_{\mathrm{n}}|} = -1\,, \tag{20.42}$$

$$\bar{\mathbf{v}}_{ti} \equiv \frac{\rho|v_{\mathrm{n}}|}{B_{\mathrm{n}}^2}\,\mathbf{v}_{ti}\,, \qquad \bar{\mathbf{B}}_{ti} \equiv \frac{\mathbf{B}_{ti}}{|B_{\mathrm{n}}|}\,, \qquad \bar{p}_i \equiv \frac{p_i}{B_{\mathrm{n}}^2} \qquad (i = 1, 2)\,, \tag{20.43}$$

where $\bar{B}_{ti}$ is directly related to the angle ϑ_i between the magnetic field and the shock normal, and $\bar{p}_i$ is related to the ratio β_{ni} between the pressure of the plasma and that of the normal magnetic field:

$$\tan\vartheta_i \equiv B_{ti}/|B_{\mathrm{n}}| \equiv \bar{B}_{ti}\,, \qquad \beta_{ni} \equiv 2p_i/B_{\mathrm{n}}^2 \equiv 2\bar{p}_i\,. \tag{20.44}$$

Since the direction of the flow is opposite to the normal and the sign of the magnetic field is not relevant for the shock conditions, in Eq. (20.42) we have assumed both $\rho v_{\mathrm{n}} < 0$ and $B_{\mathrm{n}} < 0$ (as illustrated in Fig. 20.7).

With this normalization, the jump conditions become:

$$[\![\bar{\mathbf{v}}_{\mathrm{t}}]\!] = [\![\bar{\mathbf{B}}_{\mathrm{t}}]\!] \qquad \left(\Rightarrow \bar{\boldsymbol{\omega}}^* = \mathbf{n} \times [\![\bar{\mathbf{v}}_{\mathrm{t}}]\!] = \bar{\mathbf{j}}^* = \mathbf{n} \times [\![\bar{\mathbf{B}}_{\mathrm{t}}]\!] \right)\,, \tag{20.45}$$

$$[\![(M^2 - 1)\bar{\mathbf{B}}_{\mathrm{t}}]\!] = 0\,, \tag{20.46}$$

$$[\![M^2 + \bar{p} + \tfrac{1}{2}\bar{B}_{\mathrm{t}}^2]\!] = 0\,, \tag{20.47}$$

$$\left[\!\left[\frac{\gamma}{\gamma - 1}\,\bar{p}M^2 + \tfrac{1}{2}(1 + \bar{B}_{\mathrm{t}}^2)M^4 \right]\!\right] = 0\,, \tag{20.48}$$

$$[\![\bar{p}M^{2\gamma}]\!] \leq 0\,. \tag{20.49}$$

Apart from the degree of freedom of the rotational discontinuities, discussed below, the system (20.45)–(20.48) completely determines the jumps across the discontinuity for given values of the upstream parameters, whereas the inequality (20.49) determines whether such a jump qualifies as a shock or not.

Here again two possibilities arise, according to whether the thermodynamic variables jump (genuine shock) or not across the discontinuity. The latter possibility leads to discontinuities without a gas dynamic analog, viz. *rotational discontinuities* (paragraph *(c)* below), which may be considered as the finite amplitude manifestation of the Alfvén wave. The former possibility leads to *magneto-acoustic shocks* (paragraph *(d)* below), which come in the three kinds of *slow, intermediate and fast* (Section 20.2.4). They are associated with the two magneto-acoustic waves whereas the Alfvén speed plays a central role.

We now discuss the two discontinuities with mass flow separately.

(c) Rotational (or Alfvén) discontinuities If the density is continuous across the discontinuity, $[\![\rho]\!] = 0$, the normal velocity is also continuous, $[\![v_{\mathrm{n}}]\!] = 0$, so that

$$M_1^2 = M_2^2 = 1\,, \qquad [\![\bar{p}]\!] = 0\,, \qquad [\![\bar{B}_{\mathrm{t}}^2]\!] = 0\,, \tag{20.50}$$

$$[\![\bar{\mathbf{v}}_{\mathrm{t}}]\!] = [\![\bar{\mathbf{B}}_{\mathrm{t}}]\!] \neq 0\,. \tag{20.51}$$

Hence, all thermodynamic variables (p, ρ, e and S) and the magnitude of $\mathbf{B}$ are continuous, but the tangential magnetic field $\mathbf{B}_{\mathrm{t}}$ turns through an angle φ about the normal $\mathbf{n}$ (proportional to the vortex and current sheets given inside the brackets of Eq. (20.45)). Moreover, the normal velocity and the jump in the tangential velocity are equal to their respective Alfvén velocities. This discontinuity is properly called a *rotational, or Alfvén, discontinuity* since a rotational jump of the tangential magnetic field about the normal propagates with the jump of tangential Alfvén speed.

Rotational discontinuities are essentially determined by prescribing $M^2 = 1$ and choosing values for the three free parameters,

$$\varphi \equiv 2\arcsin\left(\tfrac{1}{2}\,|[\![\bar{\mathbf{B}}_{\mathrm{t}}]\!]|/\bar{B}_{\mathrm{t}}\right)\,, \qquad \beta_{\mathrm{n}} \equiv 2\bar{p}\,, \qquad \vartheta \equiv \arctan\bar{B}_{\mathrm{t}}\,, \tag{20.52}$$

which describe the change of the direction of the tangential magnetic field, the fixed value of the pressure, and the fixed angle of the magnetic field with the normal. Note that, whereas genuine shocks are (nearly) completely determined by the values of the upstream parameters, the free parameter φ of rotational discontinuities is not determined at all by the values of the upstream parameters.

(d) Magneto-acoustic shocks If the density is discontinuous across the discontinuity, $[\![\rho]\!] \neq 0$, the normal velocity is also discontinuous, $[\![v_{\mathrm{n}}]\!] \neq 0$, so that the full system of jump conditions (20.45)–(20.49) is needed to fix the parameters. The essential features are given by

$$M_1^2 \neq M_2^2\,, \qquad [\![\bar{p}]\!] \neq 0\,, \qquad [\![\bar{B}_{\mathrm{t}}^2]\!] \neq 0\,, \tag{20.53}$$

$$[\![\bar{\mathbf{v}}_{\mathrm{t}}]\!] = [\![\bar{\mathbf{B}}_{\mathrm{t}}]\!] \parallel \bar{\mathbf{B}}_{\mathrm{t}1} \parallel \bar{\mathbf{B}}_{\mathrm{t}2}\,, \qquad (M_1^2 - 1)\bar{B}_{\mathrm{t}1} = (M_2^2 - 1)\bar{B}_{\mathrm{t}2}\,. \tag{20.54}$$

Hence, all thermodynamic variables (p, ρ, e and S) and the magnitude of $\bar{\mathbf{B}}$ are discontinuous, but the vectors $\bar{\mathbf{B}}_{\mathrm{t}1}$, $\bar{\mathbf{B}}_{\mathrm{t}2}$, $\mathbf{n}$ and $[\![\bar{\mathbf{v}}_{\mathrm{t}}]\!]$ all lie in the same plane. This discontinuity is called a *magnetohydrodynamic shock*. It is a genuine generalization of the gas dynamic shocks for magnetized plasmas.

MHD shocks are essentially determined by fixing $\varphi = 0$ and choosing values for the three free upstream parameters

$$M_1^2 \equiv -\bar{v}_{\mathrm{n}1} \equiv 1/\bar{\rho}_1 \,, \quad \beta_{\mathrm{n}1} \equiv 2\bar{p}_1 \,, \quad \vartheta_1 \equiv \arctan \bar{B}_{\mathrm{t}1} \,, \qquad (20.55)$$

which determine the values of the downstream parameters M_2^2, $\beta_{\mathrm{n}2}$ and ϑ_2 according to the jump conditions (20.45)–(20.47). The substantial algebra involved in the reduction of these conditions will be completed in Section 20.3.

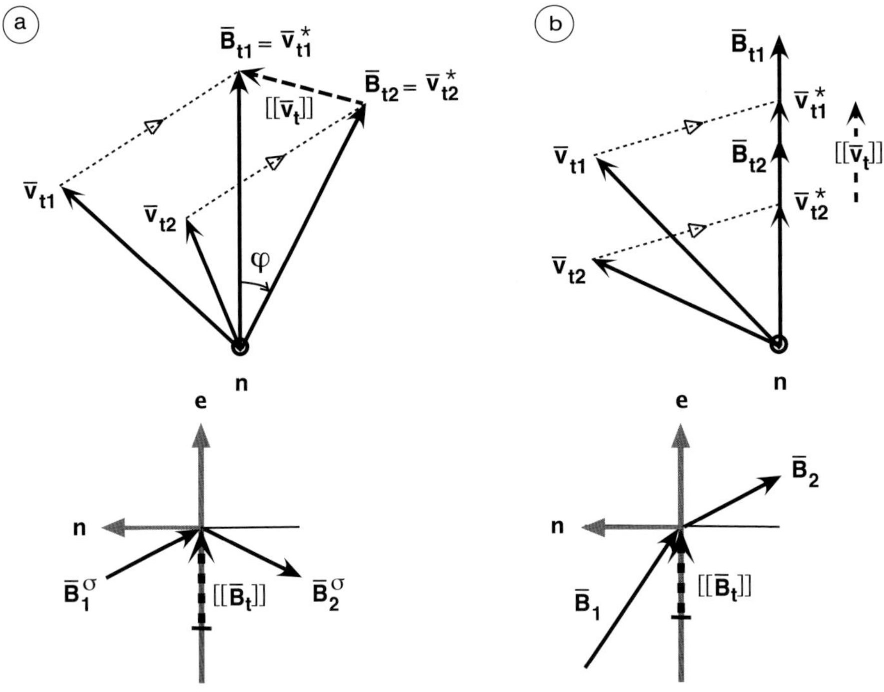

Fig. 20.7 Top: tangential magnetic fields and velocities in the de Hoffmann–Teller frame for (a) rotational discontinuities and (b) shocks; $\bar{\mathbf{v}}_{ti}^* = M_i^2 \bar{\mathbf{B}}_{ti}$ ($i = 1, 2$). Bottom: projection (on a different scale) in the e–n plane, with $\mathbf{e} \parallel [\![\bar{\mathbf{v}}_{\mathrm{t}}]\!] = [\![\bar{\mathbf{B}}_{\mathrm{t}}]\!]$. For (a), the constant part $\bar{\mathbf{B}}^\tau$ of $\bar{\mathbf{B}}_{ti} \perp \mathbf{e}$ does not show up in this projection, where $\bar{\mathbf{B}}_i = \bar{\mathbf{B}}_i^\sigma + \bar{\mathbf{B}}^\tau$, $\bar{\mathbf{B}}_i^\sigma = \pm\bar{B}_{\mathrm{t}} \sin\frac{1}{2}\varphi \, \mathbf{e} + \bar{B}_{\mathrm{n}}\mathbf{n}$, $\bar{B}_{\mathrm{n}} = -1$, $\bar{\mathbf{B}}^\tau = \bar{B}_{\mathrm{t}} \cos\frac{1}{2}\varphi \, \mathbf{e} \times \mathbf{n}$.

The geometric meaning of the jump conditions for the MHD discontinuities becomes much clearer when the tangential velocities and magnetic fields are aligned by means of a transformation to a tangentially moving frame in which the normal electric field vanishes. Since there are infinitely many of such tangential transformations, one can find a particular one involving the normal velocities as well such that the complete electric field vanishes, $\mathbf{E}_i^* = \mathbf{v}_i^* \times \mathbf{B}_i = 0$, both in front and behind the shock. Since, by definition, $\bar{v}_{\mathrm{n}i} \equiv -M_i^2$, this alignment of the transformed velocities $\mathbf{v}_i^*$ with the magnetic field $\mathbf{B}_i$ is obtained by transformation to the *de Hoffmann–Teller frame* [108],

$$\bar{\mathbf{v}}_i \to \bar{\mathbf{v}}_i^* \equiv M_i^2 \bar{\mathbf{B}}_i \qquad (i = 1, 2). \tag{20.56}$$

The boost $\bar{\mathbf{w}} \equiv \bar{\mathbf{v}}^* - \bar{\mathbf{v}}$ effecting this is the same in front and behind the shock, as follows from the jump conditions:

$$[\![\bar{\mathbf{w}}_t]\!] \equiv [\![\bar{\mathbf{v}}_t^* - \bar{\mathbf{v}}_t]\!] \equiv [\![M^2 \bar{\mathbf{B}}_t - \bar{\mathbf{v}}_t]\!] \overset{(20.45)}{=} [\![(M^2 - 1)\bar{\mathbf{B}}_t]\!] \overset{(20.46)}{=} 0,$$

$$\bar{w}_{\mathrm{n}i} \equiv \bar{v}_{\mathrm{n}i}^* - \bar{v}_{\mathrm{n}i} \equiv -M_i^2 - \bar{v}_{\mathrm{n}i} \equiv 0 \qquad \Rightarrow \qquad [\![\bar{w}_{\mathrm{n}}]\!] = 0. \tag{20.57}$$

The first part of the last line shows that the boost is actually only tangential. The de Hoffmann–Teller transformation also holds for relativistic shocks. As noted by the authors in Ref. [108], it fails for perpendicular shocks ($\vartheta = \pi/2$) since then $B_{\mathrm{n}} = 0$ but $v_{\mathrm{n}} \neq 0$. In that case, M_i^2 and the normalizations (20.44) become undefined. This defect will be cured in Section 20.3.3 by our final representation (20.93) for the velocities in the de Hoffmann–Teller frame.

In conclusion, in the de Hoffmann–Teller frame, *the tangential vectors $\bar{\mathbf{v}}_t^*$ and $\bar{\mathbf{B}}_t$ rotate with constant amplitude over the same angle φ for rotational (Alfvén) discontinuities* (Fig. 20.7(a)), whereas *they have constant direction but changing amplitudes for shocks* (Fig. 20.7(b)). The directional changes of $\bar{\mathbf{B}}$ and $\bar{\mathbf{v}}^*$ in the plane through the jumps and the normal (bottom panels of Fig. 20.7) alone do not permit us to distinguish between rotational discontinuities and shocks because intermediate shocks show similar behavior to rotational discontinuities in that plane (see Fig. 20.8(b) of the next sub-section). Those two kinds of discontinuity are distinguished by the presence or absence of a constant contribution of $\bar{\mathbf{B}}_{ti}$ in the direction orthogonal to that plane.

20.2.4 Slow, intermediate and fast shocks

Restricting the discussion now to MHD shocks, it may appear surprising that they come in three, rather than two, flavors. Anticipating the analysis of Section 20.3, where we will show that the entropy condition only permits jumps with $M_1^2 \geq M_2^2$, those three arise due to the relationship (20.54) between $\bar{B}_{t1}$ and $\bar{B}_{t2}$:

$$\begin{aligned}
M_2^2 \leq M_1^2 \leq 1 \quad &\Rightarrow \quad |\bar{B}_{t1}| \geq |\bar{B}_{t2}| \qquad &\text{(slow shocks)}, \\
M_2^2 \leq 1 \leq M_1^2 \quad &\Rightarrow \quad \bar{B}_{t1}/\bar{B}_{t2} < 0 \qquad &\text{(intermediate shocks)}, \\
1 \leq M_2^2 \leq M_1^2 \quad &\Rightarrow \quad |\bar{B}_{t1}| \leq |\bar{B}_{t2}| \qquad &\text{(fast shocks)}.
\end{aligned} \tag{20.58}$$

The magnetic fields $\mathbf{B}_{t1}$ and $\mathbf{B}_{t2}$, and hence the velocities $\mathbf{v}_{t1}^*$ and $\mathbf{v}_{t2}^*$, have the same directions for slow and fast shocks, with breaking towards the normal for slow shocks (Fig 20.8(a)) and away from it for fast shocks (Fig 20.8(c)), but they have opposite directions for intermediate shocks (Fig 20.8(b)).

When the angles ϑ_i of the directions of the magnetic fields are varied, several significant *limiting cases* are encountered.

- **Perpendicular shocks** $(\vartheta_1 = \vartheta_2 = \pi/2)$ In this limit, the normal Alfvén Mach number is not defined, so that the normalizations (20.40) and (20.42) can no longer be exploited and one should go back to the original jump conditions (20.34)–(20.37) with $B_\mathrm{n} = 0$ and $\rho v_\mathrm{n} \neq 0$. Equation (20.35) then yields $[\![\mathbf{B}_\mathrm{t}/\rho]\!] = 0$, so that $\varphi = 0$ and a rotational counterpart of the tangential discontinuity illustrated in Fig. 20.6(b) of Section 20.2.2 becomes impossible. However, perpendicular shocks are perfectly valid generalizations of tangential discontinuities with $\varphi = 0$ and jumps in p and $\bar{B}_\mathrm{t}$ such that $[\![p + \frac{1}{2}B_\mathrm{t}^2]\!] = 0$. The shock relations for this case are easily obtained from the general relations derived by replacing the variables M_1^2 and M_2^2 by v_n1 and v_n2, and properly taking the limit $B_\mathrm{n} \to 0$.

- **Parallel shocks** $(\vartheta_1 = \vartheta_2 = 0)$ When the magnetic field is parallel to the normal, both upstream and downstream (Fig 20.8(d)), and hence also the velocities $\mathbf{v}_i^*$, the relation (20.54) is trivially satisfied: $\bar{B}_\mathrm{t1} = \bar{B}_\mathrm{t2} = 0$. In that case, the jump conditions (20.45)–(20.48) reduce to the hydrodynamic jump conditions (20.13)–(20.15). This follows from the relationship between the hydrodynamical (sound) Mach number and the magnetohydrodynamical (normal Alfvén) Mach number,

$$\rho_i v_{\mathrm{n}i}^2 \equiv \gamma p_i M_i^2\big|_{\mathrm{HD}} \equiv B_\mathrm{n}^2 M_i^2\big|_{\mathrm{MHD}} . \tag{20.59}$$

Therefore, these shocks are sometimes called *hydrodynamic shocks*. Nevertheless, the normal Alfvén Mach number is defined so that one can still distinguish between fast and slow parallel shocks.

- **Switch-on shocks** $(\vartheta_1 = 0, \vartheta_2 \neq 0)$ When the upstream magnetic field is parallel to the normal $(\bar{B}_\mathrm{t1} = 0)$, as in a parallel shock, the downstream magnetic field may have a different direction (it is switched on: $\bar{B}_\mathrm{t2} \neq 0$, as illustrated in Fig 20.8(e)). According to Eq. (20.54), this happens when $M_2^2 = 1$ so that $M_1^2 > 1$. Accordingly, the switch-on shock should be termed intermediate or fast.

- **Switch-off shocks** $(\vartheta_1 \neq 0, \vartheta_2 = 0)$ The relation (20.54) between $\bar{B}_\mathrm{t1}$ and $\bar{B}_\mathrm{t2}$ is also satisfied in the opposite case $(\bar{B}_\mathrm{t1} \neq 0, \bar{B}_\mathrm{t2} = 0)$, when the magnetic field is switched off (Fig 20.8(f)). This happens for $M_1^2 = 1$ so that $M_2^2 < 1$: switch-off shocks should be termed intermediate or slow.

This appears to complete the classification of the different kinds of MHD shock. However, we still have to demonstrate that the *complete* set of jump conditions, including the ones associated with energy conservation and entropy production, can be satisfied for all those cases. This will be done in the next section, which will shed light on why and when the values of the downstream parameters may not be completely determined by the values of the upstream parameters. Thus, the

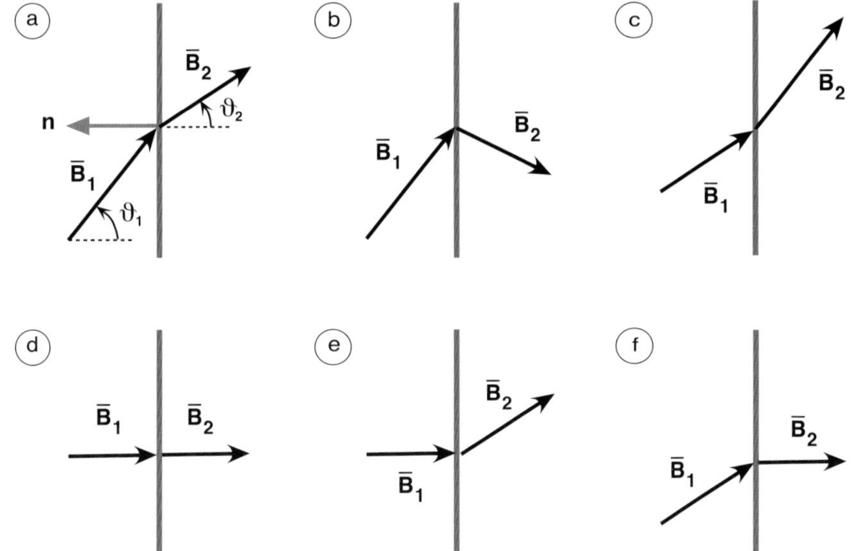

Fig. 20.8 MHD shocks: (a) slow shock, (b) intermediate shock, (c) fast shock. Limiting cases: (d) parallel ("HD") shock, (e) switch-on shock, (f) switch-off shock. The normalization is given by $\bar{B}_{ti} \equiv B_{ti}/|B_n| = \tan\vartheta_i$, $\bar{B}_n \equiv B_n/|B_n| = -1$.

distinction between fast, intermediate and slow shocks will be uniquely established. In doing so, quite a number of additional distinctions come into view.

20.3 Classification of MHD shocks

20.3.1 Distilled shock conditions

We will now complete the reduction of the MHD shock conditions (20.45)–(20.49) such that they can be solved uniquely, yielding expressions for $\bar{v}_{t2}, \bar{B}_{t2}, \bar{p}_2, M_2^2$ in terms of $\bar{v}_{t1}, \bar{B}_{t1}, \bar{p}_1, M_1^2$.

The first condition (20.45) has actually been superseded by the de Hoffman–Teller transformation (20.56) since it provides all relevant relationships between the downstream and upstream velocities in that preferred frame:

$$\left\{ \begin{array}{l} \bar{v}_{n1}^* = M_1^2 \bar{B}_n = -M_1^2 \\ \bar{v}_{t1}^* = M_1^2 \bar{B}_{t1} = M_1^2 \tan\vartheta_1 \end{array} \right. \Rightarrow \left\{ \begin{array}{l} \bar{v}_{n2}^* = M_2^2 \bar{B}_n = -M_2^2 \\ \bar{v}_{t2}^* = M_2^2 \bar{B}_{t2} = M_2^2 \tan\vartheta_2 \end{array} \right. . \quad (20.60)$$

However, the tangential velocity parameters $\bar{v}_{t1}^*$ and $\bar{v}_{t2}^*$ do not appear in the final form of the distilled shock conditions that we are going to derive so that we can defer discussion of the implications of Eqs. (20.60) for the velocity to Section 20.3.3, after we have completed the solution of the shock conditions.

The next two relations (20.46) and (20.47) determine $\bar{B}_{t2}$ and $\bar{p}_2$:

$$\bar{B}_{t2} = \frac{M_1^2 - 1}{M_2^2 - 1}\,\bar{B}_{t1}\,, \tag{20.61}$$

$$\bar{p}_2 = \bar{p}_1 + (M_1^2 - M_2^2)\left[1 - \tfrac{1}{2}\bar{B}_{t1}^2\,\frac{M_1^2 + M_2^2 - 2}{(M_2^2 - 1)^2}\right]. \tag{20.62}$$

Substituting these expressions into the jump condition (20.48) yields, after dividing out a factor $\sim (M_2^2 - M_1^2)$ (i.e. eliminating the trivial solutions without jumps), the *distilled energy jump condition*:

$$f = f(M_2^2; M_1^2, \bar{B}_{t1}^2, \bar{p}_1) = 0 \tag{20.63}$$

$$\equiv \tfrac{1}{2}(M_2^2 - 1)^2\left\{(\gamma + 1)M_2^2 - (\gamma - 1)M_1^2 - 2\gamma\bar{p}_1\right\}$$

$$+ \tfrac{1}{2}\bar{B}_{t1}^2\left\{(\gamma - 1)(M_2^2 - 1)(M_1^2 - M_2^2) - M_2^2(M_1^2 + M_2^2 - 2)\right\}.$$

Similarly, the inequality (20.49) for entropy increase across the shock yields the *distilled entropy condition*:

$$g = g(M_2^2; M_1^2, \bar{B}_{t1}^2, \bar{p}_1) \geq 0 \tag{20.64}$$

$$\equiv (M_2^2 - 1)^2\left\{M_1^2 - M_2^2 - \bar{p}_1\left[(M_1^2/M_2^2)^\gamma - 1\right]\right\}$$

$$- \tfrac{1}{2}\bar{B}_{t1}^2(M_1^2 - M_2^2)(M_1^2 + M_2^2 - 2).$$

Solutions of the distilled energy jump condition are permitted jumps, but they only qualify as shocks when they also satisfy the distilled entropy inequality.

It is straightforward to plot these conditions for particular parameter values: Figs. 20.9–20.11. For definiteness, we here present solutions demonstrating the wide variety of MHD shocks before the analysis actually has been completed. Consequently, explanation of the symbols along the curves and of the parameter regions I, II and III has to await the exposition of Section 20.3.2. Also notice that $\bar{B}_{ti}$ and $\bar{p}_i$ have been replaced by the parameters ϑ_i and β_i, defined there.

In Fig. 20.9, the complete jump curve $M_2^2 = M_2^2(M_1^2)$ is shown without suppressing the entropy-forbidden parts, as done in Fig. 20.5 of Section 20.2.1 for the gas dynamic jumps and shocks. In addition to the advantages of duality, exposed in the next sub-section, keeping them reveals the true character of the intermediate jumps/shocks as *a continuous transition in parameter space between the slow and fast jumps/shocks*. The S-shaped jump curve intersects the diagonal $M_1^2 = M_2^2$ at the points labeled S, A and F, where also the entropy-permitted (dotted) and the entropy-forbidden (white) areas interchange position. As in gas dynamics (Fig. 20.5), precisely at those points, the jump curve leaves or enters the dotted areas, distinguishing entropy-forbidden jumps from entropy-permitted

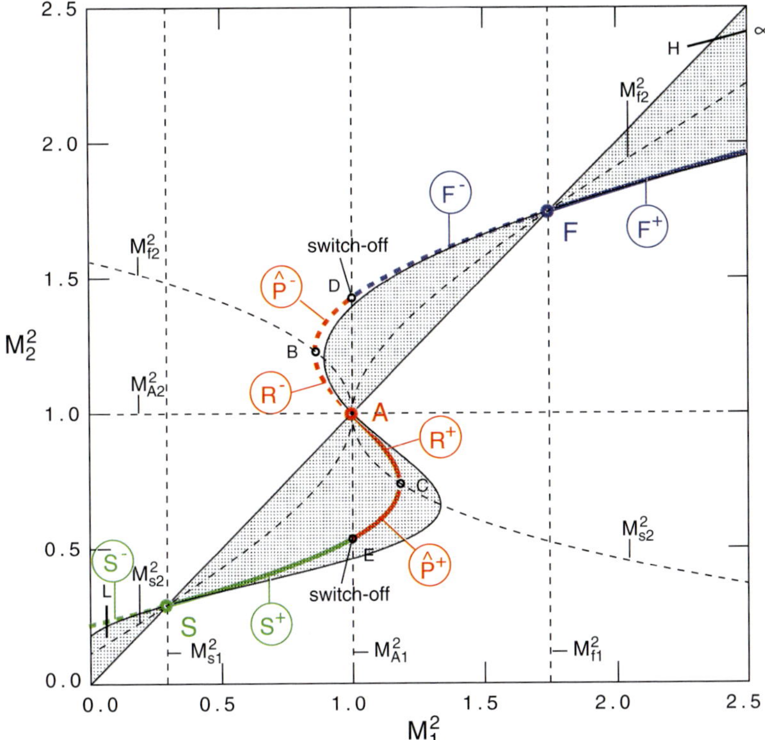

Fig. 20.9 Jump curve for parameter region II ($\vartheta_1 = 0.2\pi$, $\beta_1 = 0.4$): slow (green), intermediate (red) and fast (blue) entropy-forbidden jumps (dashed) as well as entropy-permitted shocks (drawn) are shown. The latter occur when the jump curve intersects the dotted areas, where the entropy increases downstream. The points S, A and F indicate slow, Alfvén and fast waves ($=$ weak discontinuities).

shocks. The latter only occur for values of M_1^2 and M_2^2 below the diagonal. According to Eq. (20.41), this implies that *MHD shocks are compressive.*

Going downward, the overall trend of the jump curve is *prograde*, i.e. decreasing with respect to decreases of both M_1^2 and M_2^2, up to the point B and past the point C, but it is *retrograde*, i.e. increasing with respect to M_1^2 and decreasing with respect to M_2^2 in between B and C. The implications for the intermediate jumps and shock are analyzed in depth in Section 20.3.2. As a mnemonic, the labels B–A–C have been chosen to indicate the *backward* (retrograde) part of the jump curve and the labels D–E to *demarcate* the boundaries of the intermediate jumps.

The corresponding solutions $\vartheta_2 = \vartheta_2(M_1^2)$ and $\beta_2 = \beta_2(M_1^2)$ of Eqs. (20.61) and (20.62) are shown in Fig. 20.10 for the same values of ϑ_1 and β_1 as in Fig. 20.9. The first plot clearly shows the distinct directional differences between the different kinds of shock described by Eq. (20.58): $\vartheta_2 \leq \vartheta_1$ for slow shocks, $\vartheta_2 \leq 0$ for

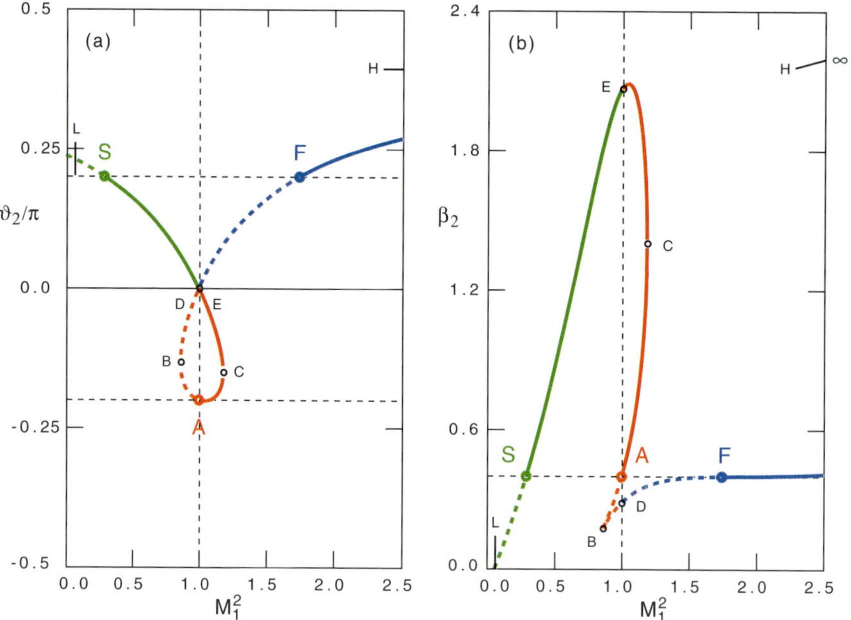

Fig. 20.10 (a) Downstream angle $\vartheta_2(M_1^2)$ and (b) relative pressure $\beta_2(M_1^2)$ of MHD jumps/shocks for the parameter values as Fig. 20.9 ($\vartheta_1 = 0.2\pi$, $\beta_1 = 0.4$). For $M_1^2 < L$, the jump curve is unphysical because $\beta_2 < 0$. The symbol H indicates the asymptote of the curves for $M_1^2 \to \infty$.

intermediate shocks and $\vartheta_2 \geq \vartheta_1$ for fast shocks. The second plot shows the enormous variation of the downstream pressure when the upstream Alfvén Mach number is varied. Similar to the situation for gas dynamic jumps and shocks, the downstream value of the pressure $\bar{p}_2$ (or β_2) from Eq. (20.62) cannot be guaranteed to be positive for arbitrary values of the upstream parameters. Consequently, the parts of the jump curve where $\beta_2 < 0$ should be discarded as being unphysical. For the parameters of Figs. 20.9 and 20.10, this occurs for $0 \leq M_1^2 < L$.

The jump curve of Fig. 20.9 intersects different regions of the M_1^2–M_2^2 plane characterizing the upstream and downstream states of the flow in terms of the characteristic speeds v_A, v_f and v_s of the Alfvén, fast and slow waves of Eqs. (20.1):

$$M_\mathrm{A}^2 \equiv 1, \qquad M_\mathrm{f,s}^2 \equiv \frac{v_\mathrm{f,sn}^2}{v_\mathrm{An}^2} = \tfrac{1}{2}(\gamma\bar{p} + \bar{B}^2) \pm \tfrac{1}{2}\sqrt{(\gamma\bar{p} + \bar{B}^2)^2 - 4\gamma\bar{p}}. \qquad (20.65)$$

The shocks may be classified according to the ranges of M_1^2 and M_2^2 with respect to these transition values. Before we proceed to compartmentalize the M_1^2–M_2^2 plane to that end, we call the attention of the reader to the most significant behavior of the jump curve when it crosses the diagonal at the points S, A and F. This is most effectively discussed in terms of the *shock strength,* $\sigma \equiv 1 - M_2^2/M_1^2$. Expressing the distilled jump condition (20.63) in terms of σ instead of M_2^2, and then taking the limit of infinitesimal shock strength, we obtain the following expression:

$$f_0 \equiv \lim_{\sigma \to 0} f(\sigma; M_1^2, \bar{B}_{t1}^2, \bar{p}_1) = (M^2 - M_S^2)(M^2 - 1)(M^2 - M_F^2) = 0, \quad (20.66)$$

where M_S^2 indicates the intersection of the curves $M_1^2 = M_{s1}^2$ and $M_2^2 = M_{s2}^2$ defined below, and similarly for M_F^2. Hence, the slow, Alfvén and fast waves of linear MHD may be considered as shocks of infinitesimal strength. (However, note the caveat in Section 20.3.3 on the relation of Alfvén waves to "weak" intermediate shocks.) In this way, the three-fold dynamics of linear MHD returns in the classification of the nonlinear shocks. Note that the distilled entropy condition (20.64) again (as in gas dynamics, see Section 20.2.1) eliminates the possibility of a finite amplitude analog of the non-propagating entropy wave of Eq. (20.1).

The upstream states of the shock separate in four by the three vertical lines $M_1^2 = M_{s1}^2 \equiv M_s^2(\bar{B}_{t1}^2, \bar{p}_1)$ (≤ 1), $M_1^2 = M_{A1}^2$ ($\equiv 1$) and $M_1^2 = M_{f1}^2 \equiv M_f^2(\bar{B}_{t1}^2, \bar{p}_1)$ (≥ 1). The downstream transition curves are more involved, except for the Alfvén line $M_2^2 = M_{A2}^2$ ($\equiv 1$), as they require solving $M_2^2 = M_{s,f2}^2 \equiv M_{f,s}^2(\bar{B}_{t2}^2, \bar{p}_2)$ with $\bar{B}_{t2}^2$ and $\bar{p}_2$ expressed in $(M_1^2, M_2^2; \bar{B}_{t1}^2, \bar{p}_1)$ through Eqs. (20.61) and (20.62). This yields a quartic for the *downstream magneto-sonic transition values*:

$$h = h(M_2^2; M_1^2, \bar{B}_{t1}^2, \bar{p}_1) = 0 \qquad (20.67)$$

$$\equiv (M_2^2 - 1)^3 \left\{ \gamma M_1^2 - (\gamma + 1)M_2^2 + \gamma \bar{p}_1 \right\}$$

$$+ \bar{B}_{t1}^2 \left\{ M_2^2(M_1^2 - 1)^2 - \tfrac{1}{2}\gamma(M_2^2 - 1)(M_1^2 - M_2^2)(M_1^2 + M_2^2 - 2) \right\},$$

determining the slow and fast transition curves labeled M_{s2}^2 and M_{f2}^2 in Fig. 20.9. One of the two curves touches the central Alfvén point $M_1^2 = M_2^2 = 1$ (viz. the fast curve for the parameters of Fig. 20.9) whereas the other one stays at a finite distance from it (viz. the slow curve, which leaves a tiny gap that is hardly visible). Together with the Alfvén line $M_2^2 = M_{A2}^2 \equiv 1$, this again yields three (dashed) curves separating the four kinds of downstream states.

The dashed transition curves divide the M_1^2–M_2^2 plane into pairs of upstream and downstream flow regimes which traditionally are labeled from 1 to 4:

1 – *super-fast,* for $M^2 > M_f^2$ (i.e. $|v_n| > |v_{fn}|$);
2 – *sub-fast, super-Alfvénic,* for $M_A^2 < M^2 < M_f^2$ (i.e. $|v_{An}| < |v_n| < |v_{fn}|$);
3 – *super-slow, sub-Alfvénic,* for $M_s^2 < M^2 < M_A^2$ (i.e. $|v_{sn}| < |v_n| < |v_{An}|$);
4 – *sub-slow,* for $M^2 < M_s^2$ (i.e. $|v_n| < |v_{sn}|$).

Accordingly, the fast shocks shown in Fig. 20.9 are 1–2 (super-fast $\to$ sub-fast) transitions, the intermediate shocks are 2–3 (super-Alfvénic $\to$ super-slow) and

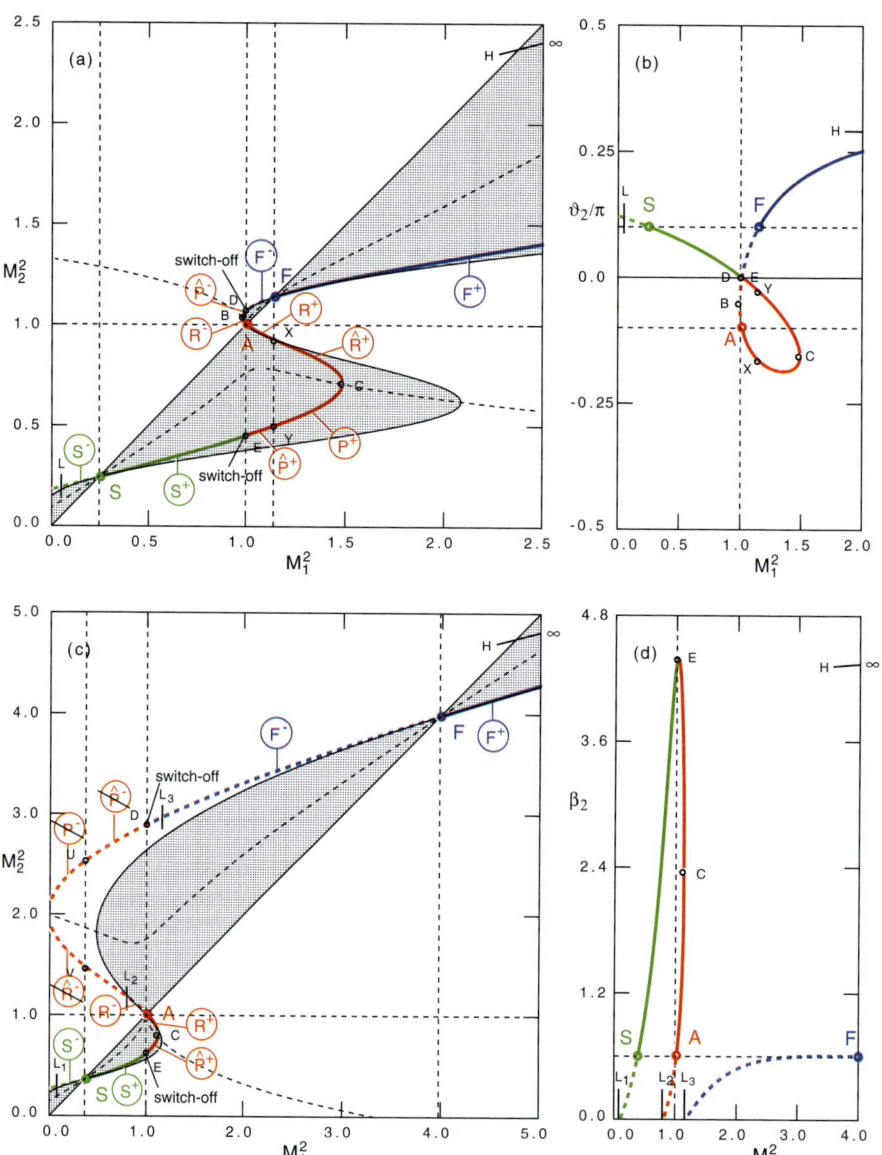

Fig. 20.11 (a) Jump curve and (b) downstream angle for parameter region I $(\vartheta_1 = 0.1\pi, \beta_1 = 0.3)$; jumps are unphysical to the left of L on the jump curve. (c) Jump curve and (d) pressure for parameter region III $(\vartheta_1 = 0.3\pi, \beta_1 = 0.6)$; jumps are unphysical to the left of L_1 and between L_2 and L_3 on the jump curve.

2–4 (super-Alfvénic $\rightarrow$ sub-slow) transitions, whereas the slow shocks are 3–4 (super-slow $\rightarrow$ sub-slow) transitions. For different parameter choices, the intermediate part of the shock curve (labeled A–X–C–Y–E now) protrudes into the super-fast regime 1, so that also 1–3 and 1–4 intermediate shocks appear (Fig. 20.11(a), parameter region I). Similarly, the entropy-forbidden part of the jump curve (labeled D–U–B–V–A, not all inside the frame) protrudes into the sub-slow regime 4, admitting 4–1 and 4–2 intermediate jumps (Fig. 20.11(b), parameter region III).

All of these solutions were obtained from the distilled jump condition (20.63), which is a cubic equation in M_2^2. Extensive literature exists on the MHD jump conditions based on quite different representations with a cubic equation, see e.g. Refs. [306], [253] and [482]. One can show these to be equivalent, but there is surprisingly little overlap with the results obtained.

In conclusion, anticipating the terminology of the next section, we encounter the following succession of entropy-forbidden jumps $(-)$ and permitted shocks $(+)$ along the jump curves, separated by the indicated transition points:

$$
\mathrm{F}_{12}^{+}, \mathrm{F}_{21}^{-}, \widehat{\mathrm{P}}_{31}^{-}, \left[\mathrm{P}_{41}^{-}, \widehat{\mathrm{R}}_{42}^{-},\right] \mathrm{R}_{32}^{-}, \mathrm{R}_{23}^{+}, \left\{\widehat{\mathrm{R}}_{13}^{+}, \mathrm{P}_{14}^{+},\right\} \widehat{\mathrm{P}}_{24}^{+}, \mathrm{S}_{34}^{+}, \mathrm{S}_{43}^{-} .
$$
$$
\quad \mathrm{H} \quad \mathrm{F} \quad \mathrm{D} \quad\quad \mathrm{U} \quad \mathrm{B} \quad \mathrm{V} \quad\quad \mathrm{A} \quad\quad \mathrm{X} \quad\quad \mathrm{C} \quad\quad \mathrm{Y} \quad\quad \mathrm{E} \quad\quad \mathrm{S} \quad\quad \mathrm{L} \tag{20.68}
$$

The discontinuities in square brackets are missing in parameter regions I and II, and the ones in curly brackets are missing in parameter regions II and III.

20.3.2 Time reversal duality

The examples of Section 20.3.1 clearly exhibit the central role of the Alfvén point A (analogous to the sonic point S in gas dynamics), and of the intermediate shocks about that point, in the theory of MHD shocks. Hence, we again define parameters

$$
\Delta_i \equiv M_i^2 - 1 \quad\quad (-1 \leq \Delta_i \leq \infty), \tag{20.69}
$$

measuring the "distances" to A, and replace $\bar{B}_{ti}$ and $\bar{p}_i$ by ϑ_i and β_{ni}, according to the definitions (20.44). Also, since scaling considerations have been sufficiently implemented now, we replace β_{ni} by the more practical parameters $\beta_i \equiv 2p_i/B_i^2$:

$$
\beta_{ni} = \beta_i(1 + \tan^2\vartheta_i). \tag{20.70}
$$

By means of the parameter sets Δ_i, ϑ_i, β_i, the jump conditions (20.61)–(20.64) acquire their most compact expression.

The distilled energy jump condition becomes the following cubic:

$$
\Delta_2^3 + p\,\Delta_2^2 + q\,\Delta_2 + r = 0 \quad\quad \Rightarrow \quad\quad \Delta_2 = \Delta_2(\Delta_1, \vartheta_1, \beta_1), \tag{20.71}
$$

$$
p \equiv \frac{2 - (\gamma - 1)\Delta_1 - \gamma\beta_1 - \gamma(1 + \beta_1)\tan^2\vartheta_1}{\gamma + 1},
$$

$$
q \equiv -\frac{[1 + (2 - \gamma)\Delta_1]\tan^2\vartheta_1}{\gamma + 1}, \quad\quad r \equiv -\frac{\tan^2\vartheta_1}{\gamma + 1}.
$$

Once this equation is solved, ϑ_2 and β_2 are found by substitution:

$$\vartheta_2 = \arctan\left(\frac{\Delta_1}{\Delta_2}\tan\vartheta_1\right) \quad\Rightarrow\quad \vartheta_2 = \vartheta_2(\Delta_1, \vartheta_1, \beta_1)\,, \qquad (20.72)$$

$$\beta_2 = \frac{\beta_1 + 2(\Delta_1 - \Delta_2) + (1 + \beta_1 - \Delta_1^2/\Delta_2^2)\tan^2\vartheta_1}{1 + (\Delta_1^2/\Delta_2^2)\tan^2\vartheta_1}$$

$$\Rightarrow\quad \beta_2 = \beta_2(\Delta_1, \vartheta_1, \beta_1)\,. \qquad (20.73)$$

The equations (20.71)–(20.73) constitute the *distilled ideal MHD jump problem*. With the addition of the distilled entropy condition,

$$\Delta_2 \leq \Delta_1\,, \qquad (20.74)$$

this becomes the *distilled MHD shock problem*.

The original six free downstream parameters ρ_1, p_1, v_{n1}, v_{t1}, B_{n1}, B_{t1} of this problem have been reduced to the three basic parameters $\Delta_1, \vartheta_1, \beta_1$ by exploitation of the scale-independence of the MHD equations, eliminating two parameters (e.g. ρ_1 and B_{n1}), and of the de Hoffmann–Teller transformation, eliminating one of the two velocity components. Consequently, *MHD shocks are essentially a three-parameter family* of solutions. Even after this maximal reduction, the MHD shock problem is much more complex than the gas dynamic one, as is manifest from the solutions of Figs. 20.9–20.11. However, we may still apply the principle of time reversal duality, formulated in Section 20.2.1, if we succeed in properly distinguishing the different sub-classes of solutions. This is the goal of the present section.

Since the distilled jump condition (20.71) is cubic in Δ_2, but linear in Δ_1, it is more easily solved from the inverse jump relation for Δ_1:

$$\Delta_1 = \frac{(\gamma + 1)\Delta_2^2 + [2 - \gamma\beta_1 - \gamma(1 + \beta_1)\tan^2\vartheta_1]\Delta_2 - \tan^2\vartheta_1}{(\gamma - 1)\Delta_2^2 + (2 - \gamma)\tan^2\vartheta_1\Delta_2 + \tan^2\vartheta_1}\Delta_2\,. \qquad (20.75)$$

The different limits of this relation exhibit the peculiarities of the jump curve.

– Around the Alfvén point $M_1^2 = M_2^2 = 1$ (or $\Delta_1 = \Delta_2 = 0$), the jump curve is *retrograde*, i.e. M_1^2 increases when M_2^2 decreases:

$$|\Delta_1| \sim |\Delta_2| \ll 1 \quad\Rightarrow\quad \Delta_1 \approx -\Delta_2\,. \qquad (20.76)$$

– For large M_1^2 and M_2^2, the jump curve has a *constant slope* $(= 4$ for $\gamma = 5/3)$:

$$\Delta_1 \sim \Delta_2 \gg 1 \quad\Rightarrow\quad \Delta_1 \approx \frac{\gamma + 1}{\gamma - 1}\Delta_2\,. \qquad (20.77)$$

– For $\vartheta_1 = 0$, when $\vartheta_2 = 0$ as well (according to Eq. (20.72)), the jump curve degenerates into a straight line:

$$(\gamma - 1)\Delta_1 - (\gamma + 1)\Delta_2 = 2 - \gamma\beta_1\,. \qquad (20.78)$$

This yields the *HD limit of parallel shocks* (Figs. 20.8(d)). The expression (20.78) may be transformed into the HD expression (20.23) by converting the Alfvén Mach number into the HD Mach number through Eq. (20.59). This shows that the striking difference of decreasing HD shock curves (Fig. 20.5) and overall increasing MHD shock curves (Fig. 20.9) is not deep, just results from different definitions of the Mach numbers.

– The expressions for *perpendicular shocks* may also be obtained from Eq. (20.75), or (20.71), in the limit $\vartheta_1 \to \pm\pi/2$. This is left as an exercise for the reader.

– The denominator of the expression (20.75) never vanishes for physical values of the parameters (since $\Delta_2 \geq -1$), except when both $\vartheta_1 = 0$ and $\Delta_2 = 0$. The numerator then also vanishes, so that the limits $\vartheta_1 \to 0$ and $\Delta_2 \to 0$ may be taken such that $\vartheta_2 \neq 0$ according to Eq. (20.72). This yields *switch-on shocks* (Figs. 20.8(e)).

Similarly, the quartic (20.67) for the magneto-sonic transition values $M_{\mathrm{f,s2}}^2(M_1^2)$ is more easily solved from the inverse relation $\Delta_{\mathrm{f,s1}}(\Delta_2)$, which is a quadratic:

$$\Delta_{\mathrm{f,s1}} = -s \pm \sqrt{s^2 - t}, \qquad s \equiv \frac{\gamma\Delta_2^3}{[2 + (2-\gamma)\Delta_2]\tan^2\vartheta_1}, \qquad (20.79)$$

$$t \equiv -\frac{[2 - \gamma\beta_1 - \gamma(1+\beta_1)\tan^2\vartheta_1 + 2(\gamma+1)\Delta_2]\Delta_2^3}{[2 + (2-\gamma)\Delta_2]\tan^2\vartheta_1}.$$

The slow and fast transition curves, labeled M_{s2}^2 and M_{f2}^2 in Fig. 20.9, are obtained by inverting these solutions again (a trivial plotting operation), where the two roots labeled $\pm$ correspond to the $\Delta_1 \geq 0$ and $\Delta_2 \leq 0$ parts of each curve.

We now analyze the consequences of the S-shape of the distilled jump curve, associated with the possibility of multiple solutions of the cubic $\Delta_2 = \Delta_2(\Delta_1)$, where intermediate jumps/shocks occur together with slow or fast jumps/shocks for the same values of the upstream parameters Δ_1, ϑ_1, β_1. This is actually not so exceptional since rotational/Alfvén discontinuities even have infinite multiplicity, labeled by the angle φ defined in Eq. (20.52). However, the indeterminacy of the downstream state in the range of intermediate shocks has frequently been associated with lack of evolutionarity of these shocks and, consequently, abandoned. We will come back to this at the end of this section, but for now note that discarding these solutions, unfortunately, has seriously hampered the development of the structure of the MHD shock relations as presented here. This negative role of evolutionarity is extensively discussed by Kennel, Blandford and Coppi [253].

Since the original cubic (20.71) is a multi-valued function of Δ_1, but the inverse jump relation (20.75) is a single-valued function of Δ_2, the jump curve along the points H, ..., L depicted in Fig. 20.9 is monotonically decreasing in Δ_2, but either decreasing (along H–F–D–B and C–E–S–L) or increasing (along B–A–C) in Δ_1. We call the corresponding intermediate jumps in these ranges *prograde* (indicated by P) when Δ_1 and Δ_2 decrease/increase together and *retrograde* (indicated by

R) when they have opposite monotonicity. The additional distinctions of *quasi-prograde* (indicated by $\widehat{P}$) and *quasi-retrograde* (indicated by $\widehat{R}$) encountered in Figs. 20.9–20.11 are explained below, when the time reversed problem is considered. We already notice that the downstream parameters have an advantage over the upstream ones in that the jump curve is monotonic in Δ_2, but not in Δ_1.

To complete the classification, we need to introduce **eleven transition points**, delimiting the twelve different kinds of MHD jumps and shocks that (may) occur on the jump curve. They are all determined by simple algebraic equations, only involving the parameters ϑ_1 and β_1.

– *Alfvén, fast and slow points* A, F *and* S,

$$\Delta_A = 0, \qquad \Delta_{F,S} = -a \pm \sqrt{a^2 - b}, \tag{20.80}$$

$$a \equiv \tfrac{1}{2}(1 - \tfrac{1}{2}\gamma\beta_1) - \tfrac{1}{2}(1 + \tfrac{1}{2}\gamma\beta_1)\tan^2\vartheta_1, \qquad b \equiv -\tan^2\vartheta_1.$$

– *Intermediate demarcation points* D *and* E,

$$\Delta_{1D,E} = 0, \qquad \Delta_{2D,E} = -c \pm \sqrt{c^2 - d}, \tag{20.81}$$

$$c \equiv \frac{1 - \tfrac{1}{2}\gamma\beta_1 - \tfrac{1}{2}\gamma(1 + \beta_1)\tan^2\vartheta_1}{\gamma + 1}, \qquad d \equiv -\frac{\tan^2\vartheta_1}{\gamma + 1}.$$

– *Intermediate turning points* B *and* C, *determined by a quartic for* $\Delta_{2B,C}$,

$$\Delta_2^4 + e\Delta_2^3 + f\Delta_2^2 + g\Delta_2 + h = 0, \qquad e \equiv \frac{2(2 - \gamma)}{\gamma - 1}\tan^2\vartheta_1, \tag{20.82}$$

$$f \equiv \frac{4\gamma + 2 + (2 - \gamma)\left[2 - \gamma\beta_1 - \gamma(1 + \beta_1)\tan^2\vartheta_1\right]}{(\gamma + 1)(\gamma - 1)}\tan^2\vartheta_1,$$

$$g \equiv \frac{2\left[2 - \gamma\beta_1 - \gamma(1 + \beta_1)\tan^2\vartheta_1\right]}{(\gamma + 1)(\gamma - 1)}\tan^2\vartheta_1, \qquad h \equiv -\frac{\tan^4\vartheta_1}{(\gamma + 1)(\gamma - 1)},$$

with corresponding $\Delta_{1B,C}$ determined by the inverse jump condition (20.75).

– *Region* I *transition points* X *and* Y, *and region* III *transition points* U *and* V,

$$\Delta_{1X,Y} = \Delta_F, \qquad \Delta_{2X,Y} = -u \pm \sqrt{u^2 - w}, \tag{20.83}$$

$$\Delta_{1U,V} = \Delta_S, \qquad \Delta_{2U,V} = -v \pm \sqrt{v^2 - w}, \tag{20.84}$$

$$u \equiv \frac{\Delta_F + 1 - \tfrac{1}{2}\gamma\beta_1 - \tfrac{1}{2}(1 + \beta_1)\tan^2\vartheta_1}{\gamma + 1},$$

$$v \equiv \frac{\Delta_S + 1 - \tfrac{1}{2}\gamma\beta_1 - \tfrac{1}{2}(1 + \beta_1)\tan^2\vartheta_1}{\gamma + 1}, \qquad w \equiv \frac{\tan^2\vartheta_1}{\gamma + 1}.$$

This enables the construction of all the points of the sequence (20.68) that appear in our new terminology for the different MHD jumps and shocks.

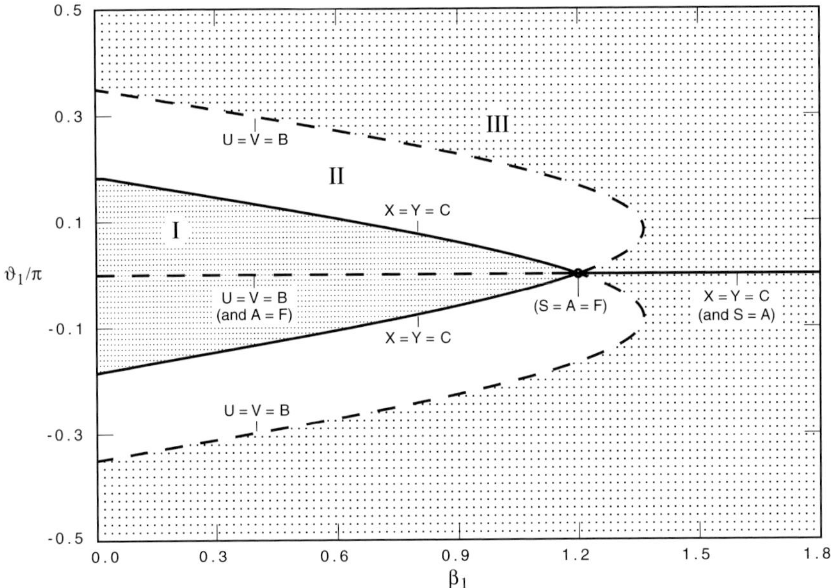

Fig. 20.12 Regions I, II and III of the β_1–ϑ_1 parameter plane determining the different shapes of the jump curves, i.e. the permitted intermediate jumps and shocks. Parameter regions I and II apply for $\Delta_1 \geq 0$, regions II and III apply for $\Delta_1 \leq 0$.

The points H and L refer to the high and low Δ_1 extremes of the jump curve. In Figs. 20.9–20.11, the label H indicates either the limiting value of the asymptote (horizontal dash) or its slope (inclined dash with ∞ on the right) for $\Delta_1 \rightarrow \infty$, exploiting Eq. (20.77). The low side is either limited by $\Delta_1 = -1$ or by L, corresponding to vanishing numerator of the expression (20.73) for β_2. This may also happen in the middle of the jump curve (e.g. in between L_2 and L_3 in Fig. 20.11(c), where three of the intermediate jumps have been struck through), so that the parts corresponding to $\beta_2 < 0$ have to be eliminated. There is nothing disturbing about this (it also occurs in HD, see Fig. 20.5), but, unfortunately, the criterion $\beta_2 \geq 0$ can only be applied after the shock problem has been solved, not before.

The parameter regions I, II and III can be determined before solving the shock problem. They just depend on the values of the two parameters ϑ_1 and β_1: see Fig. 20.12. The criteria for parameter regions I and III directly follow from the discriminants of the expressions (20.83) and (20.84):

$$D_{\mathrm{I}} \equiv u^2 - w \geq 0, \qquad D_{\mathrm{III}} \equiv v^2 - w \geq 0, \tag{20.85}$$

whereas parameter region II occurs when neither of these criteria is satisfied. Parameter region I refers to the presence ($D_{\mathrm{I}} \geq 0$) or absence ($D_{\mathrm{I}} < 0$) of the transition points C, X and Y of the intermediate shocks to the right ($\Delta_1 \geq 0$) of the Alfvén point. For $D_{\mathrm{I}} = 0$, the points C, X and Y coalesce, so that for $D_{\mathrm{I}} < 0$, the intermediate shocks $\widehat{\mathrm{R}}^+$ and P^+ (shown in Fig. 20.11(a)) disappear. Parameter region III refers to the presence ($D_{\mathrm{III}} \geq 0$) or absence ($D_{\mathrm{III}} < 0$) of the transition points B, U and V of the intermediate jumps to the left ($\Delta_1 \leq 0$) of the Alfvén point. For $D_{\mathrm{III}} = 0$, the points B, U and V coalesce, so that for $D_{\mathrm{III}} < 0$, the intermediate jumps $\widehat{\mathrm{R}}^-$ and P^- (shown in Fig. 20.11(c), although unphysical for the particular parameter values chosen) disappear. Consequently, in parameter region II (Fig. 20.9) none of the labels X, Y and U, V appear, and only the intermediate shocks R^+ and $\widehat{\mathrm{P}}^+$ and jumps R^- and $\widehat{\mathrm{P}}^-$ remain.

Let us now consider the *time reversed* problem, analogous to our exposition in Section 20.2.1 of HD jumps and shocks. Reversing the direction of time, i.e. reversing the flow, does not alter the distilled ideal MHD jump problem (20.71)–(20.73) since the Alfvén Mach number only occurs squared there. This is why jumps and shocks could be treated on an equal footing until now. Only when the additional entropy condition (20.74) is considered, i.e. when downstream entropy increase is used to fix the arrow of time for the enlarged dissipative system, are jumps and shocks discriminated. For the complete distilled MHD shock problem (20.71)–(20.74), the shocks survive and the jumps should be discarded.

An alternative point of view is to interpret the time reversal operation as just interchanging the roles of upstream and downstream states. The different jumps and shocks obtained in Figs. 20.9–20.11 then exchange position. Instead of jump curves $M_2^2 = M_2^2(M_1^2; \vartheta_1, \beta_1)$, one obtains jump curves $M_1^2 = M_1^2(M_2^2; \vartheta_2, \beta_2)$. These remain S-shaped, but they connect the different jumps and shocks in a completely different order, as illustrated by Fig. 20.13(b). For example, the intermediate shock R_{23}^+ of Fig. 20.13(a) becomes the intermediate jump R_{23}^- of Fig. 20.13(b). These may be considered as the solutions of two different physical problems, but they may also be considered as solutions of the same problem described by two different representations. This permits us to classify the MHD shocks in a new way, viz. by noting that some of the intermediate (IM) jumps/shocks remain retrograde and prograde with respect to M_1^2–M_2^2 dependences, viz. $\mathrm{R}^\pm \to \mathrm{R}^\mp$ and $\mathrm{P}^\pm \to \mathrm{P}^\mp$, but others get the opposite sense of monotonicity, viz. $\widehat{\mathrm{P}}^\pm \to \widehat{\mathrm{R}}^\mp$ and $\widehat{\mathrm{R}}^\pm \to \widehat{\mathrm{P}}^\mp$. We call the latter *quasi-retrograde* and *quasi-prograde*.

Time reversal duality between entropy-forbidden jumps and entropy-permitted shocks in MHD is now expressed by the following correspondence:

$$
\left\{
\begin{aligned}
\Delta_2 &= \Delta_2^{\alpha\pm}(\Delta_1, \vartheta_1, \beta_1) \\
\vartheta_2 &= \vartheta_2^{\alpha\pm}(\Delta_1, \vartheta_1, \beta_1) \\
\beta_2 &= \beta_2^{\alpha\pm}(\Delta_1, \vartheta_1, \beta_1)
\end{aligned}
\right.
\quad \Longleftrightarrow \quad
\left\{
\begin{aligned}
\Delta_1 &= \Delta_1^{\alpha\mp}(\Delta_2, \vartheta_2, \beta_2) \\
\vartheta_1 &= \vartheta_1^{\alpha\mp}(\Delta_2, \vartheta_2, \beta_2) \\
\beta_1 &= \beta_1^{\alpha\mp}(\Delta_2, \vartheta_2, \beta_2)
\end{aligned}
\right. , \quad (20.86)
$$

$$
\alpha^+ \equiv (\mathrm{F}^+, \widehat{\mathrm{R}}^+, \mathrm{R}^+, \mathrm{P}^+, \widehat{\mathrm{P}}^+, \mathrm{S}^+), \quad \alpha^- \equiv (\mathrm{F}^-, \widehat{\mathrm{P}}^-, \mathrm{R}^-, \mathrm{P}^-, \widehat{\mathrm{R}}^-, \mathrm{S}^-).
$$

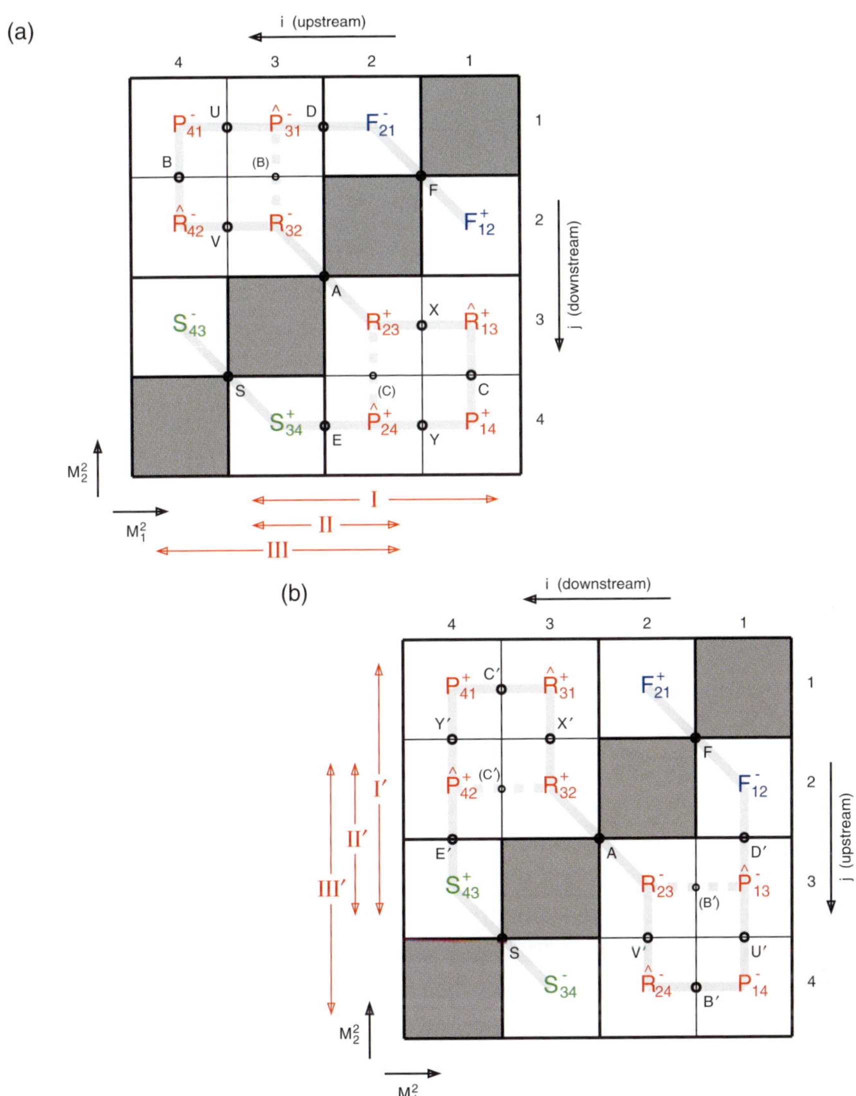

Fig. 20.13 Time reversal duality between entropy-forbidden jumps and entropy-permitted shocks in MHD and classification of discontinuities: (a) jumps and shocks connected by the forward jump curve (schematic, in grey), (b) jumps and shocks connected by the inverse jump curve. Parts of the curves are missing (connected by the dashed short cuts) for the parameter regions indicated.

With the expressions (20.71)–(20.73) of the distilled jump conditions, this scheme not only yields a straightforward prescription for the construction of shocks at each point of a transonic plasma, but it also permits us to uniquely classify the MHD jumps and shocks by means of the monotonicity properties of the jump curves.

The following unique characterization of the *six dual pairs* of entropy-allowed shocks (bold italic[+]) and entropy-forbidden jumps (italic[−]) in magnetohydrodynamics emerges:

- ***Fast shocks*** $\quad \mathrm{F}_{12}^{+}: \quad \Delta_{\mathrm{F}} \leq \Delta_1 < \infty, \quad \Delta_{\mathrm{F}} \leq \Delta_2 < \infty;$
- *Fast jumps* $\quad \mathrm{F}_{21}^{-}: \quad 0 \leq \Delta_1 \leq \Delta_{\mathrm{F}}, \quad \Delta_{2\mathrm{D}} \leq \Delta_2 \leq \Delta_{\mathrm{F}};$
- ***Quasi-retrograde IM shocks*** $\widehat{\mathrm{R}}_{13}^{+}: \quad \Delta_{1\mathrm{C}} \geq \Delta_1 > \Delta_{1\mathrm{X/C}}, \; \Delta_{2\mathrm{C}} \leq \Delta_2 < \Delta_{2\mathrm{X/C}};$
- *Quasi-prograde IM jumps* $\widehat{\mathrm{P}}_{31}^{-}: \Delta_{1\mathrm{U/B}} \leq \Delta_1 \leq 0, \quad \Delta_{2\mathrm{U/B}} \leq \Delta_2 \leq \Delta_{2\mathrm{D}};$
- ***Retrograde IM shocks*** $\quad \mathrm{R}_{23}^{+}: \Delta_{1\mathrm{X/C}} \geq \Delta_1 \geq 0, \quad \Delta_{2\mathrm{X/C}} \leq \Delta_2 \leq 0;$
- *Retrograde IM jumps* $\quad \mathrm{R}_{32}^{-}: \quad 0 \geq \Delta_1 \geq \Delta_{1\mathrm{V/B}}, \quad 0 \leq \Delta_2 \leq \Delta_{2\mathrm{V/B}};$
- ***Prograde IM shocks*** $\quad \mathrm{P}_{14}^{+}: \Delta_{1\mathrm{Y/C}} < \Delta_1 \leq \Delta_{1\mathrm{C}}, \; \Delta_{2\mathrm{Y/C}} < \Delta_2 \leq \Delta_{2\mathrm{C}};$
- *Prograde IM jumps* $\quad \mathrm{P}_{41}^{-}: \quad \Delta_{1\mathrm{B}} \leq \Delta_1 < \Delta_{1\mathrm{U/B}}, \; \Delta_{2\mathrm{B}} \leq \Delta_2 < \Delta_{2\mathrm{U/B}};$
- ***Quasi-prograde IM shocks*** $\widehat{\mathrm{P}}_{24}^{+}: \quad 0 \leq \Delta_1 \leq \Delta_{1\mathrm{Y/C}}, \; \Delta_{2\mathrm{E}} \leq \Delta_2 \leq \Delta_{2\mathrm{Y/C}};$
- *Quasi-retrograde IM jumps* $\widehat{\mathrm{R}}_{42}^{-}: \Delta_{1\mathrm{V/B}} > \Delta_1 \geq \Delta_{1\mathrm{B}}, \; \Delta_{2\mathrm{V/B}} < \Delta_2 \leq \Delta_{2\mathrm{B}};$
- ***Slow shocks*** $\quad \mathrm{S}_{34}^{+}: \quad \Delta_{\mathrm{S}} \leq \Delta_1 \leq 0, \quad \Delta_{\mathrm{S}} \leq \Delta_2 \leq \Delta_{2\mathrm{E}};$
- *Slow jumps* $\quad \mathrm{S}_{43}^{-}: \quad -1 \leq \Delta_1 \leq \Delta_{\mathrm{S}}, \quad -1 \leq \Delta_2 \leq \Delta_{\mathrm{S}}.$

$$\tag{20.87}$$

Here, for the sake of comparison with the standard terminology, the redundant notation in terms of $\alpha^{\pm}$ as well as the indices ij is presented. Of course, exploiting either one is enough. Due to the peculiar properties of the intermediate jumps and shocks, the jump curve folds over in the Alfvénic range. As a result, the indicated parameter ranges from the relations (20.80)–(20.84) overlap in Δ_1 but not in Δ_2, so that any shock or jump is uniquely described by the latter value, i.e. by M_2^2. One also recognizes the associated sub-structure: the "core" IM discontinuities $\mathrm{R}^{\pm}$, $\mathrm{P}^{\pm}$ (on the anti-diagonals of Figs. 20.13) are separated from the fast discontinuities $\mathrm{F}^{\pm}$ by the quasi-retrograde/prograde IM discontinuities $\widehat{\mathrm{R}}^{+}$, $\widehat{\mathrm{P}}^{-}$, and from the slow discontinuities $\mathrm{S}^{\pm}$ by the quasi-retrograde/prograde IM discontinuities $\widehat{\mathrm{P}}^{+}$, $\widehat{\mathrm{R}}^{-}$.

20.3.3 Angular dependence of MHD shocks

The problems of parameter reduction by the distilled jump conditions and of the classification of MHD shocks being completely solved now, it remains to show how the downstream velocities of the different discontinuities depend on the angle of incidence ϑ_1 of the upstream velocity. This will complement the provisional illustrations of Fig. 20.8 with a more quantitative description. Since the three values M_{S}, M_{A} ($\equiv 1$) and M_{F} of the normal Alfvén Mach number corresponding to the three linear MHD waves play a central role in the classification of shocks, a generalization of the Friedrichs diagram (see Figs. 5.5 and 5.13[1]) would appear to give the most logical angular representation. In the present context, this

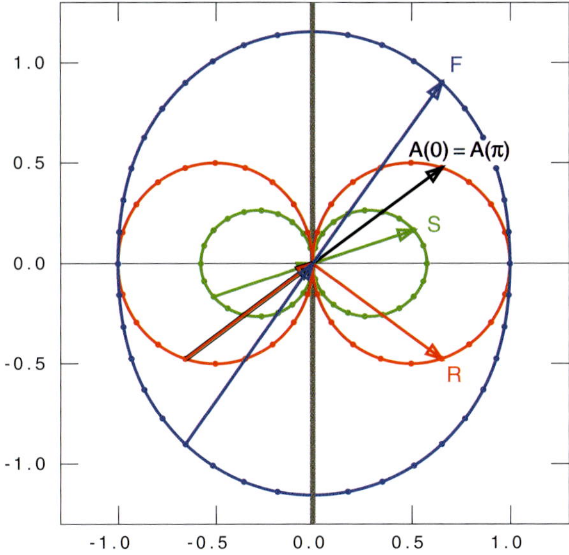

Fig. 20.14 Friedrichs-type diagram: "weak" shocks for $\beta_1 = 0.4$. In the limit $\sigma \to 0$, slow and fast shocks become magneto-acoustic waves, but intermediate shocks never (except for $\vartheta_1 \to 0$) become weak: $\vartheta_2 \neq \vartheta_1$. Instead, Alfvén waves are obtained from the rotational (or Alfvén) discontinuity in the limit $\varphi \to 0$ or π. Dots correspond to a discrete set of angles $\vartheta_{1k} = \pm(k/20)\,\pi$, with $k = 0, 1, \ldots, 10$. For distinction, the velocity vectors are shown at $k = 2$ for slow shocks, at $k = 4$ for intermediate shocks and rotational discontinuities, and at $k = 6$ for fast shocks.

becomes a superposition of the three pairs of equal upstream and downstream velocities at $M = M_{\rm S}$, $M_{\rm A}$ and $M_{\rm F}$ for all angles $|\vartheta_1| = |\vartheta_2| \leq \pi/2$, as represented in Fig. 20.14. Two caveats are in order here.

First, the shock strength, defined as $\sigma \equiv 1 - M_2^2/M_1^2$, and leading to the expression (20.66) for "weak" shocks, is actually misleading with respect to the intermediate shocks and their relationship with Alfvén waves. In the limit $\sigma \to 0$, when both $M_1^2 \to 1$ and $M_2^2 \to 1$ along the jump curve, but from opposite sides of the Alfvén point according to Eq. (20.76), the retrograde intermediate jumps R^- and shocks R^+ coalesce, but they do not become weak (hence the quotation marks above). In general (except for $\vartheta_1 \to 0$), a finite jump of the directions of the magnetic field and velocity remains, since $\vartheta_1 = -\vartheta_2$. Hence, *intermediate shocks almost never become weak*, and, in the limit $\sigma \to 0$, *they do not become Alfvén waves*. Instead, Alfvén waves proper emerge from the rotational discontinuities (appropriately called Alfvén discontinuities) in the limit $\varphi \to 0$ or π. Whereas the central Alfvén point $M = M_{\rm A} = 1$ and the intermediate jumps and shocks around it clearly mold the shape of the jump curve, intermediate discontinuities remain magneto-acoustic phenomena, with jumps restricted to the plane through the magnetic field and the normal, whereas Alfvén discontinuities have jumps in a plane perpendicular to that. The two kinds of discontinuity may be superposed at a particular point in time, but in the subsequent nonlinear dynamics the orientations of those planes will change and a complicated, non-constant, mix of Alfvén-type and magneto-sonic type of discontinuity will emerge.

Second, the Friedrichs diagram is actually an awkward representation for the angular dependence of MHD shocks since it presents the phase speeds of the different MHD waves in the direction of the normal $\mathbf{n}$. According to Eqs. (5.79) and (5.82)[1], this yields the expressions

$$(v_{\mathrm{ph}})_{\mathrm{A}} \ = \ \frac{B}{\sqrt{\rho}} \cdot \cos\vartheta, \tag{20.88}$$

$$(v_{\mathrm{ph}})_{\mathrm{F,S}} \ = \ \frac{B}{\sqrt{\rho}} \cdot \sqrt{\tfrac{1}{2}(1 + \tfrac{1}{2}\gamma\beta) \pm \tfrac{1}{2}\sqrt{(1 + \tfrac{1}{2}\gamma\beta)^2 - 2\gamma\beta\cos^2\vartheta}},$$

exhibiting the well known feature that the Alfvén and slow waves do not propagate in the perpendicular directions $\vartheta = \pm\pi/2$. Exploiting such a representation for strong shocks is undesirable since it requires separate treatment for the perpendicular shocks. In fact, the representation of Fig. 20.14 is a superposition of three plots, corresponding to the three special values (20.66) of the Alfvén Mach number,

$$M_{\mathrm{A}} \ = \ 1, \tag{20.89}$$

$$M_{\mathrm{F,S}} \ = \ \frac{1}{|\cos\vartheta|} \sqrt{\tfrac{1}{2}(1 + \tfrac{1}{2}\gamma\beta) \pm \tfrac{1}{2}\sqrt{(1 + \tfrac{1}{2}\gamma\beta)^2 - 2\gamma\beta\cos^2\vartheta}},$$

where the latter two even cover a range of values when ϑ is varied. This is not what we want. To study the effect of the angle of incidence of a shock, we wish to keep the value of the Alfvén Mach number M_1 (and also β_1) fixed, and just vary ϑ_1. Note that, in contrast to the phase speeds (20.88), the values for the Alfvén Mach numbers (20.89) do not vanish for $\vartheta = \pm\pi/2$. Clearly, dividing by $|\cos\vartheta|$ makes a crucial difference. We will make use of this.

We now return to the expressions (20.60) for the velocities $\bar{\mathbf{v}}_1^*$ and $\bar{\mathbf{v}}_2^*$. For fixed M_1^2, but varying ϑ_1, the left expression for $\bar{\mathbf{v}}_1^*$ corresponds to a straight line so that perpendicular shocks are properly described in the limit $\vartheta_1 \to \pm\pi/2$, but $|\bar{\mathbf{v}}_1^*| \to \infty$ there. This may be remedied by exploiting a powerful property of the de Hoffmann–Teller transformation by which the expressions for the velocities can be simplified even further. Since $\mathbf{v}_i^*$ and $\mathbf{B}_i$ are parallel, the squares of the upstream and downstream normal Alfvén Mach numbers may be written as

$$M_i^2 \ \equiv \ \frac{\rho_i v_{\mathrm{n}i}^2}{B_{\mathrm{n}}^2} \ = \ \frac{\rho_i v_i^{*2}}{B_i^2} \quad (i = 1, 2). \tag{20.90}$$

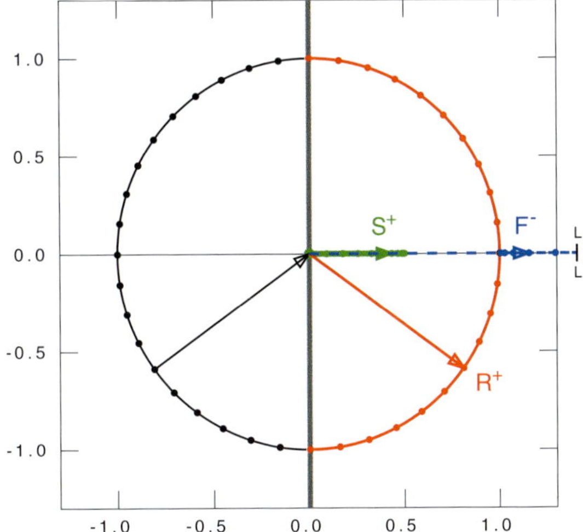

Fig. 20.15 Polar plot of the velocities $\mathbf{v}^*_{1,2}(\vartheta_1)$ in the de Hoffmann–Teller frame for $M_1 = 1$ and $\beta_1 = 0.4$. Vectors for the retrograde IM shock, the switch-off slow shock and the switch-off fast jump are shown at $k = 4$ ($\vartheta_1 = 0.2\pi$). The fast jump becomes unphysical beyond L (corresponding to $\vartheta_1 > 0.279\pi$) since $\beta_2 < 0$ there.

In other words: *for the velocities in de Hoffmann–Teller frame, the normal Alfvén Mach number coincides with the Alfvén Mach number itself!* This means that the Alfvén Mach number M_i can be exploited as a measure for the total velocity, not just for the normal component as implied by the expressions (20.60).

Recall that the bars on the velocities $\bar{\mathbf{v}}^*_1$ and $\bar{\mathbf{v}}^*_2$ come from the normalization (20.42)–(20.43) that we exploited so far, where the constancy of the factors ρv_n and B_n across the surface of discontinuity was utilized to symmetrize the jump conditions as far as possible with respect to upstream and downstream values of the parameters. This having been accomplished, we may now return to the real world, so to speak, and normalize the velocities in a more practical way, based on the upstream value $v_{\mathrm{A}1} = B_1/\sqrt{\rho_1}$ of the Alfvén speed:

$$\bar{\mathbf{v}}^*_1 \equiv \frac{\rho|v_\mathrm{n}|}{B_\mathrm{n}^2}\,\mathbf{v}^*_1 = \frac{M_1}{\cos\vartheta_1}\,\widetilde{\mathbf{v}}^*_1 , \qquad \bar{\mathbf{v}}^*_2 \equiv \frac{\rho|v_\mathrm{n}|}{B_\mathrm{n}^2}\,\mathbf{v}^*_2 = \frac{M_2}{\lambda\cos\vartheta_2}\,\widetilde{\mathbf{v}}^*_2 , \tag{20.91}$$

where $\widetilde{\mathbf{v}}^*_i \equiv \mathbf{v}^*_i/v_{\mathrm{A}1}$ and the factor λ converts the downstream velocity into the upstream normalization:

$$\lambda \equiv \frac{v_{\mathrm{A}2}}{v_{\mathrm{A}1}} \equiv \frac{B_2}{B_1}\sqrt{\frac{\rho_1}{\rho_2}} = \frac{\cos\vartheta_1}{\cos\vartheta_2}\frac{M_2}{M_1}. \tag{20.92}$$

Dropping the tildes, as usual, this yields the following asymmetric expressions for the upstream and downstream velocities in the de Hoffmann–Teller frame:

$$\begin{cases} v_{n1}^* = -M_1 \cos \vartheta_1 \\ v_{t1}^* = M_1 \sin \vartheta_1 \end{cases} \Rightarrow \begin{cases} v_{n2}^* = -\lambda M_2 \cos \vartheta_2 \\ v_{t2}^* = \lambda M_2 \sin \vartheta_2 \end{cases}. \tag{20.93}$$

Now, the upstream Alfvén Mach number M_1 and the effective downstream Alfvén Mach number λM_2 are just the normalized *total* upstream and downstream velocities. For fixed M_1 and β_1, the locus of the endpoints of the velocity $\mathbf{v}_1^*(\vartheta_1)$ is a half-circle and the corresponding locus for $\mathbf{v}_2^*(\vartheta_1)$ may be computed from the explicit solutions of the distilled jump problem (20.71)–(20.73) obtained. Hence, with this normalization, a much more effective representation of the polar plots is obtained than that resulting from either the Friedrichs representation or the one given by the original expressions (20.60). This is illustrated by Fig. 20.15, which shows the velocities of the three jumps/shocks that occur at $M_1 = 1$.

Since the Alfvén point A of the jump curve of Fig. 20.9 is a fixed point ($M_1 = M_2$, irrespective of the values of ϑ_1 and β_1) of the transformation $M_1 \to M_2$, the absolute magnitude of the velocity of the retrograde intermediate shocks R^+ (limit $M_1 \downarrow 1$) is the same upstream and downstream. Hence, the polar plot for R^+ is a circle in Fig. 20.15. With respect to the angular dependence $\vartheta_2 = \vartheta_2(\vartheta_1)$ for the retrograde IM shocks, close to the point A, it follows from Eqs. (20.76) and Eq. (20.72) that

$$\vartheta_2 \approx -\vartheta_1, \tag{20.94}$$

thus producing a second reason for the terminology *retrograde*. At A, this becomes $\vartheta_2 = -\vartheta_1$ so that the directions of the upstream and downstream velocities for the $M = 1$ retrograde IM shocks on the circle of Fig. 20.15 continuously change from $\vartheta_1 = \vartheta_2 = 0$ (parallel shock) to $\vartheta_1 = -\vartheta_2 = \pi/2$ (perpendicular shock). This shows, once more, the central role of the intermediate shocks for the description of MHD discontinuities. In contrast, for the points D and E of Fig. 20.9, which demarcate the transitions from intermediate to fast and slow jumps or shocks, the downstream Mach number M_2 depends on ϑ_1, so that the magnitudes of the downstream velocities vary but, since $M_1 = 1$, the downstream angle $\vartheta_2 = 0$ according to Eq. (20.72). Hence, these points correspond to *switch-off* fast–intermediate jumps F^- (limit $M_1 \uparrow 1$) and *switch-off* slow–intermediate shocks S^+ (limit $M_1 \downarrow 1$) for all values of the upstream angle ϑ_1. Recall that, by the principle of time reversal duality, corresponding *switch-on* fast–intermediate shocks F^+ and *switch-on* slow–intermediate jumps S^- may be obtained from these solutions by just interchanging the upstream and downstream values of the parameters.

We now complete the description of the angular dependence of MHD shocks by discussing the rather complex behavior for sub- and super-Alfvénic values of M_1, where we have selected just four representative cases in Fig. 20.16. These pictures

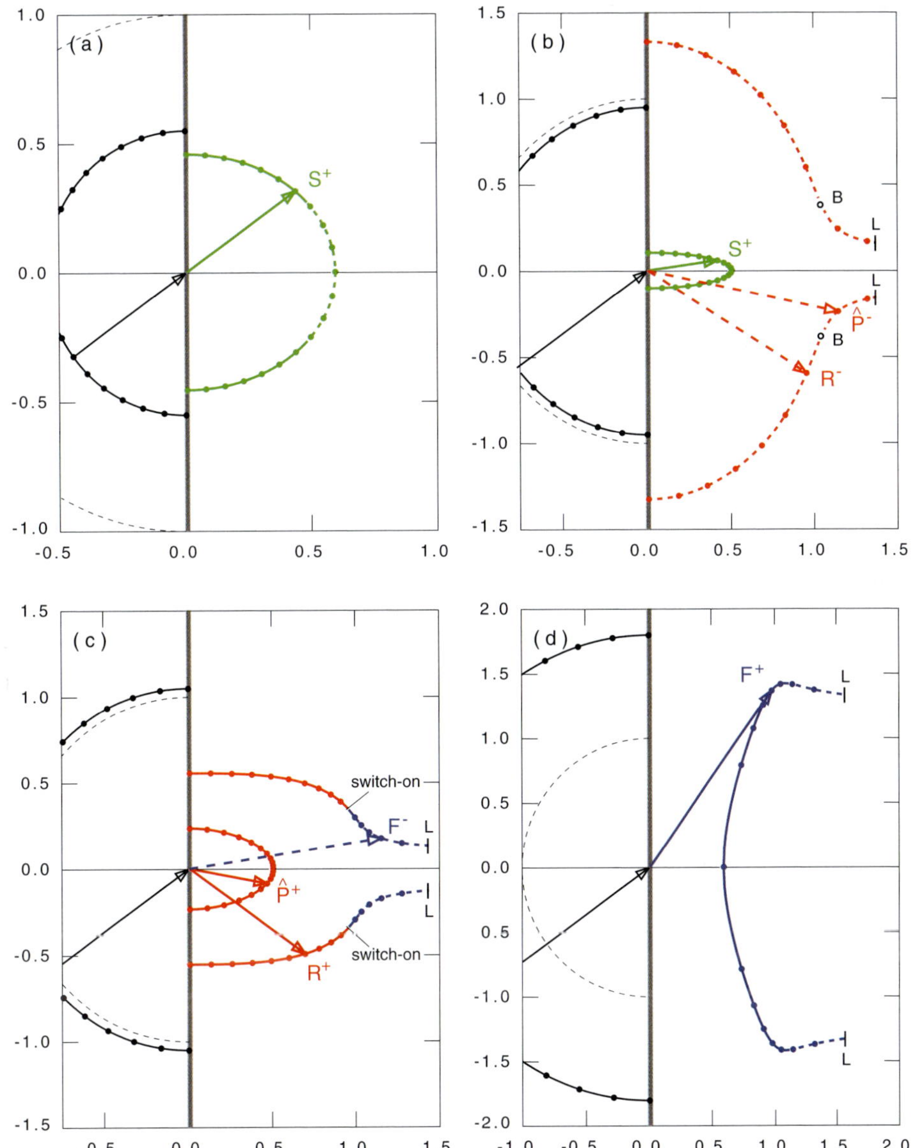

Fig. 20.16 Polar plot of the velocities $\mathbf{v}^*_{1,2}(\vartheta_1)$ in the de Hoffmann–Teller frame for $\beta_1 = 0.4$ and increasing Alfvén Mach number: (a) $M_1 = 0.55$, (b) $M_1 = 0.95$, (c) $M_1 = 1.05$, (d) $M_1 = 1.8$. Velocity vectors are shown at $k = 4$ ($\vartheta_1 = 0.2\pi$). The dashed circle on the upstream side indicates the Alfvén speed, $M_1 = 1$.

should be considered as mere illustrations of the different kinds of shocks since the actual solution of the distilled MHD shock problem has already been obtained in Section 20.3.2. Also note that, for proper interpretation of the polar plots of

Figs. 20.15 and 20.16, one would need the corresponding dependence of $\beta_2(\vartheta_1)$, which is not shown, with possible crossings to negative values like in Figs. 20.10(b) and 20.11(d). For example, the switch-off fast jumps F^- of Fig. 20.15 only exist for incident angles $0 \leq \vartheta_1 \leq 0.279\pi$. For larger angles they become unphysical since $\beta_2 < 0$ then: there is simply no discontinuity possible for those values of the upstream parameters.

In Figs. 20.16(a) and (b), the polar velocity plots are shown for *sub-Alfvénic* values of M_1. For those values, parameter region II (jump curves as in Fig. 20.9) applies for smaller ϑ_1 and parameter region III (jump curves as in Fig. 20.11(c)) applies for larger angles. The resulting polar plot of slow jumps and shocks for $M_1 = 0.55$ is shown in Fig. 20.16(a). Since M_1 is far below the Alfvénic value 1, the intermediate jumps (which would have had much larger values of M_2 than the slow discontinuities) are eliminated because they are all unphysical ($\beta_2 < 0$). The remaining slow discontinuities are S^- jumps for $\vartheta_1 < 0.161\pi$ and S^+ shocks for $\vartheta_1 \geq 0.161\pi$. Hence, in contrast to the Friedrichs diagram, weak slow shocks now only occur for the single angle $\vartheta_1 = \vartheta_2 = 0.161\pi$, which separates the slow jumps from the slow shocks. The polar plot for a near-Alfvénic value, $M_1 = 0.95$, is shown in Fig. 20.16(b). The slow jumps S^- have disappeared now and the slow shocks S^+ are focused close to the parallel direction (anticipating the switch-off behavior of Fig. 20.15). Since M_1 is now close to 1, the intermediate jumps have returned: retrograde IM jumps R^- and quasi-prograde IM jumps $\widehat{P}^-$ appear for $\vartheta_1 \geq 0.178\pi$. Starting from the turning point B (where $\vartheta_1 = 0.178$ and $\vartheta_2 = -0.111\pi$), for increasing upstream angle ϑ_1, the downstream angle ϑ_2 monotonically decreases towards $\vartheta_2 = -\pi/2$ (perpendicular jump) for R^-, whereas ϑ_1 monotonically increases towards $\vartheta_2 = -0.037\pi$ (point L, at $\vartheta_1 = 0.263\pi$) for P^-, where this jump becomes unphysical again.

In Figs. 20.16(c) and (d), the polar plots are shown for *super-Alfvénic* values of M_1. Now, parameter region I (jump curves as in Fig. 20.11(a)) applies for smaller ϑ_1 and parameter region II (jump curves as in Fig. 20.9) applies for larger angles. In the polar plot for the near-Alfvénic value $M_1 = 1.05$ shown in Fig. 20.16(c), the intermediate discontinuities have become genuine shocks. For increasing upstream angle ϑ_1, retrograde R^+ shocks switch-on at $\vartheta_1 = 0$ (where $\vartheta_2 = -0.110\pi$), while ϑ_2 monotonically decreases towards $-\pi/2$ (perpendicular R^+ shock). Quasi-prograde $\widehat{P}^+$ shocks also appear for all upstream angles, but their ϑ_2 monotonically decreases from $\vartheta_2 = 0$ towards $-\pi/2$ (perpendicular $\widehat{P}^+$ shock). For this near-Alfvénic value of M_1, fast F^+ shocks only occur for a narrow range of ϑ_1: they switch-on at $\vartheta_1 = 0$ (where $\vartheta_2 = 0.110\pi$), their value of ϑ_2 monotonically decreases to $\vartheta_1 = \vartheta_2 = 0.081\pi$, where they turn into fast F^- jumps, which become unphysical at $\vartheta_1 = 0.292\pi$ ($\vartheta_2 = 0.028\pi$). For the large super-Alfvénic value $M_1 = 1.8$ illustrated in Fig. 20.16(d), only fast discontinuities survive. Fast F^+ shocks occur for the rather wide range $0 \leq \vartheta_1 \leq 0.289\pi$ and they turn into fast F^- jumps at $\vartheta_1 = 0.289\pi$ (where $\vartheta_2 = 0.288\pi$), which become unphysical beyond $\vartheta_1 = 0.386\pi$ (where $\vartheta_2 = 0.224\pi$).

A final cautionary remark on the angular dependence $\vartheta_2(\vartheta_1)$ of the polar plots is in order. For fixed Δ_1 and β_1, this dependence follows from the misleadingly simple relationship (20.72),

$$\Delta_2 \tan \vartheta_2 = \Delta_1 \tan \vartheta_1 \,, \tag{20.95}$$

i.e. it involves the solution $\Delta_2(\vartheta_1)$ of the distilled jump condition (20.71). To avoid possible misunderstanding with respect to our terminology, which is based on the monotonicity of the jump condition with respect to the Alfvén Mach numbers: in general, because of the dependence $\Delta_2(\vartheta_1)$, the angular dependence $\vartheta_2(\vartheta_1)$ is *not monotonic*. For example, for the fast shocks and jumps of Fig. 20.16(d) just discussed, ϑ_2 rapidly increases for small ϑ_1 (a remnant of the switch-on behavior at $M_1 = 1$), becomes rather flat and reaches a maximum ($\vartheta_2 = 0.304\pi$ at $\vartheta_1 = 0.188\pi$), and then decreases again.

20.3.4 Observational considerations of MHD shocks

We have now discussed the implications of the distilled jump conditions (20.71)–(20.73) from a wide variety of view points, with the time reversal duality (20.86) and the associated classification (20.87) of shocks as the most prominent ones. They imply that a transonic plasma is uniquely described by assigning values to the three upstream parameters Δ_1, ϑ_1, β_1 at each point,[2] at shock fronts complemented with the values of the three downstream parameters Δ_2, ϑ_2, β_2. As we have seen, the relationship between these two parameter spaces is a very intricate one, where the downstream parameters are not uniquely determined by the upstream ones since the distilled energy jump condition (20.71) is a cubic equation, having one, two or three solutions at each point of the shock front, dependent on the value of Δ_1. Moreover, some of the values for the upstream parameters have to be excluded a posteriori since they correspond to negative values of the downstream pressure. Having arrived at this point of our exposition, the reader may wonder how such a rich structure, with far-reaching physical implications, could have emerged from just a set of algebraic relations (20.3)–(20.9), not even involving partial differential equations. The reason is that these relations express, at each point of the transonic plasma, the laws of conservation of mass, momentum, energy and magnetic flux of the ideal MHD model, supplemented with a prescription for the entropy change.

---

[2] Recall, though, that we have eliminated the two parameters ρ_1 and B_1 of scale independence and the two parameters of the de Hoffmann–Teller boost $\mathbf{v}^* - \mathbf{v}$. Of course, in a full 3D calculation of all seven primitive variables, these four need to be computed as well since they have different values at different positions.

In a certain sense, these equations bring together all the physical properties of the model of nonlinear magnetohydrodynamics, discussed so far, at a single point! It thus focuses on the nonlinearity of these equations, which, by definition, does not relate to the differential equations but to the algebraic aspects of the model.

Which of the multifarious MHD shocks may be realized in actual stationary or time-dependent transonic MHD flows will be determined by the dissipative and nonlinear processes acting in such an inhomogeneous configuration. The ones that survive are called *evolutionary*. Since intermediate shocks precisely occur when the distilled shock problem is multi-valued in terms of the upstream Alfvén Mach number M_1, such shocks typically have bifurcations and may split up into other types of discontinuity. For dissipative models in simple geometries in the limit of vanishing dissipation, the intermediate shocks were shown not to be stable against certain perturbations. Traditionally (Akhiezer *et al.* [6, 5], Germain [155], Jeffrey and Taniuti [250]), this failure of the test of evolutionarity has been used to rule out the physical reality of intermediate shocks. However, in 1988, Wu [487] showed intermediate shocks arising in a numerical solution of the dissipative MHD equations through nonlinear steepening from a continuous wave. This claim was further substantiated by calculations by Kennel *et al.* [254] for a dissipative model problem and by De Sterck *et al.* [112] on transonic flows in complex geometries like those encountered at solar wind–magnetosphere boundaries.

After Wu's paper, a flurry of numerical and observational papers appeared on the formation of intermediate shocks [488, 144, 490, 491, 202], on their astrophysical implications [123, 85], on their relationship with magnetic reconnection [451, 291], on their significance for the Riemann problem [344, 345, 346, 437, 438], on their occurrence in bow-shock flows [113, 114] and on their possible breakup [482]. On the other hand, Falle and Komissarov [132] presented arguments to stick to the traditional view point, subjecting numerical calculations to strict rules on what to call a shock and what a transient feature. However, as noted by Kennel *et al.* [253], and reiterated in Section 20.3.2, this view point has also hampered the development of the theory of MHD shock relations, as demonstrated by the fact that the solutions of this section were obtained only recently [172]. For the purpose of understanding how the MHD jump conditions permit a continuous description in parameter space from slow to fast shocks through the intermediate ones, the problem of evolutionarity of the intermediate shocks can be deferred to a later stage, when conclusive observational and numerical evidence has been collected. It may be noted that such a strategy is customary (because effective) in experimental fusion research when slow dissipative tokamak evolution is analyzed by means of the two-fold arbitrary family of equilibrium solutions of the Grad–Shafranov equation, and the decision on which of these equilibria is actually chosen by the experiment is left to evidence obtained from the different diagnostics. A glance at the list of plasma transport coefficients of Section 3.3.2 [1], and the realization that these just present the classical picture, which nearly always needs to be modified by anomalous corrections of even the orders of magnitude, suffices to appreciate the practical impossibility of predictive transport calculations of the dynamics of magnetized plasmas. Similarly, in the related subjects of helioseismology and MHD spectroscopy [180], discussed in Section 7.2.4 [1], and extended to accretion disks [255, 51] and general transonic flows [173, 169, 52], the full variety of MHD waves for all possible equilibrium configurations is simply cataloged and the final decision on which of those is realized in nature is left to data obtained from observations by telescopes (in astrophysics) or from the various diagnostics (in the laboratory). In physics, the difference between theoretical truth and prejudice is eventually decided by empirical evidence: nature has all the answers, let us consult her.

20.4 Stationary transonic flows

Thus far in this chapter, our exposition concentrated on the various discontinuities that can occur in stationary MHD. Magnetohydrodynamic shocks play a prominent role in the wide variety of transonic laboratory and astrophysical plasmas. Concrete examples range from those exploited in the early pinch implosion experiments [281], to more recent inertial confinement fusion experiments [99], over bow shocks encountered at planetary magnetospheres or at the heliosphere–interstellar medium boundary, up to shocks found in supersonic accretion flows onto neutron stars [395] and black holes [151]. As briefly mentioned in Chapter 14, reconnection processes [387] may also create standing magnetohydrodynamic shock fronts. In turn, steady, transient or recurring shock fronts aid in the acceleration of particles. In many astrophysical outflows, like in extragalactic jets [11, 259], shock-accelerated particles can easily reach relativistic speeds [278]. Knowledge of the MHD shock relations, extended up to relativistic MHD as introduced in Chapter 21, is thus a vital ingredient to our understanding of transonic flows and the physical processes involved in particle acceleration.

Explicit analysis, as given above for the algebraic problem of classifying all the different MHD shocks, is also important for the numerical solution of the nonlinear evolution problem by means of characteristics. In two-dimensional equilibria with flow, the characteristics exhibit both spatial (equilibrium) and temporal (wave) features such that equilibrium and waves appear to be no longer separate issues, as already pointed out in the introduction of this chapter. Some of these features of transonic MHD flow have been analyzed by Goedbloed and Lifschitz [181, 309] for a special class of ("self-similar") solutions that permits in-depth explicit analysis of the transitions and shocks of the flow, by means of a system of coupled PDEs and an algebraic Bernoulli equation. For non-planar flows, in particular MHD flows in spatially two-dimensional axi-symmetric stationary equilibria [181], a further division of the flow regimes discussed in Section 20.2.4 occurs due to the occurrence of *limiting line characteristics* in the hyperbolic magneto-sonic regions (as already shown for gas dynamics by Courant and Friedrichs [97]) and of flows that are disconnected at the Alfvén speed by a *forbidden flow regime due to the constraint of constant Bernoulli function*. These additional divisions make the computation of transonic MHD flows particularly complicated.

In the remainder of this chapter, we will discuss selected examples of continuously varying transonic flows. The solar wind is perhaps the best known example of such a "smooth" transonic flow, on top of which the complex shock patterns associated with CMEs develop. In essence, the solar wind is just one manifestation of a smooth transonic flow whose acceleration is mainly thermally driven, while magnetized stellar winds or outflows emanating from accretion disks can also reach high speeds by e.g. magneto-centrifugal mechanisms. In young star environments where circumstellar accretion disks prevail, strong stellar magnetic fields may even deviate the accreting matter to form transonic "funnel" flows. We briefly discuss the intricacies of transonic solar wind solutions in what follows, and then continue with exemplary stationary astrophysical flows.

20.4.1 Modeling the solar wind–magnetosphere boundary

The existence of intermediate shocks in numerical solutions of the dissipative MHD equations for transonic flows of the type encountered at the solar wind–magnetosphere boundary by De Sterck, Low and Poedts [112] nicely illustrates the surprising complexity of ideal MHD shock-dominated flow patterns. To model the interaction of the solar wind with a "magnetosphere", or to study the shock fronts associated with CMEs traversing the solar wind, the authors considered uniform magnetized flow around perfectly conducting obstacles. When the upstream plasma β is smaller than $2/\gamma$, while the inflow is super-Alfvénic and characterized by an Alfvén Mach number $M = v/v_{\mathrm{An}}$ such that, in agreement with Eq. (20.74) in the limit $\vartheta_1 = 0$, $\Delta_2 = 0$,

$$ 1 \leq M \leq \sqrt{\frac{\gamma(1 - \beta) + 1}{\gamma - 1}}, \tag{20.96} $$

fast switch-on shocks can exist [112] and the shock structure in the flow around the obstacle can become quite complicated. De Sterck *et al.* [112] obtained steady shock structures numerically, using shock-capturing schemes of the type discussed in Chapter 19. Figure 20.17 shows a global view and a detail of the converged bow shock solution for an initially uniform field-aligned flow with $\beta = 0.4$ and $M = 1.5$ around a cylinder in the switch-on regime. The steady shock front was

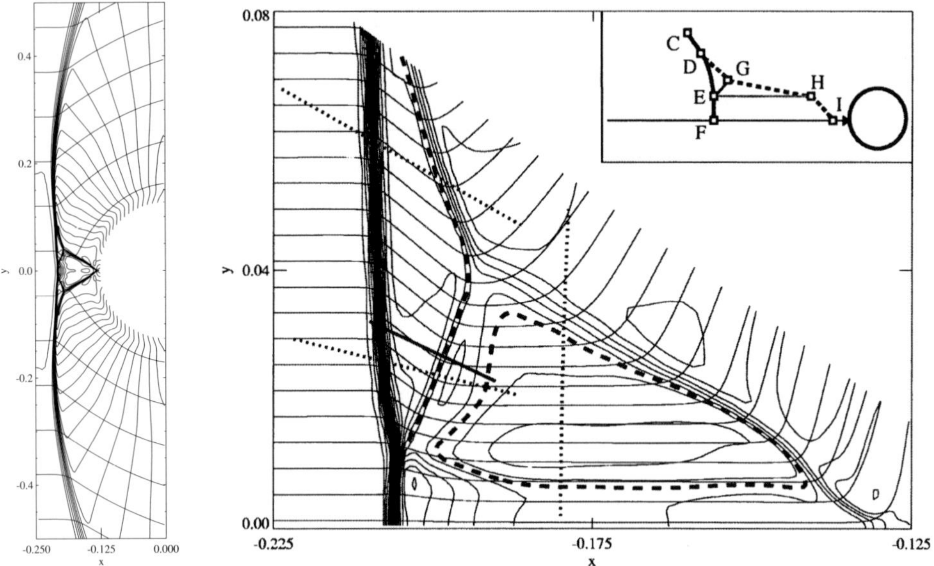

Fig. 20.17 Interacting MHD shocks in a homogeneous flow around a perfectly conducting cylinder. Global view (left) and detail of the flow in the half plane above the stagnation streamline (right). (From De Sterck *et al.* [112].)

found to consist of multiple shock segments of various types. The shock D–E is a fast shock and almost a fast switch-on shock, as $B_{y,1}$ is almost vanishing upstream. The leading shock front part E–F is a hydrodynamic shock and E–G is an intermediate shock as well as D–G–H–I. The latter one is almost a slow switch-off shock. The part indicated with E–H is a tangential discontinuity and other tangential discontinuities stretch out from points D, G and H along the streamlines to infinity. Notice that, in this inflow regime, the shock front shape is not parabolic (the shape expected in a hydrodynamic supersonic inflow about the cylinder), but shows a clear dimple in the leading shock front. This simulation may thus provide an explanation for CME related shocks, which are observed to have such a dimple.

20.4.2 Modeling the solar wind by itself

The solar wind, predicted analytically by Parker [364], is a clear example of a stationary, "smooth" transonic MHD flow. While the original analytic prediction of its existence was made under the assumption of a spherically symmetric, isothermal, unmagnetized, non-rotating solar wind (see Chapter 8 [1]), the true transonic nature of the solar wind is better described in MHD models which relax some of these assumptions. The simplest of these is the 1D Weber–Davis solution [475] which incorporates both rotation and magnetic fields. In spherical coordinates (Appendix A.2.3), it represents a steady-state ($\partial/\partial t = 0$) axi-symmetric ($\partial/\partial\phi = 0$) wind for the equatorial plane only, where the polar angle $\theta = \pi/2$ (assuming $v_\theta = 0$ and $B_\theta = 0$), and it adopts a polytropic relation between density and pressure. The transonic solution for given rotation rate Ω_*, magnetic field strength, and base temperature and density passes through the slow and fast magneto-sonic critical points of the radial momentum equation. The Alfvén point r_A, where the radial velocity equals the radial Alfvén speed, plays an important role in quantifying the angular momentum flux of the solar wind. This flux is due to the advection of angular momentum by the plasma flow and due to magnetic tension. This flux is constant as a function of radius and equal to minus the total rate of change of the angular momentum of the Sun. For the 1D Weber–Davis solution this gives

$$\frac{dJ_z}{dt} = -\frac{2}{3}\left(rv_\phi - \frac{rB_\phi B_r}{\rho v_r}\right)4\pi r^2 \rho v_r = \frac{2}{3}\Omega_* r_A^2 \frac{dM}{dt}\,, \qquad (20.97)$$

where $dM/dt = -4\pi r^2 \rho v_r$ is the solar mass loss rate. The second equality shows that the rate of angular momentum change dJ_z/dt is directly related to the Alfvén radius. The factor 2/3 comes from a moments of inertia calculation. Typical values for a 1D solar-type Weber–Davis wind solution result in a solar mass loss rate of $dM/dt = -2.94 \times 10^{-14} M_\odot yr^{-1}$. The 1D Weber–Davis solar solution is unrealistic in the sense that the actual solar wind has open field lines along the poles and closed field lines about the equator.

In a sequence of papers, Keppens and Goedbloed [256, 257] initiated a gradual approach towards dynamic solar/stellar wind numerical simulations. In [256], the authors proceeded from pure hydrodynamic Parker winds to arrive at axi-symmetric, polytropic, magnetized, rotating models. These 2D generalizations of the analytical Weber–Davis wind solution can contain both a "wind" and a "dead" zone. For these axi-symmetric, steady-state solutions, one can use the flux functions introduced in Chapter 18 to verify the physical correctness of the numerical solutions. Using explicit and implicit time-marching procedures to obtain steady-state stellar wind solutions, converged steady-state solutions for magnetized, rotating winds containing helmet streamers were obtained. An example solution is shown in Fig. 20.18. The polytropic solutions generalize the iso-thermal model by Pneuman and Kopp [372] of a mixed open–closed magnetic wind structure by including the effects of rotation. The stellar rotation causes a toroidal winding of the field lines and the resulting azimuthal field component B_φ is shown in color scale in the figure. Note the way in which the outflow accelerates to super-fast flow. This is characterized by the critical surfaces where the poloidal flow exceeds the poloidal slow, Alfvén and fast velocities. The poloidal streamlines and field lines are necessarily parallel in these stationary solutions of the flux function $\psi(R, Z)$ (exploiting the cylindrical coordinates of Appendix A.2.4 now). In ideal MHD,

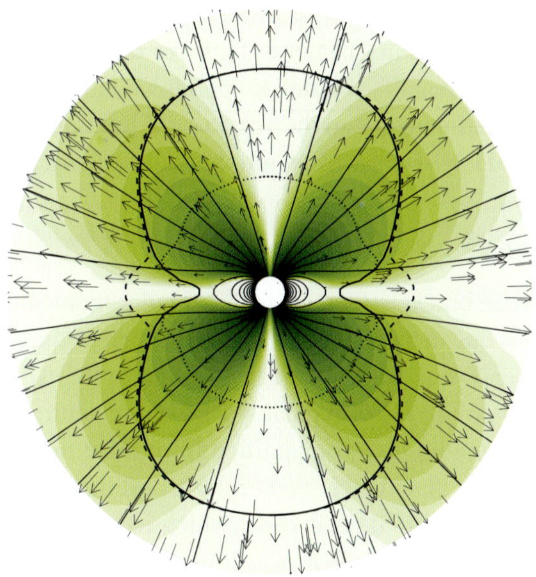

Fig. 20.18 Axi-symmetric magnetized wind with a "wind" and a "dead" zone. The poloidal magnetic field lines (solid lines) and flow vectors are shown. The colors indicate the azimuthal magnetic field component, which is non-zero in the wind zone only. The slow (dotted), Alfvén (solid) and fast (dashed) critical transitions of the solar wind are also drawn. (From Keppens and Goedbloed [257].)

one obtains various flux functions conserved along the streamlines, like $\Lambda(\psi) = Rv_\varphi - RB_\varphi B_\mathrm{p}/\rho v_\mathrm{p}$, where ρ is the density and B_p (v_p) indicates the poloidal magnetic field (velocity). The polytropic condition implies that the entropy is constant, $S = 1/\gamma$. One can further verify $\Omega(\psi) - [v_\varphi - (v_\mathrm{p}/B_\mathrm{p})B_\varphi]/R$ and the Bernoulli function $H(\psi) = \left[v_R^2 + v_Z^2 + v_\varphi^2\right]/2 + \rho^{\gamma-1}\gamma S/(\gamma-1) - GM_\odot/r - R\Omega v_\varphi$. These flux functions are specific to stationary, axi-symmetric ideal MHD flows, and they can be used to verify the correctness of transonic numerical solutions.

In [257], it was shown how reasonable changes in the coronal magnetic field strength and topology alter the detailed acceleration behavior of the solar wind during a solar cycle. Larger dead zones cause effective, fairly isotropic, acceleration to super-Alfvénic velocities as the polar open field lines are forced to fan out rapidly with radial distance. In the ecliptic, the wind outflow is modulated by the extent of the dead zone. Boundary conditions at the stellar surface imposed the mass flux and ensured the correct rotational coupling of velocity and magnetic field by prescribing the flux function $\Omega(\psi)$ at the base. The resulting–calculated–pole-to-pole variation of the base number density has a clear imprint of the dead zone, with higher densities about the equator consistent with observations. Also the density fall-off with height at the poles agrees well with the observed variation. This confirms that the macroscopic behavior of the solar wind can be described by the MHD equations, as discussed in Chapter 8 [1]. From these axi-symmetric transonic solutions, one can again quantify the stellar spin-down rate, which will be different from a Weber–Davis estimate due to the variation of the transonic transitions with polar angle. For a 2.5D axi-symmetric wind this can be written as

$$\frac{dJ_z}{dt} = -2\pi \int_0^\pi r^3 \sin^2 \theta (\rho v_r v_\phi - B_r B_\phi) d\theta . \tag{20.98}$$

The first term between brackets represents the angular momentum flux due to advection, the second is due to the magnetic tension. Whereas the angular momentum loss of the 1D Weber–Davis wind is directly related to the Alfvén radius, for the 2.5D case this is less obvious. For typical solar wind conditions, it was found that the total angular momentum flux as quantified from these models is about one order of magnitude smaller than the 1D Weber–Davis estimate. This reduction is due to the presence of the equatorial dead zone, which reduces the effective magnetic lever arm. Near the solar surface, the flux is dominated by magnetic tension, so that the angular momentum loss of the Sun via the solar wind is basically magnetic.

20.4.3 Example astrophysical transonic flows

Other transonic, magnetized plasma flows, collimated over very large distances from their source region, occur throughout our observable Universe. Astrophysical jets on light year scales are associated with young stellar objects (YSO) and the fossils of dead stars: neutron stars and stellar mass black holes in X-ray binary systems (XRB). Jets up to several millions of light year long occur in association with active galactic nuclei (AGN), containing a supermassive black hole. Many of these jets likely involve equipartition magnetic field strengths where the thermal and the magnetic energy contents are comparable. Observational and theoretical arguments favor such dynamically strong magnetic fields, which must play a role in jet launching and propagation, and in the termination of the jet through the interaction with the surrounding medium. Moreover, to explain their remarkable collimation, magnetic hoop stresses are needed to counteract the tendency of the outflow to widen by centrifugal and pressure effects. Another ingredient, common to the YSO, XRB and AGN type systems, is the presence of an accretion disk. Observational links have been established in all cases between accretion disk luminosity and jet emission, highly suggestive of a unifying jet launch scenario. The most promising scenario to explain the ubiquitous jet phenomenon relies on the interaction of a large-scale magnetic field with the accretion disk in order to give birth to bipolar self-collimated jets. The mass loaded onto the jet is then a nearly constant fraction of the mass accretion rate of the system, and its acceleration to highly super-fast magneto-sonic speeds is realized magneto-centrifugally. The presence of a structured global magnetic field configuration in accretion disk–jet systems is important in all three aspects of the astrophysical jet phenomenon: it plays a role in realizing a magneto-centrifugal jet launching, provides a natural way for jet collimation by magnetic tension, and modifies the jet linear and nonlinear stability properties against perturbations.

(a) Jets launched from magnetized accretion disks. A magnetic field threading an accretion disk can brake rotating matter, allowing accretion, and can act to transfer (part of the) angular momentum into a jet. A seminal MHD model by Blanford and Payne [49] (with transonic flows in a cold plasma) has laid the foundations for many sophisticated analytical and numerical investigations of magnetized jet dynamics. In equipartition thin accretion disks (with $\beta \simeq 1$), it is possible to realize sufficiently bent magnetic field configurations in the inner disk regions needed for magneto-centrifugal acceleration of jet material. To reach a stationary configuration in the simulations, one must model the disk internal regions in a resistive MHD framework, as material should be allowed to accrete without dragging in magnetic field lines. The anomalous resistivity mimics the effect of the magneto-turbulent nature of the inner disk plasma (with the turbulence presumably originating from magneto-rotational instabilities, or any of the transonic MHD instabilities discussed in Chapters 13 and 18). At the same time, the jet regions and the surrounding medium are adequately modeled in ideal MHD. Most simulations of astrophysical jets treat the dynamics of the disk as a mere boundary condition, and focus on acceleration and collimation. This ignores how the presence of a jet alters the magnetic and thermodynamic conditions within the jet launch region.

More recent work has numerically demonstrated the continuous launching of trans-magneto-sonic collimated jets from resistive accretion disks threaded by open large-scale magnetic fields. In particular, Casse and Keppens [80, 81] performed MHD simulations where the disk launches a non-transient ideal MHD jet accelerated to super-fast magneto-sonic velocities. In [80], the authors still assumed a simple polytropic relation between pressure and density profiles, and were able to evolve the magnetized disk over many rotational time scales. Figure 20.19 gives a 3D impression of the numerical end result, showing that a bipolar pair of self-collimated cool jets forms and is persistently ejected. The dominant part of disk matter is effectively accreted. A constant fraction reaches the inner disk surface while it accretes, where the pressure gradient lifts the matter to be propelled in the jet. Jet material is then magneto-centrifugally accelerated to reach super-fast magneto-sonic speeds. Note how the collimation is already complete near the top of the domain. In a follow-up paper [81], the energetics of the flow (without radiative losses) were accounted for as well, and the simulations produced hot jets with

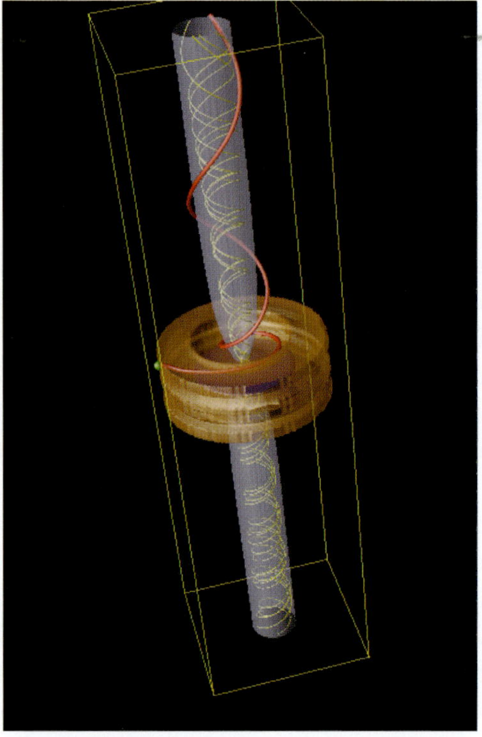

Fig. 20.19 A 3D impression of a transonic jet launched from an accretion disk. A high density iso-surface (brown) indicates the disk location, while selected magnetic field lines are shown in yellow. A fraction of the accreting disk matter gets lifted out of the disk and forms magneto-centrifugally accelerated jet matter: an indicative particle streamline (red) shows how spiraling fall-in motion gets diverted into a collimated jet. (From Casse and Keppens [80, 81].)

equipartition internal magnetic fields. In a fully self-consistent manner, the magnetized accretion–ejection structures (MAES), obtained numerically, accounted for the puzzling observation that many systems display bright collimated jets, together with under-luminous accretion disks. These and similar MHD models then quantified how the energy released by accretion can be mainly sent into the jet, naturally explaining the disk's low radiative efficiency. Note that, to make the problem computationally tractable, the simulations were performed in a 2.5D framework, where the flow and magnetic field are fully three-dimensional, under the restriction of axi-symmetry about the jet axis. This in effect eliminates potential shear flow as well as current-driven kink-mode perturbations deforming the jet, and precludes the development of non-axi-symmetric perturbations in the magnetized accretion disk. Various ongoing research efforts have started to relax these assumptions exploiting advanced computational tools, such as those mentioned in Chapter 19.

(b) Funnel accretion flows onto magnetized young stars. Our final example of a more continuous transonic flow configuration in astrophysics zooms in on the central region of the YSO-accretion disk system. While the previous example showed that jets can be launched from the inner equipartition disk regions, the fate of the major accreting fraction of the disk material was treated in a rudimentary fashion: a "sink" region allowed matter to escape the simulation box near the origin. For a YSO system, this inner region would have a typical size of about 0.1 AU, while the disk and jet dynamics were covering a region extending to several AU. The jet launch simulations thus illustrated how self-collimated, hollow jets can form within 2–3 AU. The central accreting object was, in the MAES configuration, merely important from its gravitational influence on the disk matter. We now shift our view to the star–disk dynamics happening within a range of a few stellar radii, accounting for the observational fact that young solar-type stars can harbor dynamically important global magnetic fields of strengths within several 100 G to a few kG. In reality, the very central accretion dynamics can be very complex, as the young star will rotate as it has gone through a contraction phase, and its main dipole moment does not need to be aligned with the rotation axis. We will once more avoid these complications, and discuss a computational example of magnetized star–disk interaction, which assumes axi-symmetry (with aligned magnetic dipole moment and rotation axis) and focuses on the near-stellar accretion disk dynamics.

In the presence of a stellar magnetic field, the accretion flow onto the star can be significantly different from purely equatorial accretion. The stellar field can truncate the disk at an inner truncation radius r_t and, as pointed out by Romanova *et al.* [394] and consecutively refined by Bessolaz *et al.* [37], dynamical arguments can be used to estimate this truncation radius for a given accretion rate $\dot{M}_a$, stellar field strength B_*, and other parameters like its radius R_* and mass M_*. Bessolaz *et al.* [37] showed that, once again, equipartition fields with $\beta \sim 1$, together with the requirement that the accretion ram pressure ρv_r^2 is balanced by the magnetic poloidal pressure, provides an estimate for the truncation radius given by

$$\frac{r_t}{R_*} \simeq 2\, m_s^{2/7} \left(\frac{B_*}{140\,\text{G}}\right)^{4/7} \left(\frac{\dot{M}_a}{10^{-8}\,M_\odot/\text{yr}}\right)^{-2/7} \left(\frac{M_*}{0.8\,M_\odot}\right)^{-1/7} \left(\frac{R_*}{2\,R_\odot}\right)^{5/7}.$$

(20.99)

The parameter $m_s \simeq 1$ denotes the sonic Mach number for the radial accretion flow at the disk midplane. The estimate assumed for simplicity that the real base of the funnel flow, located at r_{bf} where both $\beta \sim 1$ and $m_s \sim 1$, is close to the truncation radius r_t where by definition the radial motion is halted and diverted along a stellar field line. Under typical T-Tauri star parameters as indicated in the scalings of Eq. (20.99), the disk is thus truncated within a few stellar radii. A numerical study by Romanova *et al.* [394] was the first to convincingly demonstrate the for-

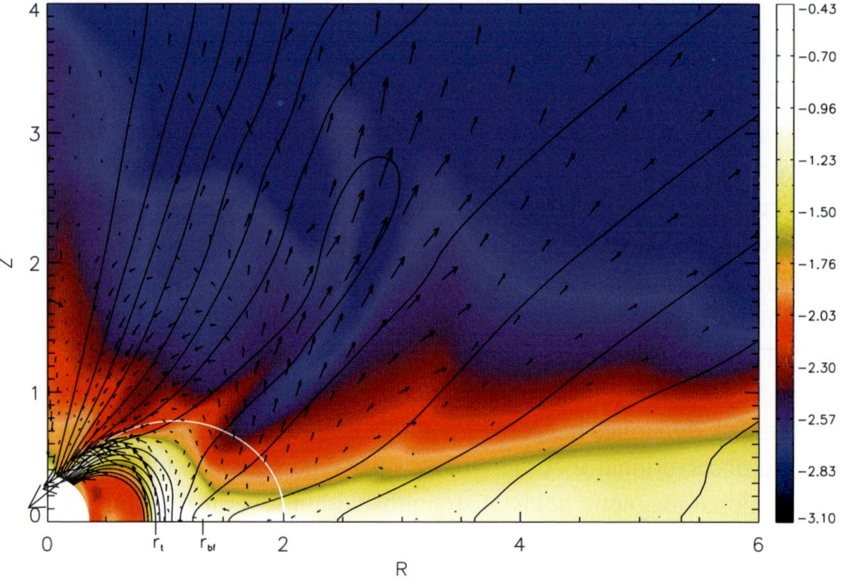

Fig. 20.20 In an axi-symmetric computation of accretion onto a young magnetized star, the accretion disk (the high density matter seen in yellow) is magnetically halted within a disk truncation radius r_t. The infalling matter gets diverted into a transonic funnel flow along the dipolar magnetic field lines (black lines) connecting star and disk. The base of the funnel is located at r_bf. (From Bessolaz *et al.* [37].)

mation of accretion "funnels" near the star, as material gets lifted out of the disk due to a vertical plasma pressure gradient and loaded onto the stellar field lines. While the original results were obtained for kG stellar fields, Fig. 20.20 shows a later result by Bessolaz *et al.* [37] which extended the findings to lower stellar field strengths of about 140 G, at even lower accretion rates of order $10^{-9}\,M_\odot/\mathrm{yr}$. The disk region was treated resistively, in a manner analogous to the jet launch studies mentioned above. The accretion funnels develop self-consistently from initial conditions where only a truncated accretion disk is embedded in a dipolar stellar magnetosphere. The end state shown in Fig. 20.20 can be considered as a stationary transonic flow configuration: when analyzing the funnel flow in detail, it is found to cross the slow magneto-sonic point, reaching sonic Mach numbers of up to 3.5. However, the funnel flow found remained sub-Alfvénic throughout. The main forces acting within the funnel are the thermal pressure gradient at the base, as matter accumulates there, while eventually gravity overwhelms near the stellar radius: matter reaches the star with approximately free-fall velocities. While these funnel flows seem to be a common ingredient to accreting magnetized stars, the stellar parameters may bring the YSO system to completely different transonic flow regimes, such as the "propeller" flow predicted in Romanova *et al.* [395].

(c) Ongoing research trends. While the previous examples clearly illustrate that transonic MHD flows are ubiquitous in astrophysics, they concentrate on highly simplified configurations to isolate one aspect of what happens in reality (be it the stellar wind, the jet launch, or the near-stellar accretion dynamics). Continued progress is made on each aspect separately, e.g. by tackling 3D accretion flows onto young stars with unaligned rotation and magnetic axes, and/or considering higher order multi-pole fields, as suggested by polarimetric measurements of starspots. While the basic funnel flow physics is well understood from the axi-symmetric simulations, the bewildering variety of multi-column accretion funnels, and the possibility of intermittent accretion tongues, forming as a result of Rayleigh–Taylor instabilities developing at the inner truncation radius [289], illustrates that the observational task of inverting from spectroscopic data, with coarse spatial (and limited temporal) resolution to obtain the ongoing thermo-dynamics is simply daunting. Furthermore, the supersonic accretion columns themselves must ultimately end on the stellar surface, and thus involve a standing shock front, from which much of the radiative losses occur. This aspect needs to be studied within the global disk dynamics simulations. Finally, the angular momentum regulation in true YSO systems is still posing many puzzles to the emerging theoretical picture: a large fraction of classical T-Tauri stars showing circumstellar disks rotate much slower than expected from pre-main-sequence evolutionary models. Due to the fact that these stars contract and hence should spin up significantly, a regulating role of star–disk coupling was typically invoked to explain the observed slow rotation rates (see e.g. Keppens *et al.* [258]). Under typical YSO parameters, Eq. (20.99) predicts that the funnel flows and disk truncation happen inwards from the co-rotation radius, which separates disk regions where the magnetic coupling would act to spin up (inwards) versus spin down (outwards) the central star. Current trends investigate the role of stellar winds in shedding the young star angular momentum (combining the disk, accretion funnel and stellar outflow problem), as well as the possibility that coherent external disk magnetic fields coexist with the stellar magnetic field component (bringing together the MAES jet launch and central accretion studies). Due to the different range in spatial and temporal scales, the full 3D demonstration of stellar outflows, (shocked) multicolumn accretion flows and disk jet/wind launching is a research item that is still challenging to the current generation of even grid-adaptive, massively parallel software tools. Nevertheless, it is guaranteed to be at the forefront of ongoing research efforts. A related challenge is the study of jet launch, propagation, stability and ultimate shock-dominated termination by interaction with the interstellar medium, for systems ranging from the YSO to the AGN category. We will give a modern example of the latter in the next chapter, but this requires an appropriate treatment of relativistically flowing magnetized plasmas. How this is done is the subject of Chapter 21.

20.5 Literature and exercises

Notes on literature

Classic treatments of gas dynamic and MHD shocks

- A classic treatment of gas dynamic shocks is *Supersonic Flow and Shock Waves* by Courant & Friedrichs [97]. Equally classic, but much shorter, are the chapters on gas dynamic shocks and MHD shocks in *Course of Theoretical Physics* (Volume 6, Chapter IX, and Volume 8, Chapter VIII) by Landau & Lifshitz [294, 295]. An in-depth treatment of MHD charateristics may be found in *Magnetohydrodynamics* by Kulikovskij & Lyubimov [288].

Textbook chapters on MHD shocks

- A number of textbooks on plasma physics have useful chapters on MHD shocks, e.g. *Magnetohydrodynamics* (Chapter VI) by Jeffrey [249]; *Solar Magnetohydrodynamics* (Chapter 5) by Priest [386]; *The Physics of Plasmas* (Section 5.6) by Boyd & Sanderson [64]; *Introduction to Plasma Physics* (Chapter 7) by Gurnett & Bhattacharjee [201].

MHD shock relations

- The material on shock relations presented in this chapter is based on the paper 'Time reversal duality of magnetohydrodynamical shocks' by Goedbloed [172]. It adds the view point of scale-independence of the MHD equations to the powerful tangential transformation that can be found in the classical, and still very readable, paper 'Magneto-hydrodynamic shocks' by de Hoffmann and Teller [108].

Exercises

[20.1] *Entropy*

In this chapter, it is stated that the entropy has to increase across a shock front.

- What is entropy? Explain why it has to increase across a shock front.

[20.2] *Essential steps*

What are the essential steps in deriving the distilled shock conditions from the original ones? Explain for each step why it is important.

[20.3] *Distilled energy jump condition*

In this exercise, you are going to reproduce the S-shape curve of Figs. 20.9 and 20.10. We assume that you use IDL, but you can also use Matlab, Maple or any other similar program. Whatever program you use, make sure that Laguerre's method [385] is implemented.

- Write $\bar{p}_1$ and $\bar{B}_{t1}$ in terms of the angle ϑ_1 and the plasma beta β_1.

- Now, write an IDL program which reproduces the S-curve shown in Fig. 20.9. Create a file called `quantities.pro` in which you put two functions, one for the upstream pressure and one computing the tangential magnetic field. Both functions use the angle ϑ_1 and the plasma beta β_1 as input parameters.

- Write the distilled energy jump condition (20.63) as a polynomial in the squared Alfvén Mach number M_2^2. How many solutions are there and what are the expressions for each of the coefficients?

- Next, create a function computing the coefficients of the energy jump condition for given ratio of specific heats γ, pressure $\bar{p}_1$, tangential magnetic field $\bar{B}_{t1}$ and squared Alfvén Mach number M_1^2. Put this function in the file shockconditions.pro.

- Next, write a generic function computing the roots of a polynomial using Laguerre's method, which is part of the IDL library. Create the file solvers.pro which creates the function rootfinder with input parameters γ, $\bar{p}_1$, $\bar{B}_{t1}$, "points" and $\max(M_1^2)$. With "points" the number of points to be calculated between 0 and $\max(M_1^2)$ is meant. At each point, you should compute the coefficients of the energy jump condition needed to compute the roots with Laguerre's method. Store only the appropriate solutions. What kind of test can you design to check if a solution is appropriate or not? How do you sort the solutions? The generic function should return the appropriate solutions as a two-dimensional array with the values for M_1^2 and M_2^2.

- Make a plot of the roots found for the given parameters of Fig. 20.9.

[20.4] *Downstream magneto-sonic transition function*

The main purpose of this exercise is to make a plot of the downstream magneto-sonic transition function as a function of the upstream Alfvén Mach number. Similarly to the previous exercise, we assume that you use IDL. If you want to use another program, like Matlab, this is no problem, but make sure that you can use Laguerre's method [385].

- Write the downstream magneto-sonic transition equation (20.67) as a polynomial in M_2^2. What are its coefficients?

- Add a function to the file shockconditions.pro which computes these coefficients. If you did not do the previous exercise, then create this file.

- Extend the function rootfinder so that it can also compute the roots of the downstream magneto-sonic transition function. Again, if you did not do the previous exercise, make a file solvers.pro creating the function rootfinder. What this function should do is specified in the last but one question of the previous exercise.

- Make a plot of the curves $M_{s2,A2,f2}^2$. If you did the previous exercise, you can just add the curves to the previously created plot.

- Finally, also add the curve $M_1^2 = M_{s1,A1,f1}^2$ to the plot.

[20.5] *Distilled entropy condition*

In this exercise, you are going to make a plot of the distilled entropy condition (20.64). We assume again that you use IDL, but you may use a similar program. Make sure you are able to use the van Wijngaarden–Dekker–Brent (WDB) method [385]. This is an advanced method to compute the zero of a monotonic function on a given domain.

- Explain why you cannot plot the distilled entropy condition (20.64) using Laguerre's method.

- For a numerical reason, you have to multiply the entropy condition by $M_2^{2\gamma}$. What is that reason precisely?

- The entropy condition will be plotted using the WDB method. Add a function computing it for given M_2^2, γ, $\bar{p}_1$, $\bar{B}_{t1}$, M_1^2 to the file shockconditions.pro. Make sure that the first input parameter is the squared Alfvén Mach number M_2^2. If you did not do one of the two previous exercises, then create this file.

- Download the file zbrent.pro from the internet. This is a part of the IDL Astronomy User's Library (idlastro.gsfc.nasa.gov) which contains the WDB method. Modify the function zbrent such that it accepts four additional parameters.

- Make a plot of the entropy condition for ϑ_1, $\beta_1 = 0.4$, M_1^2 and $M_2^2 = [0, 2.5]$. What do you observe and what does it mean for the use of the WDB method?

- What method can be used to solve the problem of the previous question?

- Add a function to the file `solvers.pro` (create it if it does not exist), which computes where the entropy condition is zero for given γ, $\bar{p}_1$, $\bar{B}_{t1}$, "points" and $\max(M_1^2)$. The zeros should be computed using the WDB method. By "points" the number of points between zero and the maximum of M_1^2 is meant. The function should return a 2D array containing the values M_1^2 and M_2^2 for which the entropy jump is zero.

- Make a plot of the solutions found. If you did one of the two previous exercises, you can just add the solutions to the existing plot.

[20.6] *Weber–Davis trans-magneto-sonic wind solution*

We here investigate the trans-magneto-sonic properties of a stellar wind for a star of mass M_* under a number of approximations according to Weber and Davis [475].

- Write down the ideal MHD equations in spherical coordinates including a spherically symmetric external gravity due to the stellar mass M_*. Drop the energy equation and assume a polytropic relation $p = p_0 (\rho/\rho_0)^\gamma$. We will look for a stellar wind solution which is stationary and axisymmetric. Apply this assumption to the equations.

- The analysis will be restricted to the equatorial plane $\theta = \pi/2$. At this equatorial plane we assume that $v_\theta = B_\theta = 0$ and that all other physical quantities ρ, v_r, v_ϕ, B_r, B_ϕ depend only on the radius r. Write out the remaining forms for mass conservation, induction equation and the $\nabla \cdot \mathbf{B} = 0$ equation.

- Show that the toroidal component of the momentum equation can be written as

$$ r v_\phi - \frac{r B_r B_\phi}{\rho v_r} = L \, , $$

where the constant L is related to the angular momentum flux.

- Show that

$$ r \left(v_\phi B_r - v_r B_\phi \right) = \Omega r^2 B_r \, , $$

where Ω is the stellar angular rotation rate.

- Express the toroidal velocity v_ϕ in terms of r, L, Ω and the radial Alfvén Mach number defined from $M_{\mathrm{Ar}}^2 \equiv \rho v_r^2 / B_r^2$. Furthermore, derive an expression for the constant L by evaluating expressions at the Alfvén point $M_{\mathrm{Ar}} = 1$.

- Express the density ρ, toroidal velocity v_ϕ, radial magnetic field B_r and toroidal magnetic field B_ϕ in terms of the radius r and radial velocity v_r. Show that the radial component of the momentum equation can be written as

$$ \frac{d}{dr} \left[\frac{1}{2} \left(v_r^2 + v_\phi^2 \right) + \frac{\gamma p}{(\gamma - 1)\rho} - \frac{GM_*}{r} - \frac{B_r B_\phi}{\rho} \frac{\Omega r}{v_r} \right] = 0 \, . $$

- Show that the radial component of the momentum equation can also be written as

$$ \frac{d v_r}{dr} = \frac{v_r}{r} \left[\frac{(v_r^2 - A_r^2)(2c_{\mathrm{s}}^2 + v_\phi^2 - GM_*/r) + 2v_r v_\phi A_r A_\phi}{(v_r^2 - A_r^2)(v_r^2 - c_{\mathrm{s}}^2) - v_r A_\phi^2} \right] \, , $$

where the (squared) sound speed $c_{\mathrm{s}}^2 \equiv \gamma p/\rho$, the radial Alfvén speed $A_r \equiv B_r/\sqrt{\rho}$ and the toroidal Alfvén speed $A_\phi \equiv B_\phi/\sqrt{\rho}$. Discuss the critical points of this ODE for the radial velocity v_r.

21

Ideal MHD in special relativity

We have seen that the MHD description for the macroscopic dynamics of plasmas offers a uniquely powerful, unifying, viewpoint on both laboratory and astrophysical plasmas. The applicability of the MHD viewpoint was discussed previously in Volume [1], along with the various approximations made to arrive at the MHD equations from first principles. For most laboratory plasmas, the single fluid ideal or resistive MHD model eventually needs to be extended towards a multi-fluid model and by including important kinetic effects, since its continuum approach to plasma modeling neglected, e.g., Landau damping as well as many other velocity-space dependent physical phenomena. For many astrophysical plasmas, we face yet another shortcoming of the MHD model used thus far, namely that we restricted all attention to non-relativistic plasma velocities. This is perfectly adequate for most of the plasma found in our own solar system. However, astronomical observations indicate that, e.g., the extragalactic jets associated with Active Galactic Nuclei clearly harbor dynamically important magnetic fields and relativistically flowing plasmas. In order to model these plasmas in a continuum model, the restriction on the plasma velocities must be alleviated, by revisiting the ideal MHD equations in a frame-invariant relativistic formulation within four-dimensional space-time. In this chapter, we present such a formulation, restricting our attention to special relativity where we still have a "flat" geometry. In recent years, modern computational techniques such as those discussed in Chapter 19 have started to be used in this more demanding relativistic MHD regime. Since such activities are necessarily still maturing, we only summarize the numeric algorithmic challenges posed by the ideal MHD model in special relativity. We end this chapter with selected example applications. In particular, we discuss insights gained from recent computations of astrophysical jets with speeds approaching the speed of light.

543

21.1 Four-dimensional space-time: special relativistic concepts

Relativity implies that physical laws do not depend on the chosen reference frame. In special relativity, the speed of light c is explicitly recognized as the maximal speed with which information can travel between different spatial locations. Since this maximal speed c is the same for all observers in uniform relative motion with respect to each other, we must abandon the familiar notion of Galilean invariance. Indeed, under Galilean invariance between two such inertial reference frames, speeds expressed in the two frames transform by adding the speed of the relative motion. In order to account for an upper limit c in propagation speed, which is the same for two inertial systems moving with relative velocity $\mathbf{v}$, one must relax their notion of simultaneity. What appears simultaneous in one frame will occur at different times in any other frame, and as a result, the temporal duration of a physical event will differ from frame to frame. This leads to a viewpoint where time is an extra coordinate, augmenting the three spatial coordinate directions, to describe physical events in a four-dimensional space-time continuum. The four coordinates associated with different inertial space-time reference frames are related by the Lorentz transformation. In the following, we introduce these and other basic concepts leading to a special relativistic formulation of nonlinear compressible gas dynamics. Section 21.2 extends this with the inclusion of electromagnetic fields, and discusses how one successively obtains the ideal MHD model in special relativity. The material in this introductory treatment towards special relativistic MHD benefited from material presented in the more broadly oriented textbook by Blandford and Thorne [50], while much more complete, but also more technical, treatments can be found in earlier monographs by Anile [8] and Lichnerowicz [307].

21.1.1 Space-time coordinates and Lorentz transformations

In Section 2.2.2 [1], we presented *the Lorentz transformation* of two inertial frames moving with relative velocity $\mathbf{v}$, given by

$$\begin{cases} \mathbf{x}' = \mathbf{x} + \dfrac{\Gamma - 1}{v^2}\, \mathbf{v}\, \mathbf{v} \cdot \mathbf{x} - \Gamma \mathbf{v}\, t, \\[2mm] t' = \Gamma\left(t - \dfrac{1}{c^2}\mathbf{v} \cdot \mathbf{x}\right), \qquad \Gamma \equiv \dfrac{1}{\sqrt{1 - v^2/c^2}}. \end{cases} \tag{21.1}$$

The Lorentz factor Γ depends nonlinearly on the relative velocity $\mathbf{v}$, which is still a vector with three components (v_1, v_2, v_3) along the three spatial (orthogonal) coordinate directions $\mathbf{x} \equiv (x_1, x_2, x_3)^{\mathrm{T}}$. We will write a Latin index $i = 1, 2, 3$ when a spatial coordinate direction is meant, e.g. in x_i or v_i. However, the Lorentz transformation (21.1) must rather be interpreted as a transformation between the four-dimensional space-time coordinates

$$\mathbf{X} \equiv \begin{pmatrix} ct \\ \mathbf{x} \end{pmatrix}, \qquad \mathbf{X}' \equiv \begin{pmatrix} ct' \\ \mathbf{x}' \end{pmatrix}, \tag{21.2}$$

when we write it as

$$\mathbf{X}' = \begin{pmatrix} \Gamma & -\Gamma\dfrac{v_1}{c} & -\Gamma\dfrac{v_2}{c} & -\Gamma\dfrac{v_3}{c} \\[2mm] -\Gamma\dfrac{v_1}{c} & 1+(\Gamma-1)\dfrac{v_1^2}{v^2} & (\Gamma-1)\dfrac{v_1 v_2}{v^2} & (\Gamma-1)\dfrac{v_1 v_3}{v^2} \\[2mm] -\Gamma\dfrac{v_2}{c} & (\Gamma-1)\dfrac{v_1 v_2}{v^2} & 1+(\Gamma-1)\dfrac{v_2^2}{v^2} & (\Gamma-1)\dfrac{v_2 v_3}{v^2} \\[2mm] -\Gamma\dfrac{v_3}{c} & (\Gamma-1)\dfrac{v_1 v_3}{v^2} & (\Gamma-1)\dfrac{v_2 v_3}{v^2} & 1+(\Gamma-1)\dfrac{v_3^2}{v^2} \end{pmatrix} \mathbf{X}. \tag{21.3}$$

The space-time coordinates $\mathbf{X}$ are specific to a pre-chosen inertial Lorentzian reference frame. If we indicate the 4×4 matrix occurring in (21.3) with $L_\alpha^{\alpha'}$, and denote the four coordinate entries as X^α and $X^{\alpha'}$, with Greek index $\alpha = 0, 1, 2, 3$, we have the compact notation

$$X^{\alpha'} = L_\alpha^{\alpha'} X^\alpha. \tag{21.4}$$

This formula implicitly assumes a choice of two Lorentzian reference frames L' with space-time coordinates $\mathbf{X}'$, and L with coordinates $\mathbf{X}$. The inertial frame L' is moving at speed $\mathbf{v}$ with respect to frame L, and Eq. (21.4) merely states how the coordinates for a unique point in four-dimensional space-time, also referred to as an "event", relate between the two selected frames. Obviously, all symbols appearing in Eq. (21.4) are frame-dependent. The inverse transformation, written symbolically as

$$X^\alpha = (L^{-1})_{\alpha'}^\alpha X^{\alpha'}, \tag{21.5}$$

can be computed by noting that the matrix $(L^{-1})_{\alpha'}^\alpha = (L_\alpha^{\alpha'})^{-1}$. As expected on physical grounds, the matrix $(L^{-1})_{\alpha'}^\alpha$ turns out to be found from $L_\alpha^{\alpha'}$ by merely replacing $\mathbf{v}$ by $-\mathbf{v}$. Note also that the Lorentz transformation matrix $L_\alpha^{\alpha'}$ (and thus its inverse) is symmetric.

Pure Lorentz boost, length contraction and time dilation We can specify the Lorentz transformation to a case where frame L' is moving with a velocity $\mathbf{v}$ directed along the x_1 coordinate axis of frame L. Then, we speak of a pure Lorentz boost with direct transformation given by

$$\begin{pmatrix} ct' \\ x_1' \\ x_2' \\ x_3' \end{pmatrix} = \begin{pmatrix} \Gamma & -\Gamma v_1/c & 0 & 0 \\ -\Gamma v_1/c & \Gamma & 0 & 0 \\ 0 & 0 & 1 & 0 \\ 0 & 0 & 0 & 1 \end{pmatrix} \begin{pmatrix} ct \\ x_1 \\ x_2 \\ x_3 \end{pmatrix}. \tag{21.6}$$

The inverse transformation replaces $\mathbf{X}$ by $\mathbf{X}'$ and v_1 by $-v_1$. A length interval Δx_1 along the x_1 axis of frame L is measured in the L' system at a fixed time t'. The inverse transformation yields immediately that, for the moving observer in frame L', the length appears to contract since

$$\Delta x_1' = \Delta x_1/\Gamma. \tag{21.7}$$

Since $\Gamma \geq 1$, lengths in the inertial frame L along the direction of motion of a moving observer in L' appear shorter to him, which is the well-known length contraction effect. Note that in the two directions perpendicular to the motion, no difference occurs, so volume measures will only differ by the single factor Γ^{-1}. For a comparison of time intervals, we use the direct transformation to compare time intervals as determined at fixed position ($\Delta x_1 = 0$) in frame L. We find similarly that

$$\Delta t' = \Gamma \Delta t, \tag{21.8}$$

telling that time in the inertial frame L appears to run slow to the moving observer in L'. Of course, when we change view point from L' to L, the same conclusions will be reached: L is also moving with velocity $-\mathbf{v}$ along the x_1' coordinate axis of frame L'. We will reach the (apparently paradoxical) conclusions that then Eqs. (21.7)–(21.8) will hold in reverse, i.e. when replacing t by t' and x_1' by x_1. This is just what relativity is all about: both observers come to the same conclusion. The paradoxical nature is a mere consequence of the fact that simultaneity has obtained a different meaning in both systems.

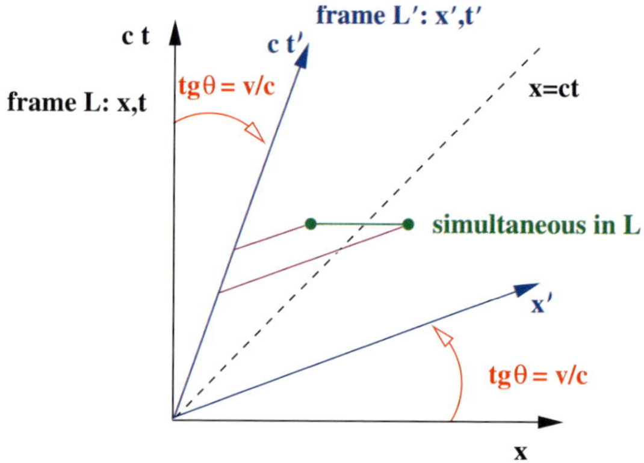

Fig. 21.1 Pure Lorentz boost and loss of simultaneity.

For the pure boost of Eq. (21.6), we can graphically illustrate the different parameterization of space-time associated with both (x_1, t) and (x_1', t') systems. When frame L' moves at speed v in frame L along x_1, and we draw the axes of

L' at the moment when origins of L and L' coincide, the t' axis is given by the straight line $ct = (c/v)x_1$. Similarly, the x_1'-axis given by $t' = 0$ points is the line $ct = (v/c)x_1$. This is illustrated in Fig. 21.1. It is then clearly seen that simultaneous events in frame L are no longer simultaneous to frame L'. Such space-time diagrams can also be used to prove in a more graphical sense the mentioned effects of length contraction and time dilation. Note finally that from Eq. (21.6) (as well as from Eq. (21.3)), we find the relation

$$- c^2t^2 + x_1^2 + x_2^2 + x_3^2 = -c^2t'^2 + x_1'^{\,2} + x_2'^{\,2} + x_3'^{\,2}, \qquad (21.9)$$

which will return in the discussion of the metric associated with flat space-time.

21.1.2 Four-vectors in flat space-time and invariants

We now consider the path traced in space-time by a single particle, which we will call the world line of this particle. Time progression as experienced by this particle is referred to as its "proper time". This proper time τ is measured by an ideal clock in the particle's local rest frame, i.e. in a reference frame moving along with the particle. Introducing a local Lorentzian reference frame with orthogonal coordinates X^α, the particle four-velocity is then given by

$$U^\alpha \equiv \frac{dX^\alpha}{d\tau}, \qquad (21.10)$$

where $dX^\alpha \equiv (c\,dt, dx_1, dx_2, dx_3)^{\mathrm{T}}$ measures the "distance" along each coordinate direction traversed in the proper time interval $d\tau$. These concepts are graphically shown in Fig. 21.2, depicting space-time as an (x, t) plane where two of the spatial directions are conveniently omitted.

The four components U^α in Eq. (21.10), as well as the coordinates X^α, are in fact contravariant components of physically meaningful four-vectors in four-dimensional space-time. The four-position vector $\mathbf{X} \equiv X^\alpha \mathbf{e}_\alpha$ connects the origin of our reference frame with the space-time point with coordinates X^α, whereby the basis vectors $\mathbf{e}_\alpha$ point along the four coordinate axes. The transformations given by Eqs. (21.4) and (21.5) can thus be understood as follows: two reference frames L and L' each introduce their set of basis vectors $\mathbf{e}_\alpha$ and $\mathbf{e}_{\alpha'}$, respectively. The matrix $L_\alpha^{\alpha'}$ quantifies how each basis vector $\mathbf{e}_\alpha$ can be written as a linear combination of the other set of basis vectors, namely $\mathbf{e}_\alpha = L_\alpha^{\alpha'} \mathbf{e}_{\alpha'}$. Note that the double appearance of the Greek index α' implies a summation, and we will from now on adopt this Einstein convention. The frame-independent four-position vector $\mathbf{X}$ is then clearly

$$\mathbf{X} = X^\alpha \mathbf{e}_\alpha = X^\alpha L_\alpha^{\alpha'} \mathbf{e}_{\alpha'} = X^{\alpha'} \mathbf{e}_{\alpha'}, \qquad (21.11)$$

which directly leads to the transformation as given in Eq. (21.4).

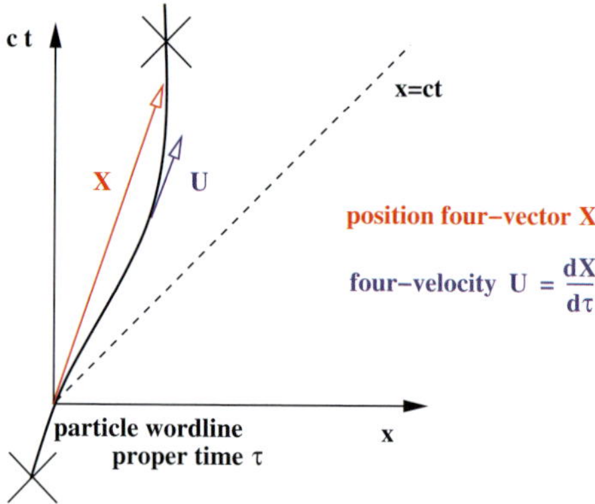

Fig. 21.2 World line of a particle, four-position vector and four-velocity.

Once a set of basis vectors $\mathbf{e}_\alpha$ is given, we can in general introduce a set of dual or reciprocal basis vectors which obey $\mathbf{e}_\alpha \cdot \mathbf{e}^\beta = \delta_\alpha^\beta$, where δ_α^β is the Kronecker delta. Covariant vector components appear when working with this dual basis, e.g. for the four-position vector $\mathbf{X} = X_\alpha \mathbf{e}^\alpha$. These dual basis vectors $\mathbf{e}^\beta$ can then be used to express the two sets of basis vectors in terms of each other, namely

$$\mathbf{e}_\alpha = \mathbf{e}_\alpha \cdot \mathbf{e}_\beta \mathbf{e}^\beta \equiv g_{\alpha\beta} \mathbf{e}^\beta \qquad \text{or} \qquad \mathbf{e}^\alpha = \mathbf{e}^\alpha \cdot \mathbf{e}^\beta \mathbf{e}_\beta \equiv g^{\alpha\beta} \mathbf{e}_\beta . \tag{21.12}$$

This introduces the 4×4 components of the metric tensor $g_{\alpha\beta}$ and its inverse $g^{\alpha\beta}$. They generally allow one to transform from contravariant components X^α to covariant components X_α by $X_\alpha = g_{\alpha\beta} X^\beta$. The metric tensor components associated with flat space-time, i.e. in a Lorentzian reference frame equipped with space-time coordinates, are given by the Minkowski metric

$$g_{\alpha\beta} = g^{\alpha\beta} = \begin{pmatrix} -1 & 0 & 0 & 0 \\ 0 & 1 & 0 & 0 \\ 0 & 0 & 1 & 0 \\ 0 & 0 & 0 & 1 \end{pmatrix} . \tag{21.13}$$

Note that the four basis vectors for this flat space-time are orthogonal to each other. Also, the three space-like directions form an orthonormal set with unit lengths since $\mathbf{e}_i \cdot \mathbf{e}_i = 1$. The squared length of the temporal basis vector $\mathbf{e}_0 \cdot \mathbf{e}_0$ has the negative value -1. For flat space-time, we then conclude that the dual basis vector $\mathbf{e}^0 = -\mathbf{e}_0$, while all spatial directions have basis vectors identical to their dual vector $\mathbf{e}^i = \mathbf{e}_i$. This simplifies the tensor calculus considerably, and one only needs to remember the sign reversal for any temporal tensor component.

In general, the squared length of a four-vector represents a frame-independent scalar invariant upon which different observers will agree. It can be evaluated by means of the metric since for a four-vector $\mathbf{X}$ we find

$$\mathbf{X} \cdot \mathbf{X} = (X^\alpha \mathbf{e}_\alpha) \cdot (X^\beta \mathbf{e}_\beta) = g_{\alpha\beta} X^\alpha X^\beta = X^\alpha X_\alpha . \tag{21.14}$$

In flat space-time, and for a differential vector $d\mathbf{X}$, this then is simply

$$ds^2 \equiv d\mathbf{X} \cdot d\mathbf{X} = -c^2\, dt^2 + dx_1^2 + dx_2^2 + dx_3^2 , \tag{21.15}$$

where we introduced the (square of the) line element ds. A four-vector with negative square length is said to separate time-like events in space-time, while a four-vector with a positive squared length separates space-like events. A zero squared length is a light-like separation. World lines of particles will always have time-like tangent four-vectors in space-time, while light (a photon) travels along a path with light-like tangents.

From (21.15), we can deduce several useful relations connected to the proper time τ and a particle's four-velocity as introduced in Eq. (21.10). It is clear that, for a stationary particle, the proper time τ will be equal to the time coordinate t of an inertial frame attached to the particle. We then find from (21.15) that $d\tau^2 = -ds^2/c^2$. Since ds is an invariant quantity, this relation between proper time intervals and the (time-like) line element tangent to the world line of a particle will always be true. One then uses this equality to find the general relation

$$- c^2 = -c^2 \frac{dt^2}{d\tau^2} + \frac{dx_1^2}{d\tau^2} + \frac{dx_2^2}{d\tau^2} + \frac{dx_3^2}{d\tau^2} . \tag{21.16}$$

Noting further that

$$\frac{dx_i}{d\tau} = \frac{dx_i}{dt} \frac{dt}{d\tau} , \qquad \frac{dx_i}{dt} = v_i , \tag{21.17}$$

where v_i is a component of the spatial three-velocity $\mathbf{v}$, one finds that

$$\frac{dt}{d\tau} = \Gamma . \tag{21.18}$$

This is again the time-dilation effect, expressing the fact that the proper time associated with a moving particle will always appear to run slow from a stationary viewpoint, since $\Gamma \geq 1$. As a result of Eq. (21.18), the contravariant components of the four-velocity can be written as

$$U^\alpha = (c\Gamma, \Gamma\mathbf{v})^{\mathrm{T}} . \tag{21.19}$$

Using the general relation (21.14), we can then verify the invariance of

$$U^\alpha U_\alpha = -c^2 . \tag{21.20}$$

This confirms the time-like nature of the four-velocity, which is the tangent vector to the particle world line. For a particle with rest mass m_0, the four-momentum is then the four-vector

$$P^\alpha = m_0 U^\alpha. \tag{21.21}$$

In Chapter 2 [1], we used the notation $E = \Gamma m_0 c^2$ and $\mathbf{p} = \Gamma m_0 \mathbf{v}$ to indicate the relativistic energy and (three-)momentum, respectively. This allows us to write

$$P^\alpha = \left(\frac{E}{c}, \mathbf{p}\right)^{\mathrm{T}}, \qquad E^2 = m_0^2 c^4 + p^2 c^2, \tag{21.22}$$

where the latter relation comes from evaluating the invariant $P^\alpha P_\alpha$. Hence, energy includes the rest mass contribution $m_0 c^2$, and momentum and energy need to be treated on the same footing.

Particle dynamics and forces in space-time The dynamics of a particle is governed by forces exerted on it, and this requires the generalization of the force concept from a three-vector $\mathbf{F}$ to its four-vector equivalent. We can expect these two concepts to coincide in the rest frame of the particle, such that we write the contravariant components of the four-vector in this frame by $F^{\alpha'} = (0, \mathbf{F})^{\mathrm{T}}$. When we then transform to the reference frame which observes the particle moving with velocity $\mathbf{v}$, we use the inverse Lorentz transform to compute

$$F^\alpha = (L^{-1})^\alpha_{\alpha'} F^{\alpha'} = \left(\Gamma \frac{\mathbf{v} \cdot \mathbf{F}}{c}, \Gamma \mathbf{F}\right)^{\mathrm{T}}, \tag{21.23}$$

where we used the fact that the velocity $\mathbf{v}$ will typically be aligned with $\mathbf{F}$. The equation of motion is then

$$\frac{dP^\alpha}{d\tau} = F^\alpha, \tag{21.24}$$

which can be decomposed into its temporal and spatial parts as

$$\frac{dE}{dt} = \mathbf{v} \cdot \mathbf{F}, \tag{21.25}$$

$$\frac{d\mathbf{p}}{dt} = \mathbf{F}. \tag{21.26}$$

We find familiar results: Eq. (21.25) states that work done by a force changes the energy, and Eq. (21.26) is the equation of motion, expressed in three-vectors. Note that the latter contains the relativistic three-momentum $\mathbf{p} = \Gamma m_0 \mathbf{v}$. This is sometimes written as $\mathbf{p} = m(v) \mathbf{v}$, where the particle "mass" $m(v) = \Gamma m_0$ increases from its rest mass value m_0 when the particle velocity approaches the speed of light.

Three-velocity addition law As stated before, since we need to use the Lorentz transformation to transform between inertial reference frames, three-velocities will no longer merely add up. We can find the special relativistic rule for three-velocity addition from the following consideration. Suppose that a particle is observed in a frame L to move with three-velocity $\mathbf{v}$. In frame L' co-moving with this particle, a signal propagates away from the particle at the speed $\mathbf{w}$ (with respect to L'). Obviously, when we then denote the Lorentz factor evaluated for this three-velocity as $\Gamma_w = \left(1 - w^2/c^2\right)^{-1/2}$, frame L' will ascribe a four-velocity $U^{\alpha'} = (c\Gamma_w, \Gamma_w \mathbf{w})^{\mathrm{T}}$ to that signal. We can then use the inverse Lorentz transform $U^\alpha = (L^{-1})^\alpha_{\alpha'} U^{\alpha'}$, to obtain the corresponding four-velocity in frame L as

$$
\begin{pmatrix} c\Gamma_u \\ \Gamma_u \mathbf{u} \end{pmatrix} = \begin{pmatrix} \Gamma & \Gamma\dfrac{\mathbf{v}}{c} \\ \Gamma\dfrac{\mathbf{v}}{c} & \mathbf{I} + (\Gamma - 1)\dfrac{\mathbf{vv}}{v^2} \end{pmatrix} \begin{pmatrix} c\Gamma_w \\ \Gamma_w \mathbf{w} \end{pmatrix} , \tag{21.27}
$$

where $\mathbf{I}$ is the 3×3 identity matrix. It is then found easily that

$$
\mathbf{u} = \frac{\mathbf{v}\left[\Gamma + (\Gamma - 1)\mathbf{v}\cdot\mathbf{w}/v^2\right] + \mathbf{w}}{\Gamma\left(1 + \mathbf{v}\cdot\mathbf{w}/c^2\right)} . \tag{21.28}
$$

It is a matter of algebra to show that $\Gamma_u = \Gamma\Gamma_w\left(1 + \mathbf{v}\cdot\mathbf{w}/c^2\right)$ is the Lorentz factor for the velocity given by this Eq. (21.28). One can reorganize this expression to one giving the three-speed $\mathbf{w}$ in frame L', obtained from adding velocities $\mathbf{u}$ and $\mathbf{v}$ known in L, namely

$$
\mathbf{w} = \frac{\mathbf{u} - \mathbf{v}\left[\Gamma - (\mathbf{u}\cdot\mathbf{v}/c^2)\Gamma^2/(\Gamma + 1)\right]}{\Gamma\left(1 - \mathbf{u}\cdot\mathbf{v}/c^2\right)} . \tag{21.29}
$$

21.1.3 Relativistic gas dynamics and stress–energy tensor

We can now formulate the governing conservation laws in four-dimensional space-time. We begin with particle conservation. The proper density ρ is the mass per unit volume as seen in the rest frame of the gas. When we indicate the number density per unit volume in this frame by n_0, we have

$$
\rho = m_0 n_0 . \tag{21.30}
$$

In the inertial Lorentzian frame where the gas is seen to move with velocity $\mathbf{v}$, volumes differ due to the length contraction effect. As a result, the number density will be $n = \Gamma n_0$. A convenient variable to quantify the "density" in this "laboratory" frame is $D = \Gamma m_0 n_0 = m_0 n = \Gamma\rho$. It should be noted that the actual lab frame density is ΓD, due to the mass increase $m(v) = \Gamma m_0$, but we will loosely refer to D as the density. Particle number conservation is then generally expressed by the vanishing divergence of the four-vector ρU^α, namely

$$\partial_\alpha (\rho U^\alpha) = 0 . \tag{21.31}$$

Written out in terms of the coordinates $(ct, \mathbf{x})$, we then find

$$\frac{\partial D}{\partial t} + \nabla \cdot (D\mathbf{v}) = 0 . \tag{21.32}$$

It is easily seen how the classical limit $\Gamma \to 1$ indeed reduces to the familiar mass conservation equation, since $D \to \rho$.

Because energy and momentum need to be treated as a single physical entity, the classical (three-)momentum and energy conservation laws will be unified in a conservation law in four dimensions, involving the vanishing divergence of a four-tensor. This four-tensor is the stress–energy tensor which we will denote in contravariant components with $T^{\alpha\beta}$. Its components contain

$$\begin{pmatrix} T^{00} & T^{0i} \\ T^{i0} & T^{ij} \end{pmatrix} = \begin{pmatrix} \textit{energy density} & \textit{energy flux} \\ \textit{momentum flux} & \textit{stresses} \end{pmatrix} . \tag{21.33}$$

In the rest frame of the fluid, explicit expressions for its components are easily given, since fluxes vanish there. The energy density $T^{0'0'}$ contains both a rest mass and an internal energy contribution. Writing the specific internal energy in the fluid frame as ϵ, the expression for the energy density is

$$T^{0'0'} = \rho c^2 + \rho \epsilon . \tag{21.34}$$

When the pressure in the fluid frame is denoted by p, isotropic pressure corresponds to stresses

$$T^{i'j'} = p\,\mathbf{I} . \tag{21.35}$$

When we now transform back to the laboratory frame where the gas moves with velocity $\mathbf{v}$, we compute

$$T^{\alpha\beta} = (L^{-1})^\alpha_{\alpha'} (L^{-1})^\beta_{\beta'} T^{\alpha'\beta'} , \tag{21.36}$$

which yields the expression

$$T^{\alpha\beta} = \left(\rho c^2 + \rho\epsilon + p\right) \frac{U^\alpha U^\beta}{c^2} + p g^{\alpha\beta} . \tag{21.37}$$

The tensor is symmetric, $T^{\alpha\beta} = T^{\beta\alpha}$, and scalar invariants computed from this stress–energy tensor, upon which all observers will agree, are its trace defined as

$$T^\alpha_\alpha = T^{\alpha\beta} g_{\alpha\beta} = 3p - \rho c^2 - \rho\epsilon , \tag{21.38}$$

and

$$T^{\alpha\beta}T_{\alpha\beta} = \left(\rho c^2 + \rho\epsilon\right)^2 + 3p^2 \,. \tag{21.39}$$

We can write this tensor in the form of Eq. (21.33), and introduce some convenient notation as

$$\begin{pmatrix} T^{00} & T^{0i} \\ T^{i0} & T^{ij} \end{pmatrix} = \begin{pmatrix} \left(\rho c^2 + \rho\epsilon + p\right)\Gamma^2 - p & \left(\rho c^2 + \rho\epsilon + p\right)\Gamma^2 \mathbf{v}/c \\ \left(\rho c^2 + \rho\epsilon + p\right)\Gamma^2 \mathbf{v}/c & \rho\Gamma^2 \mathbf{vv} + p\mathbf{I} + \left(\rho\epsilon + p\right)\Gamma^2 \mathbf{vv}/c^2 \end{pmatrix}$$

$$\equiv \begin{pmatrix} \tau_g + Dc^2 & \mathbf{S}_g/c \\ \mathbf{S}_g/c & \mathbf{S}_g \mathbf{v}/c^2 + p\mathbf{I} \end{pmatrix}. \tag{21.40}$$

In terms of the variables τ_g (the total energy density in the lab frame, minus the rest mass contribution), and the three-vector $\mathbf{S}_g$ (the relativistic energy flux), the divergence of the stress–energy tensor is then written as follows. The temporal part, i.e. the equation $\partial_\alpha T^{0\alpha} = 0$, will yield

$$\frac{\partial}{\partial t}\left(\tau_g + Dc^2\right) + \nabla \cdot \mathbf{S}_g = 0 \,. \tag{21.41}$$

We can combine this relation with particle conservation expressed by Eq. (21.32) to obtain

$$\frac{\partial \tau_g}{\partial t} + \nabla \cdot \left((\tau_g + p)\mathbf{v}\right) = 0 \,. \tag{21.42}$$

The spatial part $\partial_\alpha T^{i\alpha} = 0$ works out to be

$$\frac{\partial \mathbf{S}_g}{\partial t} + \nabla \cdot \left(\mathbf{S}_g \mathbf{v} + pc^2 \mathbf{I}\right) = 0 \,. \tag{21.43}$$

Note that in these variables, the classical Newtonian limits are directly obtained, since in that limit we find

$$\mathbf{S}_g \overset{\Gamma \to 1}{\to} c^2 \rho \mathbf{v} \,,$$
$$\tau_g \overset{\Gamma \to 1}{\to} \rho\frac{v^2}{2} + \rho\epsilon \,. \tag{21.44}$$

This set of equations still needs a specification of an equation of state, relating specific energy ϵ to the gas pressure p and proper density ρ.

Thermodynamics and special relativity The combination of mass energy, internal energy and pressure appearing in both the energy density τ_g and energy flux vector $\mathbf{S}_g$ is known as the relativistic enthalpy. We can introduce the specific enthalpy h by

$$\rho h \equiv \rho c^2 + \rho\epsilon + p \equiv \rho(c^2 + h_g) \,, \tag{21.45}$$

where $h_{\mathrm{g}} = \epsilon + p/\rho$ is the specific enthalpy of the gas as used in non-relativistic treatments. With these definitions, we can now work out the following scalar identity

$$U_\alpha \partial_\beta T^{\alpha\beta} = 0 \,, \tag{21.46}$$

which must hold since we already had vanishing $\partial_\beta T^{\alpha\beta}$ for each $\alpha = 0, 1, 2, 3$. In doing so, we can use relation (21.20) to find that for all $\beta = 0, 1, 2, 3$ we have $U^\alpha \partial_\beta U_\alpha = 0$. The temporal $\beta = 0$ relation gives an alternative expression for $\partial \Gamma / \partial t$, while the spatial ones yield expressions for $\partial \Gamma / \partial x_i$. When we also use the particle conservation law (21.32), we can rework Eq. (21.46) to

$$\rho \Gamma \left\{ \frac{\partial h_{\mathrm{g}}}{\partial t} + \mathbf{v} \cdot \nabla h_{\mathrm{g}} \right\} - \Gamma \left\{ \frac{\partial p}{\partial t} + \mathbf{v} \cdot \nabla p \right\} = 0 \,. \tag{21.47}$$

Note that $\Gamma \left\{ \partial/\partial t + \mathbf{v} \cdot \nabla \right\} \equiv U^\alpha \partial_\alpha$ is the derivative along the world line.

When we now specify the discussion to an ideal gas law equation of state, we make a commonly encountered simplification with constant polytropic index γ appearing as

$$\rho \epsilon = \frac{p}{\gamma - 1} \,. \tag{21.48}$$

It is then a matter of algebra to manipulate Eq. (21.47) to one in terms of the entropy related quantity $S = p\rho^{-\gamma}$, written as

$$\Gamma \left\{ \partial/\partial t + \mathbf{v} \cdot \nabla \right\} S = 0 \,. \tag{21.49}$$

Hence, as in the non-relativistic case, entropy is advected with the fluid. Furthermore, it is possible to show (see Sec. 21.1.4) that the relativistic expression for the sound speed c_{g} in the polytropic case becomes

$$\frac{c_{\mathrm{g}}}{c} = \sqrt{\frac{\gamma p}{\rho h}} \,. \tag{21.50}$$

The non-relativistic limit clearly reduces to $c_{\mathrm{g}} = \sqrt{\gamma p / \rho}$, as expected. Furthermore, this relation also shows that $c_{\mathrm{g}}^2 \leq (\gamma - 1)c^2$, so that there is an upper limit to sound propagation speeds.

$\triangleright$ **The Synge gas** The constant γ in Eq. (21.48) is in practice always taken in the range $\gamma \in [4/3, 5/3]$. In fact, for a perfect gas law where $p = k_{\mathrm{B}} \rho T / m_0$, the lower value $\gamma = 4/3$ is applicable at relativistic internal energies, i.e. $k_{\mathrm{B}} T \geq m_0 c^2$, while the non-relativistic limit for a monatomic gas is known to be $5/3$. A proper generalization of the perfect gas law to relativistic regimes needs to start from a relativistic kinetic plasma theory and the relativistic counterpart of the Boltzmann equation. For an equilibrium distribution function known as the Jüttner distribution function (the relativistic counterpart of the Maxwell–Boltzmann distribution), one can again deduce the perfect gas law relation $p = k_{\mathrm{B}} \rho T / m_0$.

Moreover, it is then found that

$$\rho c^2 + \rho \epsilon + p = \rho c^2 \frac{K_3(z)}{K_2(z)}, \tag{21.51}$$

where K_n denotes the modified Bessel function of the second kind, with the argument $z \equiv m_0 c^2 / k_{\rm B} T = \rho c^2 / p$. If we then write in analogy with the polytropic case

$$\hat{\gamma} \equiv 1 + \frac{p}{\rho \epsilon}, \tag{21.52}$$

we find the relation $1 + \hat{\gamma}/z(\hat{\gamma} - 1) = K_3(z)/K_2(z)$. To find the effective polytropic index $\hat{\gamma}(z)$ at low temperature ($z \to \infty$) and high temperature $z \to 0$, we can use the asymptotic expansions at small and large argument for the ratio

$$\frac{K_3(z)}{K_2(z)} \stackrel{z \to \infty}{\approx} 1 + \frac{5}{2z},$$

$$\frac{K_3(z)}{K_2(z)} \stackrel{z \to 0}{\approx} \frac{4}{z}. \tag{21.53}$$

The actual polytropic index thus increases smoothly between $4/3$ at ultra-relativistic internal energies and its non-relativistic value $5/3$, when plotted as function of z. This is shown in Fig. 21.3, and this relativistically correct perfect gas description is known as the Synge gas [428]. Writing the modified Bessel ratio as $G(z)$, it has a sound speed given by

$$\frac{c_{\rm g}^2}{c^2} = \frac{G'}{GzG' + G/z}. \tag{21.54}$$

The derivative $G' = G^2 - 1 - 5G/z$, due to recurrence relations. For computational approaches, one can use a convenient approximation introduced by Mathews [327], which avoids the (costly) Bessel function evaluations and sets

$$c^2 + \epsilon = c^2 \left\{ \frac{p}{(\gamma - 1)\rho c^2} + \sqrt{\left(\frac{p}{(\gamma - 1)\rho c^2} \right)^2 + 1} \right\},$$

$$h = \frac{1}{2} \left[(\gamma + 1)(c^2 + \epsilon) - (\gamma - 1)\frac{c^4}{c^2 + \epsilon} \right]. \tag{21.55}$$

In this approximation, the sound speed is given by

$$\frac{c_{\rm g}^2}{c^2} = \frac{p}{\rho h} \left[\frac{\gamma + 1}{2} + \frac{\gamma - 1}{2} \frac{c^4}{(c^2 + \epsilon)^2} \right], \tag{21.56}$$

while the locally effective polytropic index is then given by

$$\hat{\gamma} = \gamma - \frac{\gamma - 1}{2} \left(1 - \frac{c^4}{(c^2 + \epsilon)^2} \right). \tag{21.57}$$

In these expressions, setting the parameter $\gamma = 5/3$ yields an excellent approximation to the Synge gas variation, as shown in Fig. 21.3, while non-adiabatic effects can be approximated using different γ values. ◁

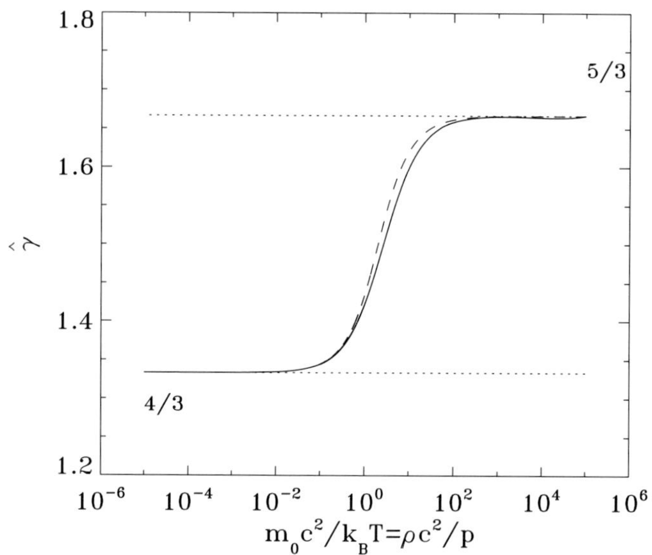

Fig. 21.3 The variation of the effective polytropic index with (inverse) temperature for the Synge gas. The dashed line represents the Mathews approximation.

21.1.4 Sound waves and shock relations in relativistic gases

To obtain the relativistic expression for the sound wave speed, one must perform the usual linearization of the relativistic hydrodynamic equations about an equilibrium configuration. We will do this for a homogeneous gas, and adopt a strategy which rewrites the governing covariant equations in four-dimensional space-time to equivalent expressions in a $3 + 1$ formalism in terms of a convenient set of primitive variables. The $3 + 1$ formalism splits temporal and spatial derivatives in a fixed Lorentzian reference frame, and shows clearly how the equations become more involved with respect to their Galilean versions used in all previous chapters. Also, the $3 + 1$ formalism is the one adopted in modern numerical approaches, where one fixes a laboratory frame and uses shock-capturing schemes to solve the governing conservation laws. We choose as primitive variables the entropy S, the rest frame proper density ρ and the velocity $\mathbf{v}$ in our laboratory frame. For simplicity, we restrict the discussion here to a constant polytropic equation of state, where $S = p\rho^{-\gamma}$ and note that we already have Eq. (21.49) for the entropy. In a similar fashion, one can (after some algebra) obtain the following set of equations:

$$\frac{\partial S}{\partial t} + \mathbf{v} \cdot \nabla S = 0 \,,$$

$$\frac{\partial \rho}{\partial t} + \mathbf{v} \cdot \nabla \rho + \frac{\rho h}{u} \nabla \cdot \mathbf{v} - \frac{1}{u\Gamma^2} \mathbf{v} \cdot \nabla \left(S\rho^\gamma \right) = 0 \,,$$

$$\frac{\partial \mathbf{v}}{\partial t} + (\mathbf{v} \cdot \nabla)\,\mathbf{v} + \frac{c^2}{\rho h \Gamma^2} \nabla\,(S\rho^\gamma)$$

$$- \mathbf{v}\,(\nabla \cdot \mathbf{v})\left(1 - \frac{yc^2}{u}\right) - \mathbf{v}\,\frac{yc^2}{u\rho h \Gamma^2}\,\mathbf{v} \cdot \nabla\,(S\rho^\gamma) = 0\,. \qquad (21.58)$$

In these equations, we introduced

$$u = h - \frac{v^2}{c^2}\gamma S\rho^{\gamma-1} \qquad \left[\to c^2\right],$$

$$y = \frac{2h}{c^2} - 1 - \frac{\gamma^2 S\rho^{\gamma-1}}{c^2(\gamma-1)} \qquad \left[\to 1\right],$$

$$h = c^2 + \frac{\gamma S\rho^{\gamma-1}}{\gamma-1} \qquad \left[\to c^2\right], \qquad (21.59)$$

where the expressions between brackets denote the Galilean limits when $\Gamma \to 1$. In this form, no approximations have been made yet, and we can clearly identify all terms denoting relativistic corrections. We will next use this form as a starting point to obtain linear wave speeds in two complementary approaches.

Waves in a static homogeneous gas The equations (21.58) can easily be linearized about a static $\mathbf{v}_0 = 0$, uniform gas with constant entropy and density S_0, ρ_0. As before, one conveniently assumes a plane wave variation $\exp(-i\omega t + i\mathbf{k} \cdot \mathbf{x})$ of all linear quantities $S_1, \rho_1, \mathbf{v}_1$ to arrive at

$$\omega S_1 = 0\,,$$

$$\omega \rho_1 = \rho_0 \mathbf{k} \cdot \mathbf{v}_1\,,$$

$$\omega \mathbf{v}_1 = \frac{c^2}{\rho_0 h_0}\mathbf{k}\left(S_0\gamma\rho_0^{\gamma-1}\rho_1 + \rho_0^\gamma S_1\right)\,. \qquad (21.60)$$

As already known from the non-relativistic case, this system admits five solutions, where three wave modes are at marginal frequency $\omega = 0$. These are the entropy wave with arbitrary S_1 but without density or velocity perturbation, together with the two transverse translations (shear waves), already encountered in Section 5.2.2 [1]. The physically more interesting modes are compressible perturbations with $\mathbf{k} \cdot \mathbf{v}_1 \neq 0$ and have the dispersion relation

$$\frac{\omega^2}{k^2 c^2} = \frac{\gamma S_0 \rho^{\gamma-1}}{h_0} = \frac{\gamma p_0}{\rho_0 h_0}\,. \qquad (21.61)$$

This shows that the sound speed c_{g} for the polytropic case is indeed given by Eq. (21.50).

Characteristic speeds A second approach to using the equations (21.58) follows the technique discussed in Section 19.1.1, noting that this form allows us to read off the components of the 5×5 coefficient matrix $\mathbf{W}$ in the quasi-linear form from Eq. (19.3), namely

$$\frac{\partial \mathbf{V}}{\partial t} + \mathbf{W}\frac{\partial \mathbf{V}}{\partial x} = 0 \,, \tag{21.62}$$

where the primitive variables are $\mathbf{V} = (S, \rho, v_x, v_y, v_z)^{\mathrm{T}}$. When we compute the five eigenvalues λ of the $\mathbf{W}$ matrix, we obtain the characteristic equation

$$(\lambda - v_x)^3 \left[\lambda^2 - 2\lambda v_x \frac{1 - c_{\mathrm{g}}2/c^2}{1 - v^2 c_{\mathrm{g}}^2/c^4} + \frac{v_x^2 \left(1 - c_{\mathrm{g}}^2/c^2\right) - c_{\mathrm{g}}^2 \left(1 - v^2/c^2\right)}{1 - v^2 c_{\mathrm{g}}^2/c^4} \right] = 0 \,. \tag{21.63}$$

Hence, the characteristic speeds either take the value $\lambda = v_x$ (which obviously correspond to the entropy and shear waves from above), or the value for the sound waves this time found from a more complicated quadratic expression. Naturally, both approaches must agree. The key observation is that by computing the characteristic speeds from the $\mathbf{W}$ matrix, we in fact linearized the equations about a moving plasma, and that we need to consider how plane waves in the gas rest frame transform relativistically to a moving reference frame. We therefore address how plane waves behave under the Lorentz transformation.

▷ **Exercise** Consider the purely 1D case where $v = v_x$, and show that the quadratic expression in Eq. (21.63) corresponds to relativistic speed addition between v_x and c_{g}. ◁

Phase and group diagrams in special relativity As usual, we will consider frame L' with coordinates $(ct', \mathbf{x}')$ to move with respect to frame L with a velocity $\mathbf{v}$. Equation (21.1) then relates the time and space coordinates, and we momentarily assume that in L' we have a plane wave with variation $\exp[-\mathrm{i}(\omega' t' - \mathbf{k}' \cdot \mathbf{x}')]$. Anticipating Doppler shifts as well as a change in the wave vector, we indicate the frequency and wave vector to be specific to L'. One then finds directly that frame L will still observe a plane wave with variation $\exp[-\mathrm{i}(\omega t - \mathbf{k} \cdot \mathbf{x})]$ with frequency and wave vector given by

$$\omega = \Gamma \left(\omega' + \mathbf{k}' \cdot \mathbf{v}\right) \,, \qquad \mathbf{k} = \mathbf{k}' + \mathbf{v} \left[\frac{\omega' \Gamma}{c^2} + (\mathbf{k}' \cdot \mathbf{v}) \frac{\Gamma - 1}{v^2} \right] \,. \tag{21.64}$$

These expressions quantify the relativistic Doppler effect (i.e. the change in frequency) and show that the wave vector changes direction when viewed from a moving vantage point. The latter effect is known as relativistic wave aberration. The inverse formulas (which are identical with $\omega' \leftrightarrow \omega$, $\mathbf{k}' \leftrightarrow \mathbf{k}$ and a sign change for $\mathbf{v} \leftrightarrow -\mathbf{v}$) then allow us to find the phase speed for frame L given by $v_{\mathrm{ph}} = \omega/k$

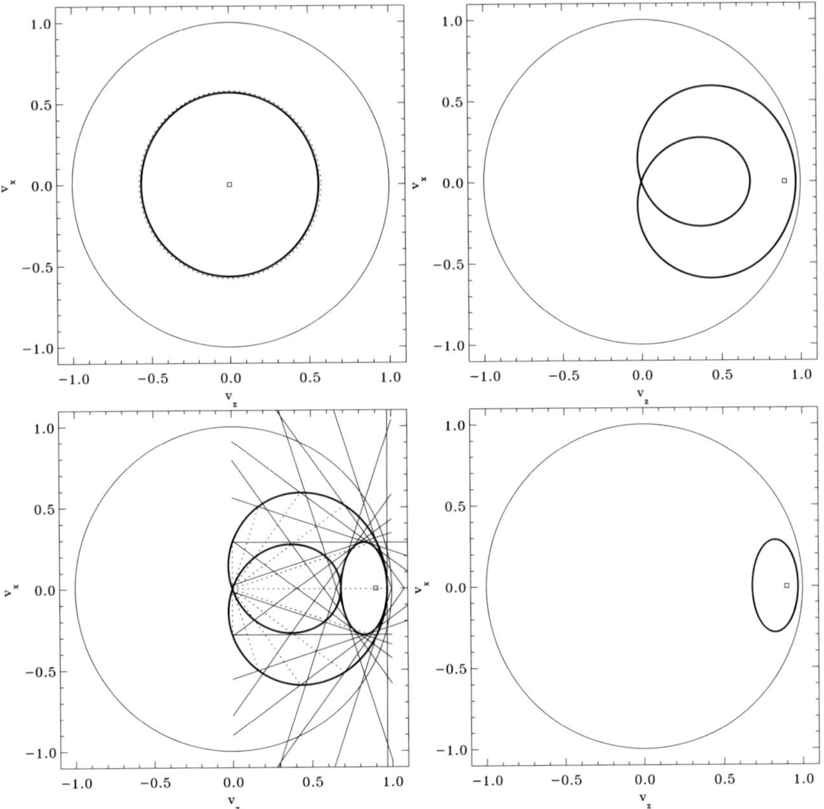

Fig. 21.4 Phase and group speed diagram in the gas rest frame (top left), compared to the phase speed in a frame where the source moves at $0.9c$ along the z-axis (top right). The bottom left panel shows the Huygens construction, which can be used to obtain the (bottom right) group speed diagram in the same lab frame.

from the formula

$$\frac{v'^2_{\mathrm{ph}}}{c^2} = \frac{\Gamma^2 \left(v_{\mathrm{ph}} - \mathbf{n} \cdot \mathbf{v}\right)^2}{c^2 + \Gamma^2 \left(v_{\mathrm{ph}} - \mathbf{n} \cdot \mathbf{v}\right)^2 - v^2_{\mathrm{ph}}}. \tag{21.65}$$

We introduced, in analogy with the notation of Section 5.4.3 [1], the unit vector $\mathbf{n} = \mathbf{k}/k$. When we invert this formula to find the phase speed in frame L, we get

$$v^2_{\mathrm{ph}} - 2v_{\mathrm{ph}}(\mathbf{n} \cdot \mathbf{v})\frac{1 - v'^2_{\mathrm{ph}}/c^2}{1 - v^2 v'^2_{\mathrm{ph}}/c^4} + \frac{(\mathbf{n} \cdot \mathbf{v})^2 \left(1 - v'^2_{\mathrm{ph}}/c^2\right) - v'^2_{\mathrm{ph}} \left(1 - v^2/c^2\right)}{1 - v^2 v'^2_{\mathrm{ph}}/c^4} = 0. \tag{21.66}$$

This is directly seen to agree with expression (21.63) for the sound waves in the moving reference frame. To complete this discussion on plane wave propagation in uniform gases for special relativity, we can draw the phase and group speed diagrams in different reference frames. This is done in Fig. 21.4, where the $z - x$ plane is drawn. We show at top left the phase $v_{\mathrm{ph}}\mathbf{n}/c$ and group speed $\mathbf{v}_{\mathrm{gr}} = \partial\omega/\partial\mathbf{k}$ in the gas rest frame, where we find isotropic sound wave propagation in all directions, and where phase and group speed diagrams overlap. The sound wave diagram will always be interior to the inner dashed circle which corresponds to the upper limit $\sqrt{1/3}c$, while the outer circle indicates the light limit. In this diagram, we actually adopted the Mathews prescription described above, and took $\rho_0 = 1$, $p_0 = 1$ (a relativistically hot gas, making the sound speed very close to the upper limit) and scaled with $c = 1$. In the right panel, the phase speed diagram is plotted as seen in a reference frame where the plane wave emitter is moving at velocity $\mathbf{v} = 0.9c\mathbf{e}_z$ along the horizontal z-axis. The wave aberration effects deformed the single circle to a kind of double loop form. In the bottom panels, we indicated how a Huygens construction then yields the corresponding group speed diagram in the same reference frame. As before, the group diagram is what evolves from a point perturbation in a finite time. The final panel just shows this group speed diagram in that frame, and demonstrates how the wave front gets "beamed" into an anti-symmetric (about the position of the point source) oval shape.

▷ **Exercise** Obtain the analytic expressions for the phase and group speed diagrams for sound waves in both reference frames, and show that the group speed diagram clearly follows the speed addition rule from Eq. (21.169) (see Exercise [21.2]). ◁

Gas dynamic shock relations To discuss gas dynamic shocks in special relativity, we need to go back to the actual conservation laws in four-dimensional space-time, and consider the limits where discontinuities occur across moving manifolds in space-time. The shock front is generally a surface in space-time, which is given by an equation $\phi(ct, \mathbf{x}) = d$ where d is some constant. The normal to the shock front is a space-like four-vector $\mathbf{l}$ (i.e. one whose invariant $l^\alpha l_\alpha > 0$), whose components are given by $l_\alpha = \partial_\alpha\phi$. We are free to normalize this such that $l^\alpha l_\alpha = 1$. The Rankine–Hugoniot conditions always follow from the conservation laws across the manifold, expressed generally as

$$[\![\rho U^\alpha]\!]l_\alpha = 0, \tag{21.67}$$

$$[\![T^{\alpha\beta}]\!]l_\alpha = 0. \tag{21.68}$$

These expressions can be written out in various reference frames, and can be used to quantify shock compression ratios, or be manipulated to identify various shock invariants. In analogy with Chapter 20, we will write out the expressions in the shock rest frame (SRF), where by definition the four-velocity of the shock is $\mathbf{U}^s = (c, 0)$. Without loss of generality, we can assume the shock normal to be oriented

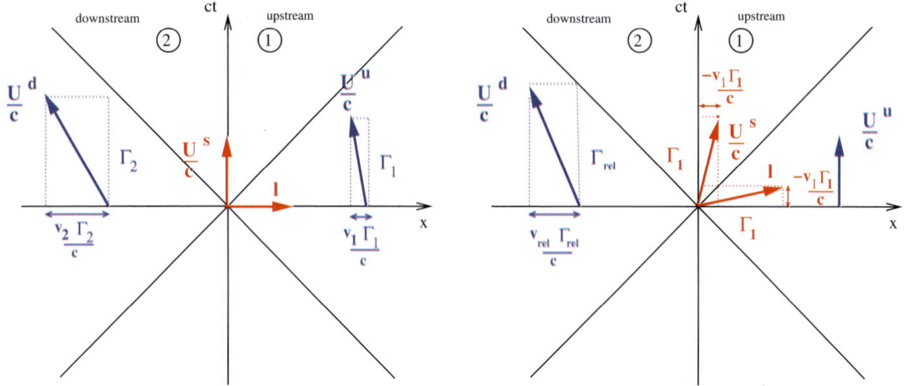

Fig. 21.5 The space-time diagrams appropriate for analysis of the Rankine–Hugoniot shock relations in the shock rest frame (SRF, left panel), and in the upstream rest frame (URF, right panel).

along the x-axis, such that $\mathbf{l} = (0, \mathbf{e}_x) \equiv (0, \mathbf{n})$, where the space part is then indicated by a three-vector $\mathbf{n}$. The four-velocities at right and left of the shock front are denoted in this SRF as

$$\mathbf{U}^u = (c\Gamma_1, \Gamma_1 \mathbf{v}_1) ,$$

$$\mathbf{U}^d = (c\Gamma_2, \Gamma_2 \mathbf{v}_2) , \tag{21.69}$$

where the index 1 is for the upstream state, while 2 denotes the downstream state.

A graphical representation of the various four-vectors in the SRF is given in the left panel of Fig. 21.5 (the right panel shows the same quantities in the upstream rest frame). Besides the three-velocities $\mathbf{v}_1$ and $\mathbf{v}_2$, which indicate the velocity of the upstream (downstream) gas with respect to the shock, a third relative velocity can be identified: the relative velocity of the upstream gas with respect to the downstream gas $\mathbf{v}_{\rm rel}$ (its explicit formula can be obtained from velocity addition rules). In any case, when we write out the jump relations (21.67)–(21.68) in the SRF, we find that

$$\rho_1 \Gamma_1 v_{1n} = \rho_2 \Gamma_2 v_{2n} ,$$

$$\rho_1 h_1 \Gamma_1^2 \mathbf{v}_1 v_{1n} + c^2 p_1 \mathbf{n} = \rho_2 h_2 \Gamma_2^2 \mathbf{v}_2 v_{2n} + c^2 p_2 \mathbf{n} ,$$

$$h_1 \Gamma_1 = h_2 \Gamma_2 . \tag{21.70}$$

From the second equation, it follows that the tangential velocities do not jump $[\![\mathbf{v}_t]\!] = 0$, where $\mathbf{t} \perp \mathbf{n}$ is a unit three-vector perpendicular to $\mathbf{n}$ in the Euclidean sense. We can thus Lorentz transform to a frame which is moving with this constant, purely tangential, three-velocity. Denoting this reference frame by tangential reference frame or TRF, the gas dynamic shock relations are then reduced to 1D relations as all tangential velocities vanish, and we find

$$\rho_1 \Gamma_1 v_1 = \rho_2 \Gamma_2 v_2 \,,$$

$$\rho_1 h_1 \Gamma_1^2 v_1^2 + c^2 p_1 = \rho_2 h_2 \Gamma_2^2 v_2^2 + c^2 p_2 \,,$$

$$h_1 \Gamma_1 = h_2 \Gamma_2 \,. \tag{21.71}$$

Note that although we used the same symbols to denote the velocities up and downstream, these in Eq. (21.71) are in the TRF, while in Eq. (21.70) we have velocities as observed in the SRF, with the two related by a Lorentz transformation involving the tangential velocity seen in the SRF. Further discussion of these shock relations is then possible for various physically interesting limiting cases, and we refer to the topical review by Kirk and Duffy [279] to find some quantitative expressions. As a final note on gas dynamic shocks, one can manipulate the expressions to the following generalization of the Hugoniot adiabat from Eq. (20.18), which is known as the Taub adiabat:

$$h_1^2 - h_2^2 + (p_2 - p_1) \left(\frac{h_1}{\rho_1} + \frac{h_2}{\rho_2} \right) = 0 \,. \tag{21.72}$$

This is an invariant across the shock, and thus holds in any reference frame.

▷ **Exercise** Verify that the coefficient of the term proportionate to c^2 in expression (21.72) reduces to the Hugoniot adiabat from Eq. (20.18). ◁

Example application: Fanaroff–Riley type I jet deceleration As a representative application of the relativistic hydro equations, we summarize findings from a modern relativistic hydro computation with shock-governed dynamics. The application focuses on active galactic nuclei (AGN) jet dynamics. Such jets typically demonstrate Lorentz factors of about $\Gamma \sim 5 - 30$, and the Fanaroff–Riley (FR) classification distinguishes between type I and FR II radio galaxies, according to the power of the jet and the corresponding accretion rate in their galactic centre. In contrast to FR II galaxies where the jet remains relativistic and narrow on all scales, the jet in FR I sources is relativistic on the parsec scale and in many cases sub-relativistic and diffuse on kparsec scales. To explain this sudden deceleration, Meliani *et al.* [330] explore a model where high Lorentz factor jets encounter density discontinuities as they propagate through the interstellar medium. In Fig. 21.6, two jet morphologies from their ten model computations are displayed for jets that start with inlet Lorentz factor 10, and are characterized by kinetic energy luminosities of about 10^{46} (top) versus 10^{43}ergs/s (bottom). The study demonstrated that as long as the jet propagates through uniform media, the density contrast between jet

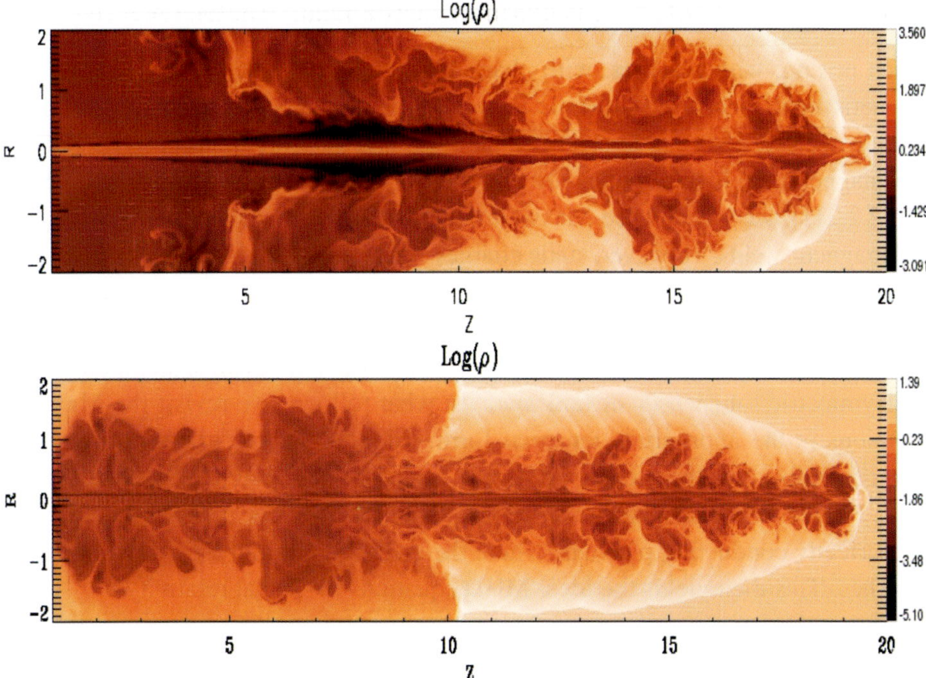

Fig. 21.6 Relativistic hydro simulations, where Lorentz factor 10 jet beams propagate through ISM regions of varying density. The two cases differ in beam kinetic luminosity (highest at top), and display the logarithm of the proper density after about 800 light crossing times of the jet beam radius. (From Meliani *et al.* [330]).

beam and ISM sets most of the propagation characteristics, fully consistent with previous modeling efforts. When the jet runs into a denser medium, the authors found a clear distinction in the deceleration of high-energy jets depending on the encountered density jump. For fairly high-density contrast, the jet becomes destabilized and can decelerate strongly, even to sub-relativistic speeds. If the density contrast is too weak, the high-energy jets continue with FR II characteristics. The trend is similar for the low-energy jet models, which start as under-dense jets from the outset, and decelerate by entrainment into the lower region as well. Another aspect of these high-resolution grid-adaptive simulations (employing an HLLC type shock-capturing solver and using the Mathews equation of state) is that they provide insight into dynamical details, like the Richtmyer–Meshkov instabilities developing at the original contact interface. This instability occurs as a shock passes a contact discontinuity, depositing vorticity on the shocked contact, eventually leading to Kelvin–Helmholtz roll-up. This is clearly seen in the density maps displayed in Fig. 21.6, where the original interface is strongly deformed and situated midway along the simulated distance at $Z \simeq 10$.

21.2 Electromagnetism and special relativistic MHD

When dealing with plasmas, with charged particle constituents, the electric and magnetic fields generated by static as well as moving charged particles introduce particle accelerations due to electrostatic fields and the Lorentz force. The electric and magnetic fields themselves are governed by the Maxwell equations. In Section 2.2.2[1], we wrote these Maxwell equations in terms of three-vector electric and magnetic fields. However, static charges which act as sources for electric fields in one Lorentzian reference frame will be moving charges, i.e. currents, leading to magnetic fields in another frame in relative motion. The combined six components of both three-vectors will in fact form the six independent tensor components of an asymmetric four-tensor of rank two. In what follows, we revisit the Maxwell equations in the appropriate tensorial formulation applicable in space-time. We then generalize the stress–energy tensor to include electromagnetic field contributions. Finally, we specify the governing equations for relativistic plasma dynamics for perfectly conducting plasmas, where the electric field in the co-moving frame vanishes. This then yields the ideal MHD equations in special relativity.

21.2.1 Electromagnetic field tensor and Maxwell's equations

In analogy with Eqs. (21.25)–(21.26), the relativistically correct equation of motion for a particle of charge q, written in three-vector notation, can be expected to be

$$\frac{d\mathbf{p}}{dt} = q\left(\mathbf{E} + \mathbf{v} \times \mathbf{B}\right). \tag{21.73}$$

Since the Lorentz force is orthogonal to the spatial three-velocity, the change in relativistic energy $E = \Gamma m_0 c^2$ is

$$\frac{dE}{dt} = \mathbf{v} \cdot (q\mathbf{E}). \tag{21.74}$$

As the Lorentz force involves the fluid velocity, we need to generalize Eq. (21.24) to a four-momentum formulation with a four-force involving the four-velocity as in

$$\frac{dP^\alpha}{d\tau} = qF^{\alpha\beta}U_\beta. \tag{21.75}$$

The tensor components of the occurring electromagnetic field tensor $F^{\alpha\beta}$ can in fact be computed from requiring the spatial and temporal components of this expression (21.75) to reduce to Eq. (21.73) and Eq. (21.74), respectively. The end result is the anti-symmetric tensor with contravariant components given by

$$
F^{\alpha\beta} = \begin{pmatrix} 0 & E_1/c & E_2/c & E_3/c \\ -E_1/c & 0 & B_3 & -B_2 \\ -E_2/c & -B_3 & 0 & B_1 \\ -E_3/c & B_2 & -B_1 & 0 \end{pmatrix}.
\tag{21.76}
$$

The components of the electric field $\mathbf{E}$ and magnetic field $\mathbf{B}$ as observed in this particular reference frame thus form the six independent entries of this tensor. Associated with this tensor, we can compute the invariant scalar quantity

$$
F^{\alpha\beta} F_{\alpha\beta} = 2 \left(\mathbf{B} \cdot \mathbf{B} - \frac{\mathbf{E} \cdot \mathbf{E}}{c^2} \right).
\tag{21.77}
$$

All observers will thus agree on the relative magnitude of electric $E^2 = \mathbf{E} \cdot \mathbf{E}$ and magnetic field vectors $B^2 = \mathbf{B} \cdot \mathbf{B}$. In particular, for electromagnetic (plane) waves in vacuum where $|\mathbf{E}| = c|\mathbf{B}|$, all observers will find a zero value.

When we change from one Lorentzian frame to another by means of the Lorentz transformation (21.3), the components of the electromagnetic field tensor transform in the usual manner. When changing from frame L to frame L' moving with velocity $\mathbf{v}$ with respect to L, we thus find from

$$
F^{\alpha'\beta'} = \begin{pmatrix} 0 & E_1'/c & E_2'/c & E_3'/c \\ -E_1'/c & 0 & B_3' & -B_2' \\ -E_2'/c & -B_3' & 0 & B_1' \\ -E_3'/c & B_2' & -B_1' & 0 \end{pmatrix} = L_\alpha^{\alpha'} L_\beta^{\beta'} F^{\alpha\beta}
\tag{21.78}
$$

that the Lorentz transformation in terms of the three-vectors for electric and magnetic fields implies

$$
\begin{cases}
\mathbf{E}' = \Gamma \left(\mathbf{E} + \mathbf{v} \times \mathbf{B} \right) - \dfrac{\Gamma^2}{(\Gamma + 1)\, c^2}\, \mathbf{v}\, \mathbf{v} \cdot \mathbf{E}, \\[3mm]
\mathbf{B}' = \Gamma \left(\mathbf{B} - \dfrac{1}{c^2} \mathbf{v} \times \mathbf{E} \right) - \dfrac{\Gamma^2}{(\Gamma + 1)\, c^2}\, \mathbf{v}\, \mathbf{v} \cdot \mathbf{B},
\end{cases}
\tag{21.79}
$$

as already given in Section 2.2.2 [1].

We can introduce a four-vector equivalent of the electric field $\mathbf{E}$, when we write the governing equation of motion (21.75) as

$$
\frac{dP^\alpha}{d\tau} = q F^{\alpha\beta} U_\beta = q e^\alpha.
\tag{21.80}
$$

The contravariant components e^α then quantify the electric field measured in the local rest frame of the moving plasma (analogous to $q\mathbf{E}$ for electrostatic acceleration of a static charge q in three-vector form). We can write this four-vector as

$$
e^\alpha = \left[\Gamma \mathbf{v} \cdot \mathbf{E}/c, \Gamma (\mathbf{E} + \mathbf{v} \times \mathbf{B}) \right]^{\mathrm{T}}.
\tag{21.81}
$$

Note that this four-vector vanishes when $\mathbf{E} = -\mathbf{v} \times \mathbf{B}$, and this observation will lead us in Section 21.2.3 to the ideal MHD limit in special relativity. Generally, though, the electric four-vector is orthogonal (in a space-time sense) to the four-velocity, since it is easily shown that

$$e^{\alpha} U_{\alpha} = 0 \,. \tag{21.82}$$

An invariant quantity is then also

$$e^{\alpha} e_{\alpha} = \Gamma^2 \left[(\mathbf{E} + \mathbf{v} \times \mathbf{B}) \cdot (\mathbf{E} + \mathbf{v} \times \mathbf{B}) - (\mathbf{v} \cdot \mathbf{E}/c)^2 \right] \,. \tag{21.83}$$

To arrive at Maxwell's equations in four-dimensional space-time notation, we will need to handle the source terms appearing in the non-homogeneous equations

$$\begin{cases} \nabla \cdot \mathbf{E} = c^2 \sigma \,, \\ \nabla \times \mathbf{B} = \mathbf{j} + \dfrac{1}{c^2} \dfrac{\partial \mathbf{E}}{\partial t} \,. \end{cases} \tag{21.84}$$

The charge density is given by $\sigma = qn = q\Gamma n_0$, where the charged particle number density is n (n_0 in the local rest frame). The current density three-vector is $\mathbf{j} = qn\mathbf{v}$. Note that we used mks units, setting $\mu_0 = 1$ for convenience (and equivalently setting the permittivity $\epsilon_0 = 1/c^2$). It is then clear how, in analogy with four-velocity or four-momentum, we introduce a four-current $J^{\alpha} = q n_0 U^{\alpha}$ such that

$$J^{\alpha} = (c\sigma, \mathbf{j})^{\mathrm{T}} \,. \tag{21.85}$$

Its invariant is $J^{\alpha} J_{\alpha} = -c^2 q^2 n_0^2$. Charge conservation is expressed as $\partial_{\alpha} J^{\alpha} = 0$ or

$$\frac{\partial \sigma}{\partial t} + \nabla \cdot \mathbf{j} = 0 \,. \tag{21.86}$$

The non-homogeneous Maxwell equations (21.84) are then unified in the following law in terms of the electromagnetic field tensor:

$$\partial_{\beta} F^{\alpha\beta} = J^{\alpha} \,. \tag{21.87}$$

Also, with the aid of this four-current J^{α}, we can combine the Lorentz force, and its contribution to the energy, for an ensemble of particles with number density n. The equivalent of the single particle expression Eq. (21.75) writes as

$$F^{\alpha\beta} J_{\beta} = F^{\alpha}_{\ \beta} J^{\beta} = \begin{pmatrix} nq\mathbf{v} \cdot \mathbf{E}/c \\ nq(\mathbf{E} + \mathbf{v} \times \mathbf{B}) \end{pmatrix} \,. \tag{21.88}$$

As a consequence, the divergence of the stress–energy tensor (21.40) for a plasma now equals

$$\partial_{\beta} T^{\alpha\beta}_{\mathrm{pl}} = F^{\alpha}_{\ \beta} J^{\beta} \,. \tag{21.89}$$

▷ **Exercise** Alternatively, we can arrive at the expression for the four-current (21.85) by noting that in the local rest frame it must boil down to $(cqn_0, \mathbf{0})^{\mathrm{T}}$, and by performing the inverse Lorentz transformation to the frame where the plasma moves at velocity $\mathbf{v}$.   ◁

For any anti-symmetric second rank tensor in space-time, such as the electromagnetic field tensor $F^{\alpha\beta}$, we can define a related anti-symmetric tensor by making use of the Levi-Civita symbol. In four-dimensional space-time, the Levi-Civita symbol is defined as

$$
\epsilon_{\mu\nu\alpha\beta} = \begin{cases} +1 & \text{for any even permutation of } 0,1,2,3 , \\ -1 & \text{for uneven permutations of } 0,1,2,3 , \\ 0 & \text{for any case with repeated indices.} \end{cases} \tag{21.90}
$$

▷ **Parity of permutations and Levi-Civita symbols** The parity of a permutation of $0,1,2,3$ is, by definition, given by the parity of the amount of numbers that appear in a non-increasing order, when going from left to right. As an example, the permutation $P = \{0,1,2,3\}$ has zero numbers appearing in non-increasing order, hence its parity is even. However, the cyclic permutation of P given by $Q = \{3,0,1,2\}$ has three numbers appear "in the wrong order", namely the sequence $0,1,2$ following 3. Therefore, its parity is odd. Permuting Q in cyclic fashion once more, we get $R = \{2,3,0,1\}$ where four numbers are non-increasing: $0,1$ follow both 2 and 3. Hence R is even again. This means that the Levi-Civita symbol in four-dimensional form changes sign under cyclic permutations of its indices, a result which is different from the perhaps more familiar 3D case. In flat space-time, we also find easily that

$$
\epsilon_{\mu\nu\alpha\beta} = -\epsilon^{\mu\nu\alpha\beta} . \tag{21.91}
$$

Finally, we note the following equalities (e.g. see [429]):

$$
\begin{aligned}
\epsilon^{\alpha\beta\mu\nu}\epsilon_{\alpha\gamma\lambda\kappa} = & -\delta^{\beta}_{\ \gamma}\delta^{\mu}_{\ \lambda}\delta^{\nu}_{\ \kappa} + \delta^{\beta}_{\ \gamma}\delta^{\mu}_{\ \kappa}\delta^{\nu}_{\ \lambda} + \delta^{\beta}_{\ \lambda}\delta^{\mu}_{\ \gamma}\delta^{\nu}_{\ \kappa} - \delta^{\beta}_{\ \lambda}\delta^{\mu}_{\ \kappa}\delta^{\nu}_{\ \gamma} \\
& -\delta^{\beta}_{\ \kappa}\delta^{\mu}_{\ \gamma}\delta^{\nu}_{\ \lambda} + \delta^{\beta}_{\ \kappa}\delta^{\mu}_{\ \lambda}\delta^{\nu}_{\ \gamma} = -\delta^{\beta\mu\nu}_{\gamma\lambda\kappa} = -\begin{vmatrix} \delta^{\beta}_{\gamma} & \delta^{\beta}_{\lambda} & \delta^{\beta}_{\kappa} \\ \delta^{\mu}_{\gamma} & \delta^{\mu}_{\lambda} & \delta^{\mu}_{\kappa} \\ \delta^{\nu}_{\gamma} & \delta^{\nu}_{\lambda} & \delta^{\nu}_{\kappa} \end{vmatrix} , \\
\epsilon^{\alpha\beta\mu\nu}\epsilon_{\alpha\beta\lambda\kappa} = & -2\delta^{\mu\nu}_{\lambda\kappa} , \\
\epsilon^{\alpha\beta\mu\nu}\epsilon_{\alpha\beta\mu\kappa} = & -6\delta^{\nu}_{\kappa} , \tag{21.92}
\end{aligned}
$$

where we introduced the generalized Kronecker delta symbols, such as $\delta^{\beta\mu\nu}_{\gamma\lambda\kappa}$ which has components that vanish unless $\beta\mu\nu$ are mutually distinct *and* $\gamma\lambda\kappa$ is a permutation of $\beta\mu\nu$. When $\gamma\kappa\lambda$ is an even permutation of $\beta\mu\nu$, its value is $+1$, while it is -1 for an uneven permutation.   ◁

With the aid of this Levi-Civita symbol, we define the dual electromagnetic tensor to be

$$
F^*_{\mu\nu} = -\frac{1}{2c}\epsilon_{\mu\nu\alpha\beta}F^{\alpha\beta} . \tag{21.93}
$$

Written out in components and transforming to contravariant notation, we get

$F^{*\alpha\beta} = g^{\alpha\mu}g^{\beta\nu}F^*_{\mu\nu}$, or

$$F^{*\alpha\beta} = \begin{pmatrix} 0 & B_1/c & B_2/c & B_3/c \\ -B_1/c & 0 & -E_3/c^2 & E_2/c^2 \\ -B_2/c & E_3/c^2 & 0 & -E_1/c^2 \\ -B_3/c & -E_2/c^2 & E_1/c^2 & 0 \end{pmatrix}. \tag{21.94}$$

We can now compute another invariant from the combination of

$$F^{*\alpha\beta}F_{\alpha\beta} = -\frac{4}{c^2}\mathbf{E}\cdot\mathbf{B}. \tag{21.95}$$

Hence, when the three-vector electric and magnetic fields are orthogonal in one frame, they will be orthogonal (in the three-space sense) in all other inertial frames. The homogeneous Maxwell equations $\nabla\cdot\mathbf{B} = 0$ and $\nabla\times\mathbf{E} = -\partial\mathbf{B}/\partial t$ are then the temporal and spatial components of the four-law

$$\partial_\beta F^{*\alpha\beta} = 0. \tag{21.96}$$

▷ **Exercise** It is of course also possible to write the homogeneous Maxwell equations directly in terms of the tensor $F^{\alpha\beta}$. Show that alternative forms are

$$\epsilon^{\alpha\beta\lambda\kappa}\partial_\beta F_{\lambda\kappa} = 0. \tag{21.97}$$

This can be rewritten, using the anti-symmetry of the tensor $F_{\lambda\kappa}$, as

$$\partial_\beta F_{\lambda\kappa} + \partial_\lambda F_{\kappa\beta} + \partial_\kappa F_{\beta\lambda} = 0. \tag{21.98}$$

Note the cyclic permutation of the indices appearing in Eq. (21.98). ◁

In complete analogy with the four-electric field, we can also consider the four-vector given by

$$b^\alpha = F^{*\alpha\beta}U_\beta. \tag{21.99}$$

Splitting into temporal and spatial components, we get

$$b^\alpha = \left[\Gamma\mathbf{v}\cdot\mathbf{B}/c, \Gamma(\mathbf{B} - \mathbf{v}\times\mathbf{E}/c^2)\right]^{\mathrm{T}}. \tag{21.100}$$

This four-vector is again found to be orthogonal in space-time to the local four-velocity since $b^\alpha U_\alpha = 0$, and is then identified as the four-magnetic field measured in the local rest frame. The invariant associated with this field is

$$b^\alpha b_\alpha = \Gamma^2\left[(\mathbf{B} - \mathbf{v}\times\mathbf{E}/c^2)\cdot(\mathbf{B} - \mathbf{v}\times\mathbf{E}/c^2) - (\mathbf{v}\cdot\mathbf{B}/c)^2\right]. \tag{21.101}$$

▷ **Alternative expressions for the electromagnetic field tensor** The four-vectors for the electric field e^α and magnetic field b^β can also be used to express the electromagnetic field tensor as follows. First note that

$$b_\alpha = F^*_{\alpha\beta}U^\beta. \tag{21.102}$$

Using this to work out an alternative form for the components of $\epsilon^{\alpha\beta\mu\nu} b_\alpha U_\beta$, and manipulating the expression to one in terms of $F^{\lambda\kappa}$ with the aid of Eq. (21.91)–(21.92), one finds

$$F^{\alpha\beta} = -\frac{1}{c}\epsilon^{\mu\nu\alpha\beta} b_\mu U_\nu - \frac{1}{c^2}\left(e^\alpha U^\beta - e^\beta U^\alpha\right),$$

$$F^{*\alpha\beta} = \frac{1}{c^2}\left(U^\alpha b^\beta - U^\beta b^\alpha\right) + \frac{1}{2c^3}\epsilon^{\alpha\beta\mu\nu}\left(e_\mu U_\nu - e_\nu U_\mu\right). \qquad (21.103)$$

These expressions simplify considerably for vanishing e^α, as will be the case in ideal relativistic MHD. $\qquad\qquad\triangleleft$

21.2.2 Stress–energy tensor for electromagnetic fields

The electromagnetic field also has an energy density $\frac{1}{2}(B^2 + E^2/c^2)$, with $B^2 = \mathbf{B} \cdot \mathbf{B}$ and $E^2 = \mathbf{E} \cdot \mathbf{E}$. The energy flux associated with the electromagnetic field is given by the Poynting flux three-vector

$$\mathbf{S}_{\text{em}} = \mathbf{E} \times \mathbf{B}. \qquad (21.104)$$

The full Maxwell stress tensor involving both electric and magnetic field contributions is further given in three-vector notation by

$$\tfrac{1}{2}\left(B^2 + E^2/c^2\right)\mathbf{I} - \mathbf{EE}/c^2 - \mathbf{BB}. \qquad (21.105)$$

Hence, the stress–energy tensor for the electromagnetic field[1] will, in analogy with the gas dynamic case from Eq. (21.40), write as

$$T_{\text{em}}^{\alpha\beta} = \begin{pmatrix} \frac{1}{2}(B^2 + E^2/c^2) & \mathbf{S}_{\text{em}}/c \\ \mathbf{S}_{\text{em}}/c & \frac{1}{2}(B^2 + E^2/c^2)\mathbf{I} - \mathbf{EE}/c^2 - \mathbf{BB} \end{pmatrix}. \qquad (21.106)$$

Using the four-tensor for the electromagnetic field introduced in the previous section, it is possible to write it as

$$T_{\text{em}}^{\alpha\beta} = F^\alpha_{\ \gamma} F^{\beta\gamma} - \tfrac{1}{4}g^{\alpha\beta} F^{\gamma\delta} F_{\gamma\delta}. \qquad (21.107)$$

This stress–energy tensor for the electromagnetic field is symmetric and traceless, i.e. $T^\alpha_\alpha = 0$. When we compute the divergence of this electromagnetic stress–energy tensor from the above expression, we find after some algebra

$$\partial_\beta T_{\text{em}}^{\alpha\beta} = -F^\alpha_{\ \gamma} J^\gamma. \qquad (21.108)$$

---

[1] This is actually the expression for electromagnetic fields in vacuum, but this restriction will be sufficient for our purposes.

In the derivation, one uses the inhomogeneous Maxwell equation (21.87), the anti-symmetry of $F^{\beta\mu}$ to write $2F^{\beta\mu}\partial_\beta F_{\nu\mu} = F^{\beta\mu}(\partial_\beta F_{\nu\mu} - \partial_\mu F_{\nu\beta})$, and the homogeneous Maxwell equation in the form (21.98). In combination with the stress–energy of the plasma given by Eq. (21.89), the governing conservation equation then considers the combination of plasma stress–energy and electromagnetic field stress–energy, such that

$$\partial_\beta \left(T_{\text{pl}}^{\alpha\beta} + T_{\text{em}}^{\alpha\beta} \right) = 0 \,. \tag{21.109}$$

▷ **Entropy conservation in a relativistic plasma** Due to the conservation law (21.109), we again can write

$$U_\alpha \partial_\beta \left(T_{\text{pl}}^{\alpha\beta} + T_{\text{em}}^{\alpha\beta} \right) = 0 \,. \tag{21.110}$$

However, the electromagnetic part in this double summation also vanishes separately, since

$$U_\alpha \partial_\beta T_{\text{em}}^{\alpha\beta} = -U_\alpha F^\alpha_{\ \gamma} J^\gamma = q n_0 e^\beta U_\beta = 0 \,. \tag{21.111}$$

Therefore, the derivation for the entropy equation (21.49) given for a relativistic gas also applies in the plasma case. The entropy $S = p\rho^{-\gamma}$ thus obeys

$$U^\alpha \partial_\alpha S = 0 \,, \tag{21.112}$$

and this can be combined with particle number conservation (21.32) to yield a conservation law for $DS = \Gamma \rho S$ since

$$\frac{\partial (DS)}{\partial t} + \nabla \cdot (DS\mathbf{v}) = 0 \,. \tag{21.113}$$

This equation can then be used instead of the energy conservation law. ◁

21.2.3 *Ideal MHD in special relativity*

To arrive at the ideal MHD limit, we now only need to consider the consequences of all the above, when the electric field in the co-moving frame vanishes. We already mentioned that this occurs when in the lab frame $\mathbf{E} = -\mathbf{v} \times \mathbf{B}$, or equivalently $e^\alpha = 0$ with e^α expressed by Eq. (21.81). Under this assumption, several expressions given earlier simplify. For example, the invariant associated with the magnetic field four-vector is then

$$b^\alpha b_\alpha = \frac{\mathbf{B} \cdot \mathbf{B}}{\Gamma^2} + \frac{(\mathbf{v} \cdot \mathbf{B})^2}{c^2} \,. \tag{21.114}$$

This is proportional to the magnetic pressure and we write $b^\alpha b_\alpha \equiv 2p_{\text{mag}}$.

▷ **Exercise** Verify that we get the same result from the transformation laws for electric and magnetic field three-vectors given by Eq. (21.79). Note that $\mathbf{E} = -\mathbf{v} \times \mathbf{B}$ will yield $\mathbf{E}' = 0$, and that Eq. (21.114) is obtained from $\mathbf{B}' \cdot \mathbf{B}'$. ◁

Revisiting the general stress–energy tensor for electromagnetic fields, we can now write for the case where co-moving electric fields vanish (i.e. $e^\alpha = 0$)

$$T_{\mathrm{em}}^{\alpha\beta} = 2p_{\mathrm{mag}}\frac{U^\alpha U^\beta}{c^2} + p_{\mathrm{mag}}g^{\alpha\beta} - b^\alpha b^\beta\,.$$

(21.115)

In a similar form to the general case given by (21.106), we get for vanishing co-moving electric fields the following expression for $T_{\mathrm{em}}^{\alpha\beta}$:

$$\begin{pmatrix} \frac{1}{2}B^2 + \frac{1}{2c^2}[B^2 v^2 - (\mathbf{v}\cdot\mathbf{B})^2] & \dfrac{\mathbf{S}_{\mathrm{em}}}{c} \\[2ex] \dfrac{\mathbf{S}_{\mathrm{em}}}{c} & \dfrac{\mathbf{S}_{\mathrm{em}}\mathbf{v}}{c^2} + p_{\mathrm{mag}}\mathbf{I} - \dfrac{\mathbf{B}\mathbf{B}}{\Gamma^2} - \dfrac{(\mathbf{v}\cdot\mathbf{B})\mathbf{v}\mathbf{B}}{c^2} \end{pmatrix}\,.$$

(21.116)

In this expression, the Poynting flux three-vector now writes as

$$\mathbf{S}_{\mathrm{em}} = B^2\mathbf{v} - (\mathbf{v}\cdot\mathbf{B})\mathbf{B}\,.$$

(21.117)

We can now write the total energy density for the gas plus electromagnetic field in the lab frame as $\tau_\mathcal{H} + Dc^2$, which separates off the rest mass contribution. Hence, we have the relations

$$\tau_\mathcal{H} = \tau_{\mathrm{g}} + \frac{1}{2}B^2 + \frac{1}{2c^2}[B^2 v^2 - (\mathbf{v}\cdot\mathbf{B})^2]\,,$$

(21.118)

where the two additions to τ_{g} represent magnetic and electric field energy densities, respectively. It is then seen from Eqs. (21.116) and (21.40) that the temporal component of the conservation law (21.109) is written as

$$\frac{\partial}{\partial t}\left(\tau_\mathcal{H} + Dc^2\right) + \nabla\cdot(\mathbf{S}_{\mathrm{g}} + \mathbf{S}_{\mathrm{em}}) = 0\,.$$

(21.119)

When we write the total energy flux $\mathbf{S}_{\mathrm{tot}} = \mathbf{S}_{\mathrm{g}} + \mathbf{S}_{\mathrm{em}}$ as the added relativistic energy flux of the gas and the Poynting flux, we can combine this relation with particle conservation $\partial D/\partial t + \nabla\cdot(D\mathbf{v}) = 0$ to obtain

$$\frac{\partial\tau_\mathcal{H}}{\partial t} + \nabla\cdot[(\tau_\mathcal{H} + p_{\mathrm{tot}})\mathbf{v} - (\mathbf{v}\cdot\mathbf{B})\mathbf{B}] = 0\,.$$

(21.120)

In so writing, the total pressure is introduced as

$$p_{\mathrm{tot}} = p + p_{\mathrm{mag}} = p + \frac{1}{2}\left[\frac{\mathbf{B}\cdot\mathbf{B}}{\Gamma^2} + \frac{(\mathbf{v}\cdot\mathbf{B})^2}{c^2}\right]\,.$$

(21.121)

The equation governing the momentum/energy flux evolution is obtained from the spatial part of (21.109), and is then recognized to become

$$\frac{\partial\mathbf{S}_{\mathrm{tot}}}{\partial t} + \nabla\cdot\left[\mathbf{S}_{\mathrm{tot}}\mathbf{v} + p_{\mathrm{tot}}c^2\mathbf{I} - c^2\frac{\mathbf{B}\mathbf{B}}{\Gamma^2} - (\mathbf{v}\cdot\mathbf{B})\mathbf{v}\mathbf{B}\right] = 0\,.$$

(21.122)

The Newtonian limits again reduce to the familiar non-relativistic ideal MHD equations, since

$$\mathbf{S}_{\mathrm{tot}} \overset{\Gamma \to 1}{\longrightarrow} c^2 \rho \mathbf{v}\,, \qquad \tau_{\mathcal{H}} \overset{\Gamma \to 1}{\longrightarrow} \tfrac{1}{2}\rho v^2 + \rho\epsilon + \tfrac{1}{2}\mathbf{B}\cdot\mathbf{B}\,, \qquad p_{\mathrm{tot}} \overset{\Gamma \to 1}{\longrightarrow} p + \tfrac{1}{2}\mathbf{B}\cdot\mathbf{B}\,. \tag{21.123}$$

$\triangleright$ **Exercise** Verify that the expressions for the spatial flux three-vectors in Eq. (21.120) and (21.122) indeed reduce to their non-relativistic counterparts for $\Gamma \to 1$. $\triangleleft$

To get a closed set of equations governing special relativistic MHD, we now need to combine the full set of Maxwell equations with particle conservation as in Eq. (21.32), and energy–momentum conservation. The latter splits in the fixed Lorentzian reference frame into Eq. (21.120) and Eq. (21.122). As for the gas dynamic case, we additionally need to provide an equation of state, such as the polytropic one given in Eq. (21.48). However, due to the vanishing electric field in the co-moving frame, it was possible to write energy–momentum conservation solely in terms of the magnetic field three-vector $\mathbf{B}$ and three-velocity $\mathbf{v}$. Hence, just as in the non-relativistic ideal MHD limit, we can close the system (mathematically speaking) by the homogeneous Maxwell equations alone, since Eq. (21.96) can use the identity

$$F^{*\alpha\beta} = (U^\alpha b^\beta - b^\alpha U^\beta)/c^2\,. \tag{21.124}$$

Written out in spatial and temporal parts, Eq. (21.96) turns into the familiar set

$$\begin{cases} \nabla \cdot \mathbf{B} = 0\,, \\[2mm] \dfrac{\partial \mathbf{B}}{\partial t} - \nabla \times (\mathbf{v} \times \mathbf{B}) = 0\,. \end{cases} \tag{21.125}$$

In contrast to the classical ideal MHD formulation, the full non-homogeneous Maxwell equations (21.84) need to be used now for computing the lab frame charge density σ and current density three-vector $\mathbf{j}$. In particular, this latter three-vector has a contribution from the displacement current $c^{-2}\partial\mathbf{E}/\partial t$, which was appropriately neglected in the non-relativistic regime.

21.2.4 Wave dynamics in a homogeneous plasma

To obtain the propagation speeds for linear waves in relativistic MHD, the governing conservation laws in tensorial form can be linearized in space-time. The algebra involved can be substantial, even in the case of linearizing about a stationary, homogeneous plasma. This is partly because of the wave aberration effects, which we mentioned already for the relativistic gas dynamic case. The analysis is tractable for the special case of the plasma rest frame, and a very elegant means to obtain the relativistic variants of the slow, Alfvén and fast wave speeds can be found in the appendix from Komissarov [283].

We note that it is possible to write down the equivalent set of equations in a $3 + 1$ formalism for the primitive variables $(S, \rho, \mathbf{v}, \mathbf{B})$, analogous to the equations (21.58) for gas dynamics. The equation for the entropy (21.49) was already shown to be identical, while Eqs. (21.125) for the magnetic field are familiar from the non-relativistic case. The continuity as well as the momentum equation become fairly cumbersome expressions, and we only mention what results from them, after linearizing with plane waves $\exp[-i(\omega t - \mathbf{k} \cdot \mathbf{x})]$ about the homogeneous plasma rest frame. Indicating as usual the background quantities by $S_0, \rho_0, \mathbf{B}_0$, and the linear variables by $S_1, \rho_1, \mathbf{B}_1, \mathbf{v}_1$, we get

$$\omega S_1 = 0 \,,$$

$$\omega \rho_1 = \rho_0 \mathbf{k} \cdot \mathbf{v}_1 \,,$$

$$\omega \mathbf{B}_1 = \mathbf{B}_0 (\mathbf{k} \cdot \mathbf{v}_1) - \mathbf{v}_1 (\mathbf{k} \cdot \mathbf{B}_0) \,, \qquad \mathbf{k} \cdot \mathbf{B}_1 = 0 \,,$$

$$\omega \mathbf{v}_1 = \frac{c^2}{w_0} \mathbf{k} \left(S_0 \gamma \rho_0^{\gamma-1} \rho_1 + \rho_0^\gamma S_1 \right) + \frac{c^2}{w_0} \left[\mathbf{k}(\mathbf{B}_0 \cdot \mathbf{B}_1) - \mathbf{B}_1 (\mathbf{k} \cdot \mathbf{B}_0) \right]$$

$$+ \frac{c^2 (\mathbf{k} \cdot \mathbf{B}_0)}{w_0 \rho_0 h_0} \mathbf{B}_0 \left(S_0 \gamma \rho_0^{\gamma-1} \rho_1 + \rho_0^\gamma S_1 \right) \,. \tag{21.126}$$

We here adopted a polytropic equation of state, where $h_0 = c^2 + [\gamma/(\gamma-1)]S_0\rho_0^{\gamma-1}$, and introduced the quantity

$$w_0 = \rho_0 h_0 + B_0^2 \,. \tag{21.127}$$

One can directly compare these expressions with the non-relativistic expressions given in Section 5.2 [1], and note that only the momentum equation yields an extra term (the last term is purely relativistic, and the coefficients for the other terms are changed to involve w_0). Not surprisingly then, the seven wave solutions return in slightly modified form. The marginal entropy mode is identical, being the solution at $\omega = 0$ for which $S_1 \neq 0$ only. The Alfvén waves return in a virtually unmodified form: they represent solutions with $\mathbf{v}_1 \neq 0$ and $\mathbf{B}_1 \neq 0$ while

$$\rho_1 = S_1 = \mathbf{k} \cdot \mathbf{v}_1 = \mathbf{k} \cdot \mathbf{B}_1 = \mathbf{B}_0 \cdot \mathbf{B}_1 = \mathbf{B}_0 \cdot \mathbf{v}_1 = 0 \,, \tag{21.128}$$

this time given by the dispersion relation

$$\omega^2 = c^2 \frac{(\mathbf{k} \cdot \mathbf{B}_0)^2}{w_0} \,. \tag{21.129}$$

They retain their field-sampling property familiar from non-relativistic MHD, and we can express their phase $\mathbf{v}_{\mathrm{ph}}$ and group velocity $\mathbf{v}_{\mathrm{gr}}$ as follows. For that purpose, assuming $\mathbf{n} = \mathbf{k}/k$ and denoting by ϑ the angle between $\mathbf{n}$ and $\mathbf{B}_0$, we find

$$\frac{\mathbf{v}_{\mathrm{ph}}}{c} = \frac{B_0 \cos \vartheta}{\sqrt{w_0}} \mathbf{n}, \qquad \frac{\mathbf{v}_{\mathrm{gr}}}{c} = \frac{B_0}{\sqrt{w_0}}. \tag{21.130}$$

The compressible modes are obtained from straightforward algebraic manipulations on Eqs. (21.126) to the dispersion relation

$$\omega^4 - \omega^2 \left[\frac{k^2 c^2}{w_0} \left(\rho_0 h_0 \frac{c_g^2}{c^2} + B_0^2 \right) + c_g^2 \frac{(\mathbf{k} \cdot \mathbf{B}_0)^2}{w_0} \right] + k^2 c^2 c_g^2 \frac{(\mathbf{k} \cdot \mathbf{B}_0)^2}{w_0} = 0. \tag{21.131}$$

Here we purposely wrote this again in terms of the squared sound speed c_g^2, while we can now also introduce the squared Alfvén speed $v_A^2 = B_0^2 c^2 / w_0$. The expressions are then generally valid, with the expressions for specific enthalpy h_0 and sound speed c_g depending on the equation of state. It is then left as an exercise to the reader that their phase speeds are found from

$$\mathbf{v}_{\mathrm{ph}}/c = (v_{\mathrm{ph}}/c)\mathbf{n} = \mathbf{n}\sqrt{\tfrac{1}{2}[(\rho_0 h_0/w_0)c_g^2 + v_A^2]/c^2} \sqrt{1 + \delta \cos^2 \vartheta \pm a}. \tag{21.132}$$

Here, the symbols δ and σ (no longer the charge density!) express the following dimensionless ratios

$$\delta = \frac{c_g^2 v_A^2}{\left[(\rho_0 h_0/w_0)c_g^2 + v_A^2 \right] c^2}, \qquad \sigma = \frac{4 c_g^2 v_A^2}{\left[(\rho_0 h_0/w_0)c_g^2 + v_A^2 \right]^2}. \tag{21.133}$$

The symbol a follows from

$$a^2 = (1 + \delta \cos^2 \vartheta)^2 - \sigma \cos^2 \vartheta. \tag{21.134}$$

Noting that $\rho_0 h_0/w_0 = 1 - v_A^2/c^2$, the phase speed for purely parallel propagation reduces to the same expression found in non-relativistic MHD, where we have

$$v_{\mathrm{ph},\parallel}/c = \sqrt{\tfrac{1}{2} \left(c_g^2/c^2 + v_A^2/c^2 \right) \left[1 \pm \sqrt{1 - 4 c_g^2 v_A^2/(c_g^2 + v_A^2)^2} \right]}. \tag{21.135}$$

The group speed is then written in terms of the orthogonal directions $\mathbf{n} = \mathbf{k}/k$ and $\mathbf{t} = [(\mathbf{B}_0/B_0) \times \mathbf{n}] \times \mathbf{n}$ as

$$\frac{\mathbf{v}_{\mathrm{gr}}}{c} = \frac{v_{\mathrm{ph}}}{c} \left[\mathbf{n} \pm \mathbf{t} \frac{[\sigma \mp 2\delta (a \pm (1 + \delta \cos^2 \vartheta))] \sin \vartheta \cos \vartheta}{2 (1 + \delta \cos^2 \vartheta \pm a) a} \right]. \tag{21.136}$$

These can be compared directly with the non-relativistic expressions, and we can note that all relativistic effects are due to the parameter δ, together with the fact that both sound and Alfvén speeds get relativistic corrections.

As representative examples, we show in Fig. 21.7 the phase and group diagrams for three cases with (again with units making $c = 1$)

- $\rho_0 = 1$, $p_0 = 0.1$, $B_0 = 0.3$, for which $c_{\mathrm{g}} = 0.354$ and $v_{\mathrm{A}} = 0.258$;
- $\rho_0 = 1$, $p_0 = 0.1$, $B_0 = 0.5$, for which $c_{\mathrm{g}} = 0.354$ and $v_{\mathrm{A}} = 0.406$;
- $\rho_0 = 0.01$, $p_0 = 0.001$, $B_0 = 1$, for which $c_{\mathrm{g}} = 0.354$ and $v_{\mathrm{A}} = 0.99$.

These assume a Mathews equation of state. In Fig. 21.8, we also show the group diagram for the first and the third case, plotted in a frame which saw the point perturbation pass by along the field lines at a velocity $0.9c$. (Also see Ref. [260]).

Characteristic speeds for relativistic MHD For the relativistic ideal MHD equations, expressions for the characteristic speeds can be derived which correspond to the phase speeds as seen from the laboratory frame (as opposed to the plasma rest frame discussed thus far). There will again be seven signal speeds, where one is related to the entropy equation (21.113). Entropy is passively advected with the characteristic speed λ_{E} in the i-coordinate direction given by

$$\lambda_{\mathrm{E}}/c = v_i/c. \tag{21.137}$$

Similar to the non-relativistic case (cf. Eqs. (5.120), (7.161)[1] and (13.22)), the characteristic speeds are again ordered according to

$$- c \leq \lambda_{\mathrm{F}}^- \leq \lambda_{\mathrm{A}}^- \leq \lambda_{\mathrm{S}}^- \leq \lambda_{\mathrm{E}} \leq \lambda_{\mathrm{S}}^+ \leq \lambda_{\mathrm{A}}^+ \leq \lambda_{\mathrm{F}}^+ \leq c. \tag{21.138}$$

In contrast to the non-relativistic case, the forward and backward wave speeds of a given family (fast, Alfvén, slow) are no longer necessarily symmetric about $\lambda_{\mathrm{E}} = v_i$. The Alfvén wave speeds are found from

$$\frac{\lambda_{\mathrm{A}}^{\pm}}{c} = \frac{v_i}{c} \pm \frac{1}{\Gamma^2} \frac{B_i}{\sqrt{\rho h_{\mathrm{tot}}} \pm \mathbf{B} \cdot \mathbf{v}/c}. \tag{21.139}$$

We thereby introduced the total specific enthalpy, where

$$h_{\mathrm{tot}} = h + 2p_{\mathrm{mag}}/\rho. \tag{21.140}$$

The fast and slow characteristic speeds can most conveniently be found from the following quartic polynomial:

$$\rho h \left(c^2 - c_{\mathrm{g}}^2 \right) \Gamma^4 \left(\frac{\lambda}{c} - \frac{v_i}{c} \right)^4 - \left(1 - \frac{\lambda^2}{c^2} \right) \left\{ \Gamma^2 \left(\rho h c_{\mathrm{g}}^2 + 2p_{\mathrm{mag}} c^2 \right) \left(\frac{\lambda}{c} - \frac{v_i}{c} \right)^2 \right.$$

$$\left. - c_{\mathrm{g}}^2 \left[\Gamma \left(\frac{\mathbf{v}}{c} \cdot \mathbf{B} \right) \left(\frac{\lambda}{c} - \frac{v_i}{c} \right) - \frac{B_i}{\Gamma} \right]^2 \right\} = 0. \tag{21.141}$$

As before, these expressions can in fact be obtained from the rest frame expressions given earlier. They become more involved due to relativistic aberration, and will

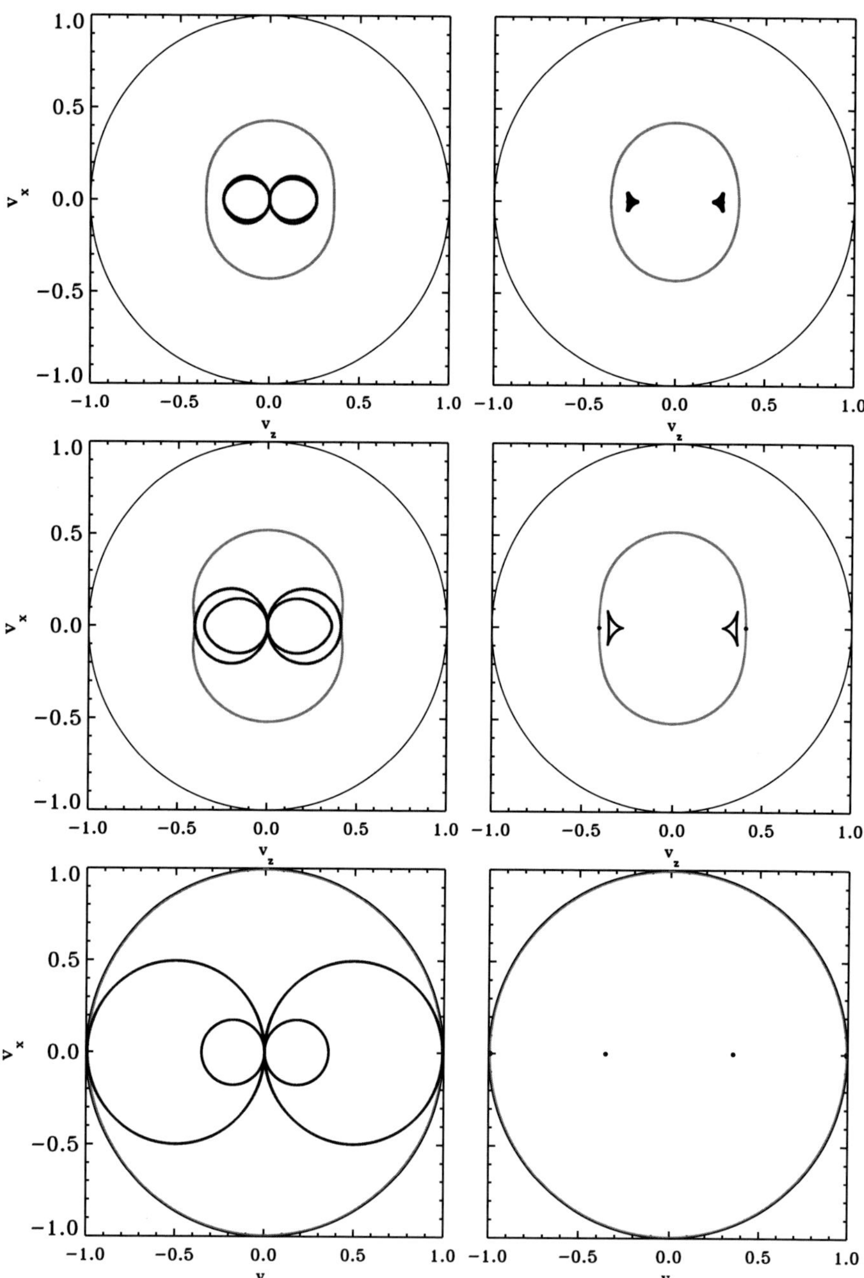

Fig. 21.7 Phase (left) and group speed (right) diagrams in the gas rest frame for three representative cases with uniform horizontal magnetic fields. Top to bottom changes the thermodynamic quantities such that $c_\mathrm{g} = 0.354 > v_\mathrm{A} = 0.258$ (top), $c_\mathrm{g} = 0.354 < v_\mathrm{A} = 0.406$ (middle), and $c_\mathrm{g} = 0.354 < v_\mathrm{A} = 0.99$ (bottom).

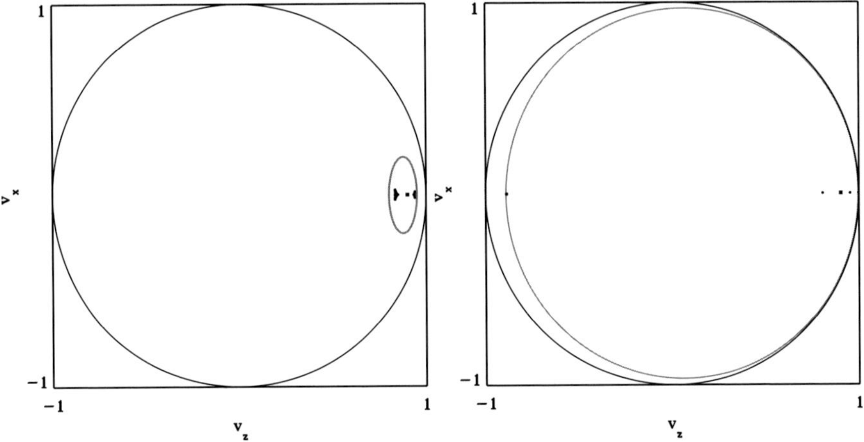

Fig. 21.8 For the case $c_{\mathrm{g}} = 0.354 > v_{\mathrm{A}} = 0.258$ (left) and the case with $c_{\mathrm{g}} = 0.354 < v_{\mathrm{A}} = 0.99$ (right), group diagrams as seen from a frame where the source moves at $0.9c$ along the z-axis aligned with the horizontal magnetic field.

involve the Lorentz transformation for the magnetic field as well, as the magnetic field component normal to the wave front will look different from differing inertial reference frames.

21.2.5 Shock conditions in relativistic MHD

The conservation laws of relativistic MHD can again be studied in the case of discontinuous jump relations across surfaces in space-time. Completely analogous to the relativistic hydro case, we write

$$[\![\rho U^{\alpha}]\!] l_{\alpha} = 0 \,, \tag{21.142}$$

$$[\![T^{\alpha\beta}]\!] l_{\alpha} = 0 \,, \tag{21.143}$$

$$[\![U^{\alpha} b^{\beta} - b^{\alpha} U^{\beta}]\!] l_{\alpha} = 0 \,. \tag{21.144}$$

where l^{α} is the space-like normal to the shock surface. The analysis of these relations can be rather complicated in general, and their precise algebraic form in a $3 + 1$ manner depends on selecting a specific Lorentzian reference frame, as usual in relativistic settings. We already noted this for relativistic hydro, where we mentioned the shock rest frame (SRF), upstream rest frame (URF), downstream rest frame (DRF) and the tangential reference frame (TRF). We will mention the significance of the important de Hoffman–Teller (or HTF) frame further on, which, like the TRF, involves a transformation with a certain tangential velocity. If we, for now, consider the relations (21.142)–(21.144) in the shock rest frame, where we denote the shock normal as $\mathbf{l} = (0, \mathbf{n})$ with $\mathbf{n}$ the normal three-vector, we find

$$[\![\rho\Gamma v_{\mathrm{n}}]\!] = 0 \,, \tag{21.145}$$

$$[\![B_{\mathrm{n}}]\!] = 0 \,, \tag{21.146}$$

$$\left[\!\!\left[\left(\rho h\Gamma^2 + B^2\right)\frac{\mathbf{v}_{\mathrm{t}} v_{\mathrm{n}}}{c^2} - \frac{(\mathbf{v}\cdot\mathbf{B})}{c^2}\left(\mathbf{B}_{\mathrm{t}} v_{\mathrm{n}} + \mathbf{v}_{\mathrm{t}} B_{\mathrm{n}}\right) - \frac{\mathbf{B}_{\mathrm{t}} B_{\mathrm{n}}}{\Gamma^2}\right]\!\!\right] = 0 \,, \tag{21.147}$$

$$\rho\Gamma v_{\mathrm{n}}\left[\!\!\left[\frac{\mathbf{B}_{\mathrm{t}}}{\rho\Gamma}\right]\!\!\right] = B_{\mathrm{n}}[\![\mathbf{v}_{\mathrm{t}}]\!] \,, \tag{21.148}$$

$$\left[\!\!\left[\left(\rho h\Gamma^2 + B^2\right)\frac{v_{\mathrm{n}}^2}{c^2} + p + p_{\mathrm{mag}} - 2\frac{(\mathbf{v}\cdot\mathbf{B})}{c^2}v_{\mathrm{n}}B_{\mathrm{n}} - \frac{B_{\mathrm{n}}^2}{\Gamma^2}\right]\!\!\right] = 0 \,, \tag{21.149}$$

$$\rho\Gamma v_{\mathrm{n}}\left[\!\!\left[h\Gamma + \frac{B^2}{\rho\Gamma}\right]\!\!\right] = B_{\mathrm{n}}[\![\mathbf{v}\cdot\mathbf{B}]\!] \,, \tag{21.150}$$

where we wrote the equations purposely in the same order as Eqs. (20.3)–(20.8) from Section 20.2, where we find the non-relativistic variants. It is a manner of algebra to show that this limit agrees, while the relativistic hydro limit is obtained for vanishing magnetic field and agrees with Eqs. (21.70).

▷ **Exercise** Verify these statements. ◁

The occurrence of the Lorentz factor in many terms of these expressions makes a discussion of the general case beyond the scope of this textbook. These relations have been studied in some detail by de Hoffmann and Teller [108], Majorana and Anile [323], Appl and Camenzind [11], and the procedure to obtain the downstream from the upstream state variables is briefly recapitulated in the appendix to the review by Kirk and Duffy [279]. The books by Lichnerowicz [307] and Anile [8] contain a precise mathematical treatment. We here only mention the important limiting case of normal perpendicular shocks, and point out the means to obtain various shock invariants from the relations (21.142)–(21.144), as already discussed by Anile [8].

Normal perpendicular shocks This limit case assumes that in the SRF, where the relations (21.145)–(21.150) hold, the normal magnetic fields vanish completely, $B_{\mathrm{n}} = 0$ (making the field purely tangential to the shock front, hence the name "perpendicular"). Moreover, we assume purely normal velocities for which $\mathbf{v}_{\mathrm{t}} = 0$ as well (hence the name "normal"). It is then easy to verify that we are left with

$$[\![\rho\Gamma v_{\mathrm{n}}]\!] = 0 \,, \tag{21.151}$$

$$[\![\mathbf{B}_{\mathrm{t}} v_{\mathrm{n}}]\!] = 0 \,, \qquad [\![\Gamma\mathbf{B}_{\mathrm{t}}' v_{\mathrm{n}}]\!] = 0 \,, \tag{21.152}$$

$$\left[\!\!\left[\left(\rho h\Gamma^2 + B_{\mathrm{t}}^2\right)\frac{v_{\mathrm{n}}^2}{c^2} + p + \frac{B_{\mathrm{t}}^2}{2\Gamma^2}\right]\!\!\right] = 0 \,, \qquad [\![\rho h_{\mathrm{tot}}\Gamma^2 v_{\mathrm{n}}^2 + p_{\mathrm{tot}}]\!] = 0 \,, \tag{21.153}$$

$$\llbracket \rho h \Gamma^2 v_{\mathrm{n}} + B_{\mathrm{t}}^2 v_{\mathrm{n}} \rrbracket = 0 \,, \qquad \llbracket \rho h_{\mathrm{tot}} \Gamma^2 v_{\mathrm{n}} \rrbracket = 0 \,. \tag{21.154}$$

The relations on the right express these SRF relations in terms of the plasma rest frame magnetic fields at left and right of the shock front (those we have in the DRF and URF, respectively), which in this normal perpendicular case are $\mathbf{B}' = \mathbf{B}/\Gamma$. Using them, the total enthalpy and pressure are written as $\rho h_{\mathrm{tot}} = \rho h + B_{\mathrm{t}}'^2$ and $p_{\mathrm{tot}} = p + B_{\mathrm{t}}'^2$. In this way, a rather direct generalization of the pure gas dynamic case is obtained, replacing pressure and enthalpy with their total counterparts.

▷ **Relativistic MHD shock invariants.** For completeness, we now list various relativistic MHD shock invariants, as described and proven already in Anile [8]. They are most conveniently found from the general expressions (21.142)–(21.144), while we will give expressions as obtained in the SRF for some of them. Of course, many more combinations may be found, but we here list the following.

(i) Normal particle flux

$$\mathcal{M} = \rho U^\alpha l_\alpha \ \left[= \rho \Gamma v_{\mathrm{n}} \right] \,. \tag{21.155}$$

(ii) Introducing $V^\beta = \left(U^\alpha b^\beta - b^\alpha U^\beta \right) l_\alpha$, we have the invariant $\mathcal{H} = -V^\alpha V_\alpha / \mathcal{M}^2$
or

$$\mathcal{H} = \frac{c^2 \left(b^\mu l_\mu \right)^2}{\mathcal{M}^2} - \frac{b^\beta b_\beta}{\rho^2} \left[= \frac{c^2}{(\rho \Gamma v_{\mathrm{n}})^2} \left(\frac{B_{\mathrm{n}}}{\Gamma} + \frac{\Gamma(\mathbf{v} \cdot \mathbf{B}) v_{\mathrm{n}}}{c^2} \right)^2 - \frac{2 p_{\mathrm{mag}}}{\rho^2} \right] \,. \tag{21.156}$$

(iii) Writing $W^\beta = T^{\alpha\beta} l_\alpha$ we have

$$\mathcal{B} = \frac{V_\alpha W^\alpha}{\mathcal{M}} = h b^\alpha l_\alpha \ \left[= h \left(\frac{B_{\mathrm{n}}}{\Gamma} + \frac{\Gamma(\mathbf{v} \cdot \mathbf{B}) v_{\mathrm{n}}}{c^2} \right) \right] \,. \tag{21.157}$$

(iv) The quantity $\mathcal{E}$ found from

$$\mathcal{E} = W^\mu l_\mu \ \left[= p + p_{\mathrm{mag}} + (\rho \Gamma v_{\mathrm{n}})^2 \left(\frac{h}{\rho c^2} - \frac{\mathcal{H}}{c^2} \right) \right] \,. \tag{21.158}$$

(v) Introducing $X^\beta = W^\beta - (W^\alpha l_\alpha) l^\beta$, we have the invariance of

$$\mathcal{K} = -\frac{X^\alpha X_\alpha}{\mathcal{M}^2} = \frac{h^2}{c^2} + \frac{\mathcal{M}^2 h^2}{\rho^2 c^4} + \left(\frac{2h}{\rho c^2} - \frac{\mathcal{H}}{c^2} \right) \left(2 p_{\mathrm{mag}} - \frac{\mathcal{M}^2 \mathcal{H}}{c^2} \right) \,. \tag{21.159}$$

(vi) Also combinations of the above, such as

$$\mathcal{L} = -\mathcal{K} \mathcal{H} + \frac{\mathcal{B}^2}{\mathcal{M}^2} = c^2 \left(2 p_{\mathrm{mag}} - \frac{\mathcal{M}^2 \mathcal{H}}{c^2} \right) \left(\frac{h}{\rho c^2} - \frac{\mathcal{H}}{c^2} \right)^2 \,. \tag{21.160}$$

With the knowledge of all the above invariants, one can obtain the relativistic MHD generalization of the Taub adiabat, called the Lichnerowicz adiabat and given by

$$\frac{h_2^2}{c^2} - \frac{h_1^2}{c^2} - \left(\frac{h_1}{\rho_1 c^2} + \frac{h_2}{\rho_2 c^2} \right) (p_2 - p_1) + \frac{1}{2} \left(\frac{h_2}{\rho_2 c^2} - \frac{h_1}{\rho_1 c^2} \right) (\chi_1 + \chi_2) + \alpha_2 \chi_2 - \alpha_1 \chi_1 = 0. \tag{21.161}$$

We introduced the quantities $\chi = 2 p_{\mathrm{mag}} - \mathcal{M}^2 \mathcal{H}/c^2$ as well as $\alpha = h/(\rho c^2) - \mathcal{H}/c^2$. This invariant then can be shown to reduce to the various limit cases we encountered thus far, namely the non-relativistic hydrodynamic Hugoniot adiabat (20.18), the relativistic Taub adiabat (21.72) and the classical MHD relation (20.39). ◁

▷ **Exercise** Verify the various limits of Eq. (21.161). ◁

We conclude the discussion of the relativistic MHD shock relations with pointing out that, in analogy with the discussion given in Chapter 20, it is likely that a more insightful classification for relativistic MHD shocks can be found than the one currently scattered throughout the literature. Such a discussion will best be done in the de Hoffmann–Teller reference frame, which writes the equations in a frame where the total electric field at left and right of the shock vanishes. This involves a Lorentz transformation from the SRF to a frame moving with a certain tangential speed with respect to the SRF. The existence of this frame was already pointed out by de Hoffmann and Teller [108], and we saw that the non-relativistic case then became most insightful when we expressed all relations using the Alfvén normal Mach number in this reference frame. This is still to be pursued in future research for the relativistic case. Note as well that we thus far did not mention the extra inequality of entropy increase across a shock front, which obviously still acts to select physically realizable shocks.

21.3 Computing relativistic magnetized plasma dynamics

We demonstrated that relativistic magneto-fluids obey, in a fixed "laboratory" Lorentz frame, a set of conservation laws given by Eqs. (21.32), (21.120), (21.122) and the familiar set of equations for the magnetic field given by Eqs. (21.125). As an alternative, also Eq. (21.113) can be used instead of (21.32) or (21.120). In any case, it is clear that we end up with a nonlinear system of conservation laws, which can be integrated numerically using shock-capturing techniques such as those presented in Chapter 19. A discussion of numerical relativistic hydrodynamics, including extensions to general relativistic hydrodynamics, can be found in the book by Wilson and Mathews [484]. Relativistic MHD poses its own challenges, as can be expected from the algorithmic complexity already needed when going from non-relativistic hydro to MHD. As an early example, Dubal [126] adopted an FCT algorithm to special relativistic MHD, where 1D Riemann problems of modest Lorentz factors (up to $\Gamma \sim 3$) were adequately resolved: Riemann invariants through rarefactions remained constant to within 1%, and even 2D spherical blast waves in initially uniform magnetic fields could be computed. Another approach was followed by van Putten [465], where a pseudo-spectral method in combination with a leapfrog scheme was exploited, demonstrating that the compound waves found in coplanar non-relativistic MHD Riemann problems persist in (numerical) relativistic MHD. As mentioned further on, more recent efforts successfully exploit approximate Riemann solvers or TVDLF schemes in challenging applications involving high Lorentz factor plasma flows. These are gaining popularity, although they still face various complications, which call for suitable algorithmic improvements.

Conservative to primitive variable transformation The most obvious complication in relativistic ideal MHD is that we have truly non-trivial relations between the primitive variables $(\rho, \mathbf{v}, p, \mathbf{B})$ and the conserved variables, since

$$
\begin{aligned}
D &= \Gamma\rho\,, \\
\mathbf{S}_{\text{tot}} &= \rho h \Gamma^2 \mathbf{v} + B^2 \mathbf{v} - (\mathbf{v}\cdot\mathbf{B})\mathbf{B}\,, \\
\tau_{\mathcal{H}} &= \rho h \Gamma^2 - p - \Gamma\rho c^2 + \tfrac{1}{2}B^2 + \frac{B^2 v^2 - (\mathbf{v}\cdot\mathbf{B})^2}{2c^2}\,.
\end{aligned}
\tag{21.162}
$$

Especially due to the occurrence of the Lorentz factor Γ, the conversion from conservative to primitive variables, which is needed to evaluate the flux expressions, is non-algebraic and must be handled numerically. Defining an auxiliary variable $\xi = \rho h \Gamma^2$, we can note that $\mathbf{S}_{\text{tot}}\cdot\mathbf{B} = \xi\mathbf{v}\cdot\mathbf{B}$ and thus we get

$$
\mathbf{v} = \frac{\mathbf{S}_{\text{tot}} + \mathbf{S}_{\text{tot}}\cdot\mathbf{B}\,\mathbf{B}/\xi}{\xi + B^2}\,.
\tag{21.163}
$$

The Lorentz factor can then be computed from ξ and the conservative variables $\mathbf{S}_{\text{tot}}$ and $\mathbf{B}$, since $\Gamma^{-2} = 1 - v^2/c^2$. This in turn means that the defining equation for the energy variable $\tau_{\mathcal{H}}$ in expressions (21.162) provides a nonlinear relation for ξ as being the zero of

$$
\xi - p - Dc^2 + B^2 - \tfrac{1}{2}\left[\frac{B^2}{\Gamma^2} + \frac{(\mathbf{S}_{\text{tot}}\cdot\mathbf{B})^2}{\xi^2 c^2}\right] - \tau_{\mathcal{H}}\,.
\tag{21.164}
$$

For a constant polytropic index, we find also

$$
p = \frac{\gamma - 1}{\gamma}\frac{\xi - \Gamma Dc^2}{\Gamma^2}\,,
\tag{21.165}
$$

so one can determine ξ from a given set of conserved variables $(D, \mathbf{S}_{\text{tot}}, \tau_{\mathcal{H}}, \mathbf{B})$ using root finding algorithms on Eq. (21.164).

The Riemann problem and modern solver strategies Just as in non-relativistic MHD, knowledge of the solution of the 1D Riemann problem in special relativistic MHD is extremely useful to test the various flavors of shock-capturing schemes. It is surprising that the exact solution for the general case where all seven nonlinear waves are accounted for has only recently been obtained [156], while the development of shock-capturing solvers for relativistic MHD problems started in earnest in the late nineties and beginning of the twenty-first century. A first Godunov-type method for relativistic MHD, exploiting an approximate Riemann solver, was presented by Komissarov [283], and his paper collects all details needed for its implementation, such as the eigenvalue and eigenvector pairs, the means to handle degenerate cases, and how to convert from primitive to conservative variables. Using a staggered representation where the magnetic field components are defined on cell interfaces, a constrained transport type approach was taken to handle the solenoidal field constraint. This allowed multi-dimensional simulations involving considerably higher Lorentz factors than those achieved in early works [126, 465]. Following this, improved, but purely 1D, approximate Riemann solver implementations are found in [22] and in [282], where the former is of TVD type. Multi-dimensional relativistic MHD schemes exploiting more central approximations to the Riemann fan (HLL and HLLC) emerged in [110] and [332]. A TVDLF type method which uses only the fastest propagation speed is exploited in [456, 457]. The latter uses parabolic source term treatments for handling $\nabla \cdot \mathbf{B} = 0$ in a grid-adaptive framework. There is currently a quickly growing research community focused on numerical MHD solvers for relativistic regimes, already involving general relativistic MHD simulations in dynamically evolving space-times [157].

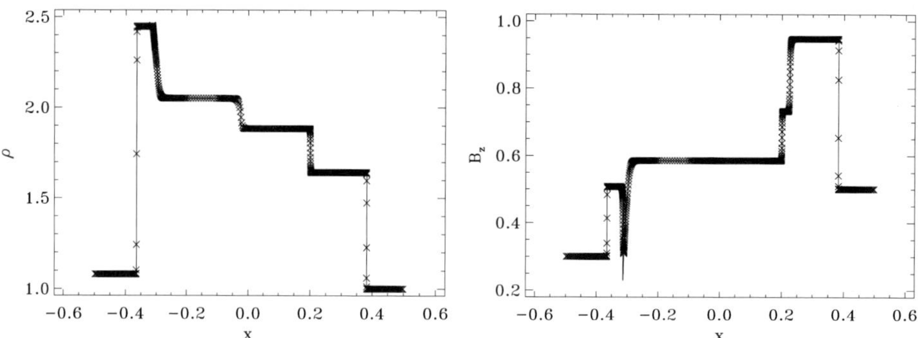

Fig. 21.9 The exact (solid line) versus numerically obtained solution of a relativistic MHD Riemann problem. Shown is the proper density and a tangential field component B_z. (From van der Holst *et al.* [457].)

An example taken from [457] compares, in Fig. 21.9, the exact solution with a grid-adaptive numerical solution of a Riemann problem, which was first presented in [22] and subsequently analytically solved in [156]. The left state has

$$(\rho, \mathbf{v}, p, \mathbf{B})_{\mathrm{L}} = (1.08, 0.4, 0.3, 0.2, 0.95, 2, 0.3, 0.3)$$

adjacent to $(1, -0.45, -0.2, 0.2, 1, 2, -0.7, 0.5)$ and constant polytropic index $\gamma = 5/3$. It is seen that the grid-adaptive result nicely recovers the solution containing a left-going fast shock, a left-going Alfvén discontinuity, a slow rarefaction, a contact discontinuity, as well as the right-going slow shock, Alfvén discontinuity and fast shock. However, close scrutiny reveals that the separation between left-going Alfvén signal and slow rarefaction is still marginally resolved. The need for grid-adaptive simulations in relativistic MHD is then already obvious for correctly solving even 1D Riemann problems.

21.3.1 Numerical challenges from relativistic MHD

In this paragraph, we merely point out various challenges associated with numerical, relativistic MHD. Some already led to clever algorithmic approaches, while others may need new ideas for further progress. A fairly straightforward observation is that the equations of relativistic MHD contain in principle the physical constants, such as the speed of light c (in mks units related to permeability μ_0 and permittivity ϵ_0 through $c^2 = 1/\mu_0\epsilon_0$) and the particle rest mass m_{p} (for protons). Obviously, in numerics one always exploits a proper scaling, which for relativistic MHD sets $c = 1$ (and thus $\mu_0 = 1 = \epsilon_0$) and, together with a reference distance and number density, all other quantities are measured in this scaled unit system. This was already the case in non-relativistic MHD, where we noted scale-independence in Section 4.1.2 and stated that units are set by choosing a length, magnetic field strength and density at a reference position. One of these latter three can now no longer be independently fixed in space-time, since c is the reference speed.

In the previous paragraph, we pointed out that the conservative to primitive variable computation is now a numerical problem by itself. It must be stressed that the accuracy with which this problem is solved numerically is a crucial element of modern solvers. Indeed, we have the obvious physical restrictions that $v < c$, which means $\Gamma \geq 1$, while $p > 0$, $\rho > 0$, and we also want the electric field density $E^2 > 0$, where $E^2 \sim B^2 v^2 - (\mathbf{v} \cdot \mathbf{B})^2$. The latter is obviously true analytically, but numerical precision is finite. All these constraints must be consistent with $\tau_{\mathcal{H}} > 0$ and $D \geq \rho$, and it helps to take explicit account of the numerical accuracy with which factors like $1 - v^2/c^2$ can actually be distinguished from zero. This can be done, by building in an appropriate "upper limit" on velocities, up to which conversions are computationally feasible. Similar tricks can be employed to guarantee consistent conversions from primitive to conservative variables, while ensuring a lower limit on attainable densities and pressures. This is similar to classical MHD, where this conversion is algebraic and trivial, but still requires that all contributions to total energy $\mathcal{H}$ are positive separately, which may introduce numerical inaccuracies in regions of very low plasma beta. This problem returns in augmented form in relativistic settings, since rest mass, internal energy, kinetic energy, magnetic and electric field densities all may dominate in localized regions of the computational domain.

When using finite volume like treatments, we discussed in Chapter 19 how limited reconstructions from cell center to cell edge are needed for raising the spatial order of accuracy, while avoiding spurious oscillations. This can still be used in relativistic MHD, but it then pays off to perform the limiting on quantities without physical bounds: one uses in practice the spatial part of the four-velocity U^α, i.e. $\Gamma \mathbf{v}$ to perform reconstructions, instead of the velocity itself $\mathbf{v}$, which must obey $v < c$. Similarly, when computing the roots of the quartic polynomial (21.141), which are needed for all shock-capturing methods exploiting knowledge of the characteristic speeds (even TVDLF which uses the maximal speed alone), one can better solve for $\Gamma(\lambda - v_i)$, or some other suitably scaled variable. In [457], a Laguerre method was then used to compute all four roots of this polynomial accurately. Also, this process is a non-trivial exercise in handling numerical accuracy, since its roots λ may come very close to each other and to 1 (the light speed in the scaled system).

Finally, just as in the non-relativistic case, the magnetic field must be solenoidal. For multi-dimensional computations, a suitable strategy must therefore be incorporated. It is noteworthy that the constrained transport idea [130] was from the beginning designed for (even general) relativistic MHD computations. Many variants now exist, and some have been applied in relativistic applications. On the other hand, since the induction equation appears identical to the non-relativistic case, also the more straightforward source terms treatments (Powell source, or parabolic cleaning, both limited to the induction equation) have been used successfully [456]. In practice, the schemes in use today may still benefit from further algorithmic improvements, although robust relativistic MHD solvers exist in modern code developments [456, 333, 157].

21.3.2 Example astrophysical applications

The current suite of shock-capturing, high resolution, schemes for relativistic MHD has already been applied to a fair variety of astrophysical problems. The study of pulsar wind nebulae is one such problem, where computational relativistic MHD has led to novel insights. In the particular case of the Crab nebula, recent observations revealed a peculiar jet–torus structure at the center (see Figure 21.10). This precise shaping of the circum-pulsar environment has been shown to relate to the shock-governed interaction of an anisotropic, toroidally magnetized, relativistic pulsar wind, interacting with the slowly expanding ejecta of the supernova [285] which created the pulsar. A separate research line focuses on dynamics in the vicinity of massive black holes to unravel the (general) relativistic processes, which must play a role in their accretion disks, and in the launch and acceleration conditions for Active Galactic Nuclei jets. In what follows, we present one example of special relativistic jet modeling, which investigates the shock-dominated processes in AGN jets at large distance from their source, such that a flat space-time description can be adopted. Such studies then address the possibility of knot formation in relativistic jet beams, deceleration processes when encountering denser interstellar medium regions, and morphological characteristics of magnetized jets, depending on the prevailing jet beam parameters.

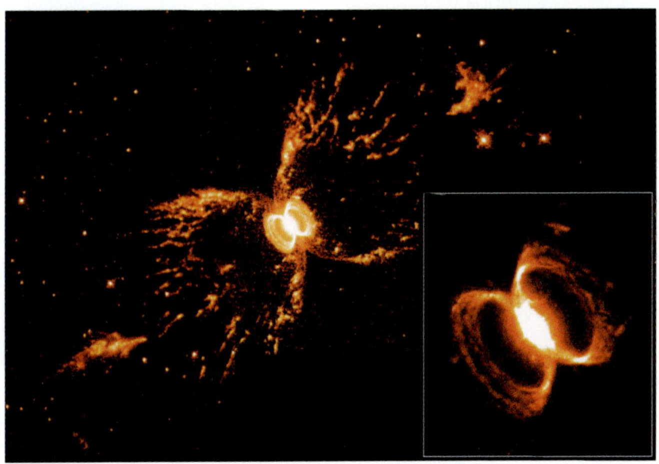

Fig. 21.10 Crab nebula. (From website `nssdc.gsfc.nasa.gov/image/astro/` `hst_southern_crab_9332.jpg`.)

Relativistic magnetized jet modeling The study of jet propagation and morphology using numerical simulations was pioneered by van Putten [466], who simulated impulsively injected toroidally magnetized jets, propagating in uniform unmagnetized environments. Using axi-symmetry, fairly mildly relativistic (Lorentz factor 2.46), light supersonic jets were shown to develop a Mach disk (shock), where the beam matter gets thermalized and subsequently splits into a backflow surrounding the jet beam (cocoon) and a recollimated nose cone. In this nose cone, a new propagating Mach disk develops, and a nozzle flow configuration was observed to form in between the stagnation point ahead of the first Mach disk and the propagating disk. The toroidal magnetic field pinched the beam and shocked beam matter, and led to distinct hot spots of compressed material. The simulation exploited 256×2048 grid points, and gave detailed insight into the early stages of relativistic jet propagation.

In a follow-up study, Komissarov [284] demonstrated that the morphology of light, highly relativistic, jets with purely toroidal magnetic fields largely depends on two characteristic parameters. While the initial magnetization of the beam is

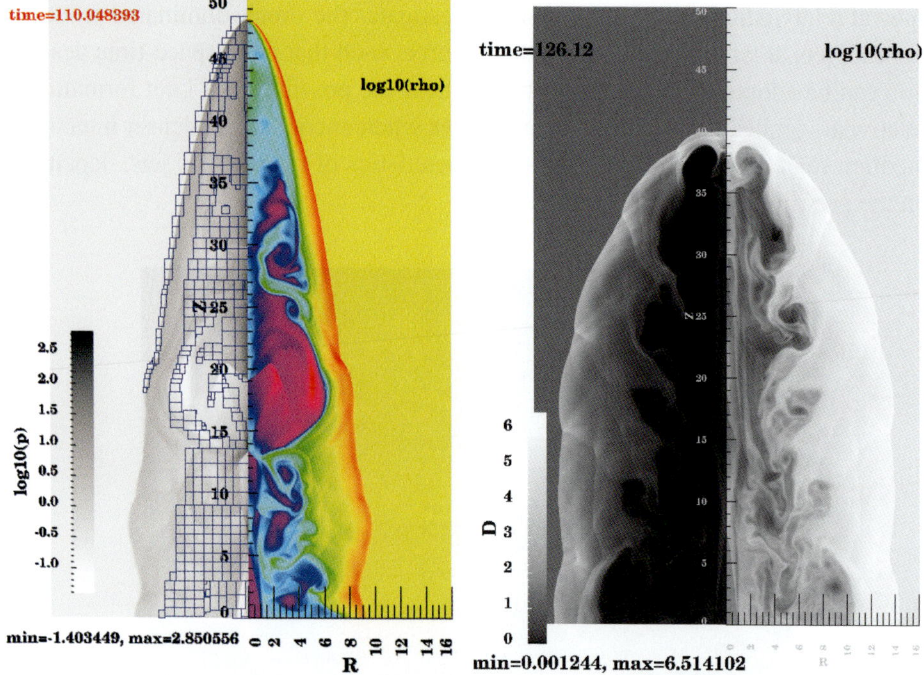

Fig. 21.11 An AMRVAC reproduction of the Poynting flux dominated jet with purely toroidal field from Komissarov [284] (left panel), demonstrating the formation of a magnetically pinched nose cone. Right panel: a purely poloidal field case as in Leismann *et al.* [300], showing no nose cone and rich cocoon dynamics.

certainly important, and expressed quantitatively as a reciprocal plasma beta parameter

$$\beta_r = p_{\mathrm{mag}}/p \,, \tag{21.166}$$

another defining parameter is the ratio of magnetic to rest mass energy density,

$$\sigma = 2p_{\mathrm{mag}}/\rho \,. \tag{21.167}$$

Conspicuous nose cones form when strongly magnetized (high β_r), Poynting flux dominated (high σ) inlet conditions prevail, in which case a strong toroidal field accumulates in the region of shocked beam plasma ahead of the Mach disk, where magnetic pinching causes the extended "nose". This was further explained using the relations for perpendicular (only B_ϕ and v_z components) shocks in relativistic MHD. An example is shown in Figure 21.11, which is a reproduction of the Komissarov result using AMRVAC [456]. In [284], only two jets were simulated, both at Lorentz factor 10, but differing mainly in their inlet σ value. The second jet was kinetic energy dominated, and formed an extensive turbulent cocoon from

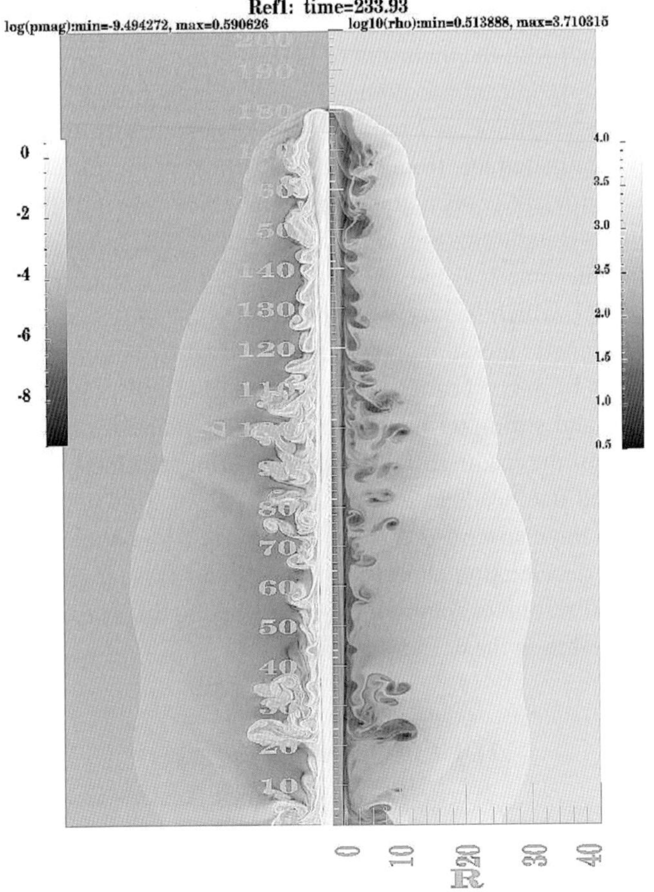

Fig. 21.12 A helically magnetized relativistic jet, taken from Keppens *et al.* [259], where a high axial beam Lorentz factor of 22 prevails.

the backflows generated at the leading Mach disk. This backflow surrounded the beam with toroidal field, aiding in the jet-cocoon confinement.

A more comprehensive study of relativistic jet morphologies was more recently presented by Leismann *et al.* [300]. These authors varied the magnetic field topology as well, still restricting the simulations to axi-symmetric conditions and focused on light relativistic jets, with relatively low Poynting flux contributions (low σ). The main innovation from this study was the systematic comparison between purely toroidal magnetic field configurations and purely poloidal field ones. In the right panel of Figure 21.11, a reproduction of a pure poloidal field model from [300] with AMRVAC is shown, and it is seen that no nose cone develops, while rich cocoon structure is evident. For purely toroidal cases with $\sigma > 0.01$ and $\beta_r \geq 1$, the nose cone morphologies were once more confirmed.

A final example, taken from a recent study by Keppens *et al.* [259], continued this line of research with the inclusion of helical field topologies. The inlet conditions now showed significant variation in Lorentz factor (from 22 on axis to about 4 at the jet radius, averaging $\Gamma = 7$ radially), and systematically explored the case of kinetic energy dominated jets from toroidal to helical to mostly poloidal magnetic fields. Using grid-adaptive techniques, it became possible to follow jet propagation over up to 140 light crossing times of the jet radius, as effective resolutions of 3200×8000 became feasible. Figure 21.12 shows a snapshot of a reference helical field model, which had an average $\sigma \approx 0.0064$ and $\beta_r \approx 0.29$. The figure quantifies the magnetic pressure and proper density distribution. It was found that the magnetization of the jet beam, and its helicity, get fairly effectively transported down the beam, and strong toroidal field regions remain localized in the vortical patterns shed from the leading Mach disk. The study retained the assumption of axi-symmetry, and one needs to investigate the fate of truly helically magnetized relativistic jets in three space dimensions to properly address their overall stability, propagation and deceleration aspects in scale-encompassing simulations. Given the computational resources required for such studies, this is a likely area of continued active research in the decades ahead.

21.4 Literature and exercises

Notes on literature

Relativistic MHD theory

– The monograph by Anile, *Relativistic fluids and magnetofluids* [8], is still the reference work on relativistic fluid treatments, with a clear mathematical orientation. This mathematical rigor is shared with the older work by Lichnerowicz, *Relativistic hydrodynamics and magnetohydrodynamics* [307].

Relativistic MHD and numerics

– The (web-based) review article by Martí and Müller, 'Numerical hydrodynamics in special relativity' [326], represents an updated account of modern numerical methods and their applications for relativistic hydrodynamics. Numerical relativistic magneto-hydrodynamics making use of Riemann solver methodology emerged in earnest with the work by Komissarov, 'A Godunov-type scheme for relativistic magnetohydrody-namics' [283].

Exercises

[21.1] *Transformation laws for tensors*

In this exercise we derive the transformation laws for the components of first rank tensors (i.e. four-vectors) and second rank tensors within flat space-time.

- Find, in analogy with (21.11), but now exploiting *covariant* components and *dual* basis vectors, the transformation for *covariant* components of a four-vector in two sets of dual basis vectors $\mathbf{e}^\alpha$ and $\mathbf{e}^{\alpha'}$.

- Use the defining relations $\mathbf{e}_\alpha \cdot \mathbf{e}^\beta = \delta_\alpha^\beta$, as well as the symmetry in the Lorentz transformation matrix $L_\alpha^{\alpha'}$, to show that, when these two sets of dual basis vectors are associated with basis vectors related by $\mathbf{e}_\alpha = L_\alpha^{\alpha'}\mathbf{e}_{\alpha'}$, they necessarily obey $\mathbf{e}^\alpha = (L^{-1})_{\alpha'}^\alpha \mathbf{e}^{\alpha'}$.

- Find as well that the Lorentz transformation leaves the components of the Minkowski metric tensor unchanged, by working out the relations $g_{\alpha'\beta'} = (L^{-1})_{\alpha'}^\alpha (L^{-1})_{\beta'}^\beta g_{\alpha\beta}$ and $g^{\alpha'\beta'} = L_\alpha^{\alpha'} L_\beta^{\beta'} g^{\alpha\beta}$. These laws for component transformations apply generally to any second rank tensor.

[21.2] *Three-velocity addition law*

In this exercise, we revisit the three-velocity addition, written in various equivalent forms.

- Work out the velocity addition law given in Eq. (21.29), starting from the four-velocity formulation in Eq. (21.27). Prove that when only one spatial direction is involved, the addition law can be written as

$$u = \frac{v + w}{1 + vw/c^2} , \tag{21.168}$$

which clearly reduces to its Galilean $u = v+w$ counterpart for non-relativistic speeds.

- Also verify that we can write Eq. (21.29) as

$$w_\| = \frac{u_\| - v}{1 - u_\| v/c^2} , \quad \text{and} \quad \mathbf{w}_\perp = \frac{\mathbf{u}_\perp}{\Gamma\left(1 - u_\| v/c^2\right)} . \tag{21.169}$$

by decomposing $\mathbf{u} = \mathbf{u}_\| + \mathbf{u}_\perp$, where $\mathbf{u}_\| \parallel \mathbf{v}$.

[21.3] *Relativistic Doppler effect, aberration and beaming for light waves*

The transformation formulas from Eq. (21.64) are perhaps best known in their specific application to light waves in vacuum. Plane wave solutions to the Maxwell equations for empty space obey $\mathbf{k} \cdot \mathbf{B} = \mathbf{k} \cdot \mathbf{E} = \mathbf{E} \cdot \mathbf{B} = 0$, have dispersion relation $\omega^2 = k^2 c^2$, and electric and magnetic field (complex) wave amplitude vectors are related by $\mathbf{B} = \pm(\mathbf{n} \times \mathbf{E})/c$ and $\mathbf{E} = \mp c(\mathbf{n} \times \mathbf{B})$.

- Using the Lorentz transformation for electromagnetic fields given by Eq. (21.79), prove that the amplitude of the light waves as observed in frames in relative motion to each other changes in accord with

$$(B')^2 = \Gamma^2 B^2 \left(1 \mp \mathbf{n} \cdot \mathbf{v}/c\right)^2 . \tag{21.170}$$

This change in the intensity of the light is referred to as relativistic beaming.

- The relativistic Doppler effect can be written in terms of the frequency $\nu = \omega/2\pi$ for light waves as

$$\nu = \nu'\Gamma\left[1 \pm (v/c)\cos\phi'\right] , \tag{21.171}$$

where we denote the angle between $\mathbf{k}'$ and $\mathbf{v}$ by ϕ'.

– Aberration is then also expressed for light waves as

$$\cos \phi = \frac{\cos \phi' \pm v/c}{1 \pm (v/c)\cos \phi'} \, .$$

(21.172)

Derive all these expressions yourself.

[21.4] *Relativistic hydro shock relations*

When discussing the gas dynamic shock relations, we wrote the jump relations (21.67)–(21.68) out for the shock rest frame (SRF). Find the expressions for the four-vectors $\mathbf{U}^{\mathrm{u}}$, $\mathbf{U}^{\mathrm{d}}$, $\mathbf{U}^{\mathrm{s}}$, 1 in reference frames in which the upstream gas is at rest (upstream rest frame or URF), and in which the downstream gas is at rest (downstream rest frame or DRF). Make a graphical representation of the various reference frames using space-time diagrams as shown in Fig. 21.5. For simplicity, assume vanishing tangential velocities. Write down and analyze the shock relations in these reference frames.

[21.5] *Ohm's law in special relativity*

Note that we wrote the four-current as $J^\alpha = (c\sigma, \mathbf{j})^{\mathrm{T}} = (c\sigma, \sigma\mathbf{v})^{\mathrm{T}}$, where σ is the charge density.

– Verify the invariant $U_\alpha J^\alpha$.

– Find that as a result

$$J^\alpha + c^{-2}(U_\beta J^\beta)U^\alpha = 0 \, .$$

(21.173)

This is actually a simplified form of Ohm's law in covariant form, where more generally a right hand side $\sigma_{\mathrm{c}} e^\alpha$ should be written, with σ_{c} the conductivity parameter.

– Find that, instead of $\mathbf{j} = \sigma\mathbf{v}$, we must then write

$$\mathbf{j} = \sigma_{\mathrm{c}}\Gamma \left[\mathbf{E} + \mathbf{v} \times \mathbf{B} - \mathbf{v}\mathbf{v} \cdot \mathbf{E}/c^2 \right] + \sigma\mathbf{v} \, .$$

(21.174)

In the plasma rest frame, we have then indeed $\mathbf{j}' = \sigma_{\mathrm{c}}\mathbf{E}'$.

[21.6] *MHD waves dynamics*

In this exercise, we expand on the relations between the rest frame expressions for the characteristic speeds, and the expressions as given in the lab frame. For a more detailed discussion of the answers to the points below, we refer to [260].

– Show that the Alfvén wave expression (21.139) indeed corresponds with transforming from the rest frame expression, where the Alfvén phase speed is found from $v_{\mathrm{ph}}'^2 = (\mathbf{B}' \cdot \mathbf{k}')^2 c^2/(\rho h + B'^2)$.

– Verify that the expressions for the characteristic speeds from Eq. (21.141) reduce to their non-relativistic counterparts for $\Gamma \to 1$. Show that Eq. (21.141) yields the relativistic hydro result from Eq. (21.63) when the magnetic field vanishes. Furthermore, check the cold plasma limit where $c_{\mathrm{g}}^2 = 0$.

– Transform the rest frame expression (21.131) to the lab frame result (21.141). Note that you will need to combine the phase speed relation given by Eq. (21.65) with the Lorentz formula (21.79) (in ideal relativistic MHD) for the magnetic field, together with the wave vector (and hence the wave front normal) relation from Eq. (21.64).

Appendix A
Vectors and coordinates

A.1 Vector identities

A list of the most frequently exploited identities:

$$\mathbf{a} \cdot (\mathbf{b} \times \mathbf{c}) = \mathbf{c} \cdot (\mathbf{a} \times \mathbf{b}) = \mathbf{b} \cdot (\mathbf{c} \times \mathbf{a}) \,, \tag{A.1}$$

$$\mathbf{a} \times (\mathbf{b} \times \mathbf{c}) = \mathbf{a} \cdot \mathbf{c}\,\mathbf{b} - \mathbf{a} \cdot \mathbf{b}\,\mathbf{c} \,, \qquad (\mathbf{a} \times \mathbf{b}) \times \mathbf{c} = \mathbf{a} \cdot \mathbf{c}\,\mathbf{b} - \mathbf{b} \cdot \mathbf{c}\,\mathbf{a} \,; \tag{A.2}$$

$$\nabla \times \nabla \Phi = 0 \,, \tag{A.3}$$

$$\nabla \cdot (\nabla \times \mathbf{a}) = 0 \,, \tag{A.4}$$

$$\nabla \times (\nabla \times \mathbf{a}) = \nabla \nabla \cdot \mathbf{a} - \nabla^2 \mathbf{a} \,; \tag{A.5}$$

$$\nabla \cdot (\Phi \mathbf{a}) = \mathbf{a} \cdot \nabla \Phi + \Phi \nabla \cdot \mathbf{a} \,, \tag{A.6}$$

$$\nabla \times (\Phi \mathbf{a}) = \nabla \Phi \times \mathbf{a} + \Phi \nabla \times \mathbf{a} \,, \tag{A.7}$$

$$\mathbf{a} \times (\nabla \times \mathbf{b}) = (\nabla \mathbf{b}) \cdot \mathbf{a} - \mathbf{a} \cdot \nabla \mathbf{b} \,, \tag{A.8}$$

$$(\mathbf{a} \times \nabla) \times \mathbf{b} = (\nabla \mathbf{b}) \cdot \mathbf{a} - \mathbf{a} \nabla \cdot \mathbf{b} \,, \tag{A.9}$$

$$\nabla (\mathbf{a} \cdot \mathbf{b}) = (\nabla \mathbf{a}) \cdot \mathbf{b} + (\nabla \mathbf{b}) \cdot \mathbf{a}$$
$$= \mathbf{a} \cdot \nabla \mathbf{b} + \mathbf{b} \cdot \nabla \mathbf{a} + \mathbf{a} \times (\nabla \times \mathbf{b}) + \mathbf{b} \times (\nabla \times \mathbf{a}) \,, \tag{A.10}$$

$$\nabla \cdot (\mathbf{a} \mathbf{b}) = \mathbf{a} \cdot \nabla \mathbf{b} + \mathbf{b} \nabla \cdot \mathbf{a} \,, \tag{A.11}$$

$$\nabla \cdot (\mathbf{a} \times \mathbf{b}) = \mathbf{b} \cdot \nabla \times \mathbf{a} - \mathbf{a} \cdot \nabla \times \mathbf{b} \,, \tag{A.12}$$

$$\nabla \times (\mathbf{a} \times \mathbf{b}) = \nabla \cdot (\mathbf{b} \mathbf{a} - \mathbf{a} \mathbf{b})$$
$$= \mathbf{a} \nabla \cdot \mathbf{b} + \mathbf{b} \cdot \nabla \mathbf{a} - \mathbf{b} \nabla \cdot \mathbf{a} - \mathbf{a} \cdot \nabla \mathbf{b} \,; \tag{A.13}$$

$$\iiint \nabla \cdot \mathbf{a}\, d\tau = \oiint \mathbf{a} \cdot \mathbf{n}\, d\sigma \quad \textit{(Gauss)}, \tag{A.14}$$

$$\mathbf{a} \to \mathbf{a} \times \mathbf{c}(\text{onst}) \quad \Rightarrow \quad \iiint \nabla \times \mathbf{a}\, d\tau = \oiint \mathbf{n} \times \mathbf{a}\, d\sigma, \tag{A.15}$$

$$\mathbf{a} \to \Phi\,\mathbf{c}(\text{onst}) \quad \Rightarrow \quad \iiint \nabla \Phi\, d\tau = \oiint \Phi\, \mathbf{n}\, d\sigma, \tag{A.16}$$

$$\mathbf{a} \to \Phi \nabla \Psi - \Psi \nabla \Phi \Rightarrow$$

$$\iiint (\Phi \nabla^2 \Psi - \Psi \nabla^2 \Phi)\, d\tau = \oiint (\Phi \nabla \Psi - \Psi \nabla \Phi) \cdot \mathbf{n}\, d\sigma \quad \textit{(Green)}; \tag{A.17}$$

$$\iint (\nabla \times \mathbf{a}) \cdot \mathbf{n}\, d\sigma = \oint \mathbf{a} \cdot d\mathbf{l} \quad \textit{(Stokes)}, \tag{A.18}$$

$$\mathbf{a} \to \mathbf{a} \times \mathbf{c}(\text{onst}) \quad \Rightarrow \quad \iint (\mathbf{n} \times \nabla) \times \mathbf{a}\, d\sigma = \oint d\mathbf{l} \times \mathbf{a}, \tag{A.19}$$

$$\mathbf{a} \to \Phi\,\mathbf{c}(\text{onst}) \quad \Rightarrow \quad \iint \mathbf{n} \times \nabla \Phi\, d\sigma = \oint \Phi\, d\mathbf{l}. \tag{A.20}$$

A.2 Vector expressions in orthogonal coordinates

Considering the position vector as a function of orthogonal coordinates x_i,

$$\mathbf{r} = \mathbf{r}(x_1, x_2, x_3) \quad \Longleftrightarrow \quad \begin{cases} x = x(x_1, x_2, x_3) \\ y = y(x_1, x_2, x_3) \\ z = z(x_1, x_2, x_3) \end{cases}, \tag{A.21}$$

the following geometric quantities are generated:

$$h_i \equiv |\partial \mathbf{r}/\partial x_i| \qquad\qquad \textit{(scale factors)}, \tag{A.22}$$

$$\mathbf{e}_i \equiv (1/h_i)\,\partial \mathbf{r}/\partial x_i, \qquad \mathbf{e}_i \cdot \mathbf{e}_j = \delta_{ij} \quad \textit{(dimensionless unit vectors)}, \tag{A.23}$$

$$d\ell = \sqrt{\sum_i (h_i dx_i)^2} \qquad\qquad \textit{(line element)}, \tag{A.24}$$

$$d\tau = h_1 h_2 h_3\, dx_1 dx_2 dx_3 \qquad\qquad \textit{(volume element)}. \tag{A.25}$$

Vector representation:

$$\mathbf{V} = \sum_i \hat{V}_i\, \mathbf{e}_i \quad (\hat{V}_i - \textit{physical components}, \text{ same dimension as } \mathbf{V}). \tag{A.26}$$

Products:

$$\mathbf{A} \cdot \mathbf{B} = \sum_i \hat{A}_i \hat{B}_i \qquad \textit{(inner product)}, \qquad (A.27)$$

$$\mathbf{A} \times \mathbf{B} = \sum_i \sum_j \sum_k \epsilon_{ijk} \hat{A}_j \hat{B}_j \, \mathbf{e}_i \qquad \textit{(vector product)}, \qquad (A.28)$$

$$\epsilon_{ijk} \equiv \begin{cases} 1 & \text{if } ijk \text{ even permutation of 123} \\ -1 & \text{if } ijk \text{ odd permutation of 123} \\ 0 & \text{otherwise} \end{cases} \quad \textit{(permutation symbol)}, \qquad (A.29)$$

$$\epsilon_{ijk}\epsilon_{ilm} = \delta_{jl}\delta_{km} - \delta_{jm}\delta_{kl} \,. \qquad (A.30)$$

Differential operators:

$$\nabla\psi = \sum \frac{1}{h_i} \frac{\partial \psi}{\partial x_i} \, \mathbf{e}_i \,, \qquad (A.31)$$

$$\nabla^2\psi = \frac{1}{h_1 h_2 h_3} \left[\frac{\partial}{\partial x_1}\left(\frac{h_2 h_3}{h_1} \frac{\partial \psi}{\partial x_1} \right) + \frac{\partial}{\partial x_2}\left(\frac{h_1 h_3}{h_2} \frac{\partial \psi}{\partial x_2} \right) \right.$$
$$\left. + \frac{\partial}{\partial x_3}\left(\frac{h_1 h_2}{h_3} \frac{\partial \psi}{\partial x_3} \right) \right], \qquad (A.32)$$

$$\nabla \cdot \mathbf{A} = \frac{1}{h_1 h_2 h_3} \left[\frac{\partial}{\partial x_1}(h_2 h_3 \hat{A}_1) + \frac{\partial}{\partial x_2}(h_1 h_3 \hat{A}_2) + \frac{\partial}{\partial x_3}(h_1 h_2 \hat{A}_3) \right], \qquad (A.33)$$

$$\nabla \times \mathbf{A} = \frac{1}{h_2 h_3} \left[\frac{\partial}{\partial x_2}(h_3 \hat{A}_3) - \frac{\partial}{\partial x_3}(h_2 \hat{A}_2) \right] \mathbf{e}_1$$
$$+ \frac{1}{h_1 h_3} \left[\frac{\partial}{\partial x_3}(h_1 \hat{A}_1) - \frac{\partial}{\partial x_1}(h_3 \hat{A}_3) \right] \mathbf{e}_2$$
$$+ \frac{1}{h_1 h_2} \left[\frac{\partial}{\partial x_1}(h_2 \hat{A}_2) - \frac{\partial}{\partial x_2}(h_1 \hat{A}_1) \right] \mathbf{e}_3 \,. \qquad (A.34)$$

Derivatives of the unit vectors:

$$\frac{\partial \mathbf{e}_1}{\partial x_1} = -\frac{1}{h_2}\frac{\partial h_1}{\partial x_2}\mathbf{e}_2 - \frac{1}{h_3}\frac{\partial h_1}{\partial x_3}\mathbf{e}_3 \,, \qquad \frac{\partial \mathbf{e}_2}{\partial x_1} = \frac{1}{h_2}\frac{\partial h_1}{\partial x_2}\mathbf{e}_1 \,, \qquad \frac{\partial \mathbf{e}_3}{\partial x_1} = \frac{1}{h_3}\frac{\partial h_1}{\partial x_3}\mathbf{e}_1 \,,$$

$$\frac{\partial \mathbf{e}_1}{\partial x_2} = \frac{1}{h_1}\frac{\partial h_2}{\partial x_1}\mathbf{e}_2 \,, \qquad \frac{\partial \mathbf{e}_2}{\partial x_2} = -\frac{1}{h_1}\frac{\partial h_2}{\partial x_1}\mathbf{e}_1 - \frac{1}{h_3}\frac{\partial h_2}{\partial x_3}\mathbf{e}_3 \,, \qquad \frac{\partial \mathbf{e}_3}{\partial x_2} = \frac{1}{h_3}\frac{\partial h_2}{\partial x_3}\mathbf{e}_2 \,,$$

$$\frac{\partial \mathbf{e}_1}{\partial x_3} = \frac{1}{h_1}\frac{\partial h_3}{\partial x_1}\mathbf{e}_3 \,, \qquad \frac{\partial \mathbf{e}_2}{\partial x_3} = \frac{1}{h_2}\frac{\partial h_3}{\partial x_2}\mathbf{e}_3 \,, \qquad \frac{\partial \mathbf{e}_3}{\partial x_3} = -\frac{1}{h_1}\frac{\partial h_3}{\partial x_1}\mathbf{e}_1 - \frac{1}{h_2}\frac{\partial h_3}{\partial x_2}\mathbf{e}_2 \,.$$

$$(A.35)$$

Hence,

$$
\begin{aligned}
\mathbf{A} \cdot \nabla \mathbf{B} = {} & \left[\frac{\hat{A}_1}{h_1}\left(\frac{\partial \hat{B}_1}{\partial x_1} + \frac{\partial h_1}{\partial x_2}\frac{\hat{B}_2}{h_2} + \frac{\partial h_1}{\partial x_3}\frac{\hat{B}_3}{h_3} \right) + \frac{\hat{A}_2}{h_2}\left(\frac{\partial \hat{B}_1}{\partial x_2} - \frac{\partial h_2}{\partial x_1}\frac{\hat{B}_2}{h_1} \right) \right. \\
& \left. + \frac{\hat{A}_3}{h_3}\left(\frac{\partial \hat{B}_1}{\partial x_3} - \frac{\partial h_3}{\partial x_1}\frac{\hat{B}_3}{h_1} \right) \right]\mathbf{e}_1 \\
& + \left[\frac{\hat{A}_1}{h_1}\left(\frac{\partial \hat{B}_2}{\partial x_1} - \frac{\partial h_1}{\partial x_2}\frac{\hat{B}_1}{h_2} \right) + \frac{\hat{A}_2}{h_2}\left(\frac{\partial \hat{B}_2}{\partial x_2} + \frac{\partial h_2}{\partial x_1}\frac{\hat{B}_1}{h_1} + \frac{\partial h_2}{\partial x_3}\frac{\hat{B}_3}{h_3} \right) \right. \\
& \left. + \frac{\hat{A}_3}{h_3}\left(\frac{\partial \hat{B}_2}{\partial x_3} - \frac{\partial h_3}{\partial x_2}\frac{\hat{B}_3}{h_2} \right) \right]\mathbf{e}_2 \\
& + \left[\frac{\hat{A}_1}{h_1}\left(\frac{\partial \hat{B}_3}{\partial x_1} - \frac{\partial h_1}{\partial x_3}\frac{\hat{B}_1}{h_3} \right) + \frac{\hat{A}_2}{h_2}\left(\frac{\partial \hat{B}_3}{\partial x_2} - \frac{\partial h_2}{\partial x_3}\frac{\hat{B}_2}{h_3} \right) \right. \\
& \left. + \frac{\hat{A}_3}{h_3}\left(\frac{\partial \hat{B}_3}{\partial x_3} + \frac{\partial h_3}{\partial x_1}\frac{\hat{B}_1}{h_1} + \frac{\partial h_3}{\partial x_2}\frac{\hat{B}_2}{h_2} \right) \right]\mathbf{e}_3 \, .
\end{aligned}
\tag{A.36}
$$

▷ **Notation** The awkward hat, used until here to avoid conflict with the covariant components of non-orthogonal coordinate systems (Section A.3), is dropped in the explicit expressions for the different coordinate systems below by writing A_{x_i} instead of $\hat{A}_i$. ◁

A.2.1 *Cartesian coordinates (x, y, z)*

$$
x \equiv x_1, \quad y \equiv x_2, \quad z \equiv x_3 \qquad \Rightarrow \qquad h_1 = h_2 = h_3 = 1 \, .
\tag{A.37}
$$

$$
\nabla \psi = \frac{\partial \psi}{\partial x}\mathbf{e}_x + \frac{\partial \psi}{\partial y}\mathbf{e}_y + \frac{\partial \psi}{\partial z}\mathbf{e}_z \, ,
\tag{A.38}
$$

$$
\nabla^2 \psi = \frac{\partial^2 \psi}{\partial x^2} + \frac{\partial^2 \psi}{\partial y^2} + \frac{\partial^2 \psi}{\partial z^2} \, ,
\tag{A.39}
$$

$$
\nabla \cdot \mathbf{A} = \frac{\partial A_x}{\partial x} + \frac{\partial A_y}{\partial y} + \frac{\partial A_z}{\partial z} \, ,
\tag{A.40}
$$

$$
\nabla \times \mathbf{A} = \left(\frac{\partial A_z}{\partial y} - \frac{\partial A_y}{\partial z} \right)\mathbf{e}_x + \left(\frac{\partial A_x}{\partial z} - \frac{\partial A_z}{\partial x} \right)\mathbf{e}_y + \left(\frac{\partial A_y}{\partial x} - \frac{\partial A_x}{\partial y} \right)\mathbf{e}_z \, .
\tag{A.41}
$$

▷ **Note** The vector identities of Section A.1, in particular the complicated ones involving vector products and curls, are most easily derived in Cartesian coordinates, exploiting Eqs. (A.31)–(A.36) with $h_i = 1$ (see, e.g., Goldston & Rutherford [185], p. 481). ◁

A.2.2 Cylindrical coordinates (r, θ, z)
(Fig. A.1)

Scale factors and derivatives of the unit vectors:

$$\begin{cases} x = r\cos\theta \\[2mm] y = r\sin\theta \\[2mm] z = z \end{cases} \quad\Rightarrow\quad h_1 = 1\,, \quad h_2 = r\,, \quad h_3 = 1\,, \tag{A.42}$$

$$\frac{\partial \mathbf{e}_r}{\partial \theta} = \mathbf{e}_\theta\,, \quad \frac{\partial \mathbf{e}_\theta}{\partial \theta} = -\mathbf{e}_r \quad (\text{only derivatives} \neq 0)\,. \tag{A.43}$$

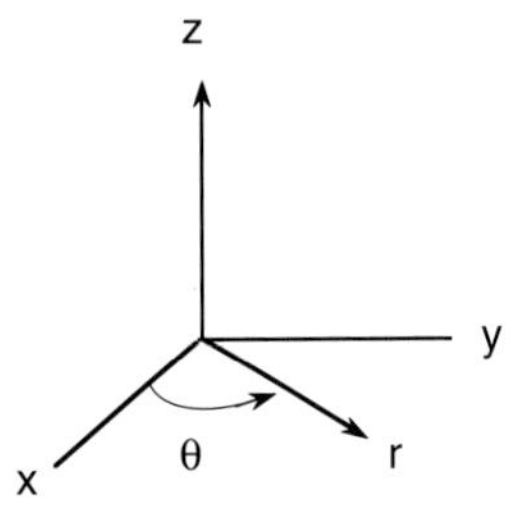

Fig. A.1

Differential operators:

$$\nabla\psi = \frac{\partial\psi}{\partial r}\,\mathbf{e}_r + \frac{1}{r}\frac{\partial\psi}{\partial\theta}\,\mathbf{e}_\theta + \frac{\partial\psi}{\partial z}\,\mathbf{e}_z\,, \tag{A.44}$$

$$\nabla^2\psi = \frac{1}{r}\frac{\partial}{\partial r}\left(r\frac{\partial\psi}{\partial r}\right) + \frac{1}{r^2}\frac{\partial^2\psi}{\partial\theta^2} + \frac{\partial^2\psi}{\partial z^2}\,, \tag{A.45}$$

$$\nabla\cdot\mathbf{A} = \frac{1}{r}\frac{\partial(rA_r)}{\partial r} + \frac{1}{r}\frac{\partial A_\theta}{\partial\theta} + \frac{\partial A_z}{\partial z}\,, \tag{A.46}$$

$$\nabla\times\mathbf{A} = \left(\frac{1}{r}\frac{\partial A_z}{\partial\theta} - \frac{\partial A_\theta}{\partial z}\right)\mathbf{e}_r$$

$$+ \left(\frac{\partial A_r}{\partial z} - \frac{\partial A_z}{\partial r}\right)\mathbf{e}_\theta + \left(\frac{1}{r}\frac{\partial(rA_\theta)}{\partial r} - \frac{1}{r}\frac{\partial A_r}{\partial\theta}\right)\mathbf{e}_z\,, \tag{A.47}$$

$$\nabla^2\mathbf{A} = \left(\nabla^2 A_r - \frac{1}{r^2}A_r - \frac{2}{r^2}\frac{\partial A_\theta}{\partial\theta}\right)\mathbf{e}_r$$

$$+ \left(\nabla^2 A_\theta - \frac{1}{r^2}A_\theta + \frac{2}{r^2}\frac{\partial A_r}{\partial\theta}\right)\mathbf{e}_\theta + \nabla^2 A_z\,\mathbf{e}_z\,, \tag{A.48}$$

$$\nabla\times\nabla\times\mathbf{A} = \left[-\frac{1}{r^2}\frac{\partial^2 A_r}{\partial\theta^2} - \frac{\partial^2 A_r}{\partial z^2} + \frac{1}{r^2}\frac{\partial^2(rA_\theta)}{\partial\theta\,\partial r} + \frac{\partial^2 A_z}{\partial z\partial r}\right]\mathbf{e}_r$$

$$+ \left[\frac{\partial}{\partial r}\left(\frac{1}{r}\frac{\partial A_r}{\partial\theta}\right) - \frac{\partial}{\partial r}\left(\frac{1}{r}\frac{\partial(rA_\theta)}{\partial r}\right) - \frac{\partial^2 A_\theta}{\partial z^2} + \frac{1}{r}\frac{\partial^2 A_z}{\partial z\partial\theta}\right]\mathbf{e}_\theta$$

$$+ \left[\frac{1}{r}\frac{\partial}{\partial r}\left(r\frac{\partial A_r}{\partial z}\right) + \frac{1}{r}\frac{\partial^2 A_\theta}{\partial\theta\partial z} - \frac{1}{r}\frac{\partial}{\partial r}\left(r\frac{\partial A_z}{\partial r}\right) - \frac{1}{r^2}\frac{\partial^2 A_z}{\partial\theta^2}\right]\mathbf{e}_z\,, \tag{A.49}$$

$$\mathbf{A} \cdot \nabla \mathbf{B} = \left[A_r \frac{\partial B_r}{\partial r} + \frac{A_\theta}{r} \left(\frac{\partial B_r}{\partial \theta} - B_\theta \right) + A_z \frac{\partial B_r}{\partial z} \right] \mathbf{e}_r$$

$$+ \left[A_r \frac{\partial B_\theta}{\partial r} + \frac{A_\theta}{r} \left(B_r + \frac{\partial B_\theta}{\partial \theta} \right) + A_z \frac{\partial B_\theta}{\partial z} \right] \mathbf{e}_\theta$$

$$+ \left[A_r \frac{\partial B_z}{\partial r} + \frac{A_\theta}{r} \frac{\partial B_z}{\partial \theta} + A_z \frac{\partial B_z}{\partial z} \right] \mathbf{e}_z . \tag{A.50}$$

A.2.3 Spherical coordinates $(r,\ \theta,\ \phi)$
(Fig. A.2)

Scale factors and derivatives of the unit vectors:

$$\begin{cases} x = R \cos \phi , \quad R = r \sin \theta \\[2mm] y = R \sin \phi \\[2mm] z = r \cos \theta \end{cases}$$

$$\Rightarrow \quad h_1 = 1 , \quad h_2 = r , \quad h_3 = r \sin \theta . \tag{A.51}$$

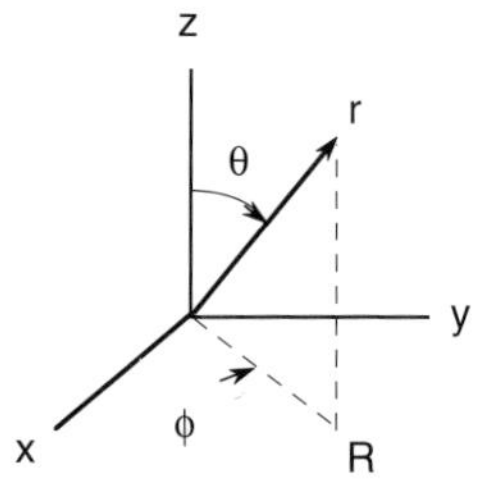

Fig. A.2

$$\frac{\partial \mathbf{e}_r}{\partial \theta} = \mathbf{e}_\theta , \qquad \frac{\partial \mathbf{e}_\theta}{\partial \theta} = -\mathbf{e}_r ,$$

$$\frac{\partial \mathbf{e}_r}{\partial \phi} = \sin \theta \, \mathbf{e}_\phi , \qquad \frac{\partial \mathbf{e}_\theta}{\partial \phi} = \cos \theta \, \mathbf{e}_\phi , \qquad \frac{\partial \mathbf{e}_\phi}{\partial \phi} = - \sin \theta \, \mathbf{e}_r - \cos \theta \, \mathbf{e}_\theta . \tag{A.52}$$

Differential operators:

$$\nabla \psi = \frac{\partial \psi}{\partial r} \mathbf{e}_r + \frac{1}{r} \frac{\partial \psi}{\partial \theta} \mathbf{e}_\theta + \frac{1}{r \sin \theta} \frac{\partial \psi}{\partial \phi} \mathbf{e}_\phi , \tag{A.53}$$

$$\nabla^2 \psi = \frac{1}{r^2} \frac{\partial}{\partial r} \left(r^2 \frac{\partial \psi}{\partial r} \right) + \frac{1}{r^2 \sin \theta} \frac{\partial}{\partial \theta} \left(\sin \theta \frac{\partial \psi}{\partial \theta} \right) + \frac{1}{r^2 \sin^2 \theta} \frac{\partial^2 \psi}{\partial \phi^2} , \tag{A.54}$$

$$\nabla \cdot \mathbf{A} = \frac{1}{r^2} \frac{\partial}{\partial r} (r^2 A_r) + \frac{1}{r \sin \theta} \frac{\partial}{\partial \theta} (\sin \theta A_\theta) + \frac{1}{r \sin \theta} \frac{\partial A_\phi}{\partial \phi} , \tag{A.55}$$

$$\nabla \times \mathbf{A} = \frac{1}{r \sin \theta} \left[\frac{\partial}{\partial \theta} (\sin \theta A_\phi) - \frac{\partial A_\theta}{\partial \phi} \right] \mathbf{e}_r$$

$$+ \frac{1}{r} \left[\frac{1}{\sin \theta} \frac{\partial A_r}{\partial \phi} - \frac{\partial}{\partial r} (r A_\phi) \right] \mathbf{e}_\theta + \frac{1}{r} \left[\frac{\partial}{\partial r} (r A_\theta) - \frac{\partial A_r}{\partial \theta} \right] \mathbf{e}_\phi , \tag{A.56}$$

$$\nabla^2 \mathbf{A} = \left[(\nabla^2 A_r) - \frac{2}{r^2} A_r - \frac{2}{r^2 \sin\theta} \frac{\partial}{\partial\theta} (\sin\theta A_\theta) - \frac{2}{r^2 \sin\theta} \frac{\partial A_\phi}{\partial\phi} \right] \mathbf{e}_r$$

$$+ \left[(\nabla^2 A_\theta) + \frac{2}{r^2} \frac{\partial A_r}{\partial\theta} - \frac{1}{r^2 \sin^2\theta} A_\theta - \frac{2\cos\theta}{r^2 \sin^2\theta} \frac{\partial A_\phi}{\partial\phi} \right] \mathbf{e}_\theta$$

$$+ \left[(\nabla^2 A_\phi) + \frac{2}{r^2 \sin\theta} \frac{\partial A_r}{\partial\phi} + \frac{2\cos\theta}{r^2 \sin^2\theta} \frac{\partial A_\theta}{\partial\phi} - \frac{1}{r^2 \sin^2\theta} A_\phi \right] \mathbf{e}_\phi ,$$

$$(A.57)$$

$$\nabla \times \nabla \times \mathbf{A} = \frac{1}{r^2 \sin\theta} \left[-\frac{\partial}{\partial\theta} \left(\sin\theta \frac{\partial A_r}{\partial\theta} \right) - \frac{1}{\sin\theta} \frac{\partial^2 A_r}{\partial\phi^2} \right.$$

$$\left. + \frac{\partial}{\partial\theta} \left(\sin\theta \frac{\partial(r A_\theta)}{\partial r} \right) + \frac{\partial^2 (r A_\phi)}{\partial\phi\,\partial r} \right] \mathbf{e}_r$$

$$+ \frac{1}{r^2} \left[r \frac{\partial^2 A_r}{\partial r \partial\theta} - \frac{\partial}{\partial r} \left(r^2 \frac{\partial A_\theta}{\partial r} \right) - \frac{1}{\sin^2\theta} \frac{\partial^2 A_\theta}{\partial\phi^2} \right.$$

$$\left. + \frac{1}{\sin^2\theta} \frac{\partial^2 (\sin\theta A_\phi)}{\partial\phi\,\partial\theta} \right] \mathbf{e}_\theta$$

$$+ \frac{1}{r^2} \left[\frac{r}{\sin\theta} \frac{\partial^2 A_r}{\partial r \partial\phi} + \frac{\partial}{\partial\theta} \left(\frac{1}{\sin\theta} \frac{\partial A_\theta}{\partial\phi} \right) \right.$$

$$\left. - \frac{\partial}{\partial r} \left(r^2 \frac{\partial A_\phi}{\partial r} \right) - \frac{\partial}{\partial\theta} \left(\frac{1}{\sin\theta} \frac{\partial(\sin\theta A_\phi)}{\partial\theta} \right) \right] \mathbf{e}_\phi ,$$

$$(A.58)$$

$$\mathbf{A} \cdot \nabla \mathbf{B} = \left[A_r \frac{\partial B_r}{\partial r} + \frac{A_\theta}{r} \left(\frac{\partial B_r}{\partial\theta} - B_\theta \right) + \frac{A_\phi}{r} \left(\frac{1}{\sin\theta} \frac{\partial B_r}{\partial\phi} - B_\phi \right) \right] \mathbf{e}_r$$

$$+ \left[A_r \frac{\partial B_\theta}{\partial r} + \frac{A_\theta}{r} \left(B_r + \frac{\partial B_\theta}{\partial\theta} \right) + \frac{A_\phi}{r} \left(\frac{1}{\sin\theta} \frac{\partial B_\theta}{\partial\phi} - \cot\theta\, B_\phi \right) \right] \mathbf{e}_\theta$$

$$+ \left[A_r \frac{\partial B_\phi}{\partial r} + \frac{A_\theta}{r} \frac{\partial B_\phi}{\partial\theta} + \frac{A_\phi}{r} \left(B_r + \cot\theta\, B_\theta + \frac{1}{\sin\theta} \frac{\partial B_\phi}{\partial\phi} \right) \right] \mathbf{e}_\phi .$$

$$(A.59)$$

A.2.4 Cylindrical coordinates for toroidal problems (R, Z, φ)
(Fig. A.3)

These coordinates are an intermediate step to the toroidal coordinates (φ ignorable) of Sections A.2.5–A.2.7 and A.3.1, where R and Z provide Cartesian coordinates of the poloidal plane. From now on, we exploit capitals X, Y, Z for the Cartesian coordinates to enable later use of the lower case x and y as scaled coordinates in the poloidal plane. Scale factors and unit vector derivatives then become:

$$\begin{cases} x \rightarrow X = R\sin\varphi, \\ y \rightarrow Y = R\cos\varphi, \\ z \rightarrow Z \end{cases}$$

$$\Rightarrow \quad h_1 = 1, \quad h_2 = 1, \quad h_3 = R. \tag{A.60}$$

$$\frac{\partial \mathbf{e}_R}{\partial \varphi} = \mathbf{e}_\varphi, \qquad \frac{\partial \mathbf{e}_\varphi}{\partial \varphi} = -\mathbf{e}_R. \tag{A.61}$$

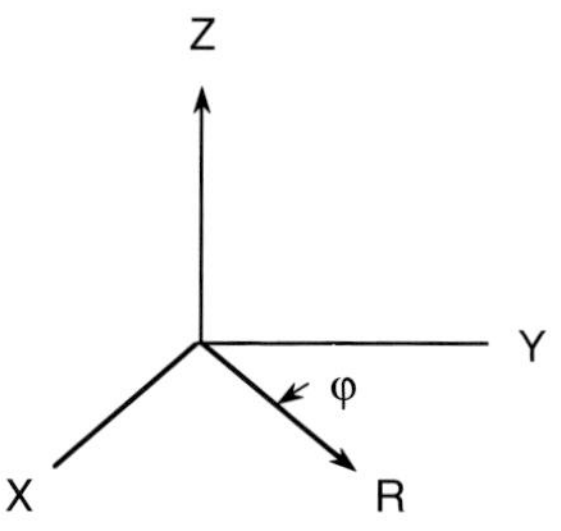

Fig. A.3

Differential operators are obtained from Eqs. (A.44)–(A.50) of Section A.2.2 with the following replacements:

$$r \rightarrow R, \qquad \theta \rightarrow \pi/2 - \varphi, \qquad z \rightarrow Z,$$

$$\partial/\partial\theta \rightarrow -\partial/\partial\varphi, \qquad \mathbf{e}_\theta \rightarrow -\mathbf{e}_\varphi, \tag{A.62}$$

$$A_r \rightarrow A_R, \qquad A_\theta \rightarrow -A_\varphi, \qquad A_z \rightarrow A_Z, \quad \text{etc}.$$

The crucial difference from ordinary cylindrical coordinates is the order R, Z, φ.

A.2.5 *Toroidal polar coordinates* (r, θ, φ)
(Fig. A.4)

Scale factors:

$$\begin{cases} X = R\sin\varphi, \quad R = R(r,\theta) = R_0 + r\cos\theta \\ Y = R\cos\varphi \\ Z = r\sin\theta \end{cases}$$

$$\Rightarrow \quad h_1 = 1, \quad h_2 = r, \quad h_3 = R = R_0 + r\cos\theta. \tag{A.63}$$

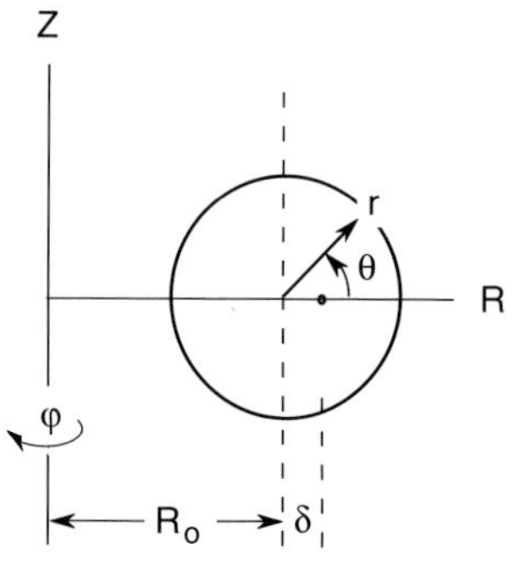

Fig. A.4

These coordinates may be used to describe global toroidal geometrical features, like vacuum fields outside the plasma, but, since they do not incorporate the shift of the magnetic axis, they are usually not appropriate to describe toroidal plasma equilibrium. For that purpose, the Shafranov shifted circle coordinates should be used; see Section 16.2.2, Eqs. (16.85)–(16.87).

A.2.6 Toroidal–conformal coordinates (s, t, φ)
(Fig. A.5)

Moebius transformation of the poloidal plane,

$$z = x + \mathrm{i}y \;\Rightarrow\; w = u + \mathrm{i}v:$$

$$\begin{cases} X = R\sin\varphi, \quad R = R(r,\theta) = R_0 + x(u,v) \\ Y = R\cos\varphi \\ Z = y(u,v), \end{cases}$$

$$s = \sqrt{u^2 + v^2}, \quad t = \arctan(v/u)$$

$$\Rightarrow \quad h = h_1 = h_2 = \frac{1 - \delta^2}{1 + 2\delta s \cos t + \delta^2 s^2},$$
$$h_3 = R. \tag{A.64}$$

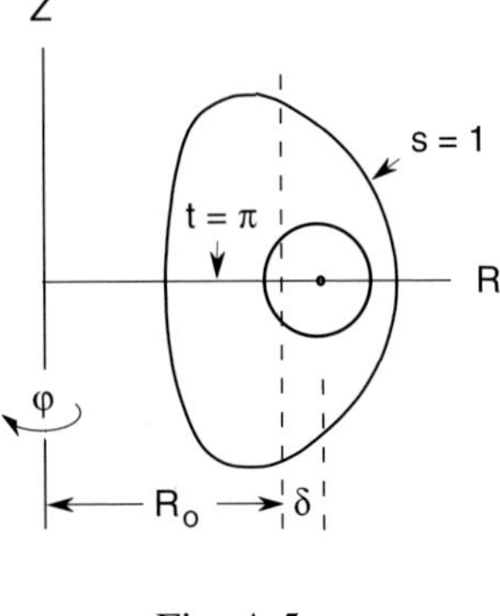

Fig. A.5

These are *not flux coordinates*, but they can be used to describe toroidal equilibrium since they do incorporate the shift of the magnetic axis; Section 16.3.3 and [162].

A.2.7 Orthogonal flux coordinates (Ψ, χ, φ)
(Fig. A.6)

These coordinates are nearly exclusively used for analytical theory since their explicit numerical construction suffers from logarithmic singularities in the distribution of the angle χ for elongated cross-sections. Their exploitation presupposes the explicit solution of the equilibrium equation for the poloidal flux Ψ and the construction of a poloidal angle χ such that $\nabla\Psi \cdot \nabla\chi = 0$, so that $\Psi(R, Z)$ and $\chi(R, Z)$ are supposed to be known. Formal inversion of these expressions provides the defining equations of the coordinates:

$$\begin{cases} X = R(\Psi, \chi)\sin\varphi, \\ Y = R(\Psi, \chi)\cos\varphi, \\ Z = Z(\Psi, \chi) \end{cases}$$

$$\Rightarrow \quad h_1 = \frac{1}{RB_\mathrm{p}}, \quad h_2 = J_\mathrm{o}B_\mathrm{p}, \quad h_3 = R. \tag{A.65}$$

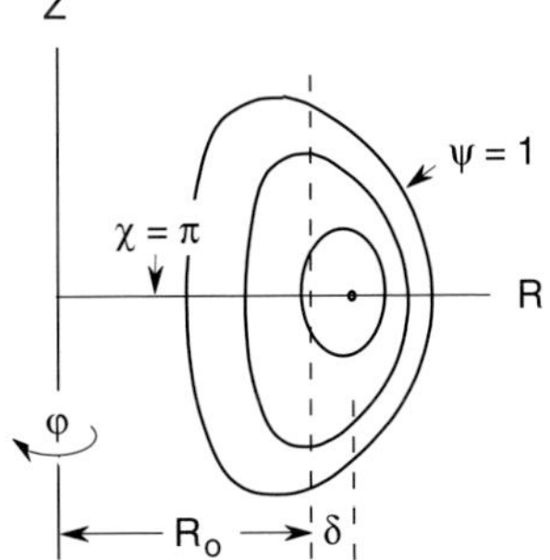

Fig. A.6

The index o on J_o distinguishes this Jacobian from that of the non-orthogonal flux coordinates of Section A.3.1. See also Section 16.1.2, Eqs. (16.16)–(16.19).

A.3 Vector expressions in non-orthogonal coordinates

Considering the position vector as a function of curvilinear coordinates x^i,

$$\mathbf{r} = \mathbf{r}(x^1, x^2, x^3) \qquad \Longleftrightarrow \qquad \begin{cases} x = x(x^1, x^2, x^3) \\ y = y(x^1, x^2, x^3) \\ z = (x^1, x^2, x^3) \end{cases}, \qquad (A.66)$$

and exploiting the shorthand notation $\partial_i \equiv \partial/\partial x^i$ for derivatives, the following geometric quantities are generated:

$$\mathbf{a}_i \equiv \partial_i \mathbf{r}, \quad \mathbf{a}^i \equiv \nabla x^i, \quad \mathbf{a}_i \cdot \mathbf{a}^j = \delta_i{}^j \qquad \text{(basis vectors)}, \qquad (A.67)$$

(in general, the basis vectors are not dimensionless!)

$$g_{ij} = \mathbf{a}_i \cdot \mathbf{a}_j, \quad g^{ij} = \mathbf{a}^i \cdot \mathbf{a}^j, \quad g_{ij} g^{jk} = \delta_i{}^k \qquad \text{(metric tensor)}, \qquad (A.68)$$

$$\mathbf{a}_1 \times \mathbf{a}_2 = J \mathbf{a}^3, \qquad \mathbf{a}^1 \times \mathbf{a}^2 = \frac{1}{J} \mathbf{a}_3 \quad \text{(cyclic)}, \qquad (A.69)$$

$$J \equiv \frac{\partial(x, y, z)}{\partial(x^1, x^2, x^3)} \equiv \begin{vmatrix} \partial_1 x & \partial_1 y & \partial_1 z \\ \partial_2 x & \partial_2 y & \partial_2 z \\ \partial_3 x & \partial_3 y & \partial_3 z \end{vmatrix} = \sqrt{\det(g_{ij})} \quad \text{(Jacobian)}, \qquad (A.70)$$

($J > 0$ for right-handed coordinate systems, i.e. excluding inversion)

$$d\ell = \sqrt{g_{ij} dx^i dx^j} \qquad \text{(line element)}, \qquad (A.71)$$

(sum over repeated indices, unless stated otherwise)

$$d\tau = J \, dx^1 dx^2 dx^3 \qquad \text{(volume element)}. \qquad (A.72)$$

Vector representations (Fig A.7):

$$\mathbf{V} = V^i \mathbf{a}_i = V_i \mathbf{a}^i \qquad (V^i - \text{contravariant}, V_i - \text{covariant}), \qquad (A.73)$$

(in general, these components have non-physical dimensions !)

$$V^i = g^{ij} V_j, \qquad V_i = g_{ij} V^j \qquad \text{(raising and lowering indices)}. \qquad (A.74)$$

▷ **Note** For orthogonal coordinates,

$$g_{ij} = (h_i)^2 \delta_{ij}, \quad h_i = |\mathbf{a}_i| = |\mathbf{a}^i|^{-1} \quad \Rightarrow \quad \mathbf{e}_i = (h_i)^{-1} \mathbf{a}_i = h_i \mathbf{a}^i$$

$$\Rightarrow \quad \hat{V}_i \equiv \mathbf{V} \cdot \mathbf{e}_i = (h_i)^{-1} V_i = h_i V^i \qquad \text{(not summing over } i). \qquad (A.75)$$

Hence, the hat on the physical components $\hat{V}_i$ in Section A.2. ◁

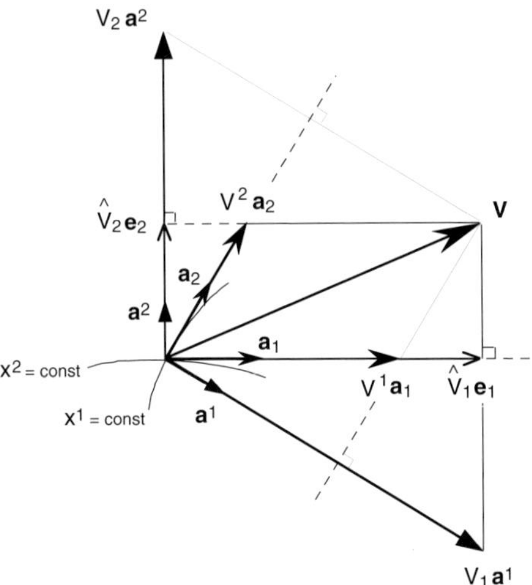

Fig. A.7 Geometrical representation of covariant (V_i), contravariant (V^i) and physical ($\hat{V}_i$) vector components in two dimensions. (Of course, the choice of the unit vector $\mathbf{e}_1 \parallel \mathbf{a}_1$ is arbitrary.)

Products:

$$\mathbf{A} \cdot \mathbf{B} = A_i B^i = g_{ij} A^i B^j = A^i B_i = g^{ij} A_i B_j \qquad \textit{(inner product)}\,, \tag{A.76}$$

$$(\mathbf{A} \times \mathbf{B})_i = J \epsilon_{ijk} A^j B^k \,, \quad (\mathbf{A} \times \mathbf{B})^i = \frac{1}{J} \epsilon^{ijk} A_j B_k \quad \textit{(vector product)}\,, \tag{A.77}$$

with the Levi-Civita symbols

$$\epsilon_{ijk} - \epsilon^{ijk} = \begin{cases} 1 & \text{if } ijk \text{ is even permutation of 123} \\ -1 & \text{if } ijk \text{ is odd permutation of 123} \,, \\ 0 & \text{otherwise} \end{cases} \tag{A.78}$$

$$\epsilon^{ijk} \epsilon_{ilm} = \delta^j_l \delta^k_m - \delta^j_m \delta^k_l \,. \tag{A.79}$$

▷ **Note** (1) The Levi-Civita symbols ϵ_{ijk} and ϵ^{ijk} are tensor densities since it requires the Jacobian to turn them into the tensors $J\epsilon_{ijk}$ and $J^{-1}\epsilon^{ijk}$. Because of the presence of J, the latter are further classified as pseudo-tensors since they change sign upon coordinate inversion ($J < 0$). Accordingly, the vector product is a pseudo-vector.
(2) Cyclic permutation of indices ($123 \to 312$, etc.) is an *even* operation in the 3D case, but in the 4D counterpart (21.90) cyclic permutation ($0123 \to 3012$, etc.) is *odd!*   ◁

Differential operators:

$$\nabla \psi = (\partial_i \psi)\, \mathbf{a}^i \,, \tag{A.80}$$

$$\nabla^2 \psi = \frac{1}{J} \partial_i (J g^{ij} \partial_j \psi) \,, \tag{A.81}$$

$$\nabla \cdot \mathbf{A} = \frac{1}{J} \partial_i (J A^i) , \tag{A.82}$$

$$(\nabla \times \mathbf{A})^i = \frac{1}{J} \epsilon^{ijk} \partial_j A_k . \tag{A.83}$$

Derivatives of the basis vectors,

$$\partial_i \mathbf{a}_j = \Gamma_{ij}^k \, \mathbf{a}_k , \qquad \partial_i \mathbf{a}^k = -\Gamma_{ij}^k \, \mathbf{a}^j , \tag{A.84}$$

involve *the Christoffel symbols*:

$$\Gamma_{ij}^k = \Gamma_{ji}^k \equiv \mathbf{a}^k \cdot \partial_i \mathbf{a}_j = -\mathbf{a}_j \cdot \partial_i \mathbf{a}^k = \tfrac{1}{2} g^{kl} (\partial_i g_{lj} + \partial_j g_{il} - \partial_l g_{ij}) , \tag{A.85}$$

$$\partial_i g_{jk} = g_{kl} \Gamma_{ij}^l + g_{jl} \Gamma_{ik}^l , \qquad \partial_i g^{jk} = -g^{jl} \Gamma_{il}^k - g^{kl} \Gamma_{il}^j , \tag{A.86}$$

$$\partial_i J = J \Gamma_{ij}^j . \tag{A.87}$$

Hence,

$$(\mathbf{A} \cdot \nabla \mathbf{B})^j = A^i \left(\partial_i B^j + \Gamma_{ik}^j B^k \right) . \tag{A.88}$$

Coordinate transformations,

$$x^j = x^j(\bar{x}^1, \bar{x}^2, \bar{x}^3) \quad \Rightarrow \quad dx^j = \alpha_i{}^j d\bar{x}^i , \quad \alpha_i{}^j \equiv \frac{\partial x^j}{\partial \bar{x}^i} , \quad \det(\alpha_i{}^j) \neq 0 , \tag{A.89}$$

involve transformations

$$\bar{\mathbf{a}}_i \equiv \frac{\partial \mathbf{r}}{\partial \bar{x}^i} = \alpha_i{}^j \, \mathbf{a}_j \qquad \textit{(of the basis)} , \tag{A.90}$$

$$V^j = \alpha_i{}^j \, \bar{V}^i \qquad \textit{(of vector components: contragredient)} , \tag{A.91}$$

$$T^{kl} = \alpha_i{}^k \alpha_j{}^l \, \bar{T}^{ij} , \quad \text{etc.} \quad \textit{(of tensor components)} . \tag{A.92}$$

A.3.1 Non-orthogonal flux coordinates $(\Psi, \vartheta, \varphi)$

As in Section A.2.4, we exploit capitals X, Y, Z for the Cartesian coordinates and lower case x and y for the normalized coordinates in the poloidal plane:

$$\begin{cases} X = R \sin \varphi , & R = R_0 + x(\Psi, \vartheta) , \\ Y = R \cos \varphi , \\ Z = y(\Psi, \vartheta) \end{cases} \tag{A.93}$$

The poloidal coordinate dependence follows from inversion of the equilibrium solution:

$$\text{equilibrium} \;\Rightarrow\; \begin{cases} \Psi = \Psi(x, y) \\ \vartheta = \vartheta(x, y) \end{cases}, \qquad \text{inversion} \;\Rightarrow\; \begin{cases} x = x(\Psi, \vartheta) \\ y = y(\Psi, \vartheta) \end{cases}. \tag{A.94}$$

Metric:

$$g_{ij} = \begin{pmatrix} g_{11} & g_{12} & 0 \\ g_{12} & g_{22} & 0 \\ 0 & 0 & g_{33} \end{pmatrix}, \qquad \begin{aligned} g_{11} &= (x_\Psi)^2 + (y_\Psi)^2 \,, \\ g_{12} &= x_\Psi\, x_\vartheta + y_\Psi\, y_\vartheta \,, \\ g_{22} &= (x_\vartheta)^2 + (y_\vartheta)^2 \,, \\ g_{33} &= R^2 = (R_0 + x)^2 \,. \end{aligned} \tag{A.95}$$

Jacobian:

$$J = (\nabla\Psi \times \nabla\vartheta \cdot \nabla\varphi)^{-1} = R\,\sqrt{g_{11}g_{22} - (g_{12})^2} = R\,(x_\Psi\, y_\vartheta - x_\vartheta\, y_\Psi)\,. \tag{A.96}$$

For further specification to *straight field line coordinates* (with Jacobian $\mathcal{J}$) see Section 16.1.2, Eqs. (16.20)–(16.26), and the relations of Section 17.1.3.

Notes on literature

Vector expressions and coordinate systems

– *NLR Plasma Formulary* by Book [58] has been in use for decades by plasma physicists to look up any of the standard formulas of vector analysis, systems of units, plasma parameters, transport coefficients, etc.
– *Flux Coordinates and Magnetic Field Structure* by D'haeseleer, Hitchon, Callen & Shohet [119] is a monograph entirely devoted to magnetic flux geometry, containing derivations of all coordinate systems in use for 2D and 3D geometries, like Hamada coordinates [206] (straight field and current lines) and Boozer coordinates [59].
– *Plasma Physics and Fusion Energy* (Appendix C) by Freidberg [141] has an extensive derivation of Boozer coordinates and an application to guiding center orbit motion.

References

[1] Previous volume: J. P. Goedbloed and S. Poedts, *Principles of Magnetohydrodynamics* (Cambridge, Cambridge University Press, 2004).

[2] Present volume: J. P. Goedbloed, R. Keppens and S. Poedts, *Advanced Magnetohydrodynamics* (Cambridge, Cambridge University Press, 2010).

---

[3] M. Abramowitz and I. A. Stegun, *Handbook of Mathematical Functions with Formulas, Graphs, and Mathematical Tables* (Washington, DC, National Bureau of Standards, 1964).

[4] Y. Z. Agim and J. A. Tataronis, 'General two-dimensional magnetohydrodynamic equilibria with mass flow', *J. Plasma Physics* **34** (1985), 337–360.

[5] A. I. Akhiezer, I. A. Akhiezer, P. V. Polovin, A. G. Sitenko and K. N. Stepanov, *Plasma Electrodynamics, Vol. 1, Linear Theory* (Oxford, Pergamon Press, 1975).

[6] A. I. Akhiezer, G. J. Lubarski and P. V. Polovin, 'The stability of shock waves in magnetohydrodynamics', *Soviet Phys.–JETP* **35** (1959), 507–511.

[7] F. Alladio and F. Crisanti, 'Analysis of MHD equilibria by toroidal multipolar expansions', *Nucl. Fusion* **26** (1986), 1143–1164.

[8] A. M. Anile, *Relativistic Fluids and Magnetofluids* (Cambridge, Cambridge University Press, 1989).

[9] K. Appert, D. Berger, R. Gruber and J. Rappaz, 'A new finite element approach to the normal mode analysis in magnetohydrodynamics', *J. Comp. Physics* **18** (1975), 284–299.

[10] K. Appert, R. Gruber and J. Vaclavik, 'Continuous spectra of a cylindrical magnetohydrodynamic equilibrium', *Phys. Fluids* **17** (1974), 1471–1472.

[11] S. Appl and M. Camenzind, 'Shock conditions for relativistic MHD jets', *Astron. Astrophys.* **206** (1988), 258–268.

[12] T. D. Arber, A. W. Longbottom, C. L. Gerrard and A. M. Milne, 'A staggered grid, Lagrangian–Eulerian remap code for 3-D MHD simulations', *J. Comp. Physics* **171** (2001), 151–181.

[13] M. Ariola and A. Pironti, *Magnetic Control of Tokamak Plasmas* (London, Springer-Verlag, 2008).

[14] V. I. Arnold, *Mathematical Methods of Classical Mechanics* (New York, Springer-Verlag, 1980).

[15] W. Arnoldi, 'The principle of minimized iterations in the solution of the matrix eigenvalue problem', *Quart. Appl. Math.* **9** (1951), 17–29.

[16] L. A. Artsimovich, G. A. Bobrovsky, E. P. Gorbunov, D. P. Ivanov, V. D. Kirillov,

E. I. Kuznetsov, S. V. Mirnov, M. P. Petrov, K. A. Razumova, V. S. Strelkov and D. A. Shcheglov, 'Experiments in tokamak devices', in *Plasma Physics and Controlled Nuclear Fusion Research*, Proc. Third Intern. IAEA Conf., 1–7 August 1968, Novosibirsk, Vol. **I** (Vienna, IAEA, 1969), 157–173. [English translation: *Nucl. Fusion Suppl.* 1969, pp. 17–24.]

[17] L. A. Artsimovich, 'Tokamak devices', *Nucl. Fusion* **12** (1972), 215–252.

[18] S. A. Balbus and J. F. Hawley, 'A powerful local shear instability in weakly magnetized disks. I. Linear analysis', *Astrophys. J.* **376** (1991), 214–222.

[19] S. A. Balbus and J. F. Hawley, 'Instability, turbulence, and enhanced transport in accretion disks', *Rev. Modern Physics* **70** (1998), 1–53.

[20] R. Balescu, *Transport Processes in Plasmas* (Amsterdam, North Holland, 1988).

[21] R. Balescu, *Aspects of Anomalous Transport in Plasmas* (Bristol, Institute of Physics, 2005).

[22] D. Balsara, 'Total variation diminishing scheme for relativistic magnetohydrodynamics', *Astrophys. J. Suppl. Ser.* **132** (2001), 83–101.

[23] R. Barrett, M. Berry, T. Chan, J. Demmel, J. Donato, J. Dongarra, V. Eijkhout, R. Pozo, C. Romine and H. van der Vorst, *Templates for the Solution of Linear Systems: Building Blocks for Iterative Methods* (SIAM, Philadelphia, 1994).

[24] E. M. Barston, 'Eigenvalue problem for Lagrangian systems', *J. Math. Phys.* **8** (1967), 523–532.

[25] H. Bateman, *Higher Transcendental Functions, Volume I*, eds. A. Erdélyi, W. Magnus, F. Oberhettinger and F. G. Tricomi (New York, McGraw-Hill Book Company, 1953).

[26] H. Baty and J. Heyvaerts, 'Electric current concentration and kink instability in line-tied coronal loops', *Astron. Astrophys.* **308** (1996), 335–350.

[27] H. Baty and R. Keppens, 'Kelvin–Helmholtz disruptions in extended magnetized jet flows', *Astron. Astrophys.* **447** (2006), 9–22.

[28] H. Baty, E. R. Priest and T. G. Forbes, 'Effect of nonuniform resistivity in Petschek reconnection', *Phys. Plasmas* **13** (2006), 022312, 1–7.

[29] A. J. C. Beliën, M. A. Botchev, J. P. Goedbloed, B. van der Holst and R. Keppens, 'FINESSE: Axisymmetric MHD equilibria with flow', *J. Comp. Physics* **182** (2002), 91–117.

[30] A. J. C. Beliën, M. A. Poedts and J. P. Goedbloed, 'Two-dimensional equilibrium in coronal magnetostatic flux tubes: an accurate equilibrium solver', *Comput. Physics Commun.* **106** (1997), 21–38.

[31] R. B. Bellan, *Spheromaks* (London, Imperial College Press, 2000).

[32] P. M. Bellan, *Fundamentals of Plasma Physics* (Cambridge, Cambridge University Press, 2006).

[33] C. M. Bender and S. Orszag, *Advanced Mathematical Methods for Scientists and Engineers* (Tokyo, McGraw-Hill International Book Company, 1978).

[34] M. J. Berger and P. Colella, 'Local adaptive mesh refinement for shock hydrodynamics', *J. Comp. Physics* **82** (1989), 64–84.

[35] I. B. Bernstein, E. A. Frieman, M. D. Kruskal and R. M. Kulsrud, 'An energy principle for hydromagnetic stability problems', *Proc. Roy. Soc. (London)* **A244** (1958), 17–40.

[36] I. B. Bernstein, E. A. Frieman, M. D. Kruskal and R. M. Kulsrud, Appendix to Ref. [35], Project Matterhorn Report NYO-7896 (Princeton, 1957).

[37] N. Bessolaz, C. Zanni, J. Ferreira, R. Keppens and J. Bouvier, 'Accretion funnels onto weakly magnetized young stars', *Astron. Astrophys.* **478** (2008), 155–162.

[38] R. Betti and J. P. Freidberg, 'Ellipticity induced Alfvén eigenmodes', *Phys. Fluids*

B3 (1991), 1865–1870.

[39] R. Betti and J. P. Freidberg, 'Stability analysis of resistive wall kink modes in rotating plasmas', *Phys. Rev. Lett.* **74** (1995), 2949–2952.

[40] S. N. Bhattacharyya and A. Bhattacharjee, 'Effect of the flow continuum on magnetohydrodynamic stablity', *Phys. Plasmas* **4** (1997), 3744–3748.

[41] L. Biermann, K. Hain, K. Jörgens and R. Lüst, 'Axialsymmetrische Lösungen der magnetohydrostatischen Gleichung mit Oberflächenströmen', *Z. Naturforsch.* **12a** (1957), 826–832.

[42] J. Birn, J. F. Drake, M. A. Shay, B. N. Rogers, R. E. Denton, M. Hesse, M. Kuznetsova, Z. W. Ma, A. Bhattacharjee, A. Otto and P. L. Pritchett, 'Geospace Environmental Modeling (GEM) Magnetic Reconnection Challenge', *J. Geophys. Res.* **106** (2001), 3715–3719.

[43] J. Birn, K. Galsgaard, M. Hesse, M. Hoshino, J. Huba, G. Lapenta, P. L. Pritchett, K. Schindler, L. Yin, J. Büchner, T. Neukirch and E. R. Priest, 'Forced magnetic reconnection', *Geophys. Res. Lett.* **32** (2005) L06105.

[44] J. Birn and M. Hesse, 'Geospace Environmental Modeling (GEM) Magnetic Reconnection Challenge: resistive tearing, anisotropic pressure and Hall effects', *J. Geophys. Res.* **106** (2001), 3737–3750.

[45] D. Biskamp, *Nonlinear Magnetohydrodynamics* (Cambridge, Cambridge University Press, 1993).

[46] D. Biskamp, *Magnetic Reconnection in Plasmas* (Cambridge, Cambridge University Press, 2000).

[47] D. Biskamp, *Magnetohydrodynamic Turbulence* (Cambridge, Cambridge University Press, 2003).

[48] D. Biskamp and H. Welter, 'Theory of sawtooth disruption', in *Plasma Physics and Controlled Fusion Research 1986*, Vol. **2** (Vienna, IAEA, 1987), 11–16.

[49] R. D. Blandford and D. G. Payne, 'Hydromagnetic flows from accretion discs and the production of radio jets', *Monthly Not. Roy. Astron. Soc.* **199** (1982), 883–903.

[50] R. D. Blandford and K. S. Thorne, *Applications of Classical Physics* (Pasadena, California Institute of Technology, 2004).

[51] J. W. S. Blokland, R. Keppens and J. P. Goedbloed, 'Unstable magnetohydrodynamical continuous spectrum of accretion disks: a new route to magnetohydrodynamical turbulence in accretion disks', *Astron. Astrophys.* **467** (2007), 21–35.

[52] J. W. S. Blokland, B. van der Holst, R. Keppens and J. P. Goedbloed, 'PHOENIX: MHD spectral code for rotating laboratory and gravitating astrophysical plasmas', *J. Comp. Physics* **226** (2007), 509–533.

[53] J. W. S. Blokland, E. van der Swaluw, R. Keppens and J. P. Goedbloed, 'Magneto-rotational overstability in accretion disks', *Astron. Astrophys.* **444** (2005), 337–346.

[54] P. Bodenheimer, G. P. Laughlin, M. Rozyczka and H. W. Yorke, *Numerical Methods in Astrophysics* (New York, Taylor & Francis, 2007).

[55] S. Bogovalov and K. Tsinganos, 'On the magnetic acceleration and collimation of astrophysical outflows', *Monthly Not. Roy. Astron. Soc.* **305** (1999), 211–224.

[56] A. Bondeson, R. Iacono and A. H. Bhattacharjee, 'Local magnetohydrodynamic instabilities of cylindrical plasma with sheared equilibrium flows', *Phys. Fluids* **30** (1987), 2167–2180.

[57] A. Bondeson and D. J. Ward, 'Stabilization of external modes in tokamaks by resistive walls and plasma rotation', *Phys. Rev. Lett.* **72** (1994), 2709–2712.

[58] D. Book, *NLR Plasma Formulary* (Washington, Naval Research Laboratory,

1980).

[59] A. H. Boozer, 'Guiding center drift equations', *Phys. Fluids* **23** (1980), 904–908.

[60] A. H. Boozer, 'Physics of magnetically confined plasmas', *Rev. Modern Physics* **76** (2004), 1071–1141.

[61] A. Borba, K. S. Riedel, W. Kerner, G. T. A. Huysmans, M. Ottaviani and P. J. Schmid, 'The pseudospectrum of the resistive magnetohydrodynamic operator: resolving the resistive Alfvén paradox', *Phys. Plasmas* **1** (1994), 3151–3160.

[62] J. P. Boris and D. L. Book, 'Flux-Corrected Transport – I. SHASTA, a fluid transport algorithm that works', *J. Comp. Physics* **11** (1973), 38–69.

[63] J. P. Boris and D.L. Book, 'Flux-Corrected Transport – III. Minimal-error FCT algorithms', *J. Comp. Physics* **20** (1976), 397–431.

[64] T. J. M. Boyd and J. J. Sanderson, *The Physics of Plasmas* (Cambridge, Cambridge University Press, 2003).

[65] B. J. Braams, 'The interpretation of tokamak magnetic diagnostics', *Plasma Physics & Controlled Fusion* **33** (1991), 715–748.

[66] C. M. Braams and P. E. Stott, *Nuclear Fusion: Half a Century of Magnetic Confinement Fusion Research* (Bristol, Institute of Physics Publishing, 2002).

[67] J. U. Brackbill, 'FLIP MHD – a particle-in-cell method for magnetohydrodynamics', *J. Comp. Physics* **96** (1991), 163–192.

[68] J. U. Brackbill and D. C. Barnes, 'The effect of non-zero $\nabla \cdot \mathbf{B}$ on the numerical solution of the magnetohydrodynamic equations', *J. Comp. Physics* **35** (1980), 426–430.

[69] S. Briguglio, F. Zonca and G. Vlad, 'Hybrid magnetohydrodynamic–particle simulation of linear and nonlinear evolution of Alfvén modes in tokamaks', *Phys. Plasmas* **5** (1998), 3287–3301.

[70] M. Brio and C. C. Wu, 'An upwind differencing scheme for the equations of ideal magnetohydrodynamics', *J. Comp. Physics* **75** (1988), 400–422.

[71] M. N. Bussac and K. Lerbinger, 'Skin currents and internal kinks in tokamaks', *Phys. Rev. Lett.* **121** (1986), 327–341.

[72] M. N. Bussac, R. Pellat, E. Edery and J. L. Soule, 'Internal kink modes in toroidal plasmas with circular cross sections', *Phys. Rev. Lett.* **35** (1975), 1638–1641.

[73] A. C. Calder, B. Fryxell, T. Plewa, R. Rosner, L. J. Dursi, V. G. Weirs, T. Dupont, H. F. Robey, J. O. Kane, B. A. Remington, R. P. Drake, G. Dimonte, M. Zingale, F. X. Timmes, K. Olson, P. Ricker, P. MacNeice and H. M. Tufo, 'On validating an astrophysical simulation code', *Astrophys. J. Suppl. Ser.* **143** (2002), 201–229.

[74] P. S. Cally and T. J. Bogdan, 'Simulation of f- and p-mode interactions with a stratified magnetic field concentration', *Astrophys. J.* **486** (1997), L67–70.

[75] C. Canuto, M. Y. Hussaini, A. Quarteroni and T. A. Zang, *Spectral Methods in Fluid Dynamics* (Berlin, Springer-Verlag, 1988).

[76] P. Cargo and G. Gallice, 'Roe matrices for ideal MHD and systematic construction of Roe matrices for systems of conservation laws', *J. Comp. Physics* **136** (1997), 446–466.

[77] K. M. Case, 'Stability of inviscid plane Couette flow', *Phys. Fluids* **3** (1960), 143–148.

[78] K. M. Case, 'Stability of an idealized atmosphere. I. Discussion of results', *Phys. Fluids* **3** (1960), 149–154.

[79] K. M. Case, 'Taylor instability of an inverted atmosphere', *Phys. Fluids* **3** (1960), 366–368.

[80] F. Casse and R. Keppens, 'Magnetized accretion-ejection structures: 2.5 D MHD simulations of continuous ideal jet launching from resistive accretion disks',

Astrophys. J. **581** (2002), 988–1001.

[81] F. Casse and R. Keppens, 'Radiatively inefficient magnetohydrodynamic accretion-ejection structures', *Astrophys. J.* **601** (2004), 90–103.

[82] M. S. Chance, J. M. Greene, R. C. Grimm and J. L. Johnson, 'Study of the MHD spectrum of an elliptic plasma column', *Nucl. Fusion* **17** (1977), 65–83.

[83] S. Chandrasekhar, 'The stability of non-dissipative Couette flow in hydromagnetics', *Proc. Nat. Acad. Sci. USA* **46** (1960), 253–257.

[84] S. Chandrasekhar, *Hydrodynamic and Hydromagnetic Stability* (Oxford, Clarendon Press, 1961).

[85] J. K. Chao, L. H. Lyu, B. H. Wu, A. J. Lazarus, T. S. Chang and R. P. Lepping, 'Observations of an intermediate shock in interplanetary space', *J. Geophys. Res.* **98** (1993), 17443–17450.

[86] C. Z. Cheng, 'Alpha particle destabilization of the toroidicity-induced Alfvén eigenmodes', *Phys. Fluids* **B3** (1991), 2463–2471.

[87] C. Z. Cheng and M. S. Chance, 'Low-n shear Alfvén spectra in axisymmetric toroidal plasmas', *Phys. Fluids* **29** (1986), 3695–3701.

[88] C. Z. Cheng and M. S. Chance, 'NOVA: a nonvariational code for solving the MHD stability of axisymmetric toroidal plasmas', *J. Comp. Physics* **71** (1987), 124–146.

[89] J. Christensen-Dalsgaard, *Stellar Oscillations* (Aarhus, Lecture Notes, Astronomisk Institut, Aarhus Universitet, 1989).

[90] J. W. Connor, R. J. Hastie and J. B. Taylor, 'Shear, periodicity, and plasma ballooning modes', *Phys. Rev. Lett.* **40** (1978), 396–399.

[91] J. W. Connor, R. J. Hastie and J. B. Taylor, 'High mode number stability of an axisymmetric toroidal plasma', *Proc. Roy. Soc. (London)* **A365** (1979), 1–17.

[92] B. Coppi, 'On the stability of hydromagnetic systems with dissipation', in *Propagation and Instabilities in Plasmas*, ed. W. I. Futterman (Stanford, Stanford University Press, 1963), pp. 70–86.

[93] B. Coppi, 'Topology of ballooning modes', *Phys. Rev. Lett.* **39** (1977), 939–942.

[94] B. Coppi, A. Ferreira, J. W.-K. Mark and J. J. Ramos, 'Ideal-MHD stability of finite-beta plasmas', *Nucl. Fusion* **19** (1979), 715–725.

[95] M. Cotsaftis, 'Formulation Lagrangienne des équations de al magnétohydrodynamique appliquée à létude de la stabilité', *Nucl. Fusion*, 1962 Suppl., Part **2** (1962), 447–450.

[96] R. Courant, K. O. Friedrichs and H. Lewy, 'Uber die Partiellen Differenzengleichungen der Mathematischen Physik', *Math. Ann.* **100** (1928), 32–74. [Trans.: 'On the Partial Difference Equations of Mathematical Physics', *IBM Journal* **11** (1967), 215–234.]

[97] R. Courant and K. O. Friedrichs, *Supersonic Flow and Shock Waves* (New York, Interscience Publishers, 1948).

[98] R. Courant and D. Hilbert, *Methods of Mathematical Physics I* (New York, Interscience, 1953).

[99] C. A. Coverdale, C. Deeney, A. L. Velikovich, R. W. Clark, Y. K. Chong, J. Davis, J. Cittenden, C. L. Ruiz, G. W. Cooper, A. J. Nelson, J. Franklin, P. D. LePell, J. P. Apruzese, J. Levine, J. Banister and N. Qi, 'Neutron production and implosion characteristics of a deuterium gas-puff Z pinch', *Phys. Plasmas* **14** (2007), 022706, 1–7.

[100] A. J. Cunningham, A. Frank, A. C. Quillen and E. G. Blackman, 'Outflow-driven cavities: numerical simulations of intermediaries of protostellar turbulence', *Astrophys. J.* **653** (2006), 416–424.

[101] R. B. Dahlburg, S. K. Antiochos and D. Norton, 'Magnetic flux tube tunneling', *Phys. Rev. E* **56** (1997), 2094–2103.

[102] R. B. Dahlburg and G. Einaudi, 'The compressible plane current-vortex sheet', *Phys. Plasmas* **7** (2000), 1356–1365.

[103] W. Dai and P. R. Woodward, 'An approximate Riemann solver for ideal magnetohydrodynamics', *J. Comp. Physics* **111** (1994), 354–372.

[104] W. Dai and P. R. Woodward, 'Extension of the piecewise parabolic method to multidimensional ideal magnetohydrodynamics', *J. Comp. Physics* **115** (1994), 485–514.

[105] H. J. de Blank and T. J. Schep, 'Theory of the $m = 1$ kink mode in toroidal plasma', *Phys. Plasmas* **B3** (1991), 1136–1151.

[106] P. de Bruyne, M. Velli and A. W. Hood, 'The ideal MHD stability of line-tied coronal loops: a truncated Fourier series approach', *Comput. Physics Commun.* **59** (1990), 55–59.

[107] A. Dedner, F. Kemm, D. Kröner, C.-D. Munz, T. Schnitzer and M. Wesenberg, 'Hyperbolic divergence cleaning for the MHD equations', *J. Comp. Physics* **175** (2002), 645–673.

[108] F. de Hoffmann and E. Teller, 'Magneto-hydrodynamic shocks', *Phys. Rev.* **80** (1950), 692-703.

[109] P. J. Dellar, 'A note on magnetic monopoles and the one-dimensional MHD Riemann problem', *J. Comp. Physics* **172** (2001), 392–398.

[110] L. Del Zanna, N. Bucciantini and P. Londrillo, 'An efficient shock-capturing central-type scheme for multidimensional relativistic flows', *Astron. Astrophys.* **400** (2003), 397–413.

[111] H. De Sterck, 'Hyperbolic theory of the "shallow water" magnetohydrodynamics equations', *Phys. Plasmas* **8** (2001), 3293–3305.

[112] H. De Sterck, B. C. Low and S. Poedts, 'Complex MHD bow shock topology in field-aligned low-β flow around a perfectly conducting cylinder', *Phys. Plasmas* **5** (1998), 4015–4027.

[113] H. De Sterck and S. Poedts, 'Intermediate shocks in three-dimensional magnetohydrodynamic bow-shock flows with multiple interacting shock fronts', *Phys. Rev. Lett.* **84** (2000), 5524–5527.

[114] H. De Sterck and S. Poedts, 'Disintegration and reformation of intermediate-shock segments in three-dimensional MHD bow shock flows', *J. Geophys. Res.* **106** (2001), 30023-30037.

[115] C. R. DeVore, 'Flux-Corrected Transport techniques for multidimensional compressible magnetohydrodynamics', *J. Comp. Physics* **92** (1991), 142–160.

[116] C. R. DeVore and S. K. Antiochos, 'Dynamical formation and stability of helical prominence magnetic fields', *Astrophys. J.* **539** (2000), 954–963.

[117] R. L. Dewar, 'Spectrum of the ballooning Schrödinger equation', *Plasma Physics & Controlled Fusion* **39** (1997), 453–470.

[118] R. L. Dewar and A. H. Glasser, 'Ballooning mode spectrum in general toroidal systems', *Phys. Fluids* **26** (1983), 3038–3052.

[119] W. D. D'haeseleer, W. N. G. Hitchon, J. D. Callen and J. L. Shohet, *Flux Coordinates and Magnetic Field Structure* (Berlin, Springer Verlag, 1991).

[120] D. A. D'Ippolito, J. P. Freidberg, J. P. Goedbloed and J. Rem, 'High-beta tokamaks surrounded by force-free fields', *Phys. Fluids* **21** (1978), 1600–1616.

[121] D. A. D'Ippolito and J. P. Goedbloed, 'Mode coupling in a toroidal sharp-boundary plasma, I Weak-coupling limit', *Plasma Physics* **22** (1980), 1091–1107; 'II Strong-coupling limit', *Plasma Physics* **25** (1983), 537–550.

[122] A. J. H. Donné, A. L. Rogister, R. Koch and H. Soltwisch (eds.), *Proc. Second Carolus Magnus Summer School on Plasma Physics*, Aachen, 1995; *Transactions of Fusion Technology* **29** (1996), 1–432.

[123] B. T. Draine and C. F. McKee, 'Theory of interstellar shocks', *Ann. Rev. Astron. Astrophys.* **31** (1993), 373–432.

[124] P. G. Drazin and W. H. Reid, *Hydrodynamic Stability*, 2nd edition (Cambridge, Cambridge University Press, 2004).

[125] J. Dreher and R. Grauer, 'Racoon: a parallel mesh-adaptive framework for hyperbolic conservation laws', *Par. Comput.* **31** (2005), 913–932.

[126] M. R. Dubal, 'Numerical simulations of special relativistic, magnetic gas flows', *Comput. Physics Commun.* **64** (1991), 221–234.

[127] B. Dubrulle and E. Knobloch, 'On instabilities in magnetized accretion disks', *Astron. Astrophys.* **274** (1993), 667–674.

[128] F. J. Dyson, 'Stability of an idealized atmosphere. II. Zeros of the confluent hypergeometric function', *Phys. Fluids* **3** (1960), 155–157.

[129] E. M. Edlund, M. Porkolab, G. J. Kramer, L. Lin, Y. Lin and S. J. Wukitch, 'Observation of reversed shear Alfvén eigenmodes between sawtooth crashes in the Alcator C-Mod tokamak', *Phys. Rev. Lett.* **102** (2009), 165003-1–4.

[130] C. R. Evans and J. F. Hawley, 'Simulation of magnetohydrodynamic flows: a constrained transport method', *Astrophys. J.* **332** (1988), 659–677.

[131] S. A. E. G. Falle, 'Rarefaction shocks, shock errors and low order of accuracy in ZEUS', *Astrophys. J.* **577** (2002), L123–L126.

[132] S. A. E. G. Falle and S. S. Komissarov, 'On the inadmissibility of non-evolutionary shocks', *J. Plasma Physics* **65** (2001), 29–58.

[133] A. Ferriz-Mas and M. Núñez, *Advances in Nonlinear Dynamos; the Fluid Dynamics of Astrophysics and Geophysics* (New York, Springer, 2003).

[134] G. B. Field, 'Thermal instability', *Astrophys. J.* **142** (1965), 531–567.

[135] C. H. Finan and J. Killeen, 'Solution of the time-dependent, three-dimensional resistive magnetohydrodynamic equations', *Comput. Physics Commun.* **24** (1981), 441–463.

[136] N. J. Fisch, 'Theory of current drive in plasmas', *Rev. Modern Physics* **59** (1987), 175–234.

[137] R. Fitzpatrick, 'A simple model of the resistive wall mode in tokamaks', *Phys. Plasmas* **6** (2002), 3459–3569.

[138] R. Fitzpatrick and E. P. Yu, 'Feedback stabilization of resistive shell modes in a reversed field pinch', *Phys. Plasmas* **6** (1999), 3536–3547.

[139] J. Frank, A. King and D. Raine, *Accretion Power in Astrophysics*, 3rd edition (Cambridge, Cambridge University Press, 2002).

[140] J. P. Freidberg, *Ideal Magnetohydrodynamics* (New York, Plenum Press, 1987).

[141] J. P. Freidberg, *Plasma Physics and Fusion Energy* (Cambridge, Cambridge University Press, 2007)

[142] J. P. Freidberg and J. P. Goedbloed, 'Equilibrium and stability of a diffuse high-beta tokamak', in *Proc. Third Topical Conf. on Pulsed High Beta Plasmas*, 9–12 Sept. 1975, Culham, ed. D. E. Evans (Oxford, Pergamon Press) 117–121 (1976).

[143] J. P. Freidberg and F. A Haas, 'Kink instabilities in a high-β tokamak', *Phys. Fluids* **16** (1973), 1909–1916.

[144] H. Freistühler, 'Some remarks on the structure of intermediate magnetohydrodynamic shocks', *J. Geophys. Res.* **96** (1991), 3825–3827.

[145] K. Fricke, 'Stability of rotating stars; II The influence of toroidal and poloidal

magnetic fields', *Astron. Astrophys.* **1** (1969), 388–398

[146] E. A. Frieman, J. M. Greene, J. L. Johnson and K. E. Weimer, 'Toroidal effects on magnetohydrodynamic modes in tokamaks', *Phys. Fluids* **16** (1973), 1108–1125.

[147] E. Frieman and M. Rotenberg, 'On the hydromagnetic stability of stationary equilibria', *Rev. Modern Physics* **32** (1960), 898–902.

[148] S. Fromang, P. Hennebelle and R. Teyssier, 'A high order Godunov scheme with constrained transport and adaptive mesh refinement for astrophysical magnetohydrodynamics', *Astron. Astrophys.* **457** (2006), 371–384

[149] G. Y. Fu and J. W. Van Dam, 'Excitation of the toroidicity-induced shear Alfvén eigenmode by fusion alpha particles in an ignited tokamak', *Phys. Fluids* **B1** (1989), 1949–1952.

[150] M. Fukugita, 'Cosmic matter distribution: Cosmic baryon budget revisited', in *International Astronomical Union Symposium* **220**, eds. S. D. Ryder, D. J. Pisano, M. A. Walker and K. C. Freeman, 21–25 July, Sydney (San Fransico, 2003), 227–332; arXiv:astro-ph/0312517v1.

[151] K. Fukumura, M. Takahashi and S. Tsuruta, 'Magnetohydrodynamic shocks in nonequitorial plasma flows around a black hole', *Astrophys. J.* **657** (2007), 415–427.

[152] H. P. Furth, J. Killeen and M. N. Rosenbluth, 'Finite resistivity instabilities of a sheet pinch', *Phys. Fluids* **6** (1963), 459–484.

[153] R. M. O. Galvão, J. P. Goedbloed, J. Rem, P. H. Sakanaka, T. J. Schep and M. Venema, 'Global kink and ballooning modes in high-beta systems and stability of toroidal drift modes', *Proc. Ninth Intern. IAEA Conf. on Plasma Physics and Controlled Nuclear Fusion Research*, 1–8 Sept. 1982, Baltimore (IAEA, Vienna) Vol. **III**, 3–16 (1983).

[154] T. A. Gardiner and J. M. Stone, 'An unsplit Godunov method for ideal MHD via constrained transport', *J. Comp. Physics* **205** (2005), 509–539.

[155] P. Germain, 'Shock waves and shock-wave structure in magneto-fluid dynamics', *Rev. Modern Physics* **32** (1960), 951–958.

[156] B. Giacomazzo and L. Rezzolla, 'The exact solution of the Riemann problem in relativistic MHD', *J. Fl. Mech.* **562** (2006), 223–259.

[157] B. Giacomazzo and L. Rezzolla, 'WhiskyMHD: a new numerical code for general relativistic magnetohydrodynamics', *Classical and Quantum Gravity* **24** (2007), S235–S258.

[158] A. H. Glasser, J. M. Greene and J. L. Johnson, 'Resistive instabilities in general toroidal plasma configurations', *Phys. Fluids* **18** (1975), 875–888.

[159] S. K. Godunov, 'Symmetric form of the equations for magnetohydrodynamics', in *Numerical Methods for Mechanics of Continuous Media*, Vol. **1** (1972), 26–31, in Russian; translation by T. Linde (*Report of the Computer Centre of the Siberian branch of the USSR Academy of Sciences*, 1972).

[160] J. P. Goedbloed, 'Stabilization of magnetohydrodynamic instabilities by force-free magnetic fields – I. Plane plasma layer', *Physica* **53** (1971), 412–444; 'II. Linear pinch', *Physica* **53** (1971), 501–534; 'III. Shearless magnetic fields', *Physica* **53** (1971), 535–570; 'IV. Boundary conditions for plasma–plasma interface', *Physica* **100C** (1980), 273–275.

[161] J. P. Goedbloed, 'Spectrum of ideal magnetohydrodynamics of axisymmetric toroidal systems', *Phys. Fluids* **18** (1975), 1258–1268.

[162] J. P. Goedbloed, 'Conformal mapping methods in two-dimensional magnetohydrodynamics', *Comput. Physics Commun.* **24** (1981), 311–321.

[163] J. P. Goedbloed, 'Free-boundary high-beta tokamaks. I. Free-boundary equilibrium', *Phys. Fluids* **25** (1982), 852–868; 'II. Mathematical intermezzo: Hilbert transforms and conformal mapping', *Phys. Fluids* **25** (1982), 2062–2072; 'III. Free-boundary stability', *Phys. Fluids* **25** (1982), 2073–2088.

[164] J. P. Goedbloed, 'Some remarks on computing axisymmetric equilibria', *Comput. Physics Commun.* **31** (1984), 123–135 [Erratum: **41** (1986), 196].

[165] J. P. Goedbloed, 'MHD waves in thermonuclear and solar plasmas', in *Trends in Physics 1991*, Eighth General Conference EPS, Amsterdam, 4–8 Sept. 1990, ed. J. Kaczér, Vol. **III** (Prague, EPS,1991) 827–844

[166] J. P. Goedbloed, 'Once more: the continuous spectrum of ideal magnetohydrodynamics', *Phys. Plasmas* **5** (1998), 3143–3154.

[167] J. P. Goedbloed, 'Expansion functions for two-dimensional incompressible fluid flow in arbitrary domains', *J. Comp. Physics* **160** (2000), 283–297.

[168] J. P. Goedbloed, 'Transonic magnetohydrodynamic flows in laboratory and astrophysical plasmas', Physica Scripta **T98** (2002), 43–47.

[169] J. P. Goedbloed, 'Variational principles for stationary one- and two-fluid equilibria of axisymmetric laboratory and astrophysical plasmas', *Phys. Plasmas* **11** (2004), L81–L84.

[170] J. P. Goedbloed, 'Response to "Comment on 'Variational principles for stationary one- and two-fluid equilibria of axisymmetric laboratory and astrophysical plasmas'" [*Phys. Plasmas* **12**, 064701 (2005)]', *Phys. Plasmas* **12** (2005), 064702–4.

[171] J.P. Goedbloed, 'Alternatives and paradoxes in rotational and gravitational instabilities', in *Proc. International Workshop on Collective Phenomena in Macroscopic Systems*, 4–6 December 2006, Como, ed. G. Bertin, R. Pozzoli, and K.R. Sreenivasan (World Scientific, Singapore, 2007), 127–136.

[172] J. P. Goedbloed, 'Time reversal duality of magnetohydrodynamical shocks', *Phys. Plasmas* **15** (2008), 062101, 1–19.

[173] J. P. Goedbloed, A. J. C. Beliën, B. van der Holst and R. Keppens, 'Unstable continuous spectra of transonic axisymmetric plasmas', *Phys. Plasmas* **11** (2004), 28–54.

[174] J. P. Goedbloed, A. J. C. Beliën, B. van der Holst and R. Keppens, 'No additional flow continua in magnetohydrodynamics', *Phys. Plasmas* **11** (2004), 4332–4340.

[175] J. P. Goedbloed and R. Y. Dagazian, 'Kinks and tearing modes in simple configurations', *Phys. Rev.* **A4** (1971), 1554–1560.

[176] J. P. Goedbloed and D. A. D'Ippolito, 'RF stabilization of external kink modes in the presence of a resistive wall', *Phys. Fluids* **B2** (1990), 2366–2372.

[177] J. P. Goedbloed and H. J. L. Hagebeuk, 'Growth rates of instabilities of a diffuse linear pinch', *Phys. Fluids* **15** (1972), 1090–1101.

[178] J. P. Goedbloed and G. Halberstadt, 'Magnetohydrodynamic waves in coronal flux tubes', *Astron. Astrophys.* **286** (1994), 275–301.

[179] J. P. Goedbloed, G. M. D. Hogeweij, R. Kleibergen, J. Rem, R. M. O. Galvão and P. H. Sakanaka, 'Investigation of high-beta tokamak stability with the program HBT', *Proc. Tenth Intern. IAEA Conf. on Plasma Physics and Controlled Nuclear Fusion Research*, 12–19 Sept. 1984, London (IAEA, Vienna) Vol. **II**, 165–172 (1985).

[180] J. P. Goedbloed, G. T. A. Huysmans, H. A. Holties, W. Kerner and S. Poedts, 'MHD spectroscopy: free boundary modes (ELMs) and external excitation of TAE modes', *Plasma Physics & Controlled Fusion* **35** (1993), B277–292.

[181] J. P. Goedbloed and A. E. Lifschitz, 'Stationary symmetric magnetohydrodynamic

flows', *Phys. Plasmas* **4** (1997), 3544–3564.

[182] J. P. Goedbloed, D. Pfirsch and H. Tasso, 'Instability of a pinch surrounded by a resistive wall', *Nucl. Fusion* **12** (1972), 649–657.

[183] J. P. Goedbloed and P. H. Sakanaka, 'New approach to magnetohydrodynamic stability. I. A practical stability concept', *Phys. Fluids* **17** (1974), 908–918.

[184] H. Goldstein, *Classical Mechanics*, 2nd edition (Reading, Addison Wesley, 1980).

[185] R. J. Goldston and P. H. Rutherford, *Introduction to Plasma Physics* (Bristol, Institute of Physics Publishing, 1995).

[186] T. I. Gombosi, K. G. Powell and D. De Zeeuw, 'Axisymmetric modelling of cometary mass loading on an adaptively refined grid, MHD results', *J. Geophys. Res.* **99** (1994), 525–539.

[187] H. Grad, 'Magnetofluid-dynamic spectrum and low shear stability', *Proc. Natl. Acad. Sci. USA* **70** (1973), 3277–3281.

[188] H. Grad and H. Rubin, 'Hydromagnetic equilibria and force-free fields', in *Proc. Second UN Intern. Conf. on Peaceful Uses of Atomic Energy* **31** (New York, Columbia University Press, 1959), 190–197.

[189] I. S. Gradshsteyn and I. M. Ryzhik, *Table of Integrals, Series, and Products*, Corrected and enlarged edition prepared by A. Jeffrey (New York, Academic Press, 1980)

[190] M. Goossens and S. Poedts, 'Linear resistive MHD computations of resonant absorption of acoustic oscillations in sunspots', *Astrophys. J.* **384** (1982), 348–360.

[191] J. M. Greene, private communication (unpublished, 1974).

[192] J. M. Greene, J. L. Johnson and K. E. Weimer, 'Tokamak equilibrium', *Phys. Fluids* **49** (1971), 671–683.

[193] R. C. Grimm, R. L. Dewar and J. Manickam, 'Ideal MHD stability calculations in axisymmetric toroidal coordinate systems', *J. Comp. Physics* **49** (1983), 94–117.

[194] R. C. Grimm, J. M. Greene and J. L. Johnson, 'Computation of the MHD spectrum in axisymmetric toroidal confinement systems', in *Methods in Computational Physics*, ed. J. Killeen, Vol. **16** (New York, Academic Press, 1976), 253–280.

[195] R. J. Groebner, 'An emerging understanding of H-mode discharges in tokamaks', *Phys. Fluids* **B5** (1993), 2343–2354.

[196] R. Gruber, 'Finite hybrid elements to compute the ideal MHD spectrum of an axisymmetric plasma', *J. Comp. Physics* **26** (1978), 379–389.

[197] R. Gruber and J. Rappaz, *Finite Element Methods in Linear Ideal Magnetohydrodynamics* (New York, Springer, 1985).

[198] R. Gruber, F. Troyon, D. Berger, L. C. Bernard, S. Rousset, R. Schreiber, W. Kerner, W. Schneider and K. V. Roberts, 'ERATO stability code', *Comput. Physics Commun.* **21** (1981), 323–371.

[199] L. Guazzotto, J. P. Freidberg and R. Betti, 'A general formulation of magnetohydrodynamic stability including flow and a resistive wall', *Phys. Plasmas* **15** (072503-1–9), 2008.

[200] K. G. Guderley, *The Theory of Transonic Flow* (Oxford, Pergamon Press, 1962; translated from the German edition of 1957 by J. R. Moszynski).

[201] D. A. Gurnett and A. Bhattacharjee, *Introduction to Plasma Physics* (Cambridge, Cambridge University Press, 2005).

[202] T. Hada, 'Evolutionarity conditions in the dissipative MHD system: stability of intermediate MHD shock waves', *Geophys. Res. Lett.* **21** (1994), 2275–2278.

[203] K. Hain, R. Lüst and A. Schlüter, 'Zur Stabilität eines Plasmas', *Z. Naturforsch.* **12a** (1957), 833–941.

[204] K. Hain and R. Lüst, 'Zur Stabilität zylinder-symmetrischer Plasmakonfigurationen mit Volumenströmen', *Z. Naturforsch.* **13a** (1958), 936–940.

[205] G. Halberstadt and J. P. Goedbloed, 'Alfvén wave heating of coronal loops: photospheric excitation', *Astron. Astrophys.* **301** (1995), 559–576.

[206] S. Hamada, 'Hydromagnetic equilbria and their proper coordinates', *Nucl. Fusion* **2** (1962), 23–37.

[207] E. Hameiri, PhD thesis New York University (Courant Institute of Mathematical Sciences Report MF-45, COO-3077-123, 1976).

[208] E. Hameiri, 'Shear stabilization of the Rayleigh–Taylor modes', *Phys. Fluids* **22** (1979), 89–98.

[209] E. Hameiri, 'Spectral estimates, stability conditions, and the rotating screw-pinch', *J. Math. Phys.* **22** (1981), 2080–2088.

[210] E. Hameiri, 'The equilibrium and stability of rotating plasmas', *Phys. Fluids* **26** (1983), 230–237.

[211] E. Hameiri, 'Dynamically accessible perturbations and magnetohydrodynamic stability', *Phys. Plasmas* **10** (2003), 2643–2648.

[212] E. Hameiri, 'The complete set of Casimir constants of the motion in magnetohydrodynamics', *Phys. Plasmas* **11** (2004), 3423–3431.

[213] E. Hameiri and J. H. Hammer, 'Unstable continuous spectrum in magnetohydrodynamics', *Phys. Fluids* **22** (1979), 1700–1706.

[214] D. S. Harned and W. Kerner, 'Semi-implicit method for three-dimensional compressible magnetohydrodynamics simulation', *J. Comp. Physics* **60** (1985), 62–75.

[215] D. S. Harned and D.D. Schnack, 'Semi-implicit method for long time scale magnetohydrodynamic computations in three dimensions', *J. Comp. Physics* **65** (1986), 57–70.

[216] A. Harten, 'High resolution schemes for hyperbolic conservation laws', *J. Comp. Physics* **49** (1983), 357–393.

[217] A. Harten, P. D. Lax and B. van Leer, 'On upstream differencing and Godunov-type schemes for hyperbolic conservation laws', *SIAM Rev.* **25** (1983), 35–61.

[218] J. F. Hawley and S. A. Balbus, 'The dynamical structure of nonradiative black hole accretion flows', *Astrophys. J.* **573** (2002), 738–748.

[219] J. F. Hawley, C. F. Gammie and S. A. Balbus, 'Local three-dimensional simulations of accretion disks', *Astrophys. J.* **440** (1995), 742–763.

[220] R. D. Hazeltine and W. A. Newcomb, 'Inversion of the ballooning transformation', *Phys. Fluids* **B2** (1990), 7–10.

[221] R. D. Hazeltine and J. D. Meiss, *Plasma Confinement* (Redwood City, Addison Wesley Publishing Company, 1992).

[222] T. A. K. Hellsten and G. O. Spies, 'Continuous magnetohydrodynamic spectrum in axially symmetric rotating plasmas', *Phys. Fluids* **22** (1979), 743–746.

[223] (a) P. Henrici, 'Fast Fourier methods in computational complex analysis', *SIAM Rev.* **21** (1979), 481–527.
(b) P. Henrici, *Applied and Computational Complex Analysis*, Volume 3: *Discrete Fourier Analysis, Cauchy Integrals, Construction of Conformal Maps, Univalent Functions* (New York, Wiley, 1993).

[224] J. Heyvaerts and C. A. Norman, 'Collimation of MHD outflows', in *Solar and Astrophysical Magnetohydrodynamic Flows*, ed. K. C. Tsinganos (Dordrecht, Kluwer Academic Publishers, 1996), 459–474.

[225] J. Heyvaerts and E. R. Priest, 'Coronal heating by phase-mixed shear Alfvén waves', *Astron. Astrophys.* **117** (1983), 220–234.

[226] C. Hirsch, *Numerical Computation of Internal and External Flow, Vol.1: Fundamentals of Numerical Discretization* (Chichester, John Wiley & Sons, 1988).

[227] D. D. Holm, J. E. Marsden, T. Ratiu and A. Weinstein, 'Nonlinear stability of fluid and plasma equilibria', *Physics Reports* **123** (1985), 1–116.

[228] H. A. Holties, G. T. A. Huysmans, W. Kerner, J. P. Goedbloed, F. X. Söldner and V. V. Parail, 'MHD stability of advanced tokamak scenarios', *Proc. 21st Eur. Conf. on Controlled Fusion and Plasma Physics*, Vol. **I** (Montpellier, EPS, 1994), 234–237.

[229] H. A. Holties, A. Fasoli, J. P. Goedbloed, G. T. A. Huysmans and W. Kerner, 'Determination of local tokamak parameters by magnetohydrodynamic spectroscopy', *Phys. Plasmas* **4** (1997), 709–719.

[230] H. A. Holties, G. T. A. Huysmans, J. P. Goedbloed, W. Kerner, V. V. Parail and F. X. Söldner, 'Stability of infernal and ballooning modes in advanced tokamak scenarios', *Nucl. Fusion* **36** (1996), 973–986.

[231] K. I. Hopcraft, 'Magnetohydrodynamcs', in *Plasma Physics: an Introductory Course*, ed. R. Dendy (Cambridge, Cambridge University Press, 1993), pp. 77–101.

[232] T. Y. Hou and P. G. Le Floch, 'Why nonconservative schemes converge to wrong solutions: error analysis', *Math. of Computation* **62** (1994) 497–530.

[233] L. N. Howard, 'Note on a paper by John W. Miles', *J. Fluid Mech.* **10** (1961) 509–512.

[234] D. Hutsemékers, R. Cabanac, H. Lamy and D. Sluse, 'Mapping extreme-scale alignments of quasar polarization vectors', *Astron. Astrophys.* **441** (2005), 915–930; 'Large-scale alignments of quasar polarization vectors: Observational evidence and possible implications', in *Conference on Outstanding Questions for the Standard Cosmological Model*, March 26–29, 2007 (Imperial College, London, 2007).

[235] G. T. A. Huysmans, 'External kink (peeling) modes in x-point geometry', *Plasma Physics & Controlled Fusion* **47** (2005), 2107–2121.

[236] G. T. A. Huysmans, 'ELMs: MHD instabilities at the transport barrier', *Plasma Physics & Controlled Fusion* **47** (2005), B165–B178.

[237] G. T. A. Huysmans and O. Czarny, 'MHD stability in X-point geometry: simulation of ELMs', *Nucl. Fusion* **47** (2007), 659–666.

[238] G. T. A. Huysmans, J. P. Goedbloed and W. Kerner, 'Isoparametric bicubic Hermite elements for solution of the Grad–Shafranov equation', in *Proc. Second International Conference on Computational Physics*, Amsterdam, 10–14 September 1990, ed. A. Tenner (Singapore, World Scientific, 1991), 371–376.

[239] G. T. A. Huysmans, J. P. Goedbloed and W. Kerner, 'Free boundary resistive modes in tokamaks', *Phys. Fluids* **B5** (1993), 1545–1558.

[240] G. T. A. Huysmans, W. Kerner, D. Borba, H. A. Holties and J. P. Goedbloed, 'Modeling the excitation of global Alfvén modes by an external antenna in the Joint European Tokamak', *Phys. Plasmas* **2** (1995), 1605–1613.

[241] G. T. A. Huysmans, W. Kerner, J. P. Goedbloed, D. Borba and F. Porcelli, 'MHD spectroscopy: modelling the excitation of TAE modes by an external antenna', *Proc. 20th Eur. Conf. on Controlled Fusion and Plasma Physics*, Vol. **I** (Lisbon, EPS, 1993), 187–190.

[242] I. V. Igumenshchev, R. Narayan and M. A. Abramovicz, 'Three-dimensional

MHD simulations of radiatively inefficient accretion flows', *Astrophys. J.* **592** (2003), 1042–1059.

[243] E. L. Ince, *Ordinary Differential Equations* (New York, Dover Publications 1956).

[244] R. C. Ireland, R. A. M. Van der Linden, A. W. Hood and M. Goossens, 'The thermal continuum in coronal loops: the influence of finite resistivity on the continuous spectrum', *Solar Phys.* **142** (1992), 265–289.

[245] S-I. Itoh, K. Itoh, M. Sasaki, A. Fujisawa, T. Ido and Y. Nagashima, 'Geodesic acoustic mode spectroscopy', *Plasma Physics & Controlled Fusion* **11** (2007), L7–L10.

[246] R. Izzo, D. A. Monticello, W. Park, J. Manickam, H. R. Strauss, R. Grimm and K. McGuire, 'Effects of toroidicity on resistive tearing modes', *Phys. Fluids* **26** (1983), 2240–2246.

[247] J. D. Jackson, *Classical Electrodynamics*, 3rd edition (New York, John Wiley & Sons, 1999).

[248] P. Janhunen, 'A positive conservative method for magnetohydrodynamics based on HLL and Roe methods', *J. Comp. Physics* **160** (2000), 649–661.

[249] A. Jeffrey, *Magnetohydrodynamics* (Edinburgh, Oliver & Boyd, 1966).

[250] A. Jeffrey and T. Taniuti, *Non-linear wave propagation. With applications to physics and magnetohydrodynamics* (New York, Academic Press, 1964).

[251] J. S. Kaastra, T. Tamura, J. R. Peterson, J. A. M. Bleeker, C. Ferrigno, S. M. Kahn, F. B. S. Paerels, R. Piffaretti, G. Branduardi-Raymont and J. Böhringer, 'Spatially resolved X-ray spectroscopy of cooling clusters of galaxies', *Astron. Astrophys.* **413** (2004), 415–439.

[252] B. B. Kadomtsev, 'Hydromagnetic stability of a plasma', in *Reviews of Plasma Physics, Vol. 2*, ed. M. A. Leontovich (New York, Consultants Bureau, 1966) pp. 153–199.

[253] C. F. Kennel, R. D. Blandford and P. Coppi, 'MHD intermediate shock discontinuities. Part 1. Rankine–Hugniot conditions', *J. Plasma Physics* **42** (1989), 299–319.

[254] C. F. Kennel, R. D. Blandford and C. C. Wu, 'Structure and evolution of small-amplitude intermediate shock waves', *Phys. Fluids* **B2** (1990), 253–269.

[255] R. Keppens, F. Casse and J. P. Goedbloed, 'Waves and instabilities in accretion disks: magnetohydrodynamic spectroscopic analysis', *Astrophys. J.* **569** (2002), L121–L126.

[256] R. Keppens and J. P. Goedbloed, 'Numerical simulations of stellar winds: polytropic models', *Astron. Astrophys.* **343** (1999), 251–260.

[257] R. Keppens and J. P. Goedbloed, 'Stellar winds, dead zones, and coronal mass ejections', *Astrophys. J.* **530** (2000), 1036–1048.

[258] R. Keppens, K. B. MacGregor and P. Charbonneau, 'On the evolution of rotational velocity distributions for solar-type stars', *Astron. Astrophys.* **294** (1995), 469–487.

[259] R. Keppens, Z. Meliani, B. van der Holst and F. Casse, 'Extragalactic jets with helical magnetic fields: relativistic MHD simulations', *Astron. Astrophys.* **486** (2008), 663–678.

[260] R. Keppens and Z. Meliani, 'Linear wave propagation in relativistic magnetohydrodynamics', *Phys. Plasmas* **15** (2008), 102103-1–11.

[261] R. Keppens, M. Nool, G. Tóth and J. P. Goedbloed, 'Adaptive Mesh Refinement for conservative systems: multi-dimensional efficiency evaluation', *Comput. Physics Commun.* **153** (2003), 317–339.

[262] R. Keppens, S. Poedts, P. M. Meijer and J. P. Goedbloed, 'A data parallel

pseudo-spectral semi-implicit MHD code', in *Lecture Notes in Computer Sciences*, eds. B. Hertzberger and P. Sloot, Vol. **1225** (1997), 190–199.

[263] R. Keppens, S. Poedts and J. P. Goedbloed, 'Data parallel simulations of the magnetohydrodynamics of plasma loops', in *Lecture Notes in Computer Sciences*, eds. P. Sloot, M. Bubak and B. Hertzberger, Vol. **1401** (1998), 233–241.

[264] R. Keppens and G. Tóth, 'Simulating magnetized plasma with the Versatile Advection Code', in *Lecture Notes in Computer Sciences*, eds. J. M. L. M. Palma, J. Dongarra and V. Hernandez, Vol. **1573** (1999), 680–690.

[265] R. Keppens and G. Tóth, 'Non-linear dynamics of Kelvin–Helmholtz unstable magnetized jets: three-dimensional effects', *Phys. Plasmas* **6** (1999), 1461–1469.

[266] R. Keppens, G. Tóth, M. A. Botchev and A. van der Ploeg, 'Implicit and semi-implicit schemes: algorithms', *Intern. Journ. for Numer. Meth. in Fluids*, **30** (1999), 335–352.

[267] R. Keppens, G. Tóth, R. H. J. Westermann and J. P. Goedbloed, 'Growth and saturation of the Kelvin–Helmholtz instability with parallel and antiparallel magnetic fields', *J. Plasma Physics* **61** (1999), 1–19.

[268] W. Kerner, 'Large-scale complex eigenvalue problems', *J. Comp. Physics* **85** (1989), 1–85.

[269] W. Kerner, 'Algorithms and software for linear and nonlinear MHD simulations', *Comput. Physics Commun.* **12** (1990), 135–175.

[270] W. Kerner, 'Equilibrium and stability of tokamaks', *Int. J. of Numerical Methods in Fluids* **11** (1990), 791–809.

[271] W. Kerner, D. Borba, G. T. A. Huysmans, F. Porcelli, S. Poedts, J. P. Goedbloed and R. Betti, 'Stability of global Alfvén waves (TAE, EAE) in JET tritium discharges', Proc. 20th Eur. Conf. on *Controlled Fusion and Plasma Physics*, Vol. **I** (Lisbon, EPS, 1993), 183–186.

[272] W. Kerner, J. P. Goedbloed, G. T. A. Huysmans, S. Poedts and E. Schwarz, 'CASTOR: Normal-mode analysis of resistive MHD plasmas', *J. Comp. Physics* **142** (1998), 271-303.

[273] W. Kerner, A. Jakoby and K. Lerbinger, 'Finite element semi discretization of linearized compressible resistive MHD', *J. Comp. Physics* **66** (1986), 332–355.

[274] W. Kerner, K. Lerbinger, R. Gruber and T. Tsunematsu, 'Normal mode analysis for resistive cylindrical plasmas', *Comput. Physics Commun.* **36** (1985), 225–240.

[275] C. E. Kieras and J. A. Tataronis, 'The shear Alfvén continuous spectrum of axisymmetric toroidal equilibria in the large aspect ratio limit', *J. Plasma Physics* **28** (1982), 395–414.

[276] W-T. Kim and E. C. Ostriker, 'Magnetohydrodynamic instabilities in shearing, rotating, stratified winds and disks', *Astrophys. J.* **540** (2000), 372–403.

[277] J. Kim, D. Ryu, T. W. Jones and S. S. Hong, 'A multi-dimensional code for isothermal magnetohydrodynamic flows in astrophysics', *Astrophys. J.* **514** (1999), 506–519.

[278] J. G. Kirk, 'Particle acceleration in relativistic current sheets', *Phys. Rev. Lett.* **92** (2004), 181101, 1–4.

[279] J. G. Kirk and P. Duffy, 'Topical Review. Particle acceleration and relativistic shocks', *J. Phys. G* **25** (1999) R163–R194.

[280] R. Kleibergen and J. P. Goedbloed, 'Alfvén wave spectrum of an analytic high-beta tokamak equilibrium', *Plasma Physics & Controlled Fusion* **30** (1988), 1961–1987.

[281] A. C. Kolb, 'Magnetic compression of plasmas', *Rev. Modern Physics* **95** (1960) 748–757.

[282] A. V. Koldoba, O. A. Kuznetsov and G. V. Ustyugova, 'An approximate Riemann solver for relativistic magnetohydrodynamics', *Monthly Not. Roy. Astron. Soc.* **333** (2002), 932–942.

[283] S. S. Komissarov, 'A Godunov-type scheme for relativistic magnetohydrodynamics', *Monthly Not. Roy. Astron. Soc.* **303** (1999), 343–366.

[284] S. S. Komissarov, 'Numerical simulations of relativistic magnetized jets', *Monthly Not. Roy. Astron. Soc.* **308** (1999), 1069–1076.

[285] S. S. Komissarov and Y. E. Lyubarsky, 'The origin of peculiar jet-torus structure in the Crab nebula', *Monthly Not. Roy. Astron. Soc.* **344** (2003), L93–L96.

[286] D. Kössl, E. Müller and W. Hillebrandt, 'Numerical simulations of axially symmetric magnetized jets. I. The influence of equipartition magnetic fields', *Astron. Astrophys.* **229** (1990), 378–396.

[287] M. D. Kruskal and R. M. Kulsrud, 'Equilibrium of a magnetically confined plasma in a toroid', *Phys. Fluids* **1** (1958), 265–274.

[288] A. G. Kulikovskij and G. A. Lyubimov, *Magnetohydrodynamics* (Reading, Addison Wesley, 1965).

[289] A. K. Kulkarni and M. M. Romanova, 'Accretion to magnetized stars through the Rayleigh–Taylor instability: global 3D simulations', *Monthly Not. Roy. Astron. Soc.* **386** (2008), 673–687.

[290] K. Kurihara, 'Improvement of tokamak plasma shape identification with a Legendre–Fourier expansion of the vacuum poloidal flux function', *Fusion Technology* **22**, 334–349 (1992).

[291] H. La Belle-Hamer, A. Otto and L. C. Lee, 'Magnetic reconnection in the presence of sheared plasma flow: intermediate shock formation', *Phys. Plasmas* **1** (1994), 706–713.

[292] K. Land and J. Magueijo, 'Examination of evidence for a preferred axis in the cosmic radiation anisotropy', *Phys. Rev. Lett.* **95** (2005), 071301, 1–4.

[293] L. D. Landau, 'On the vibrations of the electronic plasma', *J. Phys. USSR* **10** (1946), 25. [Trans.: *JETP* **16** (1946), 574; or *Collected Papers of L.D. Landau*, ed. D. ter Haar (Oxford, Pergamom Press, 1965) pp. 445–460.]

[294] L. D. Landau and E. M. Lifshitz, *Course of Theoretical Physics, Vol. 6: Fluid Mechanics*, 2nd edition (Oxford, Pergamon Press, 1987).

[295] L. D. Landau and E. M. Lifshitz, *Course of Theoretical Physics, Vol. 8: Electrodynamics of Continuous Media*, 2nd edition (Oxford, Pergamon Press, 1984).

[296] G. Lapenta, 'Self-feeding turbulent magnetic reconnection on macroscopic scales', *Phys. Rev. Lett.* **100** (2008), 235001, 1–4.

[297] S. Larsson and V. Thomée, *Partial Differential Equations with Numerical Methods* (Berlin, Springer-Verlag, 2003).

[298] G. Laval, C. Mercier and R. M. Pellat, 'Necessity of the energy principle for magnetostatic stability', *Nucl. Fusion* **5** (1965), 156–158.

[299] G. Laval, R. M. Pellat and J. S. Soule, 'Hydromagnetic stability of a current-carrying pinch with noncircular cross section', *Phys. Fluids* **17** (1974), 835–845.

[300] T. Leismann, L. Antón, M. A. Aloy, E. Müller, J. M. Martí, J. A. Miralles and J. M. Ibánez, 'Relativistic MHD simulations of extragalactic jets', *Astron. Astrophys.* **436** (2005), 503–526.

[301] K. Lerbinger and J. F. Luciani, 'A new semi-implicit method for MHD computations', *J. Comp. Physics* **97** (1991), 444–459.

[302] R. J. LeVeque, *Numerical Methods for Conservation Laws* (Berlin, Birkhäuser

Verlag, 1990).

[303] R. J. LeVeque, *Finite Volume Methods for Hyperbolic Problems* (Cambridge, Cambridge University Press, 2002).

[304] R. J. LeVeque, D. Mihalas, E. A. Dorfi and E. Müller, *Computational Methods for Astrophysical Fluid Flow* (Berlin, Springer Verlag, 1998).

[305] S. Li, 'An HLLC Riemann solver for magneto-hydrodynamics', *J. Comp. Physics* **203** (2005), 344–357.

[306] M. A. Liberman and A. L. Velikovich, *Physics of Shock Waves in Gases and Plasmas* (Berlin, Springer-Verlag, 1986).

[307] A. Lichnerowicz, *Relativistic Hydrodynamics and Magnetohydrodynamics* (New York, Benjamin, 1967).

[308] A. E. Lifschitz, *Magnetohydrodynamics and Spectral Theory* (Dordrecht, Kluwer Academic Publishers, 1989).

[309] A. Lifschitz and J. P. Goedbloed, 'Transonic magnetohydrodynamic flows', *J. Plasma Physics* **58** (1997), 61–99.

[310] C. C. Lin, *The Theory of Hydrodynamic Stability* (Cambridge, Cambridge University Press, 1955).

[311] T. J. Linde, 'A practical, general-purpose, two-state HLL Riemann solver for hyperbolic conservation laws', *Int. J. Numer. Meth. Fluids* **40** (2002), 391–402.

[312] R. Lionello, M. Velli, G. Einaudi and Z. Mikić, 'Nonlinear magnetohydrodynamic evolution of line-tied coronal loops', *Astrophys. J.* **494** (1998), 840–850.

[313] D. H. Liu and A. Bondeson, 'Improved poloidal convergence of the MARS code for MHD stability analysis', *Comput. Physics Commun.* **116** (1999), 55–64.

[314] X. Llobet, K. Appert, A. Bondeson and J. Vaclavik, 'On spectral pollution', *Comput. Physics Commun.* **59** (1990), 199–216.

[315] P. Londrillo and L. Del Zanna, 'High-order upwind schemes for multidimensional magnetohydrodynamics', *Astrophys. J.* **530** (2000), 508–524.

[316] M. J. Longo, 'Does the Universe have a handedness?', arXiv:astro-ph/0703325v3 (2007); 'Is the cosmic "axis of evil" due to a large-scale magnetic field?', arXiv:astro-ph/0703694v2 (2007); 'Evidence for a prefered handedness of spiral galaxies', arXiv:astro-ph/0904.2529v1 (2009), submitted to *Astrophys. J. Lettters*.

[317] R. V. E. Lovelace, C. Mehanian, C. M. Mobarry and M. E. Sulkanen, 'Theory of axisymmetric magnetohydrodynamic flows: disks', *Astrophys. J. Suppl. Ser.* **62** (1986), 1–37.

[318] B. C. Low and M. Zhang, 'Magnetostatic structures of the solar corona. III. Normal and inverse quiescent prominences', *Astrophys. J.* **609** (2004), 1098-1111.

[319] R. Lüst and A. Schlüter, 'Axialsymmetrische magnetohydrodynamische Gleichgewichtskonfigurationen', *Z. Naturforsch.* **12a** (1957), 850–854.

[320] H. Lütjens, A. Bondeson and A. Roy, 'Axisymmetric MHD equilibrium solver with bicubic Hermite elements', *Comput. Physics Commun.* **69** (1992), 287–298.

[321] H. Lütjens, A. Bondeson and O. Sauter, 'The CHEASE code for toroidal MHD equilibria', *Comput. Physics Commun.* **97** (1996), 219–260.

[322] M. G. Macaraeg, C. L. Streett and M. Y. Hussaini, 'A spectral collocation solution to the compressible stability eigenvalue problem', *NASA Technical Paper No. 2858* (NASA, Washington DC, 1988).

[323] A. Majorana and A. M. Anile, 'Magnetoacoustic shock waves in a relativistic gas', *Phys. Fluids* **30** (1987), 3045–3049.

[324] W. B. Manchester IV, T. I. Gombosi, I. Roussev, A. Ridley, D. L. De Zeeuw, I. V. Sokolov and K. G. Powell, 'Modeling a space weather event from the Sun to the

Earth: CME generation and interplanetary propagation', *J. Geophys. Res.* **109** (2004), A02107, 1–15.

[325] J. Manickam, N. Pomphrey and A. M. M. Todd, 'Ideal MHD stability properties of pressure driven modes in low shear tokamaks', *Nucl. Fusion* **27** (1987), 1461–1472.

[326] J. M. Martí and E. Müller, 'Numerical hydrodynamics in special relativity', *Living Rev. Relativity* **6** (2003) 7, www.livingreviews.org/lrr-2003-7.

[327] W. G. Mathews, 'The hydromagnetic free expansion of a relativistic gas', *Astrophys. J.* **165** (1971), 147–164.

[328] K. G. McClements and A. Thyagaraja, 'Azimuthally symmetric magnetohydrodynamic and two-fluid equilibria with arbitrary flows', *Monthly Not. Roy. Astron. Soc.* **323** (2001), 733–742.

[329] B. F. McMillan, R. L. Dewar and R. Storer, 'A comparison of incompressible limits for resistive plasmas', *Plasma Physics & Controlled Fusion* **46** (2004), 1027–1038.

[330] Z. Meliani, R. Keppens and B. Giacomazzo, 'Faranoff-Riley type I jet deceleration at density discontinuities', *Astron. Astrophys.* **491** (2008), 321–337.

[331] C. Mercier, 'Un critére necessaire de stabilité hydromagnétique pour un plasma en symmetrie de revolution', *Nucl. Fusion* **1** (1960), 47–53.

[332] A. Mignone and G. Bodo, 'An HLLC Riemann solver for relativistic flows. II. Magnetohydrodynamics', *Monthly Not. Roy. Astron. Soc.* **368** (2006), 1040–1054.

[333] A. Mignone, G. Bodo, S. Massaglia, T. Matsakos, O. Tesileanu, C. Zanni and A. Ferrari, 'PLUTO: a numerical code for computational astrophysics', *Astrophys. J. Suppl. Ser.* **170** (2007), 228–242.

[334] K. Miyamoto, *Plasma Physics for Nuclear Fusion*, revised edition (Cambridge, Massachusetts Institute of Technology Press, 1989).

[335] (a) K. Miyamoto, *Fundamentals of Plasma Physics and Controlled Fusion* NIFS-PROC-48 (Toki, National Institute for Fusion Science, 2001);
(b) K. Miyamoto, *Plasma Physics and Controlled Fusion* (Berlin, Springer, 2005).

[336] T. Miyoshi and K. Kusano, 'A multi-state HLL approximate Riemann solver for ideal magnetohydrodynamics', *J. Comp. Physics* **208** (2005), 315–344.

[337] H. K. Moffatt, *Magnetic Field Generation in Electrically Conducting Fluids* (Cambridge, Cambridge University Press, 1978).

[338] A. I. Morozov and L .S. Soloviev, 'The structure of magnetic fields', in *Reviews of Plasma Physics, Vol. 2*, ed. M. A. Leontovich (New York, Consultants Bureau, 1966) pp. 1–101.

[339] A. I. Morozov and L .S. Soloviev, 'Steady-state plasma flow in a magnetic field', in *Reviews of Plasma Physics, Vol. 8*, ed. M. A. Leontovich (New York, Consultants Bureau, 1980; original Russian edition, 1974) pp. 1–103.

[340] P. J. Morrison and J. M. Greene, 'Noncanonical Hamiltonian density formulation of hydrodynamics and ideal magnetohydrodynamics', *Phys. Rev. Lett.* **45** (1980), 790–794 [Erratum: **48** (1982), 569].

[341] P. M. Morse and H. Feshbach, *Methods of Theoretical Physics, Part II* (New York, McGraw-Hill, 1953).

[342] V. S. Mukhovatov and V. D. Shafranov, 'Plasma equilibrium in a tokamak', *Nucl. Fusion* **11** (1971), 605–633.

[343] K. Murawski, *Analytical and Numerical Methods for Wave Propagation in Fluid Media* (Singapore, World Scientific, 2002).

[344] R. S. Myong, 'Analytical results on MHD intermediate shocks', *Geophys. Res.*

Lett. **24** (1997), 2929–2932.

[345]   R. S. Myong and P. L. Roe, 'Shock waves and rarefaction waves in magnetohydrodynamics; 1. A model system ', *J. Plasma Physics* **58** (1997), 485–519.

[346]   R. S. Myong and P. L. Roe, 'Shock waves and rarefaction waves in magnetohydrodynamics; 2. The MHD system', *J. Plasma Physics* **58** (1997), 521–552.

[347]   W. A. Newcomb, 'Hydromagnetic stability of a diffuse linear pinch', *Ann. Phys. (New York)* **10** (1960), 232–267.

[348]   W. A. Newcomb, 'Convective instability induced by gravity in a plasma with a frozen-in magnetic field', *Phys. Fluids* **4** (1961), 391–396.

[349]   W. A. Newcomb, 'Lagrangian and Hamiltonian methods in magnetohydrodynamics', *Nucl. Fusion*, 1962 Suppl., Part **2** (1962), 451–463.

[350]   W. A. Newcomb, *Notes on Magnetohydrodynamics* (unpublished, Lawrence Livermore Laboratory, 1972).

[351]   R. J. Nijboer and J. P. Goedbloed, 'Mode coupling in two-dimensional magnetohydrodynamic flows', *J. Plasma Physics* **61** (1999), 241–262.

[352]   R. J. Nijboer, A. E. Lifschitz and J. P. Goedbloed, 'Spectrum and stability of rigidly rotating compressible plasma', *J. Plasma Physics* **58** (1997), 101–121.

[353]   R. J. Nijboer, B. van der Holst, S. Poedts and J. P. Goedbloed, 'Calculating magnetohydrodynamic flow spectra', *Comput. Physics Commun.* **106** (1997), 39–52.

[354]   M. Nool and A. van der Ploeg, 'A parallel Jacobi–Davidson-type method for solving large generalized eigenvalue problems in magnetohydrodynamics', *SIAM J. Sci. Comput.* **22** (2000), 95–112.

[355]   L. Ofman and J. M. Davila, 'Nonlinear resonant absorption of Alfvén waves in three dimensions, scaling laws, and coronal heating', *J. Geophys. Res.* **100** (1995), 23427–23441.

[356]   D. P. O'Leary, *Scientific Computing with Case Studies* (Philadelphia, SIAM Press, 2009).

[357]   S. Ortolani and D. D, Schnack, *Magnetohydrodynamics of Plasma Relaxation* (Singapore, World Scientific, 1993).

[358]   Y. P. Pao, 'Instabilities in toroidal plasmas', *Phys. Fluids* **11** (1974), 1192–1196.

[359]   Y. P. Pao, 'The continuous MHD spectrum in toroidal geometries', *Nucl. Fusion* **15** (1975), 631–635.

[360]   Y. P. Pao and W. Kerner, 'Analytic theory of stable resistive magnetohydrodynamics modes', *Phys. Fluids* **15** (1985), 287–293.

[361]   J. C. B. Papaloizou and J. E. Pringle, 'The dynamical stability of differentially rotating discs with constant specific angular momentum', *Monthly Not. Roy. Astron. Soc.* **208** (1984), 721–750.

[362]   E. N. Parker, 'Instability of thermal fields', *Astrophys. J.* **117** (1953), 431–436.

[363]   E. N. Parker, 'Sweet's mechanism for merging magnetic fields in conducting fluids', *J. Geophys. Res.* **62** (1957), 509–520.

[364]   E. N. Parker, 'Dynamics of the interplanetary gas and magnetic fields', *Astrophys. J.* **128** (1958), 664–676.

[365]   E. N. Parker, 'The dynamical state of the interstellar gas and field', *Astrophys. J.* **145** (1966), 811–833.

[366]   B. N. Parlett, *The Symmetric Eigenvalue Problem* (Berlin, Springer-Verlag, 1971).

[367]   F. Pegoraro and T. J. Schep, 'Low-frequency modes with high toroidal mode numbers: A general formulation', *Phys. Fluids* **24** (1981), 478–497.

[368] J. R. Peterson and A. C. Fabian, 'X-ray spectroscopy of cooling clusters', *Phys. Rep.* **427** (2006), 1–39.

[369] G. D. J. Petrie. J. W. S. Blokland and R. Keppens, 'Magnetohydrostatic solar prominences in near-potential coronal magnetic field', *Astrophys. J.* **865** (2007), 830–845.

[370] H. E. Petschek, 'Magnetic field annihilation', in *Physics of Solar Flares*, ed. W. N. Hess (Washington, DC, NASA SP-50) (1964), 425–439.

[371] D. Pfirsch and H. Tasso, 'A theorem on MHD instability of plasmas with resistive walls', *Nucl. Fusion* **11** (1971), 259–260.

[372] G. W. Pneuman and R. A. Kopp, 'Gas–magnetic field interactions in the solar corona', *Solar Phys.* **18** (1971), 258–270.

[373] S. Poedts, 'MHD instabilities: numerical methods and results', *Transaction of Fusion Technology* **29** (1996), 133–142.

[374] S. Poedts and G. C. Boynton, 'Nonlinear MHD of footpoint driven coronal loops', *Astron. Astrophys.* **306** (1996), 610–620.

[375] S. Poedts and J. P. Goedbloed, 'Nonlinear wave heating of solar coronal loops', *Astron. Astrophys.* **321** (1997), 935–944.

[376] S. Poedts, M. Goossens and W. Kerner, 'Numerical simulation of coronal heating by resonant absorption of Alfvén waves', *Solar Phys.* **123** (1989), 83–115.

[377] S. Poedts, D. Hermans and M. Goossens, 'The continuous spectrum of an axisymmetric self-gravitating and static equilibrium with a mixed poloidal and toroidal magnetic field', *Astron. Astrophys.* **151** (1985), 16–27.

[378] S. Poedts and W. Kerner, 'Ideal quasi-modes reviewed in resistive MHD', *Phys. Rev. Lett.* **66** (1991), 2871–2874.

[379] S. Poedts, W. Kerner, J. P. Goedbloed, B. Keegan, G. T. A. Huysmans and E. Schwarz, 'Damping of global Alfvén waves in tokamaks due to resonant absorption', *Plasma Physics & Controlled Fusion* **34** (1992), 1397–1422.

[380] S. Poedts, G. Tóth, A. J. C. Beliën and J. P. Goedbloed, 'Nonlinear MHD simulations of wave dissipation in flux tubes', *Solar Phys.* **172** (1997), 45–52.

[381] S. Poedts and E. Schwarz, 'Computation of the ideal MHD continuous spectrum in axisymmetric plasmas', *J. Comp. Physics* **105** (1993), 165–168.

[382] O. P. Pogutse and E. I. Yurchenko, 'Ballooning effects and plasma stability in tokamaks', in *Reviews of Plasma Physics, Vol. 11*, ed. M. A. Leontovich (New York, Consultants Bureau, 1982) pp. 65–151.

[383] K. G. Powell, 'An approximate Riemann solver for magnetohydrodynamics (that works in more than one dimension)', *ICASE Report No 94-24* (Langley VA, 1994).

[384] K. G. Powell, P. L. Roe, T. J. Linde, T. I. Gombosi and D. L. De Zeeuw, 'A solution-adaptive upwind scheme for ideal magnetohydrodynamics', *J. Comp. Physics* **154** (1999), 284–309.

[385] W. H. Press, B. P. Flannery, S. A. Teukolsky and W. T. Vettering, *Numerical Recipes* (Cambridge, Cambridge University Press, 1989).

[386] E. R. Priest, *Solar Magnetohydrodynamics* (Dordrecht, Reidel, 1984).

[387] E. R. Priest and T. G. Forbes, *Magnetic Reconnection; MHD Theory and Applications* (Cambridge, Cambridge University Press, 2000).

[388] J. Pringle and A. King, *Astrophysical Flows* (Cambridge, Cambridge University Press, 2007).

[389] J. J. Quirk, 'A contribution to the great Riemann solver debate', *Int. J. Num. Meth. in Fluids* **18** (1994), 555–574.

[390] J. Rappaz, 'Approximation of the spectrum of a noncompact operator given by the magnetohydrodynamic stability of plasma', *Num. Math.* **28** (1977), 15–24.

[391] K. S. Riedel, 'The spectrum of resistive viscous magnetohydrodynamics', *Phys. Fluids* **29** (1986), 1093–1104.

[392] P. L. Roe, 'Characteristic-based schemes for the Euler equations', *Ann. Rev. Fluid Mech.* **18** (1986), 337–365.

[393] P. L. Roe and D. S. Balsara, 'Notes on the eigensystem of magnetohydrodynamics', *SIAM J. Appl. Math.* **56** (1996), 57–67.

[394] M. M. Romanova, G. V. Ustyugova, A. V. Koldoba and R. V. E. Lovelace, 'Magnetohydrodynamic simulations of disk-magnetized star interactions in the quiescent regime: funnel flows and angular momentum transport', *Astrophys. J.* **578** (2002), 420-438.

[395] M. M. Romanova, O. D. Toropina, Yu. M. Toropin and R. V. E. Lovelace, 'Magnetohydrodynamic simulations of accretion onto a star in the "propeller" regime', *Astrophys. J.* **588** (2003), 400-407.

[396] M. N. Rosenbluth, R. Y. Dagazian and P. H. Rutherford, 'Nonlinear properties of the internal $m = 1$ instability in the cylindrical tokamak', *Phys. Fluids* **16** (1973), 1894–1902.

[397] G. Rüdiger and R. Hollerbach, *The Magnetic Universe; Geophysical and Astrophysical Dynamo Theory* (Weinheim, Wiley-VCH, 2004).

[398] V. V. Rusanov, 'The calculation of the interaction of non-stationary shock waves and obstacles', *USSR Comp. Math. and Math. Phys.* **1** (1961), 304–320.

[399] D. Ryu and T. W. Jones, 'Numerical magnetohydrodynamics in astrophysics: algorithm and tests for one-dimensional flow', *Astrophys. J.* **442** (1995), 228–258.

[400] D. Ryu, T. W. Jones and A. Frank, 'Numerical magnetohydrodynamics in astrophysics: algorithm and tests for multidimensional flow', *Astrophys. J.* **452** (1995), 785–796.

[401] Y. Saad and M. H. Schultz, 'GMRES: a generalized minimal residual algorithm for solving nonsymmetric linear systems', *SIAM J. Sci. Statist. Comput.* **7** (1986), 856–869.

[402] G. Schmidt, *Physics of High Temperature Plasmas*, 2nd edition (New York, Academic Press, 1979).

[403] D. D. Schnack, D. C. Barnes, D. P. Brennan, C. C. Hegna, E. Held, C. C. Kim, S. E. Kruger, A. Y. Pankin and C. R. Sovinec, 'Computational modeling of fully ionized magnetized plasmas using the fluid approximation', *Phys. Plasmas* **13** (2006), 058103-1–058103-21.

[404] D. D. Schnack and J. Killeen, 'Nonlinear, two-dimensional magnetohydrodynamic calculations', *J. Comp. Physics* **35** (1980), 110–145.

[405] L. I. Sedov, *Two-dimensional Problems in Hydrodynamica and Aerodynamics* (New York, Interscience Publishers, 1965; translated from the Russian edition of 1950 and edited by C. K. Chu, H. Cohen and B. Seckler).

[406] V. D. Shafranov, 'On magnetohydrodynamical equilibrium configurations', *Soviet Phys.–JETP* **33** (1957) 710–722.

[407] V. D. Shafranov, 'Equilibrium of a plasma toroid in a magnetic field', *Soviet Phys.–JETP* **37** (1960) 775–779.

[408] V. D. Shafranov, 'Equilibrium of a toroidal plasma in a magnetic field', *Journal of Nuclear Energy, Part C,* **5** (1963), 251–258. [*Atomnaya Energiya* **13** (1962) 521.]

[409] V. D. Shafranov, 'Plasma equilibrium in a magnetic field', in *Reviews of Plasma Physics, Vol. 2*, ed. M. A. Leontovich (New York, Consultants Bureau, 1966) pp. 103–151.

[410] N. I. Shakura and R. A. Sunyaev, 'Black holes in binary systems. Observational appearance', *Astron. Astrophys.* **24** (1973), 337–355.

[411] S. E. Sharapov, D. Testa, B. Alper, D. N. Borba, A. Fasoli, N. C. Hawkes, R. F. Heeter, M. Mantsinen, M. G. Von Hellerman and contributors to the EFDA-JET work-programme, 'MHD spectroscopy through detecting toroidal Alfvén eigenmodes and Alfvén wave cascades', *Physics Letters* **A289** (2001) 127–134.

[412] G. L. G. Sleijpen and H. A. van der Vorst, 'A Jacobi–Davidson iteration method for linear eigenvalue problems, *SIAM J. Matrix Anal. Appl.* **17** (1996), 401–425.

[413] L .S. Soloviev, 'Hydromagnetic stability of closed plasma configurations', in *Reviews of Plasma Physics, Vol. 6*, ed. M. A. Leontovich (New York, Consultants Bureau, 1975; original Russian edition, 1972) pp. 239–331.

[414] L .S. Soloviev and B. B. Shafranov, 'Plasma confinement in closed magnetic systems', in *Reviews of Plasma Physics, Vol. 5*, ed. M. A. Leontovich (New York, Consultants Bureau, 1970) pp. 1–247.

[415] C. R. Sovinec, A. H. Glasser, T. A. Gianakon, D. C. Barnes, R. A. Nebel, S. E. Kruger, D. D. Schnack, S. J. Plimpton, A. Tarditi and M. S. Chu, the NIMROD Team, 'Nonlinear magnetohydrodynamics simulation using high-order finite elements', *J. Comp. Physics* **195** (2004), 355–386.

[416] G. O. Spies, 'Magnetohydrodynamic spectrum of instabilities due to plasma rotation', *Phys. Fluids* **21** (1978), 580–587.

[417] H. C. Spruit, T. Matsuda, M. Inoue and K. Sawada 'Spiral shocks and accretion in discs', *Monthly Not. Roy. Astron. Soc.* **229** (1987), 517–527.

[418] K. Stasiewicz, 'Nonlinear Alfvén, magnetosonic, sound, and electron inertial waves in fluid formalism', *J. Geophys. Res.* **110** (2005), A03220, 1–9.

[419] O. Steiner, U. Grossmann-Doerth, M. Knölker and M. Schüssler, 'Simulation of the interaction of convective flow with magnetic elements in the solar atmosphere', *Reviews in Modern Astronomy*, ed. G. Klara, Vol. **8** (Hamburg, 1995) pp. 81–102.

[420] O. Steiner, M. Knölker and M. Schüssler, 'Dynamic interaction of convection with magnetic flux sheets: first results of a new MHD code', in *Solar Surface Magnetism*, eds. R. J. Rutten and C. J. Schrijver (Dordrecht, Kluwer Academic Publishers, 1994) pp. 441–471.

[421] J. M. Stone, J. F. Hawley, C. R. Evans and M. L. Norman, 'A test suite for magnetohydrodynamical simulations', *Astrophys. J.* **388** (1992), 415–437.

[422] J. M. Stone, J. F. Hawley, C. F. Gammie and S. A. Balbus, 'Three-dimensional magnetohydrodynamical simulations of vertically stratified accretion disks', *Astrophys. J.* **463** (1996), 656–673.

[423] J. M. Stone and M. L. Norman, 'ZEUS-2D: a radiation magnetohydrodynamics code for astrophysical flows in two space dimensions. II. The magnetohydrodynamic algorithms and tests', *Astrophys. J. Suppl. Ser.* **80** (1992), 791–818.

[424] G. Strang and G. J. Fix, *An Analysis of the Finite Element Method* (Englewood, NJ, Prentice Hall, 1973).

[425] H. R. Strauss, 'Dynamics of high β tokamaks', *Phys. Fluids* **20** (1977), 1354–1360.

[426] P. A. Sturrock, *Plasma Physics: an Introduction to the Theory of Astrophysical, Geophysical, and Laboratory Plasmas* (Cambridge, Cambridge University Press, 1994).

[427] B. R. Suydam, 'Stability of a linear pinch', in *Proc. Second UN Intern. Conf. on Peaceful Uses of Atomic Energy* **31** (New York, Columbia University Press, 1959), 157–159.

[428] J. L. Synge, *The Relativistic Gas* (Amsterdam, North Holland, 1957).

[429] J. L. Synge, *Relativity: the General Theory* (Amsterdam, North Holland, 1960).

[430] T. Tajima, *Computational Plasma Physics, with Applications to Fusion and Astrophysics* (Redwood City, Addison-Wesley Publishing Company, 1989).

[431] T. Tanaka, 'Finite volume TVD scheme on an unstructured grid system for three-dimensional MHD simulation of inhomogeneous systems including strong background potential fields', *J. Comp. Physics* **111** (1994), 381–389.

[432] T. S. Taylor, E. J. Strait, L. L. Lao, M. Mauel, A. D. Turnbull, K. H. Burrell, M. S. Chu, J. R. Ferron, R. J. Groebner, R. J. La Haye, B. W. Rice, R. T. Snyder, S. J. Thompson, D. Wròblewski and D. J. Lightly, 'Wall stabilization of high beta plasmas in DIII-D', *Phys. Plasmas* **2** (1995), 2390–2396.

[433] C. Terquem and J. C. B. Papaloizou, 'On the stability of an accretion disc containing a toroidal magnetic field', *Monthly Not. Roy. Astron. Soc.* **279** (1996), 767–784.

[434] J. Terradas, R. Oliver and J. L. Ballester, 'Damped coronal loop oscillations: time-dependent results', *Astrophys. J.* **642** (2006), 533–540.

[435] A. Thyagaraja and K. G. McClements, 'Comment on "Variational principles for stationary one- and two-fluid equilibria of axisymmetric laboratory and astrophysical plasmas" [Phys. Plasmas 11, L81 (2004)]', *Phys. Plasmas* **12** (2005), 064701–2.

[436] E. F. Toro, *Riemann Solvers and Numerical Methods for Fluid Dynamics* (Berlin, Springer-Verlag, 1997).

[437] M. Torrilhon, 'Uniqueness conditions for Riemann problems of ideal magnetohydrodynamics', *J. Plasma Physics* **69** (2003), 253–276.

[438] M. Torrilhon, 'Non-uniform convergence of finite volume schemes for Riemann problems of ideal magnetohydrodynamics', *J. Comp. Physics* **192** (2003), 73–94.

[439] G. Tóth, 'A general code for modeling MHD flows on parallel computers: Versatile Advection Code', *Astrophys. Lett. Commun.* **34** (1966), 245–250.

[440] G. Tóth, 'The $\nabla \cdot \mathbf{B} = 0$ constraint in shock-capturing magnetohydrodynamics codes', *J. Comp. Physics* **161** (2000), 605–652.

[441] G. Tóth, 'Conservative and orthogonal discretization for the Lorentz force', *J. Comp. Physics* **182** (2002), 346–354.

[442] G. Tóth, D. L. De Zeeuw, T. I. Gombosi and K. G. Powell, 'A parallel explicit/implicit time stepping scheme on block-adaptive grids', *J. Comp. Physics* **217** (2006), 722–758.

[443] G. Tóth, R. Keppens and M. A. Botchev, 'Implicit and semi-implicit schemes in the Versatile Advection Code: numerical tests', *Astron. Astrophys.* **332** (1998), 1159–1170.

[444] G. Tóth and D. Odstrčil, 'Comparison of some flux corrected transport and total variation diminishing numerical schemes for hydrodynamic and magnetohydrodynamic problems', *J. Comp. Physics* **128** (1996), 82–100.

[445] G. Tóth, I. V. Sokolov, T. I. Gombosi, D. R. Chesney, C. R. Clauer, D. L. De Zeeuw, K. C. Hansen, K. J. Kane, W. B. Manchester, R. C. Oehmke, K. G. Powell, A. J. Ridley, I. I. Roussev, Q. F. Stout, O. Volberg, R. A. Wolf, S. Sazykin, A. Chan, B. Yu and J. Kóta, 'Space weather modeling framework: a new tool for the space science community', *J. Geophys. Res.* **110** (2005), A12226, 1–21.

[446] L. N. Trefethen and M. Embree, *Spectra and Pseudospectra: the Behavior of Nonnormal Matrices and Operators* (Princeton, Princeton University Press, 2005).

[447] F. Troyon, R. Gruber, H. Saurenmann, S. Semenzato and S. Succi, 'MHD limits to plasma confinement', *Plasma Physics & Controlled Fusion* **26** (1984), 209–215.

[448] A. D. Turnbull, M. S. Chu, M. S. Chance. J. M. Greene, L. L. Lao and E. J. Strait,

'Second-order toroidicity-induced Alfvén eigenmodes and mode splitting in a low-aspect-ratio tokamak', *Phys. Fluids* **B14** (1992), 3451–3453.

[449] A. D. Turnbull, E. J. Strait, W. W. Heidbrink, M. S. Chu, H. H. Duong, J. M. Greene, L. L. Lao, T. S. Taylor and S. J. Thompson, 'Global Alfvén modes. Theory and experiment', *Phys. Fluids* **B5** (1993), 2546–2553.

[450] C. Uberoi, 'Alfvén waves in inhomogeneous magnetic fields', *Phys. Fluids* **15** (1972), 1673–1675.

[451] M. Ugai and T. Shimizu, 'Computer studies of noncoplanar slow and intermediate shocks associated with the sheared fast reconnection mechanism', *Phys. Plasmas* **1** (1994), 296–307.

[452] G. V. Ustyugova, A. V. Koldoba, M. M. Romanova, V. M. Chechetkin and R. V. E. Lovelace, 'Magnetocentrifugally driven winds: comparison of MHD simulations with theory', *Astrophys. J.* **516** (1999), 221–235.

[453] B. van der Holst, A. J. C. Beliën and J. P. Goedbloed, 'New Alfvén continuum gaps and global modes induced by toroidal flow', *Phys. Rev. Lett.* **84** (2000), 2865–2868.

[454] B. van der Holst, A. J. C. Beliën and J. P. Goedbloed, 'Low frequency Alfvén waves induced by toroidal flows', *Phys. Plasmas* **7** (2000), 4208–4222.

[455] B. van der Holst, R. J. Nijboer and J. P. Goedbloed, 'Magnetohydrodynamic spectrum of gravitating plane plasmas with flow', *J. Plasma Physics* **61** (1999), 221–240.

[456] B. van der Holst and R. Keppens, 'Hybrid block-AMR in cartesian and curvilinear coordinates: MHD applications', *J. Comp. Physics* **226** (2007), 925–946.

[457] B. van der Holst, R. Keppens and Z. Meliani, 'A multidimensional grid-adaptive relativistic magnetofluid code', *Comput. Physics Commun.* **179** (2008), 617–627.

[458] R. A. M. Van der Linden and M. Goossens, 'The thermal continuum in coronal loops: instability criteria and the influence of perpendicular thermal conduction', *Solar Phys.* **134** (1991), 247–273.

[459] R. A. M. Van der Linden, M. Goossens and J. P. Goedbloed, 'On the existence of a thermal continuum in non-adiabatic magnetohydrodynamic spectra', *Phys. Fluids* **B3** (1991), 866–868.

[460] R. A. M. Van der Linden, M. Goossens and A. W. Hood, 'The relevance of the ballooning approximation for magnetic, thermal, and coalesced magnetothermal instabilities', *Solar Phys.* **140** (1992), 317–342.

[461] R. A. M. Van der Linden, M. Goossens and W. Kerner, 'A combined finite element/Fourier series method for the numerical study of the stability of line-tied magnetic plasmas', *Comput. Physics Commun.* **59** (1990), 61–73.

[462] H. A. van der Vorst, 'Bi-CGSTAB: A fast and smoothly converging variant of Bi-CG for the solution of nonsymmetric linear systems', *SIAM J. Sci. Statist. Comput.* **13** (1992), 631–644.

[463] H. A. van der Vorst, *Iterative Krylov Methods for Large Linear Systems* (Cambridge, Cambridge University Press, 2003)

[464] T. Van Doorsselare and S. Poedts, 'Modifications to the resistive MHD spectrum due to changes in the equilibrium', *Plasma Physics & Controlled Fusion* **49** (2007), 261–271.

[465] M. H. P. M. van Putten, 'A numerical implementation of MHD in divergence form', *J. Comp. Physics* **105** (1993), 339–353.

[466] M. H. P. M. van Putten, 'Knots in simulations of magnetized relativistic jets', *Astrophys. J.* **467** (1996), L57–L60.

[467] E. P. Velikhov, 'Stability of an ideally conducting liquid flowing between cylinders

rotating in a magnetic field', *Soviet Phys.–JETP Lett.* **36** (1959) 995–998.

[468] M. Velli, G. Einaudi and A. W. Hood, 'Ideal kink instabilities in line-tied coronal loops: growth rates and geometrical properties', *Astrophys. J.* **350** (1990), 428–436.

[469] A. Vögler, S. Shelyag, M. Schüssler, F. Cattaneo, T. Emonet and T. Linde, 'Simulations of magneto-convection in the solar photosphere. Equations, methods, and results of the MURaM code', *Astron. Astrophys.* **429** (2005), 335–351.

[470] D. Voslamber and D. K. Callebaut, 'Stability of force-free magnetic fields', *Phys. Rev.* **128** (1962), 2016–2021.

[471] F. Wagner, G. Becker, K. Behringer, D. Campbell, A. Eberhagen, W. Engelhardt, G. Fussmann, O. Gehre, J. Gernhardt, G. v. Gierke, G. Haas, M. Huang, F. Karger, M. Keilhacker, O. Klüber, M. Kornherr, K. Lackner, G. Lisitano, G. G. Lister, H. M. Mayer, D. Meisel, E. R. Müller, H. Murmann, H. Niedermeyer, W. Poschenrieder, H. Rapp, H. Röhr, F. Schneider, G. Siller, E. Speth, A. Stäbler, K. H. Steuer, G. Venus, O. Vollmer and Z. Yü, 'Regime of improved confinement and high beta in neutral-beam-heated divertor discharges of the ASDEX tokamak', *Phys. Rev. Lett.* **49** (1982), 1408–1412.

[472] C. Wang, J. W. S. Blokland, R. Keppens and J. P. Goedbloed, 'Local analysis of MHD spectra for cylindrical plasmas with flows', *J. Plasma Physics* **70** (1964), 651–669.

[473] Z. X. Wang and D. R. Guo, *Special Functions* (Singapore, World Scientific, 1989).

[474] A. A. Ware, 'Role of compressibility in the magnetohydrodynamic stability of the diffuse pinch discharge', *Phys. Rev. Lett.* **12** (1964), 439–441.

[475] E. J. Weber and L. Davis Jr., 'The angular momentum of the solar wind', *Astrophys. J.* **148** (1967), 217–227.

[476] S. Wedemeyer-Böhm, I. Kamp, J. Bruls and B. Freytag, 'Carbon monoxide in the solar atmosphere. I. Numerical method and two-dimensional models', *Astron. Astrophys.* **438** (2005), 1043–1057.

[477] J. F. Wendt (ed.), *Computational Fluid Dynamics* (Berlin, Springer-Verlag, 1992).

[478] N. Werner, A. Finoguenov, J. S. Kaastra, A. Simionescu, J. P. Dietrich, J. Vink and H. Böhringer, 'Detection of hot gas in the filament connecting the clusters of galaxies Abell 222 and Abel 223', *Astron. Astrophys. Lett.* **482** (2008), L29–L33.

[479] P. Wesseling, *Principles of Computational Fluid Dynamics* (Berlin, Springer-Verlag, 2000).

[480] J. A. Wesson, 'Magnetohydrodynamic stability of tokamaks', *Nucl. Fusion* **18** (1978), 87–132.

[481] J. Wesson, *Tokamaks*, 3rd edition (Oxford, Clarendon Press, 2004).

[482] V. Wheatley, D. I. Pullin and R. Samtaney, 'Regular shock refraction at an oblique planar density interface in magnetohydrodynamics', *J. Fluid Mech.* **522** (2005), 179–214.

[483] R. B. White, *Theory of Toroidally Confined Plasmas*, 2nd edition (London, Imperial College Press, 2001).

[484] J. R. Wilson and G. J. Mathews, *Relativistic Numerical Hydrodynamics*, (Cambridge, Cambridge University Press, 2003).

[485] N. Winsor, J. L. Johnson and J. M. Dawson, 'Geodesic acoustic waves in hydromagnetic systems', *Phys. Fluids* **11** (1968), 2448–2450.

[486] L. Woltjer, 'Hydromagnetic equilibrium. IV. Axisymmetric compressible media', *Astrophys. J.* **130** (1959), 405–413.

[487] C. C. Wu, 'The MHD intermediate shock interaction with an intermediate wave: are intermediate shocks physical?', *J. Geophys. Res.* **93** (1988), 987–990.

[488] C. C. Wu, 'Formation, structure, and stability of MHD intermediate shocks', *J. Geophys. Res.* **95** (1990), 8149–8175.

[489] C. C. Wu, 'New theory of MHD shock waves', in *Viscous Profiles and Numerical Methods for shock Waves*, Chapter 17, SIAM Proceedings Series, ed. M. Shearer (Philadelphia, SIAM, 1991), 209–236.

[490] C. C. Wu and T. Hada, 'Formations of intermediate shock in both two-fluid and hybrid models', *J. Geophys. Res.* **96** (1991), 3769–3778.

[491] C. C. Wu and C. F. Kennel, 'Structure and evolution of time-dependent intermediate shocks', *Phys. Rev. Lett.* **68** (1992), 56–59.

[492] A. Yoshizawa, S.-I. Itoh and K. Itoh, *Plasma and Fluid Turbulence, Theory and Modelling* (Bristol, Institute of Physics, 2003).

[493] A. L. Zachary and P. Colella, 'A higher-order Godunov method for the equations of ideal magnetohydrodynamics', *J. Comp. Physics* **99** (1992), 341–347.

[494] L. E. Zakharov and V. D. Shafranov, 'Equilibrium of current-carrying plasmas in toroidal configurations', in *Reviews of Plasma Physics, Vol. 11*, ed. M. A. Leontovich (New York, Consultants Bureau, 1982) pp. 153–302.

[495] H. P. Zehrfeld and B. J. Green, 'Stationary toroidal equilibria at finite beta', *Nucl. Fusion* **12** (1972), 569–575.

[496] R. Zelazny, R. Stankiewicz, A. Galkowski and S. Potempski, 'Solutions to the flow equilibrium problem in elliptic regions', *Plasma Physics & Controlled Fusion* **35** (1993), 1215–1227.

[497] U. Ziegler, 'Self-gravitational adaptive mesh magnetohydrodynamics with the NIRVANA code', *Astron. Astrophys.* **435** (2005), 385–395.

Index

Note: *italic page numbers* indicate a main section on the subject.

advanced tokamak scenario, 345
Alfvén and slow continua in tokamak plasma
  ellipticity induced gap, 322
  governing ODEs, 320
  toroidicity induced gap, 322
  triangularity induced gap, 322
Alfvén wave cascades, 351
alternator, 68
angular momentum flux, 532
angular momentum flux tensor, 358
anomalous resistivity, 168, 172
apparent and spurious singularities, 53
astrophysical objects
  Active Galactic Nuclei, 543, 585
  Pinwheel Galaxy M101, 3
  pulsar wind nebulae, 584
  T-Tauri star, 537
axi-symmetric static equilibria
  average β, 251, 267
  average poloidal β_p, 264, 266, 275
  diamagnetic equilibria, 276
  distribution parameters, 265
  equilibrium-stability limit on β, 268
  flux functions, 270, 271
  geometry parameters, 265
  global confinement parameters, 265
  governing equations, 247
  Grad–Shafranov equation, 269
  high field side, 261
  high-β tokamak, 260, 267
  hoop force, 262
  internal self-inductance ℓ_i, 262, 275
  inverse aspect ratio, 251, 284
  low field side, 261
  low-β tokamak, 260, 267
  modified safety factor q^*, 266
  orthogonal flux coordinates, 252
  orthogonal toroidal coordinates, 278
  paramagnetic equilibria, 276
  poloidal currents in external toroidal field coils, 260
  poloidal magnetic flux, 249, 270

rational magnetic field lines, 254
rational magnetic surface, 254
resonant field lines/surfaces, 255, 326
safety factor, 250
safety factor in straight field line coordinates, 310
scaling parameters, 265
Shafranov shift, 260, 272
straight field line coordinates, 253, 308
toroidal current in external vertical field coils, 262
toroidal current in primary windings, 261
toroidal magnetic flux, 249
toroidal plasma current, 261
vacuum field equations, 278
vacuum field scalar potential, 280
vertical field, 262
axi-symmetric stationary equilibria
  Bernoulli equation for squared poloidal Alfvén
    Mach number, 360
  core equations, 362
  $\Delta(r)$ shift of the flux surfaces, 369
  flat accretion disk, 373
  magnetic/flow surfaces, 357
  nonlinear PDE for poloidal flux, 360
  poloidal magnetic flux function, 357
  poloidal velocity stream function, 357
  scaled flux functions, 359
  small inverse aspect ratio expansion, *366–370*
  thick accretion disk, 373
  tokamak, 373
  toroidal rescalings, 361
  toroidally rotating plasmas, 299
  variational principle, 360

ballooning modes
  Connor–Hastie–Taylor equation, 331
  governing ODEs, 330
ballooning transformation, 326
Bernoulli function, 358
β-induced Alfvén eigenmode (BAE), 349
boundary layer analysis, 143
Brunt–Väisäläa frequency, 8

centrifugal acceleration, 15
characteristic speeds, 488
classification of MHD shocks, *507–520*
Clebsch potentials, 326
conformal mapping, 293
continua for axi-symmetric stationary plasmas
    2×2 matrix reduction, 376
    Alfvén continua, 381
    Coriolis coupling factor, 379
    dispersion equation for trans-slow Alfvén
        continuum modes, 388, 389
    Eulerian entropy continua, 382
    governing ODEs, 378
    mode locking, 397
    six-mode interaction, 384, 388
    slow continua, 381
    trans-slow continua, 385
cosmic plasmas, 4
curvature vector of field line, 311

discretization
    accuracy, 181
    consistency, 181
    efficiency, 182
    global truncation error, 180
    numerical stability, 181
    residual, 181
    round-off error, 181
    total error, 181
disk truncation radius, 537
$\nabla \cdot \mathbf{B}$ treatments, *455–460*
    constrained transport, 458
    field interpolated central difference scheme, 459
    hyperbolic cleaning, 459
    method of characteristics–constrained transport
        (MOC-CT), 467
    parabolic cleaning, 459
    Powell's source terms, 457
    projection scheme, 456
    vector potential, 456
Doppler shift, 18, 32

eigenvalue problem of resistive MHD, 338
elliptic and hyperbolic flow regimes, 365
ellipticity induced Alfvén eigenmode (EAE), 349
enthalpy, 493
entropy waves, 499
epicyclic frequency, 114
epicyclic modes, 117
ϵ-stability of stationary plasmas, 34
equilibrium for tokamak plasmas, *247–269*
equilibrium for transonic plasmas, *357–365*
essential spectra, 61
Eulerian entropy continuum, 16
extended MHD, *171–174*
    collisionless reconnection, 174
    generalized Ohm's law, 173
    Hall current, 174
    Hall MHD, 172, 173
    whistler wave, 172

fast elliptic flow regime, 366

finite difference methods
    first-order accuracy, 182
    first-order backward difference, 182
    first-order forward difference, 182
    grid points, 182
    mesh points, 182
    second-order central difference, 182
finite element method
    basis functions, 186
    essential boundary conditions, 188
    Galerkin method, 187
    natural boundary conditions, 188
    residual, 187
    shape functions, 186
    weak formulation, 188
    weight functions, 187
    weighted residual formulation, 187
Fjørtoft's theorem, 78
force-free magnetic field, 6, 128
Frieman–Rotenberg formalism, *16–22*
    average Doppler–Coriolis shift, 33
    Doppler shift operator, 21
    generalized force operator, 21
    in straight-field-line coordinates, 374
    kinematic transformation of line/surface/volume
        elements, 18
    Newcomb's expression for gradient operator, 18
    parallel field operator, 375
    parallel flow operator, 375
    pre-self-adjointness relation, 26
    quasi-Lagrangian representation, 17
    standard force operator, 21
funnel flows, 537

Galerkin method, 298, 338
gas dynamic shocks, *492–498*
    distilled energy jump condition, 495
    distilled entropy condition, 497
    entropy condition, 493
    entropy-forbidden shocks, 497
    entropy-permitted shocks, 497
    Hugoniot adiabatic, 493
    shock strength, 495
    time reversal duality, 498
geodesic acoustic mode (GAM), 314, 321
geodesic curvature of magnetic field line, 312
ghost cells, 202
global tokamak modeling, 482
Grad–Shafranov equation, *269–283*
    dimensionless flux coordinate, 284
    dimensionless flux function profiles, 285
    dimensionless poloidal flux, 284
    large aspect ratio expansion, 271
    low-β tokamak approximation, 273
    non-orthogonal shifted circle coordinate system,
        272
    numerical solution methods, 293
    poloidal current stream function, 270
    poloidal field stream function, 270
    poloidal flux scaling, 287, 288
    scaled, core form, 286

unit profiles, 286
gravitating plasma equilibria, 301
gravity dominated accretion disks, 396
gravity-driven g-modes, 64

H-mode, 340
Hamiltonian formulation
   canonical momentum, 28
   Hamilton's principle for linear ideal MHD, 28
   Hamiltonian of linear perturbations, 28
Harris sheet, 162
helioseismology, 348
helmet streamer, 532
Howard's criterion, 85

ideal relativistic MHD, *570–572*
   characteristic speeds, 575
   conservative to primitive variable transformation,
      581
   de Hoffman–Teller frame, 580
   Friedrichs diagrams, 575
   Lichnerowicz adiabat, 579
   linear waves, *572–575*
   magnetic pressure invariant, 570
   normal perpendicular shocks, 578
   numerical challenges, *583–584*
   1D Riemann problem, 581
   shock conditions, *577–580*
instabilities
   ballooning instabilities, 314
   ballooning modes, 7
   coalescence instability, 166
   edge-localised modes (ELM), 340
   external kink, 7, 258
   gravitational interchange, 8
   infernal mode, 344
   interchanges, 7, 8, 314, 333
   internal kink, 7, 259
   Kelvin–Helmholtz instability, 170
   magneto-rotational instability (MRI), 7, 535
   Mercier criterion, 9
   neo-classical tearing mode, 7
   Parker instability, 7, 9
   quasi-interchange, 8
   quasi-Parker instability, 10
   Rayleigh–Taylor instability, 539
   resistive gravitational interchange mode, 149
   resistive wall mode, 152
   Schwarzschild criterion, 8
   Suydam criterion, 9
   tearing mode, 135, 344
   trans-slow Alfvén continuum (TSAC) modes, 390
ion whistler wave, 174
iso-parametric mapping, 297
iso-thermal MHD, *411–415*
   magneto-acoustic Riemann invariant, 414, 445
ITER, 3, 248

Kelvin–Helmholtz instability, *55–58, 76–85*
   dispersion equation for interface plasmas, 58
   for fluids, 76

in interface plasmas, 56
Kruskal–Shafranov limit, 258

Lagrangian representation, 16
local dispersion equation, 102, 117
logarithmic derivatives of magnetic field perturbation,
   142

Mach number, 495
magnetic axis, 249
magnetic helicity, 129
magnetic reconnection, *162–174*
   GEM challenge, 162
   Newton challenge, 162, 171
   self-feeding turbulent reconnection, 166
   stationary Petschek reconnection, 167
   Sweet–Parker reconnection, 166
magnetic Reynolds number, 135, 139
magnetic shear, 49
magnetic surface symmetry triad, 311
magnetic surface/field line triad, 374
magnetic/flow surfaces
   Gaussian curvature, 378
   geodesic curvature, 377
magnetically dominated accretion tori, 356
magnetically modified Brunt–Väisäläa frequency, 8
magnetized accretion–ejection structure (MAES), 536
magneto-centrifugal jet launching, 535
magneto-rotational instability, 112
maximal ordering
   for thick accretion disk, 362
   for thin accretion disk, 362
   for tokamak, 362
   modified Bernoulli function, 362
Mercier criterion, 332
MHD discontinuities
   contact discontinuity, 498, 499
   magneto-acoustic shock, 500, 503
   rotational or Alfvén discontinuity, 500, 503
   tangential discontinuity, 498, 500
MHD flow regimes
   sub-fast, super-Alfvénic, 511
   sub-slow, 511
   super-fast, 511
   super-slow, sub-Alfvénic, 511
MHD shock conditions, *490–507*
MHD shocks
   de Hoffmann–Teller frame, 505
   distilled energy jump condition, 508
   distilled entropy condition, 508
   distilled ideal MHD jump problem, 514
   distilled MHD shock problem, 514
   eleven transition points, 516
   fast shocks, 505
   4–1 and 4–2 jumps, 513
   hydrodynamic shocks, 506
   intermediate shocks, 505
   jump conditions in shock frame, 491
   magneto-sonic transition values, 511
   1–2, 1–3, 1–4, 2–3, 2–4 and 3–4 shocks, 511
   parallel shocks, 506, 515

perpendicular shocks, 506, 515
prograde intermediate shocks, 509, 515
quasi-prograde intermediate shocks, 516
quasi-retrograde intermediate shocks, 516
retrograde intermediate shocks, 509, 514, 515
shock evolutionarity, 515, 528
shock strength, 511
six dual pairs of entropy-allowed shocks and
    entropy-forbidden jumps, 520
slow shocks, 505
switch-off fast–intermediate jumps, 524
switch-off shocks, 506
switch-off slow–intermediate shocks, 524
switch-on fast–intermediate shocks, 524
switch-on shocks, 506, 515
switch-on slow–intermediate jumps, 524
time reversal duality, 513, 518
MHD spectroscopy for tokamak plasmas, 303, 349,
    350
MHD spectroscopy of galactic plasmas, 12
mixed field line/magnetic surface triad, 312
mode coupling, 382

non-holonomic initial data, 16
non-orthogonal eigenfunctions, 41
nonlinear conservation laws, *408–411*
    characteristic speeds, 409
    characteristic variables, 411
    conservative variables, 408
    flux Jacobian, 408
    generalized Riemann invariants, 411
    genuinely nonlinear wave field, 419
    Hugoniot locus, 417
    linearly degenerate wave field, 419
    method of characteristics, 410
    overcompressive shock, 418
    primitive variables, 408
    quasi-linear form, 408
    Riemann invariants, 410
    simple wave, 410
    strictly hyperbolic system, 409
    structure coefficient, 419
normal Alfvén Mach number, 501
normal curvature of magnetic field line, 312
numerical methods
    adaptive mesh refinement (AMR), 462
    alternating direction implicit (ADI), 474
    anti-diffusion, 425
    approximate Riemann solver, 446
    arbitrary Lagrangian–Eulerian (ALE), 470
    conservative scheme, 421
    dimensional splitting strategy, 455
    $\nabla \cdot \mathbf{B}$ treatments, 455
    entropy fix, 448
    FCT anti-diffusion coefficient, 426
    FCT diffusion coefficient, 426
    finite volume method, 430
    flux corrected transport (FCT), 425
    flux limiter, 426
    Gibbs phenomenon, 424
    Godunov method, 434
    Godunov splitting, 454
    high resolution method, 429
    HLL solver, 450
    HLLC solver, 450
    hybrid scheme, 429
    hyper-diffusion, 471
    Lax–Friedrichs scheme, 421
    linear reconstruction, 441
    local Lax–Friedrichs method, 441
    MacCormack scheme, 422
    minmod limiter, 442
    monotonicity preserving scheme, 424
    monotonized central-difference limiter, 442
    pseudo-convergence, 453
    pseudo-spectral method, 469
    Richtmyer two-step Lax–Wendroff scheme, 422
    Roe average, 447
    Roe matrix, 447
    Roe solver, 446
    semi-implicit methods, 475
    shock capturing scheme, 430
    slope limiting, 442
    Strang splitting, 454
    total variation diminishing (TVD), 427
    TVDLF method, 440
numerical MHD
    double umbilic point, 438
    quintuple umbilic point, 438
    shearing box model, 468
    solar magneto-convection, 462, 471
    Tanaka's splitting strategy, 460
    triple umbilic point, 438

oscillation theorem $\mathcal{C}$ for complex eigenvalues, *91–92*
oscillation theorem $\mathcal{R}$ for real eigenvalues, *59–64*
overstable mode, 22, 52

parallel gradient operator, 308
Picard iteration, 293
Poisson adiabatic, 494
polar plots in de Hoffmann–Teller frame for MHD
    shocks, *520–527*
poloidal Alfvén Mach number, 359
    critical cusp value, 379
poloidal curvature of magnetic surface, 311
poloidal vorticity–current density stream function,
    358
predictor–corrector method, 232
pressure-driven p-modes, 64

quadratic eigenvalue problem, 22
quasi-interchange modes, 111
quiescent prominence, 301

rarefaction wave, *418–419*
    centered simple wave, 419
    integral curve, 419
Rayleigh's circulation criterion, 114
Rayleigh's discriminant, 114
Rayleigh's inflection point theorem, 78
Rayleigh–Taylor instabilities for magnetized plasmas,
    73–76

Rayleigh–Taylor instability, 56
relativistic gas dynamic waves
    characteristic speeds, 558
    entropy waves, 557
    Huygens construction for sound waves, 560
    linear sound waves, 557
    phase and group speed diagrams, 558
    shock relations, 560
    Taub adiabat, 562
    3 + 1 formalism, 556
relativistic gas dynamics
    particle number conservation, 552
    stress–energy tensor, 552
relativistic jet simulations
    kinetic energy dominated jets, 588
    nose cone, 586
    Poynting flux dominated jets, 586
relativity and electromagnetism
    electric four-vector, 565
    electromagnetic field tensor, 564
    electromagnetic stress–energy, 569
    four-current, 566
    Lorentz transformation for $\mathbf{E}$ and $\mathbf{B}$, 565
    magnetic four-vector, 568
    Maxwell equations, 564
relativity and thermodynamics
    effective polytropic index, 555
    entropy, 554, 570
    Mathews approximation, 555
    polytropic equation of state, 554
    relativistic enthalpy, 553
    Synge gas, 554
resistive gravitational interchange mode
    growth rate, 150
    resistive layer width, 150
    stability criterion, 150
resistive MHD spectrum, *150–160*
    Alfvén dispersion equation for homogeneous
        incompressible plasma, 157
    ideal quasi-mode, 160
    spectrum for homogeneous compressible plasma,
        157
    spectrum for inhomogeneous compressible plasma,
        158
resistive normal mode analysis, 135
resistive wall mode, *150–155*
Richardson number, 85
Riemann problem for 1D MHD, *435–437*
    iso-thermal MHD, 443
rigidly rotating incompressible plasmas, *104–112*
Runge–Kutta methods, 232

safety factor, 250, 365
saturation of ideal internal kink, 481
scalar conservation law, *415–420*
    compound wave, 419
    convex flux, 419
    Godunov theorem, 429
    integral form, 430
    inviscid Burgers equation, 415
    Lax entropy condition, 417

Rankine–Hugoniot relation, 417
rarefaction wave, 418
Riemann problem, 417
scale independence of MHD, 3
Schwarzschild–Suydam stability criterion, 87
self-similar transonic flows, 529
semi-discretization, 231
Seret–Frenet triad for field lines, 311
Shafranov shifted circle approximation, 366
Shakura and Sunyaev α-parameter, 119
slow elliptic flow regime, 366
    gravitational interaction coefficient, 371, 387
    trans-slow poloidal flow ordering, 371, 377
Soloviev equilibrium, *289–292*
solution paths for stationary plasmas, *35–46*
    Doppler–Coriolis indefinite range, 43
    marginal stability transition, 43
    path of stable solutions, 37
    path of unstable solutions, 37
    sloshing energy, 39
    solution path topology, 81
    solution-averaged Doppler–Coriolis shifted
        frequency, 37
space weather, 464
space-time characteristics, 488
special relativity
    four-dimensional space-time, 544
    four-velocity, 547
    length contraction, 546
    light-like four-vector, 549
    Lorentz boost, 545
    Lorentz transformation, 544
    Minkowski metric, 548
    proper time, 547
    relativistic beaming, 589
    relativistic Doppler effect, 558
    relativistic three-momentum, 550
    relativistic wave aberration, 558
    space-like four-vector, 549
    space-time diagram, 547
    space-time event, 545
    three-velocity addition, 551
    time dilation, 546
    time-like four-vector, 549
spectral method
    Chebyshev polynomials, 200
    collocation approach, 200
    Legendre polynomials, 199
    non-Galerkin approach, 200
    tau approach, 200
spectral theory for stationary plasmas
    alternator, 68
    alternator for middle path, 69
    cluster criteria and local gravitational interchanges,
        86
    cusp value for shear Alfvén Mach number, 100
    Doppler–Coriolis indefinite range, 110
    flow continua, 59
    forward and backward Alfvén and slow continua,
        55, 60, 95

forward and backward apparent fast and slow singularities, 96

forward and backward local Doppler shifted Alfvén and slow frequencies, 54

forward and backward turning point frequencies, 55, 95

generalized Suydam criterion for cylindrical plasma, 100

gravito-MHD wave equation for plane plasma flow, 52

incompressible limit of spectral equation, 104

local Coriolis shift, 93

local Doppler shift, 50

local Doppler shifted frequency, 50

middle solution path, 68

non-holonomic Eulerian entropy continua, 60

oscillation theorem C for complex eigenvalues, 91

oscillation theorem $\mathcal{R}$ for real eigenvalues, 62

shear Alfvén Mach number, 99

solution averaged Doppler–Coriolis shift, 51, 98

sub-paths, 81

transfer of cluster sequences, 102

static tokamak plasmas
  Alfvén and slow continuum, 319
  spectral variational principle, 318
  spectral wave equation, 316

steady-state problem, 179

stellar spin-down rate, 534

stellarator, 261

straight cylinder with elliptical cross-section, 266

straight field line coordinates
  Christoffel symbols and curvature expressions, 313

straight tokamak approximation, 256

Sturm–Liouville equation, 178

sub-slow elliptic flow regime, 366

tearing induced by Kelvin–Helmholtz, 169

tearing mode
  asymptotic analysis, 139
  constant Ψ approximation, 147
  Δ' jump of logarithmic derivative, 148
  Furth–Killeen–Rosenbluth approximate solution, 147
  growth rate, 148
  inner resistive layer, 140
  matching of logarithmic derivatives, 145
  matching to resistive layer solution, 141
  regularity boundary conditions, 145
  resistive layer width, 148
  scaling of resistive layer equations, 144

tearing mode in incompressible resistive MHD, *138–147*

thermodynamical variables, 493

θ-pinch, 256

tokamak, 248

toroidal Alfvén eigenmode (TAE), 325, 347

toroidal bootstrap current, 261

toroidal curvature of magnetic surface, 311

toroidal flow Alfvén eigenmode (TFAE), 349

total toroidal current, 258, 276

trans-slow poloidal flow ordering, 371, 387, 401

transition from ellipticity to hyperbolicity, 366, 370, 489

transonic MHD flows, 529
  forbidden flow regimes, 530
  limiting line characteristics, 530
  Weber–Davis wind solution, 531

transonic transitions, 490

transonically rotating axi-symmetric plasmas, 355

Troyon limit, 334

vector potential, 129, 337

velocity shear, 50

von Neumann method, 226

wave equation for incompressible plasmas, 56

Wesson profiles, 277

WKB analysis, 102

z-pinch, 256